The Textbook of Nanoneuroscience and Nanoneurosurgery

The Textbook of

Nanoneuroscience and Nanoneurosurgery

Edited by

Babak Kateb • John D. Heiss

CRC Press
Taylor & Francis Group
Boca Raton London New York

CRC Press is an imprint of the
Taylor & Francis Group, an **informa** business

CRC Press
Taylor & Francis Group
6000 Broken Sound Parkway NW, Suite 300
Boca Raton, FL 33487-2742

Printed on acid-free paper
Version Date: 20130606

International Standard Book Number-13: 978-1-4398-4941-5 (Hardback)

Library of Congress Cataloging-in-Publication Data

Kateb, Babak.
The Textbook of nanoneuroscience and nanoneurosurgery / Babak Kateb and John D. Heiss.
p. ; cm.
Includes bibliographical references and index.
ISBN 978-1-4398-4941-5 (alk. paper)
I. Heiss, John D. II. Title.
[DNLM: 1. Nanotechnology--methods. 2. Neurosurgical Procedures. WL 368]

RD593
617.4'8--dc23 2013002513

Visit the Taylor & Francis Web site at
http://www.taylorandfrancis.com

and the CRC Press Web site at
http://www.crcpress.com

Dedication

We dedicate this textbook to our spouses, parents, family, mentors, students, and medical and scientific colleagues. We appreciate their support for our efforts.

On the Cover

Front cover: Depicts the current invasive neurosurgical approach. Comparatively, in the future non- or minimally-invasive nanoneurosurgery will be much more individualized, targeted therapy performed by smart self-assembling molecules and nanoplatforms. (Pictures are courtesy of Babak Kateb and Jean Paul Allain.)

Back cover: A view of the cortical surface of a college student with new-onset partial motor seizures affecting the left face and findings of a right frontal lobe mass on MRI. An image-guided right frontotemporal craniotomy was performed and the dura was opened. The inferoposterior part of the frontal lobe is seen on the left (anterior anatomical) side of the photograph. The superior temporal gyrus is seen on the top (inferior anatomical) portion of the photograph. The Sylvian fissure separates the frontal and temporal lobes. Note the paleness and expansion of the frontal lobe gyrus in the center of the exposure that contains the tumor. No seizures occurred after complete excision of a dysembryoplastic neuroepithelial tumor. (Picture courtesy of John D. Heiss.)

Contents

Mission Statements

SBMT MISSION STATEMENT

The Society for Brain Mapping and Therapeutics (SBMT) is a nonprofit society organized for the purpose of encouraging basic and clinical scientists who are interested in the areas of brain mapping, engineering, stem cells, nanotechnology, imaging, and medical devices to improve the diagnosis, treatment, and rehabilitation of patients afflicted with neurological disorders.

This society promotes the public welfare and improves patient care through the translation of new technologies/therapies into life-saving diagnostic and therapeutic procedures. The society is committed to excellence in education and scientific discovery. The society achieves its mission through multidisciplinary collaborations with government agencies, patient advocacy groups, educational institutes, and industry, as well as philanthropic organization.

BMF MISSION STATEMENT

Brain Mapping Foundation (BMF) is a nonprofit 501c3 charitable organization, which is established for the purpose of facilitating multidisciplinary brain and spinal cord research and expediting integration and translation of cutting-edge technologies into the field of neuroscience. BMF is focused on translating state-of-the-art technologies from space and defense industries into neuroscience in order to bring the most advanced medicine to wounded warriors as well as civilians.

NCNBE MISSION STATEMENT

The NCNBE mission is to establish collaborating research laboratories and networks throughout the state of California and beyond in order to rapidly develop solutions for neurological disorders employing advances in nanotechnology, stem cell research, and medical devices (nanobioelectronics) while fostering biotech spinoffs for the purpose of job creation.

NCNBE promotes the public welfare and improves patient care through the translation of new technologies into life-saving diagnostic and therapeutic procedures. The center is committed to excellence in education and scientific discovery. The NCNBE achieves its mission through multidisciplinary collaborations/consortia with government agencies, patient advocacy groups, educational institutions, private sector, industry, and philanthropic organizations.

Preface

Neuroscience, nanotechnology, cellular therapy, imaging, and medical devices have undergone remarkable advances in recent years. The capability to track cells in vivo with nanoplatforms, guide cells with nanowires, assess vital processes with nanosensors and treat neurological disorders such as TBI, epilepsy, Alzheimer's Disease, and brain cancer is no longer science fiction. Translation of such technologies from neuroscience to clinical applications in neurology and neurosurgery is imminent and depends on closing gaps of knowledge between disciplines in biology, physical science and medicine. By integrating nanotechnology, material science, biomedical engineering, electrical engineering, and neuroscience with neurosurgery, novel approaches can be developed to diagnose and treat a variety of neurological disorders. The application of nanotechnology in neurosurgery is known as "Nanoneurosurgery," and its application in neuroscience in general is considered "nanoneuroscience." While nanotechnology is still a developing field, we have witnessed an amazing level of scientific innovation and discovery emerge from integration of nanotechnology, imaging, devices, cellular and stem cell therapy into "Nanobioelectronics." We strongly believe that nanoneuroscience, nanoneurosurgery and nanobioelectronics will revolutionize medicine and remarkably improve the prevention, diagnosis and treatment of neurological disorders over the next 10–20 years.

Successful development of nanoneurosurgery requires carefully regulated clinical studies that validate the accuracy of diagnostic tests and verify the safety and efficacy of therapies that utilize nanotechnology, including drugs, devices, cellular therapy, and their combinations. Regulatory guidelines must be adhered to as nanoneurosurgical diagnostic techniques and therapeutic modalities are translated from the laboratory to clinical trials and to commercial medical products. This process is essential for better drug and device development, which could improve healthcare and new legislation and government-based initiatives are sorely needed to facilitate this process.

This textbook reviews the latest regulatory guidelines which influence the translation of nanotechnological research from the laboratory to the clinic as well as the most recent information on biodevices and pharmaceutical spinoffs. It highlights Presidential and Congressional initiatives and programs that may significantly impact the field in the near future. We discuss state-of-the-art science and technologies, which are applied toward diagnosis and treatment of neurological disorders, as well as regulatory issues that impact product development.

This volume also describes state-of-the-art nanotechnological advances that have already been translated or have significant potential for future application in nanoneurosurgery. The contributions of 140 researchers together with more than 3000 references are contained in this book. The potential impact of nanoneuroscisnce, nanoneurosurgery, and nanobioelectronic is described throughout this textbook. This broad coverage of nanotechnological applications in diverse areas of neuroscience, neurosurgery, neuroradiology, neurology, and the role of FDA regulation and healthcare policy provides a foundation to support and impact the future of personalized targeted therapy of neurological disorders. This textbook is devoted to implementing new ways to perform less radical procedures (surgical and non-surgical), eventually leading to approaches with minimal or no invasiveness to treat nervous system disorders including brain tumors, brain and spinal cord injuries, vascular malformations, Alzheimer's disease, and Parkinson's disease. We hope that this book provides scientists and physicians with helpful references as they devise successful treatments for these presently incurable neurological disorders.

We would like to thank all contributors for their scientific contributions to this remarkable textbook. Editors and authors of this magnificent and one of the kind textbook would like to thank CRC/Taylor & Francis Publishing: Lance Wobus (Senior Editor), Barbara Norwitz (Executive Editor),

Jessica Vakili (Project Coordinator), Jennifer Stair (Project Editor), Amor Nanas (Project Manager), Nally Dookwah-Abrams (Marketing Manager), Janet Gibbons (Editorial Assistant), Sarah A. Sporn (Copywriter), Kate Newton (Copywriter), and Shayna Murry (Graphic Designer). We strongly believe their technical insight has significantly contributed to the success of this scientific master piece. We thank the Brain Mapping Foundation, National Center for NanoBioElectronics, and Society for Brain Mapping and Therapeutics (SBMT) for their generous support and endorsement of this truly unique textbook. Finally we thank our parents, Jabbar Kateb, Maliheh Samsami Shirzad, John F. Heiss (deceased), and Joann Heiss, for their years of support and personal guidance. We trust that readers will take the ideas and methods described in this textbook to formulate novel and more effective treatments of presently incurable diseases.

Respectfully yours,

Babak Kateb and John D. Heiss

Editors

Babak Kateb
Editor of *The Textbook of Nanoneuroscience and Nanoneurosurgery*
Founding Chairman of the Board of SBMT and Brain Mapping
Foundation, Scientific Director of SBMT, and
President of Brain Mapping Foundation,
West Hollywood, CA, USA
Director of the National Center for Nano-Bio-Electronics,
NASA Research Park, Moffet, CA, USA
Senior Editor of SBMT-PLoSOne-NeuroMapping & Therapeutics, and Research Scientist,
Department of Neurosurgery, Cedars-Sinai Medical Center, Los Angeles, CA, USA
Visiting Scientist, NASA/Jet Propulsion Laboratory (JPL), Pasadena, CA, USA

John D. Heiss
Editor of *The Textbook of Nanoneuroscience and Nanoneurosurgery*
Chair, Surgical Neurology Branch,
National Institute of Neurological Disorders and Stroke,
National Institutes of Health,
Bethesda, MD, USA

Devera G. Schoenberg
Assistant Editor of *The Textbook Nanoneuroscience and Nanoneurosurgery*
National Institute of Neurological Disorders and Stroke,
National Institutes of Health,
Bethesda, MD, USA

Contributors

Michele R. Aizenberg| Division of Neurosurgery| University of Nebraska Medical Center| Nebraska Medical Center| Omaha| Nebraska
Director| Brain and Spine Cancer Center| Nebraska Medical Center| Omaha| Nebraska
Board Member| Society for Brain Mapping and Therapeutics (SBMT)| West Hollywood| California

Jean Paul Allain| College of Engineering and Birck Nanotechnology Center| Purdue University| West Lafayette| Indiana

Rina Amatya| Department of Neurosurgery| Cedars-Sinai Medical Center| Los Angeles| California

Daniel Angeles| Division of Neurosurgery| City of Hope National Medical Center| Duarte| California

Rocco Armonda| National Capital Neurosurgery Consortium at Walter Reed Military Medical Center| Bethesda| Maryland

Meghedi Babakhanian| Graduate Student Researcher| Biomedical Engineering Interdepartmental Program| Center for Advanced Surgical and Interventional Technology| University of California| Los Angeles| Los Angeles| California

Houman Badrzadeh| Western University of Health Sciences| Pomona| California

Krystof S. Bankiewicz| Department of Neurosurgery| Molecular Therapeutics Laboratory| University of California| San Francisco| San Francisco| California

Peter J. Basser| Senior Investigator| Director| Program on Pediatric Imaging and Tissue Sciences Chief| Section on Tissue Biophysics and Biomimetics (STBB)| Bethesda| Maryland
Eunice Kennedy Shriver| National Institute of Child Health and Human Development| Bethesda| Maryland

Mitchel S. Berger| Professor and Chairman| Department of Neurological Surgery| University of California, San Francisco| San Francisco| California
Kathleen M. Plant Distinguished Professor
Director| Brain Tumor Surgery Program
Director| Neurosurgical Research Centers| Brain Tumor Research Center
Board Member| Society for Brain Mapping and Therapeutics (SBMT)| West Hollywood| California

Ellen Bible| Department of Neuroscience| Institute of Psychiatry| King's College London| London| United Kingdom

Zev A. Binder| Department of Neurosurgery| The Johns Hopkins Hospital| Baltimore| Maryland

Thomas Birngruber| Senior Investigator Cerebral PK/PD| Health Institute of Biomedicine and Health Sciences| Joanneum Research| Graz| Austria

Keith L. Black| Chairman and Professor| Department of Neurosurgery| Cedars-Sinai Medical Center| Los Angeles| California
Director| Maxine Dunitz Neurosurgical Institute
Director| Johnnie L. Cochran, Jr. Brain Tumor Center
Ruth and Lawrence Harvey Chair| Neuroscience| Cedars-Sinai Medical Center| Los Angeles| California
Editorial Board Member| Society for Brain Mapping and Therapeutics (SBMT)| West Hollywood| California

Miriam S. Boer| Department of Biochemistry| University of Maryland| College Park| Maryland

Roscoe O. Brady| Developmental and Metabolic Neurology Branch| National Institute of Neurological Disorders and Stroke| National Institutes of Health| Bethesda| Maryland

Paul R. Carney| B.J. and Eve Wilder University Professor
Professor of Neurology, Pediatrics, Neuroscience, and Biomedical Engineering| Departments of Pediatrics, Neurology, and Neuroscience| J. Crayton Pruitt Family Department of Biomedical Engineering| McKnight Brain Institute| University of Florida| Gainesville| Florida

David Chau| Division of Advanced Drug Delivery| Centre for Biomolecular Sciences| School of Pharmacy| University of Nottingham| Nottingham| United Kingdom

Joseph T. Cheatle| Division of Neurosurgery| University of Nebraska Medical Center| Nebraska Medical Center| Omaha| Nebraska

Mike Y. Chen| Division of Neurosurgery| City of Hope National Medical Center| Duarte| California

Xiaoyuan Chen| Chief and Senior Investigator| Laboratory of Molecular Imaging and Nanomedicine| National Institute of Biomedical Imaging and Bioengineering| National Institutes of Health| Bethesda| Maryland

Robin P. Choudhury| Professor of Cardiovascular Medicine| Wellcome Trust Senior Research Fellow in Clinical Science| Clinical Director Oxford Acute Vascular Imaging Centre and Honorary Consultant Cardiologist| Oxford| United Kingdom

Esther Chung| Department of Neurosurgery| Cedars-Sinai Medical Center| Los Angeles| California

Heather A. Clark| Associate Professor| Bioengineering Program and Department of Pharmaceutical Sciences| Northeastern University| Boston| Massachusetts

Martin O. Culjat| Adjunct Assistant Professor| Departments of Bioengineering and Surgery| Center for Advanced Surgical and Interventional Technology (CASIT)| University of California| Los Angeles| Los Angeles| California

Antonio De Salles| Department of Neurosurgery| David Geffen School of Medicine| University of California| Los Angeles| Los Angeles| California

J. Matthew Dubach| Graduate Student| Bioengineering Program and Department of Pharmaceutical Sciences| Northeastern University| Boston| Massachusetts

Rutledge Ellis-Behnke| Medical Faculty Mannheim of the University of Heidelberg| Department of Opthalmology| Mannheim| Germany
Massachusetts Institute of Technology| Department of Brain and Cognitive Sciences| Cambridge| Massachusetts

Richard E. Fan| Postdoctoral Researcher| Department of Surgery| Center for Advanced Surgical and Interventional Technology (CASIT)| University of California| Los Angeles| Los Angeles| California

Massimo S. Fiandaca| Department of Neurosurgery| Molecular Therapeutics Laboratory| University of California| San Francisco| San Francisco| California

Christy Franco| Department of Bioengineering| Rice University| Houston| Texas

Joseph A. Frank| Senior Investigator–Chief| Frank Laboratory| National Institutes of Health| Radiology and Imaging Sciences Department| Bethesda| Maryland
Director| ISTP| National Institutes of Health| National Institute of Biomedical Imaging and Bioengineering| Bethesda| Maryland

Eleonore Fröhlich| Head Core Facility Microscopy| Center for Medical Research| Medical University of Graz| Graz| Austria

Tatsuhiro Fujii| Division of Neurosurgery| City of Hope National Medical Center| Duarte| California

Greg A. Gerhardt| RCTF Professor| Department of Anatomy and Neurobiology| Center for Microelectrode Technology| Morris K. Udall Parkinson's Disease Research Center of Excellence| University of Kentucky| Lexington| Kentucky
Director| Morris K. Udall Parkinson's Disease Research Center of Excellence| University of Kentucky| Lexington| Kentucky

Mansur Ghani| Department of Neurosurgery| Cedars-Sinai Medical Center| Los Angeles| California

Natalia Gorenkova| Department of Neuroscience Institute of Psychiatry| King's College London| London| United Kingdom

Alessandra Gorgulho| Department of Neurosurgery| David Geffen School of Medicine| University of California| Los Angeles| Los Angeles| California

J. Peter Gruen| Department of Neurological Surgery| University of Southern California| Los Angeles| California

Warren S. Grundfest| Professor| Departments of Bioengineering, Electrical Engineering, and Surgery| Center for Advanced Surgical and Interventional Technology| University of California| Los Angeles| Los Angeles| California

Zachary D. Guss| Medical Student| Johns Hopkins University School of Medicine| Baltimore| Maryland

Christopher J. Hanson| Department of Neurosurgery| Cedars-Sinai Medical Center| Los Angeles| California

J. Todd Hastings| Gill Associate Professor| Department of Electrical and Computer Engineering| Center for Nanoscale Engineering| University of Kentucky| Lexington| Kentucky

Torsten Hegmann| Associate Professor| Chemical Physics Interdisciplinary Program| Liquid Crystal Institute| Kent State University| Kent| Ohio

John D. Heiss| Chair| Surgical Neurology Branch| National Institute of Neurological Disorders and Stroke| National Institutes of Health| Bethesda| Maryland

Tuan Hoang| Lecturer and Project Administrator| Department of Biomedical Engineering| Viterbi School of Engineering| University of Southern California| Los Angeles| California

Michael E. Hoenk| Senior Technical Staff| Analytical Instruments Group| Jet Propulsion Laboratory| California Institute of Technology| Pasadena| California

Evangelia Hondroulis| Nanobioengineering/Bioelectronics Laboratory| Department of Biomedical Engineering| Florida International University| Miami| Florida

Li-Chen Hsu| Division of Neurosurgery| City of Hope National Medical Center| Duarte| California

Peter Huettl| Scientist 3| Department of Anatomy and Neurobiology| Center for Microelectrode Technology| Morris K. Udall Parkinson's Disease Research Center of Excellence| University of Kentucky| Lexington| Kentucky

Dwain K. Irvin| Department of Neurosurgery| Cedars-Sinai Medical Center| Los Angeles| California

Steven Jacobson| Viral Immunology Section| Neuroimmunology Branch| National Institute of Neurological Disorders and Stroke| National Institutes of Health| Bethesda| Maryland

Rahul Jandial| Division of Neurosurgery| City of Hope National Medical Center| Duarte| California

Babak Kateb| Founding Chairman of the Board of Directors| CEO and Scientific Director| Society for Brain Mapping and Therapeutics (SBMT)| West Hollywood| California
Director| National Center for Nano-Bio-Electronics (NCNBE)| NASA Research Park| Moffet Field| California
Research Scientist| Maxine Dunitz Neurosurgical Institute| Department of Neurosurgery| Cedars-Sinai Medical Center| Los Angeles| California
Senior Editor| SBMT–PLoSOne Neuromapping and Therapeutics| West Hollywood| California
President and Scientific Director| Brain Mapping Foundation| West Hollywood| California

Robert F. Keating| Department of Neurosurgery| Children's National Medical Center| Washington| District of Columbia

Ayaz M. Khawaja| Surgical Neurology Branch| National Institute of Neurological Disorders and Stroke| National Institutes of Health| Bethesda| Maryland

Michel Kliot| Professor of Clinical Neurosurgery| Director for Center for Management and Surgery of Peripheral Nerve Disorders| University of California| San Francisco| California

Michelle Knosp| Western University of Health Sciences| Pomona| California

Bong-Seop Lee| Department of Neurosurgery| Cedars-Sinai Medical Center| Los Angeles| California

Seulki Lee| Group Leader| Laboratory of Molecular Imaging and Nanomedicine| National Institute of Biomedical Imaging and Bioengineering| National Institutes of Health| Bethesda| Maryland

Emily C. Leibovitch| Viral Immunology Section| Neuroimmunology Branch| National Institute of Neurological Disorders and Stroke| National Institutes of Health| Bethesda| Maryland

Lucien M. Levy| Professor| Department of Neuroradiology| George Washington University Hospital| Washington| District of Columbia
Senior Editor| American Journal of Neuroradiology| George Washington University Medical Center| Washington| District of Columbia
Board Member| Society for Brain Mapping and Therapeutics (SBMT)| West Hollywood| California

Chen-Zhong Li| Principal Investigator| Nanobioengineering/Bioelectronics Laboratory| Department of Biomedical Engineering| Florida International University| Miami| Florida

Michael Lim| Assistant Professor of Neurosurgery, Oncology and Institute of NanoBiotechnology| The Johns Hopkins Hospital| Baltimore| Maryland
Director of the Brain Tumor Immunotherapy Program| The Johns Hopkins Hospital| Baltimore| Maryland
Director of the Metastatic Brain Tumor Center| The Johns Hopkins Hospital| Baltimore| Maryland

Geoffrey S. F. Ling| Professor of Neurology| Uniformed Services University of the Health Science| Bethesda| Maryland
Deputy Director| Defense Sciences Office Defense Advanced Research| Projects Agency| Arlington| Virginia

Karlo J. Lizarraga| Department of Neurosurgery| David Geffen School of Medicine| University of California| Los Angeles| Los Angeles| California

Florian Mansfeld| Professor| The Mork Family Department of Chemical Engineering and Materials Science| University of Southern California| Los Angeles| California

Martina A. McAteer| Senior Research Scientist| Department of Cardiovascular Medicine| University of Oxford| Oxford| United Kingdom

William P. Melega| Professor| Department of Molecular and Medical Pharmacology| David Geffen School of Medicine| University of California| Los Angeles| Los Angeles| California

Ellis Meng| Associate Professor| Department of Biomedical Engineering| Viterbi School of Engineering| University of Southern California| Los Angeles| California

Donald Miller| Associate Professor| Department of Pharmacology and Therapeutics| Faculty of Medicine| University of Manitoba| Winnipeg| Canada

Robbin Miranda| Research Scientist/Engineer| Schafer Corporation| Technology Management Division| Arlington| Virginia

David Mishelevich| Vice President| Engineering and Chief Medical Officer| Aubrey Group, Inc.| Irvine| California

Michel Modo| Department of Neuroscience Institute of Psychiatry| King's College London| London| United Kingdom

Bryn S. Moore| Weis Center for Research| Geisinger Health Systems| Danville| Pennsylvania

David F. Moore| Professor| Department of Neurology| Tulane University| New Orleans| Louisiana

Amit P. Mulgaonkar| Graduate Student Researcher| Biomedical Engineering Interdepartmental Program| Center for Advanced Surgical and Interventional Technology| University of California| Los Angeles| Los Angeles| California

Jairo Nelson| Nanobioengineering/Bioelectronics Laboratory| Department of Biomedical Engineering| Florida International University| Miami| Florida

Francesca Nicholls| Department of Neuroscience Institute of Psychiatry| King's College London| London| United Kingdom

Shouleh Nikzad| Society for Brain Mapping and Therapeutics (SBMT)| West Hollywood| California
Brain Mapping Foundation| West Hollywood| California
National Center for Nano-Bio-Electronics (NCNBE)| NASA Research Park| Moffett Field| California
National Aeronautics and Space Administration (NASA)| Jet Propulsion Laboratory| Pasadena| California

Gang Niu| Staff Scientist| Laboratory of Molecular Imaging and Nanomedicine| National Institute of Biomedical Imaging and Bioengineering| National Institutes of Health| Bethesda| Maryland

Alessandro Olivi| Department of Neurosurgery| The Johns Hopkins Hospital| Baltimore| Maryland

Grant Palmer| Vice President| Quality and Regulatory| Aubrey Group, Inc.| Irvine| California

Chirag G. Patil| Director| Center for Neurosurgical Outcomes Research| Associate Program Director| Neurosurgery Residency Program| Assistant Professor of Neurosurgery| Cedars-Sinai Medical Center| Los Angeles| California

Marc Pelletier| Adjunct Professor| Department of Physiology and Biophysics| Case Western Reserve University| Cleveland| Ohio
Chief Scientific Officer| Aeromics| LLC| Cleveland| Ohio

Artin Petrossians| Research Fellow| The Mork Family Department of Chemical Engineering and Materials Science| University of Southern California| Los Angeles| California

Patrick Pezeshkian| Department of Neurosurgery| David Geffen School of Medicine| University of California| Los Angeles| Los Angeles| California

Martin Pham| Department of Neurosurgery| Keck School of Medicine| University of Southern California| Los Angeles| California

Thomas R. Pieber| Professor of Medicine and Head of Division| Division of Endocrinology and Metabolism| Medical University of Graz| Graz| Austria
Head| Health Institute of Biomedicine and Health Sciences| Joanneum Research| Graz| Austria

Jay J. Pillai| Director of Functional MRI| Associate Professor| Neuroradiology Division| The Russell H. Morgan Department of Radiology and Radiological Science| Johns Hopkins University School of Medicine & The Johns Hopkins Hospital| Baltimore| Maryland

Francois Pomerleau| Scientist| Department of Anatomy and Neurobiology| Center for Microelectrode Technology| Morris K. Udall Parkinson's Disease Research Center of Excellence| University of Kentucky| Lexington| Kentucky

Gill Pratt| Defense Sciences Office| Defense Advanced Research Projects Agency| Arlington| Virginia

David A. Price| Postdoctoral Fellow| Department of Anatomy and Neurobiology| Center for Microelectrode Technology| Morris K. Udall Parkinson's Disease Research Center of Excellence| University of Kentucky| Lexington| Kentucky

Qimeng Quan| Visiting Fellows| Laboratory of Molecular Imaging and Nanomedicine| National Institute of Biomedical Imaging and Bioengineering| National Institutes of Health| Bethesda| Maryland

Ali C. Ravanpay| Resident| Department of Neurological Surgery| University of Washington| Seattle| Washington

Ali R. Rezai| Department of Neurosurgery| Center for Neuromodulation| Ohio State University| Columbus| Ohio

Allyson C. Rosen| Director of Dementia Education| VA Medical Center| Palo Alto| California
Clinical Associate Professor| Stanford University School of Medicine| Stanford| California

Michael Roy| Professor and Director of International Medicine Department of Uniformed Services University of the Health Sciences| Bethesda| Maryland
Director of Recruitment at CDRM| Maryland

Raphael Schiffmann| Medical Director| Institute of Metabolic Disease| Baylor University Medical Center at Dallas| Dallas| Texas

Cody D. Schlaff| Radiation Oncology Branch| National Institute of Cancer| National Institutes of Health| Bethesda| Maryland

Kevin Shakesheff| Division of Advanced Drug Delivery| Centre for Biomolecular Sciences| School of Pharmacy| University of Nottingham| Nottingham| United Kingdom

Kuldip Sidhu| Stem Cells| UNSW Medicine| School of Psychiatry| University of New South Wales| Sydney| Australia
President| Society for Brain Mapping and Therapeutics (SBMT)| West Hollywood| California

Rahul S. Singh| Adjunct Assistant Professor| Departments of Bioengineering and Surgery| Center for Advanced Surgical and Interventional Technology| University of California| Los Angeles| Los Angeles| California

Uttam Sinha| Associate Professor| Keck School of Medicine| University of Southern California| Los Angeles| California

Frank Sinner| Deputy Head| Health Institute of Biomedicine and Health Sciences| Joanneum Research| Graz| Austria
Division of Endocrinology and Metabolism| Medical University of Graz| Graz| Austria

Gary K. Steinberg| Bernard and Ronni Lacroute-William Randolph Hearst Professor of Neurosurgery and the Neurosciences
Director| Stanford Institute for Neuro-Innovation and Translational Neurosciences| Stanford| California
Chairman| Department of Neurosurgery| Stanford University School of Medicine| Stanford| California
Board Member| Society for Brain Mapping and Therapeutics (SBMT)| West Hollywood| California

Giancarlo Suffredini| Surgical Neurology Branch| National Institute of Neurological Disorders and Stroke| National Institutes of Health| Bethesda| Maryland

Joel C. Sunshine| Department of Biomedical Engineering| Johns Hopkins School of Medicine| Baltimore| Maryland

Alexander Taghva| Department of Neurosurgery Keck School of Medicine University of Southern California| Los Angeles| California
Department of Neurosurgery| Center for Neuromodulation| Ohio State University| Columbus| Ohio

Jian Tajbakhsh| Program Leader Translational Cytomics| Department of Surgery| Division Molecular Oncology and Imaging| Cedars-Sinai Medical Center| Los Angeles| California
Assistant Professor Department of Surgery| Division Molecular Oncology and Imaging| Cedars-Sinai Medical Center| Los Angeles| California

Pooja M. Talauliker| Postdoctoral Fellow| Department of Anatomy and Neurobiology| Center for Microelectrode Technology| Morris K. Udall Parkinson's Disease Research Center of Excellence| University of Kentucky| Lexington| Kentucky

Peyman Tashkandi| Western University of Health Sciences| Pomona| California

Charlie Teo| Conjoint Associate Professor| University of New South Wales| Sydney| Australia
Director| Centre for Minimally Invasive Neurosurgery| POW Private Hospital| Randwick| Australia
Board Member| Society for Brain Mapping and Therapeutics (SBMT)| West Hollywood| California

Teodoro Tigno| National Capital Neurosurgery Consortium at Walter Reed Military Medical Center| Bethesda| Maryland

Christos Tolias| Department of Neurosurgery| King's College Hospital| King's College London| London| United Kingdom

Steven A. Toms| Department of Neurosurgery| Geisinger Health Systems| Danville| Pennsylvania

Aimee Two| Department of Neurological Surgery| University of Southern California| Los Angeles| California

James Weiland| Associate Professor| Department of Ophthalmology and Biomedical Engineering| University of Southern California| Los Angeles| California

Jennifer West| Department of Bioengineering| Rice University| Houston| Texas

Robert T. Wicks| Department of Neurosurgery| The Johns Hopkins Hospital| Baltimore| Maryland

Jun Wu| Division of Comparative Medicine| Beckman Research Institute of the City of Hope| Duarte| California

Vicky Yamamoto| Research Fellow| Department of Otolaryngology/Head and Neck Surgery| Keck School of Medicine of University of Southern California
Board Member| Society for Brain Mapping and Therapeutics (SBMT)| West Hollywood| California

Vinith Yathindranath| University of Manitoba| Winnipeg| Canada

Allen Yen| Department of Neurosurgery| Cedars-Sinai Medical Center| Los Angeles| California

John S. Yu| Department of Neurosurgery| Cedars-Sinai Medical Center| Los Angeles| California

Gabriel Zada| Department of Neurological Surgery| University of Southern California| Los Angeles| California

Zion Zibly| Surgical Neurology Branch| National Institute of Neurological Disorders and Stroke| National Institutes of Health| Bethesda| Maryland

1 From Nanotechnology to Nanoneuroscience/ Nanoneurosurgery and Nanobioelectronics

A Historical Review of Milestones

Vicky Yamamoto, Giancarlo Suffredini, Shouleh Nikzad, Michael E. Hoenk, Miriam S. Boer, Charlie Teo, John D. Heiss, and Babak Kateb

CONTENTS

INTRODUCTION

Nanoscale technology refers to structures having a dimension of 1–100 nm. Objects smaller than 100 nm possess unique chemical and physical features that change their optical, catalytic, electrochemical, and magnetic properties, melting point, reactivity, and conductivity. At this scale, these fundamental properties have a different relationship to size, shape, and composition than is present with larger objects. The threshold where properties change from bulk- to size-dependent is distinctive for each material and property. Novel tools and substances are being developed based on the special properties of nanoscale objects that potentially could provide new solutions to current technological and health-care dilemmas (Table 1.1; Figure 1.1).

Nanomedicine refers to the diagnostic and therapeutic application of nanotechnology to medicine (Table 1.2). Diagnostic applications of nanotechnology currently in development include sensors that are millions of times more sensitive than conventional macroscale sensors and that have the ability to screen for hundreds of diseases simultaneously. Applications of nanotechnology to medicine show promise for detecting novel biomarkers that may aid in diagnosing diseases such as Alzheimer's and certain types of cancer that are undetectable by current methodologies. Nanoscale coating or etching of a surface can alter its topography and enhance interaction and integration with surrounding living tissue. Stem cell differentiation can be particularly sensitive to nanoscale topography. Scaffold topography altered by nanoetching has shown promise in inducing neuronal growth from stem cells (Migliorini et al. 2011). The most obvious application of nanotechnology to medicine is in the design of nanoparticles that carry drugs or imaging markers to disease sites (Hu et al. 2007). Nanoparticle drug carriers can be engineered with targeting ligands that recognize specific cell receptors, increasing target specificity and decreasing systemic side effects. Reformulations of approved drugs in nanoparticle vehicles, such as paclitaxel in protein-bound nanoparticles (Abraxane, Abraxis BioScience, LLC, Bridgewater, NJ), increased therapeutic effect, decreased side effects, and improved patient outcomes compared to unbound drugs (Robidoux et al. 2010).

This chapter is a synopsis of the history of nanotechnology and the emergence of nanomedicine and nanoneurosurgery. The entire textbook will provide a review of the state-of-the-art nanoplatforms and their applications in neuroscience, neurology, and neurosurgery. The volume is also focused on highlighting the strengths and shortcomings of the field and tries to address some of the weaknesses by introducing solutions and novel approaches.

TABLE 1.1
Nanotechnology Timeline

19th century	Ceramic filters used to separate viruses from cells (nanoscale porosity).
1905	Estimation of sugar molecule diameter (1 nm) by Einstein.
1931	Discovery of electron microscope by Max Knoll and Ernst Ruska.
1936	Erwin Müller invented the field emission microscope, allowing near-atomic-resolution images of materials (Nano.gov 2011).
1947	John Bardeen, William Shockley, and Walter Brattain at Bell Labs discovered the semiconductor transistor (Nano.gov 2011).
1940s	Patch clamp technique uses nanoscale capillary probes to examine individual cell surface receptors (Tibbals 2011).
	Electron microscope and x-ray crystallography enabled observation of nanoscale structures (Tibbals 2011).
	Emission microscopy enabled the observation of individual atoms on surfaces (not suitable for macromolecules) (Tibbals 2011).
1950	Victor La Mer and Robert Dinegar developed the theory and a process for growing monodisperse colloidal materials. Controlled ability to fabricate colloids enables myriad industrial uses such as specialized papers, paints, and thin films, even dialysis treatments (Nano.gov 2011).
1951	Erwin Müller pioneered the field ion microscope, a means to image the arrangement of atoms at the surface of a sharp metal tip; he first imaged tungsten atoms (Nano.gov 2011).
1956	Arthur von Hippel at MIT introduced many concepts of, and coined the term, "molecular engineering" as applied to dielectrics, ferroelectrics, and piezoelectrics (Nano.gov 2011).
1958	Jack Kilby of Texas Instruments originated the concept of, designed, and built the first integrated circuit, for which he received the Nobel Prize in 2000 (Nano.gov 2011).
1959	Richard Feynman of the California Institute of Technology gave what is considered to be the first lecture on technology and engineering at the atomic scale, "There's Plenty of Room at the Bottom," at an American Physical Society meeting at Caltech (Nano.gov 2011).
1965	Intel cofounder Gordon Moore described the trend now known as "Moore's Law." This described the density of transistors on an integrated chip (IC) doubling every 12 months (later amended to every 2 years) (Nano.gov 2011).
1974	It is generally acknowledged that the term "nanotechnology" was coined by Professor N. Taniguchi of the Tokyo Science University in a paper, "On the Basic Concept of 'Nanotechnology'" in 1974, which stated that nanotechnology mainly consists of "the processing of separation, consolidation, and deformation of materials by one atom or one molecule" (Nano.gov 2011; Tibbals 2011).
	Aviram and Pattner started to develop molecular electronics.
1976	Erwin Neher and Bert Sakmann at the Max Planck Institute in Goettingen devised a patch clamp technique able to measure the currents flowing through single channels.
1979	Donald Tomalia is credited with creating dendrimers (Donald 2013).
	Butten and Hopkins used colloidal gold nanoparticles as a form of electron dense probes in immunohistochemistry and electron microscopy.
1981	Scanning tunneling microscopy (STM) invented by Gerd Karl Binnig, Heinrich Rohrer, and James Gimzewski, who pioneered research on electrical contacts with single atoms and molecules and light emission using STM. Binnig and Rohrer won the Nobel Prize for this discovery in 1986 (Nano.gov 2011).
	Drexler conceptualized the design molecular machines.
1985	Rice University researchers Harold Kroto, Sean O'Brien, Robert Curl, and Richard Smalley discovered the C60 allotrope of carbon, named the "buckminsterfullerene." The team was awarded the 1996 Nobel Prize in Chemistry for their roles in this discovery and that of the fullerene class of molecules more generally (Nano.gov 2011).
	Discovery of bulky balls by Robert Curl, Richard Smalley, and Harold Kroto, which culminated in 1996 Nobel Prize.
	Bell Labs' Louis Brus discovered colloidal semiconductor nanocrystals (quantum dots), for which he shared the 2008 Kavli Prize in Nanotechnology (Nano.gov 2011).

(*continued*)

TABLE 1.1 (Continued)

Nanotechnology Timeline

1986	Calvin Quate and Christoph Gerber at the IBM research laboratory in Zurich collaborated with Gerd Karl Binnig to invent atomic force microscopy (AFM), which provided the ability to measure and manipulate nanoscale structures, down to atomic scales (Nano.gov 2011).
1987	Douglas et al. demonstrated cancer targeting with nanoparticles coated with monoclonal antibody.
	Dexler published his book on nanotechnology: *Engines of Creation.*
1988	Construction of artificial smart molecules that could interact with each other by Lehn (Nobel Prize).
1989	IBM researchers Don Eigler and Erhard Schweizer spell out the company logo using 35 Xenon atoms, demonstrating how nanoparticles can be manipulated (Nano.gov 2011).
	Scanning probe microscope (SPM) enabled observation and manipulation of individual atoms on solid surfaces.
	Early nanotechnology companies began to operate, e.g., Nanophase Technologies in 1989, Helix Energy Solutions Group in 1990, Zyvex in 1997, Nano-Tex in 1998 (Nano.gov 2011).
1990	Atoms visualized by Eigler and Schweizer (Nobel Prize).
1991	Sumio Iijima of Nippon Electric Company (NEC) is credited with discovering the carbon nanotube. Iijima shared the Kavli Prize in Nanoscience in 2008 for this and other advances in the field (Nano.gov 2011).
1993	Moungi Bawendi of MIT invented a method for controlled synthesis of nanocrystals (quantum dots), paving the way for applications ranging from computing to biology to high-efficiency photovoltaics and lighting (Nano.gov 2011).
1994	Nanoparticle-based drug delivery by Kreuter.
1995	FDA approved Doxil and Abelcet.
1997	First nanotechnology company, Zybex, established.
1998	The Interagency Working Group on Nanotechnology (IWGN) was formed under the National Science and Technology Council to investigate the state of the art in nanoscale science and technology and to forecast possible future developments. The IWGN's study and report, "Nanotechnology Research Directions: Vision for the Next Decade" (1999) defined the vision for and led directly to formation of the US National Nanotechnology Initiative in 2000 (Nano.gov 2011).
	Bruchez et al. demonstrated the first use of nanocrystals as biological labels.
	Trung-Lee et al. used DNA-gelatin nanospheres for controlled gene delivery.
	Freitas used the term "nanomedicine" in a publication.
1999	Cornell University researchers Wilson Ho and Hyojune Lee probed secrets of chemical bonding by assembling a molecule [iron carbonyl $Fe(CO)_2$] from constituent components [iron (Fe) and carbon monoxide (CO)] with a scanning tunneling microscope.
1999	Chad Mirkin at Northwestern University invented dip-pen nanolithography (DPN), leading to manufacturable, reproducible "writing" of electronic circuits as well as patterning biomaterials for cell biology research, nanoencryption, and other applications (Nano.gov 2011).
1999–early 2000s	Consumer products making use of nanotechnology began appearing in the marketplace (Nano.gov 2011).
2001	National Nanotechnology Initiative.
2003	Naomi Halas, Jennifer West, Rebekah Drezek, and Renata Pasqualin at Rice University developed gold nanoshells, which, when "tuned" in size to absorb near-infrared light, serve as a platform for the integrated discovery, diagnosis, and treatment of breast cancer without invasive biopsies, surgery, or systemically destructive radiation or chemotherapy (Nano.gov 2011).
2005	Erik Winfree and Paul Rothemund from the California Institute of Technology developed theories for *DNA-based computation* and "algorithmic self-assembly" in which computations are embedded in the process of nanocrystal growth (Nano.gov 2011).
2010	IBM used a silicon tip measuring only a few nanometers at its apex (similar to the tips used in atomic force microscopes) to chisel away material from a substrate to create a complete nanoscale 3-D relief map of the world one-thousandth the size of a grain of salt—in 2 min, 23 s. This activity demonstrated a powerful patterning methodology for generating *nanoscale patterns and structures as small as 15 nm* at greatly reduced cost and complexity, opening up new prospects for fields such as electronics, optoelectronics, and medicine (Nano.gov 2011).

Source: Jain, K.K., *The Handbook of Nanomedicine*, 1–6, 2008.

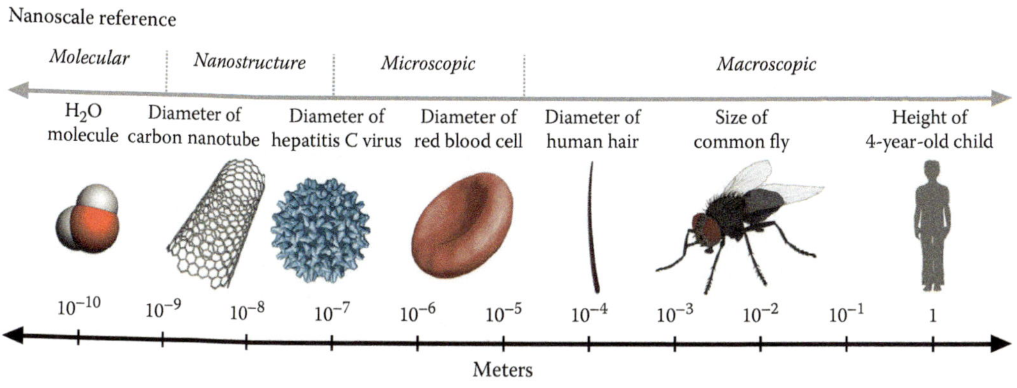

FIGURE 1.1 Size and scale of nanostructures relative to commonly known objects. (Reprinted from Kateb, B. et al., *Neuroimage*, 54, Suppl 1, 2011. With permission.)

WHAT IS NANONEUROSURGERY AND NANONEUROSCIENCE?

Nanoneurosurgery is defined as subcellular, nanoscale manipulation of the nervous system, which consists of the central nervous system (CNS) and the peripheral nervous system (PNS), using nanotechnology combined with molecular and cellular biology. Nanoneuroscience is defined as application of nanotechnology in neuroscience. The purposes of using nanotechnology in neurosurgery are to deliver nucleic acids (such as DNA, RNA, siRNA, and shRNA), pharmaceuticals, labeling markers, and other macromolecules to targeted nervous tissues/cells more precisely or to create a better interface between a device and the neurosurgical patient. Thus, nanoneurosurgery has its own particular scope, application, and focus in nanomedicine. In other words, nanomedicine refers to a much broader application of nanotechnology in medicine, while nanoneurosurgery specifically focuses on targeted and precision guided therapeutics in the nervous system, and nanoneuroscience refers to study of the subcellular and nanoscale levels in the nervous system. The nanoneuroscience and nanoneurosurgery fields in the last 10 years alone have emerged as new frontiers in science, leading to astonishing advancements in biomedical science and clinical practice and the emerging field of next-generation brain mapping. Brain mapping is defined as "The study of the anatomy and function of the brain and spinal cord through the use of imaging (including intra-operative, microscopic, endoscopic and multi-modality imaging), immunohistochemistry, molecular & optogenetics, stem cell and cellular biology, engineering (material, electrical and biomedical), neurophysiology and nanotechnology."

NANOMATERIALS IN ANCIENT TIMES

Nanotechnology emerged as a field of study in the 20th century, but the Greek philosophers Leucippus and Democritus actually proposed 25 centuries earlier that all matter was made from tiny particles, such as atoms (Vo-Dinh 2007). Ancient artisans and sword makers unknowingly utilized nanotechnology. One example is the Lycurgus Cup, created by medieval glassmakers during the Roman Period (30 BC–640 AD), and presently viewable at the British Museum in London (Cook 2005). The color of this cup is blue-green when it is externally illuminated but changes to red when internally illuminated. Modern evaluation of the cup using transmission electron microscopy revealed the presence of silver and gold nanoparticles similar in size to the wavelengths of visible light. Internal illumination of the cup with visible light (390–750 nm wavelength) results in the cup appearing red because colloidal silver and gold nanoparticles in the glass reflect the blue-green light (500 nm wavelength) but do not prevent red light (700 nm wavelength) from passing through the glass. External illumination of the cup with visible light results in the cup appearing blue-green because red light is again transmitted through the glass, while blue-green light (500 nm wavelength) is reflected back to the observer (Aronson 2001). Other optical properties of gold nanoparticles that allow them to scatter visible light (Mie scattering)

TABLE 1.2
Nanomedicine Timeline

Late 1960s	Peter Paul Speiser developed a nanoparticle for sustained delivery of vaccines (Kreuter 2007).
Late 1970s	Richard Oppenheim and Jennifer Marty created a controlled method to produce gelatin and albumin nanoparticles (Kreuter 2007).
1979	Patrick Couvreur is credited with developing the first rapidly biodegradable acrylic nanoparticles (Kreuter 2007).
1983	Illum and Davis found that they could prolong the circulation time and decrease RES recognition of intravenously delivered nanoparticles by coating the nanoparticles with polyoxyethylene and polyoxypropylene (Illum and Davis 1983).
1986	Yasuhiro Matsumura and Hiroshi Maeda of the Kumamoto University Medical School in Japan first described the enhanced permeability and retention effect (EPR) (Matsumura and Maeda 1986).
1994	Ken Audus showed that surfactant-coated nanoparticles coated with polysorbate-80 were taken up by brain capillary endothelial cells in vitro (Borchard et al. 1994).
	Nanoparticle-based drug delivery by Kreuter.
1994/1995	The benefits of the covalent attachment of PEG chains discovered independently in 1994 by Gref and in 1995 by Bazile are noted (Bazile et al. 1995; Gref et al. 1994).
1995	FDA approved Doxil and Abelcet.
1997	First nanotechnology company, Zyvex, established.
1998	The potential in luminescent semiconductor nanocrystals (quantum dots) for medical diagnostics is recognized (Bruchez et al. 1998).
	Bruchez et al. demonstrated the first use of nanocrystals as biological labels.
	Truong-Lee et al. used DNA-gelatin nanospheres for controlled gene delivery.
	Freitas used the term "nanomedicine" in a publication.
2002	In vivo targeting of peptide-coated quantum dots to vasculature of normal lungs and tumors is accomplished (Akerman et al. 2002).
2004	First Annual World Congress for Brain Mapping was held at University of Southern California (USC) Keck School of Medicine.
2005	Society for Brain Mapping and Therapeutics (previously known as International Brain Mapping and Intraoperative Surgical Planning Society-IBMISPS) and its foundation was established by Babak Kateb.
2010	Combination of active and passive targeting increased cellular uptake in tumors.
	Nanotopography induced stem cell differentiation of neuronal precursors (Migliorini et al. 2011).
	Congressional Neuroscience was formed by Congressman Earl Blumenauer, which is now chaired by him and co-chaired by Congresswoman McMoriss Rodgers.
2011	Landmark review position policy paper by Kateb et al.; nanoplatforms for constructing new approaches to cancer treatment, imaging, and drug delivery: what should be the policy?
	Brain Mapping Foundation launched the National Center for Nanobioelectronics (NCNBE).
	NCNBE formed global alliance for nanobioelectronics in partnership with US, Canadian, and Australian scientists.
2012	Congressman Fattah introduced legislation urging all federal agencies to coordinate their efforts in the area of neuroscience.
	Congressman Jim Moran, Brain Mapping Foundation, and Society for Brain Mapping & Therapeutics introduced report language into the defense appropriation bill focusing on the integration of nanotechnology, devices, imaging, and cellular/stem cell therapy.
2013	President Obama introduced a new initiative on brain mapping (BRAIN Initiative: Brain Research through Advancing Innovative Neurotechnologies) authorizing $100M for the field, which has huge implications for the field of nanoneuroscience.
	The textbook of nanoneuroscience and nanoneurosurgery published by CRC/Taylor & Francis.

and absorb light energy and resonate (surface plasmon resonance [SPR]) are currently employed to improve optical imaging techniques (Cook 2005; Gartia et al. 2013). Damascus blades produced by sword smiths in the Middle East in the 5th century had exceptional strength and flexibility. Recent examination of a Damascus blade made in the 17th century suggests that nanoscale impurities in the steel were responsible for its extraordinary mechanical properties (Reibold et al. 2006; Sachse 2008).

EARLY PIONEERS IN NANOTECHNOLOGY AND NANOMEDICINE

Feynman's Lecture

The conception of nanotechnology can be traced back to a lecture delivered by Richard Feynman, the American theoretical physicist and Nobel laureate (in Physics, 1965). Feynman delivered his famous talk entitled, "There's Plenty of Room at the Bottom," to the American Physical Society at the California Institute of Technology on December 29, 1959, stating, "What I want to talk about is the problem of manipulating and controlling things on a small scale" (Feynman 1960). Feynman went on to describe a series of extravagant ideas, such as writing the entire Encyclopedia Britannica on the head of a pin, reducing words to a binary code written in atoms, making computer components with diameters of less than 100 atoms, modeling information systems after biological systems, manufacturing extremely small devices, and manipulating individual atoms. In this speech, he reiterated that he did not know how to do these procedures, but because these procedures do not violate the laws of physics, scientists should accept the challenges and figure them out. The common belief is that Feynman's paper represents the origin of contemporary nanotechnology (Toumey 2009). Feynman's original concept of nanomedicine revolved around nanorobots performing highly specific and noninvasive surgery. Feynman reported in 1960:

> "...A friend of mine (Albert R. Hibbs) suggests a very interesting possibility for relatively small machines. He says that, although it is a very wild idea, it would be interesting in surgery if you could swallow the surgeon. You put the mechanical surgeon inside the blood vessel and it goes into the heart and 'looks' around. (Of course the information has to be fed out.) It finds out which valve is the faulty one and takes a little knife and slices it out. Other small machines might be permanently incorporated in the body to assist some inadequately-functioning organ..." (Feynman 1960).

Naming of Nanotechnology

Although the origins of nanotechnology are generally attributed to Feynman, the term "nanotechnology" was coined by the Japanese physicist Norio Taniguchi of the Tokyo Science University. In a paper entitled, "On the Basic Concept of Nanotechnology," Taniguchi writes, "...nanotechnology mainly consists of the processing of separation, consolidation, and deformation of materials by one atom or one molecule" (Nano.gov 2011; Tibbals 2011). Taniguchi proposed in the paper that the precision of material-processing technology would be on the nanometer scale by the year 2000. His description of nanotechnology is still the basic definition of nanotechnology.

Molecular Nanotechnology

Eric Drexler of the Massachusetts Institute of Technology first described the concept of molecular nanotechnology. Drexler (1981) proposed building machines that used "molecular assemblers" to manipulate individual atoms, a pioneering concept of nanotechnology. In 1986, Drexler published these ideas in a book intended for the general public entitled, *Engines of Creation: The Coming Era of Nanotechnology* (Drexler 1986), which helped popularize the term "nanotechnology." In the book, which is considered to be the first book on nanotechnology, Drexler discussed the technological significance of nanoscale phenomena. Drexler received the first doctoral degree in nanotechnology in 1991 from MIT and in 1992 published the first textbook of nanotechnology entitled, *Nanosystems: Molecular Machinery, Manufacturing and Computation* (Drexler 1992), detailing the principles and mechanisms of molecular nanotechnology.

Although naturally occurring molecular assemblers exist within the cell, such as the ribosome that reads strands of mRNA and assembles amino acids into proteins, a synthetic molecular assembler has yet to be created. Ongoing research seeks to develop a molecular assembler that can synthesize unique molecules that may act as novel pharmaceuticals or be used to manufacture innovative materials (Freitas and Merkle 2004).

Nanomedicine

The first known use of the term "nanomedicine" was described in 1991 by Eric Drexler, Chris Peterson, and Gayle Pergamit in their book entitled, *Unbounding the Future* (Drexler et al. 1991). Similar to Feynman, these authors presented many ideas for diagnostic and therapeutic applications of nanotechnology, although they did not discuss any means of engineering or manufacturing that might bring these ideas to fruition. "Modern medicine" is portrayed by the authors as crude and limited by its understanding and tools; extending surgical control to the molecular level is a goal (Drexler et al. 1991). The first nanomedicine textbook was published by Robert A. Freitas, Jr. in 1999 (Freitas 1999).

The visionary leadership of Feynman, Drexler, Freitas and others set the field of nanomedicine on a course that would eventually lead to applying nanomechanical concepts to reverse disease processes and to perform precision-guided neurosurgery or nanoneurosurgery. Nanomedicine as a whole involves using nanoscale tools to diagnose and treat disease. Since the publication of *Unbounding the Future,* nanomedicine has advanced significantly. The field of nanotechnology has now found its way to neuroscience in general and neurosurgery in particular. This textbook is the first combined textbook of nanoneuroscience and nanoneurosurgery, and will break new ground for constructing more personalized and noninvasive future treatments of neurological disorders and for spurring new discoveries in brain function and mapping.

TECHNOLOGICAL ADVANCEMENTS CRUCIAL FOR NANOTECHNOLOGY DEVELOPMENT

Electron Microscope

Accelerated electrons used in electron microscopy have wavelengths that are 100,000 times shorter than visible light, giving the electron microscope much greater resolution than the light microscope. German scientists Max Knoll and his graduate student Ernst Ruska developed a prototype of the electron microscope in 1931 (Cosslett 1962). This prototype initially led to developing an electron microscope with a resolution greater than 400 times, making it more powerful than a light microscope. The prototype eventually inspired contemporary electron microscopes, such as the field-ion microscope in 1951 and the helium ion microscope in 2006, which can magnify 2 million times and marks an important innovation in studying and characterizing nanoscale structures (Economou et al. 2012).

Scanning Tunneling Microscopy and Atomic Force Microscopy

The scanning tunneling microscope (STM) was developed by Gerd Binnig and Heinrich Rohrer at the IBM Zurich Research Laboratory in 1981 (Binnig and Rohrer 1986). Inventing STM enabled imaging a single atom because STM imaging resolution was 0.1 nm laterally and 0.01 nm deep. Binnig and Rohrer later won the Nobel Prize for this invention in 1986 (Nano.gov 2011; Tibbals 2011). STM can manipulate single atoms and molecules on surfaces (Eigler and Schweizer 1990; Stroscio and Celotta 2004). For instance, Donald Eigler and Erhard Schweizer arranged individual xenon atoms using an STM to spell their company logo, IBM (Eigler and Schweizer 1990).

In 1986, Calvin Quate and Christoph Gerber at the IBM Research Laboratory in Zurich collaborated with Gerd Binnig to invent atomic force microscopy (AFM) (Binnig et al. 1986). AFM, a more advanced version of STM, improved the ability to measure and manipulate nanoscale structures at atomic scales (Nano.gov 2011; Tibbals 2011). Such technological advances not only provided ways to visualize nanoparticles and their components but also methods to better understand their interactions with biological tissue.

Colloid/Interface Science and Surface Engineering

Using nanotechnology in medicine requires understanding and controlling the behavior of nanoparticles/platforms and their interaction with biomacromolecules in the physiological environment.

This is crucial because this interaction controls the signaling, kinetics, and biodistribution induced by nanoparticles in the body (Mout et al. 2012; Walkey and Chan 2012). In addition, it is necessary to chemically or physically modify (a process referred to as "functionalization") the surface of nanoparticles/platforms, making them multifunctional, stable, less toxic, and targeted to a specific tissue or cell type (Subbiah et al. 2010). Increased understanding and utilization of colloid/interface science and surface engineering encouraged translating nanotechnological advances to medicine.

Role of Quantum Physics in Advancement of Nanotechnology

Many of the most innovative applications of nanotechnology arise from qualitative rather than quantitative properties of matter that emerge only as length scales are reduced to less than a few nanometers. The laws of quantum mechanics predict that electrons can pass through an energy barrier (tunneling) that would be energetically insurmountable according to the laws of classical physics. Tunneling was first encountered experimentally in the study of radioactive decay in the late 19th and early 20th centuries, long before it was understood theoretically on the basis of quantum mechanics in the late 1920s (Gamow 1928). The Japanese physicist Leona Esaki showed that nanoscale materials are controlled by quantum mechanics rather than classical physics. Esaki developed the first semiconductor device that was explicitly based on electron tunneling toward the end of the 1950s (Esaki 1958). Esaki discovered that a positive–negative (p-n) diode formed in highly doped germanium exhibits negative differential resistance, which he attributed to quantum mechanical tunneling of electrons across a potential barrier at the nanoscale junction. This invention marked the beginning of nanoscale engineering of semiconductors, which today encompasses a variety of optoelectronic devices, including semiconductor transistors, detectors, lasers, and thermovoltaic devices.

Moore's (1965) law provides a heuristic scaling of semiconductor circuits to ever smaller length scales and higher densities. Yet nothing can be scaled indefinitely, and Moore's law is rapidly approaching hard limits that concern the physical properties of nanoscale materials (as opposed to the technological limits of nanofabrication technologies). A prime example of this is current leakage in transistors; as oxide forming the insulating barrier in transistor gates is reduced to nanometer thickness, the insulating properties of the oxide begin to break down, and the gate oxide loses its ability to prevent electrons from "leaking" from the gate to the semiconductor. This limitation also reflects the quantum mechanical phenomenon of tunneling.

Whereas tunneling imposes limits on the dimensions of conventional field-effect transistors, quantum mechanics also offers many opportunities for discovering, innovating, and inventing nanoscale devices. For example, tunneling has led to innovative technologies for imaging and spectroscopy, including the above-mentioned STM (Binnig and Rohrer 1983). Tunneling is also operative in nanoelectromechanical sensing (NEMS) systems that are used as pressure sensors for "nearly noninvasive" monitoring. A clear example of the NEMS application was the nanoTrek device, which was developed by the Quantum-π company. This NEMS sensor was developed to act as an implantable medical pressure and flow meter, replacing infection-prone catheters. Further applications include pressure monitoring the intraocular space, cerebral ventricles, cerebral vascular system, and vertebral disc (Michalewicz et al. 2007).

GROUNDBREAKING EVENTS IN NANOTECHNOLOGY

New Carbon Allotropes

Nanotechnology started to expand after Rice University researchers Harry Kroto, Richard Smalley, and Robert Curl discovered the carbon structure, buckminsterfullerene C60, sometimes referred to as the buckyball, in 1985 (Kroto et al. 1985). This seminal discovery resulted in the Nobel Prize for chemistry being awarded to Kroto, Smalley, and Curl in 1996 (Nano.gov 2011). Due to their unique physical, biological, and chemical properties, fullerenes are being considered for medical applications, including in neuroscience since C60 passes the blood–brain barrier (BBB) (Bakry et al. 2007; Lin and Lu 2012; Tong et al. 2011).

Sumio Iijima of Nippon Electric Company (NEC, Limited) in Japan first reported the discovery of the multiwalled carbon nanotube (MWCNT) in 1991 (Iijima 1991). Two years later, the single-walled carbon nanotube (SWCNT) was reported independently by Iijima and Donald Bethune of IBM (Bethune et al. 1993; Iijima and Ichihashi 1993). The carbon nanotube (CNT) discovery further fueled an intense race to investigate and explore nanotechnologic applications in science and medicine. CNTs are now widely studied for their potential application in nanomedicine, including neuroscience and neurosurgery (Foldvari and Bagonluri 2008a,b; Nunes-Pereira et al. 2012; Ren et al. 2012).

Smalley continued his work with fullerenes, focusing on CNTs and applying their extraordinary properties (Adams and Baughman 2005). He and his team developed the laser ablation and high-pressure carbon monoxide processes for manufacturing single-walled CNTs. He is recognized as having contributed to worldwide scientific progress by providing other scientists with access to these high-quality nanotubes through both Rice University and his company, Carbon Nanotechnologies, Inc. (Adams and Baughman 2005).

Quantum Dots

Louis Brus is credited with discovering colloidal semiconductor nanocrystals (quantum dots [QDs]) while working at Bell Laboratories in 1985, for which he shared the 2008 Kavli Prize in Nanotechnology (Nano.gov 2011). In 1993, Moungi Bawendi of MIT invented a method for controlled synthesis of nanocrystals (QDs), paving the way for applications in computing, biology, high-efficiency photovoltaics, and lighting. Within the next several years, Brus and other researchers such as Chris Murray reported additional methods for synthesizing QDs (Nano.gov 2011). QDs are effective biological labels because they exhibit size-dependent fluorescence. For example, small interfering RNAs (siRNAs) were made into siRNA–QD complexes that visualizing and quantifying transfected cells, while polyethylene glycol (PEG)–coated QDs can be used to image sentinel lymph nodes near a tumor (Ballou et al. 2007; Kobayashi et al. 2007). When injected into a tumor, QDs rapidly move to the adjacent sentinel lymph nodes, effectively tracking the flow of lymph fluid from the tumor. There are benefits of QD fluorescence not found in any other labeling technique, such as excellent photostability and a wide range of fluorescent wavelengths. Ballou et al. (2007) described the possibility of using unlabeled QDs that fluoresce at one wavelength to nonspecifically label a tumor and its lymph nodes along with a second set of QD–ligand complexes that fluoresce at an alternate wavelength to specifically label the tumor or any of its metastases. Such a system could guide operating surgeons in minimally invasive procedures. QDs can gain biologic specificity by conjugating them to a targeting ligand such as an antibody, peptide, or oligonucleotide. In 2002, Dr. Maria Akerman of the Burnham Institute in California showed that peptide-coated QDs in vivo can selectively target the vasculature of a tumor and a normal lung in mice (Akerman et al. 2002). QDs are being explored as agents to improve noninvasive diagnostic imaging of cerebral aneurysm, cerebral vasospasm, arteriovenous malformation, and cerebral ischemic disease (Taghva et al. 2010a,b).

Single Electron Tunneling Transistor

The single electron tunneling transistor first achieved the controlled movement of individual electrons through a nanoscale device. Dimitri Averin and Konstantin Likharev are credited with proposing the idea in 1991, and Theodore Fulton and Gerald Dolan from Bell Laboratories are credited with building the device. Discontinuity in a wire or semiconductor prevents electrical conduction at the macroscale, but at the nanoscale, an electron can "tunnel" through a gap (tunnel barrier) of 1 nm between conductors and maintain conduction (Lu et al. 2003). Researchers are working to reduce voltage variation in these devices and to combine the components of a single electron tunneling transistor into a single molecule. Development of reliable nanoscale transistors would support development of smaller and more complex computing devices (Ubbelohde et al. 2012).

DNA and Gold Colloids to Assemble Inorganic Materials

In 1996, Chad Mirkin and Robert Letsinger of Northwestern University described a method for attaching strands of synthetic DNA onto gold nanoparticles to direct the formation of macromolecular gold nanoparticle colloids. This strategy opened the possibility of tailoring the optical, electronic, and structural properties of colloidal aggregates by using the specificity of DNA interactions to direct the interactions between particles of different size and composition.

The light-scattering properties and SPR of gold nanoparticles, as discussed in the context of the Lycurgus Cup, are the reasons they are suitable for imaging. Gold nanoparticles can be modified with antibodies for molecular specificity and do not suffer from photobleaching effects like fluorescent probes. Additionally, they demonstrate QD properties; gold nanoparticle aggregates will emit a different signal from dispersed nanoparticles. Given the aforementioned traits: ability to be modified with antibodies for molecular specificity, a lack of photobleaching, and their QD behavior, gold nanoparticles have the potential to be a powerful tool used to observe the location of certain molecules on the cell surface and to detect whether they are clustered or dispersed (Sperling et al. 2008). Gold nanoshells are optically tunable nanoparticles composed of a silicon dielectric core coated with a thin metallic layer of gold. These nanoshells can be manufactured by adjusting the relative thickness of the core and shell, so they have peak absorption of near-infrared light, which can penetrate into tissue. In 2003, a platform was developed at Rice University for photothermal tumor ablation using near-infrared–absorbing nanoparticles (O'Neal et al. 2004). Using a mouse model, O'Neal et al. (2004) showed that systemically administered gold nanoshells (130 nm in diameter) preferentially accumulated in tumor tissue via the enhanced permeability and retention effect (EPR); illuminating the gold nanoshells with a diode laser achieved complete tumor resorption within 10 days. Gold nanoparticles with tumor-specific binding molecules were developed to bind selectively to the tumor (Sperling et al. 2008). Similarly, gold nanoparticles can be attached to polymer capsules containing anticancer drugs, and with the same technique used above, the gold nanoparticles can generate heat. The heat then opens the polymer capsules and releases the desired drug within the target tissue, simultaneously reducing systemic exposure while achieving higher concentrations within the tumor itself (Sperling et al. 2008). Targeted delivery of gold nanoparticles to tumor tissue is essential because gold nanoparticles are cytotoxic. Some studies demonstrated that they interrupt actin filaments and form complexes with DNA (Sperling et al. 2008).

RECENT MEDICAL AND NEUROSCIENTIFIC ADVANCEMENTS USING NANOTECHNOLOGY

After the discovery of carbon nanoparticles and nanotubes, there was intense scientific interest in nanotechnology, since the technology has a wide variety of potential applications in material, electrical, and optical engineering. Soon thereafter, a search for potential applications in biomedical science began. Nanotechnology can be used for drug delivery, nucleic acid/macromolecule delivery, imaging, and as an interface to treat disorders and injuries, including in the CNS. Use of nanotechnology for neurosurgical applications is covered in this textbook.

Early Dawn of Nanobased Drug Delivery

Peter Paul Speiser at the ETH in Zurich attempted to miniaturize delivery systems for controlled drug release in the mid-1950s. His goal was to achieve chronic immune stimulation with a single injection of a sustained-release preparation (Kreuter 2007). Instead of following the prescribed practice of administering multiple injections to achieve an antibody response sufficient to defend against tetanus, diphtheria, and viral infections, Speiser sought to evoke an equivalent antibody response from a single injection (Kreuter 2007). By the late 1960s, he developed the first nanoparticle that could be used for drug delivery and vaccines (Kreuter 2007). In the late 1970s, Richard Oppenheim and Jennifer Marty could produce gelatin and albumin nanoparticles. Their method became standard in many laboratories and subsequently led to several important nanoparticle developments (Kreuter 2007).

In 2000, Jorg Kreuter modified his process to enable nanoparticles to transport oligonucleotides, peptides, nucleic acids, and genes across cell membranes (Kreuter 2007). Cell targeting using these nanoparticles was further improved by the covalent attachment of antibodies, allowing highly specific targeting (Wartlick et al. 2004), and covalent attachment of apolipoprotein E, facilitating transport across the BBB (Michaelis et al. 2006). Patrick Couvreur is credited with developing the first rapidly biodegradable acrylic nanoparticles in 1979, which were made of polymethylcyanoacrylate and polyethylcyanoacrylate. Couvreur also demonstrated that nanoparticles could increase the effectiveness and decrease the toxicity of cancer therapy (Brasseur et al. 1980; Couvreur et al. 1982). Today, there are numerous nanoplatforms in development and in clinical use with core structures consisting of silver, gold, iron oxide, liposomes, QDs, proteins, polymers, or dendrimers (Figure 1.2; Tables 1.3 and 1.4).

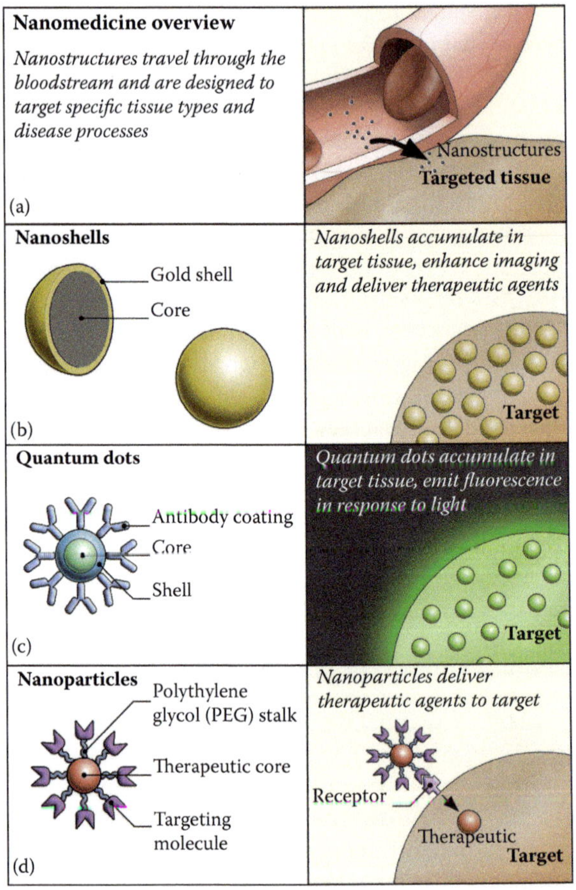

FIGURE 1.2 (a) One pathway for nanostructures to reach target tissues is via the bloodstream. (b) Nanoshells can be used in imaging, as well as drug delivery. The core can be loaded with therapeutic agents that affect pathophysiological processes in the target tissue. (c) QDs include a semiconductor nanocrystal core, which emits fluorescence in response to light. This schematic shows a QD-based structure with an antibody coating. (d) Nanoparticle functionalized with a targeting molecule interacting with a receptor at the target site. In this way, nanoparticles can deliver therapeutics and have localized versus systemic effects. (Reprinted from Kateb, B. et al., *Neuroimage*, 54, Suppl 1, 2011. With permission.)

TABLE 1.3
Nanoscale Platforms Presently Being Developed for Use in Nanomedicine

	Complex	Definition/Possible Uses	References
Gold and Silver	AuNPs (gold nanoparticles) (Figure 1.2b)	Present AuNPs are highly stable but also toxic. They have an incredible potential to be developed for imaging and drug delivery applications such as in vivo sensors, semiconductive photoactive agents for optical imaging, contrast enhancers for computed tomographic (CT) imaging, x-ray absorbers at tumor sites, and carriers of free-radical-generating chemicals to tumor sites. As a result, the development of nontoxic AuNP vectors is very important for advances in nanomedicine. A method was described to synthesize and stabilize AuNPs within a nontoxic phytochemical gum-arabic matrix (GA-AuNPs).	Kattumuri et al. 2007; Sperling et al. 2008
	Silver nanoparticles	Nanosilver structures are used to enhance wound healing and for coating plastic catheters. Coated catheters are nontoxic, and their sustained release of silver reduces infection at the site of the implant because silver is antibacterial.	Ellis-Behnke et al. 2007; Roe et al. 2008
Liposome	Liposomes with Sialyl Lewis X (SLX)	Liposomes with SLX are not only an efficient imaging reagent but also an effective drug delivery agent. In delivering substances like fluorescent markers, proteins, genes, and chemical substances to regions of inflammation or tumor in a mouse model, the accumulation of liposomes with SLX was much higher than in controls.	Hirai et al. 2007; Hashida et al. 2008
	Lysolipid-based temperature-sensitive liposomes (LTSLs)	In conjunction with local hyperthermia, LTSLs containing chemotherapy drugs were found to increase tumor drug concentrations and improve antitumor efficacy of the drugs.	Ponce et al. 2007
	Lipid-encapsulated perfluorocarbon nanoparticle molecular imaging contrast agent	Utilizes a paramagnetic chemical exchange saturation transfer (PARACEST) chelate to target and detect antifibrin antibodies. Since PARACEST can be turned on and off by adjusting the pulse-sequence settings, this technique avoids the need for preinjection and postinjection images to define contrast agent binding. Furthermore, PARACEST nanoparticles demonstrate >10% signal enhancement and produce a contrast-to-noise ratio (CNR) of 10 at the clot surface.	Winter et al. 2006

(*continued*)

TABLE 1.3 (Continued)
Nanoscale Platforms Presently Being Developed for Use in Nanomedicine

	Complex	Definition/Possible Uses	References
	MAGfect	A novel liposome formulation containing a lipidic gadolinium contrast agent for magnetic resonance imaging (MRI), Gd-DOTA-Chol, is designed to enter and label cells. Furthermore, MAGfect is not only an effective vehicle for transporting plasmid DNA (pDNA) into cells but also an optimizing agent for cellular uptake.	Oliver et al. 2006; Terreno et al. 2008
Quantum Dots	Polyethylene glycol (PEG)-coated quantum dots (QDs), PEG-polyhexadecyl	When combined with carboxyl, amino, or methoxyl groups, PEG-conjugated QDs migrate and confine themselves to sentinel lymph nodes. Localization of the QDs within the lymphatic circulation enables easy tagging of the lymph nodes, aiding the surgeon and the pathologist. Metastasis of tumors to adjacent lymph nodes can also be visualized.	Ballou et al. 2007
	cyanoacrylate (PHDCA)	PEG-PHDCA stable imaging marker, which could be used for drug delivery in brain. PEG-treated polyalkylcyanoacrylate nanoparticles cross the blood–brain barrier (BBB).	Ellis-Behnke et al. 2007
	Small interfering RNA (siRNA)–QD complexes	Since variations in delivery of siRNA, cytotoxic effects due to delivery of siRNA, and "off-target" effects at high siRNA concentrations often blur the outcome of many studies, siRNA was coupled with semiconductor QDs that act as multicolor biological probes. Due to the inherent photostability of QDs and their variable optical properties, the QD–siRNA complex is well suited for tracking the delivery of nucleic acid, sorting cells by amount of transfection, and purifying silenced subpopulations.	Chen et al. 2005
	Annexin A5 functionalized QDs	These complexes can be bimodal contrast agents allowing anatomic and subcellular imaging of structures in the vessel wall.	Prinzen et al. 2007
Iron Oxide	Magnetism-engineered iron oxide (MEIO) nanoprobes	Compared with currently available probes, antibody-conjugated MEIO nanoprobes display enhanced MRI sensitivity in detecting tumor markers. Nanobased probes have immense potential to improve visualization of key biological events in diagnostic and therapeutic nanomedicine.	Lee et al. 2007; Lin et al. 2008; Peng et al. 2008
	Superparamagnetic magnetite nanoparticles coated with lactobionic acid	Superparamagnetic magnetite nanoparticles with surface-attached lactobionic acid allow better nanoparticle hepatocyte targeting. This could lead to their use as contrast agents in liver disease diagnoses.	Kamruzzaman et al. 2007

	Iron oxide (magnetite) nanoparticle	This nanoparticle complex is synthesized within a genetically engineered human H-chain ferritin (HFn) with a cell-specific targeting peptide (RGD-4C) that binds $\alpha_v\beta_3$-integrins upregulated on tumor vessels. This complex also bound to C32 melanoma cells in vitro, illustrating the ability of genetically modified protein cage nanoparticles to act as agents for oxide loading and cell-specific targeting.	Uchida et al. 2006
Protein	Calmodulin	By combining iron oxide nanoparticle-based contrast mechanisms with the calcium-sensing protein calmodulin and its targets, these calcium indicators for MRI can be used for functional molecular imaging of biological signaling networks in live systems.	Atanasijevic et al. 2006
	$\alpha_v\beta_3$-integrin-targeted ^{111}In integrin nanoparticles ($\alpha_v\beta_3$-targeted ^{111}In)	$\alpha_v\beta_3$-targeted ^{111}In can be used as another means for sensitively detecting angiogenesis in emerging tumors, especially when combined with high-resolution imaging techniques like MRI. The biochemical and morphological detection and characterization of angiogenesis induced by very small tumors can also be accomplished by using $\alpha_v\beta_3$-targeted ^{111}In. This allows localization of emergent cancers and phenotypic characterization of patient populations for therapy.	Hu et al. 2007; Winter et al. 2003
Polymer	Multifunctional polymeric nanoparticle	This nanoparticle complex overcomes the drawback of photodynamic therapy (PDT), prolonged cutaneous photosensitization, by encapsulating nanoparticles with a photodynamic agent. Consisting of a tumor vasculature targeting F3 peptide and encapsulated PDT and imaging agents, this complex significantly enhances MRI contrast and improves detection of tumors. As a result, this translated into a significant improvement in the survival rate of glioma-bearing rats over controls.	Reddy et al. 2006
	Fluorescein isothiocyanate conjugated glycol chitosan (FGC) nanoparticles	FGC nanoparticles show highly selective tumor localization since they can clearly define tumors from adjacent tissues, allowing more accurate cancer diagnosis. These complexes also enable better therapy by permitting the delivery of multiple therapeutic agents and imaging probes at high local concentrations.	Cho et al. 2007

TABLE 1.4
FDA-Approved Drug Products in Nanometer-Size Range

	Drug Name	Generic Name	Company	Indication	Route	Approval Date
Liposomal platforms	Diprivan	Propofol	Zeneca Pharma	Anesthesia	IV	Oct. 1989
	Doxil	Doxorubicin	Ortho Biotech Schering Plough	Antineoplastic	IV	Nov. 1995
	Abelcet	Amphotericin B lipid complex injection	Enzon Pharmaceuticals	Antifungal	IV	Nov. 1995
	DaunoXome	Daunorubicin citrate liposome injection	Diatos	Antineoplastic	IV	Apr. 1996
	Amphotec	Amphotericin B cholesteryl sulfate injection	Three Rivers Pharmaceuticals	Antifungal	IV	Nov. 1996
	AmBisome	Amphotericin B liposome injection	Astellas Pharma US, Inc.	Antifungal	IV	Aug. 1997
	DepoCyt	Cytarabine liposome injection	Enzon Pharmaceuticals	Lymphomatous meningitis	IT	Apr. 1999
	Visudyne	Verteporfin for injection	Novartis	Photodynamic therapy for age-related macular degeneration	IV	Apr. 2000
Nanocrystal platforms	Rapamune	Sirolimus	Wyeth	Immunosuppressant	Oral	Aug. 2002
	Emend	Aprepitant	Merck & Co., Inc.	Antiemetic	Oral	Mar. 2003
	Depodur	Morphine sulfate	Pacira Pharmaceuticals	Relief of postsurgical pain	Epidural	May 2004
	TriCor	Fenofibrate	Abbott Laboratories	Hypercholesterolemia and hypertriglyceridemia	Oral	Nov. 2004
	Triglide	Fenofibrate	Sciele Pharma, Inc.	Hypercholesterolemia and hypertriglyceridemia	Oral	May 2005

	Megace ES	Megestrol acetate	Strativa, subsidiary of Par Pharmaceutical, Inc.	Anorexia, cachexia, or an unexplained significant weight loss in AIDS patients	Oral	Jul. 2005
	Elestrin	Estradiol	BioSante'	Hot flushes in menopausal women	Transdermal	Dec. 2006
Other platforms	Adagen	Pegademase (adenosine deaminase)	Enzon	Enzyme replacement for severe combined immunodeficiency syndrome	IV	Mar. 1990
	Oncaspar	Pegaspargase: pegylated L-asparaginase	Enzon Pharmaceuticals	Acute lymphoblastic leukemia	IM or IV	Feb. 1994
	Feridex	Ferumoxides injectable solution (superparamagnetic iron oxide)	Bayer HealthCare Pharmaceuticals	MRI contrast agent	IV	Aug. 1996
	Copaxone	Glatiramer acetate	TEVA Pharmaceuticals	Multiple sclerosis	SC	Dec. 1996; Mar. 2009
	Renagel	Sevelemar	Genzyme	Hyperphosphatemia associated with kidney failure	PO	Oct. 1998
	Pegasys	Peginterferon alpha-2a	Nektar; Hoffman-LaRoche	Chronic hepatitis C	SC	Oct. 2002
	PEGIntron	PEGinterferon alpha-2b	Enzon, Schering-Plough	Chronic hepatitis C	SC	Jan. 2001
	Neulasta/ Neulastim	Pegfilgrastim	Amgen (Neulasta); Roche (Neulastim)	Neutropenia due to cancer chemotherapy	SC	Jan. 2002
	Somavert	Pegvisomant	Pfizer	Acromegaly	SC	Mar. 2003
	Estrasorb	Estradiol topical emulsion	Graceway Pharmaceuticals, LLC	Vasomotor symptoms associated with menopause	Transdermal	Oct. 2003
	Macugen	Pegaptanib (Pegylated aptamer)	OSI Pharmaceuticals (within USA); Pfizer (outside USA)	Anti-VEGF against age-related macular degeneration (ARMD)	Intravitreal	Dec. 2004
	Abraxane	Paclitaxel albumin-bound particles for injectable suspension	Abraxis Oncology	Metastatic breast cancer	IV	Jan. 2005

Medical Use of Nanomaterials

The first FDA-approved nanotechnology application for drugs was Doxil in 1995. Doxil is made of nanosized liposomes enclosing the chemotherapeutic drug doxorubicin and has been used to treat Kaposi's sarcoma, ovarian cancer, and multiple myeloma (Barenholz 2012). Unmodified doxorubicin itself may damage cardiac muscle. In contrast, polyethylene-glycol-coated, liposome-encapsulated doxorubicin prevents heart damage by making the drug less accessible to the heart (Barenholz 2012; O'Brien et al. 2004). Another notable nanotech reformulation of a chemotherapeutic reagent is Abraxane, which was approved by the FDA in 2005 (Petrelli et al. 2010). Abraxane is human albumin-coated paclitaxel, a mitotic inhibitor, and was initially approved to treat metastatic breast cancer. This is the first drug that utilizes nanoparticle albumin bound (nab) technology. This protein-bound formulation improved the solubility of paclitaxel, which is hydrophobic, and reduced its toxicity (Jain and Stylianopoulos 2010; Petrelli et al. 2010). Currently, there are more than 20 FDA-approved nanodrugs (Davis 2008; Jain and Stylianopoulos 2010).

Other Important Developments in Nanotechnology

Liposomes

Liposomes are spherical nanoparticles with a phospholipid bilayer surface. Alec Bangham and Robert Horne created these structures by adding lecithin, a phospholipid, to water and first observed their structure using electron microscopy. Their lamellar appearance appeared similar to that of the cell membrane (Bangham and Horne 1964). The term "liposome" was coined later (Sessa and Weissmann 1968). Liposomes show promise as carriers of both imaging agents and drugs. When combined with Sialyl Lewis X (SLX) and a fluorescent marker, liposomes accumulate in much greater concentrations in inflamed tissues and tumor sites, facilitating both imaging and drug delivery (Hirai et al. 2007). SLX is a carbohydrate that, in the context of liposome–cell interaction, is thought to prefer sites of inflammation and tumors, enhancing liposome incorporation into the tissue. Both inflamed tissue and certain tumor cells express the cell-to-cell adhesion molecule, E-selectin, which is thought to favorably interact with SLX (Hirai et al. 2007). In studies, murine models of the Ehrlich ascites tumor and induced arthritis showed enhanced liposomal uptake from the bloodstream into the tissue as illustrated by a fluorescent marker (Hirai et al. 2007). The concentration of SLX on the liposome surface influenced inclusion within the tissue; if the concentration of SLX was either too low or too high, the liposomes were not maximally absorbed into the target area (Hirai et al. 2007). Overall, liposomes with SLX showed an enhanced degree of both specificity and absorption, two characteristics that can certainly be exploited in nanomedicine. The ability to track delivery using a fluorescent marker on the liposome will help future efforts to develop more tissue-specific imaging and drug delivery. Lysolipid-based temperature-sensitive liposomes (LTSLs) in conjunction with hyperthermia may be particularly useful in localized drug delivery (Lindner et al. 2008; Negussie et al. 2010; Ponce et al. 2007). LTSLs are temperature-sensitive liposomes that can deliver a drug, such as doxorubicin, into a target region upon heating (Ponce et al. 2007). In experiments, heating appeared to play a role in how deeply doxorubicin penetrated into the tumor, prompting researchers to suggest allowing the liposomes to diffuse into a tumor's interstitial space beyond the vasculature before applying heat to increase tumor drug concentration, enhance antivascular effects, and accelerate the tumor's demise (Ponce et al. 2007). These studies are examples of a targeted drug release system independent of specific binding between the nanoparticle and tumor cell or vasculature. The researchers here used a catheter for heating, but noninvasive heating methods are also available (Dromi et al. 2007; Kateb et al. 2009; Ranjan et al. 2012). Overall, liposomes show great promise as a drug delivery tool. The techniques used in the above two examples of liposome–SLX and LTSL–doxorubicin complexes demonstrate both binding- and heat-dependent delivery, respectively. Additionally, the former example shows how liposomes may be tracked as they act, which also demonstrates their imaging capability.

Evading the Reticuloendothelial System

The body's reticuloendothelial system (RES) contains monocytes, macrophages, and phagocytes that actively remove and sequester microbes and tiny fragments of biological and synthetic materials, including nanoparticles. Phagocytic cells are found in greatest concentration in the liver, spleen, lymph nodes, bone marrow, and blood (Almeida et al. 2011). Uncoated nanoparticles are quickly opsonized and removed by macrophages of the RES. Lisbeth Illum of the Royal Danish School of Pharmacy and Stanley Davis of the University of Nottingham found that they could prolong the circulation time and decrease RES recognition of intravenously delivered nanoparticles by coating the nanoparticles with polyoxyethylene and polyoxypropylene (Illum and Davis 1983). Following this discovery, other surface attachments were investigated, most importantly the covalent attachment of PEG chains discovered independently in 1994 by Gref (Gref et al. 1994) and in 1995 by Bazile (Bazile et al. 1995). Today, PEGylation is an extremely common surface modification made to a variety of nanoparticles to increase their solubility and ability to evade the RES.

The Enhanced Permeability and Retention (EPR) Effect

In 1986, Yasuhiro Matsumura and Hiroshi Maeda of the Kumamoto University Medical School in Japan first described the EPR, a property that allows macromolecules to preferentially accumulate in tumor tissue rather than normal tissue. They found that (1) after intravenous injection, proteins tend to preferentially accumulate in tumor tissue rather than circulate in the blood; (2) proteins were retained for long periods of time in the tumor tissue; and (3) there was little lymphatic recovery of the macromolecules from the tumor tissue (Maeda et al. 2000; Matsumura and Maeda 1986). EPR results from the defective endothelial cells that line tumor vasculature. Tumor vessels exhibit a wider-than-normal lumen, wide fenestrations, an absent smooth muscle layer and innervation, and impaired angiotensin II receptors (Maeda et al. 2000). Nanoparticles administered systemically are captured preferentially by the EPR effect in the tumor. This effect was observed in xenograft brain tumor models after systemic injection of nanoparticles and may be exploited for brain tumor treatment (Ljubimova et al. 2008; Na et al. 2011).

NANOTECHNOLOGY AND THE NERVOUS SYSTEM

The FDA has not yet approved a nanomedicine for a neurological or neurosurgical application. However, a variety of nanoplatforms for biomedical application are available for research and clinical trials, including solid lipid nanoparticles, polymeric nanoparticles, dendrimers, nanogels, gold nanoparticles, and liposomes, to name a few (Re et al. 2012a,b).

For instance, CNTs are being developed to deliver drugs and genes to treat CNS malignancies (Kateb et al. 2010; VanHandel et al. 2009). CNTs, like bucky balls, are composed entirely of carbon, but in a tubular shape. They exhibit extraordinary properties in terms of strength, electrical and thermal conductivity, and other characteristics (Nano.gov 2011). CNTs may be synthesized by rolling sheets of carbon into hollow tubes that are single-walled (SWCNT, 0.4–2 nm diameter), double-walled (DWCNT, 1–3.5 nm diameter), or multiwalled (MWCNT, 2–100 nm diameter). MWCNTs, SWCNTs, and DWCNTs are used as semiconductors in electronic devices (Collins et al. 2001; Rius-Ruiz et al. 2011; Sgobba and Guldi 2009). They are also being investigated as vectors for drug delivery because their physical and chemical properties allow them to carry drugs and to be internalized by phagocytic cells, such as microglia. The in vivo toxicity of CNTs is currently under investigation. Systemic delivery of CNTs can induce oxidative stress in end organs (Aillon et al. 2009). Inhaled CNTs can result in acute lung injury, inflammation, and fibrosis (Sanchez et al. 2009). Efforts were made to attenuate the immune response to CNTs by altering their surface chemistry with heparin (Murugesan et al. 2006; Salvador-Morales et al. 2006).

Thermal ablation applications using gold nanoparticles developed for neurosurgery include specific and localized therapy for CNS malignancies (Etame et al. 2012; O'Neal et al. 2004; Sperling et al. 2008).

Targeted Drug Delivery

The ability to covalently attach targeting molecules such as monoclonal antibodies and other targeting ligands to nanoparticles introduced a new level of specificity to drug targeting. In cancer therapy, although nanosized drugs accumulate at the tumor site as a consequence of the EPR, some of these drugs may be confined to the interstitial space and not enter the intracellular compartment. In 2010, Forrest Kievit and colleagues at the University of Washington showed that attaching targeting moieties to nanovectors did not affect accumulation of nanoparticles within a xenograft brain tumor model but did specifically increase cellular uptake of the nanovector and expression of the marker gene (GFP) by the cancer cell (Kievit et al. 2010). Increasing expression of a suicide gene within tumor cells could potentially improve the effectiveness of gene therapy for glioma.

Biomaterials and Regenerative Medicine

Recent work, which will be extensively covered in other chapters in this volume, has demonstrated the unique sensitivity of stem cell precursors to topography, allowing them to be differentiated into neuronal precursors by altering the topography of their surroundings on a nanoscale (Migliorini et al. 2011). Researchers are investigating implantable, biocompatible, nanomodified scaffolds containing stem cells for their potential to regenerate tissue in the CNS (Ellis-Behnke and Schneider 2011; Liang et al. 2011, Ellis-Behnke 2007; Ellis-Behnke et al. 2006, 2007). Altering the surface topography of implantable biomaterials may dramatically increase the compatibility of implantable materials (Maclaine et al. 2012; Teo et al. 2010; Ravichandran et al. 2009; Decuzzi and Ferrari 2010; Draghi and Cigada 2007; Simon et al. 2007). However, certain nanoscale topographical features may also increase the immunogenicity of implantable biomaterials (Biggs et al. 2010).

CHALLENGES OF APPLYING NANOTECHNOLOGY TO NANONEUROSCIENCE AND NANONEUROSURGERY

Challenges of using nanomedicine in neurological disorders include the wide variety and complex nature of neurological disorders and the presence of the BBB, which poses an obstacle in delivering therapeutics. Some nanomaterials can cross the BBB if injected systemically (Albanese et al. 2012; Trickler et al. 2011, 2012), providing a way to deliver therapeutics to the brain (Jain 2012; Tosi et al. 2012). However, there are concerns of neurotoxicity (Jain 2012; Kim et al. 2007; Ulbrich et al. 2011) as well as adverse effects of nanomaterials on other organs, such as their asbestos like effect in the lung (Powell and Kanarek 2006a,b; Tsuji et al. 2006). It is necessary to conduct rigorous safety research when new products are developed. Advanced functionalization may make nanocarriers less toxic and more nervous system–specific.

The ideal nanocarrier for passage across the BBB should have the following characteristics (adapted from Jain 2012):

- Well-controlled drug delivery.
- Does not damage the BBB.
- Biodegradable.
- Nontoxic.
- Selectively crosses the BBB.
- Selectively delivered to the site of intended action in the brain.
- Therapeutic concentrations of drug/macromolecules and duration of action are adequate and optimized in the brain once the nanocarrier crosses the BBB.

In 1994, it was observed at the University of Kansas that surfactant-coated nanoparticles, specifically polybutylcyanoacrylate (PBCA) coated with polysorbate-80, were taken up by brain capillary endothelial cells in vitro (Borchard et al. 1994). Following this discovery, several researchers performed in vivo animal studies that used polysorbate-80-coated PBCA nanoparticles to carry drugs to the CNS that otherwise would not enter the CNS because they could not penetrate the BBB. A group in Moscow showed that significant brain concentrations of doxorubicin could be achieved with this delivery system (Gulyaev et al. 1999). This observation spurred subsequent research to engineer nanoparticles that can carry therapeutic agents across the BBB to noninvasively treat malignant brain tumors, CNS infections, and other CNS disorders (Jain et al. 2011; Li et al. 2011a,b,c; Montenegro et al. 2011).

ESTABLISHMENT OF THE NATIONAL NANOTECHNOLOGY INITIATIVE

The National Nanotechnology Initiative (NNI) was initially formed as the Interagency Working Group in Nanotechnology under the National Science and Technology Council in 1996 and was launched by President Clinton in 2000. The group sponsored workshops to define the state of the art in nanoscale science development and discuss possible future progress in the field. In 2000, the group submitted a budget proposal to Congress under the Clinton administration, referring to it as the NNI. The group received strong presidential and congressional support. Mihail (Mike) Roco is credited with being a key architect of the NNI, coordinating the preparation of the US National Science and Technology Council reports on "Nanotechnology Research Directions" and the "National Nanotechnology Initiative." The NNI has helped fund the development of devices to detect biomarkers for cancer and to detect misfolded proteins for prion diseases such as bovine spongiform encephalopathy (mad cow disease) (Turner et al. 2007). NNI funding has supported development of iron oxide and gadolinium nanoparticles to improve imaging primary brain tumors by increasing MRI contrast before surgery. NNI funding has also supported development of multifunctional nanoparticles that can increase the optical contrast between a brain tumor and the surrounding brain during an operative procedure and can be used for photodynamic therapy (Koo et al. 2007; Tang et al. 2005).

ROLE OF THE US CONGRESS AND THE WHITE HOUSE IN ADVANCEMENT OF NANONEUROSCIENCE AND NANONEUROSURGERY

In 2010, Congressman Blumenauer formed the United States Congressional Neuroscience Caucus. Today, the caucus is co-chaired by him and Congresswoman Cathy McMorris Rodgers. The mission of the caucus is "to build awareness of the intrinsic role brain research plays in understanding ourselves and our society, to help communicate the progress and the benefits of this research, and to help inform federal policy."

The caucus organizes events that focus on neuroscience: "developing and promoting legislation that will further advance neuroscience research; sponsoring briefings and information sessions on neuroscience research and findings; collaborating with patient advocacy, physician, and research organizations to build awareness; and distributing congressional communications and information such as Dear Colleagues and Congressional Record statements." One such legislation is the *Military Construction and Veterans Appropriations Act*, which was initiated by Congressman Blumenauer who offered and successfully passed a bipartisan amendment. The act "ensures the Veterans Administration continues to have the resources it needs to find innovative medicines and enhanced diagnostics for soldiers and veterans suffering from traumatic brain injuries and post-traumatic stress disorder. It requires that no less than $35 million in medical research goes toward post-traumatic stress disorder and traumatic brain injury research and treatment in order to expedite a cure for hundreds of thousands of active duty personnel and veterans suffering the effects of brain- and psychological-trauma incurred during service."

Therefore, there is a great need for more collaborative efforts between the US Congress and medical associations such as SBMT and nonprofit organizations such as the Brain Mapping Foundation.

Such close collaboration will result in better policies that could enable integration and translation of science and technology across disciplines and into the neuroscience field.

While the NNI initiative has been a great step toward advancing the field of nanotechnology, it has been less aggressive in supporting groundbreaking translational and integrative research, which focuses on integration of nanotechnology, stem cell/cellular therapy, medical imaging, and devices. President of Brain Mapping Foundation and Chairman of the Board of SBMT, Babak Kateb and his policy advisor, Terry Lierman in collaboration with Congressman Moran submitted an initiative to the US Congress in order to:

1. Encourage multidisciplinary research.
2. Create a National Data Repository for Neuroscience, which could help avoid duplication and identify the best practices while reducing cost by identifying the most effective technologies.
3. Rapidly introduce diagnostic and therapeutic solutions for neurological disorders by supporting integration of nanotechnology, stem cell/cellular therapy, imaging, and medical devices. These are the critical goals of the National Center for Nanobioelectronics (NCNBE), which will greatly advance nanoneuroscience and nanoneurosurgery.

In this way, policy and initiatives can continue to play a role in advancing progress in nanotechnology in general and nanoneurosurgery in particular.

The following is the exact language of the legislation:

> "The Committee encourages the Secretary of Defense to support multi-disciplinary research toward translational medicine that may provide better diagnostic tools and treatment outcomes for service members who suffer from traumatic brain injury, post-traumatic stress disorder, and other neurological diseases. With this support the Secretary of Defense will enable the creation of a nationwide scientific consortium aimed at integrating nanotechnology, stem cell, cellular therapy, medical imaging, electronic medical record, information technology, Brain Mapping, and medical device. The Committee encourages the Secretary of Defense to provide the capabilities necessary for researchers, scientists, surgeons, physicians, healthcare professionals, and patients to effectively communicate their findings and outcomes. With proper support, translational research outcomes would be augmented through real-time access to information and its integration between researchers, physicians, hospitals, and patients. The Committee encourages the Secretary of Defense to establish this program through the NAVY Bureau of Medicine and Surgery (BUMED). The Committee directs the Secretary of Defense to provide a report not later than 90 days after enactment of this act on a possible implementation of this program."

This report language helped in shaping the national debate about formation of a National Alliance for Nano-Bio-Electronics, which is aimed at fostering and encouraging integration, translation, and commercialization of technologies across disciplines of nanotechnology, stem cell/cellular therapy, imaging, and devices. This legislation is the first aimed at integration of these scientific disciplines.

In 2013, President Obama introduced a new initiative on brain mapping authorizing $100 million for the field, which has huge implications for the field of nanoneuroscience. The initiative was coordinated through the office of Science, Technology, and Policy at the White House and led by the Principal Assistant Director for Science, Office of Science and Technology Policy at the Executive Office of the President of the United States, Philip Rubin, and deputy Director for Technology and Innovation, Thomas Kalil. They successfully engaged leading industry, foundation, and government agencies to ensure that President Obama's vision will be a success.

As part of President Obama's 2013 Brain Mapping Project, Brain Mapping Foundation, NCNBE, and SBMT presented the White House with a 10-page whitepaper proposing the following:

- National Alliance for Nano-Bio-Electronics (NANBE)
- National Network for Human Brain and Specimen Banks (NNHBSN)
- National Data Repository and Analysis for Neuroscience (NDRAN)

The program would be established in order to (1) foster partnership in technology, science, and medicine through the development of national consortia; (2) systemically and actively facilitate the integration, translation, and commercialization of technologies and scientific advancement from the various physical, social, and biological sciences into clinical medicine; (3) establish consortia, scholarships, and educational programs at all educational levels in order to encourage ingenuity and cross-disciplinary interactions, which could contribute to the development of future scientists, technologists, and health professionals; and (4) promote education curricula addressing disease prevention. This novel program will contribute to US leadership in the global health-care marketplace while building long-term employment and (5) establishing brain mapping centers, minimally invasive and noninvasive technologies such as nanotechnology, biophotonics, ultralow MRI, portable CT, high-definition diffusion tensor imaging (DTI), simulations, cellular therapeutics/stem cells and devices, and other methods that could help map the brain and enable physicians to better treat neurological disorders.

CONCLUSION

In this chapter, we reviewed the history of nanotechnology and several of its milestones and outlined the importance of groundbreaking policies that played a significant role in advancing nanotechnology and nanoneurosurgery.

Recent work, which will be extensively covered in other chapters in this book, revealed the unique sensitivity of stem cell precursors to topography, allowing them to be differentiated into neuronal precursors by altering the topography of their surroundings on a nanoscale (Migliorini et al. 2011). Researchers are investigating implantable, biocompatible, nanomodified scaffolds containing stem cells for their potential to regenerate tissue in the CNS (Ellis-Behnke and Schneider 2011; Liang et al. 2011; Ellis-Behnke 2007; Ellis-Behnke et al. 2006, 2007). However, certain nanoscale topographical features may increase the immunogenicity of implantable biomaterials (Biggs et al. 2010). Altering surface topography of implantable biomaterials holds promise for dramatically increasing the compatibility of implantable materials (Maclaine et al. 2012; Teo et al. 2010; Ravichandran et al. 2009; Decuzzi and Ferrari 2010; Draghi and Cigada 2007; Simon et al. 2007).

The future of nanoneurosurgery and nanoneuroscience depends on important legislative acts such as the ones described in this chapter. Nanotechnology by itself is incapable of treating any diseases. The National Nanotechnology Initiative focuses on nanotechnology more than medical applications and has led to relatively few medical advances. The future importance of nanotechnology in medicine depends on how well we integrate this nanotechnology with cellular/stem cell therapy, imaging, and devices. Neurosurgery, neuroscience, and neurology residencies and fellowships and scientific doctoral programs should provide education in nanobioelectronics that will enable a new generation of integrative thinkers to devise new diagnostic and therapeutic solutions to presently incurable conditions.

Within the next decade, nanoneurosurgery could change the way gliomas are diagnosed and treated, neurotrauma is managed utilizing revolutionary new nanoplatforms, and neurorehabilitation is approached using axonal guidance systems and stem cell integration. Neurodegenerative disorders may also be impacted by nanotechnology leading to improved drug and gene delivery.

We are now in an era of personalized medicine and minimally invasive therapy. Nanobioelectronics, which is defined as the integration of nanotechnology, device, imaging, and cellular/genomic/stem cell therapy, will be part of a new frontier that could make many current neurosurgical procedures obsolete. Lessons learned from the history of nanotechnology, nanoneuroscience, and nanoneurosurgery will guide the continued development of nanoneurosurgery and provide a foundation for new diagnostic and therapeutic advances that will impact future medical practice and personalized medicine.

REFERENCES

Adams, W.W., Baughman, R.H. 2005. Retrospective: Richard E. Smalley (1943–2005). *Science,* 310, 1916.

Aillon, K.L., Xie, Y., El-Gendy, N. et al. 2009. Effects of nanomaterial physicochemical properties on in vivo toxicity. *Adv Drug Deliv Rev,* 61, 457–66.

Akerman, M.E., Chan, W.C., Laakkonen, P. et al. 2002. Nanocrystal targeting in vivo. *Proc Natl Acad Sci U S A,* 99, 12617–21.

Albanese, A., Tang, P.S., Chan, W.C.W. 2012. The effect of nanoparticle size, shape, and surface chemistry on biological systems. *Annu Rev Biomed Eng, Vol 14,* 14, 1–16.

Almeida, J.P., Chen, A.L., Foster, A. et al. 2011. In vivo biodistribution of nanoparticles. *Nanomedicine (Lond),* 6, 815–35.

Aronson, M. 2001. Materials Science of Art & Archeological Objects. Tucson, Arizona: University of Arizona.

Atanasijevic, T., Shusteff, M., Fam, P. et al. 2006. Calcium-sensitive MRI contrast agents based on superparamagnetic iron oxide nanoparticles and calmodulin. *Proc Natl Acad Sci U S A,* 103, 14707–12.

Bakry, R., Vallant, R.M., Najam-ul-Haq, M. et al. 2007. Medicinal applications of fullerenes. *Int J Nanomed,* 2, 639–49.

Ballou, B., Ernst, L.A., Andreko, S. et al. 2007. Sentinel lymph node imaging using quantum dots in mouse tumor models. *Bioconjug Chem,* 18, 389–96.

Bangham, A.D., Horne, R.W. 1964. Negative staining of phospholipids and their structural modification by surface-active agents as observed in the electron microscope. *J Mol Biol,* 8, 660–8.

Barenholz, Y. 2012. Doxil(R)—the first FDA-approved nano-drug: lessons learned. *J Control Release,* 160, 117–34.

Bazile, D., Prud'homme, C., Bassoullet, M.T. et al. 1995. Stealth Me. PEG-PLA nanoparticles avoid uptake by the mononuclear phagocytes system. *J Pharm Sci,* 84, 493–8.

Bethune, D.S., Kiang, C.H., Devries, M.S. et al. 1993. Cobalt-catalyzed growth of carbon nanotubes with single-atomic-layer walls. *Nature,* 363, 605–7.

Biggs, M.J., Richards, R.G., Dalby, M.J. 2010. Nanotopographical modification: a regulator of cellular function through focal adhesions. *Nanomedicine,* 6, 619–33.

Binnig, G., Quate, C.F., Gerber, C. 1986. Atomic force microscope. *Phys Rev Lett,* 56, 930–3.

Binnig, G., Rohrer, H. 1983. Scanning tunneling microscopy. *Surf Sci,* 126, 236–44.

Binnig, G., Rohrer, H. 1986. Scanning tunneling microscopy. *IBM J Res Dev,* 30, 355–69.

Borchard, G., Audus, K.L., Shi, F. et al. 1994. Uptake of surfactant-coated poly(methyl methacrylate)-nanoparticles by bovine brain microvessel endothelial cell monolayers. *Int J Pharm,* 110, 29–35.

Brasseur, F., Couvreur, P., Kante, B. et al. 1980. Actinomycin D absorbed on polymethylcyanoacrylate nanoparticles: increased efficiency against an experimental tumor. *Eur J Cancer,* 16, 1441–5.

Cho, Y.W., Park, S.A., Han, T.H. et al. 2007. In vivo tumor targeting and radionuclide imaging with self-assembled nanoparticles: mechanisms, key factors, and their implications. *Biomaterials,* 28, 1236–47.

Collins, P.G., Arnold, M.S., Avouris, P. 2001. Engineering carbon nanotubes and nanotube circuits using electrical breakdown. *Science,* 292, 706–9.

Cook, K.A. 2005. DiscoverNANO.

Cosslett, V.E. 1962. Electron microscopy: introduction. *Br Med Bull,* 18, 179–82.

Couvreur, P., Kante, B., Grislain, L. et al. 1982. Toxicity of polyalkylcyanoacrylate nanoparticles II: Doxorubicin-loaded nanoparticles. *J Pharm Sci,* 71, 790–2.

Davis, A.E., 3rd. 2008. New treatments addressing the pathophysiology of hereditary angioedema. *Clin Mol Allergy,* 6, 2.

Decuzzi, P., Ferrari, M. 2010. Modulating cellular adhesion through nanotopography. *Biomaterials,* 31, 173–9.

Donald A. Tomalia. Chemical Heritage Foundation. 2013. http://www.chemheritage.org/discover/online-resources/chemistry-in-history/themes/microelectronics-and-nanotechnology/tomalia.aspx. (accessed Apr 4, 2013).

Draghi, L., Cigada, A. 2007. Nanostructured surfaces for biomedical applications. Part I: nanotopography. *J Appl Biomater Biomech,* 5, 61–9.

Drexler, K.E. 1981. Molecular engineering: an approach to the development of general capabilities for molecular manipulation. *Proc Natl Acad Sci U S A,* 78, 5275–8.

Drexler, K.E. 1986. *Engines of Creation: The Coming Era of Nanotechnology.* New York: Anchor Books.

Drexler, K.E. 1992. *Nanosystems: Molecular Machinery, Manufacturing, and Computation.* New York: John Wiley & Sons, Inc.

Drexler, K.E., Peterson, C., Pergamit, G. 1991. *Unbounding the Future: The Nanotechnology Revolution.* New York: William Morrow and Company, Inc.

Dromi, S., Frenkel, V., Luk, A. et al. 2007. Pulsed-high intensity focused ultrasound and low temperature-sensitive liposomes for enhanced targeted drug delivery and antitumor effect. *Clin Cancer Res,* 13, 2722–7.
Economou, N.P., Notte, J.A., Thompson, W.B. 2012. The history and development of the helium ion microscope. *Scanning,* 34, 83–9.
Eigler, D.M., Schweizer, E.K. 1990. Positioning single atoms with a scanning tunneling microscope. *Nature,* 344, 524–6.
Ellis-Behnke, R. 2007. Nano neurology and the four P's of central nervous system regeneration: preserve, permit, promote, plasticity. *Med Clin North Am,* 91, 937–62.
Ellis-Behnke, R.G., Liang, Y.X., You, S.W. et al. 2006. Nano neuro knitting: peptide nanofiber scaffold for brain repair and axon regeneration with functional return of vision. *Proc Natl Acad Sci U S A,* 103, 5054–9.
Ellis-Behnke, R.G., Schneider, G.E. 2011. Peptide amphiphiles and porous biodegradable scaffolds for tissue regeneration in the brain and spinal cord. *Methods Mol Biol,* 726, 259–81.
Ellis-Behnke, R.G., Teather, L.A., Schneider, G.E. et al. 2007. Using nanotechnology to design potential therapies for CNS regeneration. *Curr Pharm Des,* 13, 2519–28.
Esaki, L. 1958. New phenomenon in narrow germanium p-n junctions. *Phys Rev,* 109, 603–4.
Etame, A.B., Diaz, R.J., O'Reilly, M.A. et al. 2012. Enhanced delivery of gold nanoparticles with therapeutic potential into the brain using MRI-guided focused ultrasound. *Nanomedicine,* 8, 1133–42.
Feynman, R.P. 1960. There's plenty of room at the bottom. *Engineering and Science,* 23, 22–36.
Foldvari, M., Bagonluri, M. 2008a. Carbon nanotubes as functional excipients for nanomedicines: I. pharmaceutical properties. *Nanomedicine-Nanotechnology Biology and Medicine,* 4, 173–82.
Foldvari, M., Bagonluri, M. 2008b. Carbon nanotubes as functional excipients for nanomedicines: II. drug delivery and biocompatibility issues. *Nanomed-Nanotechnol Biol Med,* 4, 183–200.
Freitas, R.A., Jr. 1999. *Nanomedicine, Volume I: Basic Capabilities.* Georgetown, Texas: Landes Bioscience.
Freitas, R.A.J., Merkle, R.C. 2004. *Kinematic Self-Replicating Machines.* Georgetown, TX: Landes Bioscience.
Gamow, G. 1928. Quantum theory of the atomic nucleus. *Z Phys,* 51, 1–9.
Gartia, M.R., Hsiao, A., Pokhriyal, A. et al. 2013. Colorimetric plasmon resonance imaging using nano Lycurgus cup arrays. *Adv Opt Mater,* 1, 68–76.
Gref, R., Minamitake, Y., Peracchia, M.T. et al. 1994. Biodegradable long-circulating polymeric nanospheres. *Science,* 263, 1600–3.
Gulyaev, A.E., Gelperina, S.E., Skidan, I.N. et al. 1999. Significant transport of doxorubicin into the brain with polysorbate 80-coated nanoparticles. *Pharm Res,* 16, 1564–9.
Hirai, M., Minematsu, H., Kondo, N. et al. 2007. Accumulation of liposome with Sialyl Lewis X to inflammation and tumor region: application to in vivo bio-imaging. *Biochem Biophys Res Commun,* 353, 553–8.
Hu, G., Lijowski, M., Zhang, H. et al. 2007. Imaging of Vx-2 rabbit tumors with alpha(nu)beta3-integrin-targeted 111In nanoparticles. *Int J Cancer,* 120, 1951–7.
Iijima, S. 1991. Helical microtubules of graphitic carbon. *Nature,* 354, 56–8.
Iijima, S., Ichihashi, T. 1993. Single-shell carbon nanotubes of 1-nm diameter. *Nature,* 363, 603–5.
Illum, S.L., Davis, S.S. 1983. Effect of the nonionic surfactant poloxamer 338 on the fate and deposition of polystyrene microspheres following intravenous administration. *J Pharm Sci,* 72, 1086–9.
Jain, A., Chasoo, G., Singh, S.K. et al. 2011. Transferrin-appended PEGylated nanoparticles for temozolomide delivery to brain: in vitro characterisation. *J Microencapsul,* 28, 21–8.
Jain, K.K. 2008. Chapter 1: introduction. In *The Handbook of Nanomedicine*, ed. K.K. Jain, 1–6. Totowa, NJ: Humana Press.
Jain, K.K. 2012. Nanobiotechnology-based strategies for crossing the blood–brain barrier. *Nanomedicine (Lond),* 7, 1225–33.
Jain, R.K., Stylianopoulos, T. 2010. Delivering nanomedicine to solid tumors. *Nat Rev Clin Oncol,* 7, 653–64.
Kateb, B., Yamamoto, V., Alizadeh, D. et al. 2010. Multi-walled carbon nanotube (MWCNT) synthesis, preparation, labeling, and functionalization. *Methods Mol Biol,* 651, 307–17.
Kateb, B., Yamamoto, V., Yu, C. et al. 2009. Infrared thermal imaging: a review of the literature and case report. *Neuroimage,* 47 Suppl 2, T154–62.
Kievit, F.M., Veiseh, O., Fang, C. et al. 2010. Chlorotoxin labeled magnetic nanovectors for targeted gene delivery to glioma. *ACS Nano,* 4, 4587–94.
Kim, H.R., Gil, S., Andrieux, K. et al. 2007. Low-density lipoprotein receptor-mediated endocytosis of PEGylated nanoparticles in rat brain endothelial cells. *Cell Mol Life Sci,* 64, 356–64.
Kobayashi, H., Hama, Y., Koyama, Y. et al. 2007. Simultaneous multicolor imaging of five different lymphatic basins using quantum dots. *Nano Lett,* 7, 1711–16.
Koo, Y.E., Fan, W., Hah, H. et al. 2007. Photonic explorers based on multifunctional nanoplatforms for biosensing and photodynamic therapy. *Appl Opt,* 46, 1924–30.

Kreuter, J. 2007. Nanoparticles—a historical perspective. *Int J Pharm,* 331, 1–10.
Kroto, H.W., Heath, J.R., Obrien, S.C. et al. 1985. C-60: Buckminsterfullerene. *Nature,* 318, 162–3.
Li, J., Feng, L., Fan, L. et al. 2011a. Targeting the brain with PEG-PLGA nanoparticles modified with phage-displayed peptides. *Biomaterials,* 32, 4943–50.
Li, J., Zhou, L., Ye, D. et al. 2011b. Choline-derivate-modified nanoparticles for brain-targeting gene delivery. *Adv Mater,* 23, 4516–20.
Li, J.M., Zhao, M.X., Su, H. et al. 2011c. Multifunctional quantum-dot-based siRNA delivery for HPV18 E6 gene silence and intracellular imaging. *Biomaterials,* 32, 7978–87.
Liang, Y.X., Cheung, S.W., Chan, K.C. et al. 2011. CNS regeneration after chronic injury using a self-assembled nanomaterial and MEMRI for real-time in vivo monitoring. *Nanomedicine,* 7, 351–9.
Lin, C.M., Lu, T.Y. 2012. C60 fullerene derivatized nanoparticles and their application to therapeutics. *Recent Pat Nanotechnol,* 6, 105–13.
Lindner, L.H., Hossann, M., Vogeser, M. et al. 2008. Dual role of hexadecylphosphocholine (miltefosine) in thermosensitive liposomes: active ingredient and mediator of drug release. *J Control Release,* 125, 112–20.
Ljubimova, J.Y., Fujita, M., Khazenzon, N.M. et al. 2008. Nanoconjugate based on polymalic acid for tumor targeting. *Chem Biol Interact,* 171, 195–203.
Lu W., Ji Z., Pfeiffer L. et al. 2003. Real-time detection of electron tunnelling in a quantum dot. *Nature,* 423, 422–5.
Maclaine, S.E., Gadhari, N., Pugin, R. et al. 2012. Optimizing the osteogenicity of nanotopography using block co-polymer phase separation fabrication techniques. *J Orthop Res,* 30, 1190–7.
Maeda, H., Wu, J., Sawa, T. et al. 2000. Tumor vascular permeability and the EPR effect in macromolecular therapeutics: a review. *J Control Release,* 65, 271–84.
Matsumura, Y., Maeda, H. 1986. A new concept for macromolecular therapeutics in cancer chemotherapy: mechanism of tumoritropic accumulation of proteins and the antitumor agent smancs. *Cancer Res,* 46, 6387–92.
Michaelis, K., Hoffmann, M.M., Dreis, S. et al. 2006. Covalent linkage of apolipoprotein e to albumin nanoparticles strongly enhances drug transport into the brain. *J Pharmacol Exp Ther,* 317, 1246–53.
Michalewicz, M.T., Sasse, A., Rymuza, Z. 2007. Quantum tunneling NEMS devices for bio-medical applications. In talk presented at the *Technology Commercialisation Forum, Singapore, (2007 Nov 5–6).*
Migliorini, E., Grenci, G., Ban, J. et al. 2011. Acceleration of neuronal precursors differentiation induced by substrate nanotopography. *Biotechnol Bioeng,* 108, 2736–46.
Montenegro, L., Campisi, A., Sarpietro, M.G. et al. 2011. In vitro evaluation of idebenone-loaded solid lipid nanoparticles for drug delivery to the brain. *Drug Dev Ind Pharm,* 37, 737–46.
Moore, G.E. 1965. Cramming more components onto integrated circuits. *Electronics,* 38, 114–7.
Mout, R., Moyano, D.F., Rana, S. et al. 2012. Surface functionalization of nanoparticles for nanomedicine. *Chem Soc Rev,* 41, 2539–44.
Murugesan, S., Park, T.J., Yang, H. et al. 2006. Blood compatible carbon nanotubes—nano-based neoproteoglycans. *Langmuir,* 22, 3461–3.
Na, J.H., Koo, H., Lee, S. et al. 2011. Real-time and non-invasive optical imaging of tumor-targeting glycol chitosan nanoparticles in various tumor models. *Biomaterials,* 32, 5252–61.
Nano.gov. 2011. United States Nanotechnology Initiative.
Negussie, A.H., Miller, J.L., Reddy, G. et al. 2010. Synthesis and in vitro evaluation of cyclic NGR peptide targeted thermally sensitive liposome. *J Control Release,* 143, 265–73.
Nunes-Pereira, J., Lopes, A.C., Costa, C.M. et al. 2012. Porous membranes of montmorillonite/poly(vinylidene fluoride-trifluorethylene) for Li-ion battery separators. *Electroanalysis,* 24, 2147–56.
O'Brien, M.E., Wigler, N., Inbar, M. et al. 2004. Reduced cardiotoxicity and comparable efficacy in a phase III trial of pegylated liposomal doxorubicin HCl (CAELYX/Doxil) versus conventional doxorubicin for first-line treatment of metastatic breast cancer. *Ann Oncol,* 15, 440–9.
O'Neal, D.P., Hirsch, L.R., Halas, N.J. et al. 2004. Photo-thermal tumor ablation in mice using near infrared-absorbing nanoparticles. *Cancer Lett,* 209, 171–6.
Petrelli, F., Borgonovo, K., Barni, S. 2010. Targeted delivery for breast cancer therapy: the history of nanoparticle-albumin-bound paclitaxel. *Expert Opin Pharmacother,* 11, 1413–32.
Ponce, A.M., Viglianti, B.L., Yu, D. et al. 2007. Magnetic resonance imaging of temperature-sensitive liposome release: drug dose painting and antitumor effects. *J Natl Cancer Inst,* 99, 53–63.
Powell, M.C., Kanarek, M.S. 2006a. Nanomaterial health effects—part 1: background and current knowledge. *WMJ,* 105, 16–20.
Powell, M.C., Kanarek, M.S. 2006b. Nanomaterial health effects—part 2: uncertainties and recommendations for the future. *WMJ,* 105, 18–23.

Ranjan, A., Jacobs, G.C., Woods, D.L. et al. 2012. Image-guided drug delivery with magnetic resonance guided high intensity focused ultrasound and temperature sensitive liposomes in a rabbit Vx2 tumor model. *J Control Release,* 158, 487–94.
Ravichandran, R., Liao, S., Ng, C. et al. 2009. Effects of nanotopography on stem cell phenotypes. *World J Stem Cells,* 1, 55–66.
Re, F., Gregori, M., Masserini, M. 2012a. Nanotechnology for neurodegenerative disorders. *Nanomed-Nanotechnol Biol Med,* 8, S51–8.
Re, F., Moresco, R., Masserini, M. 2012b. Nanoparticles for neuroimaging. *J Phys D Appl Phys,* 45.
Reibold, M., Paufler, P., Levin, A.A. et al. 2006. Materials: carbon nanotubes in an ancient Damascus sabre. *Nature,* 444, 286.
Ren, L., Pint, C.L., Arikawa, T. et al. 2012. Broadband terahertz polarizers with ideal performance based on aligned carbon nanotube stacks. *Nano Lett,* 12, 787–90.
Rius-Ruiz, F.X., Kisiel, A., Michalska, A. et al. 2011. Solid-state reference electrodes based on carbon nanotubes and polyacrylate membranes. *Anal Bioanal Chem,* 399, 3613–22.
Robidoux, A., Buzdar, A.U., Quinaux, E. et al. 2010. A phase II neoadjuvant trial of sequential nanoparticle albumin-bound paclitaxel followed by 5-fluorouracil/epirubicin/cyclophosphamide in locally advanced breast cancer. *Clin Breast Cancer,* 10, 81–6.
Sachse, M. 2008. *Damascus Steel: Myth, History, Technology, Applications.* Dusseldorf: Stahleisen Communications.
Salvador-Morales, C., Flahaut, E., Sim, E. et al. 2006. Complement activation and protein adsorption by carbon nanotubes. *Mol Immunol,* 43, 193–201.
Sanchez, V.C., Pietruska, J.R., Miselis, N.R. et al. 2009. Biopersistence and potential adverse health impacts of fibrous nanomaterials: what have we learned from asbestos? *Wiley Interdiscip Rev Nanomed Nanobiotechnol,* 1, 511–29.
Sessa, G., Weissmann, G. 1968. Phospholipid spherules (liposomes) as a model for biological membranes. *J Lipid Res,* 9, 310–18.
Sgobba, V., Guldi, D.M. 2009. Carbon nanotubes—electronic/electrochemical properties and application for nanoelectronics and photonics. *Chem Soc Rev,* 38, 165–84.
Simon, K.A., Burton, E.A., Han, Y. et al. 2007. Enhancing cell adhesion and confinement by gradient nanotopography. *J Am Chem Soc,* 129, 4892–3.
Sperling, R.A., Rivera Gil, P., Zhang, F. et al. 2008. Biological applications of gold nanoparticles. *Chem Soc Rev,* 37, 1896–908.
Stroscio, J.A., Celotta, R.J. 2004. Controlling the dynamics of a single atom in lateral atom manipulation. *Science,* 306, 242–7.
Subbiah, R., Veerapandian, M., Yun, K.S. 2010. Nanoparticles: functionalization and multifunctional applications in biomedical sciences. *Curr Med Chem,* 17, 4559–77.
Taghva, A., Khalessi, A.A., Kim, P.E. et al. 2010a. From atom to brain: applications of molecular imaging to neurosurgery. *World Neurosurg,* 73, 477–85.
Taghva, A., Kim, P.E., Liu, C.Y. et al. 2010b. Molecular imaging, part 1: apertures into the landscape of genomic medicine. *World Neurosurg,* 73, 307–16.
Tang, W., Xu, H., Kopelman, R. et al. 2005. Photodynamic characterization and in vitro application of methylene blue-containing nanoparticle platforms. *Photochem Photobiol,* 81, 242–9.
Teo, B.K., Ankam, S., Chan, L.Y. et al. 2010. Nanotopography/mechanical induction of stem-cell differentiation. *Methods Cell Biol,* 98, 241–94.
Tibbals, H.F. 2011. *Medical Nanotechnology and Nanomedicine.* Boca Raton: Taylor and Francis Group.
Tong, J., Zimmerman, M.C., Li, S. et al. 2011. Neuronal uptake and intracellular superoxide scavenging of a fullerene (C60)-poly(2-oxazoline)s nanoformulation. *Biomaterials,* 32, 3654–65.
Tosi, G., Badiali, L., Ruozi, B. et al. 2012. Can leptin-derived sequence-modified nanoparticles be suitable tools for brain delivery? *Nanomedicine (Lond),* 7, 365–82.
Toumey, C. 2009. Plenty of room, plenty of history. *Nat Nanotechnol,* 4, 783–4.
Trickler, W.J., Lantz, S.M., Murdock, R.C. et al. 2011. Brain microvessel endothelial cells responses to gold nanoparticles: In vitro pro-inflammatory mediators and permeability. *Nanotoxicology,* 5, 479–92.
Trickler, W.J., Lantz, S.M., Schrand, A.M. et al. 2012. Effects of copper nanoparticles on rat cerebral microvessel endothelial cells. *Nanomedicine (Lond),* 7, 835–46.
Tsuji, J.S., Maynard, A.D., Howard, P.C. et al. 2006. Research strategies for safety evaluation of nanomaterials, part IV: risk assessment of nanoparticles. *Toxicol Sci,* 89, 42–50.
Turner, N.W., Liu, X., Piletsky, S.A. et al. 2007. Recognition of conformational changes in beta-lactoglobulin by molecularly imprinted thin films. *Biomacromolecules,* 8, 2781–7.

Ubbelohde, N., Fricke, C., Flindt, C. et al. 2012. Measurement of finite-frequency current statistics in a single-electron transistor. *Nat Commun,* 3, 612.

Ulbrich, K., Knobloch, T., Kreuter, J. 2011. Targeting the insulin receptor: nanoparticles for drug delivery across the blood–brain barrier (BBB). *J Drug Target,* 19, 125–32.

VanHandel, M., Alizadeh, D., Zhang, L. et al. 2009. Selective uptake of multi-walled carbon nanotubes by tumor macrophages in a murine glioma model. *J Neuroimmunol,* 208, 3–9.

Vo-Dinh, T. 2007. Nanotechnology in biology and medicine: the new frontier. In *Nanotechnology in Biology and Medicine: Methods, Devices, and Applications,* ed. T. Vo-Dinh, 1–9. Boca Raton, FL: CRC Press, Taylor & Francis Group.

Walkey, C.D., Chan, W.C. 2012. Understanding and controlling the interaction of nanomaterials with proteins in a physiological environment. *Chem Soc Rev,* 41, 2780–99.

Wartlick, H., Michaelis, K., Balthasar, S. et al. 2004. Highly specific HER2-mediated cellular uptake of antibody-modified nanoparticles in tumour cells. *J Drug Target,* 12, 461–71.

Winter P.M., Caruthers, S.D., Kassner, A., et al. 2003. Molecular imaging of angiogenesis in nascent Vx-2 rabbit tumors using a novel alpha(nu)beta3-targeted nanoparticle and 1.5 tesla magnetic resonance imaging. *Cancer Res*, 63, 5838–43.

2 Nanoneurology and Nanoneurosurgery

Geoffrey S. F. Ling, Robbin Miranda, and Gill Pratt

CONTENTS

INTRODUCTION

Of the diseases faced by medical providers, those afflicting the central and peripheral nervous systems are among the most vexing. Most are difficult to treat, and many cannot be cured. Because of years of dedicated effort in neuroscience, there has been increased understanding of the pathobiology of these conditions, which provides the foundation for new approaches and technologies. Nanotechnology, in particular, offers an exciting unique potential since advances in this realm are on a previously unattainable scale. By adapting emerging nanotechnologies to clinical care of neurological diseases, there are now opportunities to diagnose, treat, and eventually, cure.

The application of nanotechnology to medical care is "nanomedicine" or "nanobiomedicine." Over the last 20 years, nanomedicine concepts were refined, evaluated, and transitioned into actual capabilities. This offers the potential to revolutionize a wide array of medical devices and procedures including tissue manipulation, safe and highly targeted drug delivery, improved diagnostics, medical implants, and many others. Nanomedicine also offers the potential to combine diagnosis and therapy. Some of the first medical applications of nanotechnology occurred in the early 2000s, including the use of silver nanoparticles as an antibacterial agent and the development of nanoparticle-based sunscreens. However, it is only recently that the broad potential for nanomedicine has been realized as it is applied to diseases of the nervous system.

TARGETED DELIVERY OF NANOPARTICLES

Many current medical applications for nanotechnology entail delivery of nanoparticles to target locations within the body. Nanomedical drug delivery approaches enable transport of drug molecules to target tissue. Similar delivery mechanisms underlie new imaging approaches by using nanoparticles to elucidate specific cell or tissue types. Such imaging technologies have been utilized for diagnostic purposes and to assist in surgery, for instance, by illuminating tumors.

The small size of nanoparticles confers a number of advantages over microparticles, including the ability to travel to the bloodstream without sedimentation or microvasculature blockage, as well as the ability to circulate throughout the body and adhere to or penetrate specific target cell types via natural biological mechanisms such as endocytosis or receptor binding (Khademhosseini and Langer 2006). The sections below describe some of the general mechanisms and challenges associated with delivering nanoparticles to target locations, including encapsulation of the particles, injection into the body, circulation, and target specificity.

ENCAPSULATION OF NANOPARTICLES

Delivery of therapeutic agents to the central nervous system (CNS) is complicated by having to cross the blood–brain barrier (BBB). Nanoparticles (such as small molecules, DNA, RNA, or proteins) can penetrate this vascular endothelium–glia interface. This is generally facilitated by encapsulating the particles within an aqueous core and encompassing this core with biodegradable material such as polymers, dendrimers, or liposomes (Irvine 2011; Khademhosseini and Langer 2006). Such nanoparticles can be fabricated via top-down approaches, such as those based on integrated circuit processing, or bottom-up approaches, such as the generation of tissue engineering scaffolds from the self-assembly of nanomaterials (Khademhosseini and Langer 2006). Insertion of drug particles into nanoporous silica cores via adsorption-mediated drug loading can greatly improve the drug dose per particle, compared with currently employed clinical methods (Irvine 2011).

ADMINISTRATION OF NANOPARTICLES

Nanoparticles can be administered locally to the targeted area via injection or implantation or systemically via transdermal or intravenous approaches. One transdermal administration approach uses microneedles that penetrate the outer 10–20 μm of skin (Khademhosseini and Langer 2006). This approach is painless, since sensory nerves are not activated. Additional approaches specific to delivery of nanoparticles to the brain will be discussed in more detail below.

Individual administrations of nanoparticles are limited in their ability to provide sustained release of therapeutic agents over long periods of time. An alternative method of nanoparticle delivery is being developed that utilizes preprogrammed, remotely activated implanted devices, called "NanoGlands," to chronically release therapeutic nanoparticles (Grattoni et al. 2012). These devices consist of nanochannel membranes, through which analytes are delivered via electrokinetic transport mechanisms. The implantable capsule consists of a battery and electronic components that can control time, duration, and frequency of drug administration.

CIRCULATION OF NANOPARTICLES

One of the major challenges with targeted delivery of nanoparticles is controlling interactions between the nanoparticles and nontarget cells. Such interactions can limit the number of particles that reach the targeted locations and can also result in unwanted toxicity associated with nontargeted delivery of the nanoparticles. This is especially relevant for delivering toxic cancer drugs targeted at tumors. A widely used mechanism to overcome this challenge is to coat the nanoparticles with poly(ethylene glycol) (PEG) or polysaccharides that decrease surface interactions with nontarget cells and increase the circulation time of the nanoparticles (Khademhosseini and Langer 2006). Such an approach can also protect the nanoparticle drug from chemical or enzymatic degradation, for instance, by reducing interactions of the nanoparticles with macrophages (Heidarpour et al. 2011). A complementary approach to protect nanoparticles from degradation by the immune system is to use liposome–nanoparticle assemblies, which results in the concentration of small nanoparticles (Preiss and Bothun 2011).

An additional challenge associated with systemic circulation is the cleaning of debris in the bloodstream by the body's reticuloendothelial system, causing the majority of injected nanoparticles to accumulate in the liver, spleen, and lymph nodes, rather than at the targeted location(s) (Weldon et al. 2011). Adding long PEG chains on nanoparticles helps prevent removal of the particles from systemic circulation via the reticuloendothelial system, although the resulting steric hindrance can impair receptor-mediated uptake of the nanoparticles that may be required for targeted delivery (Yang 2010). A solution to this trade-off was developed by McNeeley et al. (2009), who describe a cleavable PEG that initially inhibits reticuloendothelial clearance of the nanoparticles. Once the nanoparticles reach the target tumors, the PEG coating is cleaved to enable receptor-mediated uptake by the target cells.

TARGET SPECIFICITY OF NANOPARTICLES

The encapsulated nanoparticles can be targeted to desired tissues through physical means, based on size, charge, and hydrophobicity of the nanoparticles, or by conjugating specific targeting molecules to the surface of the encapsulating material (Khademhosseini and Langer 2006). Such targeting molecules, or ligands, include proteins, nucleic acids, or other small molecules that recognize specific structures within the target cell surface, such as receptors (Irvine 2011; Khademhosseini and Langer 2006). Including multiple copies of targeting ligands on nanoparticle surfaces was shown to increase binding with target cells; however, a trade-off of this approach is that since multiple copies of identical surface molecules resemble viral structure, such nanoparticles are more susceptible to degradation via immune response attacks (Irvine 2011).

Nanoparticles are released upon degradation of the encapsulating material (e.g., bulk erosion; diffusion; external environment triggers including changes in pH, light, temperature; or presence of analytes such as glucose) (Khademhosseini and Langer 2006). Multiple drugs can also be included within different layers of encapsulation in a single nanoparticle, allowing temporal and sequential release of the drugs. In one example of a nanoparticle targeted to and taken up by tumor cells, degradation of the outer encapsulation layer results in release of an antiangiogenesis agent, which leads to localized vascular collapse and trapping of the nanoparticles within the tumor cells (Sengupta et al. 2005). An inner encapsulation layer of the nanoparticle then degrades, releasing a chemotherapy agent within the tumor cell. A current challenge is the ability to control degradation rate to enable sustained release of the encapsulated drugs.

DELIVERY OF NANOPARTICLES TO THE CNS

Delivery of drugs to the CNS presents its own unique challenges, that is, passage across the BBB. The BBB can be bypassed by administering drugs directly via brain injection or drug reservoir implants. These approaches are limited because they require highly invasive neurosurgical procedures. A transnasal approach was suggested (Biddlestone-Thorpe et al. 2011). More commonly, drugs are delivered systemically into the blood stream, after which they gain entry into the CNS by passing through the BBB. However, the BBB's glial–vascular complex's tight junctions restrict the passage of most drugs.

Only highly lipid-soluble molecules with an atomic mass of 400–600 Da or less can pass through the BBB (Khawaja 2011; Yang 2010). Although this limits traditional medications, it offers great opportunity for nanoparticles, as the small size and targeting specificity of nanoparticles confer their ability to "hitchhike" on the backs of particles that preferentially pass through the BBB.

Mechanisms upon which substances can cross the BBB include the paracellular aqueous pathway, transcellular lipophilic pathway, transport proteins, receptor-mediated transcytosis, and adsorptive-mediated transcytosis (Yang 2010). Transport of nanoparticles across the BBB is facilitated by two of these mechanisms: receptor-mediated transcytosis and adsorptive transcytosis. One example of the use of receptor-mediated transcytosis for transporting nanoparticles across the BBB includes the OX26 monoclonal antibody (mAb), which mimics the glycoprotein transferrin and binds to transferrin receptors at the BBB (Yang 2010). Chitosan–PEG nanospheres containing OX mAb were shown to cross the BBB following systemic administration in mice (Aktaş et al. 2005). One challenge in the use of mAbs in the transfer of nanoparticles across the BBB is that most brain-targeting mAbs are species-specific and vary in efficacy across species (Yang 2010). Thus, humanized forms of mAbs will need to be tested in humans to fully assess and validate their efficacy.

Adsorptive-mediated transcytosis is facilitated by interactions between positively charged moieties on circulating proteins and negatively charged regions on the surface of brain epithelial cells, resulting in endocytosis of the proteins (Herve' et al. 2008). However, such cationized proteins also tend to be adsorbed at anionic sites of all cells, thus greatly increasing the risk of nonspecific delivery and reducing distribution of the particles to target brain tissue (Herve' et al. 2008). Despite this risk, selected cationized proteins demonstrated high levels of distribution to the brain compared to other organs (Herve' et al. 2008).

NEUROSURGICAL APPLICATIONS OF NANOTECHNOLOGY (NANONEUROSURGERY)

The ability to deliver nanoparticles to targeted cell types within the CNS enabled advancement of novel neuroimaging and drug delivery techniques, which have numerous implications in both diagnostic and neurosurgical applications (Andrews 2009; Collins et al. 2010; Elder et al. 2008; Khawaja 2011; Weldon et al. 2011).

Perhaps the most widely developed application for delivery of nanoparticles to the brain is elucidating and treating brain carcinomas. It is often the case that CNS tumors cannot be removed surgically due to location within or proximal to critical brain regions or because the cancer is too small to be detected. Resections are often incomplete. Gliomas, due to their indistinct margins, are particularly difficult to fully remove. Targeted delivery of nanoparticles to brain tumor cells can help illuminate cancerous tissue via novel neuroimaging techniques, such that they can be seen by the surgeon in the operating room. Also, such delivery can provide therapeutic agents that specifically target and thus selectively kill the cancerous cells. This could reduce the need for an open craniotomy and neurosurgical removal of tissue. In addition to imaging and treating brain cancer, nanotechnology has also been applied to diagnosing and treating other brain-specific disorders and diseases such as Alzheimer's disease.

IMAGING OF BRAIN TUMORS

Nanotechnology enables imaging at cellular and subcellular levels. Two major categories of nanoparticles used for brain imaging include (1) nanosized imaging agents and (2) nanocarriers attached to imaging agents, while potentially exhibiting additional functional properties (Yang 2010).

Quantum dots (qdots), semiconducting nanocrystals that can enhance biological imaging, became prominent in the 1980s. Qdots consist of a metalloid crystalline core that is <10 nm in diameter, such as cadmium selenide (Andrews 2007, 2009). This is surrounded by a shielding shell, for example, zinc sulfide. Qdots have been used as both imaging agents as well as therapeutic carriers and can do both functions simultaneously. Qdots were demonstrated to pass through the BBB in vitro when conjugated with transferrin (Yang 2010). One of the unique features of qdots is that they are available in a variety of forms that are each excited by a different wavelength of light (Michalet et al. 2005). Thus, qdots can be used to color-code and track different cells. Although qdots provide an optical detection up to 1000 times better than conventional dyes, use is limited due to their size and risk of toxicity.

A new technology, Cornell Dots (C dots) developed at Cornell University (Ow et al. 2005), is currently under investigation as an improved means to highlight cancer cells in positron emission tomography (PET) imaging. C dots are less than 8 nm in diameter, sufficiently small to travel through the body and be subsequently excreted through urine. They are composed of silica spheres enclosing optical dye molecules (Ow et al. 2005). To extend the circulation time of tumor targeting, C dots are coated with PEG so they will not be recognized as foreign substances, thereby evading the immune system. Under near-infrared light, the dots fluoresce much brighter than traditional dye, enabling identification and visualization of target cells during surgical treatment, nodal mapping, as well as the extent of treatment response. This brightness allows cells to be tracked for longer periods.

An investigative clinical trial with C dots is underway with melanoma patients to verify safety, movement throughout the body, and time course elimination (Benezra et al. 2011; Friedman 2011). The silica-based particles will be labeled with radioactive iodine, and PET scans will be used to track their course. As the technology becomes more widespread, it is hopeful that C dots will improve tumor targeting and disease staging and enable patient-tailored cancer treatment.

Gold nanoparticles are also being investigated as a means of imaging brain tumor cells. Seekell and associates at Duke University immersed mouse brain slices containing tumors into a solution of gold nanorods conjugated to antibodies that bind to growth factor proteins found in high concentrations on the cancer cells (Seekell et al. 2011). The nanoparticles then accumulated in locations of the tumors and were visible when illuminated with a specific frequency of light. Importantly, gold nanorods can be synthesized at various length-to-width ratios. Because different-sized nanoparticles scatter different frequencies of light, they allow flexibility in choosing a light frequency that is not readily absorbed by biological tissue. Once accumulated into the tumor bed, gold nanoparticles can be used to kill cells (Kitz et al. 2011). When irradiated by nanosecond laser pulses, vapor bubbles are formed, which induce cell damage.

Dye-loaded polyacrylamide nanoparticles are being studied to enhance tumor visualization during surgery (Collins et al. 2010). These are methylene blue-containing nanoparticles on which the surface is conjugated with targeting peptides (Tang et al. 2008). In this way, the nanoparticles bind specifically to glioma cells. The blue dye is visible under ambient light to the neurosurgeon's naked eye. The benefit of this is that the tumor margin is better defined, which will allow more complete resection (Collins et al. 2010).

CONTRAST AGENTS FOR ENHANCED NEUROIMAGING

The progression of neuroimaging technology from standard x-ray to computed tomographic (CT) and magnetic resonance imaging (MRI) has improved the ability to noninvasively visualize and assess the nervous system. Likewise, the integration of contrast imaging agents with neuroimaging technologies has further improved the ability to clearly differentiate between blood vessels and surrounding tissues to elucidate tumors and vascular malformations. However, contrast materials are not without limitations. While most radio-opaque contrast materials have a half-life of 2 h, bolus injections and gadolinium-based contrast have a half-life of 5–10 and 12 min, respectively, impeding repeat imaging of a lesion. Additionally, these agents do not cross the BBB. Most importantly, iodinated and gadolinium contrast was shown to result in long-term disability or death in patients with renal insufficiency (Winer et al. 2011).

New, biocompatible contrast agents composed of various conjugates of ultrasmall superparamagnetic iron oxide nanoparticles (USPIONs) were investigated in animal and clinical models of cancer and were shown to provide improved signal intensity over traditional agents (Collins et al. 2010; Yang 2010). USPIONs are designed with prolonged half-life, improved absorption, and the ability to target specific cells. The presence of these iron oxide nanoparticles in the brain can be detected noninvasively using MRI and near-infrared fluorescence imaging (Yang 2010). USPION ferumoxtran-10 is currently undergoing clinical trial to brain image tumors and inflammation (Collins et al. 2010).

Fullerenes are also being studied as contrast agents in neuroimaging (Collins et al. 2010; D'Aquino et al. 2006; Elder et al. 2008). Endohedral metallofullerenes are metal ions encapsulated in a fullerene cage. This complex is highly stable and maintains its integrity under various biological conditions. For MRI use, gadolinium is encapsulated. An additional benefit is obtained by functionalizing fullerenes with hydroxyl or carboxyl groups to increase bioavailability and optimize renal elimination. By doing so, toxicity from bone and reticuloendothelial uptake is minimized. When compared to traditional gadolinium contrast agents, for example, Magnevist, gadofullerenes provide superior MRI images with much less agent, almost 20- to 40-fold less (Mikawa et al. 2001). This is due to the better T1 spin-lattice relaxation rates. An exciting potential opportunity is combining radioactive ^{177}Lu with gadolinium, which would result in simultaneous imaging and brachytherapy (Zhang et al. 2010). Thus, tumors would be imaged and treated simultaneously.

GOLD NANOPARTICLES TO DETECT EARLY-STAGE ALZHEIMER'S DISEASE

Due to their unique optical properties at the nanoscale level, gold nanoparticles have been utilized in biomedical imaging. Gold nanoparticles vary in their absorption and scattering of light depending on the particles' size, shape aggregation, and local environment (Murphy et al. 2008). In 2009, a team from Jackson State University demonstrated the use of gold nanoparticles to detect tau protein, a biomarker of Alzheimer's disease (Neely et al. 2009). Tau protein is involved in microtubule stability and has been observed at significantly higher concentrations, particularly in hyperphosphorylated form, in patients with mild cognitive impairment and Alzheimer's disease compared to control populations. Conjugating gold nanoparticles with anti-tau mAb enabled detection of tau protein at concentration levels two orders of magnitude lower than the Alzheimer's disease cutoff value for total tau protein in cerebrospinal fluid (CSF). In the presence of tau protein, several gold nanoparticles bound to each protein and formed nanoparticle aggregates. Compared to the nonaggregated gold nanoparticles, the formation of nanoparticle aggregates resulted in increased absorption of the 670-nm band wavelength of light, thus causing a visible color shift from red to blue. In addition, aggregation of gold nanoparticles in the presence of tau protein resulted in an approximately 16-fold increase in two-photon Rayleigh scattering intensity, enabling detection of tau protein at levels almost three orders of magnitude less than that using a colorimetric assay. This ultrasensitive method of detecting tau protein may lead to early diagnosis of Alzheimer's disease, enabling clinicians to effectively advance the onset of drug treatments that slow down progression of dementia.

NEURAL INTERFACES

Patients such as those with intractable epilepsy undergo implantation of microelectrode arrays consisting of sensory electrodes for neurophysiological mapping. Neurosurgical implantation of microwires for monitoring and therapy has become increasingly common. These are sensory electrodes. Patients with severe Parkinson's disease or other neurodegenerative disorders may undergo implantation of microelectrodes into the subthalamic nucleus or globus pallidus for deep brain stimulation. A limitation of chronic implantation of microelectrodes is glial scar formation around the microwire, which impedes the microwire's ability to function, whether it is to sense or stimulate. Nanotechnology may offer effective solutions. One is to create nanowires with carbon nanotubes (CNTs) (Andrews 2009; Yang et al. 2007). CNTs are graphite hollow tubes only nanometers in diameter and can be either single or multiple walled. They are durable and able to flex and conduct a high current density. Because of these features, they can be implanted without breaking, move when the brain moves, which reduces local tissue injury, and are small enough to minimize host inflammatory response. Furthermore, CNTs have the potential to be functionalized with bioactive molecules, thus providing potential for additional therapeutic roles (Collins et al. 2010; Elder et al. 2008).

NEUROPROTECTION

An especially frustrating area of neurotherapeutics is finding the most appropriate neuroprotection agent. In spite of decades of intense effort, there is still no effective neuroprotection medication for humans in clinical practice. Nanotechnology is offering new and different hope. Fullerenes are being developed as free-radical scavengers (Elder et al. 2008). In particular, hydroxyl-functionalized C-60 fullerenes called fullerenols have shown neuroprotection efficacy in preclinical models through glutamate receptor blockade and antioxidation activity (Silva 2005). The glutamate receptor antagonism leads to less glutamate-induced increase in intracellular calcium, which results in less excitotoxic and apoptotic cell death. The antioxidation effect is achieved through potentiation of the degradation of superoxide anions to oxygen and hydrogen peroxide. The peroxide is further degraded by endogenous peroxidases (Collins et al. 2010).

NEURO-ONCOLOGY

Nanotechnology is enabling a new class of therapy. Nanoparticles can be constructed that will specifically bind to targeted tumor cells. Attachment of a reporter such as a dye will illuminate the tumor. Additional attachment of a tumor toxic agent will simultaneously kill the tumor. Thus, a single class of nanoparticles can specifically mark, kill, and remark tumors. One example is the application of this approach to glioblastoma multiforme (GBM), a highly malignant brain tumor (Khawaja 2011). Once diagnosed, GBM typically leads to death in less than a year. Current therapy is neurosurgical resection with radiation therapy and chemotherapy with temozolomide; however, this improves survival by only a few months. Nano-LDL (low-density lipoprotein) particles for GBM therapy are an exciting new development. GBM upregulates LDL receptors. Nanoparticles are constructed with Apo-B100 (the LDL receptor binding domain), lecithin, cholesteryl oleate, and a lipid binding domain. This construction assures that the nanoparticles will avidly bind only to GBM cells. By tagging the nanoparticles with a fluorescent marker, the tumor will be illuminated. When paclitaxel oleate is added to the nano-LDL particles, they have an added capability to kill tumor cells. Thus, nano-LDL particles can diagnose and treat at the same time—noninvasively and with the potential of minimal toxicity (Khawaja 2011).

CNS AND PERIPHERAL NERVOUS SYSTEM TISSUE REPAIR

Trauma, stroke, and degeneration are common examples of injury to neurons, glia, nerves, and support structures. Effective techniques to repair such injury would be a major clinical advance. Traditional surgical and medical approaches have had limited benefit. Self-assembling peptide nanofiber scaffolds are recent innovations and have been shown to permit axons to grow through a site of injury, thereby allowing restoration of function, such as vision (Ellis-Behnke et al. 2006). Peptide amphiphiles can be introduced into a site needing repair. There, they will self-assemble into cylindrical nanofibers, similar in dimension to endogenous extracellular matrix (Collins et al. 2010). In an ionic environment, these amphiphiles will orient such that their hydrophobic ends will be in the center and their hydrophilic ends will be on the surface (Webber et al. 2010). The hydrophilic end can be engineered with a bioactive compound such as isoleucine–lysine–valine–alanine–valine (IKVAV). IKVAV can promote neural stem cells to become neurons rather than glia (Silva et al. 2004). CNTs also can be used as scaffolds and, when combined with appropriate DNA or protein, can support tissue regeneration (Collins et al. 2010; Elder et al. 2008; Khawaja 2011).

NANOROBOTICS FOR NEUROSURGERY

The design and simulation of nanorobotics have received much focus, yet such envisioned devices have not yet come to fruition but may soon have a revolutionary impact on medicine. While not yet nanosized, several miniaturized robotic systems for image-guided surgeries are in use. The SpineAssist device, developed by Mazor Robotics Ltd., allows surgeons to plan a spinal procedure in 3-D. Designed to reduce the need for CT scan during spinal implant procedures and improve placement accuracy, the device has been used in more than 2000 surgeries worldwide and has placed tens of thousands of implants in patients. In May 2011, SpineAssist received a European CE Mark for use in brain surgeries. Likewise, the StealthStation guidance system, developed by Medtronic Surgical Navigation Technologies, collects images from traditional imaging sources (x-ray, CT, MRI, and ultrasound) and transforms them into 3-D images, which can then be used by the surgeon to plan the surgical procedure as well as monitor the surgery in real time. The StealthStation system is utilized for brain, spinal, orthopedic, and ear/nose/throat surgeries.

CONCLUSION

Nanotechnology offers significant advancements in medical diagnostics and treatments. The challenge lies in fully transitioning these applications to practical medicine as they pertain to the nervous system.

In addition to its application for imaging and treating brain cancer and various neural disorders, nanotechnology has the potential to revolutionize the diagnosis and treatment of traumatic brain injury (TBI). One of the characteristics of explosive blast-related TBI is the diffuse nature of the injuries, making it nearly impossible to detect such injuries using macroscale imaging techniques such as MRI, although various correlates of this type of TBI have been observed at the cellular level. Nanoparticles targeted to such cellular abnormalities confer the potential to image or treat neural dysfunction resulting from TBI.

Nanoscale technologies not only have potential to revolutionize diagnostic and drug therapy capabilities in neuroscience but also could be applied toward applications such as brain–machine interfaces (BMIs). In order to promote a nanotechnology-facilitated BMI that would enable neural control of complex sensorimotor or cognitive tasks, imaging must occur in real time and across large populations of neurons. Ideally, such real-time imaging of nanoparticles would occur noninvasively. This remains a challenge; even though nanoparticles have been detected noninvasively through PET and MRI, the temporal resolution of these techniques is limited because of the brain's hemodynamic response properties.

Nanomedicine will remove the traditional boundaries separating the medical specialties of surgery, medicine, and radiology. As the scale of technical manipulation is reduced, so will the need to perform large open craniotomy tissue resection procedures. As technology becomes more focused on the cellular and subcellular levels, imaging will become more functionally based. When interventions are more specific, their applications are more clearly defined. So, for neurosurgeons, neurologists, and neuroradiologists, a time may come when the three specialties become one. Perhaps they will be referred to as neuromedical physicians, who, through neuronanomedicine, can diagnose, treat, and cure—all at the same time.

DISCLAIMER

The opinions expressed herein belong solely to the authors. They do not represent and should not be interpreted as being those of or endorsed by the US Army, the Defense Advanced Research Projects Agency, the Department of Defense, or any other branch of the federal government.

DISCLOSURES

The authors have no disclosures to make.

ACKNOWLEDGMENTS

The authors gratefully acknowledge the technical and editorial help of Nicole Draghic in the preparation of this manuscript.

REFERENCES

Aktaş, Y., Yemisci, M., Andrieux, K. et al. 2005. Development and brain delivery of chitosan–PEG nanoparticles functionalized with the monoclonal antibody OX26. *Bioconjug Chem* 16: 1503–11.

Andrews, R.J. 2007. Neuroprotection at the nanolevel—Part I: Introduction to nanoneurosurgery. *Ann N Y Acad Sci* 1122: 169–84.

Andrews, R.J. 2009. Nanotechnology and neurosurgery. *J Nanosci Nanotech* 9: 5008–13.

Benezra, M., Penate-Medina, O., Zanzonico, P.B. et al. 2011. Multimodal silica nanoparticles are effective cancer-targeted probes in a model of human melanoma. *J Clin Invest* 121: 2768–80.

Biddlestone-Thorpe, L., Marchi, N., Guo, K. et al. 2011. Nanomaterial-mediated CNS delivery of diagnostic and therapeutic agents. *Adv Drug Deliv Rev* 64: 605–13.

Collins, K.I., Orringer, D.A., and Patil, P.G. 2010. Nanotechnology in neurosurgery. *J Nanotech Eng Med* 1: 034001.

D'Aquino, R., Harper, T., and Vas, C.R. 2006. Nanobiotechnology—Fulfilling the promise of nanomedicine. *Chem Eng Prog* 102: 35–7.

Elder, J.B., Liu, C.Y., and Apuzzo, M.L. 2008. Neurosurgery in the realm of 10(-9). Part 2: Applications of nanotechnology to neurosurgery—Present and future. *Neurosurgery* 62: 269–84; discussion 284–5.

Ellis-Behnke, R.G., Liang, Y.X., You, S.W. et al. 2006. Nano neuro knitting: Peptide nanofiber scaffold for brain repair and axon regeneration with functional return of vision. *Proc Natl Acad Sci U S A* 103: 5054–9.

Friedman, R. 2011. Nano dot technology enters clinical trials. *J Natl Cancer Inst* 103: 1428–9.

Grattoni, A., Tasciotti, E., Fine, D. et al. 2012. Nanotechnologies and regenerative medical approaches for space and terrestrial medicine. *Aviat Space Environ Med* 83: 1025–36.

Heidarpour, F., Mohammadabadi, M.R., Zaidul, I.S. et al. 2011. Use of prebiotics in oral delivery of bioactive compounds: a nanotechnology perspective. *Pharmazie* 66: 319–24.

Herve', F., Ghinea, N., and Scherrmann, J.M. 2008. CNS delivery via adsorptive transcytosis. *AAPS J* 10: 455–72.

Irvine, D.J. 2011. Drug delivery: One nanoparticle, one kill. *Nat Mater* 10: 342–3.

Khademhosseini, A., and Langer, R. 2006. Nanobiotechnology—Drug delivery and tissue engineering. *Chem Eng Prog* 102: 38–42.

Khawaja, A.M. 2011. The legacy of nanotechnology: Revolution and prospects in neurosurgery. *Int J Surg* 9: 608 14.

Kitz, M., Preisser, S., Wetterwald, A. et al. 2011. Vapor bubble generation around gold nano-particles and its application to damaging of cells. *Biomed Opt Express* 2: 291–304.

McNeeley, K.M., Karathanasis, E., Annapragada, A.V. et al. 2009. Masking and triggered unmasking of targeting ligands on nanocarriers to improve drug delivery to brain tumors. *Biomaterials* 30: 3986–95.

Michalet, X., Pinaud, F.F., Bentolila, L.A. et al. 2005. Quantum dots for live cells, in vivo imaging, and diagnostics. *Science* 307: 538–44.

Mikawa, M., Kato, H., Okumura, M. et al. 2001. Paramagnetic water-soluble metallofullerenes having the highest relaxivity for MRI contrast agents. *Bioconjug Chem* 12: 510–14.

Murphy, C.J., Gole, A.M., Stone, J.W. et al. 2008. Gold nanoparticles in biology: Beyond toxicity to cellular imaging. *Acc Chem Res* 41: 1721–30.

Neely, A., Perry, C., Varisli, B. et al. 2009. Ultrasensitive and highly selective detection of Alzheimer's disease biomarker using two-photon Rayleigh scattering properties of gold nanoparticle. *ACS Nano* 3: 2834–40.

Ow, H., Larson, D.R., Srivastava, M. et al. 2005. Bright and stable core-shell fluorescent silica nanoparticles. *Nano Letters* 5: 113–7.

Preiss, M.R., and Bothun, G.D. 2011. Stimuli-responsive liposome–nanoparticle assemblies. *Expert Opin Drug Deliv* 8: 1025–40.

Seekell, K., Wilson, C., Price, H. et al. 2011. Controlled Synthesis of Gold Nanorods and Application to Brain Tumor Delineation. In *Frontiers in Optics*, ed. O. S. A. T. Digest, paper FWL4.

Sengupta, S., Eavarone, D., Capila, I. et al. 2005. Temporal targeting of tumour cells and neovasculature with a nanoscale delivery system. *Nature* 436: 568–72.

Silva, G.A. 2005. Nanotechnology approaches for the regeneration and neuroprotection of the central nervous system. *Surg Neurol* 63: 301–6.
Silva, G.A., Czeisler, C., Niece, K.L. et al. 2004. Selective differentiation of neural progenitor cells by high-epitope density nanofibers. *Science* 303: 1352–5.
Tang, W., Xu, H., Park, E.J. et al. 2008. Encapsulation of methylene blue in polyacrylamide nanoparticle platforms protects its photodynamic effectiveness. *Biochem Biophys Res Commun* 369: 579–83.
Webber, M.J., Kessler, J.A., and Stupp, S.I. 2010. Emerging peptide nanomedicine to regenerate tissues and organs. *J Intern Med* 267: 71–88.
Weldon, C., Tian, B., and Kohane, D.S. 2011. Nanotechnology for surgeons. *Wiley Interdiscip Rev Nanomed Nanobiotechnol* 3: 223–8.
Winer, J.L., Kim, P.E., Law, M. et al. 2011. Visualizing the future: enhancing neuroimaging with nanotechnology. *World Neurosurg* 75: 626–37; discussion 18–9.
Yang, H. 2010. Nanoparticle-mediated brain-specific drug delivery, imaging, and diagnosis. *Pharm Res* 27: 1759–71.
Yang, W., Thordarson, P., Gooding, J. et al. 2007. Carbon nonotubes for biological and biomedical applications. *Nanotechnology* 18: 412001.
Zhang, J., Fatouros, P.P., Shu, C. et al. 2010. High relaxivity trimetallic nitride (Gd3N) metallofullerene MRI contrast agents with optimized functionality. *Bioconjug Chem* 21: 610–5.

3 Nanopolymers and Nanoconjugates for Central Nervous System Diagnostics and Therapies

Giancarlo Suffredini and Lucien M. Levy

CONTENTS

INTRODUCTION

The National Nanotechnology Initiative defines nanotechnology as "…the understanding and control of matter at dimensions between approximately 1 and 100 nanometers, where unique phenomena enable novel applications" ("National Nanotechnology Initiative" 2010). Because there has been confusion concerning the nomenclature of nanoparticles, a comprehensive definition for pharmaceutical purposes defines nanoparticles as "…solid colloidal particles ranging in size from 10 to 1000 nm (1 μm). They consist of macromolecular materials in which the active principle (drug or biologically active material) is dissolved, entrapped, encapsulated and/or to which the active principle is adsorbed or attached" (Kreuter 2007, p. 8). Nanoparticles have been further categorized into seven different platforms: polymeric nanoparticles, polymeric nanoconjugates, micelles, peptides and proteins, metal complexes, carbon nanoplatforms, and dendrimers (Kateb et al. 2011).

POLYMERIC NANOPARTICLES AND POLYMERIC NANOCONJUGATES

Polymeric Nanoparticles

The term "polymeric nanoparticle" is a collective term that describes both nanospheres and nanocapsules (Muthu and Singh 2009; Pinto Reis et al. 2006). The structure of a nanosphere consists of a matrix in which the drug is physically and uniformly dispersed, either at the sphere surface or encapsulated within the vesicle. Nanocapsules confine the drug to a central cavity consisting of a liquid core surrounded by a polymer matrix (Leary et al. 2006) (Figure 3.1). A number of nontoxic and biodegradable polymers such as polylactic acid (PLA), polylactic-*co*-glycolic acid (PLGA), poly-L-caprolactone (PCL), various polyamino acids, polyphosphazene derivatives, as well as natural polymers such as albumin, collagen, gliadin, gelatin, chitosan, pullulan, and curdlan may be used to encapsulate drugs without chemical modification (Peer et al. 2007; Zhao et al. 2009).

PLA has long been recognized to undergo acid-catalyzed homopolymerization to yield a low-molecular-weight polymer. The cyclic diester of lactic acid can polymerize using the anionic ring-opening addition mechanism, creating materials that can then be cast into strong films or spun into fibers (Kulkarni et al. 1966). PLA polymers undergo hydrolysis in vivo to form lactic acid, which is metabolized by the citric acid cycle. Drug release from a PLA polymer is controlled by diffusion of the drug through the matrix and by esterase-mediated degradation of the polymer (Shive and Anderson 1997).

PLGA is a synthetic copolymer of lactic acid (α-hydroxypropanoic acid) and glycolic acid (hydroxyacetic acid). Lactic acid and glycolic acid have nearly identical molecular structures, with lactic acid bearing an additional methylene group at the beta carbon position. The copolymer degrades by bulk erosion through hydrolysis of the ester bonds. The copolymer ratio of lactic acid to glycolic acid affects the rate of degradation; higher lactic acid concentration decreases the rate of degradation due to decreased water imbibition. The size of the spheres is also a principle determinant of the rate of polymer degradation and drug release. As the size of the particle increases, the surface area/volume ratio decreases, resulting in a slower rate of drug release. Increased molecular weight is also a factor in retarding the erosion process (Pai et al. 2009).

Both PLA and PLGA have been studied extensively and have inclusion status in the Food and Drug Administration's Generally Recognized as Safe (GRAS) list for use in medical devices and drug formulations. The typical drug release profile from PLA or PLGA polymers consists of an initial "burst" release of payload in proximity to the surface of the microsphere, followed by relatively little additional release of payload by diffusion through the microsphere pores (generated by hydrolysis of the polymer backbone) (Shive and Anderson 1997). Although this pharmacokinetic profile may not be ideal, modification via PEGylation of protein payloads carried by a PLA or PLGA depot can prevent the pharmacokinetic "burst" profile and promote sustained release of active conjugate (Pai et al. 2009).

Polymeric nanoparticles have received a great deal of attention due to their stability and ease of surface modification. The most important features of polymeric nanoparticles are their size and surface chemistry. Nanoparticle size affects the concentration profile in blood vessels, the mechanism of clearance (Table 3.1), and extravasation from blood vessels (Veiseh et al. 2010). Most endothelial barriers allow uncoated nanoparticles less than 150 nm to pass. Surface chemistry manipulation

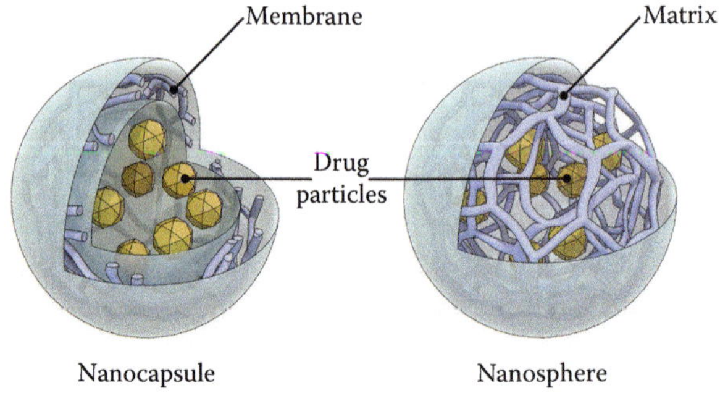

FIGURE 3.1 Nanocapsule confines the drug to a central cavity surrounded by a polymer matrix. A nanosphere consists of a matrix in which the drug is physically and uniformly dispersed, either at the sphere surface or encapsulated within the vesicle.

TABLE 3.1
Nanoparticle Clearance Mechanism by Size

Size	Clearance from Circulation
<20 nm	Renal
30–150 nm	Bone marrow, heart, stomach, kidney
150–300 nm	Liver and spleen

is necessary to cross more restrictive endothelial barriers such as the blood–brain barrier (BBB). Polysorbate coating of polymeric nanoparticles facilitates Apolipoprotein E adherence and has been reported to improve brain targeting through interactions (receptor-mediated endocytosis) with the low-density lipoprotein receptor present on BBB endothelial cells (Hasadsri et al. 2009; Steiniger et al. 2004). Furthermore, polysorbate inhibits the effects of the *p*-glycoprotein activity in glioma cell lines (Chavanpatil et al. 2007). Controlled drug release from polymeric nanoparticles is governed by surface or bulk erosion, diffusion through the polymer matrix, swelling followed by diffusion, and changes in local environment, such as changes in pH (Peer et al. 2007; Singh and Lillard 2009).

Polymeric Nanoconjugates

Polymeric nanoconjugates contain multiple biochemically active groups covalently bound to the same nanoconjugate molecule and are designed to target specific cellular and molecular markers of diseased tissues (Ljubimova et al. 2008) (Figure 3.2). The functional modules contained within and on the surface of a nanoconjugate serve to efficiently deliver, protect, and target the drug payload. Surface moieties such as monoclonal antibodies and peptides directed against cell surface antigens unique to a given cell population or overexpressed in tumors are used to drive cellular uptake via the endosomal pathway (Ljubimova et al. 2008). Poly-β-malic acid (PMLA) is a water-soluble and biodegradable polymer that has received much attention for its use in the biosynthesis of polymeric nanoconjugates. PMLA is a linear polyester containing a pendant carboxyl group available for conjugation with different functional groups. Polymeric nanoconjugates significantly differ from polymeric nanoparticles because they contain a covalently linked payload. Nanocarrier formulations that do not contain covalent linkages, such as polymeric nanoparticles, are susceptible to damage, which leads to premature leakage and release of their payload, which could potentially result in toxicity to healthy tissue (Ljubimova et al. 2008). Furthermore, polymeric nanoparticles rely on passive targeting, whereas polymeric nanoconjugates may exploit both active and passive targeting.

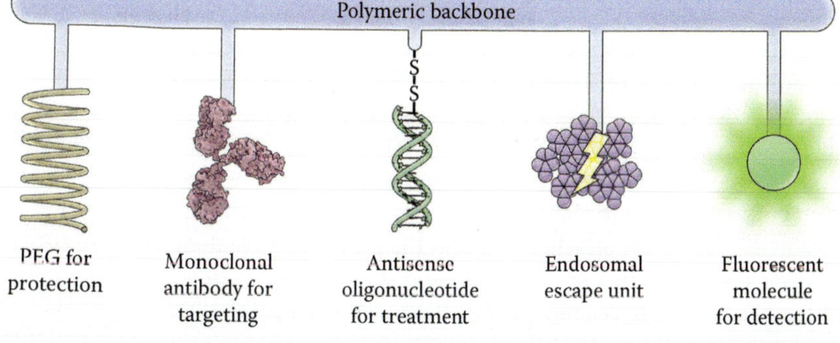

FIGURE 3.2 Polymeric nanoconjugates contain a polymeric backbone and multiple covalently bound biochemically active groups, which are designed to target specific cellular and molecular markers of diseased tissues.

NANOPARTICLE DRUG DELIVERY

Principles

The idea that targeted molecular delivery could greatly improve drug therapy is attributed to Nobel laureate Paul Ehrlich and his envisioned use of "magic bullets" for the therapy of various diseases (Kreuter 2007). Drug therapies may be targeted through active and passive mechanisms. Passive targeting relies on the enhanced permeability and retention effect (EPR) and the bystander effect. Effective delivery of nanoparticles to target tissue through the blood stream requires nanoparticles to escape the scavenging effect of the reticuloendothelial system by virtue of their small size (size evasion) and to avoid binding to plasma proteins (surface evasion). After reaching the vasculature of target tissue, nanoparticles selectively accumulate in the tumor due to unique characteristics of the tumor microenvironment, including increased microdensity of vessels in the tumor, vessels with widened gap junctions, and compromised extracellular or lymphatic drainage, factors responsible for the EPR (Maeda et al. 2003). Active targeting is achieved by attaching agents to the surface of a nanocarrier that bind to a specific cell type. Current strategies for intravascular delivery employ a system composed of the ligand or antibody as a targeting moiety, a polymer or lipid as the carrier, and a sophisticated payload delivery system that enables drug release and endosomal escape (Ljubimova et al. 2008; Peer et al. 2007). An alternative but more invasive delivery strategy is intratumoral delivery, in which a catheter is placed within the tumor, nanoparticles are infused directly into the target organ or tumor tissue, and nanoparticles spread through the extracellular space to reach the cells of interest (Sahoo et al. 2002; Singh and Lillard 2009).

Therapeutic Applications

Polymeric nanoparticles have been a focus for the treatment of a variety of neurodegenerative diseases and central nervous system (CNS) malignancies because of their high drug-loading capacity and ability to cross the BBB (Modi et al. 2010; Tosi et al. 2008). A variety of anticancer drugs including paclitaxel (Fonseca et al. 2002; Koziara et al. 2006; Petrelli et al. 2010), doxorubicin (DOX) (Alyautdin et al. 1997; Xiong et al. 2010; Yoo et al. 2000), 5-fluorouracil (5-FU) (Bhadra et al. 2003; Yan et al. 2010), and dexamethasone (Panyam and Labhasetwar 2004) have been incorporated into polymeric nanoparticles. Nanoparticle albumin bound paclitaxel (nab P) has been approved for human use. Testing of the other systems is ongoing in animal models of Alzheimer's disease and glioblastoma (Table 3.2).

DOX [(8S,10S)-10-(4-amino-5-hydroxy-6-methyl-tetrahydro-2*H*-pyran-2-yloxy)-6,8,11trihydroxy-8-(2-hydroxyacetyl)-1-methoxy-7,8,9,10-tetrahydrotetracene-5,12-dione] is synthesized from daunorubicin, a class I anthracycline antibiotic derived from *Streptomyces peucetius*. DOX disrupts DNA synthesis and transcription by intercalating into nuclear DNA via electrostatic and structural interactions and by binding to topoisomerase II. Polymeric nanocarriers of DOX have been developed that prevent the distribution of DOX to normal cells and that increase DOX levels in tumor cells by bypassing drug resistance mechanisms, such as the *p*-glycoprotein efflux transporter found in cancerous cells. Two-block copolymer–DOX nanoconjugates based on poly(ethylene oxide)-block-poly(ε-caprolactone) micelles with αvβ3 integrin-targeting ligand (receptor overexpressed on endothelial cells and certain cancer cells) on the micellar surface have been recently shown to improve the therapeutic efficacy of DOX in both DOX-sensitive and resistant human cancer cell lines (Xiong et al. 2010). The two delivery vehicles covalently attached DOX to the poly-ε-caprolactone core by either an amide or hydrazone bond. The amide linkage formulation resulted in more selective accumulation of DOX in mitochondria, whereas the hydrazone formulation increased cellular uptake of DOX, producing preferential nuclear accumulation. Both formulations offer potential solutions to treat DOX-resistant cancer cells. Cardiotoxicity is known to be associated with DOX, presumed currently to result from the generation of free oxygen radicals and the peroxidation of lipid membranes, increased calcium release through alterations in the ryanodine calcium receptor, and inhibition of topoisomerase II.

TABLE 3.2
Examples of Potential Nanodiagnostics and Nanotherapies

Disease	Therapeutic Agent	Potential Therapy
Alzheimer's disease	Clioquinol (CQ)	A quinoline derivative that is known to solubilize β-amyloid plaques in vitro and has been shown to inhibit the accumulation of β-amyloid in vivo in mice (Cherny et al. 2001). CQ-encapsulated polybutylcyanoacrylate (PBCA) delivered intravenously can cross the BBB at a higher threshold than CQ alone, making it a promising candidate to treat Alzheimer's disease.
	Thioflavin-T (Th-T)	A benzothiazole dye that exhibits enhanced fluorescence upon binding to amyloid fibrils (Khurana et al. 2005). Successful delivery of Th-T-loaded nanocapsules, prepared by the polymerization of styrene, delivered by direct intrahippocampal injection into the brains of mice has been reported (Hartig et al. 2003).The authors suggest that intravenous delivery of ThT-encapsulated PBCA has potential as a β-amyloid detection method for Alzheimer's disease.
	Rivastigmine tartrate	A reversible cholinesterase inhibitor used to treat Alzheimer's disease. Currently this drug is given orally, requiring frequent dosing because of its restricted entry into the brain due to its hydrophilicity. A variety of cholinergic side effects result from such frequent dosing, including bradycardia, nausea, dyspepsia, vomiting, and anorexia. Rivastigmine-loaded polylactic-*co*-glycolic acid (PLGA) and PBCA have recently been studied in an animal model demonstrating a significantly higher transport of the drug into the brain than rivastigmine solution (Joshi et al. 2010).
	Tacrine	A cholinesterase inhibitor to treat Alzheimer's disease. Recently, a polysorbate-coated PBCA nanoparticle formulation of tacrine was shown in animal studies to achieve significantly greater concentrations in the brain when compared to uncoated nanoparticle and free drug (Wilson et al. 2008).
Cancer	Abraxane	A cremophor-free albumin-bound nanoparticle formulation of paclitaxel that eliminates the cremophor-associated infusion toxicities and requirement for pretreatment steroid medication. The results showed that Abraxane was very well tolerated and should be further evaluated in a randomized setting in both adjuvant and neoadjuvant trials.
	Doxorubicin bound to polysorbate coated PBCA	This nanoparticle formulation of doxorubicin can cross the BBB and reach therapeutic concentrations in a rat glioma model (Alyautdin et al. 1997).
	TiO_2–DNA nanoconjugates	Titanium dioxide nanoparticles covalently bound to dopamine, which in turn is bound to a single-strand DNA oligonucleotide, are being investigated for their application in antisense oligonucleotide therapy. Photoactivation of these nanoparticles causes a charge separation within the nanoparticle, resulting in the migration of an electropositive hole from the nanoparticle onto the DNA, resulting in its scission (Thurn et al. 2009).

Cardiotoxicity of DOX could be accentuated if the amide linkage formulation of DOX was allowed to increase the level of DOX in the mitochondria of normal myocardial cells. To avoid this cardiotoxicity, targeting moieties on nanoparticles with the amide linkage formulation must avoid nonspecific binding to myocardial cells.

5-FU is a pyrimidine analogue that prevents DNA synthesis in cancer cells by noncompetitive inhibition of thymidylate synthase. Since effective suppression of mitotic activity with this agent requires prolonged inhibition of DNA synthesis, the preferred method of drug delivery for 5-FU is by low-dose continuous intravenous infusion ("Meta-Analysis Group in Cancer" 1998;

Tanaka et al. 2000). Chronic infusions are unfortunately associated with complications including catheter-related vasculitis and sepsis (Desai et al. 2002; Hansen et al. 1996). Avoidance of these complications would improve the tolerability of 5-FU in cancer patients. Recently, an injectable, biodegradable, controlled-release PEG-hydrogel formulation of 5-FU has been developed, which can be delivered subcutaneously (Yi et al. 2010). A cross-linking reaction between two PEG solutions (6-arm polyethylene glycol *N*-hydroxysuccinimidyl glutarate and 6-arm polyethylene glycol amine) forms the hydrogel structure that entraps 5-FU. At its optimal concentration within 15% PEG-hydrogel, 5-FU maintained stable serum concentrations for 4 days in a rat model. This method for providing prolonged delivery of a therapeutic agent offers an alternative approach to long-term intravenous infusion of a therapy such as 5-FU.

The nanoparticle formulation of paclitaxel serves as the paradigm for improving drug delivery of existing pharmaceuticals with nanotechnology. Paclitaxel [(2α,4α,5β,7β,10β,13α)-4,10-bis(acetyloxy)-13-{[(2*R*,3*S*)-3-(benzoylamino)-2-hydroxy-3-phenylpropanoyl]oxy}-1,7-dihydroxy-9-oxo-5,20-epoxytax-11-en-2-yl benzoate] is a clinical anticancer agent with activity against breast, lung, ovarian, and head and neck cancer. Paclitaxel binds the β subunit of tubulin, which hyperstabilizes the microtubule structure of the cell and destroys its ability to use its cytoskeleton in a flexible manner. It also binds to Bcl-2 protein and induces apoptosis (Bhalla et al. 1993; Schiff and Horwitz 1980). Paclitaxel is hydrophobic, and until recently, all available formulations required Cremophor EL (a polyoxyethylated castor oil) and an ethanol vehicle to enhance its solubility for parenteral administration. These solubilizing agents have been implicated as the cause of serious adverse side effects such as hypersensitivity reactions and peripheral neuropathy (Szebeni et al. 1998; Windebank et al. 1994).

Nanoparticle albumin-bound paclitaxel (Abraxane, Abraxis BioScience, Bridgewater, NJ) is an albumin-bound form of paclitaxel with a mean particle size of 130 nm. The nanoparticle colloidal suspension is prepared by high-pressure homogenization of paclitaxel in the presence of serum albumin (Petrelli et al. 2010). After intravenous injection, the nanoparticles dissolve into smaller particles with a mean size of 10 nm, similar in size to that of endogenous serum albumin. Current models show that nab-P has ready bioavailability after intravenous infusion and accumulates in tumors via transcytosis (Petrelli et al. 2010). Abraxane is indicated for treating breast cancer after failure of combination chemotherapy or relapse within 6 months of adjuvant chemotherapy ("Abraxane" 2008). A phase II clinical trial recently evaluated the efficacy of Abraxane as a neoadjuvant that was followed by 5-fluorouracil/epirubicin/cyclophosphamide in patients with locally advanced breast cancer. Abraxane was very well tolerated in this trial, which supports its further evaluation in a randomized setting in both adjuvant and neoadjuvant trials (Robidoux et al. 2010).

Active targeting with polymeric nanoconjugates is a promising method that can be individually tailored to patients with a variety of diseases. Polycefin is an anticancer nanoconjugate drug delivery system based on a PMLA polymer backbone. Currently, PMLA-conjugated drug systems include antisense oligonucleotides to messenger RNAs (mRNAs) of the α4 and β1 chains of laminin-411 (overexpressed in the vessel walls of high-grade glial tumors, primary breast cancers, and breast cancer metastases). An antibody to the vascular endothelial transferrin receptor (TfR) is the targeting module that facilitates passage of polycephin across the blood–tumor barrier and into the tumor. The nanoconjugate binds to the TfR on the surface of tumor endothelial cells, undergoes transcytosis, subsequently interacts with the tumor cell TfR, and enters the endosome of the tumor cell.

Passive targeting has limited drug specificity in certain tumors, such as those not exhibiting the EPR effect and those having inhomogeneous vessel permeability (Jain 1994). Active targeting increases specificity and decreases the toxic effects that conventional systemically administered drugs have on healthy tissue. In active targeting, nanocarriers are internalized through receptors on specific cells that mediate their endocytosis (Table 3.3). After entering the target cell, polymeric nanoconjugates that have been engineered to bypass multidrug resistance transporters can overcome drug resistance, a bane of conventional chemotherapy (Lee et al. 2006).

TABLE 3.3
Examples of Specific Ligands Used for Targeting

Ligand	Receptor	Target
Albumin	SPARC	Endothelial cells, tumor cells
Transferrin	Transferrin receptor (TfR)	Tumor cells
Lipoprotein B, E	Lipoprotein receptor	Brain
Folate	Folate receptor	Tumor cells
Mannose	Mannose receptor	Antigen-presenting cells
Antibody against receptor	Erb/HER	Tumor cells
	EGF-R	Tumor cells

Note: SPARC = secreted protein acidic and rich in cysteine; EGF-R = epidermal growth factor receptor; Erb/HER = type 1 transmembrane receptor tyrosine kinases.

Methods of Drug Delivery

Polymeric nanoparticles and nanoconjugates may be delivered locally or systemically. Polymeric implants slowly release drugs locally into the extracellular space of the tumor or brain. Drug distributes by diffusion around the implant, which limits therapeutic drug levels to tissues within 1–2 mm of the implant. Delivery of a therapeutic concentration of drug throughout a larger volume of target tissue can be achieved using the convection-enhanced delivery technique (Bobo et al. 1994). Co-infusion of the therapeutic drug with an imaging tracer such as gadolinium-diethylenetriamine pentaacetate (DTPA) during real-time magnetic resonance imaging allows one to monitor and adjust the volume of distribution of therapeutic drug (Croteau et al. 2005; Heiss et al. 2010; Lonser et al. 2007; Nguyen et al. 2003). Nanoparticles have been distributed widely throughout the brain using the convection-enhanced delivery technique (Hadjipanayis et al. 2010; Vinchon-Petit et al. 2010; Voges et al. 2003). Local drug delivery requires invasive methods to place a slow-release polymer or an intracerebral catheter for convection-enhanced delivery. Although entering the brain can be achieved with little risk of permanent disability, the necessity of performing a surgical procedure and of staying in the hospital during an infusion makes local brain delivery more invasive and costly than systemic delivery, in which drug(s) can be administered during an outpatient visit.

Systemic drug delivery for CNS-related diseases can be achieved through injection, inhalation, ingestion, or absorption through the skin. Nanoparticles pass the BBB or blood–tumor barrier through receptor-mediated transcytosis or adsorptive-mediated transcytosis. Loperamide (4-[4-(4-chlorophenyl)-4-hydroxypiperidin-1-yl]-*N,N*-dimethyl-2,2-diphenylbutanamide) is a μ-opioid receptor agonist that cannot readily cross the BBB. Recently, intravenous administration of a polymeric formulation of loperamide [poly(D,L-lactide-*co*-glycolide), surface-decorated with the peptide Gly-L-Phe-D-Thr-Gly-L-Phe-L-Leu-L-Ser(O-β-D-glucose)-CONH2] has been reported as being successfully delivered to the brain in a mouse model. This nanoparticle provided efficiency of biodistribution two orders of magnitude greater when compared to other nanoparticle drug carriers (Vergoni et al. 2009). The peptide used for surface decoration on the nanoparticle vehicle exhibits a sequence-specific effect that promotes successful crossing of the BBB. In vivo studies are ongoing to elucidate further details of the mechanism of this novel method of drug transport into the CNS.

Intranasal administration is an attractive drug delivery route that allows drugs to circumvent the BBB and access the brain (Lerner et al. 2004). Successful entry of drugs into the brain via intranasal administration of a liposomal formulation of rivastigmine has been reported (Arumugam et al. 2008). A multilamellar liposomal formulation of rivastigmine administered intranasally reaches concentrations in the brain that are 10 times that of free drug administered intranasally and 5 times greater than with oral delivery of rivastigmine. The pharmacokinetic profile of this liposomal formulation exhibited "burst"-type kinetics, with 56% of the rivastigmine payload being delivered within 6 h of administration. Presumably, the remainder of the payload was not released because drug concentrations equilibrated at the donor and recipient sites and prevented further drug release.

Oral administration of nanoparticles may increase drug levels by avoiding presystemic hepatic metabolism. Polymeric nanoparticles that are less than 500 nm can cross M cells in the Peyer's patches of the intestine and enter the lymphatic system rather than the portal vein (Muthu and Singh 2009). Transdermal drug delivery is a noninvasive method to provide steady and sustained drug levels (Cevc and Vierl 2010).

DIAGNOSTIC APPLICATION

Polymeric nanoparticles and nanoconjugates have the potential to lead to advances in medical diagnostics. A recent study reports that nanoconjugates composed of titanium dioxide nanoparticles, DNA oligonucleotides, and a gadolinium contrast agent achieved significantly greater intracellular accumulation, retention, and subcellular localization when compared to the gadolinium contrast agent alone (Paunesku et al. 2008). Diseased tissues should be detected with greater sensitivity than present magnetic resonance techniques allow because the nanoconjugate-bound agents increase T1 signal and contrast both immediately and on a delayed basis from prolonged retention of gadolinium. This formulation contains a large number of potential binding sites for targeting molecules, creating the possibility of tissue-specific imaging; the oligonucleotide sequence can be modified to enhance binding to various subcellular organelles. Polymeric nanoparticles may also increase the sensitivity of laboratory diagnostics. They have been engineered to exhibit fluorescence upon binding disease-specific markers. One such example is in a test to detect Alzheimer's disease in brain tissue, which depends upon thioflavin-T, a benzothiazole dye that exhibits enhanced fluorescence upon binding to amyloid fibrils (Khurana et al. 2005).

SIDE EFFECTS AND TOXICITY

Nanoparticle delivery systems aim to improve drug efficacy by decreasing systemic clearance of nanoparticles and by increasing drug uptake into target cells. To date, there have been few studies reporting specific in vivo toxicity of nanopolymers and nanoconjugates. Concerns have been raised about the effects of polymeric nanoparticles on the BBB. An in vitro model suggests that polybutylcyanoacrylate (PBCA) nanoparticles cause nonspecific opening of the BBB, presumably related to the effects of serum esterases required for their biodegradation (Olivier et al. 1999).

SUMMARY

Polymeric nanoparticles and polymeric nanoconjugates are promising platforms for drug delivery. The central goal to be achieved through these technologies is to increase the specificity of therapeutics and diagnostics and to limit toxicity in healthy tissues. This approach has the potential to accelerate developing novel treatments and imaging agents for CNS malignancies, multiple sclerosis, and neurodegenerative diseases such as Parkinson's and Alzheimer's diseases.

ACKNOWLEDGMENT

This research was supported by the Intramural Research Program of the National Institute of Neurological Disorders and Stroke at the National Institutes of Health.

REFERENCES

Abraxane (paclitaxel) indications and dosage. 2008. DrugLib.com http://www.druglib.com/druginfo/abraxane/indications_dosage/.

Alyautdin, R.N., Petrov, V.E., Langer, K. et al. 1997. Delivery of loperamide across the blood–brain barrier with polysorbate 80-coated polybutylcyanoacrylate nanoparticles. *Pharm Res* 14: 325–8.
Arumugam, K., Subramanian, G.S., Mallayasamy, S.R. et al. 2008. A study of rivastigmine liposomes for delivery into the brain through intranasal route. *Acta Pharm* 58: 287–97.
Bhadra, D., Bhadra, S., Jain, S. et al. 2003. A PEGylated dendritic nanoparticulate carrier of fluorouracil. *Int J Pharm* 257: 111–24.
Bhalla, K., Ibrado, A.M., Tourkina, E. et al. 1993. Taxol induces internucleosomal DNA fragmentation associated with programmed cell death in human myeloid leukemia cells. *Leukemia* 7: 563–8.
Bobo, R.H., Laske, D.W., Akbasak, A. et al. 1994. Convection-enhanced delivery of macromolecules in the brain. *Proc Natl Acad Sci U S A* 91: 2076–80.
Cevc, G., and Vierl, U. 2010. Nanotechnology and the transdermal route: A state of the art review and critical appraisal. *J Control Release* 141: 277–99.
Chavanpatil, M.D., Khdair, A., Gerard, B. et al. 2007. Surfactant-polymer nanoparticles overcome *p*-glycoprotein-mediated drug efflux. *Mol Pharm* 4: 730–8.
Cherny, R.A., Atwood, C.S., Xilinas, M.E. et al. 2001. Treatment with a copper–zinc chelator markedly and rapidly inhibits beta-amyloid accumulation in Alzheimer's disease transgenic mice. *Neuron* 30: 665–76.
Croteau, D., Walbridge, S., Morrison, P.F. et al. 2005. Real-time in vivo imaging of the convective distribution of a low-molecular-weight tracer. *J Neurosurg* 102: 90–7.
Desai, A.A., Vogelzang, N.J., Rini, B.I. et al. 2002. A high rate of venous thromboembolism in a multi-institutional phase II trial of weekly intravenous gemcitabine with continuous infusion fluorouracil and daily thalidomide in patients with metastatic renal cell carcinoma. *Cancer* 95: 1629–36.
Fonseca, C., Simoes, S., and Gaspar, R. 2002. Paclitaxel-loaded plga nanoparticles: Preparation, physicochemical characterization and in vitro anti-tumoral activity. *J Control Release* 83: 273–86.
Hadjipanayis, C.G., Machaidze, R., Kaluzova, M. et al. 2010. EGFRvIII antibody-conjugated iron oxide nanoparticles for magnetic resonance imaging-guided convection-enhanced delivery and targeted therapy of glioblastoma. *Cancer Res* 70: 6303–12.
Hansen, R.M., Ryan, L., Anderson, T. et al. 1996. Phase III study of bolus versus infusion fluorouracil with or without cisplatin in advanced colorectal cancer. *J Natl Cancer Inst* 88: 668–74.
Hartig, W., Paulke, B.R., Varga, C. et al. 2003. Electron microscopic analysis of nanoparticles delivering thioflavin-T after intrahippocampal injection in mouse: Implications for targeting beta-amyloid in Alzheimer's disease. *Neurosci Lett* 338: 174–6.
Hasadsri, L., Kreuter, J., Hattori, H. et al. 2009. Functional protein delivery into neurons using polymeric nanoparticles. *J Biol Chem* 284: 6972–81.
Heiss, J.D., Walbridge, S., Asthagiri, A.R. et al. 2010. Image-guided convection-enhanced delivery of muscimol to the primate brain. *J Neurosurg* 112: 790–5.
Jain, R.K. 1994. Barriers to drug delivery in solid tumors. *Sci Am* 271: 58–65.
Joshi, S.A., Chavhan, S.S., and Sawant, K.K. 2010. Rivastigmine-loaded PLGA and PBCA nanoparticles: Preparation, optimization, characterization, in vitro and pharmacodynamic studies. *Eur J Pharm Biopharm* 76: 189–99.
Kateb, B., Chiu, K., Black, K.L. et al. 2011. Nanoplatforms for constructing new approaches to cancer treatment, imaging, and drug delivery: What should be the policy? *Neuroimage* 54 Suppl 1: S106–24.
Khurana, R., Coleman, C., Ionescu-Zanetti, C. et al. 2005. Mechanism of thioflavin T binding to amyloid fibrils. *J Struct Biol* 151: 229–38.
Koziara, J.M., Whisman, T.R., Tseng, M.T. et al. 2006. In-vivo efficacy of novel paclitaxel nanoparticles in paclitaxel-resistant human colorectal tumors. *J Control Release* 112: 312–9.
Kreuter, J. 2007. Nanoparticles—a historical perspective. *Int J Pharm* 331: 1–10.
Kulkarni, R.K., Pani, K.C., Neuman, C. et al. 1966. Polylactic acid for surgical implants. *Arch Surg* 93: 839–43.
Leary, S.P., Liu, C.Y., and Apuzzo, M.L. 2006. Toward the emergence of nanoneurosurgery: Part III—nanomedicine: Targeted nanotherapy, nanosurgery, and progress toward the realization of nanoneurosurgery. *Neurosurgery* 58: 1009–26; discussion 09–26.
Lee, B.S., Fujita, M., Khazenzon, N.M. et al. 2006. Polycefin, a new prototype of a multifunctional nanoconjugate based on poly(beta-L-malic acid) for drug delivery. *Bioconjug Chem* 17: 317–26.
Lerner, E.N., van Zanten, E.H., and Stewart, G.R. 2004. Enhanced delivery of octreotide to the brain via transnasal iontophoretic administration. *J Drug Target* 12: 273–80.
Ljubimova, J.Y., Fujita, M., Ljubimov, A.V. et al. 2008. Poly(malic acid) nanoconjugates containing various antibodies and oligonucleotides for multitargeting drug delivery. *Nanomedicine (Lond)* 3: 247–65.
Lonser, R.R., Schiffman, R., Robison, R.A. et al. 2007. Image-guided, direct convective delivery of glucocerebrosidase for neuronopathic gaucher disease. *Neurology* 68: 254–61.

Maeda, H., Fang, J., Inutsuka, T. et al. 2003. Vascular permeability enhancement in solid tumor: Various factors, mechanisms involved and its implications. *Int Immunopharmacol* 3: 319–28.
Meta-analysis Group in Cancer. 1998. Efficacy of intravenous continuous infusion of fluorouracil compared with bolus administration in advanced colorectal cancer. *J Clin Oncol* 16: 301–8.
Modi, G., Pillay, V., and Choonara, Y.E. 2010. Advances in the treatment of neurodegenerative disorders employing nanotechnology. *Ann N Y Acad Sci* 1184: 154–72.
Muthu, M.S., and Singh, S. 2009. Targeted nanomedicines: Effective treatment modalities for cancer, aids and brain disorders. *Nanomedicine (Lond)* 4: 105–18.
National Nanotechnology Initiative. 2010. What is nanotechnology? http://www.nano.gov/html/facts/whatIsNano.html.
Nguyen, T.T., Pannu, Y.S., Sung, C. et al. 2003. Convective distribution of macromolecules in the primate brain demonstrated using computerized tomography and magnetic resonance imaging. *J Neurosurg* 98: 584–90.
Olivier, J.C., Fenart, L., Chauvet, R. et al. 1999. Indirect evidence that drug brain targeting using polysorbate 80-coated polybutylcyanoacrylate nanoparticles is related to toxicity. *Pharm Res* 16: 1836–42.
Pai, S.S., Tilton, R.D., and Przybycien, T.M. 2009. Poly(ethylene glycol)-modified proteins: Implications for poly(lactide-*co*-glycolide)-based microsphere delivery. *AAPS J* 11: 88–98.
Panyam, J., and Labhasetwar, V. 2004. Sustained cytoplasmic delivery of drugs with intracellular receptors using biodegradable nanoparticles. *Mol Pharm* 1: 77–84.
Paunesku, T., Ke, T., Dharmakumar, R. et al. 2008. Gadolinium-conjugated TiO_2–DNA oligonucleotide nanoconjugates show prolonged intracellular retention period and T1-weighted contrast enhancement in magnetic resonance images. *Nanomedicine* 4: 201–7.
Peer, D., Karp, J.M., Hong, S. et al. 2007. Nanocarriers as an emerging platform for cancer therapy. *Nat Nanotechnol* 2: 751–60.
Petrelli, F., Borgonovo, K., and Barni, S. 2010. Targeted delivery for breast cancer therapy: The history of nanoparticle-albumin-bound paclitaxel. *Expert Opin Pharmacother* 11: 1413–32.
Pinto Reis, C., Neufeld, R.J., Ribeiro, A.J. et al. 2006. Nanoencapsulation I. Methods for preparation of drug-loaded polymeric nanoparticles. *Nanomedicine* 2: 8–21.
Robidoux, A., Buzdar, A.U., Quinaux, E. et al. 2010. A phase II neoadjuvant trial of sequential nanoparticle albumin-bound paclitaxel followed by 5-fluorouracil/epirubicin/cyclophosphamide in locally advanced breast cancer. *Clin Breast Cancer* 10: 81–6.
Sahoo, S.K., Panyam, J., Prabha, S. et al. 2002. Residual polyvinyl alcohol associated with poly (D,L-lactide-co-glycolide) nanoparticles affects their physical properties and cellular uptake. *J Control Release* 82: 105–14.
Schiff, P.B., and Horwitz, S.B. 1980. Taxol stabilizes microtubules in mouse fibroblast cells. *Proc Natl Acad Sci U S A* 77: 1561–5.
Shive, M.S., and Anderson, J.M. 1997. Biodegradation and biocompatibility of PLA and PLGA microspheres. *Adv Drug Deliv Rev* 28: 5–24.
Singh, R., and Lillard, J.W., Jr. 2009. Nanoparticle-based targeted drug delivery. *Exp Mol Pathol* 86: 215–23.
Steiniger, S.C., Kreuter, J., Khalansky, A.S. et al. 2004. Chemotherapy of glioblastoma in rats using doxorubicin-loaded nanoparticles. *Int J Cancer* 109: 759–67.
Szebeni, J., Muggia, F.M., and Alving, C.R. 1998. Complement activation by cremophor el as a possible contributor to hypersensitivity to paclitaxel: An in vitro study. *J Natl Cancer Inst* 90: 300–6.
Tanaka, F., Fukuse, T., Wada, H. et al. 2000. The history, mechanism and clinical use of oral 5-fluorouracil derivative chemotherapeutic agents. *Curr Pharm Biotechnol* 1: 137–64.
Thurn, K.T., Paunesku, T., Wu, A. et al. 2009. Labeling TiO_2 nanoparticles with dyes for optical fluorescence microscopy and determination of TiO_2–DNA nanoconjugate stability. *Small* 5: 1318–25.
Tosi, G., Costantino, L., Ruozi, B. et al. 2008. Polymeric nanoparticles for the drug delivery to the central nervous system. *Expert Opin Drug Deliv* 5: 155–74.
Veiseh, O., Gunn, J.W., and Zhang, M. 2010. Design and fabrication of magnetic nanoparticles for targeted drug delivery and imaging. *Adv Drug Deliv Rev* 62: 284–304.
Vergoni, A.V., Tosi, G., Tacchi, R. et al. 2009. Nanoparticles as drug delivery agents specific for cns: In vivo biodistribution. *Nanomedicine* 5: 369–77.
Vinchon-Petit, S., Jarnet, D., Paillard, A. et al. 2010. In vivo evaluation of intracellular drug-nanocarriers infused into intracranial tumours by convection-enhanced delivery: Distribution and radiosensitisation efficacy. *J Neurooncol* 97: 195–205.
Voges, J., Reszka, R., Gossmann, A. et al. 2003. Imaging-guided convection-enhanced delivery and gene therapy of glioblastoma. *Ann Neurol* 54: 479–87.
Wilson, B., Samanta, M.K., Santhi, K. et al. 2008. Targeted delivery of tacrine into the brain with polysorbate 80-coated poly(*n*-butylcyanoacrylate) nanoparticles. *Eur J Pharm Biopharm* 70: 75–84.

Windebank, A.J., Blexrud, M.D., and de Groen, P.C. 1994. Potential neurotoxicity of the solvent vehicle for cyclosporine. *J Pharmacol Exp Ther* 268: 1051–6.

Xiong, X.B., Ma, Z., Lai, R. et al. 2010. The therapeutic response to multifunctional polymeric nano-conjugates in the targeted cellular and subcellular delivery of doxorubicin. *Biomaterials* 31: 757–68.

Yan, G.P., Zong, R.F., Li, L. et al. 2010. Anticancer drug-loaded nanospheres based on biodegradable amphiphilic epsilon-caprolactone and carbonate copolymers. *Pharm Res* 27: 2743–52.

Yi, H., Cho, H.J., Cho, S.M. et al. 2010. Pharmacokinetic properties and antitumor efficacy of the 5-fluorouracil loaded PEG-hydrogel. *BMC Cancer* 10: 211.

Yoo, H.S., Lee, K.H., Oh, J.E. et al. 2000. In vitro and in vivo anti-tumor activities of nanoparticles based on doxorubicin–PLGA conjugates. *J Control Release* 68: 419–31.

Zhao, Z., He, M., Yin, L. et al. 2009. Biodegradable nanoparticles based on linoleic acid and poly(beta-malic acid) double grafted chitosan derivatives as carriers of anticancer drugs. *Biomacromolecules* 10: 565–72.

4 Micelles and Liposomes
Lipid Nanovehicles for Intracerebral Drug Delivery

Massimo S. Fiandaca and Krystof S. Bankiewicz

CONTENTS

INTRODUCTION

One of the earliest nanoscale platforms developed for therapeutic delivery to the central nervous system (CNS) featured nanoparticles whose walls were composed of phospholipid moieties that form a "core-shell structure" (Blanco et al. 2009; Torchilin 2001). Formulation and chemical modification of these basic structures allow these delivery platforms to carry varying amounts and types of therapeutic substances and, in some cases, to specifically target the delivery of therapeutics to particular cell types within the CNS. The ability to traverse the blood–brain barrier (BBB) in certain cases, while avoiding clearance by the immune surveillance system, allows some nanoplatforms to enter the CNS without requiring a neurosurgical procedure. Micelles and liposomes, the topics of this chapter, together with the other nanoscale platforms described in other chapters of this book, are playing an ever-increasing role in the diagnostic and therapeutic armamentarium of the nanoneurosurgeon and nanoneuroscientist.

Within this chapter, we will assess the historical development and chemical nature of nanomicelles and nanoliposomes. Consideration will be given to the differences and similarities between these two nanocarriers. We will also review the known cellular incorporation mechanisms associated with these nanoparticles. Finally, a brief introduction to specific applications of these diagnostic and therapeutic delivery vehicles will be presented, emphasizing how their structure and function make them particularly suited for these clinical roles. Greater detail on the applications of these and other nanoplatforms in current and future clinical applications will be addressed by other chapter authors in this volume.

MICELLES AND LIPOSOMES FOR INTRACEREBRAL DRUG DELIVERY

Historical and Chemical Basis for Lipid Nanostructures

Plants and animals were the original sources of the fats (conjugated fatty acids) that were eventually developed into soaps (detergents) through reactions with strong bases (e.g., potassium or sodium hydroxide, KOH or NaOH) (Figure 4.1). The breakdown of these conjugated fatty acids resulted in a polyalcohol (e.g., glycerol) and fatty acid salts of potassium or sodium (i.e., a soap). Mixed in water, these soaps (or detergents) have acted as cleaning agents for hundreds of years. The basis for these cleaning properties of soaps (as well as other properties) was not fully elucidated until J.W. McBain's pioneering work in the early 20th century (McBain 1913), for which he was elected a Fellow of the Royal Society of London in 1923. He noted electrical conductivity in soapy solutions and coined the term "colloidal ions" as being responsible for this property. Working in parallel, additional investigation by McBain and by G.S. Hartley (Hartley 1936; Hartley and Runnicles 1938) resulted in understanding the structural basis for these properties of soaps and the conceptualization and nomenclature for the micelle. The soap molecules are amphiphilic, meaning that they have hydrophilic (water-loving) and hydrophobic (water-fearing) properties within the same molecule (Figure 4.1). The polar (hydrophilic) side of the soap attracts water molecules that dissolve away molecularly charged (water-soluble) substances. The nonpolar (hydrophobic) hydrocarbon tails of the soap are lipophilic (fat loving) and attract and dissolve noncharged hydrophobic substances (e.g., oils, grease, gasoline). These soap molecules can exist as individual monomers in water, with single-chain or double-chain hydrocarbon tails, tending to accumulate at the surface,

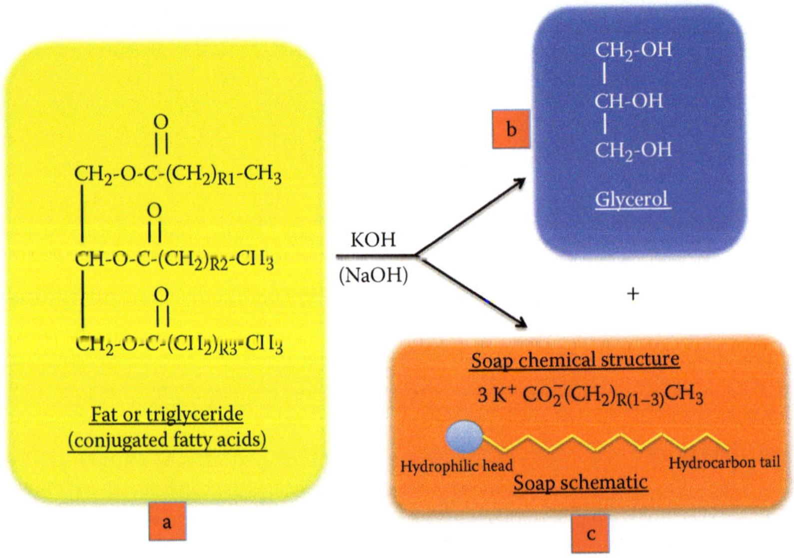

FIGURE 4.1 Basic chemical structures of fats and soaps. Conjugated fatty acids attached to glycerol are the common components of fat or triglycerides (a). Triglycerides found in nature commonly have dissimilar hydrocarbon chain lengths in each molecule, but most typically range from 16 to 20 carbon atoms. Since both plants and animals synthesize fats with acetyl CoA, which conveys 2 carbon atoms at a time to the synthesis, most of these naturally occurring triglycerides have an even number of carbons in the fatty acids. Reaction of a fat with a strong base will break down the triglyceride structure into glycerol (b) and salts of the three fatty acids (c). Note that the R groups (R1, R2, and R3) of the fat designate the different chain lengths included in a single triglyceride (a and c). Note that the chemical structure of the soaps yielded from the reaction can contain up to three different chemical structures, R(1–3) (c). The schematic representation of soap features the hydrophilic head group, made up of the charged end of the molecule and the hydrophobic hydrocarbon tail (c). This schematic depicts the typical structure of a saturated fatty acid (without double bonds).

with hydrophobic tails extending into the air above the water and the hydrophilic heads extending into the water. Above a specific concentration of monomers (critical micelle concentration [CMC]) (Baeurle and Kroener 2004), however, these individual soap molecules can aggregate to form a spherical or ellipsoidal structure (a process called micellization), with the charged (hydrophilic) end of the soap on the surface of the structure and toward the water and the hydrocarbon (hydrophobic) end within the structure. Micellization is the result of two major forces: the hydrophobic force of water on the hydrocarbon tails and the electrorepulsive forces on the polar heads of soap molecules (Tanford 1974) (Figure 4.2).

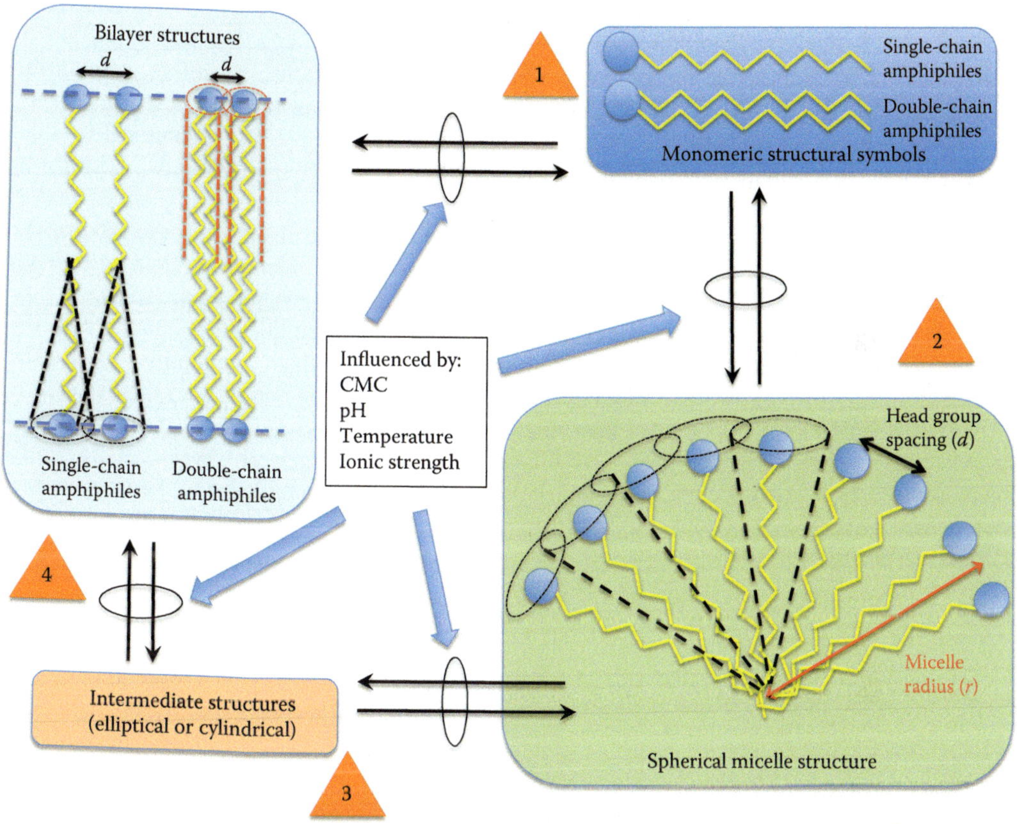

FIGURE 4.2 Schematic representation of structural equilibrium for amphiphilic molecules in aqueous solution. There is equilibrium present within a solution that allows the varying preponderance of four major structural entities, influenced by critical micellar concentration (CMC), temperature, pH, and ionic strength. (1) Monomeric structural symbols for single- and double-chain amphiphiles. The blue circle represents the polar heads, while the yellow zigzag lines represent the hydrocarbon tails (simplified to show saturated fatty acids). (2) Single-chain amphiphiles prefer to aggregate in a spherical micelle structure with a radius (*r*) that depends on the length of the hydrocarbon tails. Dashed lines and ellipses indicate the preferred steric occupancy volumes around the amphiphiles that help determine the preferred spherical micellar shape for single-chain species. Dotted ellipses around the polar head groups represent the spherical circumference of hydration shells associated with the hydrophilic end. Head group spacing (*d*) is determined by the nature of the charged group and its hydration shell and helps determine the preferred number of amphiphiles per micelle. (3) Intermediate structures, including elliptical and cylindrical configurations, are possible, based on the influential factors present within the solution. (4) Bilayer structures can also be created, either spontaneously, as with double-chain amphiphiles, or with significant influences (see center of figure) provided to the single-chain amphiphiles. Dashed (red vs. black) lines indicate differences in steric occupancy volumes between single-chain amphiphiles (black cone-shaped volumes) and double-chain amphiphiles (red cylindrical volumes). These steric differences allow decreased energy requirement for reduced head group spacing (*d*) with double-chain amphiphiles as opposed to single-chain molecules.

It is the balance of these forces (Tanford 1974), partly determined by the structural chemistry of the soap involved, together with temperature of the solution (Krafft temperature, critical micellar temperature), pH, and ionic strength, that helps determine the structure and size of the soap micelle (Ikeda et al. 1980; Sugihara et al. 1982).

Single-chain amphiphiles tend to form spherical micelles rather than bilayers, due to the steric conditions determined by the charged head groups on the surface of a sphere and the rather limited steric effects of the hydrocarbon chain (Figure 4.2). Head group spacing is maximized on a spherical shape, while it is reduced toward a minimum in a bilayer structure (Katsaras and Gutberlet 2010). Extrinsic effects can influence the equilibrium between single-chain amphiphilic structures, from the preferred spherical micelle to other structural possibilities (Figure 4.2). The alkyl group length of the single-chain amphiphile monomers determines the micelle radius. Therefore, varying the length of the hydrophobic chain most commonly alters micelle size (and radius). For a given monomer structure, there are an optimal number of amphiphiles that form a stable spherical micelle, based on the surface charge dispersion requirements (Figure 4.2) and micelle radius. Attempts to enlarge the micelle by adding more monomers destabilize the optimal structure by either forcing the charged head groups closer together (decreasing the surface charge dispersion) or expanding the micelle beyond the micelle radius, thereby undermining the central hydrophobic domain through the introduction of water. Elliptical and cylindrical structures are possible but require special conditions and additional energy; however, they permit incorporation of additional monomers into the structure. Ultimately, such environmental factors may drive single-chain amphiphiles toward the forming bilayers, which allow relatively unlimited incorporation of monomers into their structure. In contrast, double-chain amphiphiles favor the formation of bilayers (Figure 4.2), and although they can exist in other structural forms, the bilayer structure is energetically preferred. Tanford's forces (Tanford 1974) are different, with double-chain amphiphiles as opposed to the single-chain species, due to increased hydrophobic stability and steric restriction associated with the second hydrocarbon chain (Figure 4.2).

Nanomicelle and Nanoliposome Structure and Production

Nanoparticles, such as nanomicelles and nanoliposomes, typically range in size from 10 to several hundred nanometers (10^{-9} m). Phospholipid nanomicelles tend to form spontaneously and have been produced historically by placing soaps or detergents in solution and raising their concentration to or above CMC. As discussed above, micelles are typically spherical and have an outer surface of charged or hydrophilic moieties and an inner lipophilic region (Figure 4.3). Phospholipid micelles can be utilized to carry other amphiphilic molecules or, based on the micelle radius, significant amounts of hydrophobic substances within their inner lipophilic region (Kabanov et al. 1989; Mittal 1977) (Figure 4.3a). Currently, amphiphilic block copolymers are commonly used to produce clinical nanomicelles to deliver therapeutics (Jones and Leroux 1999). Polymer micelle categories include hydrophobically assembled or polyion-complex structures (Gaucher et al. 2005). The former structures are typically amphiphilic copolymers that spontaneously form nanoparticles in aqueous solution (Figure 4.3b). Tailoring the micelle structure with copolymers enables variable drug delivery kinetics and particle longevity (Moghimi and Hunter 2001; Torchilin 2001), while maintaining the basic micellar size and architecture. The nature of the hydrophobic core-forming copolymer blocks provides the basis for the delivery kinetics of these nanomicelles (Attwood et al. 2007). Using specific block copolymers featuring polyethylene glycol (PEG) as the hydrophilic block, a hydrophilic surface corona can be enhanced to reduce the affinity of opsonizing proteins, thus reducing phagocytic clearance and extending circulating half-life (Moghimi and Hunter 2001). Polyion-complex micelles use charged copolymer blocks to bind and deliver therapeutic charged proteins or nucleic acids (Oishi et al. 2005; Tian et al. 2005).

Liposomes, on the other hand, are typically composed of double-chain phospholipid amphiphiles combined with cholesterol, forming spheroidal bilayer membrane structures that encompass

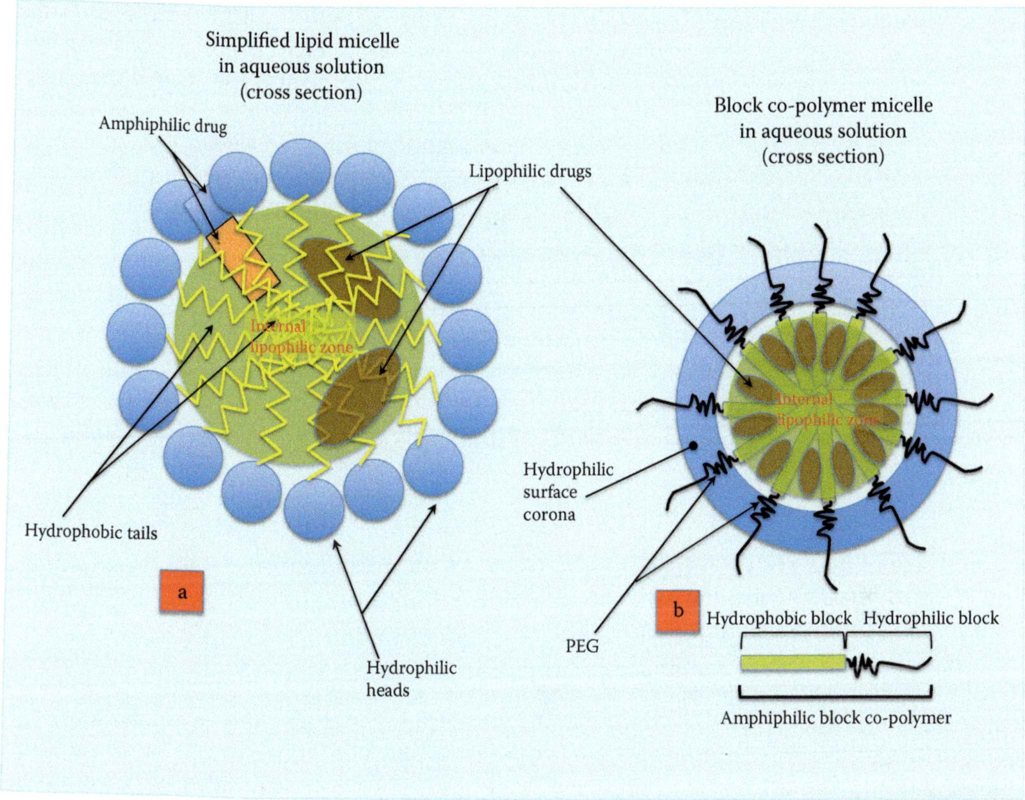

FIGURE 4.3 Schematic representation of the structure of a micelle in aqueous solution. (a) In a simplified lipid micelle, single-chain amphiphiles tend to aggregate themselves in a spherical structure in solution once the CMC is exceeded. The classic spherical lipid micelle typically contains the hydrophilic head groups on the surface and the hydrophobic tails within the center. The internal lipophilic zone acts as a reservoir for hydrophobic drugs. Certain amphiphilic substances (drugs or other compounds) can also incorporate into the substance of the micelle (blue/orange rectangle). (b) With a block copolymer micelle, an internal lipophilic zone is formed by the hydrophobic block of the copolymer, while the hydrophilic block, typically PEG, is present on the micellar surface and helps form a hydration shell, or hydrophilic surface corona, that suppresses opsonin binding and phagocytosis.

an aqueous internal domain (Torchilin 2001, 2005) (Figure 4.4). The length of the fatty acid chains and the presence or absence of double bonds within the liposome bilayer lipids affect membrane fluidity, as does the combination of different phospholipids within the membrane structure (Chapman 1975; Chapman and Benga 1984). Cholesterol moieties strengthen and stabilize the bilayer membrane (Kates and Manson 1984) and reduce cation leakage in physiologic systems (Haines 2001). Increasing the molar cholesterol content of liposomal drug carriers typically reduces the release kinetics of the therapeutic from the nanocarrier (Panwar et al. 2010). Therefore, specific liposomal properties can be tailored by the membrane component makeup (Panwar et al. 2010; Szoka and Papahadjopoulos 1980). Liposomes are typically formed by the addition of energy to amphiphilic phospholipids in aqueous solution. As discussed in the previous section on the chemical basis of lipid nanostructures, double-chain amphiphiles tend to preferentially form bilayer structures. Based on lipid monomer concentration and environmental factors, however, liposomal structures can range from long tubules to spheres, with dimensions ranging from several hundred angstroms to fractions of a millimeter (Pagano and Weinstein 1978). A prototypical liposomal vesicle has a single, closed lipid bilayer (unilamellar) confining a single internal aqueous volume (Figure 4.4). Outer bilayer

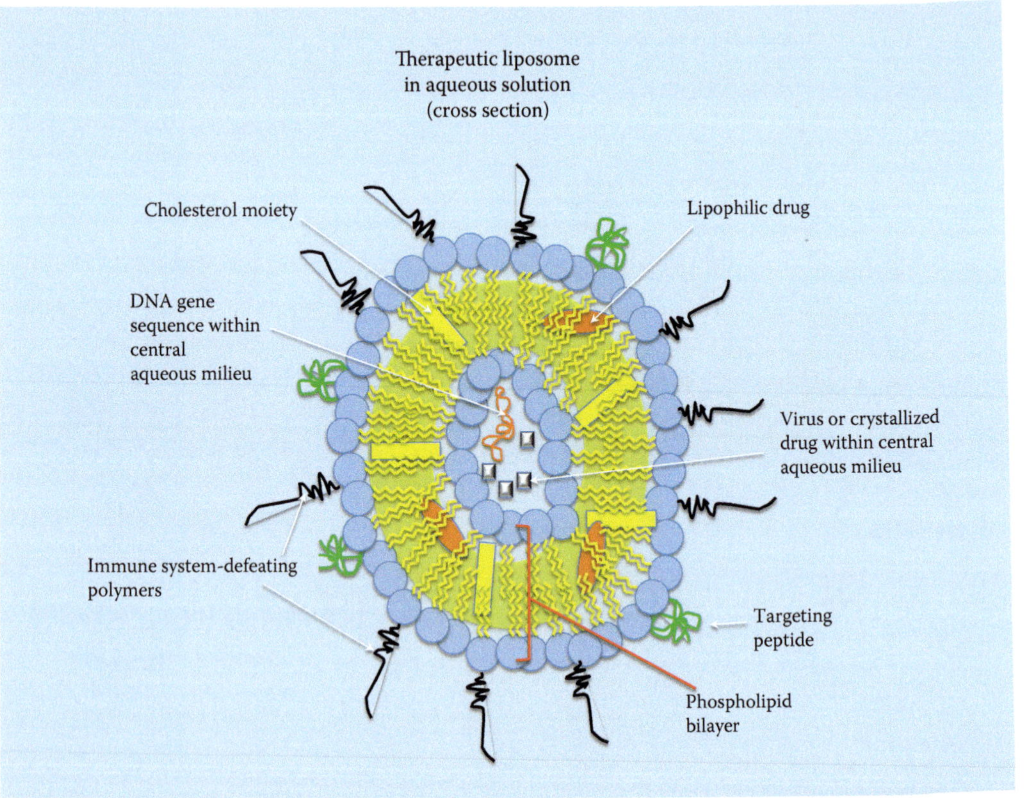

FIGURE 4.4 Schematic representation of a therapeutic liposome in aqueous solution. The double-chain amphiphiles arrange themselves in a phospholipid bilayer vesicle, with water surrounding the outside of the liposome and within the central aqueous milieu. Cholesterol within the lipid bilayer (yellow rectangles) stabilizes the lipid membrane. Complex targeting molecules (green) are shown on the outer surface of the bilayer structure and enable preferential binding of the liposome to cell surface receptors for possible receptor-dependent cellular uptake. Immune system–defeating molecules (black) on the outer surface of the bilayer structure, through their enhanced steric effects, decrease opsonization and phagocytic clearance while increasing liposome longevity and availability for cellular internalization. Within the phospholipid bilayer (red brackets), there is a lipophilic zone within which lipophilic drugs (ovoid structures) can be assimilated and transported. Finally, small and large hydrophilic molecular species can exist within the liposome's aqueous core, including proteins, crystalline drugs, genetic material, viruses, and other particulates.

membrane modifications with PEG, similar to those seen in block copolymer micelles, can be used to give the liposomal structure "stealth" characteristics (Immordino et al. 2006; Klibanov et al. 1990; Murayama 2002; Murayama et al. 1990; Torchilin et al. 1994) as well as cellular targeting options with the addition of certain surface ligands (Bareford and Swaan 2007; Hillaireau and Couvreur 2009; Marsh and Helenius 2006). The three basic types of liposomal structures (Figure 4.5) include multilamellar vesicles (MLVs, typically >500 nm), small unilamellar vesicles (SUVs, <100 nm), and large unilamellar vesicles (LUVs, ≥100 nm) (Pagano and Weinstein 1978; Samad et al. 2007; Szoka and Papahadjopoulos 1980). Sonication of phospholipids in an aqueous solution can produce liposomes (Pagano and Weinstein 1978) but, in extreme circumstances, can damage the vesicles. Low shear conditions favor development of MLVs, while increasing shear produces LUVs and, finally, SUVs. Regardless of the preparation methodologies (Deamer and Bangham 1976; Gould-Fogerite and Mannino 1993; Kikuchi et al. 1994; Lasic 1988; Lasic et al. 2001; Mozafari et al. 2002a,b, 2004; Pick 1981; Schieren et al. 1978; Szoka and Papahadjopoulos 1978; Talsma et al. 1994; Vemuri

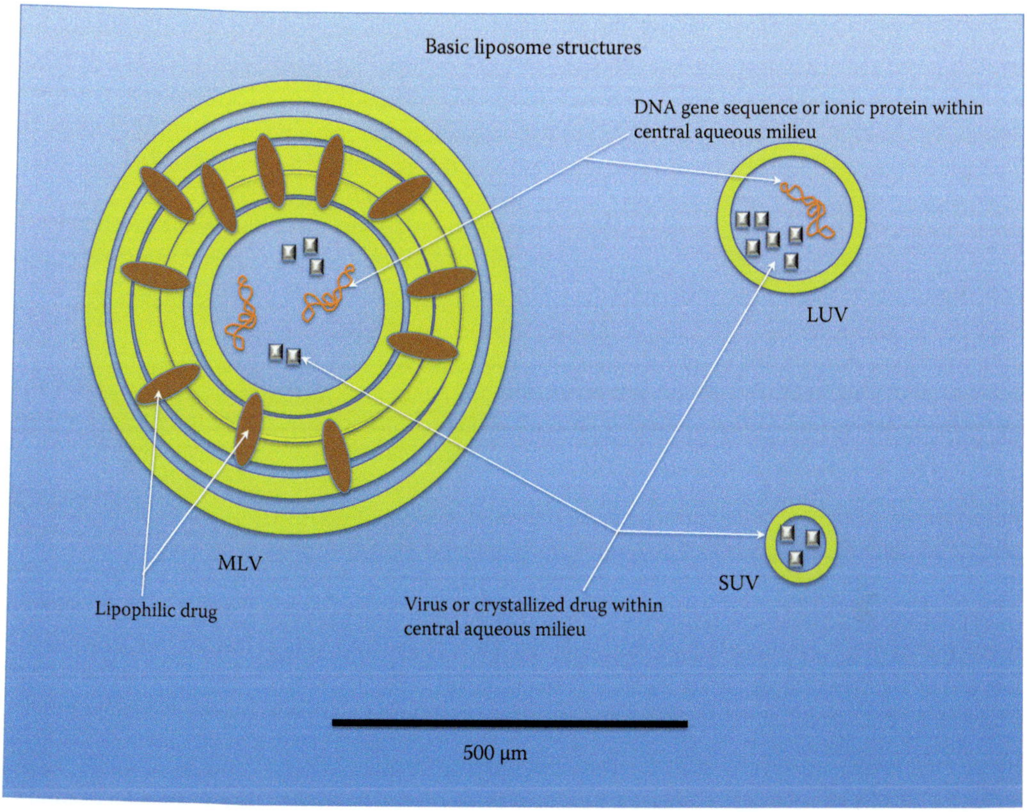

FIGURE 4.5 Schematic representation of liposomal structures and their relative sizes. Liposomes can be manufactured to contain all of the previously noted bilayer structures seen in Figure 4.4 but, in addition, can have varying sizes and numbers of bilayer membranes. *Multilamellar vesicles* (*MLVs*) are structured to have an increased lipophilic volume through the onion-like layering of multiple phospholipid membranes. With typical sizes at >500 nm, they represent some of the largest and most complex of the liposomal structures. MLVs have the greatest capacity for transporting lipophilic agents. They can also carry hydrophilic therapeutics within their central aqueous core. *Large unilamellar vesicles* (*LUVs*) are typically >100 nm in diameter and structured to contain the largest internal aqueous volume, thereby providing the most capacity for delivery of hydrophilic therapeutic agents. Their size is limited by the development of instability of the single bilayer membrane at larger volume/surface area ratios. *Small unilamellar vesicles* (*SUVs*) feature a very stable, single bilayer membrane structure <100 nm in diameter. These liposomal structures have the least volumetric capacity for distribution of therapeutic agents but, due to their size, have attractive cellular uptake and processing characteristics.

and Rhodes 1995), liposome formation results from the addition of energy (e.g., heating, sonication, homogenization, shaking, etc.) altering the tendency for lipid membranes to form a flat bilayer at an aqueous interface, which instead form bilayered vesicles (Mozafari 2005). Unfortunately, only a few of the conventional liposomal production methods can entrap large quantities of water-soluble agents (Vemuri and Rhodes 1995). Conventional liposomal production methods, such as the reverse-phase evaporation technique (Szoka and Papahadjopoulos 1978), ether injection/vaporization technique (Deamer and Bangham 1976; Schieren et al. 1978), and freeze–thaw method (Pick 1981), yield a heterogeneous mixture of LUVs and MLVs (Mozafari 2005). Production of a more homogeneous liposome mixture has been accomplished through centrifugation (Barenholz et al. 1977) or filtering methods (Hope et al. 1985; MacDonald et al. 1991). With the increased lipophilic volume associated with MLVs, lipophilic drugs can be concentrated within liposomes (Figure 4.5) as well as within micellar structures.

Lipid Nanostructure Transport into the Cell

Although the exact mechanisms for transport of nanomicelles and nanoliposomes into cells have yet to be fully elucidated, common cellular uptake mechanisms appear to be involved. For some negatively charged liposomal vesicles, fusion with the cell membrane allows contained drugs to be released directly into the cytoplasm (Ranade 1989). For the majority of both nanomicelles and nanoliposomes, however, two major pathways for cellular internalization predominate: phagocytosis and endocytosis (Hillaireau and Couvreur 2009). Phagocytosis occurs primarily in professional phagocytes (e.g., macrophages, monocytes, neutrophils, and dendritic cells) (Aderem and Underhill 1999). Fibroblasts, endothelial cells, and epithelial cells have some phagocytic capabilities but to a much lower extent (Rabinovitch 1995). Serum proteins, including immunoglobulins, complement components, laminin, fibronectin, C-reactive protein, and type-I collagen, are associated with the opsonization of the nanoparticles for phagocytosis (Owens and Peppas 2006; Vonarbourg et al. 2006). Opsonized particles specifically attach to phagocytes via receptor–ligand interactions, which trigger a signaling cascade resulting in actin-dependent pseudopodia extension and eventual engulfing of the particle, ingestion, and processing through phagolysosomes (Hillaireau and Couvreur 2009). The entire process can take 30 min to 2 h and is highly dependent on the surface properties of the ingested particle (Aderem and Underhill 1999). Particle size plays a key role in cellular phagocytosis, with the process optimized for particles in the blood or extracellular fluid (ECF) >250 nm in diameter and with smaller particles less efficiently internalized (Korn and Weisman 1967). Larger-sized lipid nanoparticles show increased opsonization by serum proteins and phagocytic clearance, a process that has been largely defeated in the peripheral circulation by using PEG coating (or PEGylation) (Hillaireau and Couvreur 2009; Otsuka et al. 2003; Torchilin et al. 1994). Additionally, nanoparticles with a significant surface charge (positive or negative) have a much higher binding affinity to phagocytes than vesicles that are neutral, while hydrophobic nanoparticles are more readily taken up than hydrophilic nonionic structures (Hillaireau and Couvreur 2009). Finally, shape and rigidity of the nanoparticles also influence the likelihood of cellular uptake via phagocytosis. Less spherical and more rigid particles directly stimulate phagocytic ingestion (Hillaireau and Couvreur 2009).

The uptake and intracellular fate of nanoparticles are highly dependent on the above-mentioned factors but are especially determined by particle size. Larger particles (and ECF volume) are taken up by phagocytosis and macropinocytosis through two different mechanisms sharing a similar intracellular fate (Figure 4.6) (Fiandaca et al. 2011). Smaller particles (<150 nm) are taken up and processed via at least three other mechanisms with variable cellular processing (Figure 4.6) (Fiandaca et al. 2011).

Clinical Applications of Nanomicelles and Nanoliposomes

This volume contains specific chapters on the diagnostic and therapeutic use of nanoparticles. This section is meant as a brief overview of some of the details within upcoming chapters in this volume and with specific introductions to how nanomicelles and nanoliposomes are finding a niche in the medical armamentarium.

Nanoparticles are currently being used to deliver various substances in the evaluation and treatment of many human diseases (Sahoo and Labhasetwar 2003). Major portals of entry into the body for these therapeutic agents include the gastrointestinal (Chang et al. 2004) and parenteral (intravascular and intranasal) pathways. Although these same pathways have been used to deliver comparable compounds to the CNS, especially via the intravascular (Gulyaev et al. 1999; Lockman et al. 2002) and intranasal routes (Dhuria et al. 2010; Kim et al. 2000; Vyas et al. 2005), providing an indirect approach to CNS delivery, both methods are limited by the presence of the BBB. Direct CNS delivery bypasses the BBB and includes the distribution of therapeutics via the cerebrospinal fluid (CSF) spaces (intrathecal subarachnoid or intraventricular) (Glantz et al. 2010) or via intraparenchymal routes (Dickinson et al. 2010; Pradilla et al. 2006).

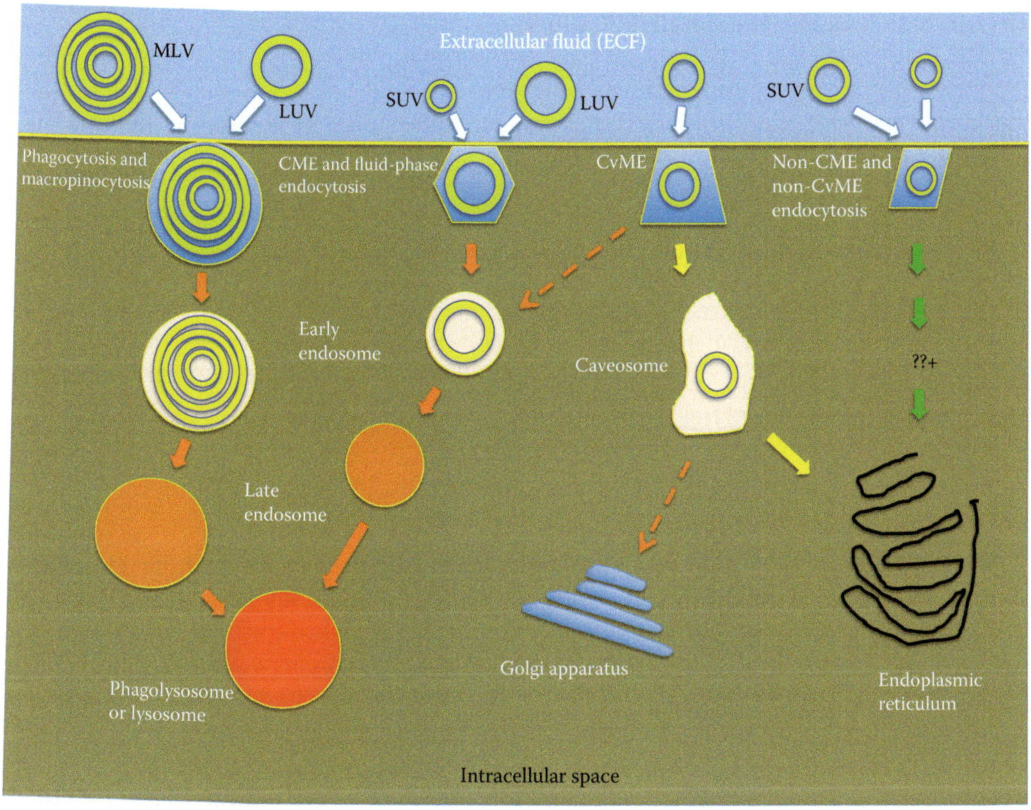

FIGURE 4.6 Schematic representation of the size-dependent cellular uptake mechanisms for lipid nanoparticles and variable cellular processing. *Phagocytosis and macropinocytosis* utilize actin-mediated membrane extensions to engulf particles that are typically >250 nm in diameter, but smaller particles can also be taken up. Phagocytosis usually depends on receptor binding kinetics and activation, while macropinocytosis does not. Intracellular processing of the engulfed particles typically goes through the early and late endosome stages, with the addition of hydrolytic enzymes and, finally, the acidified phagolysosome, for final degradation of the ingested material. *Clathrin-mediated endocytosis (CME) and fluid-phase endocytosis* occur in membrane regions rich with the cytosolic protein clathrin. Receptor-mediated and nonspecific CME occur and require dynamin-dependent formation of a clathrin-coated vacuole. Particles most commonly taken up by these mechanisms are in the 80–150 nm diameter range and limited by the size of the formed vacuoles. Intracellular processing of ingested materials is similar to that of larger particles, leading to degradative lysosomes. *Caveolin-mediated endocytosis* (*CvME*) occurs in membrane regions rich in caveolin protein. Smaller particles (<80 nm in diameter) are typically taken up through this highly regulated, membrane receptor–mediated internalization mechanism. Similar to CME, CvME requires the dynamin GTPase for fission of a caveolin-coated vesicle. The minority of these vesicles undergo processing via a lysosomal pathway, while the majority lead to formation of a caveosome for nonenzymatic processing and transport via the endoplasmic reticulum or Golgi apparatus. *Non-CME and non-CvME endocytosis* remains to be fully elucidated. Small particle (<50 nm diameter) binding to lipid-rich membrane regions has been shown as well as the avoidance of the lysosomal processing pathway.

Nanomicelles and nanoliposomes are currently being used for various imaging applications (Moffat et al. 2003; Papagiannaros et al. 2010; Peira et al. 2003), although some of the contained materials and certain nanoparticles themselves are potentially toxic (De Jong and Borm 2008) and require additional investigation before clinical translation. Iron-oxide crystals contained within the core of lipid nanoparticles have been utilized for over a decade (Moffat et al. 2003; Peira et al. 2003), having many specific advantages compared to standard magnetic resonance imaging (MRI) contrast agents, such as gadolinium (Orringer et al. 2009). Nanoliposomes carrying gadolinium, however, have been effectively used for real-time monitoring of convection-enhanced delivery to the CNS (Dickinson et al. 2008; Fiandaca et al. 2009) and may offer similar advantages to encapsulated iron oxide, if properly packaged within nanoparticles.

With the previously described particulate architecture, nanomicelles and nanoliposomes can also be formulated and targeted to deliver therapeutic agents to the periphery as well as to the CNS via passive or active targeting (Hu et al. 2010). Brain delivery of neuroleptics was noticeably enhanced when carried within lipid micelles (Kabanov et al. 1989). Polymeric micelles are effective delivery vehicles for various drugs, including antibiotics (Liu et al. 2008), cancer chemotherapeutic agents (Blanco et al. 2009; Inoue et al. 2009), antimigraine formulations (Jain et al. 2010), and gene therapy (Osada et al. 2009). Nanoliposomes have also been effectively used within the CNS to deliver antioxidants in experimental models of multiple sclerosis (Kizelsztein et al. 2009), chemotherapeutic agents in neuro-oncology (Dickinson et al. 2008; Grahn et al. 2009), and nucleic acids for gene therapy (Li and Szoka 2007; Maitani and Hattori 2009). As the reader will see in subsequent chapters, there is a rich future for innovation utilizing lipids as well as other nanoparticles to treat human diseases.

CONCLUSIONS AND FUTURE CONSIDERATIONS

This chapter is meant to provide a basis of understanding regarding the structure and function of micellar and liposomal nanoparticles for use in medicine and specifically within the clinical neurosciences. A brief introduction to the cellular processing of these nanostructures was presented to allow the reader to better appreciate the content and strategies developed in subsequent chapters of this volume. The explosion of innovations in the neurosciences and nanotechnology spheres, partly depicted within this textbook, will continue to drive investigators toward translating revolutionary ideas into practical, safe, and efficacious treatments for patients.

REFERENCES

Aderem, A., and Underhill, D. 1999. Mechanisms of phagocytosis in macrophages. *Annu Rev Immunol* 17: 593–623.

Attwood, D., Booth, C., Yeates, S.G. et al. 2007. Block copolymers for drug solubilisation: relative hydrophobicities of polyether and polyester micelle-core-forming blocks. *Int J Pharm* 345: 35–41.

Baeurle, S.J., and Kroener, J. 2004. Modeling effective interactions of micellar aggregates of ionic surfactants with the Gauss-core potential. *J Math Chem* 36: 409–21.

Bareford, L.M., and Swaan, P.W. 2007. Endocytic mechanisms for targeted drug delivery. *Adv Drug Deliv Rev* 59: 748–58.

Barenholz, Y., Gibbes, D., Litman, B.J. et al. 1977. A simple method for the preparation of homogeneous phospholipid vesicles. *Biochemistry* 16: 2806–10.

Blanco, E., Kessinger, C.W., Sumer, B.D. et al. 2009. Multifunctional micellar nanomedicine for cancer therapy. *Exp Biol Med* 234: 123–31.

Chang, S.F., Chang, H.Y., Tong, Y.C. et al. 2004. Nonionic polymeric micelles for oral gene delivery in vivo. *Hum Gene Ther* 15: 481–93.

Chapman, D. 1975. Lipid dynamics in cell membranes. In *Cell Membranes: Biochemistry, Cell Biology and Pathology*, ed. G. Weissmann and R. Claiborne, 13–22. New York: Hospital Practice.

Chapman, D., and Benga, G. 1984. Biomembrane fluidity-studies of model and natural membranes. In *Biological Membranes*, ed. D. Chapman, 1–56. London: Academic.

De Jong, W.H., and Borm, P.J. 2008. Drug delivery and nanoparticles: applications and hazards. *Int J Nanomed* 3: 133–49.

Deamer, D., and Bangham, A.D. 1976. Large volume liposomes by an ether vaporization method. *Biochim. Biophys Acta* 443: 629–34.

Dhuria, S.V., Hanson, L.R., and Frey, W.H., II. 2010. Intranasal delivery to the central nervous system: mechanisms and experimental considerations. *J Pharm Sci* 99: 1654–73.

Dickinson, P.J., LeCouteur, R.A., Higgins, R.J. et al. 2008. Canine model of convection-enhanced delivery of liposomes containing CPT-11 monitored with real-time magnetic resonance imaging: laboratory investigation. *J Neurosurg* 108: 989–98.

Dickinson, P.J., LeCouteur, R.A., Higgins, R.J. et al. 2010. Canine spontaneous glioma: a translational model system for convection-enhanced delivery. *Neuro Oncol* 12: 928–40.
Fiandaca, M.S., Berger, M.S., and Bankiewicz, K.S. 2011. The use of convection-enhanced delivery with liposomal toxins in neurooncology. *Toxins* 3: 369–7.
Fiandaca, M.S., Varenika, V., Eberling, J. et al. 2009. Real-time MR imaging of adeno-associated viral vector delivery to the primate brain. *Neuroimage* 47: T27–T35.
Gaucher, G., Dufresne, M.H., Sant, V.P. et al. 2005. Block copolymer micelles: preparation, characterization and application in drug delivery. *J Control Releasey* 109: 169–88.
Glantz, M.J., Van Horn, A., Fisher, R. et al. 2010. Route of intracerebrospinal fluid chemotherapy administration and efficacy of therapy in neoplastic meningitis. *Cancer* 116: 1947–52.
Gould-Fogerite, S., and Mannino, R.J. 1993. Preparation of large unilamellar liposomes with high entrapment yield by rotary dialysis or agarose plug diffusion. In *Liposome Technology. Liposome Preparation and Related Techniques*, ed. G. Gregoriadis, 67–79. Boca Raton: CRC Press.
Grahn, A.Y., Bankiewicz, K.S., Dugich-Djordjevic, M. et al. 2009. Non-PEGylated liposomes for convection-enhanced delivery of topotecan and gadodiamide in malignant glioma: initial experience. *J Neurooncol* 95: 185–97.
Gulyaev, A.E., Gelperina, S.E., Skidan, I.N. et al. 1999. Significant transport of doxorubicin into the brain with polysorbate 80-coated nanoparticles. *Pharm Res* 16: 1564–9.
Haines, T.H. 2001. Do sterols reduce proton and sodium leaks through lipid bilayers? *Prog Lipid Res* 40: 299–324.
Hartley, G.S. 1936. *Aqueous Solutions of Paraffin Chain Salts, a Study in Micelle Formation.* Paris: Hermann et Cie.
Hartley, G.S., and Runnicles, D.F. 1938. The determination of the size of paraffin-chain salt micelles from diffusion measurements. *Proc R Soc Lond A Math Phys Sci* 168: 420–40.
Hillaireau, H., and Couvreur, P. 2009. Nanocarriers' entry into the cell: relevance to drug delivery. *Cell Mol Life Sci* 66: 2873–96.
Hope, M.J., Bally, M.B., Webb, G. et al. 1985. Production of large unilamellar vesicles by a rapid extrusion procedure. Characterization of size distribution, trapped volume and ability to maintain a membrane potential. *Biochim Biophys Acta* 812: 55–65.
Hu, X., Liu, S., Huang, Y. et al. 2010. Biodegradable block copolymer–doxorubicin conjugates via different linkages: preparation, characterization, and in vitro evaluation. *Biomacromolecules* 11: 2094–102.
Ikeda, S., Ozeki, S., and Hayashi, S. 1980. Size and shape of charged micelles of surfactants in aqueous salt solutions. *Biophys Chem* 11: 417–23.
Immordino, M.L., Dosio, F., and Cattel, L. 2006. Stealth liposomes: review of the basic science, rationale, and clinical applications, existing and potential. *Int J Nanomed* 1: 297–315.
Inoue, T., Yamashita, Y., Nishihara, M. et al. 2009. Therapeutic efficacy of a polymeric micellar doxorubicin infused by convection-enhanced delivery against intracranial 9L brain tumor models. *Neuro Oncol* 11: 151–7.
Jain, R., Nabar, S., Dandekar, P. et al. 2010. Micellar nanocarriers: potential nose-to-brain delivery of zolmitriptan as novel migraine therapy. *Pharm Res* 27: 655–64.
Jones, M., and Leroux, J. 1999. Polymeric micelles—a new generation of colloidal drug carriers. *Eur J Pharm Biopharm* 48: 101–11.
Kabanov, A.V., Chekhonin, V.P., Alakhov, V.Y. et al. 1989. The neuroleptic activity of haloperidol increases after its solubilization in surfactant micelles. Micelles as microcontainers for drug targeting. *FEBS Lett* 258: 343–5.
Kates, M., and Manson, L.A. (eds.). 1984. *Book Membrane Fluidity. Biomembranes.* New York: Plenum.
Katsaras, J., and Gutberlet, T. (eds.). 2010. *Book Lipid Bilayers: Structure and Interactions. Biological Physics Series*, ed. E. Greenbaum. Berlin: Springer.
Kikuchi, H., Yamauchi, H., and Hirota, S. 1994. A polyol dilution method for mass production of liposomes. *J. Liposome Res.* 4: 71–91.
Kim, T.W., Chung, H., Kwon, I.C. et al. 2000. In vivo gene transfer to the mouse nasal cavity mucosa using a stable cationic lipid emulsion. *Mol Cells* 10: 142–7.
Kizelsztein, P., Ovadia, H., Garbuzenko, O. et al. 2009. PEGylated nanoliposomes remote-loaded with the antioxidant tempamine ameliorate experimental autoimmune encephalomyelitis. *J Neuroimmunol* 213: 20–5.
Klibanov, A.L., Murayama, K., Torchilin, V.P. et al. 1990. Amphipathic polyethyleneglycols effectively prolong the circulation time of liposomes. *FEBS Lett* 268: 235–7.
Korn, E.D., and Weisman, R.A. 1967. Phagocytosis of latex beads by Acanthamoeba. II. Electron microscopic study of the initial events. *J Cell Biol* 34: 219–27.

Lasic, D.D. 1988. The spontaneous formation of unilamellar vesicles. *J Colloid Interf Sci* 124: 428–35.
Lasic, D.D., Joannic, R., Keller, B.C. et al. 2001. Spontaneous vesiculation. *Adv Colloid Interfac Sci* 89–90: 337–49.
Li, W., and Szoka, F.C., Jr. 2007. Lipid-based nanoparticles for nucleic acid delivery. *Pharm Res* 24: 438–49.
Liu, L., Venkatraman, S.S., Yang, Y.Y. et al. 2008. Polymeric micelles anchored with TAT for delivery of antibiotics across the blood–brain barrier. *Biopolymers* 90: 617–23.
Lockman, P.R., Mumper, R.J., Khan, M.A. et al. 2002. Nanoparticle technology for drug delivery across the blood–brain barrier. *Drug Dev Ind Pharm* 28: 1–13.
MacDonald, R.C., MacDonald, R.I., Menco, B.P.M. et al. 1991. Small volume extrusion apparatus for preparation of large unilamellar vesicles. *Biochim Biophys Acta* 1061: 297–303.
Maitani, Y., and Hattori, Y. 2009. Oligoarginine–PEG–lipid particles for gene delivery. *Expert Opin Drug Deliv* 6: 1065–77.
Marsh, M., and Helenius, A. 2006. Virus entry: open sesame. *Cell* 124: 729–40.
McBain, J.W. 1913. Colloids and their viscosity. *Trans Faraday Soc* 9: 99–101.
Mittal, K.L. (ed.). 1977. *Book Micellization, Solubilization and Microemulsions.* New York: Plenum Press.
Moffat, B.A., Reddy, G.R., McConville, P. et al. 2003. A novel polyacrylamide magnetic nanoparticle contrast agent for molecular imaging using MRI. *Mol Imaging* 2: 324–32.
Moghimi, S.M., and Hunter, A.C. 2001. Recognition by macrophages and liver cells of opsonized phospholipid vesicles and phospholipid headgroups. *Pharm Res* 18: 1–8.
Mozafari, M.R. 2005. Liposomes: an overview of manufacturing techniques. *Cell Mol Biol Lett* 10: 711–9.
Mozafari, M.R., Reed, C.J., and Rostron, C. 2002a. Development of non-toxic liposomal formulations for gene and drug delivery to the lung. *Technol Health Care* 10: 342–4.
Mozafari, M.R., Reed, C.J., Rostron, C. et al. 2002b. Construction of stable anionic liposome-plasmid particles using the heating method: a preliminary investigation. *Cell Mol Biol Lett* 7: 923–7.
Mozafari, M.R., Reed, C.J., Rostron, C. et al. 2004. Transfection of human airway epithelial cells using a lipid-based vector prepared by the heating method. *J Aerosol Med* 17: 100.
Murayama, K. 2002. PEG–immunoliposoome. *Biosci Rep* 22: 251–66.
Murayama, K., Kennel, S.J., and Huang, L. 1990. Lipid composition is important for highly efficient target binding and retention of immunoliposomes. *Proc Natl Acad Sci U S A* 87: 5744–8.
Oishi, M., Hayama, T., Akiyama, Y. et al. 2005. Supramolecular assemblies for the cytoplasmic delivery of antisense oligodeoxynucleotide: polyion complex (PIC) micelles based on poly(ethylene glycol)-SS-oligodeoxynucleotide conjugate. *Biomacromolecules* 6: 2449–54.
Orringer, D.A., Koo, Y.E., Chen, T. et al. 2009. Small solutions for big problems: the application of nanoparticles to brain tumor diagnosis and therapy. *Clin Pharm Ther* 85: 531–4.
Osada, K., Christie, R.J., and Kataoka, K. 2009. Polymeric micelles from poly(ethylene glycol)-poly(amino acid) block copolymer for drug and gene delivery. *J R Soc Interface* 6 Suppl 3: S325–39.
Otsuka, H., Nagasaki, Y., and Kataoka, K. 2003. PEGylated nanoparticles for biological and pharmaceutical applications. *Adv Drug Deliv Rev* 55: 403–19.
Owens, D., and Peppas, N. 2006. Opsonization, biodistribution, and pharmacokinetics of polymeric nanoparticles. *Int J Pharm* 307: 93–102.
Pagano, R.E., and Weinstein, J.N. 1978. Interaction of liposomes with mammalian cells. *Ann Rev Biophys Bioeng* 7: 435–68.
Panwar, P., Pandey, B., Lakhera, P.C. et al. 2010. Preparation, characterization, and in vitro release study of albendazole-encapsulated nanosize liposomes. *Int J Nanomedicine* 5: 101–8.
Papagiannaros, A., Upponi, J., Hartner, W. et al. 2010. Quantum dot loaded immunomicelles for tumor imaging. *BMC Med Imaging* 10: 22.
Peira, E., Marzola, P., Podio, V. et al. 2003. In vitro and in vivo study of solid lipid nanoparticles loaded with superparamagnetic iron oxide. *J Drug Target* 11: 19–24.
Pick, U. 1981. Liposomes with a large trapping capacity prepared by freezing and thawing of sonicated phospholipid mixtures. *Arch Biochem Biophys* 212: 186–94.
Pradilla, G., Wang, P.P., Gabikian, P. et al. 2006. Local intracerebral administration of paclitaxel with the paclimer delivery system: toxicity study in a canine model. *J Neurooncol* 76: 131–8.
Rabinovitch, M. 1995. Professional and nonprofessional phagocytes—an introduction. *Trends Cell Biol* 5: 85–7.
Ranade, V.V. 1989. Drug delivery systems. 1. Site-specific drug delivery using liposomes as carriers. *J Clin Pharmacol* 29: 685–94.
Sahoo, S.K., and Labhasetwar, V. 2003. Nanotech approaches to drug delivery and imaging. *Drug Discov Today* 8: 1112–20.

Samad, A., Sultana, Y., and Aqil, M. 2007. Liposomal drug delivery systems: an update review. *Curr Drug Deliv* 4: 297–305.

Schieren, H., Rudolph, S., Finkelstein, M. et al. 1978. Comparison of large unilamellar vesicles prepared by a petroleum ether vaporization method with multilamellar vesicles: ESR, diffusion and entrapment analyses. *Biochim Biophys Acta* 542: 137–53.

Sugihara, G., Yamakawa, K., Murata, Y. et al. 1982. Effects of pH, pNa, and temperature on micelle formation and solubilization of cholesterol in aqueous solutions of bile salts. *J Phys Chem* 86: 2784–8.

Szoka, F.J., and Papahadjopoulos, D. 1978. Procedure for preparation of liposomes with large internal aqueous space and high capture by reverse-phase evaporation. *Proc Natl Acad Sci U S A* 75: 4194–8.

Szoka, F.J., and Papahadjopoulos, D. 1980. Comparative properties and methods of preparation of lipid vesicles (liposomes). *Ann Rev Biophys Bioeng* 9: 467–508.

Talsma, H., Van Steenbergen, M.J., Borchert, J.C.H. et al. 1994. A novel technique for the one-step preparation of liposomes and nonionic surfactant vesicles without the use of organic solvents. Liposome formation in a continuous gas stream: The bubble method. *J Pharm Sci* 83: 276–80.

Tanford, C. 1974. Thermodynamics of micelle formation: prediction of micelle size and size distribution. *Proc Natl Acad Sci U S A* 71: 1811–5.

Tian, H.Y., Deng, C., Lin, H. et al. 2005. Biodegradable cationic PEG–PEI–PBLG hyperbranched block copolymer: synthesis and micelle characterization. *Biomaterials* 26: 4209–17.

Torchilin, V.P. 2001. Structure and design of polymeric surfactant-based drug delivery systems. *J Control Release* 73: 137–72.

Torchilin, V.P. 2005. Recent advances with liposomes as pharmaceutical carriers. *Nat Rev Drug Discov* 4: 145–60.

Torchilin, V.P., Omelyanenko, V.G., Papisov, M.I. et al. 1994. Poly(ethylene glycol) on the liposome surface: on the mechanism of polymer-coated liposome longevity. *Biochim Biophys Acta* 1195: 11–20.

Vemuri, S., and Rhodes, C.T. 1995. Preparation and characterization of liposomes as therapeutic delivery systems: a review. *Pharm Acta Helv* 70: 95–111.

Vonarbourg, A., Passirani, C., Saulnier, P. et al. 2006. Parameters influencing the stealthiness of colloidal drug delivery systems. *Biomaterials* 27: 4356–73.

Vyas, T.K., Shahiwala, A., Marathe, S. et al. 2005. Intranasal drug delivery for brain targeting. *Curr Drug Deliv* 2: 165–75.

5 Peptide and Protein-Based Nanoparticles

Ayaz M. Khawaja and John D. Heiss

CONTENTS

INTRODUCTION

Nanoparticles have at least one dimension of 1–100 nm, with each particle behaving as a unit with many independent functions and properties. The basic structure of a nanoparticle consists of a core, in which a therapeutic or imaging agent can be loaded, and surface components, which modify the nanoparticle by (1) improving its stability, (2) enhancing its solubility, (3) aiding in its transport across the blood–brain barrier (BBB), and/or (4) promoting its binding to a target at the site of interest, for example, a deep-seated brain tumor. The surface of the nanoparticle can be tagged with various structures, including polyethylene glycol (PEG) to enhance its half-life and water solubility; TAT protein (trans-activator of transcription, an HIV peptide) to improve its central nervous system (CNS) bioavailability; fluorescent dyes to enhance its site-specific detection; and monoclonal antibodies, aptamers, and peptides to increase its accumulation in diseased tissue. Modification of the surface of the nanoparticle can add multivalent affinity that produces stronger anchoring, homing to multiple receptors, and activating of cell signaling pathways. For drug delivery, nanoparticles offer several advantages over conventional drug delivery, including (1) isolating the drug from the

TABLE 5.1
Nanoparticle-Based Therapeutic Agents

Agent	Type of Nanoparticle	Active Component	Clinical Use
Abraxane	Albumin-bound	Paclitaxel	Treatment of breast cancer unresponsive to conventional chemotherapy.
Copaxone	Polymeric platform of amino acids	Glutamic acid, alanine, lysine, and tyrosine	Immunomodulator that reduces severity and frequency of relapses in multiple sclerosis.
Adagel	PEGylated	Adenosine deaminase	Treatment of adenosine deaminase deficiency in those who cannot tolerate or fail bone marrow replacement.
Mucagen	PEGylated	Anti-vascular endothelial growth factor (VEGF) aptamer	Treatment of age-related macular degeneration.
Pegasys	PEGylated	α-interferon 2a	Used with ribavirin to treat chronic hepatitis C. Used alone to treat hepatitis B.
Neulasta	PEGylated	Granulocyte colony-stimulating factor (G-CSF)	Used to treat neutropenia associated with cancer chemotherapy.
Somavert	PEGylated	Growth hormone receptor antagonist	Acromegaly treatment.
Oncaspar	PEGylated	L-asparaginase	Acute lymphoblastic leukemia treatment.
Cimzia	PEGylated	Anti-tumor necrosis factor (TNF) α antibody	In phase III clinical trials, to treat rheumatoid arthritis and Crohn's disease.
Puricase	PEGylated	Uricase	In phase III clinical trials, to manage hyperuricemia and gout.
Hepacid	PEGylated	Arginine deaminase	In phase I and II clinical trials, for hepatocellular carcinoma treatment.
Camptothecin	Polyglutamate	Camptothecin	In phase I clinical trials, for ovarian and colorectal carcinoma treatment.
Paclitaxel	Polyglutamate	Paclitaxel	In phase III clinical trials, for non-small cell lung carcinoma and ovarian carcinoma.

Source: Zhang, L. et al., *Clin Pharmacol Ther* 83: 761–9, 2007.
Notes: TNF, Tumor necrosis factor; VEGF, vascular endothelial growth factor.

systemic circulation, which minimizes systemic side effects; (2) penetrating the BBB; (3) avoiding the *P*-glycoprotein (*P*-gp) efflux system; (4) retention within the site of interest; and (5) sustained release of the drug. By nature of their unique size, nanoparticles preferentially accumulate within tumors rather than in normal tissues because of a passive process known as the enhanced permeability and retention (EPR) effect. Sequestration of nanoparticles within a tumor can enable highly selective treatments.

The fundamental objective of research and development in nanomedicine is to develop more sensitive diagnostics and more effective therapeutics. According to the European Science and Technology Observatory, more than 150 companies presently are developing therapeutics involving nanotechnology (Wagner et al. 2006). Twenty-four therapeutic products are currently approved for clinical use (Table 5.1). Most of these are based on nanoparticles with liposomal or polymeric cores (Zhang et al. 2007). This chapter explores the medical applications of nanoparticles with a peptide or protein core or with other types of cores and a surface that is heavily modified by peptides or proteins.

NANOPARTICLES WITH A CORE OF PROTEIN OR PEPTIDE

Human Serum Albumin-Core Nanoparticles

Human Serum Albumin-Core Nanoparticles for Photodynamic Therapy

Human serum albumin (HSA)–based nanoparticles are conjugated with photosensitizers for use in photodynamic therapy (PDT) (Table 5.2). Photosensitizers are nontoxic molecules until activated by light. Their activation can be observed by photodynamic imaging. PDT involves cellular uptake of the photosensitizer followed by applying light that activates the photosensitizer, producing reactive oxygen

TABLE 5.2
Protein and Peptide-Based Nanoparticles

Core Nanoparticle		Targeting Component	Therapeutic Component	Imaging Component
Albumin (human or bovine)		Not used with photosensitizer	Chlorin e6 (Ce6) (photosensitizer) and LED lamp (670–690 nm)	Ce6 fluorescence
		Trans-activating transcriptor (TAT)	Saquinavir	Fluorescein
		hTRAIL	Plasmid DNA (pORF-hTRAIL)	6-coumarin-labeled CBSA–NP, EMA-labeled pDNA
Short cationic amphiphilic peptide with glycine and arginine residues		TAT	Peptide	None
Vault		Anti-epidermal growth factor (EGFR) monoclonal antibody, protein VI (from adenovirus)	Ribotoxin, pDNA, CCL21	mCherry fluorochrome, Cy3.5
Liposomes/nLDL		Apolipoprotein B-100 (targets low-density lipoprotein receptor [LDLR])	Paclitaxel oleate	Fluorescein isothiocyanate (FITC)
		Interleukin 13 (IL-13; binds IL-13R)	Doxorubicin	Rhodamine
Gold	Nanorods	Aptamer	Laser excitation of aptamer conjugated GNR	Carboxyfluorescein succinimidyl ester (CFSE)
	Nanospheres	Polyclonal antibody/ anti-amyloid-β-derived diffusible ligand (ADDL) antibody	None	Silver ions
Iron oxide nanoparticle		Anti-EGFRvIII antibody	Iron oxide NP-antibody complex	Iron oxide
		HIV TAT protein	None	Cy5.5, annexin-V
Multifunctional polymeric NP		Peptide F3	Photofrin (photosensitizer)	Alexa Fluor 594
Biopoly(β-L-malic acid)		Antinucleosome antibody C25; anti-mouse transferrin receptor antibody	Antisense oligonucleotides	Oregon green, Alexa Fluor 680

Notes: CBSA–NP, cationic bovine serum albumin conjugated with poly(ethylene glycol)–poly(lactide) nanoparticle; EMA, ethidium monoazide (a fluorescent dye); GNR, gold nanorods; LED, light emitting diode; nLDL, nano-low density lipoprotein; NP, nanoparticle; pDNA - plasmid DNA.

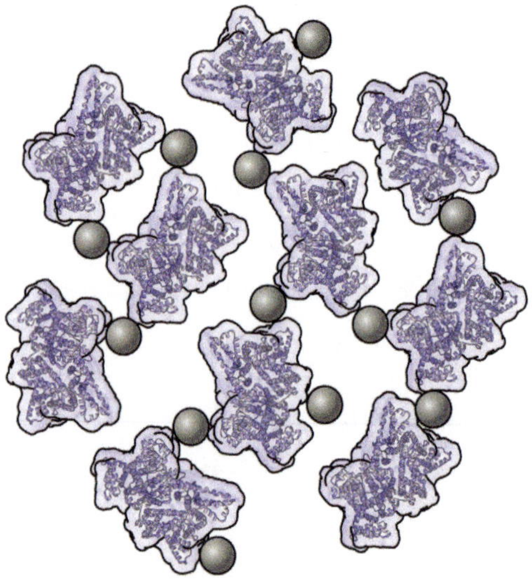

FIGURE 5.1 Albumin–chlorin e6 complex showing the nanoparticle structure with multiple albumin particles (violet) and chlorin e6 photosensitizers (gray).

species that induce apoptotic cell death. The clinical use of PDT has been limited because most photosensitizers display nonspecific targeting of tumor tissue and poor water solubility. Nanoparticles provide a method to address these limiting factors. By conjugating the photosensitizer to a nanoparticle, the EPR effect leads to preferential accumulation of these particles within the tumor site rather than in normal tissue. Nanoparticle delivery can also increase water solubility of the photosensitizer. In an in vitro study, efficient singlet oxygen generation resulted when albumin nanoparticle–photosensitizer conjugates were incubated with Jurkat cells and activated by light (Wacker et al. 2010). The porphyrin-based photosensitizer (mTHPP or mTHPC) was released from the nanoparticles after they were taken up by the cells. Chlorin e6 (Ce6) is another photosensitizer that is conjugated to HSA nanoparticles (Figure 5.1). HSA-conjugated Ce6 delivered to tumor-bearing mice for PDT reduced tumor volume by 70% compared to 34% for free Ce6 and 0% for normal saline. On histological analysis of the tumor, cytotoxicity after PDT was much more severe using HSA-Ce6 nanoparticles rather than free Ce6 (Jeong et al. 2011).

HSA-Core Nanoparticles for CNS HIV Infection

Radioactively labeled nanoparticles composed of HSA, polymethylacrylate, and polyalkylcyanoacrylate were actively phagocytosed into HIV-1-infected macrophages cultured in vitro. Macrophages are the main focus of HIV infection because they harbor the virus, allow it to replicate, and release the newly formed viral particles. Neuronal apoptosis is a proposed mechanism for HIV-induced neurotoxicity, resulting from oxidative stress caused by activating macrophages and glial cells. Effects predominate in the basal ganglia. Not surprisingly, macrophages have become the target of nanoparticles carrying antiviral drugs (Schäfer et al. 1992). Uptake of nanoparticles is greater in HIV-infected than noninfected macrophages. These nanoparticles can be loaded with anti-HIV drugs, with some showing increased efficacy against HIV compared to the free form of anti-HIV drugs. Anti-HIV drugs that have been loaded into nanoparticles include zalcitabine and saquinavir, whose free forms reduce production of HIV type 1 (HIV-1) antigen (Bender et al. 1996). The nanoparticle-bound zalcitabine has not shown superior efficacy compared to its free form. However, loading saquinavir into a nanoparticle increases its antiviral potency, lowering its minimal inhibitory concentration to 0.39 nM, compared to 4.23 nM for the free form of the drug. These nanoparticles can be conjugated further with trans-activating transcriptor (TAT) peptide (Rao et al. 2008) to bypass the efflux system of *P*-gp. TAT conjugation and nanoparticle delivery increased the CNS bioavailability of the antiviral drug, ritonavir, up to 800 times compared to the free form of the drug in vivo (Rao et al. 2008).

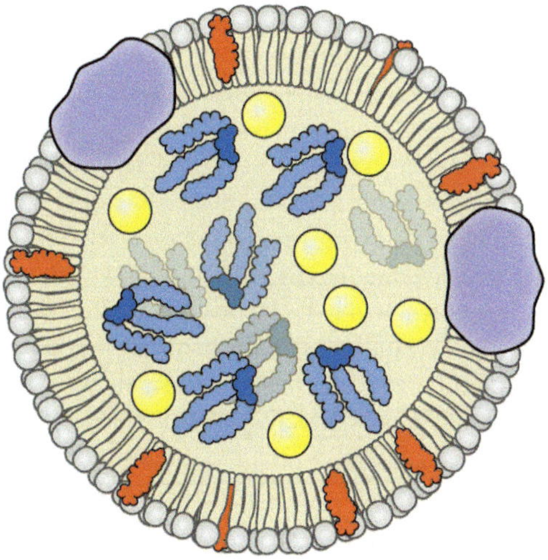

FIGURE 5.2 Drawing of the structure of a lipoprotein nanoparticle construct containing apoprotein (violet), lecithin bilayer, cholesterol (orange), triacylglycerol (blue), and incorporated paclitaxel (yellow) within the core.

Lipoprotein Core Nanoparticles

Low-density lipoprotein (LDL) and high-density lipoprotein (HDL) nanoparticles were investigated as carriers of chemotherapeutic drugs such as paclitaxel oleate (PO) (McConathy et al. 2008) (Figure 5.2). They have favorable characteristics including stability that is unaffected by ultracentrifugation, storage capability longer than 6 months, and a half-maximal inhibitory concentration that is 5%–20% of the free drug. LDL normally acts as a transport protein for cholesterol, which is otherwise insoluble in blood. LDL receptors (LDLRs) are upregulated in tumor cells, increasing their uptake of cholesterol, which is required for membrane formation and cell division (Maletínská et al. 2000). Recently, LDL nanoparticles have been studied for their ability to carry PO and provide target-specific cytotoxicity (Nikanjam et al. 2007). These particles are constructed using (1) a synthetic peptide containing a lipid binding motif and the LDLR binding domain of apolipoprotein B-100, (2) lecithin, (3) cholesteryl oleate, and (4) PO within the lipophilic core. A fluorescent marker, DiO, was added later to allow the degree of cellular uptake to be examined by fluorescence microscopy. This construct delivers hydrophobic drugs across the BBB and into the tumor using a mechanism that mimics the natural pathway of LDL endocytosis, incorporation into lysosomes, and degradation. Suramin (an inhibitor of the LDLR) blocked the action of this nanoconjugate on tumor cells, significantly improving cell survival (51%) compared to controls without suramin (12%), providing support that drug delivery with this nanoconjugates occurs via the LDLR (Nikanjam et al. 2007).

Protein-Cage Nanoparticles

A cell-specific targeting peptide (RGD-4C) was attached to a human H-chain ferritin molecule that encases an iron oxide nanoparticle This nanoconjugate can bind C32 melanoma cells in vitro (Uchida et al. 2006).

Vault Nanoparticles

Kedersha and Rome of University of California–Los Angeles (UCLA) School of Medicine first isolated major vault protein (MVP) in 1986, when they noticed structures looking like coated vesicles in rat liver microsomal extracts (Kedersha and Rome 1986). Measuring 35 × 65 nm, they were described by the investigators as "resembling multiple arches which form cathedral vaults" when examined under an electron microscope (Kedersha and Rome 1986, p. 699). Vaults are ribonucleoproteins by nature whose structure consists of several polypeptides, of which a single polypeptide species with a molecular weight (MW) of 104,000 predominates. RNA constitutes only 4.6% of the structure. These are universally present in all cell types across all species (Kedersha and Rome 1986). About 5% of vault particles are associated with nuclear pore complexes, suggesting that they are involved in nucleocytoplasmic exchange of other particles. Lung resistance-related protein (LRP), whose overexpression predicts a poor response to chemotherapy in acute myeloid leukemia and ovarian carcinoma, is a human vault protein that was postulated to be involved in multidrug resistance in various cancers (Scheffer et al. 1995). Several clinical studies indicate that MVP negatively impacts response to chemotherapy in cancer cells and affects prognosis and eventual survival. It is unclear, however, whether MVP is directly involved in resistance or is merely a marker (Mossink et al. 2003).

Vaults are naturally occurring organelles with a large internal cavity, which has been exploited and investigated for its capacity to carry biologically active materials. Vaults were conjugated with various enzymes and fluorescent molecules and exhibited enzymatic properties and fluorescence in an in vitro system using HeLa cells (Kickhoefer et al. 2005) (Figures 5.3 and 5.4). Vaults have further been engineered to allow various structures to bind on the top and bottom of the vault. These structures include immunoglobulin G (IgG)–binding peptide and epidermal growth factor (EGF). Anti-EGF receptor (EGFR) antibody was also conjugated to the vaults and was shown by fluorescent labeling (mCherry fluorochrome) to selectively target EGF-overexpressing A431 cells (an epidermal carcinoma cell line) in an in vitro system (Lai et al. 2009). To further enhance tumor cell targeting, penetration of cellular membranes, histological localization, and tumoricidal effect of recombinant vaults, they were conjugated with the membrane lytic domain of adenovirus protein VI, the Cy3.5 fluorescent molecule, and soluble ribotoxin. An in vitro system was designed using RAW

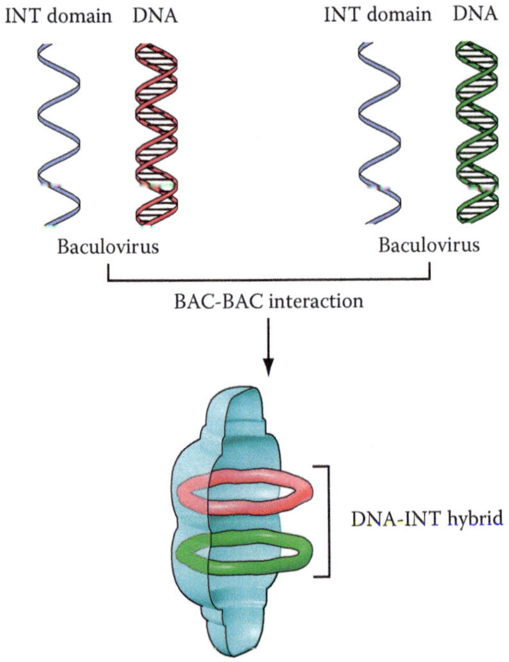

FIGURE 5.3 INT is the amino acid sequence required for interaction and incorporation of materials into vault. The red-coded DNA is the gene for luciferase enzyme, and the green-coded DNA is the gene for green fluorescent protein (GFP). Conjugation with the INT domain is important for incorporation into the vault, which is achieved with the Bac to Bac expression system (Invitrogen). The two DNAs are incorporated as rings within the belly of the vault.

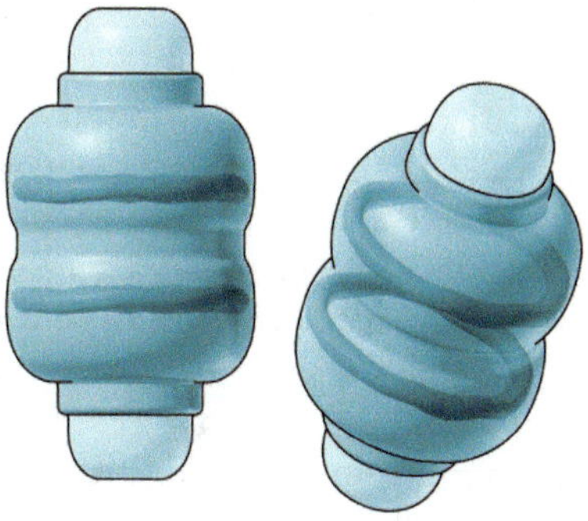

FIGURE 5.4 Incorporated DNAs appear as densities on imaging of vault nanoparticles.

264.7 mouse macrophages and A549 human epithelial cells. The vault–pVI combination in the presence of saporin exhibited cellular cytotoxicity (Lai et al. 2009). The same experimental design was used to investigate delivery of plasmid DNA into cells by vault–pVI compared to delivery of calcium phosphate precipitate DNA without vault–pVI; a sevenfold increase in delivery was observed with vault–pVI (Lai et al. 2009). Delivery of antitumor compounds was also explored by internalizing CCL21 (a chemokine) and the fluorescent molecule mCherry-INT within vaults. This construct was then injected intratumorally in mice bearing 9-day 3LL lung tumors, and its effect was compared to that of control vaults without the chemokine. There was a predominantly CD-3 T-cell infiltrate in the treatment group, with a sevenfold decrease in tumor burden, compared to control (Kar et al. 2011). Efforts are currently underway at the UCLA Brain Tumor Nanotechnology Laboratory to evaluate if bioengineered vaults can produce an immune response in brain tumors similar to that seen in the lung tumor model.

NANOPARTICLES WITH PROTEIN OR PEPTIDE SURFACE LIGANDS FOR RECEPTOR TARGETING

Proteins and peptides can be linked to (functionalize) many types of nanoparticles and alter their surface properties. Various peptides can be tagged to the surface of a nanoparticle to enable selective targeting of receptors expressed by cancer cells, tumor vasculature, and the lymphatic network, and to enable nanoparticle accumulation within the targeted tissue (Table 5.2) (Reubi and Maecke 2008).

Polymeric Nanoparticles with Surface Peptide Ligand

Multifunctional polymeric nanoparticles tagged with F3 (a peptide that targets tumor vasculature), porfimer sodium (Photofrin, AxCan Pharma, Birmingham, AL, a photosensitizer), and imaging agents were evaluated for use in PDT (Reddy et al. 2006). This study compared the outcome of PDT in mice implanted with glioma cell lines that received these nanoparticles versus those that received free Photofrin by intravenous (IV) injection. Animals receiving the nanoparticle preparation had a mean survival of 33 days compared to 8.5 days in the control group. Two of five nanoparticle-treated mice survived beyond 6 months. More selective targeting of tumor tissue with nanoparticle-conjugated photosensitizers may improve the safety, effectiveness, and usefulness of PDT in clinically managing malignant brain tumors.

Polymeric Nanoparticles with Surface Protein

The immunogenicity and side effects of gene therapy delivered by viral vectors led to developing alternative nonviral vectors. Cationic bovine serum albumin (CBSA) conjugated with poly(ethylene glycol)–poly(lactide) (PEG-PLA) nanoparticle (CBSA-NP) is a nanoparticle designed for gene delivery. A nonviral vector for gene delivery was produced by incorporating plasmid DNA (pORF-hTRAIL) into this nanoparticle (CBSA-NP-hTRAIL) (Lu et al. 2006). After administering this nanoparticle IV to mice with implanted intracerebral glioma, hTRAIL messenger RNA (mRNA) and protein were found 24 h later in normal brain and in tumor tissue. Repeated IV injections of CBSA-NP-hTRAIL resulted in apoptosis of tumor cells, inhibition of tumor growth, and prolonged survival time, suggesting that this nonviral vector could be developed for gene therapy of malignant brain tumors.

Nanoparticles with Surface Antibodies

Polymeric Nanoparticles Conjugated to Antibodies

Targeting a nanoparticle to a tumor cell can be made more specific by tagging the nanoparticle with an antibody that binds to a receptor on the tumor cell. Nanoparticles were also tagged with more than one antibody to further enhance their accumulation in tumor cells (Fujita et al. 2007). Two different monoclonal antibodies, anti-mouse transferrin receptor antibody and mouse autoimmune antinucleosome antibody 2C5, were conjugated to a polymeric nanoparticle, poly(ß-L-malic acid), that carried antisense oligonucleotides to vascular protein laminin-8. This dual-antibody nanoparticle variant of polycefin was compared to a single-antibody form. After IV injection in glioma-bearing xenogeneic animals (human glioblastoma multiforme [GBM] in athymic mice), tandem antibody tagging resulted in significantly more tumor accumulation of polycefin compared to single-antibody tagging (Fujita et al. 2007).

Liposomes Conjugated to Antibodies

There is overexpression of other cell surface receptors in glioblastoma. Nanoparticles targeting EGFR (Kuo and Liang 2011), platelet-derived growth factor receptor (PDGFR), fibroblast growth factor receptor (FGFR), and human epidermal receptor type 2 (HER2) (Cimini and Ippoliti 2011) are also being studied as agents that may be used to improve targeting of chemotherapeutic drugs to tumor tissue. Anti-HER2 immunoliposomes were conjugated with monoclonal antibody fragments that target the HER2 protein expressed by breast cancer cells. The antibody selectively binds to cancer cells and leads to internalization of the liposome within the cancer cells in vitro (Kirpotin et al. 2006).

Iron Oxide Nanoparticles Conjugated to Antibodies

Iron oxide nanoparticles provide a novel way to image the tumor margin by providing magnetic resonance imaging (MRI) contrast. Iron oxide nanoparticles with a surface coat of amphiphilic polymers were conjugated to a purified antibody that selectively binds to the EGFR deletion mutant (EGFRvIII) that is overexpressed in GBM (Figure 5.5). A human glioblastoma cell line (U87HG) was stably transfected with a plasmid to overexpress EGFRvIII or with a plasmid for wild-type EGFR (Hadjipanayis et al. 2010). In vitro, an iron oxide nanoparticle conjugated to EGFRvIII resulted in apoptosis and significantly reduced survival of the EGFRvIII-expressing glioma cells, without causing toxicity to normal astrocytes. The same set of researchers went on to intracerebrally implant human GBM xenografts expressing mutated EGFRvIII protein in athymic nude mice. Iron oxide nanoparticles with the EGFRvIII antibody were introduced into the tumors by MRI-guided convection-enhanced delivery (CED). Mice treated with the EGFRvIII antibody-conjugated nanoparticles, nanoparticles alone, and EGFRvIII alone all survived significantly longer than controls receiving Hank's buffered salt solution (HBSS) by CED (Hadjipanayis et al. 2010). EGFRvIII antibody-conjugated nanoparticles had a greater antitumor effect in tumors expressing EGFRvIII compared to tumors expressing wild-type EGFR, demonstrating the ability of the conjugated nanoparticles to target tumor cells expressing a specific tumor biomarker.

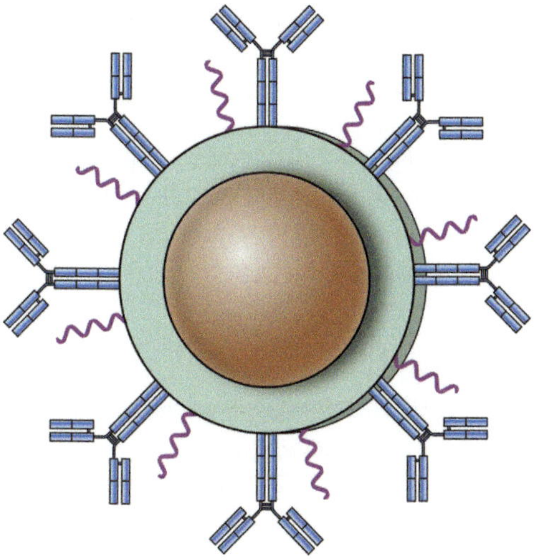

FIGURE 5.5 Iron oxide nanoparticle construct showing iron oxide in the core (light orange), polymer layer (aqua), antibodies on the surface for targeting (blue), and polyethylene glycol (violet).

Iron Oxide Nanoparticles Conjugated to Fluorescent Dyes

Achieving adequate margins of tumor resection is critical to improving long-term survival and increasing the time to tumor recurrence in primary brain neoplasms, including GBM. Accurately delineating the tumor margin can be challenging and is based on the subjective impression of the neurosurgeon and radiographic image guidance during the resection. Even after presumed total resection, residual tumor can still remain (Smith et al. 2005). Cy5.5 is a fluorescent dye of the cyanine family that has been used to produce compact discs (CDs). Cy5.5 is fluorescent in the far red light region at a wavelength of 650 to 670 nm. This dye has reactive groups on either side of nitrogen chains that enable polymerization with superparamagnetic iron oxide (SPIO) nanoparticles, proteins, and nucleic acids. SPIOs are used mainly to provide contrast on T2 MRI images and are taken up avidly by microglia (Fleige et al. 2001). These particles have long circulating times and are subsequently concentrated within the tumor. SPIOs are coated with dextran to provide groups for reaction with the substrates succinimidyl iodoacetate and R4 to form R4-SC CLIO (cross-linked iron oxide). Cy5.5 is conjugated with this product thereafter. The Cy5.5 portion on the nanoparticle emits near-infrared (NIR) fluorescence that enables a high contrast between the target and the background tissue, as normal brain tissue is minimally autofluorescent (Josephson et al. 2002). NIR fluorescence has reduced scatter and higher sensitivity than visible fluorescence. This NIR region is also compatible with using diode lasers and silicon detectors, providing a noninvasive method of dye excitation and imaging. Compared to other contrast agents, Cy5.5–CLIO is phagocytosed by the tumor cells and does not escape from them, allowing it to be administered either before or during the surgery.

The Cy5.5-R4 portion can be activated by the proteases in macrophages as they cleave the arginine bonds to make Cy5.5-RR and RR-SC-CLIO, providing increased NIR fluorescence signals. This was confirmed in animal studies by injecting Cy5.5-R4-SC-CLIO and Cy3.5-R4-SC-CLIO (which is not cleaved by the proteases) and measuring the Cy5.5-to-Cy3.5 fluorescence ratio (Kircher et al. 2002). This ratio was found to be highest in the liver, followed by the spleen and lungs. Rats implanted with gliosarcoma cell lines that expressed green fluorescent protein (GFP) were injected with CLIO–Cy5.5 via tail vein injection (Kircher et al. 2003). This was followed by MRI scans after 24 h, cryosectioning of the tumor, and evaluating for fluorescence of GFP and Cy5.5. Tumor extent measured on MRI scans correlated highly with tumor extent measured by histopathologic examination (linear regression, $R^2 = 0.996$). The border of the tumor as detected by

CLIO extended minimally beyond the GFP border (0.02 mm). This study demonstrated the extent to which Cy5.5–CLIO can accurately delineate the tumor margin.

The Cy5.5–CLIO nanoparticle can be modified in different ways. For example, it can be used to image apoptotic cells if it is conjugated with annexin-V (Schellenberger et al. 2002), which recognizes phosphatidylserine in apoptotic cells. The annexin-V–CLIO nanoparticle can identify apoptotic cells even at very low concentrations of the magnetic substrate. The Cy5.5–CLIO nanoparticle was conjugated to a TAT protein sequence (Josephson et al. 1999), which enabled a 100-fold increase in internalization by lymphocytes when compared to control particles, as measured by fluorescence microscopy. After uptake of CLIO nanoparticles, magnetically labeled cells can be tracked noninvasively by MRI.

Liposomal Nanoparticles Conjugated to Interleukin-13

Receptors for interleukin-13 (IL-13) are upregulated in glioblastoma (Pandya et al. 2012). IL-13 is a cytokine secreted primarily by helper T cells but also by other cell types. It has both proinflammatory and anti-inflammatory roles. IL-13 can induce B cells to release IgE, which can produce an antiparasitic response or an allergic response by the lung (as demonstrated by goblet cell metaplasia and excessive mucous secretion). IL-13 has two different receptors: IL-13/4R and IL-13R α (Wynn 2003). The former is present on normal cells and also binds IL-4; the latter is especially expressed by the tumor and offers a very specific target for drug delivery. As with the LDL nanoparticles mentioned above, IL-13-bound liposomes are tagged with a fluorescent marker, rhodamine, to characterize liposomal uptake with flow cytometry and fluorescence microscopy (Madhankumar et al. 2006). The IL-13-bound liposomes can be taken up in vitro by various glioma cell lines. In an in vivo study in adult female athymic nude mice with human glioma xenografts, IL-13-targeted liposomes and nontargeted liposomes were given by intraperitoneal injection over 7 weeks. The IL-13-targeted liposomes decreased tumor volume to less than one quarter of that of controls treated with nontargeted liposomes (Madhankumar et al. 2006).

Gold Nanorods Conjugated to Aptamer

An aptamer is a protein or oligonucleotide sequence that selectively binds to its target with an affinity rivaling that of antibodies. DNA and RNA aptamers consist of a short linear sequence

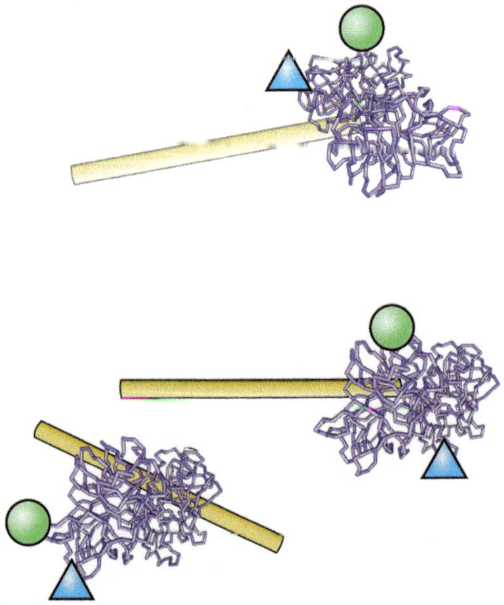

FIGURE 5.6 Complex of gold nanorods (gold) and aptamers (violet) showing the typical aptamer structure involving a variable chain of protein conjugated to protein scaffolds (blue) at either end and to fluorescent molecule (green). Aptamers provide specific targeting for the gold nanorods.

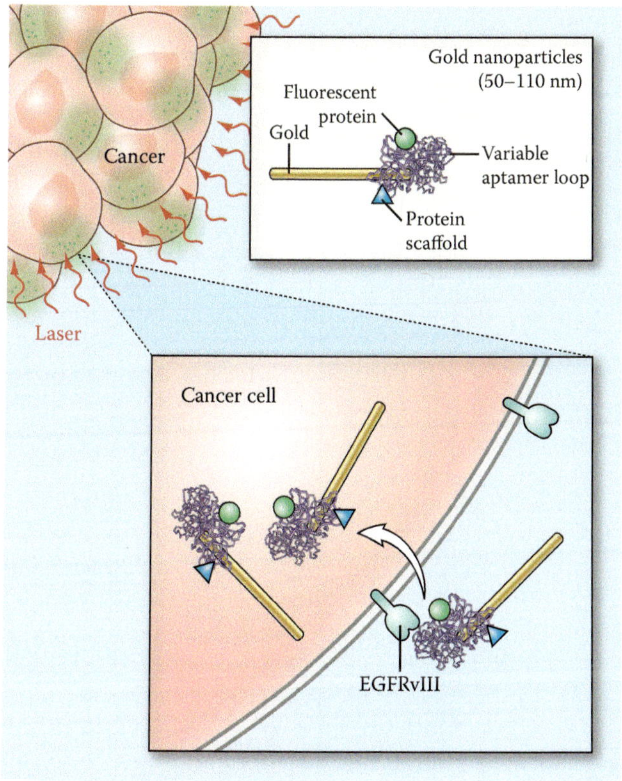

FIGURE 5.7 Multiple gold nanorod–aptamer complexes bind cancerous cells with tumor-specific targeting provided by aptamers. A fluorescent protein attached allows localization within the tumor by fluorescence microscopy. Laser application induces apoptosis in the cancerous cells by inducing release of thermogenic energy from the gold nanorods.

of nucleotides, usually about 15–40, that can assume a complex three-dimensional shape due to molecular interactions. These aptamers are engineered to bind the molecule of interest through repeated in vitro selection or systematic evolution of ligands by exponential enrichment (SELEX) and elicit negligible immunogenicity. Protein aptamers are composed of a short peptide domain that is attached on either end to a protein scaffold, which increases the binding affinity of the structure. Aptamers interfere with protein interactions within cells. They were conjugated with gold nanorods to enable the nanorods to specifically target glioma cell lines. A fluorescein molecule attached to the aptamer allowed binding affinity to be measured with fluorescent activated cell sorting (FACS). Cells bound with 99% affinity to the aptamer. Subsequent exposure to laser energy resulted in 75%–85% destruction of the tumor cells compared to little or no effect of aptamer or gold nanorod controls with and without laser (Oli 2010) (Figures 5.6 and 5.7).

Magnetic Nanoparticles Conjugated to Antibody and Gold Nanospheres Conjugated to Antibodies and DNA for Highly Sensitive Detection of Amyloid Protein in Cerebrospinal Fluid (CSF)

Amyloid-β-derived diffusible ligands (ADDLs) are present in elevated levels in the brains of patients with Alzheimer's disease (AD). They are thought to be involved in memory loss associated with the disease. Ultrasensitive assays, such as the polymerase chain reaction (PCR), are commonly used to detect nucleic acids. Assays of protein using enzyme-linked immunosorbent assay (ELISA) are, in comparison, less sensitive, although immuno-PCR (IPCR), in which a secondary antibody is

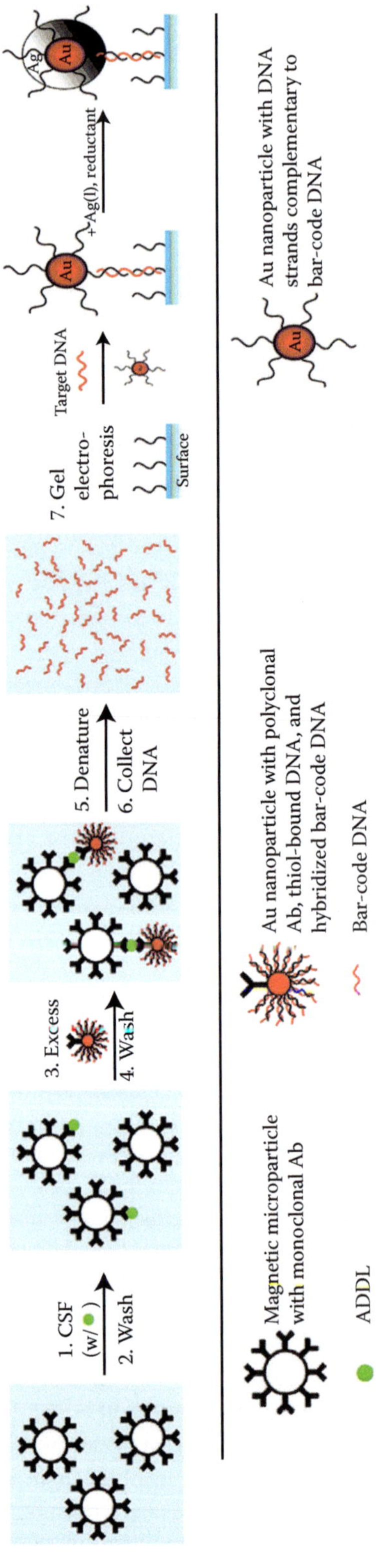

FIGURE 5.8 Magnetic nanoparticles with antibody are washed with CSF containing ADDLs. The complex is later exposed to polyclonal antibody and bio-barcode DNA-bearing gold nanoparticles. The resulting complex is denatured to release the bio-barcode DNA. (From Keating, C.D., *PNAS* 102: 2263–4, 2005. With permission.)

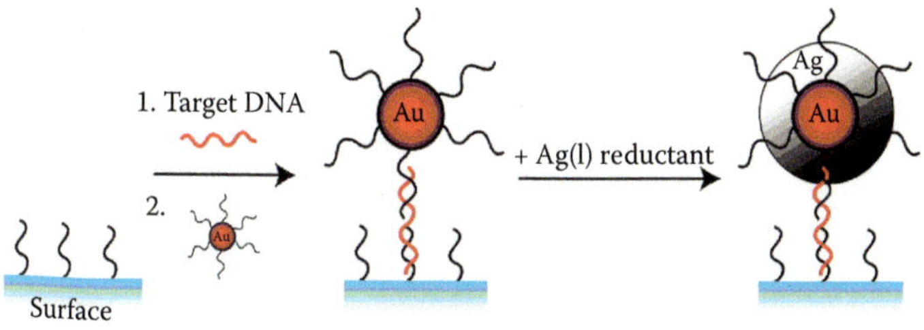

FIGURE 5.9 Target DNA is detected by gel electrophoresis. Later, gold nanospheres are bound to the target DNA and act as nucleation sites for silver. The complex is subsequently quantified using a scanner to detect the concentration of ADDLs. (From Keating, C.D., *PNAS* 102: 2263–4, 2005. With permission.)

coupled with an oligonucleotide rather than an enzyme, can increase the sensitivity of ELISA for detecting protein by about 1000-fold (Keating 2005). Georganopoulou et al. (2005) developed the DNA bio-barcode amplification method, an even more sensitive assay than IPCR for ADDLs in CSF that utilizes nanoparticles. To conduct the assay, (1) human CSF is exposed to anti-ADDL antibodies that are bound to magnetic nanoparticles; (2) these nanoparticles are then washed and later (3) exposed to the polyclonal antibody-bearing gold nanospheres, which carry covalently bound DNA, complementary barcoded DNA, and antibodies (Figure 5.8). Later the specimen is (4) denatured, and the identical barcode DNA strands are released that can then be detected by (5) gel electrophoresis and quantified using the (6) scanometric DNA detection method (Taton et al. 2000), in which detection strands bound to gold nanospheres bind the target DNA and ultimately act as nucleation sites for silver. This can be quantified with a simple desktop scanner (Figure 5.9). This method is extremely sensitive and can detect ADDLs in concentrations of only 200×10^{-18} M in normal patients, with higher levels detected in AD patients.

Short Cationic Amphiphilic Peptides Conjugated to TAT Protein and a Cholesterol Core

Short cationic amphiphilic peptides form an important part of the innate immune system by killing invading pathogens directly and by activating the adaptive immune system. They have antimicrobial properties and are present in almost every life form (Hancock and Sahl 2006). Additionally, they have an unusually broad spectrum of activity against different infectious agents, including viruses, and also exhibit tumoricidal properties. These are gene-encoded short peptide sequences with about 10–50 amino acids and an overall positive charge with a significant proportion of hydrophobic residues. This enables three-dimensional folding of the protein molecule, especially upon contact with cellular membranes. Four categories of these folded peptides can be defined: alpha-helical peptides, beta-peptide sheets, loop peptides, and extended structures rich in glycine, tryptophan, proline, histidine, and/or arginine. They may exert their action through enhancing the effects of innate immunity through immunomodulation. The mechanism involves binding the cationic portion of these peptides with the polyanionic microbial cell membrane, which results either in pore formation by disrupting the physical integrity of the membrane or in cell membrane penetration resulting in various intracellular effects such as inhibition of DNA and protein synthesis.

Liu et al. (2009) used the TAT peptide sequence as a targeting molecule. TAT protein sequence is a part of the trans-activator of transcription (TAT) protein of HIV-1 and is the minimal peptide sequence required for translocation through the cellular membrane (Liu et al. 2009) (Figure 5.10). TAT promotes increased cellular uptake of proteins, small interfering RNA (siRNA), and liposomes when it is chemically attached to them. It also enables the transfer of quantum dots and polymeric micelles across the BBB. Liu et al. designed a short amphiphilic peptide tagged with TAT and six arginine residues that increased the cationic charge on the surface for translocation. The hydrophobic portion participated in the transfer of cholesterol-containing materials. This configuration allowed self-assembly of nanoparticles with a hydrophobic inner portion (cholesterol) and a hydrophilic surface that is tagged with TAT (Liu et al. 2009). The local density of positive

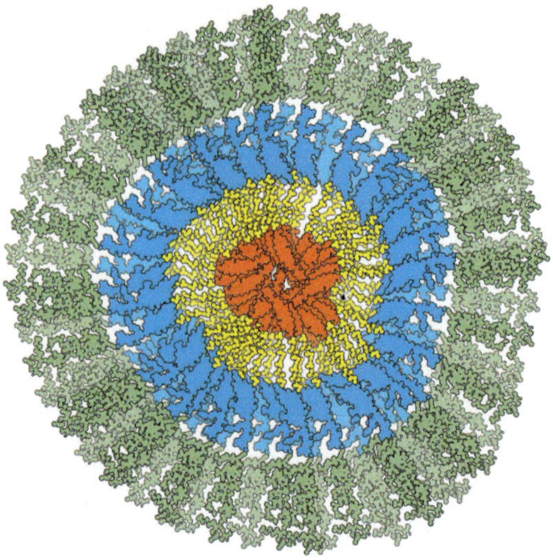

FIGURE 5.10 Amphiphilic peptide structural configuration showing a cholesterol core (orange), glycine layer (yellow), arginine layer (blue), and trans-activating transcriptor (TAT) protein surface layer (green).

charge enhances the antimicrobial properties of the peptide mass and enables it to cross the BBB. These nanoparticles were found to have a broad spectrum of antimicrobial properties, inhibiting the growth of a variety of gram-positive bacteria, fungi, and yeast. They have low minimal inhibitory concentrations and have a high therapeutic index, especially against *Staphylococcus aureus* infections, which are usually very aggressive, especially in the CNS. These nanoparticles have great potential as future antimicrobial agents and appear to have only minimal systemic toxicity. Of note, these peptides have only been recently evaluated in preclinical and early clinical trials.

CONCLUSION

Nanoparticle-based therapy has developed steadily and is beginning to have a substantial impact on clinical practice. Nanotechnology holds great potential to radically alter the treatment of patients with diseases of the CNS. It may be able to improve recognition of tumor margins and aid in resecting the nidus of primary brain tumors. Nanotechnology may enable better recognition of tumor cells invading surrounding brain by introducing new or modified imaging techniques based on iron oxide particles or quantum dots. It may also improve adjuvant therapies such as thermotherapy and PDT and extend survival of patients with primary brain tumors. Rigorous clinical trials are already underway to test nanotechnology-based drugs. Nanoparticle conjugates are being developed to improve transit of certain drugs that otherwise could not cross the BBB or enter the CNS. Preclinical and clinical testing is also being conducted to address the biodegradability and potential toxicity of nanoparticles. Nanotechnology provides new opportunities to develop novel diagnostic tests and therapeutics that may improve the treatment of presently disabling and incurable diseases. Basic, translational, and clinical researchers will drive progress in this field by persisting in their developing and testing of diagnostic agents and therapeutics that employ nanoparticles.

REFERENCES

Bender, A., von Briesen, H., Kreuter, J. et al. 1996. Efficiency of nanoparticles as a carrier system for antiviral agents in human immunodeficiency virus-infected human monocytes/macrophages in vitro. *Antimicrob. Agents Chemother.* 40: 1467–71.

Cimini, A., and Ippoliti, R. 2011. Innovative therapies against human glioblastoma multiforme. *ISRN Oncol* 2011: 787490.

Fleige, G., Nolte, C., Synowitz, M. et al. 2001. Magnetic labeling of activated microglia in experimental gliomas. *Neoplasia* 3: 489–99.

Fujita, M., Lee, B.-S., Khazenzon, N.M. et al. 2007. Brain tumor tandem targeting using a combination of monoclonal antibodies attached to biopoly(β-L-malic acid). *J Controlled Release* 122: 356–63.

Georganopoulou, D.G., Chang, L., Nam, J.-M. et al. 2005. Nanoparticle-based detection in cerebral spinal fluid of a soluble pathogenic biomarker for Alzheimer's disease. *PNAS* 7: 2273–6.

Hadjipanayis, C.G., Machaidze, R., Kaluzova, M. et al. 2010. EGFRvIII antibody-conjugated iron oxide nanoparticles for magnetic resonance imaging-guided convection-enhanced delivery and targeted therapy of glioblastoma. *Cancer Res* 70: 6303–12.

Hancock, R.E.W., and Sahl, H.-G. 2006. Antimicrobial and host-defense peptides as new anti-infective therapeutic strategies. *Nat Biotech* 24: 1551–7.

Jeong, H., Huh, M., Lee, S.J. et al. 2011. Photosensitizer-conjugated human serum albumin nanoparticles for effective photodynamic therapy. *Theranostics* 1: 230–9.

Josephson, L., Kircher, M.F., Mahmood, U. et al. 2002. Near-infrared fluorescent nanoparticles as combined MR/optical imaging probes. *Bioconj Chem* 13: 554–60.

Josephson, L., Tung, C.-H., Moore, A. et al. 1999. High-efficiency intracellular magnetic labeling with novel superparamagnetic–TAT peptide conjugates. *Biocon Chem* 10: 186–91.

Kar, U.K., Srivastava, M.K., Andersson, A. et al. 2011. Novel CCL21–vault nanocapsule intratumoral delivery inhibits lung cancer growth. *PLoS One* 6: e18758.

Keating, C.D. 2005. Nanoscience enables ultrasensitive detection of Alzheimer's biomarker. *PNAS* 102: 2263–4.

Kedersha, N.L., and Rome, L.H. 1986. Isolation and characterization of a novel ribonucleoprotein particle: large structures contain a single species of small RNA. *J Cell Biol* 103: 699–709.

Kickhoefer, V.A., Garcia, Y., Mikyas, Y. et al. 2005. Engineering of vault nanocapsules with enzymatic and fluorescent properties. *Proc Natl Acad Sci U S A* 102: 4348–52.

Kircher, M.F., Josephson, L., and Weissleder, R. 2002. Ratio imaging of enzyme activity using dual wavelength optical reporters. *Mol Imaging* 1: 89–95.

Kircher, M.F., Mahmood, U., King, R.S. et al. 2003. A multimodal nanoparticle for preoperative magnetic resonance imaging and intraoperative optical brain tumor delineation. *Cancer Res* 63: 8122–5.

Kirpotin, D.B., Drummond, D.C., Shao, Y. et al. 2006. Antibody targeting of long-circulating lipidic nanoparticles does not increase tumor localization but does increase internalization in animal models. *Cancer Res* 66: 6732–40.

Kuo, Y.-C., and Liang, C.-T. 2011. Inhibition of human brain malignant glioblastoma cells using carmustine-loaded catanionic solid lipid nanoparticles with surface anti-epithelial growth factor receptor. *Biomaterials* 32: 3340–50.

Lai, C.Y., Wiethoff, C.M., Kickhoefer, V.A. et al. 2009. Vault nanoparticles containing an adenovirus-derived membrane lytic protein facilitate toxin and gene transfer. *ACS Nano* 3: 691–9.

Liu, L., Xu, K., Wang, H. et al. 2009. Self-assembled cationic peptide nanoparticles as an efficient antimicrobial agent. *Nat Nano* 4: 457–63.

Lu, W., Sun, Q., Wan, J. et al. 2006. Cationic albumin-conjugated pegylated nanoparticles allow gene delivery into brain tumors via intravenous administration. *Cancer Res* 66: 11878–87.

Madhankumar, A.B., Slagle-Webb, B., Mintz, A. et al. 2006. Interleukin-13 receptor-targeted nanovesicles are a potential therapy for glioblastoma multiforme. *Mol Canc Ther* 5: 3162–9.

Maletínská, L., Blakely, E.A., Bjornstad, K.A. et al. 2000. Human glioblastoma cell iines: levels of low-density lipoprotein receptor and low-density lipoprotein receptor-related protein. *Cancer Res* 60: 2300–3.

McConathy, W.J., Nair, M.P., Paranjape, S. et al. 2008. Evaluation of synthetic/reconstituted high-density lipoproteins as delivery vehicles for paclitaxel. *Anti-Cancer Drugs* 19: 183–8.

Mossink, M.H., van Zon, A., Scheper, R.J. et al. 2003. Vaults: a ribonucleoprotein particle involved in drug resistance? *Oncogene* 22: 7458–67.

Nikanjam, M., Gibbs, A.R., Hunt, C.A. et al. 2007. Synthetic nano-LDL with paclitaxel oleate as a targeted drug delivery vehicle for glioblastoma multiforme. *J Control Release* 124: 163–71.

Oli, M. 2010. Aptamer conjugated gold nanorods for targeted nanothermal radiation of glioblastoma cancer cells (A novel selective targeted approach to cancer treatment). *Young Sci J* 3: 18–25.

Pandya, H., Gibo, D.M., Garg, S. et al. 2012. An interleukin 13 receptor α 2-specific peptide homes to human glioblastoma multiforme xenografts. *Neuro Oncol* 14: 6–18.

Rao, K.S., Reddy, M.K., Horning, J.L. et al. 2008. TAT-conjugated nanoparticles for the CNS delivery of anti-HIV drugs. *Biomaterials* 29: 4429–38.
Reddy, G.R., Bhojani, M.S., McConville, P. et al. 2006. Vascular targeted nanoparticles for imaging and treatment of brain tumors. *Clin Canc Res* 12: 6677–86.
Reubi, J.C., and Maecke, H.R. 2008. Peptide-based probes for cancer imaging. *J Nucl Med* 49: 1735–8.
Schäfer, V., von Briesen, H., Andreesen, R. et al. 1992. Phagocytosis of nanoparticles by human immunodeficiency virus (HIV)-infected macrophages: a possibility for antiviral drug targeting. *Pharm Res* 9: 541–6.
Scheffer, G.L., Wijngaard, P.L., Flens, M.J. et al. 1995. The drug resistance-related protein LRP is the human major vault protein. *Nat Med* 1: 578–82.
Schellenberger, E.A., Bogdanov, A., Jr., Hogemann, D. et al. 2002. Annexin V–CLIO: a nanoparticle for detecting apoptosis by MRI. *Mol Imaging* 1: 102–7.
Smith, J.S., Cha, S., Mayo, M.C. et al. 2005. Serial diffusion-weighted magnetic resonance imaging in cases of glioma: distinguishing tumor recurrence from postresection injury. *J Neurosurg* 103: 428–38.
Taton, T.A., Mirkin, C.A., and Letsinger, R.L. 2000. Scanometric DNA array detection with nanoparticle probes. *Science* 289: 1757–60.
Uchida, M., Flenniken, M.L., Allen, M. et al. 2006. Targeting of cancer cells with ferrimagnetic ferritin cage nanoparticles. *J Am Chem Soc* 128: 16626–33.
Wacker, M., Chen, K., Preuss, A. et al. 2010. Photosensitizer loaded HSA nanoparticles. I: preparation and photophysical properties. *Int J Pharmaceutics* 393: 254–63.
Wagner, V., Dullaart, A., Bock, A.-K. et al. 2006. The emerging nanomedicine landscape. *Nat Biotech* 24: 1211–7.
Wynn, T.A. 2003. IL-13 effector functions. *Ann Rev Immunol* 21: 425–56.
Zhang, L., Gu, F.X., Chan, J.M. et al. 2007. Nanoparticles in medicine: therapeutic applications and developments. *Clin Pharmacol Ther* 83: 761–9.

6 Endohedral Metallofullerenes, Iron Oxide Agents, and Gold Nanoparticles for Brain Imaging

Martina A. McAteer and Robin P. Choudhury

CONTENTS

INTRODUCTION

Considerable advances in nanotechnology have led to developing a range of nanomaterials with great promise for advancing medical diagnostic imaging and therapeutics. Current clinical brain imaging modalities include magnetic resonance imaging (MRI), x-ray computed tomography, and ultrasound. While MRI provides high-spatial-resolution images with soft tissue contrast, without using ionizing radiation, x-ray computed tomography exposes patients to ionizing radiation, and ultrasound cannot reliably penetrate the human skull. MRI has inherently low sensitivity, but this can be overcome by using MRI contrast agents. In clinical practice, gadolinium (Gd) chelates such as Gd diethylenetriaminepentaacetic acid (Gd-DTPA) are the most commonly used MRI contrast agents. For brain imaging, impairment of the blood–brain barrier (BBB) enables preferential accumulation of Gd-DTPA in regions of tumors and inflammatory lesions due to increased vascular permeability. However, Gd chelates have been reported to increase the risk of nephrogenic systemic fibrosis, a serious medical condition in patients with renal impairment (Lim et al. 2007; Stratta et al. 2008). Therefore, there is a continuing need for safer contrast agents, with higher relaxivity and molecular specificity. MRI is very responsive to molecular imaging through the use of contrast agents conjugated to specific targeting ligands such as antibodies, peptides, or aptamers (Choudhury et al. 2004). An array of nanoparticles and microparticles of iron oxide have been developed for in vivo molecular imaging in experimental systems. Gd-based metallofullerenes are also emerging as a new class of high-relaxivity MRI contrast agents, with a higher safety margin than conventional Gd chelates (MacFarland et al. 2008).

Optical imaging techniques also promise noninvasive biomedical diagnostic imaging tools, which provide high-spatial-resolution real-time images at relatively low cost and do not expose patients to ionizing irradiation. However, optical techniques such as optical coherence tomography (OCT) and two-photon microscopy cannot penetrate beyond a soft depth limit of ~1 mm in the skin (Wang 2008). This fundamental depth limitation has recently been countered by the development of photoacoustic tomography (PAT), which combines strong optical absorption contrast and high ultrasonic resolution in a single modality (Wang 2008), enabling imaging of biological tissue up to a depth of several centimeters (Debbage and Jaschke 2008). High-resolution optical images can be enhanced by using optically excitable contrast agents and quantitative detection of resulting oscillatory contrast agent movement with ultrasound. Gold nanostructures are of great interest for optical imaging due to their remarkable capacity to absorb and scatter light in both the visible and near-infrared (NIR) regions. Gold nanoparticles (GNPs) also have intense photothermal properties whereby they can convert optical energy into heat. Localized heating effects can be directed toward neoplastic tissue or other diseased tissue via targeting with GNPs (Jain et al. 2008). The wavelength of light used for photothermal therapy is invisible to most tissues, allowing tissue penetration and excitation of GNP with minimal energy transfer to surrounding structures (Collins et al. 2010).

In this chapter, we review the properties and recent applications of Gd-based metallofullerenes and iron oxide nanoparticles and microparticles for MRI and GNP for the PAT of the brain. The development of targeted molecular imaging agents, as well as multifunctional nanoprobes with combined diagnostic and therapeutic capabilities, is also discussed. Finally, the potential of these nanomaterials for clinical implementation is highlighted.

ENDOHEDRAL METALLOFULLERENES

Gd-based endohedral metallofullerenes are emerging as a promising new generation of paramagnetic MRI contrast agents (MacFarland et al. 2008). Fullerenes, also known as buckminsterfullerenes or "buckyballs," were first described by Kroto et al. (1985) as a truncated icosahedron crystal cage structure composed of 60 carbon atoms. The hollow interior of the fullerene cage can trap atoms, ions, or clusters, such as Gd ions or trimetallic nitride (Gd_3N), inside to form endohedral metallofullerenes (Fatouros et al. 2006; MacFarland et al. 2008). Most endohedral metallofullerenes have been synthesized using fullerenes with molecular weight of C80 or more due to the limited interior volume of C60 (Liang et al. 2008). The encapsulation of Gd prevents release of toxic Gd ions into surrounding tissues, providing a high biological safety margin for in vivo applications (Aloya et al. 2006; Krause and Dunsch 2005). The hydrophobic character of fullerenes is overcome by covalent conjugation of hydrophilic hydroxyl groups or polyethylene glycol (PEG) to the fullerene cage. Recently, novel Gd_3N derivatives, termed hydrochalarones, which are made water soluble by adding polar ethylene glycols (rather than hydroxyls) to the fullerene cage, have been developed (MacFarland et al. 2008).

MRI Relaxivity Properties

Gd-based endohedral metallofullerenes have significantly greater MRI relaxivity properties than conventional Gd chelates by up to 20 times (Mikawa et al. 2001) and 30 times (Fatouros et al. 2006). The relaxivity mechanism differs from the classic inner-sphere mechanism of Gd chelates, which relies on direct water to Gd^{3+} interactions. Since the fullerene cage prevents passage of water or Gd ions (Laus et al. 2005), the relaxivity mechanism is thought to be due to electronic interactions of the fullerene cage with water molecules (Liang et al. 2008). Endohedral metallofullerenes containing one Gd ion, derived by polyhydroxylation or through addition of carboxyl groups to improve water solubility, exhibit high relaxivity, although this has been attributed primarily to the formation of aggregates (Laus et al. 2005).

This problem has been overcome by adding groups on the fullerene cage that optimize magnetic coupling between unpaired electrons on the enclosed Gd^{3+} and water protons outside the cage (MacFarland et al. 2008). Trimetallic nitride metallofullerenes encapsulating three Gd^{3+} ions per molecule exhibit the highest relaxivities due to increased charge transfer between the Gd_3N cluster and the fullerene cage, producing a larger magnetic moment and longer rotational correlation time (Qian et al. 2007; Shu et al. 2009). Hydroxyl surface groups further facilitate proton exchange with surrounding water protons, transmitting paramagnetic relaxation to bulk water molecules (Shu et al. 2009).

Applications of Endohedral Metallofullerenes

Here, we highlight a number of endohedral metallofullerenes that have been applied in experimental models for MRI detection and tracking of brain tumors and their potential as multimodal imaging and therapeutic agents. The first, hydroxylated/PEGylated trimetallic nitride metallofullerene ($Gd_3N@C_{80}[DiPEG_{5000}(OH)_x]$), increased T_1 relaxivity up to 30 times compared to Gd chelates (Fatouros et al. 2006). In vivo infusion of $Gd_3N@C_{80}[DiPEG_{5000}(OH)_x]$ into rats 14 days after intracerebral implantation of glioma improved tumor delineation by MRI (2.4 T) compared to gadodiamide alone due to prolonged residence within the tumor. Shu et al. (2009) reported a novel hydroxylated and carboxylated trimetallic nitride templated (TNT) metallofullerene (TNT $Gd_3N@C_{80}$) that enabled tumor delineation at 1/50 the dose of commercial Gd chelates in a rat model of experimental brain glioma. TNT $Gd_3N@C_{80}$ enabled clear delineation of tumors up to 7 days postinjection, demonstrating its potential as a long-term diagnostic agent, due to its slow diffusion behavior (Figure 6.1). The reactive surface carboxyl terminal groups also allow ligand conjugation for future molecular imaging studies.

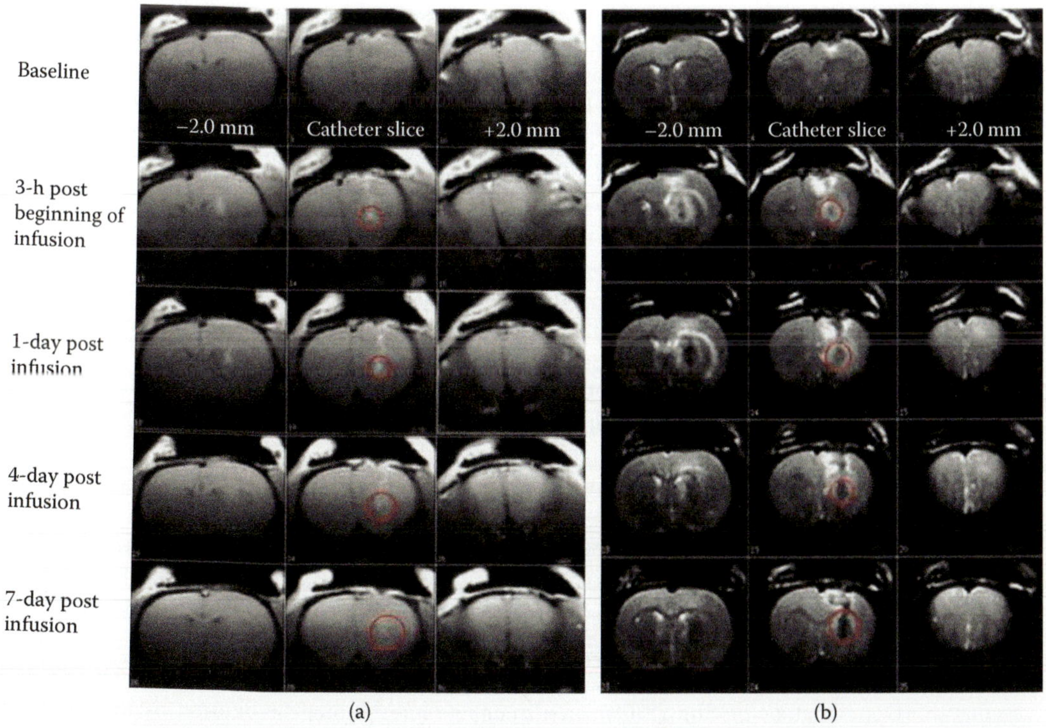

FIGURE 6.1 T_1-weighted (a) and T_2 weighted (b) images of direct infusion of TNT $Gd_3N@C_{80}$ agent into tumor-bearing rat brain. Infusion was applied for 120 min at 0.2 μL/min. (From Shu, C. et al., *Bioconjug Chem* 20: 1186–93, 2009. With permission.)

The relaxivity properties of endohedral trimetallic nitride metallofullerenes ($Gd_3N@C_{80}$ [DiPEG$(OH)_x$]) can also be influenced by the molecular weight of the PEG surface coat (350–5000 Da) (Zhang et al. 2010a). $Gd_3N@C_{80}$[DiPEG$(OH)_x$] derivatives PEGylated with 350 Da PEG exhibited the highest MRI relaxivity (2.4 T). Direct in vivo infusion of 350 Da PEGylated $Gd_3N@C_{80}$[DiPEG$(OH)_x$] into tumor-bearing rat brains demonstrated an initial uniform distribution, and the potential for effective local tumor radiation (brachytherapy) therapy, if encapsulated Gd^{3+} ions were replaced with radioactive ^{177}Lu (Zhang et al. 2010a). Multimodal single-walled carbon nanohorns (SWNHs) were also synthesized, which can be encapsulated with $Gd_3N@C_{80}$ for MRI or $Lu_3N@C_{80}$ as an x-ray contrast agent or ^{177}Lu as a radiotherapeutic agent (Zhang et al. 2010b). The SWNHs can be further conjugated with cadmium selenide/zinc sulfide quantum dots ($Gd_3N@C_{80}$@SWNH-QDs) for fluorescence confocal microscopy of brain sections. Direct in vivo infusion of $Gd_3N@C_{80}$@SWNH-QDs into U87 intracerebral tumors in mice significantly enhanced tumor detection that persisted 1 and 3 days postinfusion on T_1-weighted magnetic resonance (MR) images. The prolonged tumor residence of SWNHs may provide a useful probe for long-term diagnosis and evaluation of local drug delivery therapy. Thus, trimetallic nitride metallofullerenes can offer a multimodal platform for combining MRI or radiochemical agents, x-ray-based imaging contrast, or fluorescence (Fatouros et al. 2006).

IRON OXIDE CONTRAST AGENTS

Iron oxides are superparamagnetic agents composed of a magnetite (Fe_2O_3) and/or maghemite (Fe_3O_4) core, surrounded by a dextran or polymer coat. They are classified according to size as ultrasmall superparamagnetic particles of iron oxide (USPIOs) (20–50 nm), superparamagnetic particles of iron oxide (SPIOs) (60–250 nm), or microparticles of iron oxide (MPIOs) (0.9–4.5 μm). By MRI, iron oxide agents shorten T_2 and T_2^* relaxation times, creating hypointense signal areas that appear black on MR images. "Positive" sequences were also developed, which generate bright signals in regions containing iron oxide, to distinguish between iron contrast and normal tissue heterogeneity and susceptibility artifacts (Briley-Saebo et al. 2008; Lipinski et al. 2008). Below, we highlight properties and applications of USPIOs and MPIOs for nontargeted and targeted molecular imaging, which we have reviewed in detail elsewhere (McAteer et al. 2010; McAteer and Choudhury 2009).

Molecular MRI Using USPIOs

USPIOs, due to their long blood half-life (24 h) and small diameter, are readily phagocytosed by inflammatory cells and can detect inflammatory central nervous system (CNS) lesions by in vivo MRI both in experimental models (Brochet et al. 2006; Deloire et al. 2004; Dousset et al. 1999a,c; Floris et al. 2004; Kim et al. 2008; Kleinschnitz et al. 2005; Rausch et al. 2001, 2002, 2003; Schroeter et al. 2004; Zimmer et al. 1995) and in patients with multiple sclerosis (Dousset et al. 2006; Manninger et al. 2005), stroke (Nighoghossian et al. 2007; Saleh et al. 2004; Trivedi et al. 2006), and intracranial tumors (Neuwelt et al. 2004). Since USPIO uptake can precede and persist longer than Gd enhancement (Vellinga et al. 2008), which relies on increased permeability of impaired BBB (Dousset et al. 1999a; Floris et al. 2004; Saleh et al. 2004), it has been suggested that UPSIOs may be taken up by circulating monocyte/macrophages across an activated but intact BBB (Dousset et al. 2006; Manninger et al. 2005). USPIOs may also passively enter the CNS by transcytosis across endothelial cells (Xu et al. 1998) or via cellular incorporation by activated microglia outside the vasculature (Dousset et al. 1999b). SPIOs can also enable brain tumor delineation by in vivo MRI (Fleige et al. 2001; Moore et al. 2000; Muldoon et al. 2006) due to enhanced tumor cell internalization and prolonged clearance from the tumor site (Varallyay et al. 2002). However, this nontargeted approach cannot distinguish between malignant and nonmalignant tumors. SPIOs, conjugated to lactoferrin, whose receptor is highly expressed on the glioblastoma cell surface, significantly

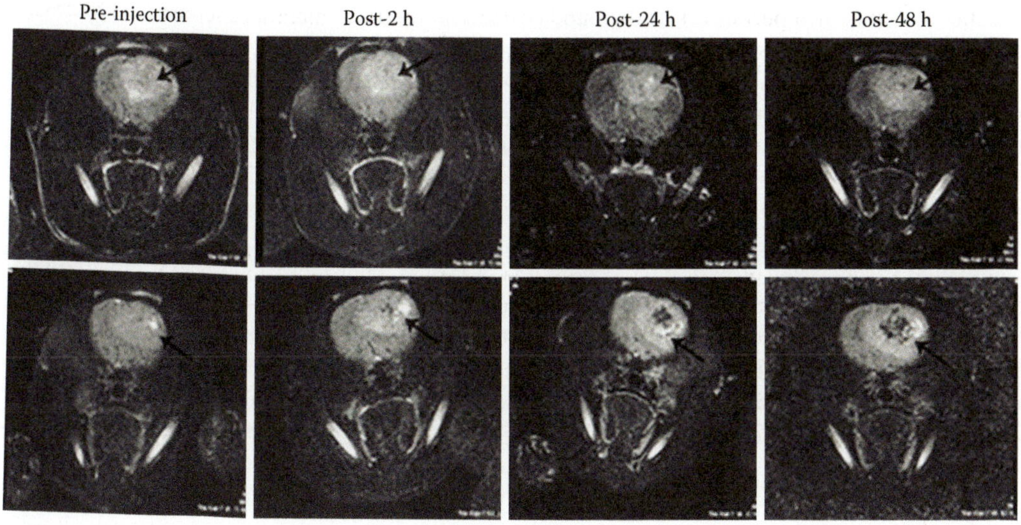

FIGURE 6.2 In vivo T_2-weighted MR images of rats bearing C6 brain gliomas. Upper row: acquired after administration of nontargeted superparamagnetic iron oxide nanoparticles (SPIONs) (12 mg/kg); lower row: acquired after administration of lactoferrin targeted SPIONs (12 mg/kg). (From Xie, H. et al., *Biomaterials* 32: 495–502, 2011. With permission.)

enhanced brain glioma detection by in vivo MRI compared to nontargeted SPIOs (Figure 6.2) (Xie et al. 2011). Multimodal nanoparticles can enable brain tumor margin delineation in rats by MRI presurgery and postsurgery as well as intraoperatively using NIR fluorescence imaging (Kircher et al. 2003). Targeted multimodal USPIOs, fluorescently labeled with Cy5.5 and covalently conjugated to a peptide chlorotoxin, which specifically binds to matrix metalloproteinase-2 (MMP-2) endopeptidase, which is upregulated on glioma cells (Meng et al. 2007; Veiseh et al. 2005), improved glioma detection both in vitro (Veiseh et al. 2005) and in mice bearing 9L xenograft tumors by in vivo MRI (Sun et al. 2008). Multimodal, fluorescent USPIOs, conjugated to cyclic arginine–glycine–aspartic acid (RGD) peptides, have also permitted in vivo fluorescence and MRI detection of $\alpha_V\beta_3$ expression, a marker of tumor angiogenesis (Montet et al. 2006). However, biological effects of high doses must be considered because high uptake of RGD-targeted USPIOs can lead to loss of cell adhesion and endocytotic activity in glioblastoma (U87MG) cells in vitro and in vivo (Kiessling et al. 2009). Theranostic nanoprobes were also developed for combined detection, diagnosis, and treatment of brain tumors in rats (Reddy et al. 2006). The photodynamic therapy agent, Photofrin, was embedded together with USPIOs inside fluorescently labeled PEGylated amine-functionalized polyacrylamide nanoparticles, conjugated with an F3 peptide, which specifically binds to tumor endothelial cells. F3-targeted, light-activated nanoparticles enhanced in vivo MRI tumor detection in rats and improved survival rates compared to Photofrin or untargeted nanoparticles.

Microparticles of Iron Oxide

MPIOs, due to their high iron content, create significantly greater hypointense contrast effects than USPIOs on T_2^*-weighted MR images that extend up to 50 times their physical diameter (Shapiro et al. 2006). The large MPIO contrast effects can enable in vivo MRI detection of single cells (Shapiro et al. 2006), such as single macrophages in the mouse brain (Heyn et al. 2006a), and cell tracking (Shapiro et al. 2005), including tracking brain cancer metastases from the single-cell stage (Heyn et al. 2006b; Ramadan et al. 2008). For endovascular imaging, the large size and incompressible nature of MPIOs make them less susceptible to nonspecific vascular egress than USPIOs (Briley-Saebo et al. 2006). MPIOs have a short circulation half-life (<2 min) (Ye et al. 2008) compared to

USPIOs (>24 h), which minimizes background contrast. Targeting MPIOs can be achieved by covalent conjugation of ligands to reactive surface groups, such as amine, carboxylic acid, and *p*-toluene sulfhydryl (tosyl) groups. Molecular targeting ligands include monoclonal or single-chain antibodies, aptamers, or small peptides (McAteer et al. 2010). Below, we highlight several targeted MPIO approaches for molecular imaging of cerebrovascular inflammation.

Molecular MRI Using MPIOs

Endothelial vascular cell adhesion molecule-1 (VCAM-1) is an attractive molecular imaging target since it is rapidly upregulated upon endothelial inflammation (Carlos et al. 1990), mediating leukocyte recruitment and subsequent lesion development (Elices et al. 1990). Initial leukocyte recruitment across an activated but intact endothelium cannot be observed by conventional MRI, which relies on increased BBB permeability in later stages of disease. We developed VCAM-1-targeted MPIOs for in vivo detection of early endothelial activation in mice (McAteer et al. 2007). MPIOs (1 μm diameter), with reactive tosyl groups, were covalently conjugated to monoclonal VCAM-1 antibodies (VCAM–MPIO) (McAteer et al. 2011). VCAM–MPIO bound in a dose-dependent manner to tumor necrosis factor alpha stimulated endothelial cells in vitro, which was blocked by the addition of soluble extracellular VCAM-1 chimeric protein (Fc-VCAM) but not

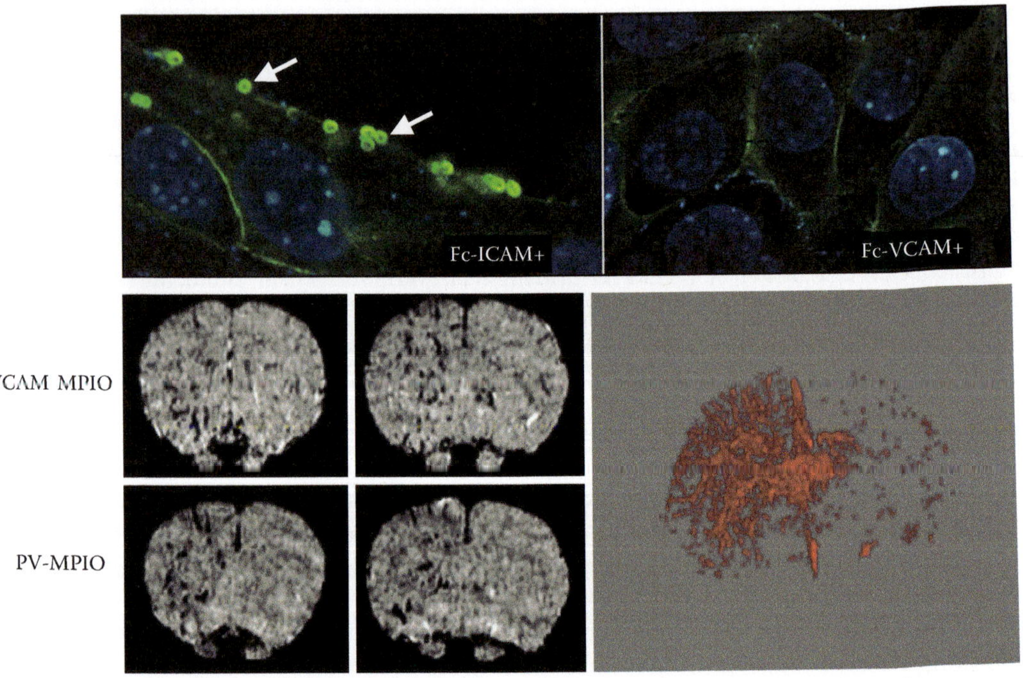

FIGURE 6.3 (a) VCAM–MPIO binding (autofluorescent spheres [see arrows]) on the surface of mouse skin endothelioma cell line sEND-1 cells, stimulated with tumor necrosis factor (TNF)-α, colocalizes with VCAM-1 immunofluorescence. Preincubation of VCAM–MPIO with Fc-VCAM-1 abolished VCAM–MPIO binding, despite cell surface expression of VCAM-1, as confirmed by immunofluorescence, while preincubation with Fc-ICAM-1 did not inhibit VCAM–MPIO binding. (b) T_2^*-weighted 3-D gradient echo MR image showing intense low-signal areas on the left IL-1β–stimulated brain hemisphere reflecting specific VCAM–MPIO with virtually absent contrast effects in the contralateral hemisphere. Similar, unilateral contrast effects were observed in mice administered dual-targeted VCAM-1 and P-selectin MPIO (PV-MPIO). (c) 3-D volumetric map of segmented VCAM–MPIO contrast delineating the architecture of inflamed cerebral vasculature in the left IL-1β-stimulated hemisphere with absence of binding on the contralateral side. (From McAteer, M.A. et al., *Nat Med* 13: 1253–8, 2007. With permission.)

by extracellular ICAM-1 (Fc-ICAM) (Figure 6.3). For in vivo studies, acute brain inflammation was induced by stereotactic injection of interleukin 1β (IL-1β) into the left cerebral hemisphere, and 3 h later, VCAM–MPIOs or control immunoglobulin G (IgG)–MPIOs were intravenously injected. In vivo T_2^*-weighted MR images showed highly specific, negative contrast effects in the IL-1β-activated hemisphere due to VCAM–MPIO binding, which delineated activated cerebral blood vessel architecture, with minimal binding in the nonstimulated hemisphere (Figure 6.3). We have also applied VCAM–MPIO for in vivo MRI of early ischemia–reperfusion injury using a mouse model of acute experimental stroke (Akhtar et al. 2010; Hoyte et al. 2010). VCAM–MPIO detected a greater volume of activated endothelium in the lesioned hemisphere compared to diffusion-weighted imaging (DWI), suggesting that VCAM–MPIO may detect potentially salvageable tissue in the penumbral region surrounding the ischemic core of the lesion, while DWI detects only the ischemic core (Hoyte et al. 2010). Following preconditioning, VCAM–MPIO binding to the lesioned hemisphere was reduced, demonstrating the protective effect of cerebral ischemic preconditioning.

In order to attenuate immunogenicity of monoclonal antibodies, several alternative ligands, including synthetic sugars and single-chain antibodies, have been investigated. Van Kasteren et al. (2009) developed iron oxide glyconanoparticles functionalized with a glycan ligand, sialyl Lewis^X (sLe^x), that binds with high affinity to E- and P-selectin. This agent enabled molecular MRI of early endothelial inflammation in mouse models of acute cerebral vascular inflammation, experimental autoimmune encephalomyelitis (EAE), and focal cerebral stroke. The sLe^X sugar ligand has the advantage of transferable cross-species utility that is difficult to achieve with antibodies. MPIOs have also been developed to detect activated platelets through conjugation with single-chain antibodies specific for ligand-induced binding sites (LIBSs) on glycoprotein (GP) IIb/IIIa receptors (LIBS–MPIO) (von Zur Muhlen et al. 2009). The LIBS–MPIOs were applied for in vivo MRI detection of platelet aggregation in murine cerebral malaria (von Zur Muhlen et al. 2008a) and carotid thrombosis (von Zur Muhlen et al. 2008c). The potential of LIBS–MPIO for human use was also investigated ex vivo in explanted symptomatic carotid artery plaque specimens (von Zur Muhlen et al. 2008c) and in human platelet-rich clots using a 3 T clinical MR scanner (von Zur Muhlen et al. 2008b).

Clinical Translation of MPIOs

For clinical application, commercially available MPIOs are not biodegradable due to their polyurethane coat. However, the synthesis of biodegradable MPIOs has been under development (Chen et al. 2005a; Hamoudeh and Fessi 2006; Hemmingsson et al. 1987; Zhu et al. 2006). The immunogenicity of monoclonal antibody ligands may be overcome by using modified or "humanized" antibody (Hwang and Foote 2005), single-chain antibody (Schwarz et al. 2006; von Zur Muhlen et al. 2009), or synthetic sugar-based ligands, such as sLe^X (van Kasteren et al. 2009). In our in vivo studies, the MPIO dose we employed closely reflects the human USPIO doses (2.6 mg iron/kg) used for human oncological MRI (Will et al. 2006). MPIOs were well tolerated in mice, with no short-term ill effects. They are rapidly sequestered by the liver and spleen, with no hemorrhage or tissue infarction (McAteer et al. 2008).

GOLD NANOPARTICLES

GNPs have become highly promising contrast agents for optical imaging because of their unique optical resonance properties, biocompatibility, and uncomplicated conjugation to biomolecules through well-established gold–thiolate bonding chemistry. GNPs exhibit strong light scattering and absorption peaks in the visible and NIR regions due to the collective excitation of their surface electrons in an electromagnetic field, known as localized surface plasmon resonance (LSPR) (Hutter and Fendler 2004). Solid spherical GNPs have the longest history in terms of their fundamental imaging capabilities, but their in vivo application is limited since absorption occurs in the visible spectrum, which cannot penetrate deep tissues and can cause photochemical damage (Jain et al. 2008). Recently, nonspherical GNPs, such as nanoshells (Melancon et al. 2008; Schwartzberg et al. 2006), nanorods (Nikoobakht and El-Sayed 2003), nanocages (Yang et al. 2007), and nanourchins (Bakr et al. 2006; Lu et al. 2008), have been synthesized that have peak absorption in the NIR range, where optical attenuation caused by blood and soft tissue is essentially negligible. Moreover, these GNPs do not cause photochemical damage (Skrabalak et al. 2007). The optical properties of nonspherical GNPs are easily tuned by adjusting the size, structure, and surface chemistry (Wang et al. 2004). Smaller GNPs, such as nanorods and nanocages (<50 nm), exhibit optical absorption and scattering coefficients an order of magnitude greater than nanoshells (>100 nm diameter) (Jain et al. 2006). Gold nanorods can be optically tuned by increasing the aspect ratio (i.e., length-to-width ratio) (Jain et al. 2008), while gold nanocages, which are synthesized through a simple galvanic replacement reaction between silver nanocubes and chloroauric acid ($AuCl_4$) (Chen et al. 2005b; Yang et al. 2007), can be tuned by controlling the molar ratio between gold and $HAuCl_4$ (Cang et al. 2005). Features such as bumps and thorns on gold nanourchins can also be used to enhance localized electromagnetic fields (Bakr et al. 2006; Hutter et al. 2010). For in vivo applications, the surface of GNPs can easily be conjugated with PEG or cetyltrimethylammonium bromide (CTAB) surface groups to enhance stability and increase the blood circulation half-life time. Below, we discuss the properties and application of nonspherical GNP for imaging the cerebral vasculature and brain tumor detection using PAT and multimodal imaging, as well as their potential for targeted photothermal therapy.

Photoacoustic Imaging and Photothermal Therapy Using GNPs

Photoacoustic imaging, also called optoacoustic imaging, is a rapidly evolving field that offers the potential to image organs, including the brain, with simultaneous high optical absorption contrast and high ultrasonic spatial resolution (Xu and Wang 2006). GNPs are promising contrast agents for PAT due to their optically tunable properties in the NIR region. Gold nanoshells can enhance in vivo PAT visualization of cerebral cortex vasculature (Wang et al. 2004) and brain tumor detection via passive extravasation of nanoshells from localized leaky vasculature and progressive accumulation in the tumor cortex (Li et al. 2009). Smaller, hollow gold nanocages (~50 nm) can improve in vivo PAT visualization of cerebral vasculature in rats by up to 81% (Yang et al. 2007). Hollow gold nanospheres (HAuNS) (40–50 nm) are also promising vascular imaging agents due to their strong NIR absorption (unlike solid spheres) and spherical shape, which is optimal for contrast delivery (Lu et al. 2010). PEGylated HAuNS enabled high resolution in vivo PAT of cerebral vasculature, depicting blood vessels as small as ~100 μm in diameter (Figure 6.4). Importantly, no histological signs of acute toxicity were observed following a single dose of PEG–HAuNS.

Gold nanorods can be applied for in vivo tumor detection, but this is based on passive accumulation due to increased tumor vasculature porosity and cannot be used for specific tumor-targeted therapy (Dickerson et al. 2008). Functionalization of GNPs with tumor-specific antibodies, such as epidermal growth factor receptor (EGFR) monoclonal antibodies, was shown to specifically detect tumor cells in vitro (Huang et al. 2006; Mallidi et al. 2007) and can also be used for selective photothermal destruction of tumor cells (Chen et al. 2007; El-Sayed et al. 2006). The feasibility of bioconjugated gold nanorods for PAT detection of early inflammatory responses ex vivo was also reported using stimulated endothelial cells labeled with nanorods conjugated to anti-intercellular adhesion molecule-1 (Kim et al. 2007). Gold nanorods, conjugated to tumor-specific peptides (EGFR, urokinase plasminogen activator receptor, and the $a_v\beta_3$ integrin receptor), were also developed (Huang et al. 2010). However, these tumor peptide-targeted nanorods showed only marginal in vivo tumor enhancement compared to nontargeted nanorods due to limited mass transport of nanorods across the tumor vasculature, which remains largely unaffected by receptor binding. Therefore, direct tumor injection of GNP may be better than intravenous injection for photothermal therapy.

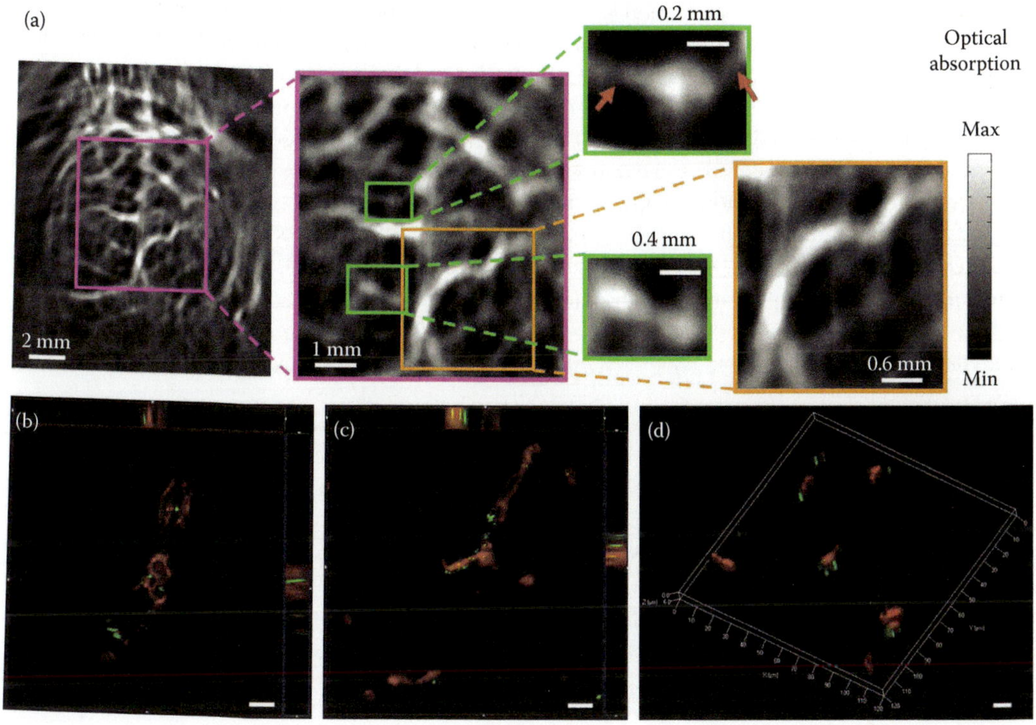

FIGURE 6.4 (a) Enhanced photoacoustic signals revealed clear and detailed structure of large (yellow-framed picture) and small (green-framed picture) blood vessels in the mouse brain at higher magnification 2 h after intravenous injection of PEG–HAuNS. Arrows represent the small blood vessels with a diameter of about 100 μm, which can be seen in the contrast-enhanced images. (b through d) Distribution of PEG–HAuNS in brain vessels 2 h after injection (bar = 10 μm). Brain vessels were stained with anti-CD31 antibody (red fluorescence), while the scattering signals of gold particles were detected under a dark field (pseudo-green). Z-stack images showed the particles located on the luminal side of brain blood vessels (b and c). In brain capillaries, 3-D reconstruction images show that the particles colocalized or stayed adjacent to the brain capillary endothelial cells (d). (From Lu, W. et al., *Biomaterials* 31: 2617–26, 2010. With permission.)

Multimodal Imaging Using GNPs

Several GNPs were developed for multimodal imaging including gold-coated quantum dots for combined optical and fluorescence imaging (Jin and Gao 2009) and gold-coated iron oxide nanoparticles for optical and MR imaging (Melancon et al. 2009, 2011). Functionalized GNPs conjugated with a multilayered shell of Gd chelates show enhanced relaxivities compared to GD^{3+} ions alone (Moriggi et al. 2009) and can enable in vivo MRI detection of brain tumors with follow-up treatment by x-ray microbeam irradiation, which improved the survival of rats (Figure 6.4) (Alric et al. 2008).

Clinical Application of GNPs

For clinical application, more data are needed on in vivo interactions of GNPs with CNS cells. In vitro experiments showed that spherical GNPs can be toxic to microglial cells, while gold nanorods and nanourchins were relatively innocuous (Hutter et al. 2010). Microglial responses to GNPs of different shapes were also tested in vivo in transgenic mice expressing luciferase controlled by the microglia-specific promoter driven by toll-like receptor-2, a marker of microglial activation. Gold nanorods and nanourchins caused transient microglial activation, whereas gold nanospheres did not. Additional in vitro studies found that the activation mechanism depended, at least in part, on

GNP shape (Hutter et al. 2010). Concerns also remain about in vivo GNP biocompatibility due to potential cytotoxicity (Liang et al. 2008) and nonspecific cellular uptake related to CTAB surfactant used during GNP synthesis (Huff et al. 2007). Partially thiolated polyamidoamine dendrimers were used to replace CTAB molecules to improve in vivo biocompatibility of gold nanorods (Li et al. 2010). Dendrimer-modified gold nanorods (dGNRs) conjugated with RGD peptides (RGD-dGNRs)] specifically bound to tumor cells in vivo and exhibited selective photothermal destruction of tumor cells under NIR laser irradiation, which extended the life span of tumor-bearing mice (Li et al. 2010). Gum Arabic matrix may also be used as a nontoxic, biocompatible alternative to stabilize GNPs (Kattumuri et al. 2007). In vivo biodistribution and pharmacokinetic studies in pigs confirmed no toxic effects in organs up to 72 h postinjection.

CONCLUSIONS AND FUTURE PERSPECTIVES

One of the current challenges in biomedicine is to develop safe and effective nanoprobes that can be applied clinically. For MRI, the high relaxivity and biological safety properties of trimetallic nitride metallofullerenes offer a potential new diagnostic tool for brain tumor detection and tracking, which through combination with radiochemical agents, x-ray-based imaging contrast, or fluorescence can offer a multimodal platform for diagnostic imaging and treatment of tumors. Iron oxide nanoparticles and microparticles were also developed for noninvasive, molecular MRI detection of neurovascular inflammation and glioma and tumor angiogenesis. For future clinical application, MPIOs offer a versatile platform for molecular endovascular imaging, which, with suitable modification using a biodegradable surface coat, may enable earlier detection of neuroinflammatory diseases and guide specific therapies. Identifying alternative ligands such as single-chain antibodies or synthetic sugar ligands may overcome immunogenicity and steric constraints of antibodies. Higher MRI field strengths (3 T) and the introduction of multifunctional and multimodal nanoparticles offer potential for integrated diagnosis, treatment, and tracking of response to therapy. GNPs are promising nanomaterials for optical PAT and multimodal imaging due to their unique physical, chemical, and optically tunable properties. Considerable progress is being made to improve biocompatibility by removing cytotoxic CTAB, which can interfere with selective cell uptake. Functionalization of GNPs for targeted tumor cell delivery, combined with photothermal therapy, offers a noninvasive alternative to surgery for future cancer treatment.

REFERENCES

Akhtar, A.M., Schneider, J.E., Chapman, S.J. et al. 2010. In vivo quantification of VCAM-1 expression in renal ischemia reperfusion injury using non-invasive magnetic resonance molecular imaging. *PLoS One* 5: e12800.

Aloya, R., Shirvan, A., Grimberg, H. et al. 2006. Molecular imaging of cell death in vivo by a novel small molecule probe. *Apoptosis* 11: 2089–101.

Alric, C., Taleb, J., Le Duc, G. et al. 2008. Gadolinium chelate coated gold nanoparticles as contrast agents for both x-ray computed tomography and magnetic resonance imaging. *J Am Chem Soc* 130: 5908–15.

Bakr, O.M., Wunsch, B.H., and Stellacci, F. 2006. High-yield synthesis of multi-branched urchin-like gold nanoparticles. *Chem. Mater.* 18: 3297–301.

Briley-Saebo, K.C., Johansson, L.O., Hustvedt, S.O. et al. 2006. Clearance of iron oxide particles in rat liver: Effect of hydrated particle size and coating material on liver metabolism. *Invest Radiol* 41: 560–71.

Briley-Saebo, K.C., Mani, V., Hyafil, F. et al. 2008. Fractionated feridex and positive contrast: In vivo mr imaging of atherosclerosis. *Magn Reson Med* 59: 721–30.

Brochet, B., Deloire, M.S., Touil, T. et al. 2006. Early macrophage mri of inflammatory lesions predicts lesion severity and disease development in relapsing eae. *Neuroimage* 32: 266–74.

Cang, H., Sun, T., Li, Z.Y. et al. 2005. Gold nanocages as contrast agents for spectroscopic optical coherence tomography. *Opt Lett* 30: 3048–50.

Carlos, T.M., Schwartz, B.R., Kovach, N.L. et al. 1990. Vascular cell adhesion molecule-1 mediates lymphocyte adherence to cytokine-activated cultured human endothelial cells. *Blood* 76: 965–70.

Chen, H.H., Le Visage, C., Qiu, B. et al. 2005a. Mr imaging of biodegradable polymeric microparticles: A potential method of monitoring local drug delivery. *Magn Reson Med* 53: 614–20.
Chen, J., Saeki, F., Wiley, B.J. et al. 2005b. Gold nanocages: Bioconjugation and their potential use as optical imaging contrast agents. *Nano Lett* 5: 473–7.
Chen, J., Wang, D., Xi, J. et al. 2007. Immuno gold nanocages with tailored optical properties for targeted photothermal destruction of cancer cells. *Nano Lett* 7: 1318–22.
Choudhury, R.P., Fuster, V., and Fayad, Z.A. 2004. Molecular, cellular and functional imaging of atherothrombosis. *Nat Rev Drug Discov* 3: 913–25.
Collins, K.L., Orringer, D.A., and Patil, P.G. 2010. Nanotechnology in neurosurgery. *J Nanotechnol Eng Med* 1: 034001.
Debbage, P., and Jaschke, W. 2008. Molecular imaging with nanoparticles: Giant roles for dwarf actors. *Histochem Cell Biol* 130: 845–75.
Deloire, M.S., Touil, T., Brochet, B. et al. 2004. Macrophage brain infiltration in experimental autoimmune encephalomyelitis is not completely compromised by suppressed t-cell invasion: In vivo magnetic resonance imaging illustration in effective anti-VLA-4 antibody treatment. *Mult Scler* 10: 540–8.
Dickerson, E.B., Dreaden, E.C., Huang, X. et al. 2008. Gold nanorod assisted near-infrared plasmonic photothermal therapy (PPTT) of squamous cell carcinoma in mice. *Cancer Lett* 269: 57–66.
Dousset, V., Ballarino, L., Delalande, C. et al. 1999a. Comparison of ultrasmall particles of iron oxide (USPIO)-enhanced T2-weighted, conventional T2-weighted, and gadolinium-enhanced T1-weighted mr images in rats with experimental autoimmune encephalomyelitis. *AJNR Am J Neuroradiol* 20: 223–7.
Dousset, V., Brochet, B., Deloire, M.S. et al. 2006. MR imaging of relapsing multiple sclerosis patients using ultra-small-particle iron oxide and compared with gadolinium. *AJNR Am J Neuroradiol* 27: 1000–5.
Dousset, V., Delalande, C., Ballarino, L. et al. 1999b. In vivo macrophage activity imaging in the central nervous system detected by magnetic resonance. *Magn Reson Med* 41: 329–33.
Dousset, V., Gomez, C., Petry, K.G. et al. 1999c. Dose and scanning delay using USPIO for central nervous system macrophage imaging. *Magma* 8: 185–9.
El-Sayed, I.H., Huang, X., and El-Sayed, M.A. 2006. Selective laser photo-thermal therapy of epithelial carcinoma using anti-EGFR antibody conjugated gold nanoparticles. *Cancer Lett* 239: 129–35.
Elices, M.J., Osborn, L., Takada, Y. et al. 1990. VCAM-1 on activated endothelium interacts with the leukocyte integrin VLA-4 at a site distinct from the VLA-4/fibronectin binding site. *Cell* 60: 577–84.
Fatouros, P.P., Corwin, F.D., Chen, Z.J. et al. 2006. In vitro and in vivo imaging studies of a new endohedral metallofullerene nanoparticle. *Radiology* 240: 756–64.
Fleige, G., Nolte, C., Synowitz, M. et al. 2001. Magnetic labeling of activated microglia in experimental gliomas. *Neoplasia* 3: 489–99.
Floris, S., Blezer, E.L., Schreibelt, G. et al. 2004. Blood–brain barrier permeability and monocyte infiltration in experimental allergic encephalomyelitis: A quantitative mri study. *Brain* 127: 616–27.
Hamoudeh, M., and Fessi, H. 2006. Preparation, characterization and surface study of poly-epsilon caprolactone magnetic microparticles. *J Colloid Interface Sci* 300: 584–90.
Hemmingsson, A., Carlsten, J., Ericsson, A. et al. 1987. Relaxation enhancement of the dog liver and spleen by biodegradable superparamagnetic particles in proton magnetic resonance imaging. *Acta Radiol* 28: 703–5.
Heyn, C., Ronald, J.A., Mackenzie, L.T. et al. 2006a. In vivo magnetic resonance imaging of single cells in mouse brain with optical validation. *Magn Reson Med* 55: 23–9.
Heyn, C., Ronald, J.A., Ramadan, S.S. et al. 2006b. In vivo MRI of cancer cell fate at the single-cell level in a mouse model of breast cancer metastasis to the brain. *Magn Reson Med* 56: 1001–10.
Hoyte, L.C., Brooks, K.J., Nagel, S. et al. 2010. Molecular magnetic resonance imaging of acute vascular cell adhesion molecule-1 expression in a mouse model of cerebral ischemia. *J Cereb Blood Flow Metab* 30: 1178–87.
Huang, X., El-Sayed, I.H., Qian, W. et al. 2006. Cancer cell imaging and photothermal therapy in the near-infrared region by using gold nanorods. *J Am Chem Soc* 128: 2115–20.
Huang, X., Peng, X., Wang, Y. et al. 2010. A reexamination of active and passive tumor targeting by using rod-shaped gold nanocrystals and covalently conjugated peptide ligands. *ACS Nano* 4: 5887–96.
Huff, T.B., Hansen, M.N., Zhao, Y. et al. 2007. Controlling the cellular uptake of gold nanorods. *Langmuir* 23: 1596–9.
Hutter, E., Boridy, S., Labrecque, S. et al. 2010. Microglial response to gold nanoparticles. *ACS Nano* 4: 2595–606.
Hutter, E., and Fendler, J.H. 2004. Exploitation of localised surface plasmon resonance. *Adv. Mater* 16: 1685–706.

Hwang, W.Y., and Foote, J. 2005. Immunogenicity of engineered antibodies. *Methods* 36: 3–10.
Jain, P.K., Huang, X., El-Sayed, I.H. et al. 2008. Noble metals on the nanoscale: Optical and photothermal properties and some applications in imaging, sensing, biology, and medicine. *Acc Chem Res* 41: 1578–86.
Jain, P.K., Lee, K.S., El-Sayed, I.H. et al. 2006. Calculated absorption and scattering properties of gold nanoparticles of different size, shape, and composition: Applications in biological imaging and biomedicine. *J Phys Chem B* 110: 7238–48.
Jin, Y., and Gao, X. 2009. Plasmonic fluorescent quantum dots. *Nat Nanotechnol* 4: 571–6.
Kattumuri, V., Katti, K., Bhaskaran, S. et al. 2007. Gum arabic as a phytochemical construct for the stabilization of gold nanoparticles: in vivo pharmacokinetics and x-ray-contrast-imaging studies. *Small* 3: 333–41.
Kiessling, F., Huppert, J., Zhang, C. et al. 2009. RGD-labeled USPIO inhibits adhesion and endocytotic activity of alpha v beta3-integrin-expressing glioma cells and only accumulates in the vascular tumor compartment. *Radiology* 253: 462–9.
Kim, J., Kim, D.I., Lee, S.K. et al. 2008. Imaging of the inflammatory response in reperfusion injury after transient cerebral ischemia in rats: Correlation of superparamagnetic iron oxide-enhanced magnetic resonance imaging with histopathology. *Acta Radiol* 49: 580–8.
Kim, K., Huang, S.W., Ashkenazi, S. et al. 2007. Photoacoustic imaging of early inflammatory response using gold nanorods. *Appl Phys Lett* 90: 223901-1–01-3.
Kircher, M.F., Mahmood, U., King, R.S. et al. 2003. A multimodal nanoparticle for preoperative magnetic resonance imaging and intraoperative optical brain tumor delineation. *Cancer Res* 63: 8122–5.
Kleinschnitz, C., Schutz, A., Nolte, I. et al. 2005. In vivo detection of developing vessel occlusion in photothrombotic ischemic brain lesions in the rat by iron particle enhanced mri. *J Cereb Blood Flow Metab* 25: 1548–55.
Krause, M., and Dunsch, L. 2005. Gadolinium nitride Gd_3N in carbon cages: The influence of cluster size and bond strength. *Angew Chem Int Ed Engl* 44: 1557–60.
Kroto, H.W., Heath, J.R., O'Brien, S.C. et al. 1985. C60: Buckminsterfullerene. *Nature* 318: 162–3.
Laus, S., Sitharaman, B., Toth, E. et al. 2005. Destroying gadofullerene aggregates by salt addition in aqueous solution of gd@c(60)(oh)(x) and gd@c(60)[c(cooh(2))](10). *J Am Chem Soc* 127: 9368–9.
Li, M.L., Wang, J.C., Schwartz, J.A. et al. 2009. In-vivo photoacoustic microscopy of nanoshell extravasation from solid tumor vasculature. *J Biomed Opt* 14: 010507.
Li, Z., Huang, P., Zhang, X. et al. 2010. RGD-conjugated dendrimer-modified gold nanorods for in vivo tumor targeting and photothermal therapy. *Mol Pharm* 7: 94–104.
Liang, X.J., Chen, C., Zhao, Y. et al. 2008. Biopharmaceutics and therapeutic potential of engineered nanomaterials. *Curr Drug Metab* 9: 697–709.
Lim, Y.L., Lee, H.Y., Low, S.C. et al. 2007. Possible role of gadolinium in nephrogenic systemic fibrosis: Report of two cases and review of the literature. *Clin Exp Dermatol* 32: 353–8.
Lipinski, M.J., Briley-Saebo, K.C., Mani, V. et al. 2008. "Positive contrast" inversion-recovery with on[corrected]-resonant water suppression magnetic resonance imaging: A change for the better? *J Am Coll Cardiol* 52: 492–4.
Lu, L.H., Ai, K., and Ozaki, Y. 2008. Environmentally friendly synthesis of highly monodisperse biocompatible gold nanoparticles with urchin-like shape. *Langmuir* 24: 1058–63.
Lu, W., Huang, Q., Ku, G. et al. 2010. Photoacoustic imaging of living mouse brain vasculature using hollow gold nanospheres. *Biomaterials* 31: 2617–26.
MacFarland, D.K., Walker, K.L., Lenk, R.P. et al. 2008. Hydrochalarones: A novel endohedral metallofullerene platform for enhancing magnetic resonance imaging contrast. *J Med Chem* 51: 3681–3.
Mallidi, S., Larson, T., Aaron, J. et al. 2007. Molecular specific optoacoustic imaging with plasmonic nanoparticles. *Opt Express* 15: 6583–8.
Manninger, S.P., Muldoon, L.L., Nesbit, G. et al. 2005. An exploratory study of ferumoxtran-10 nanoparticles as a blood-brain barrier imaging agent targeting phagocytic cells in CNS inflammatory lesions. *AJNR Am J Neuroradiol* 26: 2290–300.
McAteer, M.A., Akhtar, A.M., von Zur Muhlen, C. et al. 2010. An approach to molecular imaging of atherosclerosis, thrombosis, and vascular inflammation using microparticles of iron oxide. *Atherosclerosis* 209: 18–27.
McAteer, M.A., and Choudhury, R.P. 2009. Chapter 4—Applications of nanotechnology in molecular imaging of the brain. *Prog Brain Res* 180: 72–96.
McAteer, M.A., Schneider, J.E., Ali, Z.A. et al. 2008. Magnetic resonance imaging of endothelial adhesion molecules in mouse atherosclerosis using dual-targeted microparticles of iron oxide. *Arterioscler Thromb Vasc Biol* 28: 77–83.

McAteer, M.A., Sibson, N.R., von Zur Muhlen, C. et al. 2007. In vivo magnetic resonance imaging of acute brain inflammation using microparticles of iron oxide. *Nat Med* 13: 1253–8.
McAteer, M.A., von Zur Muhlen, C., Anthony, D.C. et al. 2011. Magnetic resonance imaging of brain inflammation using microparticles of iron oxide. *Methods Mol Biol* 680: 103–15.
Melancon, M.P., Elliott, A., Ji, X. et al. 2011. Theranostics with multifunctional magnetic gold nanoshells: Photothermal therapy and T2* magnetic resonance imaging. *Invest Radiol* 46: 132–40.
Melancon, M., Lu, W., and Li, C. 2009. Gold-based magneto/optical nanostructures: Challenges for in vivo applications in cancer diagnostics and therapy. *Mater Res Bull* 34: 415–21.
Melancon, M.P., Lu, W., Yang, Z. et al. 2008. In vitro and in vivo targeting of hollow gold nanoshells directed at epidermal growth factor receptor for photothermal ablation therapy. *Mol Cancer Ther* 7: 1730–9.
Meng, X.X., Wan, J.Q., Jing, M. et al. 2007. Specific targeting of gliomas with multifunctional superparamagnetic iron oxide nanoparticle optical and magnetic resonance imaging contrast agents. *Acta Pharmacol Sin* 28: 2019–26.
Mikawa, M., Kato, H., Okumura, M. et al. 2001. Paramagnetic water-soluble metallofullerenes having the highest relaxivity for MRI contrast agents. *Bioconjug Chem* 12: 510–4.
Montet, X., Montet-Abou, K., Reynolds, F. et al. 2006. Nanoparticle imaging of integrins on tumor cells. *Neoplasia* 8: 214–22.
Moore, A., Marecos, E., Bogdanov, A., Jr. et al. 2000. Tumoral distribution of long-circulating dextran-coated iron oxide nanoparticles in a rodent model. *Radiology* 214: 568–74.
Moriggi, L., Cannizzo, C., Dumas, E. et al. 2009. Gold nanoparticles functionalized with gadolinium chelates as high-relaxivity MRI contrast agents. *J Am Chem Soc* 131: 10828–9.
Muldoon, L.L., Tratnyek, P.G., Jacobs, P.M. et al. 2006. Imaging and nanomedicine for diagnosis and therapy in the central nervous system: Report of the Eleventh Annual Blood–Brain Barrier Disruption Consortium Meeting. *AJNR Am J Neuroradiol* 27: 715–21.
Neuwelt, E.A., Varallyay, P., Bago, A.G. et al. 2004. Imaging of iron oxide nanoparticles by MR and light microscopy in patients with malignant brain tumours. *Neuropathol Appl Neurobiol* 30: 456–71.
Nighoghossian, N., Wiart, M., Cakmak, S. et al. 2007. Inflammatory response after ischemic stroke: A USPIO-enhanced MRI study in patients. *Stroke* 38: 303–7.
Nikoobakht, B., and El-Sayed, M.A. 2003. Preparation and growth mechanism of gold nanorods (nrs) using seed-mediated growth method. *Chem Mater* 15: 1957–62.
Qian, M., Ong, S., Khanna, S. et al. 2007. Magnetic endohedral metallofullerenes with floppy interiors. *Phys Rev B* 75 (104424): 1–6.
Ramadan, S.S., Heyn, C., Mackenzie, L.T. et al. 2008. Ex-vivo cellular MRI with b-SSFP: Quantitative benefits of 3 T over 1.5 T. *Magma* 21: 251–9.
Rausch, M., Baumann, D., Neubacher, U. et al. 2002. In-vivo visualization of phagocytotic cells in rat brains after transient ischemia by USPIO. *NMR Biomed* 15: 278–83.
Rausch, M., Hiestand, P., Baumann, D. et al. 2003. MRI-based monitoring of inflammation and tissue damage in acute and chronic relapsing EAE. *Magn Reson Med* 50: 309–14.
Rausch, M., Sauter, A., Frohlich, J. et al. 2001. Dynamic patterns of USPIO enhancement can be observed in macrophages after ischemic brain damage. *Magn Reson Med* 46: 1018–22.
Reddy, G.R., Bhojani, M.S., McConville, P. et al. 2006. Vascular targeted nanoparticles for imaging and treatment of brain tumors. *Clin Cancer Res* 12: 6677–86.
Saleh, A., Schroeter, M., Jonkmanns, C. et al. 2004. In vivo MRI of brain inflammation in human ischaemic stroke. *Brain* 127: 1670–7.
Schroeter, M., Saleh, A., Wiedermann, D. et al. 2004. Histochemical detection of ultrasmall superparamagnetic iron oxide (USPIO) contrast medium uptake in experimental brain ischemia. *Magn Reson Med* 52: 403–6.
Schwartzberg, A.M., Olson, T.Y., Talley, C.E. et al. 2006. Synthesis, characterization, and tunable optical properties of hollow gold nanospheres. *J Phys Chem B* 110: 19935–44.
Schwarz, M., Meade, G., Stoll, P. et al. 2006. Conformation-specific blockade of the integrin GPIIb/IIIa: A novel antiplatelet strategy that selectively targets activated platelets. *Circ Res* 99: 25–33.
Shapiro, E.M., Sharer, K., Skrtic, S. et al. 2006. In vivo detection of single cells by MRI. *Magn Reson Med* 55: 242–9.
Shapiro, E.M., Skrtic, S., and Koretsky, A.P. 2005. Sizing it up: Cellular MRI using micron-sized iron oxide particles. *Magn Reson Med* 53: 329–38.
Shu, C., Corwin, F.D., Zhang, J. et al. 2009. Facile preparation of a new gadofullerene-based magnetic resonance imaging contrast agent with high 1H relaxivity. *Bioconjug Chem* 20: 1186–93.
Skrabalak, S.E., Chen, J., Au, L. et al. 2007. Gold nanocages for biomedical applications. *Adv Mater* 19: 3177–84.

Stratta, P., Canavese, C., and Aime, S. 2008. Gadolinium-enhanced magnetic resonance imaging, renal failure and nephrogenic systemic fibrosis/nephrogenic fibrosing dermopathy. *Curr Med Chem* 15: 1229–35.

Sun, C., Veiseh, O., Gunn, J. et al. 2008. In vivo MRI detection of gliomas by chlorotoxin-conjugated superparamagnetic nanoprobes. *Small* 4: 372–9.

Trivedi, R.A., Mallawarachi, C., JM, U.K.-I. et al. 2006. Identifying inflamed carotid plaques using in vivo USPIO-enhanced MR imaging to label plaque macrophages. *Arterioscler Thromb Vasc Biol* 26: 1601–6.

van Kasteren, S.I., Campbell, S.J., Serres, S. et al. 2009. Glyconanoparticles allow pre-symptomatic in vivo imaging of brain disease. *Proc Natl Acad Sci U S A* 106: 18–23.

Varallyay, P., Nesbit, G., Muldoon, L.L. et al. 2002. Comparison of two superparamagnetic viral-sized iron oxide particles ferumoxides and ferumoxtran-10 with a gadolinium chelate in imaging intracranial tumors. *AJNR Am J Neuroradiol* 23: 510–9.

Veiseh, O., Sun, C., Gunn, J. et al. 2005. Optical and MRI multifunctional nanoprobe for targeting gliomas. *Nano Lett* 5: 1003–8.

Vellinga, M.M., Oude Engberink, R.D., Seewann, A. et al. 2008. Pluriformity of inflammation in multiple sclerosis shown by ultra-small iron oxide particle enhancement. *Brain* 131: 800–7.

von zur Muhlen, C., Peter, K., Ali, Z.A. et al. 2009. Visualization of activated platelets by targeted magnetic resonance imaging utilizing conformation-specific antibodies against glycoprotein IIb/IIIa. *J Vasc Res* 46: 6–14.

von Zur Muhlen, C., Sibson, N.R., Peter, K. et al. 2008a. A contrast agent recognizing activated platelets reveals murine cerebral malaria pathology undetectable by conventional mri. *J Clin Invest* 118: 1198–207.

von Zur Muhlen, C., von Elverfeldt, D., Choudhury, R.P. et al. 2008b. Functionalized magnetic resonance contrast agent selectively binds to glycoprotein IIb/IIIa on activated human platelets under flow conditions and is detectable at clinically relevant field strengths. *Mol Imaging* 7: 59–67.

von Zur Muhlen, C., von Elverfeldt, D., Moeller, J.A. et al. 2008c. Magnetic resonance imaging contrast agent targeted toward activated platelets allows in vivo detection of thrombosis and monitoring of thrombolysis. *Circulation* 118(3): 258–67.

Wang, L.V. 2008. Prospects of photoacoustic tomography. *Med Phys* 35: 5758–67.

Wang, Y., Xie, X., Wang, X. et al. 2004. Photoacoustic tomography of a nanoshell contrast agent in the in vivo rat brain. *Nano Lett* 4, No. 9: 1689–92.

Will, O., Purkayastha, S., Chan, C. et al. 2006. Diagnostic precision of nanoparticle-enhanced MRI for lymph-node metastases: A meta-analysis. *Lancet Oncol* 7: 52–60.

Xie, H., Zhu, Y., Jiang, W. et al. 2011. Lactoferrin-conjugated superparamagnetic iron oxide nanoparticles as a specific MRI contrast agent for detection of brain glioma in vivo. *Biomaterials* 32: 495–502.

Xu, M., and Wang, L.V. 2006. Photoacoustic imaging in biomedicine. *Rev Sci Instrum* 77: 041101-1–01-12.

Xu, S., Jordan, E.K., Brocke, S. et al. 1998. Study of relapsing remitting experimental allergic encephalomyelitis SJL mouse model using MION-46L enhanced in vivo MRI: Early histopathological correlation. *J Neurosci Res* 52: 549–58.

Yang, X., Skrabalak, S.E., Li, Z.Y. et al. 2007. Photoacoustic tomography of a rat cerebral cortex in vivo with Au nanocages as an optical contrast agent. *Nano Lett* 7: 3798–802.

Ye, Q., Wu, Y.L., Foley, L.M. et al. 2008. Longitudinal tracking of recipient macrophages in a rat chronic cardiac allograft rejection model with noninvasive magnetic resonance imaging using micrometer-sized paramagnetic iron oxide particles. *Circulation* 118: 149–56.

Zhang, J., Fatouros, P.P., Shu, C. et al. 2010a. High relaxivity trimetallic nitride (Gd_3N) metallofullerene MRI contrast agents with optimized functionality. *Bioconjug Chem* 21: 610–5.

Zhang, J., Ge, J., Shultz, M.D. et al. 2010b. In vitro and in vivo studies of single-walled carbon nanohorns with encapsulated metallofullerenes and exohedrally functionalized quantum dots. *Nano Lett* 10: 2843–8.

Zhu, D., White, R.D., Hardy, P.A. et al. 2006. Biocompatible nanotemplate-engineered nanoparticles containing gadolinium: Stability and relaxivity of a potential MRI contrast agent. *J Nanosci Nanotechnol* 6: 996–1003.

Zimmer, C., Weissleder, R., Poss, K. et al. 1995. MR imaging of phagocytosis in experimental gliomas. *Radiology* 197: 533–8.

7 Carbon Nanoplatforms and Their Application for Neurological Disorders

Houman Badrzadeh, Peyman Tashkandi, Michelle Knosp, and Babak Kateb

CONTENTS

INTRODUCTION

Nanoparticles are entities with dimensions of 100 nm or less (Berhanu et al. 2009). Their applications have expanded in a variety of different fields such as medicine and electronics (Berhanu et al. 2009). The physicochemical properties of the particles vary according to their size, structure, and chemical makeup and determine their suitability for certain applications. In this chapter, we focus on carbon nanoplatforms including fullerenes, nanodiamonds (NDs), and carbon nanotubes (CNTs).

The fullerene or buckyball, a spherical allotrope of carbon, was discovered by Kroto et al. in 1985 (Kroto et al. 1985). In 1990, Krätschmer et al. reported mass production of fullerenes by creating an alternating current (AC) arc discharge between two graphite electrodes. Yoshinori Ando carried out direct current (DC) arc discharge between two graphite electrodes and obtained a large volume of carbon soot, including fullerenes with additional carbon deposits at the end of the negative graphite rod; this technique subsequently gave rise to the exciting discovery of the CNTs by Sumio Iijima in 1991 (Ando 2010; Iijima 1991).

CNTs are a significant part of the history of nanotechnology and one of the initial major building blocks of the nanotechnology field, with unique electrical, mechanical, and thermal properties that have provided new capabilities and potential applications in medicine, electronics, aerospace, and other industries (Lam et al. 2006). NDs are proving to be more biocompatible than other carbon nanomaterials. They also provide more stable photoluminescence, greater ease of purification, better commercial availability, and less cytotoxicity (Enoki et al. 2009; Holt 2007; Krueger 2008; Xing and Dai 2009).

CNTs AND THEIR SYNTHESIS

CNTs can be synthesized as biodegradable nanostructures by rolling sheets of carbon into hollow tubes (cylindrical buckytubes). They can be classified as single walled (0.4–2 nm diameter), double walled (1–3.5 nm diameter), or multiwalled (2–100 nm diameter) and have great potential for serving as nanovectors for drug and small interfering RNA (siRNA) delivery (Kateb et al. 2011). The first CNTs were multiwalled CNTs (MWCNTs) containing two, five, and seven coaxial cylinders (Ando 2010). CNTs are mainly produced by the (1) arc discharge method, (2) laser ablation method, and (3) chemical vapor deposition (CVD) method (Salvador-Morales et al. 2006).

The arc discharge method requires applying DC current to two pure graphite electrodes located 1–2 mm apart in a production chamber. Electrons emitted from the cathode hit the anode at high velocity. As the anode temperature rises higher than the cathode temperature, carbon atoms evaporate. The evaporated carbon atoms coalesce to CNPs such as fullerenes in the chamber. A small portion of the evaporated carbon is deposited on the adjacent cathode that has a relatively lower temperature and leads to growth of MWCNTs (Ando 2010). Single-wall (SWCNTs) and double-walled CNTs (DWCNTs) are also produced by the arc discharge method. Production conditions such as type of catalyst or gas affect the type of CNT produced. For example: SWCNTs with 1.2–1.5 nm tube diameter and 2–10 μm length can be produced by arc plasma jet (APJ) in which involves 4% Nickel (Ni) and 500 torr Helium (He). SWCNTs with same length but smaller tube diameter can be produced by ferrum-hydrogen (FH) methods involving 1% Iron (Fe) with 200 torr combined Hydrogen (H2) and Argon (Ar). A combination catalyst of 0.25% Fe, 0.9% Ni, and 0.9% Co with 500 torr combined Ambient gases of H2 and 1% H2S using normal DC arc discharge will lead to production of SWCNs and DWCNTs with 1.4–4 nm tubes and length of greater than 10 μm. MWCNTs with length tube of greater than 10 μm can be produced by DC arc discharge evaporation of a pure graphiterod in pure hydrogen gas. In this method, 30–100 torr H2 with no catalyst will result in 3–30 walled carbon nano tubes with narrow inner diameter (10–30 nm), called nanowire;

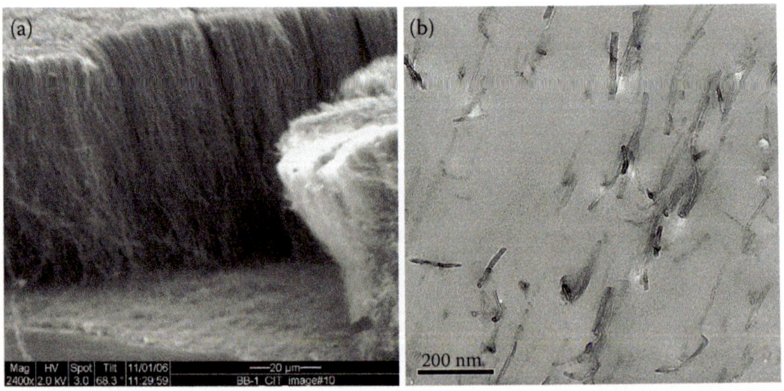

FIGURE 7.1 MWCNTs were imaged using scanning electron microscopy (SEM) and transmission electron microscopy (TEM). (a) MWCNTs were imaged using SEM before processing to observe homogenous growth and size. (b) Processed MWCNTs were imaged using TEM after processing to ascertain the consistency in length of the MWCNTs after processing. MWCNTs were determined to be between 200 and 400 μm for more than 80% of the suspension. (From Kateb, B. et al., *Neuroimage* 37 Suppl 1: S9–17, 2007. With permission. The picture is courtesy of Dr. Harish Manohara, Jet Propulsion Laboratory of NASA in Pasadena, CA, and Dr. Babak Kateb, Department of Neurosurgery, Cedars Sinai Medical Center, Beverly Hills, CA.)

However, if Scandium (Sc) or Lanthanum is used with 500 torr He, mass production of MWCNTs with 20–40 walls with wider diameter (20–40 nm) are produced (Ando 2010). The laser ablation method, a pulsed laser vaporizes graphite in a reaction chamber containing an inert gas; CNTs develop as the carbon vapor cools in the reaction chamber (Guo et al. 1995). In the CVD method, a vapor containing a carbon compound is placed over catalytic metal particles; the CNT nanotubes grow from the edges of the metal particles (Inami et al. 2007). Innovative methods were developed to purify the CNTs after they are produced (Figure 7.1).

IMMUNOGENICITY AND TOXICITY OF CNTs

CNTs can be exploited as molecular transporters to carry cargoes into cells. A fluoresceinated protein entered cells only when conjugated to a CNT. The functionalized SWCNT crosses the cell membrane and enters the human promyelocytic leukemia cell line HL60 and human T cells without resultant toxicity. Figure 7.2b and c shows how CNTs penetrate the cell membrane and enter the cell. The capability of nanoparticles to be taken up by cells enables their use as delivery vehicles to introduce therapeutic agents into diseased cells (Figure 7.2a).

Pure CNTs lack immunogenicity in contrast to common protein carriers. However, when attached to an antigen, they elicit an immune response (Salvador-Morales et al. 2006). Protein binding to CNTs is highly selective. Fibrinogen and apolipoproteins (AI, AIV, and CIII) are proteins that bind to CNTs in greatest quantity (Salvador-Morales et al. 2006). CNTs can also activate human complement and evoke the damaging effects of excessive complement activation, including inflammation and granuloma formation. SWCNTs activate the human serum complement system via the classical pathway. DWCNTs activate both the classical and alternative pathways (Salvador-Morales et al. 2006).

C1q, the multivalent protein for attaching to the complement fixation sites of immunoglobulin, binds directly to CNTs. In the classical pathway, the protein C1q recognizes activators mainly via charge and hydrophobic interactions. This activation leads to generating inflammatory peptides C3a, C4a, and C5a, a neutrophil chemotactic factor. The opsonins, C3b and iC3b, can also coat CNTs. When these nanomaterials are introduced into a mammalian host, complement activation

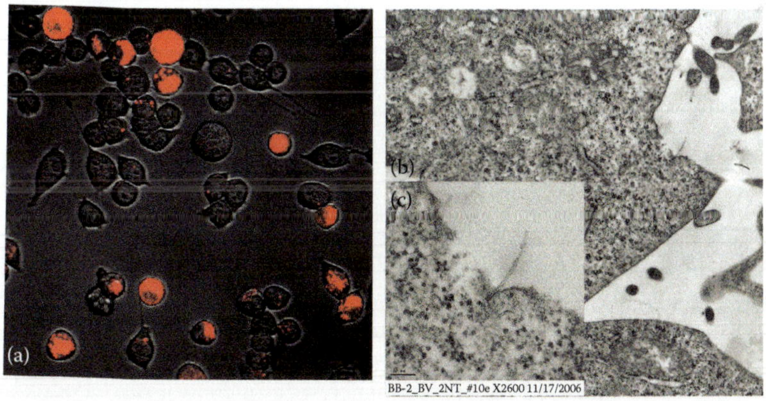

BB-2_BV_2NT_#10h X4400 11/17/2006

FIGURE 7.2 MWCNT internalization by BV2 microglia: 2.3 µg pMWCNTs-PKH (a dye for cell fluorescent labeling) was incubated with 1e6 BV2 cells for 48 h. Images were taken at 15, 24, and 48 h after pMWCNTs-PKH were added. Increasing numbers of pMWCNTs-PKH-positive cells were observed throughout this time course. (a) pMWCNTs-PKH-positive BV2 cells were imaged using an LSM 510 Meta confocal microscope at 48 h, magnification 400× (please also see the online video). (b) Transmission electron microscopy (TEM) was performed on cells incubated with MWCNTs after 2, 6, and 24 h. At 2 h, single pMWCNTs were observed penetrating the cell surface, magnification 4400×. (c) Magnified image showing MWCNTs penetrating cell surface, magnification ×26,000. (Reproduced from Kateb, B. et al., *Neuroimage* 37 Suppl 1: S9–17, 2007. With permission.)

leads to accumulation of neutrophils and adherence of phagocytic cells. Discharge of degradative enzymes by neutrophils may lead to tissue damage. Granuloma formation may also originate from macrophage adherence to the nanotubes (Salvador-Morales et al. 2006).

The activation of complement via the alternative pathway starts with binding of C3b to the surface of a pathogen or DWCNTs; subsequent events of complement activation via this pathway are similar to those of the classical pathway. CNTs, which become coated with iC3b and C3d, may act as adjuvants. The mechanism of adjuvant activity relies on complement and complement receptor interactions, such as interaction of C3d bound to the complement activator with its receptor CR2 on antigen-presenting cells. This receptor–ligand interaction can stimulate antigen presentation, B lymphocyte proliferation, and immunoglobulin secretion (Salvador-Morales et al. 2006).

MWCNTs may have no effect on proliferation or cytokine formation in vitro. However, to prevent immune reaction, it is essential to very carefully control CNT size and composition when designing nanotubes for intracellular drug delivery. Complement activation can be diminished or eliminated by functionalizing the CNTs or otherwise altering their surface chemistry to make them more biocompatible (Kateb et al. 2011). Variations in surface charge may promote binding of factor H, a downregulator of complement activation (Salvador-Morales et al. 2006). Heparinization of the nanotubes is another way to significantly enhance their biocompatibility (Kateb et al. 2011).

Current commercially available nanomaterials are characterized based on their desired properties only and are usually not tested for potential toxicity. Concern about the possible toxicity of carbon-based nanomaterials and CNTs arises from the potential presence of metal impurities within them that remain as by-products of their synthesis (Berhanu et al. 2009). Catalytic metals are often used in the synthesis of CNTs and may result in impurities in SWCNTs and MWCNTs at different scales, which may be undetectable using commercial analytical methods (Berhanu et al. 2009). Therefore, detailed characterization of engineered nanomaterials is necessary to establish if the nanomaterials are toxic (Berhanu et al. 2009). Systemic use of CNTs through Intravenous (IV) injection and inhalation seems to be much more toxic than local injection of CNTs into the brain (Kateb et al. 2007, 2011).

APPLICATIONS OF CNTs IN ALZHEIMER'S AND PARKINSON'S DISEASES

A case of what is now known as Alzheimer's disease (AD) was first reported by the German physician Alois Alzheimer in 1907 (Alzheimer 1907; Cipriani et al. 2011; Strassnig and Ganguli 2005). This progressive and degenerative brain disease results in memory loss, behavioral changes, and cognitive and linguistic dysfunction (Podolski et al. 2007). AD affects an estimated 5.4 million people in the United States (Alzheimer's Association 2010; Podolski et al. 2007). This type of dementia can be conclusively confirmed only by postmortem examination of brain tissue. Brain tissue changes that are common in people affected by AD include neurofibrillary tangles and beta-amyloid plaques (Yang et al. 2010). Groundbreaking work by Podolski et al. (2007) demonstrated that the hydrated fullerene C60 inhibited fibrillization of amyloid-beta$_{25\text{-}35}$ peptide in vitro and, when injected intracerebroventricularly in rats with cognitive impairment induced by amyloid-beta$_{25\text{-}35}$ peptide, prevented impaired task performance seen in control rats (Podolski et al. 2007). By preventing fibrillization of this peptide, accumulation of beta-amyloid plaques could be reduced. Such work encourages additional research to develop this modality as a new and promising treatment for AD.

Various drug therapies are used to reduce or reverse the mental and behavioral manifestations of AD. Reduced activity of cholinergic neurons in AD may increase cognitive impairment (Wu 2008). Patients with mild-to-moderate AD are often prescribed cholinesterase inhibitors, which inhibited the breakdown of the neurotransmitter acetylcholine, thereby increasing acetylcholine levels in the brain. Yang et al. (2010) recently studied the use of SWCNT as a drug carrier in treating a murine model of AD. Although nanoparticles may be toxic, the level of toxicity can be controlled by regulating the dosage of SWCNTs. Dosages less than 300 mg/kg were successfully taken up by lysosomes within neurons and neurites without resultant toxicity. This study demonstrated that acetylcholine loaded in SWCNTs was taken up preferentially by lysosomes at low dosages and showed that the ability to learn recovered to that of normal levels. Because free acetylcholine cannot cross the blood–brain barrier (BBB), and treatment with cholinesterase inhibitors can result in desensitization, using acetylcholine-loaded SWCNTs is potentially a future therapeutic treatment for cognitive changes resulting from AD (Yang et al. 2010).

Implementation of nanotechnology has also been explored in treating Parkinson's disease (PD), which is a chronic and progressive movement disorder (Trimmer et al. 2009). It is believed that a minimum of 1 million Americans suffer from PD (Haniu et al. 2010). Like AD, the cause of PD in a particular patient is usually uncertain, and current medical intervention only partially alleviates symptoms but does not cure the disease. As PD progresses, the level of dopamine produced by dopaminergic neurons originating in the substantia nigra diminishes, and the affected patient experiences cardinal symptoms of the disease including tremor, stiffness of the legs and trunk (dystonia), slowness of movement (bradykinesia), and instability while standing or walking. Currently, the primary treatment of PD is an oral tablet containing levodopa (L-DOPA) and carbidopa that is taken three to four times per day (Cai et al. 2008). Levodopa is absorbed from the gut, carried through the blood stream, transported across the BBB, and converted within the brain to dopamine by DOPA decarboxylase (Li and Andrews 2007).

Ali et al. (2007) used a composite of poly(anilineboronic acid) (PABA) and SWCNT functionalized with single-stranded DNA to electrochemically detect extremely low (40–1000 pM) concentrations of dopamine in solution. Creation of a biosensor that could effectively and accurately measure neurotransmitter concentration in a specific disease in which it is reduced, such as PD, could serve as a model for using molecular diagnostic tools in this and other diseases. At present, dopamine detection is accomplished by ascorbic acid oxidation of dopamine, which is relatively insensitive and nonselective. Ali et al. took a different approach and created a nonoxidative method to measure dopamine levels. The nanocomposite constructed in this study (PABA/ss-DNA/SWCNT) was deposited onto electrodes and detected very low concentrations (1 nM) of dopamine. The high sensitivity and selectivity achieved in this experiment were attributed to the large surface area of CNTs, which permitted increased boronic acid receptor density and dopamine binding (Ali et al. 2007). This electrode could be modified for in vivo use to measure dopamine levels in brain extracellular fluid. Ultimately, new treatments must be developed to prevent or reverse the underlying degeneration of dopaminergic cells in the substantia nigra, which ultimately leads to diminished release of dopamine in the basal ganglia of PD patients.

CNTs AS THERAPEUTIC AGENTS FOR STROKE

As reported by the Centers for Disease Control and Prevention, the third leading cause of death in the United States is stroke; approximately 137,000 Americans die from a stroke annually. Not all strokes result in death, and survivors can suffer from paralysis, speech difficulties, and emotional problems (Keenan and Shaw 2011). In a recent study, the effective delivery of siRNA directly to the central nervous system (CNS) normalized phenotypes in animal models of stroke and other neurological diseases. Using an endothelin-1–induced stroke model, the perilesional stereotactic administration of a caspase-3 siRNA (si Cas 3) delivered by functionalized CNTs (f-CNTs) both before and after focal ischemic damage of the rodent motor cortex reduced neurodegeneration and promoted preservation of neurological function. These observations illustrate the opportunity offered by CNT-mediated siRNA delivery and gene silencing of neuronal tissue to treat a variety of different neuropathological conditions where intervention at well-localized brain foci may produce therapeutic and functional benefits (Al-Jamal et al. 2011). This proof of principle study of in vivo gene silencing of caspase-3 in neuronal tissue suggests that siRNA-mediated gene silencing could target messenger RNA (mRNA) for other proteins that are overexpressed in stroke.

Simultaneously targeting multiple stroke pathways with siRNA may be more effective in preventing stroke-induced neurodegeneration than attacking a single pathway by administering a caspase-3 inhibitor intracerebrally or systemically. Similar inhibition of multiple pathways of neurodegeneration may be more difficult or impossible with pharmacological inhibitors, which are more limited in scope than siRNA and can target only certain cellular receptors and enzymes. The siRNA approach could improve treatment of a variety of CNS diseases by providing flexible, genetics-based combinatory modalities that can be integrated with various currently used neurosurgical interventions. In particular, they may be employed to treat neuropathologies, such as stroke and PD, for which well-localized brain foci are identified and known to be implicated in the phenotypic expression of the disease.

FULLERENES

Fullerenes are a family of carbon allotropic compounds that were predicted to exist by Osawa and his team in 1970 and officially discovered by Kroto and colleagues in 1985 (Kroto et al. 1985; Osawa 1970). Fullerenes vary in size depending on their number of carbon atoms. Discovery of C61-butylic acid led to the discovery and synthesis of other fullerene variations, such as C70, C20 (the smallest member), CNTs (elongated, tube-structured fullerene), carbon nano-onions, and nanobuds (Iijima 1991; Kateb et al. 2011; Langa and Nierengarten 2007; Nasibulin et al. 2007; Prinzbach et al. 2000; Sano et al. 2001). Synthesis of C84 fullerene was reported using the flash vacuum pyrolysis method (Amsharov and Jansen 2009).

Fullerenes can assume different forms and can encage compounds. The unique material (physical, chemical, electrical, and optical) properties of fullerenes and their derivatives led to their incorporation into new and improved devices and materials that have advanced uses in bioengineering (Avouris et al. 2007; Bakry et al. 2007; Bosi et al. 2003; Guldi and Prato 2000; Koruga et al. 1996; Zagal et al. 2009). Well-defined fullerene-containing polymers and stimuli-sensitive amphiphilic systems can be synthesized as a result of developing the azido coupling and atom transfer radical addition process (Ravi et al. 2007). Hydrated fullerenes are radioprotective. Hydrated C60 fullerene protects against damage from x-irradiation (7 Gy) in vitro and in vivo in mice by reducing the formation of reactive oxygen species (Andrievsky et al. 2009).

The poor solubility of fullerenes in aqueous solution makes them incompatible with intravenous infusion and leads to toxicity in biological systems. Research continues to increase the solubility and reduce the toxicity of fullerenes and their derived compounds (Johnston et al. 2010). The problem is being addressed by deriving functionalized fullerenes that are more soluble and less toxic than naked fullerenes. Table 7.1 shows the list of various functionalized fullerenes and their potential applications.

Partha and his colleagues restricted their review of functionalized fullerenes to those based on C60 platforms. They specifically focused on three important applications to medicine and the functionalized fullerenes that make these applications possible: 1) use of the paclitaxel-embedded buckysomes (PEBs) as drug delivery nanoparticles; 2) carboxyfullerenes to serve as neuroprotective agents against reactive oxidative species (ROS); and 3) endohedral metallofullerenes to be used as contrast agents in magnetic resonance imaging (MRI) (Partha and Conyers 2009).

PEBs are amphiphilic fullerenes that are spherical in shape and 100–200 nm in diameter. These nanoparticles have a hydrophobic core that can carry the anticancer drug paclitaxel (Taxol) and a water-soluble exterior that can solubilize it. Using PEBs to deliver this anticancer drug avoids the need for nonaqueous agents that would otherwise be required to deliver it; the nonaqueous solvent, polyoxyethylated castor oil (Cremophor EL), which is presently used to solubilize paclitaxel for clinical use, leads to allergic reactions (rarely fatal) and uncomfortable injection site side effects in many patients. Partha and Conyers (2009) suggest that PEBs could deliver more paclitaxel over time than the standard paclitaxel preparation and potentially reduce its infusion time. Additionally, PEBs were shown to be as effective as an injectable suspension of paclitaxel albumin-bound particles (Abraxane, Abraxis BioScience, Celgene Corp., Summit, NJ) in hindering MCF-7 human breast

TABLE 7.1
List of Functionalized Fullerenes and Their Potential Applications in Biomedicine

Type of Functionalized Fullerene (Key Words)	Potential Application	References
AF-1, buckysomes, PEB	Drug delivery	Partha et al. (2007)
DF-1	Radioprotection	Theriot et al. (2010)
Fullerene–paclitaxel	Cancer therapy	Zakharian et al. (2005)
Fullerene polyamine (tetra-amino fullerene)	Gene delivery, transfection	Sitharaman et al. (2008)
Aminofullerene adducts	Nonviral gene delivery	Isobe et al. (2006)
Fullerene-based amino acids and peptides	Peptide delivery	Bianco et al. (2001)
Cystine C_{60}, beta-alanine C_{60}	H_2O_2-induced apoptosis protection	Hu et al. (2007)
Fullerene lipidosome	Antiviral activity	Ji et al. (2008)
Fullerene vesicle	Oxidative stress reduction	Maeda et al. (2008)
Hydrophilic or cationic fullerenes	Photodynamic cancer therapy	Mroz et al. (2007)
C3-F-tris-MDC	Oxidative stress reduction	Dugan et al. (2001)

Source: Modified from Partha, R., and Conyers, J.L., *Int J Nanomedicine* 4: 261–75, 2009. With permission.
Note: AF-1, amphiphilic fullerene-1; PEB, paclitaxel-embedded buckysomes; DF-1, dendrofullerene-1; C3-F-tris-MDC, C3-fullero-tris-methanodicarboxylic acid.

cancer cell division in vitro. PEBs were also created with the ability to slowly release paclitaxel, which may lead to enhanced therapeutic effects in vivo (Partha et al. 2008).

Mitochondrial dysfunction was recognized in the mid-1980s as a possible cause of PD and led to the development of treatment to counteract this and slow neurologic deterioration (Ramsey and Giasson 2007). Recent studies showed that the polyhydroxyl fullerene derivative C60(OH)(24) is an antioxidant that can effectively protect human neuroblastoma cells from acute injury resulting from 1-methyl-4-phenylpyridinium (MPTP), a chemical that induces PD in experimental animals and humans (Cai et al. 2008). This report describes the potential use of fullerenes as neuroprotectants. Fullerenes might also be used as a drug carrier to treat neurodegenerative diseases such as PD.

NANODIAMONDS

Unknown to the rest of world until the 1980s, NDs were first constructed by explosive detonation in the USSR in the 1960s. The use of explosives provided a source of carbon as well as the energy necessary for the creation of these particles. When detonation occurs in a closed chamber filled with inert gas (dry synthesis) or a water coolant (wet synthesis), it yields a soot composed of diamond particles 4 to 5 nm in diameter, along with other allotropes of carbon and impurities (Mochalin et al. 2011). These methods of synthesis can generate up to 75% diamonds by weight; purification of the soot is required to remove any entrapped impurities from the particles. Currently, the most promising method of purification of soot containing NDs is gas oxidation (Mochalin et al. 2011). NDs may also be produced by the high-pressure–high-temperature (HPHT) technique. As in detonation, HPHT yields a mixture of impurities and NDs, and acid treatment is necessary for purification. As miniaturized diamonds, NDs have an assortment of unique physical and chemical properties including exceptional hardness, biocompatibility, optical properties such as fluorescence (Mochalin et al. 2011) and transparency (Zhang et al. 2010), high thermal conductivity, electrical resistivity, chemical stability, and resistance to severe environmental conditions (Mochalin et al. 2011).

FLUORESCENCE

Detonation creates NDs with diameters up to 100 nm and green fluorescence (Mamaeva et al. 2011), whereas HPHT creates NDs with diameters around 25 nm and bright red fluorescence (Vial et al.

2008). NDs may also become fluorescent upon irradiation with high-energy particles followed by vacuum annealing (heating followed by slow cooling). Vacancies within the ND structure are created by irradiation. During the annealing process, these vacancies migrate and come into proximity of the nitrogen atoms within the diamond structure, producing nitrogen-vacancy (NV) centers. At present, researchers are investigating the use of NDs and their implementation in high-resolution magnetic sensing, fluorescence resonance energy transfer, and biomedical imaging. To date, internal and size-related defects within NDs have prevented their application in imaging. An alternative method to produce NDs with fluorescent properties is to functionalize them with fluorophores via linking or adsorbing techniques. The conjugated NDs can be transported through cellular compartments of varying pH without sacrificing cell viability (Mochalin et al. 2011).

BIOCOMPATIBILITY AND TOXICITY OF NDs

Compared to their cousin the quantum dot, NDs were shown to be relatively nontoxic and therefore appear to have greater potential for in vivo applications. Both green and red fluorescent NDs (fNDs) are taken up by cells, localize in the cytoplasm, and are biocompatible with a wide variety of cell types. NDs have little to no in vitro toxicity (Sano et al. 2001). One study that instilled NDs within the trachea reported low toxicity in pulmonary epithelial tissue, a declining presence of NDs in the alveolae over time, and internalization of the NDs by macrophages observed after 28 days of exposure (Mochalin et al. 2011). However, the work of Zhang et al. (2010) yielded conflicting results. Their study reported respiratory toxicity after intratracheal instillation that ultimately led to systemic toxicity and accumulation of NDs in the liver, bone, and spleen (Zhang et al. 2010). This disparity in toxicity may have resulted from differences in experimental technique between studies. These studies emphasize that development of NDs for medical applications requires in vivo testing to evaluate their possible toxic effects on organ systems. Conjugating NDs with functional groups may increase their biocompatibility in certain applications. Graphitization of NDs creates a carbon "onion," which has potential application in energy storage, composites, and catalytic activity (Sano et al. 2001; Shevchenko et al. 2007).

BIOMEDICAL APPLICATION OF NDs

Imaging

Implementing NDs to visualize cells has important ramifications for neurosurgery. Work done by Fang et al. (2011) demonstrated successful internalization of fNDs with NV centers in three cell lines (HeLa cervical cancer cells, 3T3-L1 preadipocytes, and 489-2.1 multipotent stromal cells) via endocytosis. The nanoparticles did not affect cell growth or proliferation. Exocytosis of the NDs over 6 days was less than 15% for HeLa and 3T3-L1 cells and up to 30% for multipotent stromal cells. These in vitro findings suggest that these fNDs could be used for long-term cell tracking and cell labeling applications (Fang et al. 2011). Tetramethylrhodamine-conjugated NDs were similarly taken up by endocytosis and sequestered in the lysosomes of neuroblastoma cells with low toxicity. The rhodamine label of the NDs providing bright fluorescence and protein conjugation remained stable. Conjugating NDs with a fluorophore is another way to utilize NDs for cell labeling and tracking (Schrand et al. 2011).

Drug Delivery

For a material to be successfully developed as a platform for drug delivery, it must possess certain properties, which include biocompatibility, targeted therapy, and water solubility. A complex (NDX) of Doxorubicin (Dox) and Nanodiamonds (ND) were tested in vivo against murine liver tumor and mammary carcinoma models; NDX had less myelotoxicity than unbound doxorubicin, a popular anticancer drug. Using NDs enhanced NDX's circulation time, reduced the tumor cells' ability to expel the drug, and significantly increased apoptosis of tumor cells. Tumor growth inhibition was significantly greater for ND–Dox compared to Dox alone (Chow et al. 2011). When functionalized with polyethylenimine 800 (PEI800), coated NDs can carry and deliver nucleic acids, including siRNA (Mochalin et al. 2011).

Tissue Scaffolding and Surgical Implants

Regenerative medicine is a field of increasing interest in biomedical research that may benefit from further development of ND technology. Recent studies demonstrated that NDs in a monolayer can direct murine neuronal growth and promote formation of neural networks (Thalhammer et al. 2010). Neural regeneration is of great importance to patients with neurological disorders that result in interruptions of neural networks involved in movement, speech, vision, and cognition, including stroke, spinal cord injury, multiple sclerosis, and PD. ND layering on a smooth surface creates points of attachment for neuronal adhesion and directs neuronal growth. In cell culture, neurons developed electrochemical connections via synapses within 5 days and neural networks within 7 days (Thalhammer et al. 2010). Advancing this work from growth on a two-dimensional surface to growth in a three-dimensional tissue-engineering scaffold requires additional research. Promoting neuronal growth in medical implants may lead to improved brain–machine interfaces. NDs are also being studied for their potential to promote in-growth of bone to replace bone tissue that is removed during treatment of diseases of the spine or other skeletal structures. Adding 10% by weight of octadecylamine-functionalized ND (ND–ODA) to the biodegradable polymer, poly(L-lactic acid) (PLLA), increased its stiffness twofold and its hardness eightfold, close to the level of human cortical bone. The material supported in vitro proliferation and differentiation of murine osteoblasts and was nontoxic (Zhang et al. 2011). Successful development of safe and biocompatible bone replacement materials may reduce reliance on bone autografts and allografts in spinal and extraspinal surgery.

Conclusion

The application of carbon nanoplatforms in the diagnosis and treatment of neurological disorders is still under investigation. So far, there is encouraging evidence that carbon nanoplatforms can be developed for drug and siRNA delivery to cells affected by any number of diseases. These platforms can be functionalized to improve their targeting of diseased cells and tissues. Cellular imaging can be advanced by exploiting the intrinsic fluorescence of NDs with NV centers or by conjugating carbon nanoplatforms with fluorophores. Carbon nanoplatforms appear to have substantial potential to play a major role in developing novel therapeutic and imaging agents that will contribute to future advances in personalized medicine and targeted therapy and the field of nano neuro surgery.

References

Al-Jamal, K.T., Gherardini, L., Bardi, G. et al. 2011. Functional motor recovery from brain ischemic insult by carbon nanotube-mediated siRNA silencing. *Proc Natl Acad Sci U S A* 108: 10952–7.

Ali, S.R., Ma, Y., Parajuli, R.R. et al. 2007. A nonoxidative sensor based on a self-doped polyaniline/carbon nanotube composite for sensitive and selective detection of the neurotransmitter dopamine. *Anal Chem* 79: 2583–7.

Alzheimer, A. 1907. Ueber eine eigenartige Erkrankung der Hirnrinde [About a peculiar disease of the cerebral cortex]. *Allg Z Psychiat Med* 64: 146–8.

Alzheimer's Association. 2010. Alzheimer's disease facts and figures. *Alzheimers Dement* 6: 158–94.

Amsharov, K., and Jansen, M. 2009. Synthesis of a higher fullerene precursor-an "unrolled" C84 fullerene. *Chem Commun (Camb)* 2691–3.

Ando, Y. 2010. Carbon nanotube: the inside story. *J Nanosci Nanotechnol* 10: 3726–38.

Andrievsky, G.V., Bruskov, V.I., Tykhomyrov, A.A. et al. 2009. Peculiarities of the antioxidant and radioprotective effects of hydrated C60 fullerene nanostuctures in vitro and in vivo. *Free Radic Biol Med* 47: 786–93.

Avouris, P., Chen, Z., and Perebeinos, V. 2007. Carbon-based electronics. *Nat Nanotechnol* 2: 605–15.

Bakry, R., Vallant, R.M., Najam-ul-Haq, M. et al. 2007. Medicinal applications of fullerenes. *Int J Nanomed* 2: 639–49.

Berhanu, D., Dybowska, A., Misra, S.K. et al. 2009. Characterisation of carbon nanotubes in the context of toxicity studies. *Environ Health* 8 Suppl 1: S3.

Bianco, A., Da Ros, T., Prato, M. et al. 2001. Fullerene-based amino acids and peptides. *J Pept Sci* 7: 208–19.

Bosi, S., Da Ros, T., Spalluto, G. et al. 2003. Fullerene derivatives: an attractive tool for biological applications. *Eur J Med Chem* 38: 913–23.

Cai, X., Jia, H., Liu, Z. et al. 2008. Polyhydroxylated fullerene derivative C(60)(OH)(24) prevents mitochondrial dysfunction and oxidative damage in an MPP(+)-induced cellular model of Parkinson's disease. *J Neurosci Res* 86: 3622–34.

Chow, E.K., Zhang, X.Q., Chen, M. et al. 2011. Nanodiamond therapeutic delivery agents mediate enhanced chemoresistant tumor treatment. *Sci Transl Med* 3: 73ra21.

Cipriani, G., Dolciotti, C., Picchi, L. et al. 2011. Alzheimer and his disease: a brief history. *Neurol Sci* 32: 275–9.

Dugan, L.L., Lovett, E.G., Quick, K.L. et al. 2001. Fullerene-based antioxidants and neurodegenerative disorders. *Parkinsonism Relat Disord* 7: 243–6.

Enoki, T., Takai, K., Osipov, V. et al. 2009. Nanographene and nanodiamond; new members in the nanocarbon family. *Chem Asian J* 4: 796–804.

Fang, C.Y., Vaijayanthimala, V., Cheng, C.A. et al. 2011. The exocytosis of fluorescent nanodiamond and its use as a long-term cell tracker. *Small* 7: 3363–70.

Guldi, D.M., and Prato, M. 2000. Excited-state properties of C-60 fullerene derivatives. *Acc Chem Res* 33: 695–703.

Guo, T., Nikolaev, P., Thess, A. et al. 1995. Catalytic growth of single-walled nanotubes by laser vaporization. *Chem Phys Lett* 243: 49–54.

Haniu, H., Matsuda, Y., Takeuchi, K. et al. 2010. Proteomics-based safety evaluation of multi-walled carbon nanotubes. *Toxicol Appl Pharmacol* 242: 256–62.

Holt, K.B. 2007. Diamond at the nanoscale: applications of diamond nanoparticles from cellular biomarkers to quantum computing. *Philos Transact A Math Phys Eng Sci* 365: 2845–61.

Hu, Z., Guan, W., Wang, W. et al. 2007. Synthesis of beta-alanine C60 derivative and its protective effect on hydrogen peroxide-induced apoptosis in rat pheochromocytoma cells. *Cell Biol Int* 31: 798–804.

Iijima, S. 1991. Helical microtubules of graphitic carbon. *Nature* 354: 56–8.

Inami, N., Mohamed, M.A., Shikoh, E. et al. 2007. Synthesis-condition dependence of carbon nanotube growth by alcohol catalytic chemical vapor deposition method. *Sci Technol Adv Mater* 8: 292–5.

Isobe, H., Nakanishi, W., Tomita, N. et al. 2006. Gene delivery by aminofullerenes: structural requirements for efficient transfection. *Chem Asian J* 1: 167–75.

Ji, H., Yang, Z., Jiang, W. et al. 2008. Antiviral activity of nano carbon fullerene lipidosome against influenza virus in vitro. *J Huazhong Univ Sci Technolog Med Sci* 28: 243–6.

Johnston, H.J., Hutchison, G.R., Christensen, F.M. et al. 2010. The biological mechanisms and physicochemical characteristics responsible for driving fullerene toxicity. *Toxicol Sci* 114: 162–82.

Kateb, B., Chiu, K., Black, K.L. et al. 2011. Nanoplatforms for constructing new approaches to cancer treatment, imaging, and drug delivery: what should be the policy? *Neuroimage* 54 Suppl 1: S106–24.

Kateb, B., Van Handel, M., Zhang, L. et al. 2007. Internalization of MWCNTs by microglia: possible application in immunotherapy of brain tumors. *Neuroimage* 37 Suppl 1: S9–17.

Keenan, N.L., and Shaw, K.M. 2011. Coronary heart disease and stroke deaths—United States, 2006. *MMWR Surveill Summ* 60 Suppl: 62–6.

Koruga, D., Matija, L., Misic, N. et al. 1996. Fullerene C60: properties and possible applications. *Mater Sci Forum* 214: 49–56.

Kroto, H.W., Heath, J.R., O'Brien, S.C. et al. 1985. C-60: Buckminsterfullerene. *Nature* 318: 162–3.

Krueger, A. 2008. New carbon materials: biological applications of functionalized nanodiamond materials. *Chemistry* 14: 1382–90.

Lam, C.W., James, J.T., McCluskey, R. et al. 2006. A review of carbon nanotube toxicity and assessment of potential occupational and environmental health risks. *Crit Rev Toxicol* 36: 189–217.

Langa, F., and Nierengarten, J.-F. 2007. *Fullerenes—Principles and Applications. RSC Nanoscience & Nanotechnology*. Cambridge, UK: RSC Publishing.

Li, J., and Andrews, R.J. 2007. Trimodal nanoelectrode array for precise deep brain stimulation: prospects of a new technology based on carbon nanofiber arrays. *Acta Neurochir Suppl* 97: 537–45.

Maeda, R., Noiri, E., Isobe, H. et al. 2008. A water-soluble fullerene vesicle alleviates angiotensin II-induced oxidative stress in human umbilical venous endothelial cells. *Hypertens Res* 31: 141–51.
Mamaeva, E.S., Baron, A.V., Puzyr, A.P. et al. 2011. Modified nanodiamonds obtained by detonation synthesis in constructing biochemical indication systems (as exemplified by the glucose determination system). *Dokl Biochem Biophys* 439: 182–4.
Mochalin, V.N., Shenderova, O., Ho, D. et al. 2011. The properties and applications of nanodiamonds. *Nat Nanotechnol* 7: 11–23.
Mroz, P., Pawlak, A., Satti, M. et al. 2007. Functionalized fullerenes mediate photodynamic killing of cancer cells: type I versus type II photochemical mechanism. *Free Radic Biol Med* 43: 711–9.
Nasibulin, A.G., Pikhitsa, P.V., Jiang, H. et al. 2007. A novel hybrid carbon material. *Nat Nanotechnol* 2: 156–61.
Osawa, E. 1970. Superaromaticity. *Kagaku (Kyoto)* 25: 854–63.
Partha, R., and Conyers, J.L. 2009. Biomedical applications of functionalized fullerene-based nanomaterials. *Int J Nanomedicine* 4: 261–75.
Partha, R., Lackey, M., Hirsch, A. et al. 2007. Self assembly of amphiphilic C60 fullerene derivatives into nanoscale supramolecular structures. *J Nanobiotechnology* 5: 6.
Partha, R., Mitchell, L.R., Lyon, J.L. et al. 2008. Buckysomes: fullerene-based nanocarriers for hydrophobic molecule delivery. *ACS Nano* 2: 1950–8.
Podolski, I.Y., Podlubnaya, Z.A., Kosenko, E.A. et al. 2007. Effects of hydrated forms of C60 fullerene on amyloid 1-peptide fibrillization in vitro and performance of the cognitive task. *J Nanosci Nanotechnol* 7: 1479–85.
Prinzbach, H., Weiler, A., Landenberger, P. et al. 2000. Gas-phase production and photoelectron spectroscopy of the smallest fullerene, C20. *Nature* 407: 60–3.
Ramsey, C.P., and Giasson, B.I. 2007. Role of mitochondrial dysfunction in Parkinson's disease: implications for treatment. *Drugs Aging* 24: 95–105.
Ravi, P., Dai, S., Wang, C. et al. 2007. Fullerene containing polymers: a review on their synthesis and supramolecular behavior in solution. *J Nanosci Nanotechnol* 7: 1176–96.
Salvador-Morales, C., Flahaut, E., Sim, E. et al. 2006. Complement activation and protein adsorption by carbon nanotubes. *Mol Immunol* 43: 193–201.
Sano, N., Wang, H., Chhowalla, M. et al. 2001. Synthesis of carbon 'onions' in water. *Nature* 414: 506–7.
Schrand, A.M., Lin, J.B., Hens, S.C. et al. 2011. Temporal and mechanistic tracking of cellular uptake dynamics with novel surface fluorophore-bound nanodiamonds. *Nanoscale* 3: 435–45.
Shevchenko, V.Y., Madison, A.E., and Mackay, A.L. 2007. Coherent coexistence of nanodiamonds and carbon onions in icosahedral core-shell particles. *Acta Crystallogr A* 63: 172–6.
Sitharaman, B., Zakharian, T.Y., Saraf, A. et al. 2008. Water-soluble fullerene (C60) derivatives as nonviral gene-delivery vectors. *Mol Pharm* 5: 567–78.
Strassnig, M., and Ganguli, M. 2005. About a peculiar disease of the cerebral cortex: Alzheimer's original case revisited. *Psychiatry (Edgmont)* 2: 30–3.
Thalhammer, A., Edgington, R.J., Cingolani, L.A. et al. 2010. The use of nanodiamond monolayer coatings to promote the formation of functional neuronal networks. *Biomaterials* 31: 2097–104.
Theriot, C.A., Casey, R.C., Moore, V.C. et al. 2010. Dendro[C(60)]fullerene DF-1 provides radioprotection to radiosensitive mammalian cells. *Radiat Environ Biophys* 49: 437–45.
Trimmer, P.A., Schwartz, K.M., Borland, M.K. et al. 2009. Reduced axonal transport in Parkinson's disease cybrid neurites is restored by light therapy. *Mol Neurodegener* 4: 26.
Vial, S., Mansuy, C., Sagan, S. et al. 2008. Peptide-grafted nanodiamonds: preparation, cytotoxicity and uptake in cells. *Chembiochem* 9: 2113–9.
Wu, C.K. 2008. Mild cognitive impairment, healthy aging and Alzheimer's disease. *Med Health R I* 91: 132–3.
Xing, Y., and Dai, L. 2009. Nanodiamonds for nanomedicine. *Nanomedicine (Lond)* 4: 207–18.
Yang, Z., Zhang, Y., Yang, Y. et al. 2010. Pharmacological and toxicological target organelles and safe use of single-walled carbon nanotubes as drug carriers in treating Alzheimer disease. *Nanomedicine* 6: 427–41.
Zagal, J.H., Griveau, S., Ozoemena, K.I. et al. 2009. Carbon nanotubes, phthalocyanines and porphyrins: attractive hybrid materials for electrocatalysis and electroanalysis. *J Nanosci Nanotechnol* 9: 2201–14.
Zakharian, T.Y., Seryshev, A., Sitharaman, B. et al. 2005. A fullerene–paclitaxel chemotherapeutic: synthesis, characterization, and study of biological activity in tissue culture. *J Am Chem Soc* 127: 12508–9.
Zhang, Q., Mochalin, V.N., Neitzel, I. et al. 2011. Fluorescent PLLA-nanodiamond composites for bone tissue engineering. *Biomaterials* 32: 87–94.
Zhang, X., Yin, J., Kang, C. et al. 2010. Biodistribution and toxicity of nanodiamonds in mice after intratracheal instillation. *Toxicol Lett* 198: 237–43.

8 Dendrimers

Ayaz M. Khawaja and John D. Heiss

CONTENTS

INTRODUCTION

Dendrimers are macromolecular structures exhibiting a repetitive, highly branched, symmetrical, and regular three-dimensional structure. They have a size on the nanometer scale but differ significantly from other nanoparticles in their structures and properties (Khawaja 2011). Dendrimers derive their name from the Greek words *dendri* (treelike) and *meros* (part of) (Tomalia et al. 1985). They are also known as "arborols" and "cascade molecules." Technically, dendrimers are polymers that are monodisperse, meaning that they have the same size and molecular weight (M_w) in solution, unlike most polymers, which are polydisperse, having different sizes and M_w in solution. Dendrimers are of interest for drug delivery and imaging applications because of their (1) predictable M_w, (2) ability to include molecules within their structure, and (3) large surface area, which allows attachment of many functional groups and targeting through multivalent attachment to cell targets (Mammen et al. 1998) (Table 8.1, see page 104).

This chapter reviews investigations of dendrimers in which they (1) have acted as drug carriers for gene therapy, chemotherapy, and boron neutron capture therapy (BNCT) against in vitro and in vivo models of glioblastoma multiforme models; (2) have carried contrast agent for magnetic resonance (MR) imaging; and (3) have exerted direct actions against diseases such as HIV infection, Alzheimer's dementia, and spongiform encephalopathy. Methods to increase transport of dendrimers across the blood–brain barrier (BBB), delivery of dendrimers to the target site, and uptake of dendrimers into cells are discussed. The effects of adding functional groups to the dendrimer surface are also reported.

TABLE 8.1
List of Different Functional Groups Discussed for Dendrimer Modification

Property Imparted	Functional Group Used	Utility Demonstrated with Dendrimers in Investigations
Targeting	TAT	Used in siRNA transfer in in vivo glioma model suppressing oncoprotein expression.
	Angiopep	Targets lipoprotein-related protein 1. Used for gene therapy in an in vivo model with TRAIL, causing enhanced apoptosis and survival.
	Folate	Exhibits remarkable specificity for tissues expressing high levels of folate receptors, raising the cytotoxic profile of the drug within the dendrimer, reducing the toxicity of the dendrimer proper, enhancing AO transfer to exert gene-silencing effect in in vivo glioma animal model.
	CTX	Provides specific targeting toward receptors in an in vivo glioma model, for gene transfer.
	Monoclonal Ab	C225 targets EGFR in an experiment suppressing its expression and exhibiting remarkable specificity.
	Aptamer	A specific aptamer, GBI-10, recognizes an extracellular matrix protein with high affinity and specificity in glioma cell lines.
	RGD peptides	RGD recognizes integrins and inhibits adhesion and migration of cells in glioma spheroids.
BBB transport	Transferrin Lactoferrin Tamoxifene Glucosamine	All molecules are involved in aiding the dendrimer transport across the BBB through caveolin- and clathrin-mediated endocytic mechanism. Tamoxifene inhibits efflux, further increasing bioavailability for tumors.
Imaging	DOTA	DOTA–Gd specifically targets folate receptor-bearing tumors. More stable, affords higher relaxivity, and less toxic than DTPA.
	DTPA	A Gd chelate, along with DOTA.
	ACPP	Recognizes levels of matrix metalloproteinase in tumors and a cleavable link with dendrimers increasing ion relaxivity and signal intensity.
Fluorescence	FITC, FMNP	Both used together in an experiment to demonstrate release of FITC (model for drug release). Drug release was lower for dendrimer complex compared to FMNP alone.
	Cy5.5	Used for fluorescence with a variety of nanoparticles.
Other	PEG	Increases hydrophilicity, prolongs $T_{1/2}$, contributes resistance to biodegradability, escapes RES detection, increases drug encapsulation, attaches folate.
	Acetylation	In a siRNA delivery experiment, acetylation lowered cytotoxicity of dendrimers toward glioma cells.
	Methyl acrylate	Can modulate release of FITC through control of surface groups. Decreases FITC release.

STRUCTURE AND PROPERTIES OF DENDRIMERS

Dendrimers can be conveniently classified by generation, with generation zero (G0) being the core structure around which the dendrimer is built, and G1 and subsequent generations being more elaborate structures produced by adding identical monomers (XY_2) (Figure 8.1a). Generations can be added sequentially to produce progressively larger dendrimer structures, with increasing molecular size and weight.

Dendrimeric structure is described from inside out, with (1) an inner core, (2) an intermediate shell, which typically carries drug, and (3) a surface, to which active groups can be added to modify (functionalize) dendrimers. The shape of the core usually dictates the shape of the entire dendrimer.

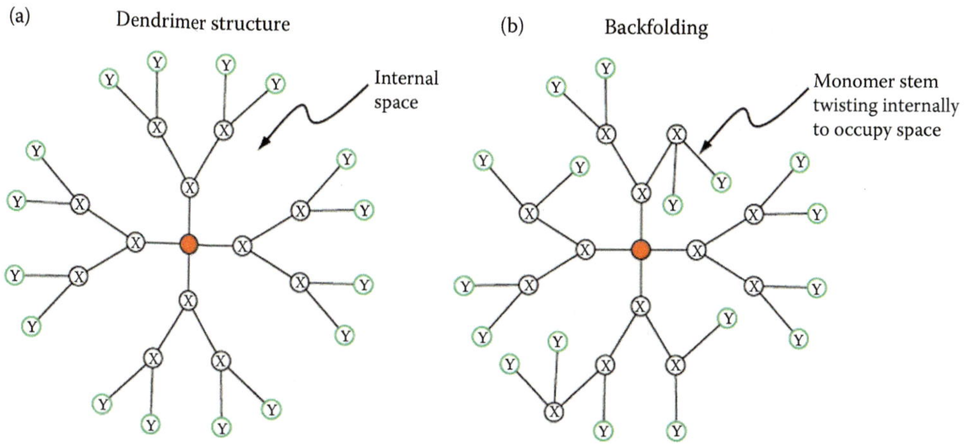

FIGURE 8.1 Typical dendrimer structure (a). The concept of backfolding within a G2 dendrimer is also shown (b). Backfolding occurs when an outwardly extending monomer twists and assumes a more random and internal position within the dendrimer.

Some researchers have produced asymmetric lower-generation dendrimers with open structures (Klajnert and Bryszewska 2001). Higher-generation dendrimers have more and longer branches and appear more globular than lower-generation structures (Caminati et al. 1990). As the number of groups increases exponentially on the dendrimer surface, steric hindrance is introduced, making the surface extremely dense and unable to accommodate further molecules (Degennes and Hervet 1983; the starburst effect, Fischer and Vogtle 1999). The density of a dendrimer decreases toward the core, introducing spaces for drug loading, although inward twisting of a monomer stem (backfolding) may obliterate these spaces (Chai et al. 2001) (Figure 8.1b). Dendrimers can depart from typical globular structure, even in higher generations, if more flexible end groups backfold, usually in response to external stimuli (Bosman et al. 1999). pH has a significant effect on the structural configuration of dendrimers due to the presence of charged groups at the surface (Lee et al. 2002).

Dendrimers also exhibit fluorescent properties. There is strong fluorescence emission from G2 and G4 NH_2-terminated PAMAM dendrimers, OH- and carboxylate-terminated PAMAM, and NH_2 polypropylenimine (PPI) dendrimers (Wang and Imae 2004). Both G2 and G4 PAMAM emit fluorescence at different levels throughout the pH range. For G4, there is little change in intensity from pH 11 to 6, but further decrease in pH dramatically increases emission intensity until it reaches a maximum at pH 2.5. For G2, intensity increases steadily from pH 11 to 4, where the maximal point is reached (Wang and Imae 2004).

SYNTHESIS OF DENDRIMERS

Dendrimers can be synthesized by a variety of mechanisms that include divergent, convergent, or "click-chemistry" methods.

Divergent

Dendrimers are formed using a step-by-step approach, which is critical to minimizing defects and impurities within the final dendrimer structure and proceeds sequentially from lower to higher generations (Figure 8.2). G0 possesses a defined number of active sites to which similar structures can be covalently attached to result in G1 (Newkome and Shreiner 2008). Additional monomers are attached to active sites on the periphery of the new G1 structure to produce a higher-molecular-weight structure (G2). The structure that is added onto the active site can be described as either XY_2 or XY_3 (Newkome and Shreiner 2008). Theoretically, this can be continued almost indefinitely. It is

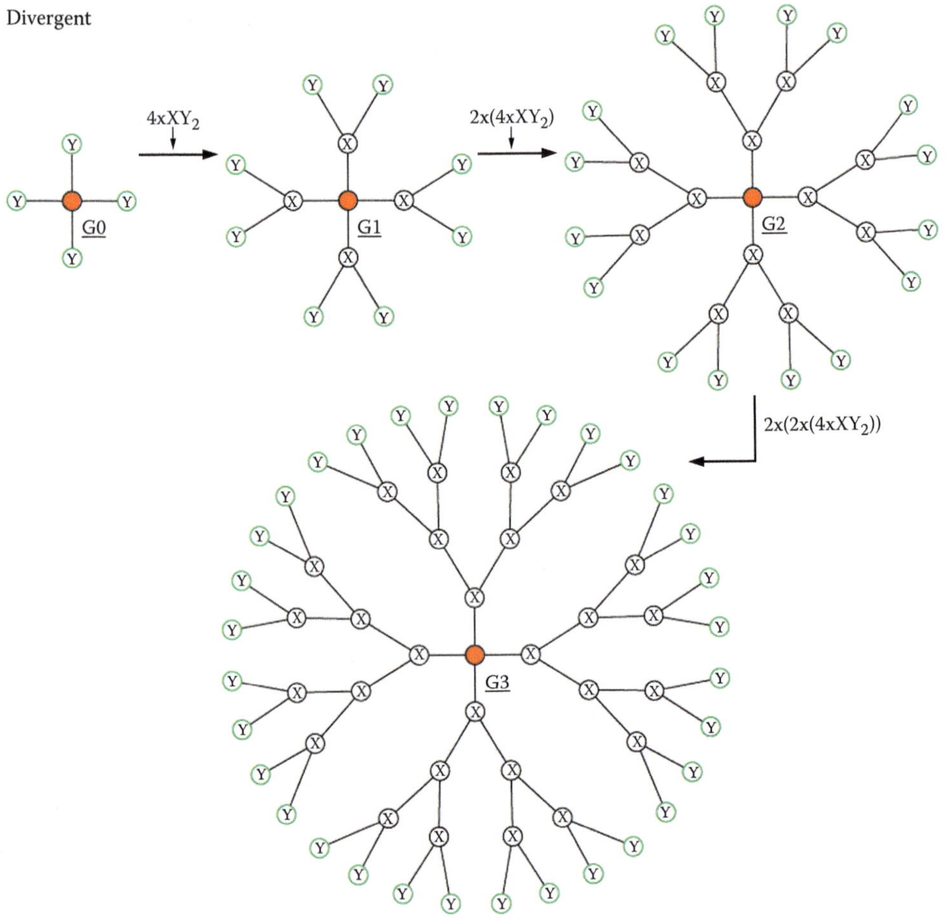

FIGURE 8.2 These drawings depict the steps in divergent synthesis. The top left drawing demonstrates that divergent synthesis begins with a core (red circle) that has multiple functional groups (Y) available for molecular bonding (four in this drawing). Proceeding to the other drawings in a clockwise manner, XY_2 monomers are added that create molecular bonds with the functional groups present on the core. The dendrimer structure subsequently grows, and a G1 dendrimer is constituted. Addition of more monomers occupies the functional groups (Y) and extends the dendrimer structure. The shape of higher generations of dendrimer is dependent on the shape of the core (spherical in this figure).

important to realize the sequential nature of the synthesis mechanism, making it cumbersome and expensive to prepare higher-generation dendrimers. Such a reaction mechanism was first outlined in 1978, when investigators prepared a heptamine structure using a sequential reaction mechanism starting from an amine and activated olefin (Buhleier et al. 1978). This was accomplished by the "Michael reaction," which consists of adding a carbon ion with an unpaired electron group to a nucleophile. The resultant structure was PPI (Tomalia et al. 1985). Newkome et al. (1985) utilized a similar approach for synthesizing dendrimers.

Convergent

Hawker and Frechet (1990) introduced the convergent method for growing dendrimers (Figure 8.3). The reaction initially forms individual dendrons, using the typical divergent method, that subsequently react with a suitable core. In the convergent method, synthesis proceeds from surface to

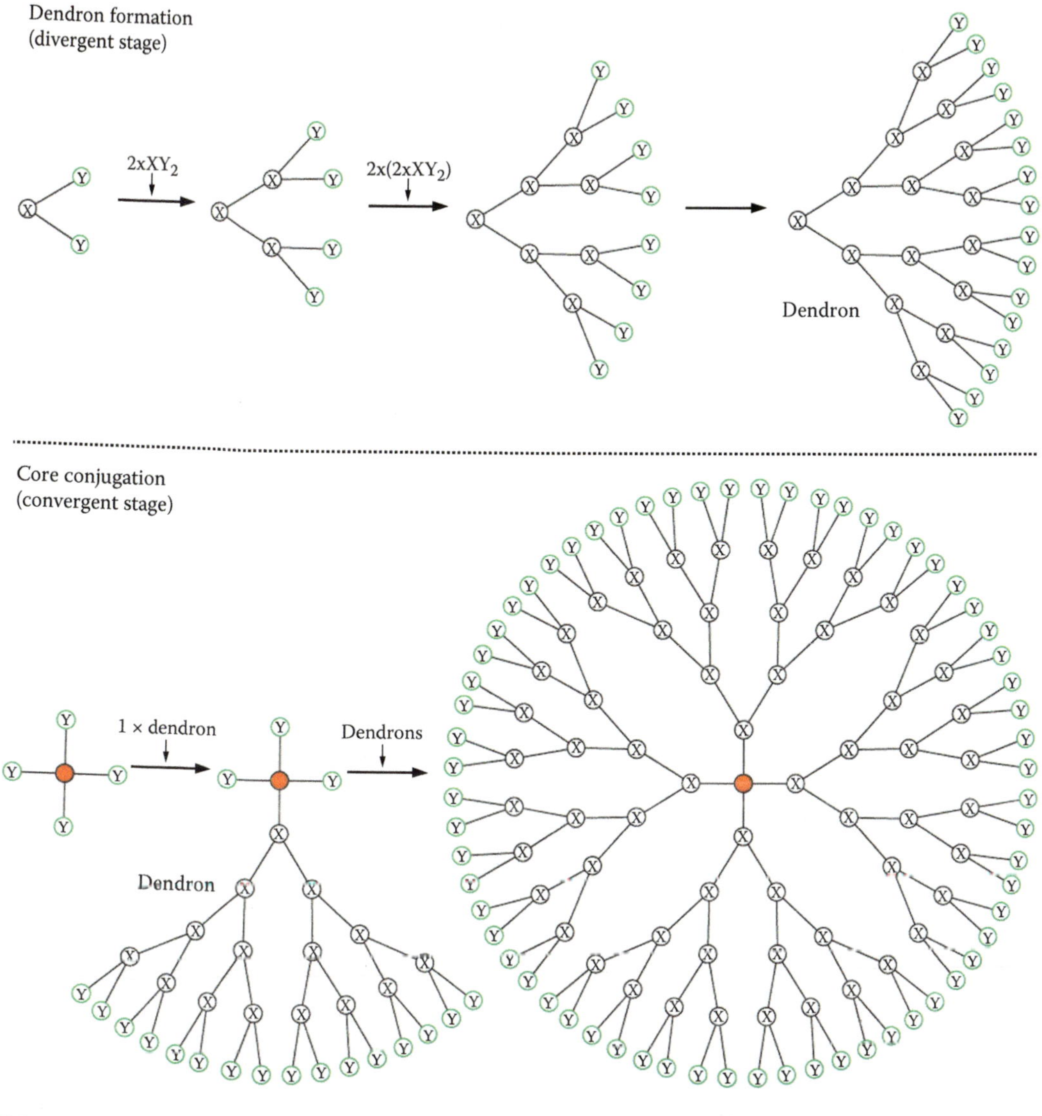

FIGURE 8.3 This illustration demonstrates that convergent synthesis begins with the formation of a single dendron (top drawing), which starts from a single XY_2 monomer whose functional groups undergo conjugation with additional monomers. The individual dendrons are subsequently conjugated with the functional groups of the core, constituting a higher-generation dendrimer (G4 in this figure).

core, rather than from core to surface as in the divergent method. The convergent method introduces fewer defects within the dendrimer than the divergent method, since the number of active sites is considerably less and only a limited number of side products can be formed (Bosman et al. 1999). Because there are fewer components in the reaction mixture, product purification is facilitated, and reactions can be completed with only a slight excess of reagents, contrary to the divergent process, where monomers have to be used in excess each time the generation number is extended (Grayson and Frechet 2001).

Click Chemistry

K. Barry Sharpless introduced the "click chemistry" methodology (Kolb et al. 2001). The basic idea revolves around forming carbon heteroatom (noncarbon atom, such as sulfur, nitrogen, or oxygen) bonds in synthesis reactions rather than carbon–carbon bonds. Sharpless notes the similarity between this approach and nature's way of creating naturally occurring structures using C–X–C bonds, with only minimal incorporation of –C–C–. The purpose is to generate structures quickly using only basic units under simple conditions, utilizing no solvent or only water, exhibiting atom economy, generating inoffensive by-products, being thermodynamically stable and stereospecific, and producing high yields (Franc and Kakkar 2008, 2009; Killops et al. 2008).

DESIGNING DENDRIMERS FOR USE AS DRUG CARRIERS

A drug-carrying structure should have a long shelf life, be able to carry an adequate number of drug molecules, be biodegradable, be able to target and be retained within the area of interest, prevent release of drugs in ectopic locations, avoid recognition by the reticuloendothelial system (RES), be detectable visually or radiographically, and exhibit minimal toxicity. Dendrimers with greater M_w (>30 kDa), a rigid globular structure, and more polymeric arms have longer half-lives and decreased renal filtration (Fox et al. 2009). Polymeric particles preferentially accumulate at tumor sites because of increased permeability of tumor vessels to particles of M_w less than 200 kDa (Tabata et al. 1998), known as the enhanced permeability and retention (EPR) effect (Tabata et al. 1998). Rigid tubular dendrimers penetrate pores in tumor vessels and accumulate in tumor tissue (Kaminskas et al. 2008; Uzgiris et al. 2004).

Pegylated dendrimers are more resistant to biodegradation in vivo compared to nonpegylated structures (Kaminskas et al. 2008) (Figure 8.4). Pegylation blocks opsonin proteins from binding to the dendrimer, prevents phagocytosis of dendrimers by the RES, and ultimately prolongs dendrimer half-life ($T_{1/2}$) (Owens and Peppas 2006). Dendrimers with linear or branched polyethylene glycol (PEG) chains have longer $T_{1/2}$ than linear PEG chains (both with similar M_w) (Bowen et al. 1999; Lim et al. 2008). Pegylation increases the hydrophilicity of a dendrimer while retaining a hydrophobic internal structure that is ideal for carrying hydrophobic compounds.

Dendrimers can use encapsulation and covalent linkage to carry and deliver drugs. Encapsulation utilizes available cavities within the dendrimer to incorporate drug molecules; covalent linkage uses the multivalent functional groups on the surface of the dendrimer to bind the drugs (Figure 8.4). Dendritic micelles are typically used for this purpose (Gillies and Frechet 2005). Pegylation increases encapsulation of anticancer drugs such as MTX and Adriamycin (Kojima et al. 2000). Covalently linked MTX–dendrimers were shown to be more stable in a phosphate-buffered saline compared to encapsulated MTX–dendrimers that release the drug readily in the phosphate-buffered saline (Patri et al. 2005). Release of encapsulated drugs can be controlled by attaching polyethylene oxide chains and pH-sensitive acetyl groups to the dendrimer surface, which undergo acetyl hydrolysis only at slightly low pH values (pH 5), disrupting the micelle and releasing the drug (Gillies et al. 2004).

Covalent modification can be achieved through (1) acid labile *cis*-aconityl or acyl hydrazone groups, (2) ester groups, and (3) disulfide groups (Mintzer and Grinstaff 2011). The acid labile groups undergo hydrolysis in an acidic environment such as within the lysosome, after cellular uptake, as shown by a *cis*-aconityl-linked DOX–dendrimer conjugate, which displayed fluorescence after cellular uptake and release from the lysosome but exhibited no fluorescence when lysosomal pH was altered (Zhu et al. 2010). Acyl hydrazone linkages were used to conjugate DOX with a bow-tie dendrimer that exhibited antitumor activity similar to commercially available liposomal-based DOX in an in vivo animal model and could be stored for a longer period of time in an external environment (Lee et al. 2006). Ester linkages were used to conjugate paclitaxel and folic acid with dendrimers (Bi et al. 2007). Conjugating MTX and folic acid with a G5 PAMAM dendrimer using ester and amide linkages, respectively, resulted in an agent with efficient targeting and cytotoxicity in in vitro cancer cell culture, less toxicity than a nontargeted dendrimer, and greater efficacy and less toxicity than free MTX in an in vivo animal model (Majoros et al. 2005). Folic acid receptors are overexpressed in many cancer cell lines, and folic acid has been used widely for conjugation with dendrimers for targeting purposes (Kono et al. 1999; Liu et al. 2000). Folic acid targeting with

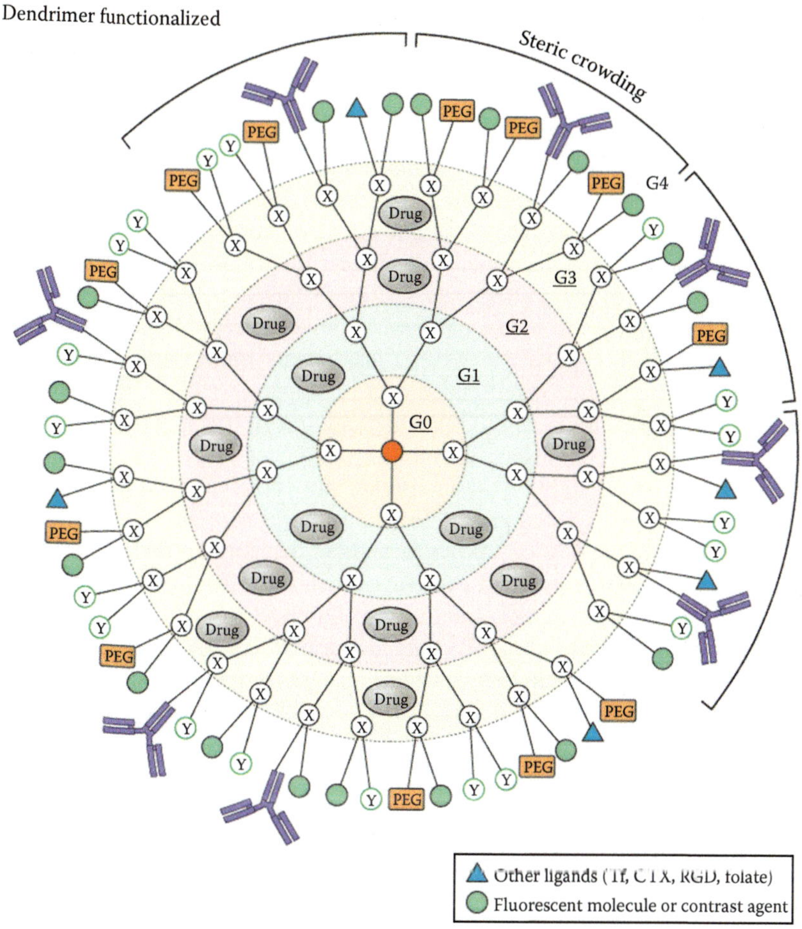

FIGURE 8.4 G4 dendrimer that functions as a drug carrier is depicted. Drug molecules occupy the internal spaces between the dendrons. The surface groups are functionalized with molecules that impart specific functions to the dendrimer. PEG groups increase the hydrophilicity of the structure. A fluorescent dye or contrast agent (green circles) enables intracellular tracking of the internalized dendrimer. Antibody molecules (purple) enable specific targeting. Ligands such as transferrin, folate, or lactoferrin (blue triangles) allow penetration of the BBB and homing to receptors at the target site complementary to the ligand. Due to the addition of multiple functional groups at the dendrimer surface, steric crowding occurs, which hinders extension into higher-generation dendrimer structures.

MTX conjugated to a G5 PAMAM dendrimer also showed specific binding to KB cells (cell line derived from keratin-forming tumor cells known as HeLa cells) and a 100-fold increased cytotoxicity compared to free MTX (Quintana et al. 2002). Attaching fluorescein-doped magnetic mesoporous silica nanoparticles (FMNPs) to the surface of amine- and ester-terminated G2 PAMAM dendrimers resulted in an agent that can be used to image glioma cells and monitor drug release (FITC) in an in vitro model of 9L glioma cell culture (Yu et al. 2009). The release rates of FITC from G2-FMNPs were lower than FMNP alone, indicating the potential use of dendrimers in controlled release systems (Yu et al. 2009).

The cationic groups present on the surface of dendrimers are cytotoxic. By allowing self-assembly of amphiphilic dendrimers, it is possible to completely eliminate the cationic charge, reducing cytotoxicity while still maintaining desired attributes (Meyers et al. 2008).

Dendrimer Modifications for Efficient Delivery across BBB

Increasing the M_w of a dendrimer reduces its renal filtration and prolongs its $T_{1/2}$ in circulation but ultimately reduces its delivery across the BBB. Delivering a dendrimer to a malignant glioma in an animal model decreased beyond 11.7 nm (Sarin et al. 2008). Uptake of a dendrimer harboring MTX is enhanced in vitro by attaching glucosamine molecules to the surface of the dendrimer, resulting in higher MTX–dendrimer concentrations, compared to dendrimer–MTX without glucosamine derivatives (Dhanikula et al. 2008). The same authors also showed, in a glioma spheroid model, that glucosylated compared to nonglucosylated dendrimers distributed more rapidly throughout the spheroid and inhibited growth more effectively and uniformly over time (Dhanikula et al. 2008). Dendrimer architecture appears to affect dendrimer delivery across the BBB. For example, G2 rhodamine-labeled polyether–copolyester (PEPE) dendrimers more rapidly penetrated the BBB through clathrin- and caveolin-mediated endocytosis, compared to G1 PEPE, saturating only at higher concentrations (Dhanikula et al. 2009). Attaching transferrin to a dendrimer used for gene transfer increased brain uptake, transfection efficiency, and gene expression compared to a nontargeted dendrimer (Huang et al. 2007). Transferrin receptors are expressed at the BBB and neuronal cell membranes, allowing targeted gene delivery to central nervous system (CNS) neurons (Li and Qian 2002). In another publication, the same authors compared a PAMAM–PEG–lactoferrin construct to a PAMAM–PEG–transferrin construct for their ability to carry and transfer genetic information in in vivo and in vitro models and found that transfection efficiency and gene expression were considerably higher for PAMAM–PEG–lactoferrin than PAMAM–PEG–transferrin (Huang et al. 2008). G4 PAMAM functionalized with PEG and transferrin on the surface and tamoxifen and DOX within the interior had greater cytotoxicity in an in vitro C6 glioma spheroid model than dendrimers without PEG and/or tamoxifen (Li et al. 2012).

POTENTIAL APPLICATIONS IN NEUROSURGERY

Gene Delivery and Other Modes of Glioma Therapy

Dendrimers were used to deliver antitumor drugs such as MTX and camptothecin to malignant glioma in animal models (Dhanikula et al. 2008; Gopin et al. 2006). They were also utilized for gene delivery using ligands such as transferrin and lactoferrin to penetrate the BBB (Huang et al. 2007, 2008). Targeting with Angiopep, the ligand of low-density lipoprotein receptor-related protein (LRP) present on the BBB (Shao et al. 2010), conjugated to pegylated PAMAM and BODIPY (boron-dipyrromethene, a fluorescent dye), was evaluated for gene delivery and expression in the brain in an in vivo Balb/c mouse model (Ke et al. 2009). Cellular uptake was dose dependent, and transfection and genetic expression efficiency were considerably greater with Angiopep-modified PAMAM than with the nonmodified dendrimer. Angiopep-2 was conjugated to PAMAM and to a therapeutic gene (DNA), which is an expression vector containing the human tumor necrosis factor–related apoptosis-inducing ligand (TRAIL) open reading frame (pORF-hTRAIL), and checked for ability to induce apoptosis in an in vivo murine glioma model (Huang et al. 2011b). This agent resulted in better biodistribution, more widespread apoptosis in the tumor, and longer median survival compared to either PAMAM–DNA or commercially available temozolomide (Huang et al. 2011b). DNA entered the nucleus through the endosome/lysosome pathway (Huang et al. 2011b). Other authors have added other targeting groups to PAMAM, including a peptide derived from rabies virus glycoprotein (RVG29), which targets metabotropic transmembrane receptors for gamma aminobutyric acid ($GABA_B$) receptors and nicotinic acetylcholine receptors (nAchR), and noted that in an in vitro model of brain capillary endothelial cells (BCECs), PAMAM–PEG–DNA–RVG29 crossed the BBB more efficiently than PAMAM–DNA (Liu et al. 2009). Chlorotoxin (CTX) was used to provide specific targeting of a pORF/TRAIL-loaded G5–PEG–PAMAM dendrimer toward the small-conductance chloride channels overexpressed on glioma cells in an in vivo murine model with stereotactically implanted C6 glioma cells (Huang et al. 2011a).

Both the distribution within glioma cells and the gene expression induced by the PAMAM–PEG–CTX–TRAIL complex were significantly better than dendrimers without CTX, indicating the specificity provided by CTX. Both the apoptotic effect and survival (59.5 days for the complex, 49 days for temozolomide) were significantly better for the CTX-modified dendrimer than for temozolomide (Huang et al. 2011a).

Antibody-targeted dendrimers were tested for their specificity and ability to suppress tumor growth in in vivo murine glioma models with intracerebral implants of either $F98_{EGFR}$ (EGFR-bearing) or $F98_{WT}$ (wild-type, non-EGFR-bearing) gliomas. Cetuximab (C225) was conjugated with G5 PAMAM bearing MTX (Wu et al. 2006). Radiolabeled ^{125}I G5–C225–MTX complex accumulated to a much greater degree (almost 5.5 times) in $F98_{EGFR}$ gliomas compared to $F98_{WT}$ gliomas, showing the specificity of targeting by C225 in the dendrimer complex. Radioactive signal was seven times greater in the tumor compared to an unaffected brain. The complex also had a significantly lower toxicity compared to free MTX, even with a higher MTX payload. However, survival times did not differ among groups of rats receiving free MTX, C225 alone, or the whole complex. The authors suggest that lowered cytotoxicity associated with MTX delivery by the dendrimer complex was due to either altered substrate binding affinity or nonrelease from the complex and that drug release could be improved using cleavable dendrimer linkages. They argue that the dendrimer complex may still be superior to free MTX due to a lower toxic profile (Wu et al. 2006). G5 PAMAM was conjugated with arginylglycylaspartic acid tripeptide (RGD) peptides that recognize integrin molecules, which are important in intercellular and cell–extracellular matrix adhesion. In an in vitro U87 glioma spheroid model, G5–RGD prevented adhesion of U87 cells to fibrinogen-coated plates (Waite and Roth 2009). Antisense-microRNA-21 [miRNA-21, an oncogene involved in cell growth, apoptosis suppression, and evolution of multiple carcinomas (Asangani et al. 2008)] and 5-fluorouracil were conjugated to G5 PAMAM and tested for a synergistic antitumor effect in vitro on the U251 human glioma cell line (Ren et al. 2010a). The dendrimer efficiently transferred both the drug and the antisense-miRNA, potentiating their effects and resulting in a dramatic increase in apoptosis and a decrease in migration of the tumor cells (Ren et al. 2010a). This complex increased the sensitivity of both U251 (PTEN-mutant) and LN229 (PTEN-wild type) glioma cell lines to paclitaxel (Taxol) by enhancing apoptosis and inhibiting cell migration (Ren et al. 2010b). This effect was independent of PTEN status. The constructs were studied for their ability to efficiently deliver siRNA to exert an antitumor effect (Morille et al. 2008). It was determined that an increasing number of RGD groups correlated well with an increasingly efficient siRNA transfer mechanism, although the actual gene-silencing effect was unremarkable (Waite and Roth 2009). To reduce the cytotoxic effect of these dendrimers, acetylation was attempted while simultaneously determining the effect of such a modification on siRNA delivery (Waite et al. 2009). It was concluded that acetylation significantly reduces the toxicity toward U87 glioma cells in an in vitro system, although it negatively impacts the delivery and gene-silencing effect of siRNA, as well as increases dendrimer–siRNA dissociation. However, 20% of the primary amines on PAMAM can be acetylated with no remarkable negative effect on siRNA delivery (Waite et al. 2009). Another novel mechanism for siRNA delivery involved conjugation of TAT protein (trans-activator of transcription) and bacterial magnetic nanoparticles (BMPs). This mechanism was investigated in the U251 glioma cell line both in vitro and in vivo in a murine model (Han et al. 2010). There were significantly decreased tumor volume, downregulated major oncoproteins, and improved survival of mice treated with TAT–BMP–PAMAM–siRNA complex compared to control groups. These results did not differ from siRNA delivery via commercially available Lipofectamine 2000 (Life Technologies, Grand Island, NY) (Han et al. 2010). Higher-generation PAMAM–siRNA complexes seem to exhibit the highest gene-silencing effect as evidenced by a G7 PAMAM–siRNA inhibition effect toward green fluorescent protein (GFP) on T98G glioma cells and J-774 macrophages (Perez et al. 2009). The uptake of dendrimer–siRNA complexes was modulated by clathrin- and caveolin-mediated endocytosis in J774 macrophages, whereas only caveolin-mediated internalization was involved with T98G glioma cells (Perez et al. 2011). Transferrin can be employed to target the dendrimer complex preferentially through the clathrin pathway to exert a better gene-silencing effect (Perez et al. 2011). Gene therapy with siRNA was also employed in in vitro HIV-infected astrocyte cultures to determine the efficiency of gene transfer and inhibition of HIV replication (Jimenez et al. 2010). Amino-terminated carbosilane dendrimers were loaded with siRNA, were minimally cytotoxic to both infected and noninfected astrocytes, and reduced replication of the HIV strains. Dendrimers that targeted folate receptors on tumor cells were used to deliver antisense oligonucleotides (AOs) that silence the EGFR (Kang et al. 2010). Folate-functionalized G5 PAMAM was complexed with AOs and delivered to in vitro C6 glioma cells, resulting in significant suppression of EGFR expression and cell growth. In rats with intracerebral glioma implants, the folate-functionalized dendrimer complex led to greater suppression of tumor growth and enhanced survival compared to the non-folate-functionalized dendrimer complex (Kang et al. 2010).

Boron Neutron Capture Therapy

PAMAM dendrimers were also used for BNCT. This therapy depends upon irradiating a stable isotope of boron (^{10}B) with low-energy thermal neutrons to produce energetic alpha (α) particles and recoiling lithium-7 nuclei, which are toxic to tumors (Barth et al. 1992; Liu et al. 1995). Successful therapy requires delivering 10^9 boron atoms per cell (Gillies and Frechet 2005), which can be accomplished by boronating and adding tumor-targeting antibodies to a dendrimer, as reported with PAMAM G2 and G4 conjugated to isocyanato polyhedral borane and IB16-6 monoclonal Antibody (Barth et al. 1994). BNCT has been used to treat glioblastoma multiforme patients in Japan since the 1960s (Hatanaka 1975; Nakagawa et al. 2003), but dendrimers have been only recently included as boron delivery agents. Since epidermal growth factor receptor (EGFR) is a frequently upregulated receptor on glial tumors, investigators conjugated epidermal growth factor (EGF) to a G4 starburst dendrimer boronated with isocyanato polyhedral borane, resulting in capturing 960 atoms of boron per EGF (Capala et al. 1996). This was then tested for targeting and accumulation in an in vitro malignant glioma (U-343MG) cell line and was localized in endosomes and lysosomes (Capala et al. 1996). Other investigators tested this system in an in vivo study, delivering boronated G4 starburst dendrimers linked to EGF either intratumorally or intravenously in Fischer 344 rats with stereotactically implanted C6 glioma cells (Yang et al. 1997). Intratumoral injection resulted in greater accumulation of boron within the tumor compared to systemically administered dendrimer, which also accumulated within the reticuloendothelial organs (Yang et al. 1997). The antitumoral effect of G5 PAMAM boronated dendrimers conjugated with C225 (monoclonal Ab targeting EGFR) versus G5 PAMAM boronated dendrimers without C225 was tested in an in vivo model of rats implanted with either the EGFR-F98 glioma line or non-EGFR-F98 glioma line (Wu et al. 2003). Intratumoral injection of the C225–dendrimer complex resulted in ~2.5 times more accumulation in EGFR-F98 glioma than in non-EGFR glioma and ~14 times more accumulation than the non-C225 dendrimer in the EGFR-F98 glioma (Wu et al. 2003).

MR Imaging and Contrast Delivery

Several studies used contrast agents conjugated to dendrimers to increase the efficiency and sensitivity of MR imaging (MRI). Lower-generation dendrimers (G2 PPI-DAB, G3-PPI-DAB, G2 PAMAM) are cleared more rapidly via renal excretion than higher-generation dendrimers (Kobayashi et al. 2003). G2–G4 PPI dendrimers exhibit more rapid clearance from circulation compared to G2–G4 PAMAM dendrimers, even if they have the same number of branches, and may be better suited for clinical use. PPI dendrimers exhibited comparable MRI efficiency for outlining hepatic anatomy to PAMAM in an in vivo animal study (Kobayashi et al. 2001a, 2003). Increasing PAMAM-chelate generations and M_w had a favorable effect on intensifying the vascular signal intensity with high vascular-to-organ signal intensity ratios and presented vasculature with high quality and detail in in vivo animal studies; however, the higher generations were retained longer (Kobayashi et al. 2001b;

Sato et al. 2001). Long retention periods pose a theoretical risk of gadolinium (Gd) dissociation from the complex and possible toxicity, such as NSF (Kobayashi et al. 2003). Gd–DOTA is a better contrast agent in terms of stability and toxicity compared to DTPA (Bousquet et al. 1988), with higher relaxivity, lower dissociation of Gd ions even at low pH values (Wang et al. 1992), and higher thermodynamic stability (Benazeth et al. 1998). Conjugating DOTA with Gd and then attaching it to the dendrimer produces a more hydrophobic agent with higher relaxivity and faster clearance from blood, compared to a postmetalation strategy, potentially decreasing the risk of NSF (Nwe et al. 2010).

Specific targeting of the dendrimer–contrast complexes can be achieved by attaching ligands to the dendrimer surface. The G4 dendrimer–chelate complex conjugated with folate on its surface was tested for specific targeting against folate receptor-positive and -negative ovarian tumors in an in vivo model using athymic mice. The folate receptor-positive tumors had significantly greater accumulation of contrast compared to receptor-negative tumors (Konda et al. 2002). Monoclonal Ab-directed accumulation was tested by conjugating G4 PAMAM with OST7Ab and DTPA-derivative chelate, which achieved a significantly greater concentration in tumors in an in vivo animal setting compared to nontargeted complexes (Kobayashi et al. 2000). If DTPA–OST7Ab was labeled with radioactive indium, additional conjugation to dendrimers increased labeling with minimal loss of immunoreactivity and quicker clearance from the circulation within the mice compared to complexes without dendrimers (Kobayashi et al. 2000). Very specific targeting toward folate receptor-bearing tumors was seen with a DOTA–Gd complex conjugated with G5 PAMAM-surface folate in an in vivo animal model of epithelial cancer (Swanson et al. 2008).

Dendrimers can be conjugated to targeting molecules and imaging labels to detect and image biomarkers in tumors. Matrix metalloproteinases (MMPs), enzymes associated with malignancy and metastasis, were targeted by conjugating activated cell-penetrating peptides (ACPPs) to dendrimers alongside Cy5.5 (fluorescent molecule) and Gd. This complex accumulated in tumors to a far greater degree than unconjugated ACPPs in an in vivo animal model. Cy5.5 detected a tumor as small as 200 μm. Gd retention allowed follow-up with MRI for several days. The link between the dendrimer and ACPP in the complex was broken by tumor-specific MMP-2 and MMP-9, resulting in this agent having four times the ion relaxivity and twice the signal intensity as a noncleavable conjugate. This complex could prove useful in aiding intraoperative localization, resection, and follow-up imaging of tumors (Olson et al. 2010).

Dendrimers were also used to enable imaging with quantum dots. The surfaces of quantum dots were modified with PAMAM dendrimer to make them water soluble and subsequently conjugated with an aptamer (GBI-10) that can recognize tenascin-C, an extracellular matrix protein, on glioma cells. High binding affinity and specificity were observed for such a construct against the U251 glioma cell line in vitro, with high quantum yield and biocompatibility compared to non-aptamer-targeted dendrimer–quantum dots (Zhiming et al. 2009).

Spongiform Encephalopathy and Alzheimer's Dementia

Studies probing other potential applications of dendrimers in neurology and neurosurgery are limited. Quaternized PEI (polyethyleneimine) and PAMAM dendrimers were tested for their potential antiprion activity in an in vitro system involving scrapie-infected N2a neuroblastoma cells (Lim et al. 2010). Quaternization of polyamine dendrimers increases charge density and reduces toxicity from the cationic surface groups. These quaternized polyamines exhibited potent antiprion activity through inhibiting conversion of normal prion protein into PrP^{Sc}; however, efficiency of inducing prion degradation was reduced due to quaternization (Lim et al. 2010). Other studies showed that dendrimers can completely clear prion proteins in cell culture. Greater positive charge on the dendrimer surface, higher generation number, and lower pH correlate with effectiveness (Cordes et al. 2007). One in vivo study also demonstrated the ability of phosphorus-containing dendrimers to decrease PrP^{Sc} by 80% in the spleen of scrapie-infected mice (Solassol et al. 2004). Dendrimers were also shown to hold promise as a treatment for Alzheimer's disease. G3–G5 PAMAM dendrimers disrupted amyloid fibrils and inhibited their aggregation in an in vitro model (Klajnert et al. 2006). This activity increased with increasing generation of dendrimers (Klajnert et al. 2006). Sialic acid–modified dendrimers were shown to sequester amyloid in solution and reduce its toxic effect on human neuroblastoma cells in culture, at much lower concentrations than sialic acid alone (Patel et al. 2006). Increasing size and generation number of the dendrimer increased binding affinity to amyloid and decreased toxicity (Patel et al. 2006). Attaching sialic acid to dendrimers through hydroxyl groups decreased amyloid toxicity at much lower concentrations than attaching sialic acid through carboxylic groups (Patel et al. 2007).

Peripheral Nerve Regeneration

The potential use of dendrimers in promoting nerve regeneration is supported by a study demonstrating the utility of YIGSR, a cell adhesion peptide conjugated to the surface of G2 PPI dendrimer–collagen hydrogel through covalent attachment, in promoting growth and neurite extension from chicken embryo dorsal root ganglia cells (Duan et al. 2007). The YIGSR–dendrimer–collagen complex also promoted the proliferation and adhesion of corneal epithelial cells on collagen hydrogel in an in vitro model of human corneal epithelial cells (Duan et al. 2007). In other studies, YIGSR promoted epithelial and neuronal growth in in vitro experiments; these implants into pigs' corneas display regeneration of the host's corneal epithelium and permit functional regeneration of nerves (Li et al. 2004; Suuronen et al. 2004). Magnaghi et al. (2011) examined PAMAM polymer hydrogels in promoting peripheral nerve regeneration. The authors discuss a method for creating hydrogel tubing from an amphoteric-based polyamidoamine polymer (AGMA1) carrying guanidine groups and tested it in vivo in rats. Rats were divided into two groups: one with sciatic nerves transected, inverted, and autografted (group 1); and another with sciatic nerves transected and tubing implanted for guiding regeneration (group 2). Nerves were both visually and morphologically analyzed through microscopy at 30, 90, and 180 days for any interruptions or neuromas. Morphological features such as fiber density, axon size, fiber and axon area, and myelin thickness increased progressively over time in both groups but significantly more in group 2 than group 1. Group 2 rats also had better behavioral and functional recovery, with almost complete behavioral and functional recovery of the sciatic nerve. These hydrogels were biocompatible, had good mechanical strength, and exhibited minimal immunogenicity, although their biodegradation was prolonged. These reports by Duan and Magnaghi show the potential of PPI dendrimers and PAMAM polymers to create biocompatible scaffolds that promote nerve regeneration.

CONCLUSIONS

Scientists can create almost limitless varieties and configurations of a dendrimer structure. Dendrimer synthesis can be controlled to produce dendrimer molecules of a particular size and M_w that are well suited to medical applications. Subsequent addition of functional groups to the dendrimer can introduce desired properties including fluorescence, targeting cell-surface receptors, and facilitating transport across the BBB. The interior structure of the dendrimer can carry chemotherapeutic agents. Evaluation of dendrimers for applications in neurosurgery and neuroscience has been limited to research studies in both in vitro and in vivo models. Clinical trials in other medical fields are underway, such as the topical use of a dendrimer (VivaGel, StarPharma, Melbourne, Australia) to prevent spread of HIV infection. Several concerns pertaining to dendrimers are limiting their translation to clinical studies, including (1) expense involved in manufacturing dendrimers due to the control and fine-tuning required at each individual step of the reaction; (2) toxicity of dendrimers that has been addressed in multiple studies by reducing the surface cationic charge, albeit with some loss of functionality; (3) limited biodegradability of dendrimers resulting in prolonged retention times; (4) significant concerns about toxicity resulting from release of functional groups at ectopic locations; and (5) uncontrolled release of active agents that may either limit usefulness due to slow release or increase cytotoxicity due to rapid release. A fine balance is required between several factors: M_w, diameter, generation number, cationic charge, renal clearance, $T_{1/2}$ in circulation, and uptake by the RES. It remains to be seen how dendrimers will compare to other existing nanoparticle constructs that are in preclinical studies, in clinical trials, or approved by the Food and Drug Administration (FDA). Experiments such as in vitro and in vivo studies of dendrimer-conjugated temozolomide against glioma demonstrated that dendrimer-conjugated drugs can be more effective than commercially available agents. The ultimate usefulness of dendrimer-based agents will not be established until they are tested and shown to be effective in clinical trials.

ABBREVIATIONS

ACPP: Activated cell-penetrating peptides
AO: Antisense oligonucleotides
BNCT: Boron neutron capture therapy
C225: Cetuximab
CTX: Chlorotoxin
DOTA: 1,4,7,10-tetraazacyclododecane-1,4,7,10-tetraacetic acid
DOX: Doxorubicin
DTPA: Diethylene triamine pentaacetic acid
EGFR: Epidermal growth factor receptor
EPR: Enhanced permeability and retention
FITC: Fluorescein isothiocyanate
G#: Generation number (1, 3, 7, etc.)
Gd: Gadolinium
miRNA: micro RNA
MTX: Methotrexate
M_w: Molecular weight
NSF: Nephrogenic systemic fibrosis
PAMAM: Polyamidoamine
PEPE: Polyether–copolyester
PPI: Polypropylenimine
PrP^{Sc}: Scrapie prion protein
RES: Reticuloendothelial system
siRNA: Small interfering RNA
$T_{1/2}$: Half-life

REFERENCES

Asangani, I.A., Rasheed, S.A., Nikolova, D.A. et al. 2008. MicroRNA-21 (miR-21) post-transcriptionally downregulates tumor suppressor Pdcd4 and stimulates invasion, intravasation and metastasis in colorectal cancer. *Oncogene* 27: 2128–36.

Barth, R.F., Adams, D.M., Soloway, A.H. et al. 1994. Boronated starburst dendrimer–monoclonal antibody immunoconjugates: evaluation as a potential delivery system for neutron capture therapy. *Bioconjug Chem* 5: 58–66.

Barth, R.F., Soloway, A.H., Fairchild, R.G. et al. 1992. Boron neutron capture therapy for cancer: realities and prospects. *Cancer* 70: 2995–3007.

Benazeth, S., Purans, J., Chalbot, M.C. et al. 1998. Temperature and pH dependence XAFS study of Gd(DOTA)(-) and Gd(DTPA)(2)(-) complexes: solid state and solution structures. *Inorg Chem* 37: 3667–74.

Bi, X., Shi, X., Majoros, I.J. et al. 2007. Multifunctional poly(amidoamine) dendrimer–taxol conjugates: synthesis, characterization and stability. *Comput Theor Nanosci* 4: 1179–87.

Bosman, A.W., Janssen, H.M., and Meijer, E.W. 1999. About dendrimers: structure, physical properties, and applications. *Chem Rev* 99: 1665–88.

Bousquet, J.C., Saini, S., Stark, D.D. et al. 1988. Gd–DOTA: characterization of a new paramagnetic complex. *Radiology* 166: 693–8.

Bowen, S., Tare, N., Inoue, T. et al. 1999. Relationship between molecular mass and duration of activity of polyethylene glycol conjugated granulocyte colony-stimulating factor mutein. *Exp Hematol* 27: 425–32.

Buhleier, E., Wehner, W., and Vogtle, F. 1978. Cascade-chain-like and nonskid-chain-like syntheses of molecular cavity topologies. *Synthesis (Stuttg)* 2: 155–8.

Caminati, G., Turro, N.J., and Tomalia, D.A. 1990. Photophysical investigation of starburst dendrimers and their interactions with anionic and cationic surfactants. *J Am Chem Soc* 112: 8515–22.

Capala, J., Barth, R.F., Bendayan, M.Ø. et al. 1996. Boronated epidermal growth factor as a potential targeting agent for boron neutron capture therapy of brain tumors. *Bioconjug Chem* 7: 7–15.

Chai, M.H., Niu, Y.H., Youngs, W.J. et al. 2001. Structure and conformation of DAB dendrimers in solution via multidimensional NMR techniques. *J Am Chem Soc* 123: 4670–8.

Cordes, H., Boas, U., Olsen, P. et al. 2007. Guanidino- and urea-modified dendrimers as potent solubilizers of misfolded prion protein aggregates under non-cytotoxic conditions. Dependence on dendrimer generation and surface charge. *Biomacromolecules* 8: 3578–83.

Degennes, P.G., and Hervet, H. 1983. Statistics of starburst polymers. *J Physique Lett* 44: L351–L60.

Dhanikula, R.S., Argaw, A., Bouchard, J.-F. et al. 2008. Methotrexate loaded polyether–copolyester dendrimers for the treatment of gliomas: enhanced efficacy and intratumoral transport capability. *Mol Pharm* 5: 105–16.

Dhanikula, R.S., Hammady, T., and Hildgen, P. 2009. On the mechanism and dynamics of uptake and permeation of polyether–copolyester dendrimers across an in vitro blood–brain barrier model. *J Pharm Sci* 98: 3748–60.

Duan, X.D., McLaughlin, C., Griffith, M. et al. 2007. Biofunctionalization of collagen for improved biological response: scaffolds for corneal tissue engineering. *Biomaterials* 28: 78–88.

Fischer, M., and Vogtle, F. 1999. Dendrimers: from design to application—a progress report. *Angew Chem (Int Ed Engl)* 38: 885–905.

Fox, M.E., Szoka, F.C., and Frechet, J.M.J. 2009. Soluble polymer carriers for the treatment of cancer: the importance of molecular architecture. *Acc Chem Res* 42: 1141–51.

Franc, G., and Kakkar, A. 2008. Dendrimer design using Cu^I-catalyzed alkyne-azide "click-chemistry." *Chem Commun* 2008: 5267–76.

Franc, G., and Kakkar, A.K. 2009. Diels–Alder "click" chemistry in designing dendritic macromolecules. *Chem Eur J* 15: 5630–9.

Gillies, E.R., and Frechet, J.M. 2005. Dendrimers and dendritic polymers in drug delivery. *Drug Discov Today* 10: 35–43.

Gillies, E.R., Jonsson, T.B., and Frechet, J.M. 2004. Stimuli-responsive supramolecular assemblies of linear-dendritic copolymers. *J Am Chem Soc* 126: 11936–43.

Gopin, A., Ebner, S., Attali, B. et al. 2006. Enzymatic activation of second-generation dendritic prodrugs: conjugation of self-immolative dendrimers with poly(ethylene glycol) via click chemistry. *Bioconjug Chem* 17: 1432–40.

Grayson, S.M., and Frechet, J.M. 2001. Convergent dendrons and dendrimers: from synthesis to applications. *Chem Rev* 101: 3819–68.

Han, L., Zhang, A., Wang, H. et al. 2010. Tat–BMPs–PAMAM conjugates enhance therapeutic effect of small interference RNA on U251 glioma cells in vitro and in vivo. *Hum Gene Ther* 21: 417–26.

Hatanaka, H. 1975. A revised boron-neutron capture therapy for malignant brain tumors. II. Interim clinical result with the patients excluding previous treatments. *J Neurol* 209: 81–94.

Hawker, C.J., and Frechet, J.M.J. 1990. Preparation of polymers with controlled molecular architecture—a new convergent approach to dendritic macromolecules. *J Am Chem Soc* 112: 7638–47.

Huang, R., Ke, W., Han, L. et al. 2011a. Targeted delivery of chlorotoxin-modified DNA-loaded nanoparticles to glioma via intravenous administration. *Biomaterials* 32: 2399–406.

Huang, R.Q., Ke, W.L., Liu, Y. et al. 2008. The use of lactoferrin as a ligand for targeting the polyamidoamine-based gene delivery system to the brain. *Biomaterials* 29: 238–46.

Huang, R.-Q., Qu, Y.-H., Ke, W.-L. et al. 2007. Efficient gene delivery targeted to the brain using a transferrin-conjugated polyethyleneglycol-modified polyamidoamine dendrimer. *FASEB J* 21: 1117–25.

Huang, S., Li, J., Han, L. et al. 2011b. Dual targeting effect of Angiopep-2-modified, DNA-loaded nanoparticles for glioma. *Biomaterials* 32: 6832–8.

Jimenez, J.L., Clemente, M.I., Weber, N.D. et al. 2010. Carbosilane dendrimers to transfect human astrocytes with small interfering RNA targeting human immunodeficiency virus. *BioDrugs* 24: 331–43.

Kaminskas, L.M., Boyd, B.J., Karellas, P. et al. 2008. The impact of molecular weight and PEG chain length on the systemic pharmacokinetics of PEGylated poly L-lysine dendrimers. *Mol Pharm* 5: 449–63.

Kang, C., Yuan, X., Li, F. et al. 2010. Evaluation of folate–PAMAM for the delivery of antisense oligonucleotides to rat C6 glioma cells in vitro and in vivo. *J Biomed Mater Res A* 93: 585–94.

Ke, W.L., Shao, K., Huang, R.Q. et al. 2009. Gene delivery targeted to the brain using an Angiopep-conjugated polyethyleneglycol-modified polyamidoamine dendrimer. *Biomaterials* 30: 6976–85.

Khawaja, A.M. 2011. The legacy of nanotechnology: revolution and prospects in neurosurgery. *Int J Surg* 9: 608–14.

Killops, K.L., Campos, L.M., and Hawker, C.J. 2008. Robust, efficient, and orthogonal synthesis of dendrimers via thiol-ene "click" chemistry. *J Am Chem Soc* 130: 5062–4.

Klajnert, B., and Bryszewska, M. 2001. Dendrimers: properties and applications. *Acta Biochim Pol* 48: 199–208.

Klajnert, B., Cortijo-Arellano, M., Cladera, J. et al. 2006. Influence of dendrimer's structure on its activity against amyloid fibril formation. *Biochem Biophys Res Commun* 345: 21–8.

Kobayashi, H., Kawamoto, S., Jo, S.K. et al. 2003. Macromolecular MRI contrast agents with small dendrimers: pharmacokinetic differences between sizes and cores. *Bioconjug Chem* 14: 388–94.

Kobayashi, H., Kawamoto, S., Saga, T. et al. 2001a. Novel liver macromolecular MR contrast agent with a polypropylenimine diaminobutyl dendrimer core: comparison to the vascular MR contrast agent with the polyamidoamine dendrimer core. *Magn Reson Med* 46: 795–802.

Kobayashi, H., Sato, N., Hiraga, A. et al. 2001b. 3D-micro-MR angiography of mice using macromolecular MR contrast agents with polyamidoamine dendrimer core with reference to their pharmacokinetic properties. *Magn Reson Med* 45: 454–60.

Kobayashi, H., Sato, N., Saga, T. et al. 2000. Monoclonal antibody–dendrimer conjugates enable radiolabeling of antibody with markedly high specific activity with minimal loss of immunoreactivity. *Eur J Nucl Med* 27: 1334–9.

Kojima, C., Kono, K., Maruyama, K. et al. 2000. Synthesis of polyamidoamine dendrimers having poly(ethylene glycol) grafts and their ability to encapsulate anticancer drugs. *Bioconjug Chem* 11: 910–17.

Kolb, H.C., Finn, M.G., and Sharpless, K.B. 2001. Click chemistry: diverse chemical function from a few good reactions. *Angew Chem (Int Ed Engl)* 40: 2004–21.

Konda, S.D., Wang, S., Brechbiel, M. et al. 2002. Biodistribution of a 153 Gd-folate dendrimer, generation = 4, in mice with folate-receptor positive and negative ovarian tumor xenografts. *Invest Radiol* 37: 199–204.

Kono, K., Liu, M., and Frechet, J.M. 1999. Design of dendritic macromolecules containing folate or methotrexate residues. *Bioconjug Chem* 10: 1115–21.

Lee, C.C., Gillies, E.R., Fox, M.E. et al. 2006. A single dose of doxorubicin-functionalized bow-tie dendrimer cures mice bearing C-26 colon carcinomas. *Proc Natl Acad Sci* 103: 16649–54.

Lee, I., Athey, B.D., Wetzel, A.W. et al. 2002. Structural molecular dynamics studies on polyamidoamine dendrimers for a therapeutic application: effects of pH and generation. *Macromolecules* 35: 4510–20.

Li, F., Carlsson, D., Lohmann, C. et al. 2004. Cellular and nerve regeneration within a biosynthetic extracellular matrix for corneal transplantation. *Am J Ophthalmol* 138: 318.

Li, H., and Qian, Z.M. 2002. Transferrin/transferrin receptor-mediated drug delivery. *Med Res Rev* 22: 225–50.

Li, Y., He, H., Jia, X. et al. 2012. A dual-targeting nanocarrier based on poly(amidoamine) dendrimers conjugated with transferrin and tamoxifen for treating brain gliomas. *Biomaterials* 33: 3899–908.

Lim, J., Guo, Y., Rostollan, C.L. et al. 2008. The role of the size and number of polyethylene glycol chains in the biodistribution and tumor localization of triazine dendrimers. *Mol Pharm* 5: 540–7.

Lim, Y.B., Mays, C.E., Kim, Y. et al. 2010. The inhibition of prions through blocking prion conversion by permanently charged branched polyamines of low cytotoxicity. *Biomaterials* 31: 2025–33.

Liu, L., Barth, R.F., Adams, D.M. et al. 1995. Bispecific antibodies as targeting agents for boron neutron capture therapy of brain tumors. *J Hematother* 4: 477–83.

Liu, M., Kono, K., and Frechet, J.M. 2000. Water-soluble dendritic unimolecular micelles: their potential as drug delivery agents. *J Control Release* 65: 121–31.

Liu, Y., Huang, R., Han, L. et al. 2009. Brain-targeting gene delivery and cellular internalization mechanisms for modified rabies virus glycoprotein RVG29 nanoparticles. *Biomaterials* 30: 4195–202.

Magnaghi, V., Conte, V., Procacci, P. et al. 2011. Biological performance of a novel biodegradable polyamidoamine hydrogel as guide for peripheral nerve regeneration. *J Biomed Mater Res A* 98A: 19–30.

Majoros, I.N.J., Thomas, T.P., Mehta, C.B. et al. 2005. Poly(amidoamine) dendrimer-based multifunctional engineered nanodevice for cancer therapy. *J Med Chem* 48: 5892–9.

Mammen, M., Choi, S.-K., and Whitesides, G.M. 1998. Polyvalent interactions in biological systems: implications for design and use of multivalent ligands and inhibitors. *Angew Chem (Int Ed Engld)* 37: 2754–94.

Meyers, S.R., Juhn, F.S., Griset, A.P. et al. 2008. Anionic amphiphilic dendrimers as antibacterial agents. *J Am Chem Soc* 130: 14444–5.

Mintzer, M.A., and Grinstaff, M.W. 2011. Biomedical applications of dendrimers: a tutorial. *Chem Soc Rev* 40: 173–90.

Morille, M., Passirani, C., Vonarbourg, A. et al. 2008. Progress in developing cationic vectors for non-viral systemic gene therapy against cancer. *Biomaterials* 29: 3477–96.

Nakagawa, Y., Pooh, K., Kobayashi, T. et al. 2003. Clinical review of the Japanese experience with boron neutron capture therapy and a proposed strategy using epithermal neutron beams. *J Neurooncol* 62: 87–99.

Newkome, G.R., and Shreiner, C.D. 2008. Poly(amidoamine), polypropylenimine, and related dendrimers and dendrons possessing different 1 -> 2 branching motifs: an overview of the divergent procedures. *Polymer* 49: 1–173.

Newkome, G.R., Yao, Z.Q., Baker, G.R. et al. 1985. Cascade molecules—a new approach to micelles. *Abst Pap Am Chem Soc* 189: 166–ORGN.

Nwe, K., Bryant, L.H., Jr., and Brechbiel, M.W. 2010. Poly(amidoamine) dendrimer based MRI contrast agents exhibiting enhanced relaxivities derived via metal preligation techniques. *Bioconjug Chem* 21: 1014–7.

Olson, E.S., Jiang, T., Aguilera, T.A. et al. 2010. Activatable cell penetrating peptides linked to nanoparticles as dual probes for in vivo fluorescence and MR imaging of proteases. *Proc Natl Acad Sci* 107: 4311–6.

Owens, D.E., 3rd, and Peppas, N.A. 2006. Opsonization, biodistribution, and pharmacokinetics of polymeric nanoparticles. *Int J Pharm* 307: 93–102.

Patel, D., Henry, J., and Good, T. 2006. Attenuation of beta-amyloid induced toxicity by sialic acid-conjugated dendrimeric polymers. *Biochim Biophys Acta* 1760: 1802–9.

Patel, D.A., Henry, J.E., and Good, T.A. 2007. Attenuation of beta-amyloid-induced toxicity by sialic-acid-conjugated dendrimers: role of sialic acid attachment. *Brain Res* 1161: 95–105.

Patri, A.K., Kukowska-Latallo, J.F., and Baker, J.R., Jr. 2005. Targeted drug delivery with dendrimers: comparison of the release kinetics of covalently conjugated drug and non-covalent drug inclusion complex. *Adv Drug Deliv Rev* 57: 2203–14.

Perez, A.P., Cosaka, M.L., Romero, E.L. et al. 2011. Uptake and intracellular traffic of siRNA dendriplexes in glioblastoma cells and macrophages. *Int J Nanomedicine* 6: 2715–28.

Perez, A.P., Romero, E.L., and Morilla, M.J. 2009. Ethylendiamine core PAMAM dendrimers/siRNA complexes as in vitro silencing agents. *Int J Pharm* 380: 189–200.

Quintana, A., Raczka, E., Piehler, L. et al. 2002. Design and function of a dendrimer-based therapeutic nanodevice targeted to tumor cells through the folate receptor. *Pharm Res* 19: 1310–6.

Ren, Y., Kang, C.S., Yuan, X.B. et al. 2010a. Co-delivery of as-miR-21 and 5-FU by poly(amidoamine) dendrimer attenuates human glioma cell growth in vitro. *J Biomater Sci Polym Ed* 21: 303–14.

Ren, Y., Zhou, X., Mei, M. et al. 2010b. MicroRNA-21 inhibitor sensitizes human glioblastoma cells U251 (PTEN-mutant) and LN229 (PTEN-wild type) to taxol. *BMC Cancer* 10: 27.

Sarin, H., Kanevsky, A.S., Wu, H.T. et al. 2008. Effective transvascular delivery of nanoparticles across the blood–brain tumor barrier into malignant glioma cells. *J Transl Med* 6: 80.

Sato, N., Kobayashi, H., Hiraga, A. et al. 2001. Pharmacokinetics and enhancement patterns of macromolecular MR contrast agents with various sizes of polyamidoamine dendrimer cores. *Magn Reson Med* 46: 1169–73.

Shao, K., Huang, R., Li, J. et al. 2010. Angiopep-2 modified PE–PEG based polymeric micelles for amphotericin B delivery targeted to the brain. *J Control Release* 147: 118–26.

Solassol, J.R.M., Crozet, C., Perrier, V.R. et al. 2004. Cationic phosphorus-containing dendrimers reduce prion replication both in cell culture and in mice infected with scrapie. *J Gen Virol* 85: 1791–9.

Suuronen, E.J., Nakamura, M., Watsky, M.A. et al. 2004. Innervated human corneal equivalents as in vitro models for nerve-target cell interactions. *FASEB J* 18: 170–2.

Swanson, S.D., Kukowska-Latallo, J.F., Patri, A.K. et al. 2008. Targeted gadolinium-loaded dendrimer nanoparticles for tumor-specific magnetic resonance contrast enhancement. *Int J Nanomedicine* 3: 201–10.

Tabata, Y., Murakami, Y., and Ikada, Y. 1998. Tumor accumulation of poly(vinyl alcohol) of different sizes after intravenous injection. *J Control Release* 50: 123–33.

Tomalia, D.A., Baker, H., Dewald, J. et al. 1985. A new class of polymers—starburst-dendritic macromolecules. *Polym J* 17: 117–32.

Uzgiris, E.E., Cline, H., Moasser, B. et al. 2004. Conformation and structure of polymeric contrast agents for medical imaging. *Biomacromolecules* 5: 54–61.

Waite, C.L., and Roth, C.M. 2009. PAMAM–RGD conjugates enhance siRNA delivery through a multicellular spheroid model of malignant glioma. *Bioconjug Chem* 20: 1908–16.

Waite, C.L., Sparks, S.M., Uhrich, K.E. et al. 2009. Acetylation of PAMAM dendrimers for cellular delivery of siRNA. *BMC Biotechnol* 9: 38.

Wang, D.J., and Imae, T. 2004. Fluorescence emission from dendrimers and its pH dependence. *J Am Chem Soc* 126: 13204–5.

Wang, X., Jin, T., Comblin, V. et al. 1992. A kinetic investigation of the lanthanide DOTA chelates. Stability and rates of formation and of dissociation of a macrocyclic gadolinium(III) polyaza polycarboxylic MRI contrast agent. *Inorg Chem* 31: 1095–9.

Wu, G., Barth, R.F., Yang, W. et al. 2003. Site-specific conjugation of boron-containing dendrimers to anti-EGF receptor monoclonal antibody cetuximab (IMC-C225) and its evaluation as a potential delivery agent for neutron capture therapy. *Bioconjug Chem* 15: 185–94.

Wu, G., Barth, R.F., Yang, W. et al. 2006. Targeted delivery of methotrexate to epidermal growth factor receptor-positive brain tumors by means of cetuximab (IMC-C225) dendrimer bioconjugates. *Mol Cancer Ther* 5: 52–9.

Yang, W., Barth, R.F., Adams, D.M. et al. 1997. Intratumoral delivery of boronated epidermal growth factor for neutron capture therapy of brain tumors. *Cancer Res* 57: 4333–9.

Yu, J., Zhao, H., Ye, L. et al. 2009. Effect of surface functionality of magnetic silica nanoparticles on the cellular uptake by glioma cells in vitro. *J Mater Chem* 19: 1265–70.

Zhiming, L., Peng, H., Rong, H. et al. 2009. Aptamer-conjugated dendrimer-modified quantum dots for glioblastoma cells imaging. *J Phys Conf Ser* 188: 012032.

Zhu, S., Hong, M., Zhang, L. et al. 2010. PEGylated PAMAM dendrimer–doxorubicin conjugates: in vitro evaluation and in vivo tumor accumulation. *Pharm Res* 27: 161–74.

9 Ligand-Based Targeting of Therapeutic and Imaging Agents for Cancer

Qimeng Quan, Seulki Lee, and Xiaoyuan Chen

CONTENTS

INTRODUCTION

The diagnosis and treatment of cancer remain a challenge despite recent improvements in surgery and systemic therapy. Molecular imaging enables visualization of specific molecular events that play key roles in disease processes, facilitating earlier diagnosis as well as monitoring therapeutic responses. Various imaging modalities, including positron emission tomography (PET), single photon emission computed tomography (SPECT), optical fluorescence imaging, magnetic resonance imaging (MRI), computed tomography, and ultrasound imaging, are all successfully employed in the field of molecular imaging (Cai and Chen 2008). Specific imaging relies substantially on contrast agents. Therefore, a variety of sophisticated targeting ligands (i.e., small molecules, peptides or peptidomimetics, proteins, and antibodies or their immunospecific fragments) have been surface functionalized with different types of imaging moieties (i.e., radioisotopes, fluorophores, and nanoparticles) that can specifically target cancerous cells and angiogenesis-associated tumor endothelial cells. Such novel designs have led to extensive interest in their potential for in vivo molecular imaging of cancer (Lee et al. 2010) and for cancer therapy.

Conventional therapeutics have several drawbacks, such as the limited ability of drug delivery to tumor tissues, requirement of high doses, undesirable cytotoxicity, drug resistance, and nonspecific targeting. Drug therapies for brain cancer have been particularly ineffective due to the blood–brain barrier (BBB) (Begley 2003; Orive et al. 2010; Yang 2010). Some ligand-based targeting probes have emerged as potential vectors for brain targeting and delivery. This platform can overcome the problems of conventional technologies and can enable tumor-specific detection, treatment, and follow-up monitoring (Das et al. 2009; Fang and Zhang 2010; Hsu et al. 2007; Rybak et al. 2007). Rapid expansion in the nanotechnology field has led to the development of an array of nanoparticle (NP)-based targeting contrast agents for application in therapy and imaging.

Design of ligand-based conjugates relies on awareness of the various molecules, so-called homing ligands that bind to the biological target of interest. These include any molecules that recognize and bind to target antigens or receptors overexpressed or selectively expressed by particular cells or tissue components. An ideal ligand for targeting and delivery would be one with high affinity and specificity of binding to the receptors. In addition, a few more characteristics are needed to improve targeting, such as the following: (1) stimulation of cellular internalization for the conjugated carrier or moiety to perform its intracellular action; (2) chemical modification capability for conjugation; (3) nontoxicity; and (4) ability for large-scale manufacture (Lu et al. 2006). Targeting ligands can significantly impact drug delivery. Agents with targeting ligands are more effective than unconjugated ones because drugs from the former group can reach their target site more efficiently (Hahn et al. 2011). A targeted drug delivery system is expected to be nontoxic to normal tissue. Numerous nanoparticle-based drug delivery systems have been developed or are currently under development. Nanoparticles are typically functionalized with biocompatible small molecules or polymers providing flexibility to add ligands, drugs, or imaging moieties as diagnostic tools, therapeutic carriers, or both (Orive et al. 2010). Such multifunctional nanoparticles have been demonstrated in vitro and in vivo to improve cell-specific binding, chemotherapy, and gene transfection. With size dimensions on the nanoscale, these particles can navigate through microvasculature and across various biological barriers to reach target tissues (Fang and Zhang 2010; Nel et al. 2009).

In this chapter, we describe several ligand–target interactions widely used for specific tumor imaging and therapy. Recent advances in their application to molecular imaging of the brain are also briefly reviewed. We focus, in particular, on multifunctional isotope-labeled probes, nanoprobes, and fluorophore probes with combined diagnostic and therapeutic capabilities. Multimodal nanoprobes that can be simultaneously analyzed using a combination of MRI and PET or SPECT, or MRI and optical imaging techniques such as near-infrared fluorescence (NIRF), are also highlighted.

LIGAND-BASED CONJUGATES FOR TUMOR IMAGING AND THERAPY

Peptide-Based Conjugates

Ideal imaging contrast agents would exhibit high binding affinity for the target, specific uptake and retention in the target, rapid clearance from nontarget tissue, adequate capillary permeability, high stability and integrity in physiological conditions, and safety. Peptide or peptidomimetic-based imaging probes have been increasingly studied in the molecular imaging field, given their distinctive advantages over other small molecules or macromolecules. Peptides can be directly or indirectly labeled with imaging moieties to provide or augment the imaging signal while maintaining their biological activity against specific receptors. Molecular modification techniques permit synthesis of a variety of bioactive peptides with chelating groups, without compromising biological properties. Various techniques have been developed that allow efficient and site-specific labeling of peptides with clinically useful radionuclides such as ^{99m}Tc, ^{123}I, ^{111}In, and ^{18}F (Okarvi 2004). Organic near-infrared (NIR) fluorophores or quantum dots (QDs) for optical imaging and magnetic nanoparticles (iron oxide nanoparticles [IONPs]) for MRI can be conjugated to peptides via organic linkers, macrocyclic or branched chelators, polymers, or nanoplatforms by well-established bioconjugation techniques (Frangioni 2003; Lee et al. 2010; Reubi and Maecke 2008). Advances in molecular cancer biology have elucidated an increasing number of tumor receptors. To date, various receptors including somatostatin (SST), integrin, gastrin-releasing peptide (GRP), cholecystokinin (CCK), R-melanocyte stimulating hormone (R-MSH), and glucagon-like peptide-1 (GLP-1) have been identified and characterized for tumor receptor imaging (Cai et al. 2008; Miao and Quinn 2007; Reubi 2003, 2007). Specific peptide molecules for these receptors represent a class of perfect ligands for cancer targeting and have accelerated interest in developing peptide-based imaging probes (Aloj and Morelli 2004; Okarvi 2004).

SSTs, also known as somatotropin release inhibiting factors (SRIFs), are a family of cyclopeptides that have inhibitory effects on the secretion of hormones such as growth hormone, insulin, and glucagon (Weckbecker et al. 2003). There are two naturally occurring active SSTs: the tetradecapeptide, SST-14, and its *N*-terminally extended form, SST-28. Five different subtypes of G-protein-coupled human SST receptors (SSTrs, SSTr1-SSTr5) have been identified (Susini and Buscail 2006). The role of SST as a neurotransmitter and neuromodulator has been addressed in electrophysiological studies (Hannon et al. 2002). The rationale for the therapeutic use of SST analogs is largely based on the expression of SSTrs in relevant normal and tumor tissues (Froidevaux and Eberle 2002; Reubi and Waser 2003). A high concentration of SSTrs can also be detected in the brain, and these are targets for developing centrally acting agents. Most tumor tissues preferentially express SSTr2; less frequently SSTr1, 3, and 5; and rarely SSTr4 (Reubi et al. 2001).

SSTrs represent one of the best examples of receptor-targeted ligand-based radiolabeled probes. A large library of SST analogs with improved metabolic stability and biological activity has been developed to overcome the therapeutic limitation of the native form of SSTs, which have a short plasma half-life of less than 3 min (Patel and Wheatley 1983). The most well-investigated peptide is the cyclic octapeptide octreotide (DPhe^1-Cys^2-Phe^3-D-Trp^4-Lys^5-Thr^6-Cys^7-Thr^8-ol) and its analogs. Octreotide is more stable than native SSTs (half-life: 90–110 min vs. <3 min).

The first Food and Drug Administration (FDA)-approved agent was [^{111}In-diethyene triamine pentaacetic acid (DTPA)]-octreotide (OctreoScan), which was originally designed for scintigraphy of neuroendocrine tumors. Octreotide analogs with a higher affinity for SSTr2 have been discovered and labeled with a 1,4,7,10-tetraazacyclododecane-N,N′,N″,N‴-tetraacetate (DOTA) chelator, which chelates ^{111}In as well as the β-emitting nuclide ^{90}Y. The resulting compound, [^{90}Y-DOTA-D-Phe^1,Tyr^3]-octreotide, formed thermodynamically and kinetically stable complexes with metals, extending its therapeutic application (Stolz et al. 1998). SST analogs have also been designed for PET imaging. Different DOTA-octreotide analogs, including [^{68}Ga-DOTA]-TOC, [^{68}Ga-DOTA]-NOC, and [^{64}Cu-DOTA]-TATE, have been reported. Their efficacy has been further confirmed in various preclinical and clinical settings (Li et al. 2002). To improve pharmacokinetic profiles of the probe for clinical application, ^{18}F-labeled carbohydrate conjugates of tyrosine-octreotide (TOC) such as [^{18}F-FP-Gluc-Lys^0-Tyr^3]-octreotide were developed. Such analogs demonstrated improved resolution in PET imaging in SSTr-positive tumors and provided superior diagnostic performance (Meisetschlager et al. 2006; Schottelius et al. 2002). Different libraries of SST analogs that were labeled with fluorophores and tested in SSTr-positive tumor-bearing mice showed enhancement in tumor detection (Kostenich et al. 2005, 2008).

Integrins are a family of heterodimeric transmembrane glycoproteins consisting of two noncovalently associated receptor subunits, α and β (Horton 1997; Ruoslahti and Pierschbacher 1987). The $\alpha_v\beta_3$ integrin is highly expressed on activated and proliferating endothelial cells during tumor angiogenesis and metastasis (Cai et al. 2008). Integrin $\alpha_v\beta_3$ has attracted much attention as a target for diagnostic cancer imaging because of its overexpression in several tumor forms including melanoma, ovarian and lung carcinoma, neuroblastoma, glioblastoma, and breast cancer. The $\alpha_v\beta_3$ receptor, integrins, and several extracellular matrix proteins such as vitronectin contain the Arg-Gly-Asp (RGD) sequence. Additionally, a significant number of small linear and cyclic RGD peptide analogs strongly bind to the $\alpha_v\beta_3$ receptor (Horton 1997).

A series of active and selective cyclic RGD analogs containing the sequence of cyclo(Arg-Gly-Asp-D-Tyr-Lys), c(RGDyK), and cyclo(Arg-Gly-Asp-D-Phe-Lys), c(RGDfK), were designed by backbone cyclization. Two different ^{18}F-labeled RGD analogs, ^{18}F-galacto-c(RGDfK), which was developed by ^{18}F labeling of the RGD-containing glycopeptides, and ^{18}F-AH111585, an ^{18}F-radiolabeled small peptide containing the RGD sequence with binding affinities to $\alpha_v\beta_3$, $\alpha_v\beta_5$, and $\alpha_{IIb}\beta_3$ integrins, are under clinical trial in patients with melanoma, sarcoma, breast cancer, and glioblastoma (Beer et al. 2007, 2008; Haubner et al. 2001; Kenny et al. 2008; Morrison et al. 2009; Schnell et al. 2009). Several different types of ^{18}F-labeled RGD dimers, multivalent dimers, and tetramers of the cyclic RGD-based peptides have been developed and labeled with various radioisotopes, such as ^{64}Cu, ^{68}Ga, and ^{99m}Tc, to improve its pharmacokinetic profiles and binding affinity (Chen et al. 2004b,c; Liu et al. 2009b; Shi et al. 2008, 2009; Wu et al. 2007; Zhang et al. 2006). In particular, ^{18}F-FPRGD$_2$ and ^{18}F-FPPRGD$_2$ were approved by the FDA for the first-in-human studies (Lee et al. 2010). The NIR fluorophore molecule can be conjugated with RGD peptide analogs for highly sensitive, in vivo optical imaging. Cy5.5 dye was conjugated to the ε-amino group of the lysine residue of c(RGDyK) or c(RGDfK) and allowed tumor imaging with high contrast in the U87MG glioblastoma and KS167 Kaposi's sarcoma xenograft models (Chen et al. 2004a; Wang et al. 2004). Similarly, dimeric and tetrameric RGD peptide analogs were labeled with NIR fluorophores including Cy5.5 and Cy7 and provided increased receptor binding affinity and imaging efficacy compared to the monomeric compound (Cheng et al. 2005; Wu et al. 2006).

RGD-based nanoconjugates are one of the most investigated ligand-based nanoparticles applied in the molecular imaging field. Nanomaterials, including magnetic IONPs, QDs, and carbon nanotubes, were labeled with RGD peptide analogs and investigated using different imaging modalities such as MRI, optical imaging, and photoacoustic imaging. For example, [Gly-Ser-Ser-Lys(FITC)-Gly-Gly-Gly]-c(RGD) was chemically conjugated to the primary amines on the surface of amino cross-linked iron oxide (CLIO) nanoparticles and demonstrated integrin-mediated cellular uptake and tumor accumulation in BT-20 tumors in vivo (Zhang 2007). The c(RGDyK) peptide was also conjugated to 4-methylcatechol-coated ultrasmall IONPs (4.5 nm) via a Mannich reaction (Jun et al. 2005) and showed increased cellular uptake and enhanced MR signal in a U87MG tumor-bearing mouse model. For dual-modality imaging, an RGD peptide-based MRI/PET probe was generated by conjugating ^{64}Cu-DOTA and c(RGDyK) on the surface of polyaspartic acid-coated IONPs (Figure 9.1) (Lee et al. 2008). The labeled RGD analog successfully delivered ^{64}Cu-DOTA–labeled IONPs to U87MG tumors and enabled dual in vivo imaging. Such probes are designed to combine the high spatial and temporal resolution and tissue penetration of MRI with the high sensitivity of PET. Similarly, different peptide-based ligands were labeled onto IONPs and were studied for multimodal imaging in in vivo models (Chen et al. 2009; Veiseh et al. 2009; Yang et al. 2009). Besides the use of IONPs, RGD peptide-coated ^{64}Cu-labeled single-walled carbon nanotubes (SWCNTs) were prepared to target integrin-positive U87MG tumors in mice, and imaging was performed by Raman and PET imaging (Liu et al. 2007). Along with Raman characteristics, SWCNTs provide

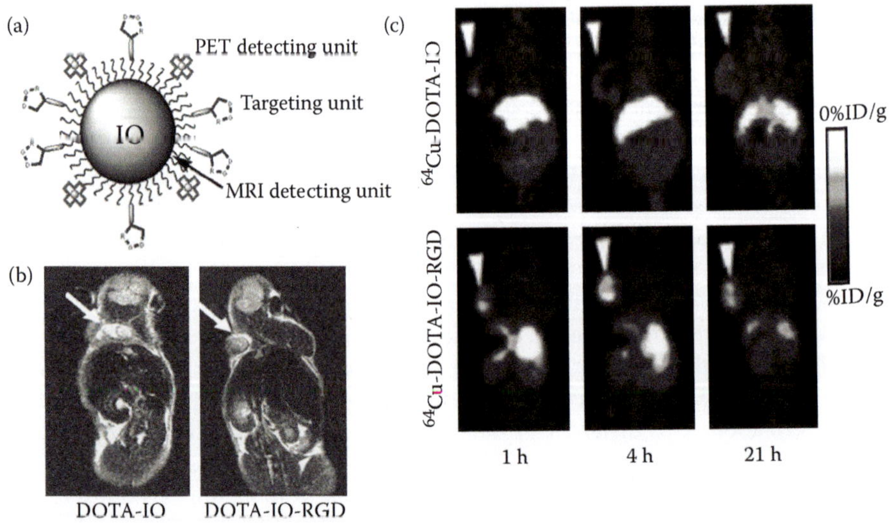

FIGURE 9.1 (a) Illustration of PET/MRI probe based on iron oxide (IO) nanoparticle. (b) T2-weighted MR images of nude mice bearing U87MG tumor at 4 h after tail-vein injection of DOTA-IO and DOTA-IO-RGD. (c) Decay-corrected whole-body coronal PET images of nude mouse bearing human U87MG tumor at 1, 4, and 21 h after injection of ^{64}Cu-DOTA-IO and ^{64}Cu-DOTA-IO-RGD. (Adapted from Lee, H.Y. et al., *J Nucl Med* 49: 1371–9, 2008. With permission.)

optical contrast with improved spatial resolution and tissue penetration by strong photoacoustic signals in a dose-dependent manner (Shan 2010).

Bombesin (BBN) is an amphiphilic homologue of mammalian GRP (pGlu^{1}-Gln^{2}-Arg^{3}-Leu^{4}-Gly^{5}-Asn^{6}-Gln^{7}-Trp^{8}-Ala^{9}-Val^{10}-Gly^{11}-His^{12}-Leu^{13}-Met^{14}-NH_2) that exhibits high affinity and specificity for the GRP receptor (GRPr) (Smith et al. 2005). A variety of cancers including small cell lung cancer and prostate, breast, pancreas, and gastrointestinal tumors were shown to overexpress GRPrs (Reubi 2003). Various BBN analogs based on their key amino acid sequence, BBN$^{7-14}$, were screened and developed as imaging probes to improve the circulation half-life of BBN. BBN analogs were developed to provide flexibility for both SPECT and PET applications. ^{99m}Tc-labeled BBN$^{7-14}$ was evaluated in mice with PC-3 tumors and in breast cancer patients (Van de Wiele et al. 2001, 2008). [^{111}In-DOTA-11-Aun]-BBN$^{7-14}$, [^{68}Ga-DOTA-PEG$_4$]-BBN$^{7-14}$, [^{64}Cu-DOTA-Pro^{1},Tyr^{4}]-BBN, [^{18}F-Lys^{3}]-BBN, and [^{64}Cu-NOTA-8-Aoc]-BBN$^{7-14}$ were developed, and their efficiency was successfully confirmed in vivo (Begley 2003; Biddlecombe et al. 2007; Hoffman et al. 2003; Prasanphanich et al. 2007; Zhang et al. 2007). Potent BBN agonists and antagonists including Demobesin-1 and Demobesin-4 were designed and characterized both in vitro and in vivo by PET imaging (Cescato et al. 2008; Chopra 2010a). As an alternative to using single or homogeneous multivalent peptides, peptide ligands that recognize both GRPr and integrin receptors were designed as a dual-receptor-targeting ligand, which significantly improved tumor targeting efficacy and pharmacokinetics in prostate and breast cancer models. For instance, a BBN-RGD heterodimer was synthesized by coupling BBN$^{7-14}$ and c(RGDyK) peptides and labeled with various radioisotopes, such as ^{18}F, ^{64}Cu, and ^{68}Ga along with a PEG$_3$ linker or chelating agents including DOTA and 1,4,7-triazacyclononane-N,N′,N″-triacetic acid (NOTA) (Figure 9.2) (Li et al. 2008; Liu et al. 2009a,c,d,e). Application of BBN peptide-based probes is considered to have a broad spectrum of specificity for various tumors.

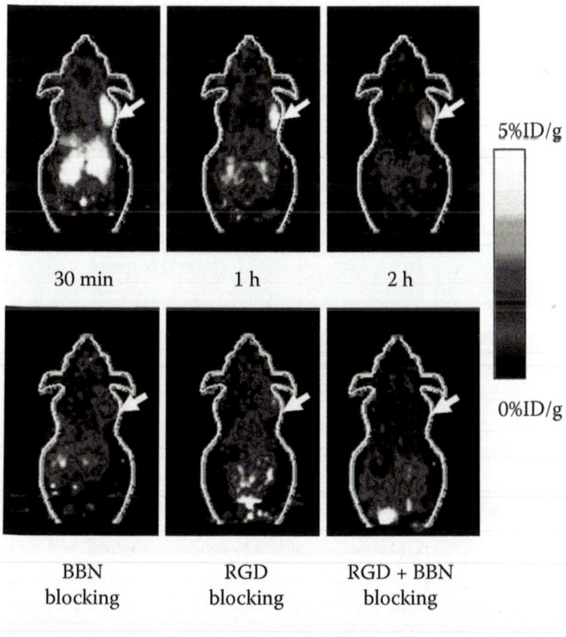

FIGURE 9.2 Coronal microPET images of PC-3 tumor-bearing mice at 30, 60, and 120 min after injection of 3.7 MBq ^{18}F-FB-PEG$_3$-Glu-RGD-BBN (top row) and the images at 1 h after injection of 3.7 MBq ^{18}F-FB-PEG$_3$-Glu-RGD-BBN and a blocking dose of c(RGDyK) (10 mg/kg of mouse body weight), BBN peptide (15 mg/kg mouse body weight), or RGD + BBN peptides (10 mg/kg for RGD and 15 mg/kg for BBN) (bottom row). Arrows indicate presence of PC-3 tumors. (Adapted from Li, Z.B. et al., *J Nucl Med* 49: 453–61, 2008. With permission.)

Chlorotoxin (CTX) has recently been suggested as a potential brain tumor ligand. CTX is a small 36-amino acid peptide purified from the venom of the giant Israeli scorpion (*Leiurus quinquestriatus*). This peptide binds with high affinity to the membrane-bound matrix metalloproteinase-2 (MMP-2), which is preferentially upregulated in glioma, medulloblastoma, and other tumors of neuroectodermal origin (Deshane et al. 2003; Kachra et al. 1999; Lampert et al. 1998). Targeted radiotherapy using ^{131}I-CTX has been promising for postsurgical treatment

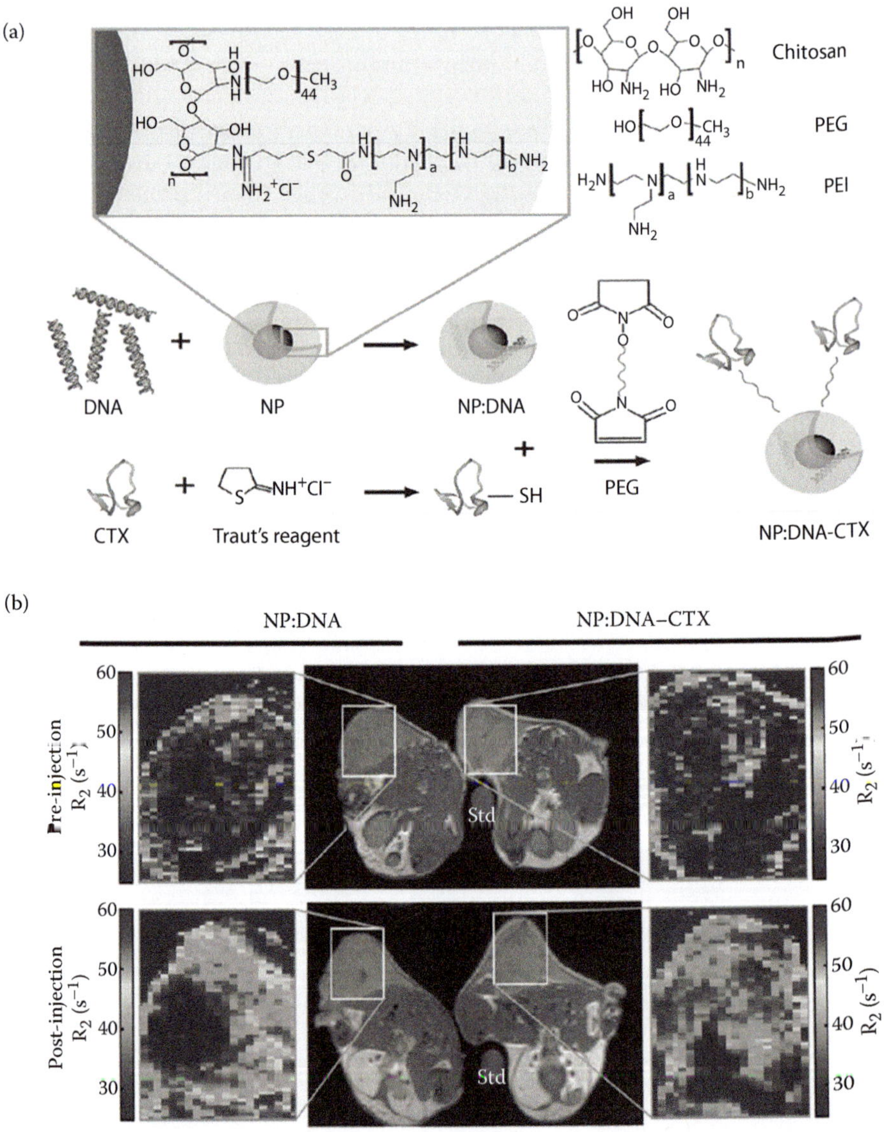

FIGURE 9.3 CTX-labeled magnetic nanovectors for targeted gene delivery to glioma. (a) The IONP core was coated with a copolymer of chitosan, poly(ethylene glycol) (PEG), and poly(ethylene imine) (PEI) to produce NP. DNA was loaded into NP through the electrostatic interaction between negatively charged DNA and positively charged NP coating to form NP:DNA. CTX was conjugated onto NP:DNA to form NP:DNA–CTX through a heterobifunctional PEG linker. (b) Nanovector delivery to C6 xenograft tumors monitored by T2-weighted MRI. Both nontargeted (NP:DNA) and targeted (NP:DNA–CTX) nanovector-treated tumors showed similar enhancement of R2 contrast. (Adapted from Kievit, F.M. et al., *ACS Nano* 4: 4587–94, 2011. With permission.)

of brain tumors (Shen et al. 2005). In combination with IONPs, CTX was also used to deliver nanoparticle-based glioma-targeting gene delivery carriers (Figure 9.3) (Huang et al. 2011; Kievit et al. 2011). Various types of peptide-based ligands were designed to target different receptors such as CCK-2/gastrin receptor (Chen et al. 1999; Good et al. 2008; Reubi 2007), melanocortin-1 receptor (MC-1r) (Chen et al. 2001; McQuade et al. 2005; Tatro et al. 1992), GLP-1 receptor (GLP-1r) (Gotthardt et al. 2002; Kieffer and Habener 1999), and vasoactive intestinal peptide (VIP) receptor (Thakur et al. 2004; Zhang et al. 2008) and were applied to imaging and therapeutic applications.

Macromolecule-Based Conjugates

Proteins that specifically interact with certain receptors overexpressed in cancer cells were successfully developed as targeting ligands for drug delivery and imaging applications (Reubi and Maecke 2008). The interaction of proteins with the cell membrane results in cellular penetration or the pore formation within the cell membrane, similar to the way peptides work. Proteins, including antibodies, enzymes, and structural proteins, have higher molecular weights, more complex three-dimensional structures, and generally more specific interaction with receptors than peptide ligands (Kirpotin et al. 2006; Sofou and Sgouros 2008). Here, we describe a few representative target receptors and ligands that were developed for therapeutic and imaging applications.

Epidermal growth factor receptors (EGFRs) are receptors that regulate the growth, survival, differentiation, and migration of cells by activating an associated intracellular tyrosine kinase (TK) signaling pathway. EGFRs are overexpressed in various malignant epithelial tumors, but particularly in glioblastoma and to a lesser extent in prostate and breast cancer (Pines et al. 2010; Spector and Blackwell 2009). The most common mutation observed in the receptor is the deletion of an extracellular domain segment of EGFR, including the ligand-binding region; this generates a variant known as EGFRvIII (Hatanpaa et al. 2010; Scott et al. 2007). EGFRvIII antibodies, such as the monoclonal antibody (mAb) L8A4, labeled with various radioisotopes are widely used as ligand-based imaging agents. Different types of radioisotope-labeled L8A4, such as [^{125}I]-SGMIB-L8A4 (labeled with ^{125}I using *N*-succinimidyl 4-guanidinomethyl-3-[^{125}I]-iodobenzoate, [^{125}I]-SGMIB) and [^{177}Lu]-CHX-A″ DTPA-L8A4, were prepared and applied in U87MG glioblastoma-bearing mice for imaging and radioimmunotherapy (Chopra 2010b,c). Two other FDA-approved anti-EGFR mAbs, cetuximab and panitumumab, were labeled with ^{99m}Tc, ^{64}Cu, and ^{18}F to image EGFR expression in tumor models (Leung 2010; Pal et al. 2011; Schechter et al. 2003; Tian et al. 2011). Alternatively, EGFR-targeted nanoparticles were designed for MRI, optical imaging, and dual-modality imaging. For example, anti-EGFR mAb-conjugated IONPs were developed for MRI-guided delivery and targeted therapy of glioblastoma and infiltrative glioblastoma (Hadjipanayis et al. 2010). For therapeutic applications, IONPs loaded with doxorubicin (DOX) and cetuximab-immunomicelles were demonstrated in vitro as MRI-visible, EGFR-targeted drug delivery agents that potentially could improve tumor imaging and therapy (Liao et al. 2011). Magnetofluorescent silica nanoparticles were conjugated with cetuximab to target and image colon cancer (Cho et al. 2010). Streptavidin-coated QDs conjugated to anti-EGFR mAb selectively bound to brain tumor cells expressing EGFR in vitro (Wang et al. 2007).

The discovery of vascular endothelial growth factors (VEGFs), especially VEGF-A, as prime mediators of angiogenesis has spawned much further investigation. VEGFs, VEGF receptors (VEGFRs), and their complexes as antigens are well reported for ligand-based targeting of tumor neovasculature (Ferrara 2002, 2004; Prewett et al. 1999). mAbs to VEGF-A (Cooke et al. 2001), VEGFR-2 (Prewett et al. 1999), and the VEGFA–VEGFR-2 complex (Brekken et al. 2000) have been used for tumor targeting. Even though these antibodies can be used by intravenous injection, the poor targeting kinetics of antibodies to tumors with low or medium abundance of antigens limited clinical application. Recently, the humanized anti-VEGF mAb bevacizumab (Avastin, Genentech) was FDA-approved for cancer treatment combined with chemotherapy (Ferrara et al.

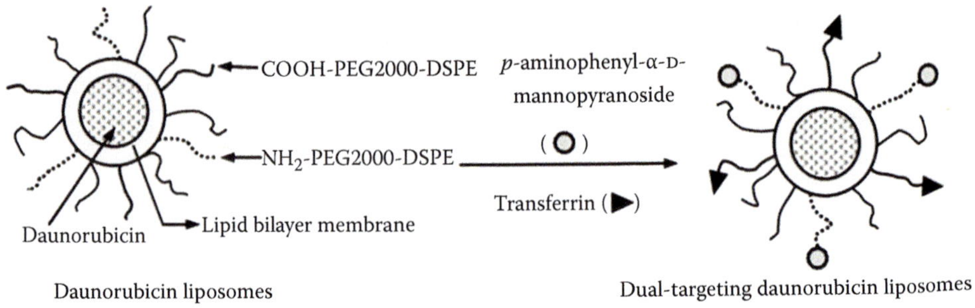

FIGURE 9.4 Overall schematic representation for the preparation of daunorubicin liposomes modified with *p*-aminophenyl-α-D-mannopyranoside (MAN) and TF. (Adapted from Kievit, F.M. et al., *ACS Nano* 4: 4587–94, 2011. With permission.)

2004). Monitoring and imaging the expression of both VEGF and VEGFRs are important for cancer diagnosis and serial assessment of therapeutic efficacy. Because of the soluble and dynamic nature of VEGF proteins, imaging of VEGF expression is challenging. Examining tumors in the same animals or cancer patients with both VEGF- and VEGFR-targeted tracers may provide important insights during cancer development and cancer therapy. Multimodal molecular imaging and accurate measurements of VEGF, VEGFR expression during angiogenesis, and VEGF–VEGFR interactions in vitro and in vivo are fully reviewed elsewhere (Cai and Chen 2007, 2008; Hsu et al. 2007; Niu and Chen).

Membrane transferrin receptor (TFr)-mediated endocytosis is an efficient cellular uptake pathway for drug delivery of anticancer drugs (Visser et al. 2004). TFrs are overexpressed in many tumors and have been widely studied (Li and Qian 2002). Since TFr is also present in the BBB, transferrin (TF)-conjugated solid lipid nanoparticles enabled enhanced delivery of quinine dihydrochloride to the brain (Gupta et al. 2007). It is reported that human serum albumin-conjugated nanoparticles (HSA-NPs) coupled to TF or TFr-mAb carried loperamide across the BBB and into the brain (Ulbrich et al. 2009). Dual-targeting liposomes modified with *p*-aminophenyl-α-D-mannopyranoside (MAN) and TF demonstrated efficient delivery of the anticancer drug daunorubicin to gliomas in the brain (Figure 9.4) (Ying et al. 2010). FC5 single-domain antibody (sdAb) is a BBB-transmigrating antibody isolated from the llama by sdAb phage-display library (Muruganandam et al. 2002). FC5 is ten times smaller in size (15 kDa) than immunoglobulin G (IgG) and highly stable and resistant to proteases and pH and temperature changes. These unique characteristics of sdAb make it an ideal ligand (Abulrob et al. 2005). A variety of engineered proteins that are screened from phage display or derived from natural binding sources have been selected and used as efficient ligands for targeted tumor imaging and therapy.

LIGAND-BASED NANOTHERAPEUTICS FOR BRAIN TUMOR

The development of imaging and therapeutic agents against neuronal targets is hampered by the limited access of probes and drugs to the central nervous system across the BBB. The BBB is responsible for transporting essential nutrients and metabolites to the brain while protecting and regulating its internal environment (Begley 2003; Iqbal et al. 2011; Yang 2010). Biological molecules, such as peptides, proteins, and oligonucleotides, that are hydrophilic and tend to have larger molecular weights have limited BBB penetration (Orive et al. 2010). Nanoparticles composed of polymeric materials, liposomes, dendrimers, IONPs, or QDs can improve BBB penetration and enhance their therapeutic and diagnostic efficacy by the following mechanisms: (1) increased retention time of the nanoparticles in the blood of brain capillaries can create a higher concentration gradient for drugs and lead to an opening of the tight junctions between the endothelial cells, which would enhance their transport across the endothelial cell layer; (2) a surfactant effect characterized by solubilization of endothelial cell membrane lipids by nanoparticles can lead to membrane fluidization and thus to enhanced drug permeability through the BBB; and (3) nanoparticles loaded with drugs can be endocytosed and transcytosed through the endothelial cells followed by the release of the drugs within the brain (Kreuter et al. 2002; Smith and Gumbleton 2006).

Nanoparticles of a certain size range (30–200 nm) can be injected interstitially to deliver novel therapeutics directly to target tissue while avoiding macrophage clearance and loss of drug through leakage into capillaries (Orive et al. 2010). Among various biomaterials, polymeric nanoparticles can improve systemic drug delivery to the tumor if they are functionalized with specific ligands that can target the tumor endothelium or tumor cells. Ligands can be immobilized in situ on nanoparticles via covalent or noncovalent interactions during synthesis of nanoparticles or via postsynthesis surface modification (Fang and Zhang 2010). Polymer-based formulations have multiple functional groups ideally suited for modification with a wide range of imaging moieties or drugs, prolonged plasma half-lives, improved stability, and enhanced targeting. To date, a few polymers have been approved for clinical applications by the FDA or are currently under clinical trials, including PEG, dextran, carboxydextran, β-cyclodextrin (βCD), poly(D,L-lactide-*co*-glycolide) (PLGA), chitosan, and hyaluronic acid (Fang and Zhang 2010). In addition to using polymer conjugates, strategies such as integration of lipophilic substances, for example, liposomes, were shown to facilitate passive diffusion of macromolecules into the brain parenchyma (Chiu et al. 2009; Kaur et al. 2008; Souto and Muller 2010). Liposomes are spherical lipid envelopes constructed from bilayer membranes, which create an effective permeability barrier between the interior liposomal aqueous compartment and the external aqueous environment. This property allows liposomes to deliver hydrophilic agents carried in their interior aqueous core by solubilizing the hydrophobic lipid bilayer.

Polymeric and liposomal nanoparticles conjugated with ligands such as apolipoproteins (Apos) (Kreuter et al. 2002), TF antibodies (Smith and Gumbleton 2006; Ulbrich et al. 2009), cell-penetrating peptides [e.g., transactivator of transcription (TAT) peptide] (Rao et al. 2008), and cationic moieties promote receptor-mediated and adsorptive-mediated transcytosis of macromolecules across the BBB. McNeeley et al. (2007) discovered that the blood circulation time of folate-carrying PEGylated liposomes was significantly reduced, counteracting the effect of facilitated receptor-mediated drug delivery to gliomas. Liu et al. (2008) fabricated biologically active polymeric core/shell nanoparticles composed of self-assembled cell-penetrating TAT peptide conjugated with PEG *b*-cholesterol (TAT–PEG-*b*-Chol) and used them as carriers to deliver the antibiotic ciprofloxacin. Their animal studies qualitatively confirmed that TAT-conjugated micelles can cross the BBB and locate around the cell nucleus of neurons. In another example, polysorbate 80-coated nanoparticles selectively adsorbed plasma proteins such as Apo E and B to promote receptor-mediated endocytosis across the BBB (Gelperina et al. 2010; Yang 2010). Proteins such as albumin covalently attached to Apo E were explored as carriers for brain drug delivery (Zensi et al. 2009). Human serum albumin-based nanoparticles associated with Apo E demonstrated significantly enhanced capillary transcytosis and accumulation in brain parenchyma compared to nanoparticles without Apo E in a mouse model. Nanoparticle-based formulations are increasingly important in the targeted delivery of therapeutics or imaging agents across the BBB and will continue to grow. Careful optimization and validation processes of such formulations in various conditions and models are important for clinical translation. In addition, there is an urgent need to verify chronic toxicities of nanoparticle-based formulations in vivo.

CONCLUSIONS

This chapter introduced and described numerous emerging approaches categorized by the term "ligand-based targeted delivery," with respect to imaging probe and drug design strategies and applications. To date, significant research efforts have been expended toward developing ligand-based imaging probes and therapeutic formulations. However, many novel probes and drugs reported in animal models have failed to translate into clinical products. The development and application of ligand-based probes and therapeutics have a long way to go; nevertheless, interdisciplinary research at the interface of molecular imaging, drug delivery, and nanotechnology will provide numerous possibilities with clinically available tools in the near future.

ACKNOWLEDGMENT

This work was supported in part by the Intramural Research Program (IRP), the National Institute of Biomedical Imaging and Bioengineering (NIBIB), and the National Institutes of Health (NIH).

REFERENCES

Abulrob, A., Sprong, H., Van Bergen en Henegouwen, P. et al. 2005. The blood–brain barrier transmigrating single domain antibody: mechanisms of transport and antigenic epitopes in human brain endothelial cells. *J Neurochem* 95: 1201–14.

Aloj, L., and Morelli, G. 2004. Design, synthesis and preclinical evaluation of radiolabeled peptides for diagnosis and therapy. *Curr Pharm Des* 10: 3009–31.

Beer, A.J., Grosu, A.L., Carlsen, J. et al. 2007. [18F]galacto-RGD positron emission tomography for imaging of alphavbeta3 expression on the neovasculature in patients with squamous cell carcinoma of the head and neck. *Clin Cancer Res* 13: 6610–6.

Beer, A.J., Niemeyer, M., Carlsen, J. et al. 2008. Patterns of alphavbeta3 expression in primary and metastatic human breast cancer as shown by 18F-Galacto-RGD PET. *J Nucl Med* 49: 255–9.

Begley, D.J. 2003. Understanding and circumventing the blood–brain barrier. *Acta Paediatr Suppl* 92: 83–91.

Biddlecombe, G.B., Rogers, B.E., de Visser, M. et al. 2007. Molecular imaging of gastrin-releasing peptide receptor-positive tumors in mice using 64Cu- and 86Y-DOTA-(Pro1,Tyr4)-bombesin(1-14). *Bioconjug Chem* 18: 724–30.

Brekken, R.A., Overholser, J.P., Stastny, V.A. et al. 2000. Selective inhibition of vascular endothelial growth factor (VEGF) receptor 2 (KDR/Flk-1) activity by a monoclonal anti-VEGF antibody blocks tumor growth in mice. *Cancer Res* 60: 5117–24.

Cai, W., and Chen, X. 2007. Multimodality imaging of vascular endothelial growth factor and vascular endothelial growth factor receptor expression. *Front Biosci* 12: 4267–79.

Cai, W., and Chen, X. 2008. Multimodality molecular imaging of tumor angiogenesis. *J Nucl Med* 49 Suppl 2: 113S–28S.

Cai, W., Niu, G., and Chen, X. 2008. Imaging of integrins as biomarkers for tumor angiogenesis. *Curr Pharm Des* 14: 2943–73.

Cescato, R., Maina, T., Nock, B. et al. 2008. Bombesin receptor antagonists may be preferable to agonists for tumor targeting. *J Nucl Med* 49: 318–26.

Chen, J., Cheng, Z., Owen, N.K. et al. 2001. Evaluation of an (111)In-DOTA-rhenium cyclized alpha-MSH analog: a novel cyclic-peptide analog with improved tumor-targeting properties. *J Nucl Med* 42: 1847–55.

Chen, J., Giblin, M.F., Wang, N. et al. 1999. In vivo evaluation of 99mTc/188Re-labeled linear alpha-melanocyte stimulating hormone analogs for specific melanoma targeting. *Nucl Med Biol* 26: 687–93.

Chen, K., Xie, J., Xu, H. et al. 2009. Triblock copolymer coated iron oxide nanoparticle conjugate for tumor integrin targeting. *Biomaterials* 30: 6912–9.

Chen, X., Conti, P.S., and Moats, R.A. 2004a. In vivo near-infrared fluorescence imaging of integrin alphavbeta3 in brain tumor xenografts. *Cancer Res* 64: 8009–14.

Chen, X., Hou, Y., Tohme, M. et al. 2004b. Pegylated Arg-Gly-Asp peptide: 64Cu labeling and PET imaging of brain tumor alphavbeta3-integrin expression. *J Nucl Med* 45: 1776–83.

Chen, X., Park, R., Shahinian, A.H. et al. 2004c. 18F-labeled RGD peptide: initial evaluation for imaging brain tumor angiogenesis. *Nucl Med Biol* 31: 179–89.

Cheng, Z., Wu, Y., Xiong, Z. et al. 2005. Near-infrared fluorescent RGD peptides for optical imaging of integrin alphavbeta3 expression in living mice. *Bioconjug Chem* 16: 1433–41.

Chiu, G.N., Wong, M.Y., Ling, L.U. et al. 2009. Lipid-based nanoparticulate systems for the delivery of anti-cancer drug cocktails: implications on pharmacokinetics and drug toxicities. *Curr Drug Metab* 10: 861–74.

Cho, Y.S., Yoon, T.J., Jang, E.S. et al. 2010. Cetuximab-conjugated magneto-fluorescent silica nanoparticles for in vivo colon cancer targeting and imaging. *Cancer Lett* 299: 63–71.

Chopra, A. 2010a. [^{111}In]-DOTA-Aminohexanoyl-[D-Phe6,Leu-NHCH2CH2CH313,desMet14] BBN[6-14] 2010 May 20 [Updated 2010 Jul 8]. In: Molecular Imaging and Contrast Agent Database (MICAD) [Internet]. Bethesda (MD): National Center for Biotechnology Information (US); 2004–2012. Available from: http://www.ncbi.nlm.nih.gov/books/NBK45125/.

Chopra, A. 2004b. [^{125}I]-Labeled monoclonal antibody L8A4 against epidermal growth factor receptor variant III (EGFRvIII) 2010 Dec 24 [Updated 2011 Feb 17]. In: Molecular Imaging and Contrast Agent Database (MICAD) [Internet]. Bethesda (MD): National Center for Biotechnology Information (US); 2004–2012. Available from: http://www.ncbi.nlm.nih.gov/books/NBK52783/.

Chopra, A. 2010c. [^{177}Lu]-Labeled α-(5-isothiocyanato-2-methoxyphenyl)-1,4,7,10-tetraazacyclododecane-1,4,7,10-tetraacetic acid (MeO-DOTA) conjugated monoclonal antibody L8A4 against epidermal growth factor receptor variant III (EGFRvIII) 2010 Dec 24 [Updated 2011 Feb 17]. In: Molecular Imaging and Contrast Agent Database (MICAD) [Internet]. Bethesda (MD): National Center for Biotechnology Information (US); 2004–2012. Available from: http://www.ncbi.nlm.nih.gov/books/NBK52780/.

Cooke, S.P., Boxer, G.M., Lawrence, L. et al. 2001. A strategy for antitumor vascular therapy by targeting the vascular endothelial growth factor: receptor complex. *Cancer Res* 61: 3653–9.

Das, M., Mohanty, C., and Sahoo, S.K. 2009. Ligand-based targeted therapy for cancer tissue. *Expert Opin Drug Deliv* 6: 285–304.

Deshane, J., Garner, C.C., and Sontheimer, H. 2003. Chlorotoxin inhibits glioma cell invasion via matrix metalloproteinase-2. *J Biol Chem* 278: 4135–44.

Fang, C., and Zhang, M. 2010. Nanoparticle-based theragnostics: integrating diagnostic and therapeutic potentials in nanomedicine. *J Control Release* 146: 2–5.

Ferrara, N. 2002. VEGF and the quest for tumour angiogenesis factors. *Nat Rev Cancer* 2: 795–803.

Ferrara, N. 2004. Vascular endothelial growth factor: basic science and clinical progress. *Endocr Rev* 25: 581–611.

Ferrara, N., Hillan, K.J., Gerber, H.P. et al. 2004. Discovery and development of bevacizumab, an anti-VEGF antibody for treating cancer. *Nat Rev Drug Discov* 3: 391–400.

Frangioni, J.V. 2003. In vivo near-infrared fluorescence imaging. *Curr Opin Chem Biol* 7: 626–34.

Froidevaux, S., and Eberle, A.N. 2002. Somatostatin analogs and radiopeptides in cancer therapy. *Biopolymers* 66: 161–83.

Gelperina, S., Maksimenko, O., Khalansky, A. et al. 2010. Drug delivery to the brain using surfactant-coated poly(lactide-co-glycolide) nanoparticles: influence of the formulation parameters. *Eur J Pharm Biopharm* 74: 157–63.

Good, S., Walter, M.A., Waser, B. et al. 2008. Macrocyclic chelator-coupled gastrin-based radiopharmaceuticals for targeting of gastrin receptor-expressing tumours. *Eur J Nucl Med Mol Imaging* 35: 1868–77.

Gotthardt, M., Fischer, M., Naeher, I. et al. 2002. Use of the incretin hormone glucagon-like peptide-1 (GLP-1) for the detection of insulinomas: initial experimental results. *Eur J Nucl Med Mol Imaging* 29: 597–606.

Gupta, Y., Jain, A., and Jain, S.K. 2007. Transferrin-conjugated solid lipid nanoparticles for enhanced delivery of quinine dihydrochloride to the brain. *J Pharm Pharmacol* 59: 935–40.

Hadjipanayis, C.G., Machaidze, R., Kaluzova, M. et al. 2010. EGFRvIII antibody-conjugated iron oxide nanoparticles for magnetic resonance imaging-guided convection-enhanced delivery and targeted therapy of glioblastoma. *Cancer Res* 70: 6303–12.

Hahn, M.A., Singh, A.K., Sharma, P. et al. 2011. Nanoparticles as contrast agents for in-vivo bioimaging: current status and future perspectives. *Anal Bioanal Chem* 399: 3–27.

Hannon, J.P., Nunn, C., Stolz, B. et al. 2002. Drug design at peptide receptors: somatostatin receptor ligands. *J Mol Neurosci* 18: 15–27.

Hatanpaa, K.J., Burma, S., Zhao, D. et al. 2010. Epidermal growth factor receptor in glioma: signal transduction, neuropathology, imaging, and radioresistance. *Neoplasia* 12: 675–84.

Haubner, R., Wester, H.J., Weber, W.A. et al. 2001. Noninvasive imaging of alpha(v)beta3 integrin expression using 18F-labeled RGD-containing glycopeptide and positron emission tomography. *Cancer Res* 61: 1781–5.

Hoffman, T.J., Gali, H., Smith, C.J. et al. 2003. Novel series of 111In-labeled bombesin analogs as potential radiopharmaceuticals for specific targeting of gastrin-releasing peptide receptors expressed on human prostate cancer cells. *J Nucl Med* 44: 823–31.

Horton, M.A. 1997. The alpha v beta 3 integrin "vitronectin receptor." *Int J Biochem Cell Biol* 29: 721–5.

Hsu, A.R., Cai, W., Veeravagu, A. et al. 2007. Multimodality molecular imaging of glioblastoma growth inhibition with vasculature-targeting fusion toxin VEGF121/rGel. *J Nucl Med* 48: 445–54.

Huang, R., Ke, W., Han, L. et al. 2011. Targeted delivery of chlorotoxin-modified DNA-loaded nanoparticles to glioma via intravenous administration. *Biomaterials* 32: 2399–406.

Iqbal, U., Abulrob, A., and Stanimirovic, D.B. 2011. Integrated platform for brain imaging and drug delivery across the blood–brain barrier. *Methods Mol Biol* 686: 465–81.

Jun, Y.W., Huh, Y.M., Choi, J.S. et al. 2005. Nanoscale size effect of magnetic nanocrystals and their utilization for cancer diagnosis via magnetic resonance imaging. *J Am Chem Soc* 127: 5732–3.

Kachra, Z., Beaulieu, E., Delbecchi, L. et al. 1999. Expression of matrix metalloproteinases and their inhibitors in human brain tumors. *Clin Exp Metastasis* 17: 555–66.

Kaur, I.P., Bhandari, R., Bhandari, S. et al. 2008. Potential of solid lipid nanoparticles in brain targeting. *J Control Release* 127: 97–109.

Kenny, L.M., Coombes, R.C., Oulie, I. et al. 2008. Phase I trial of the positron-emitting Arg-Gly-Asp (RGD) peptide radioligand 18F-AH111585 in breast cancer patients. *J Nucl Med* 49: 879–86.

Kieffer, T.J., and Habener, J.F. 1999. The glucagon-like peptides. *Endocr Rev* 20: 876–913.

Kievit, F.M., Veiseh, O., Fang, C. et al. 2011. Chlorotoxin labeled magnetic nanovectors for targeted gene delivery to glioma. *ACS Nano* 4: 4587–94.

Kirpotin, D.B., Drummond, D.C., Shao, Y. et al. 2006. Antibody targeting of long-circulating lipidic nanoparticles does not increase tumor localization but does increase internalization in animal models. *Cancer Res* 66: 6732–40.

Kostenich, G., Livnah, N., Bonasera, T.A. et al. 2005. Targeting small-cell lung cancer with novel fluorescent analogs of somatostatin. *Lung Cancer* 50: 319–28.

Kostenich, G., Oron-Herman, M., Kimel, S. et al. 2008. Diagnostic targeting of colon cancer using a novel fluorescent somatostatin conjugate in a mouse xenograft model. *Int J Cancer* 122: 2044–9.

Kreuter, J., Shamenkov, D., Petrov, V. et al. 2002. Apolipoprotein-mediated transport of nanoparticle-bound drugs across the blood–brain barrier. *J Drug Target* 10: 317–25.

Lampert, K., Machein, U., Machein, M.R. et al. 1998. Expression of matrix metalloproteinases and their tissue inhibitors in human brain tumors. *Am J Pathol* 153: 429–37.

Lee, H.Y., Li, Z., Chen, K. et al. 2008. PET/MRI dual-modality tumor imaging using arginine–glycine–aspartic (RGD)-conjugated radiolabeled iron oxide nanoparticles. *J Nucl Med* 49: 1371–9.

Lee, S., Xie, J., and Chen, X. 2010. Peptides and peptide hormones for molecular imaging and disease diagnosis. *Chem Rev* 110: 3087–111.

Leung, K. 2010. ^{64}Cu-1,4,7,10-Tetraazacyclododecane-N,N′,N″,N‴-tetraacetic acid-panitumumab. 2010 Jun 25 [Updated 2010 Sep 30]. In: Molecular Imaging and Contrast Agent Database (MICAD) [Internet]. Bethesda (MD): National Center for Biotechnology Information (US); 2004-2012. Available from: http://www.ncbi.nlm.nih.gov/books/NBK47078/.

Li, H., and Qian, Z.M. 2002. Transferrin/transferrin receptor-mediated drug delivery. *Med Res Rev* 22: 225–50.

Li, W.P., Lewis, J.S., Kim, J. et al. 2002. DOTA-D-Tyr(1)-octreotate: a somatostatin analogue for labelling with metal and halogen radionuclides for cancer imaging and therapy. *Bioconjug Chem* 13: 721–8.

Li, Z.B., Wu, Z., Chen, K. et al. 2008. 18F-labeled BBN-RGD heterodimer for prostate cancer imaging. *J Nucl Med* 49: 453–61.

Liao, C., Sun, Q., Liang, B. et al. 2011. Targeting EGFR-overexpressing tumor cells using cetuximab-immunomicelles loaded with doxorubicin and superparamagnetic iron oxide. *Eur J Radiol* 80: 699–705.

Liu, L., Guo, K., Lu, J. et al. 2008. Biologically active core/shell nanoparticles self-assembled from cholesterol-terminated PEG-TAT for drug delivery across the blood–brain barrier. *Biomaterials* 29: 1509–17.

Liu, Z., Cai, W., He, L. et al. 2007. In vivo biodistribution and highly efficient tumour targeting of carbon nanotubes in mice. *Nat Nanotechnol* 2: 47–52.

Liu, Z., Li, Z.B., Cao, Q. et al. 2009a. Small-animal PET of tumors with (64)Cu-labeled RGD-bombesin heterodimer. *J Nucl Med* 50: 1168–77.

Liu, Z., Niu, G., Shi, J. et al. 2009b. (68)Ga-labeled cyclic RGD dimers with Gly3 and PEG4 linkers: promising agents for tumor integrin alphavbeta3 PET imaging. *Eur J Nucl Med Mol Imaging* 36: 947–57.

Liu, Z., Niu, G., Wang, F. et al. 2009c. (68)Ga-labeled NOTA-RGD-BBN peptide for dual integrin and GRPR-targeted tumor imaging. *Eur J Nucl Med Mol Imaging* 36: 1483–94.

Liu, Z., Yan, Y., Chin, F.T. et al. 2009d. Dual integrin and gastrin-releasing peptide receptor targeted tumor imaging using 18F-labeled PEGylated RGD-bombesin heterodimer 18F-FB-PEG3-Glu-RGD-BBN. *J Med Chem* 52: 425–32.

Liu, Z., Yan, Y., Liu, S. et al. 2009e. (18)F, (64)Cu, and (68)Ga labeled RGD-bombesin heterodimeric peptides for PET imaging of breast cancer. *Bioconjug Chem* 20: 1016–25.

Lu, Y., Yang, J., and Sega, E. 2006. Issues related to targeted delivery of proteins and peptides. *AAPS J* 8: E466–78.

McNeeley, K.M., Annapragada, A., and Bellamkonda, R.V. 2007. Decreased circulation time offsets increased efficacy of PEGylated nanocarriers targeting folate receptors of glioma. *Nanotechnology* 18: 385101.

McQuade, P., Miao, Y., Yoo, J. et al. 2005. Imaging of melanoma using 64Cu- and 86Y-DOTA-ReCCMSH(Arg11), a cyclized peptide analogue of alpha-MSH. *J Med Chem* 48: 2985–92.

Meisetschlager, G., Poethko, T., Stahl, A. et al. 2006. Gluc-Lys([18F]FP)-TOCA PET in patients with SSTR-positive tumors: biodistribution and diagnostic evaluation compared with [111In]DTPA-octreotide. *J Nucl Med* 47: 566–73.

Miao, Y., and Quinn, T.P. 2007. Alpha-melanocyte stimulating hormone peptide-targeted melanoma imaging. *Front Biosci* 12: 4514–24.

Morrison, M.S., Ricketts, S.A., Barnett, J. et al. 2009. Use of a novel Arg-Gly-Asp radioligand, 18F-AH111585, to determine changes in tumor vascularity after antitumor therapy. *J Nucl Med* 50: 116–22.

Muruganandam, A., Tanha, J., Narang, S. et al. 2002. Selection of phage-displayed llama single-domain antibodies that transmigrate across human blood–brain barrier endothelium. *FASEB J* 16: 240–2.

Nel, A.E., Madler, L., Velegol, D. et al. 2009. Understanding biophysicochemical interactions at the nano-bio interface. *Nat Mater* 8: 543–57.

Niu, G., and Chen, X. 2010. Vascular endothelial growth factor as an anti-angiogenic target for cancer therapy. *Curr Drug Targets* 11: 1000–17.

Okarvi, S.M. 2004. Peptide-based radiopharmaceuticals: future tools for diagnostic imaging of cancers and other diseases. *Med Res Rev* 24: 357–97.

Orive, G., Ali, O.A., Anitua, E. et al. 2010. Biomaterial-based technologies for brain anti-cancer therapeutics and imaging. *Biochim Biophys Acta* 1806: 96–107.

Pal, A., Balatoni, J.A., Mukhopadhyay, U. et al. 2011. Radiosynthesis and initial in vitro evaluation of [(18)F] F-PEG (6)-IPQA—a novel PET radiotracer for imaging EGFR expression-activity in lung carcinomas. *Mol Imaging Biol* 13: 853–61.

Patel, Y.C., and Wheatley, T. 1983. In vivo and in vitro plasma disappearance and metabolism of somatostatin-28 and somatostatin-14 in the rat. *Endocrinology* 112: 220–5.

Pines, G., Kostler, W.J., and Yarden, Y. 2010. Oncogenic mutant forms of EGFR: lessons in signal transduction and targets for cancer therapy. *FEBS Lett* 584: 2699–706.

Prasanphanich, A.F., Nanda, P.K., Rold, T.L. et al. 2007. [64Cu-NOTA-8-Aoc-BBN(7-14)NH2] targeting vector for positron-emission tomography imaging of gastrin-releasing peptide receptor-expressing tissues. *Proc Natl Acad Sci U S A* 104: 12462–7.

Prewett, M., Huber, J., Li, Y. et al. 1999. Antivascular endothelial growth factor receptor (fetal liver kinase 1) monoclonal antibody inhibits tumor angiogenesis and growth of several mouse and human tumors. *Cancer Res* 59: 5209–18.

Rao, K.S., Reddy, M.K., Horning, J.L. et al. 2008. TAT-conjugated nanoparticles for the CNS delivery of anti-HIV drugs. *Biomaterials* 29: 4429–38.

Reubi, J.C. 2003. Peptide receptors as molecular targets for cancer diagnosis and therapy. *Endocr Rev* 24: 389–427.

Reubi, J.C. 2007. Targeting CCK receptors in human cancers. *Curr Top Med Chem* 7: 1239–42.

Reubi, J.C., and Maecke, H.R. 2008. Peptide-based probes for cancer imaging. *J Nucl Med* 49: 1735–8.

Reubi, J.C., and Waser, B. 2003. Concomitant expression of several peptide receptors in neuroendocrine tumours: molecular basis for in vivo multireceptor tumour targeting. *Eur J Nucl Med Mol Imaging* 30: 781–93.

Reubi, J.C., Waser, B., Schaer, J.C. et al. 2001. Somatostatin receptor sst1-sst5 expression in normal and neoplastic human tissues using receptor autoradiography with subtype-selective ligands. *Eur J Nucl Med* 28: 836–46.

Ruoslahti, E., and Pierschbacher, M.D. 1987. New perspectives in cell adhesion: RGD and integrins. *Science* 238: 491–7.

Rybak, J.N., Trachsel, E., Scheuermann, J. et al. 2007. Ligand-based vascular targeting of disease. *Chem Med Chem* 2: 22–40.

Schechter, N.R., Yang, D.J., Azhdarinia, A. et al. 2003. Assessment of epidermal growth factor receptor with 99mTc-ethylenedicysteine-C225 monoclonal antibody. *Anticancer Drugs* 14: 49–56.

Schnell, O., Krebs, B., Carlsen, J. et al. 2009. Imaging of integrin alpha(v)beta(3) expression in patients with malignant glioma by [18F] Galacto-RGD positron emission tomography. *Neuro Oncol* 11: 861–70.

Schottelius, M., Wester, H.J., Reubi, J.C. et al. 2002. Improvement of pharmacokinetics of radioiodinated Tyr(3)-octreotide by conjugation with carbohydrates. *Bioconjug Chem* 13: 1021–30.

Scott, A.M., Lee, F.T., Tebbutt, N. et al. 2007. A phase I clinical trial with monoclonal antibody ch806 targeting transitional state and mutant epidermal growth factor receptors. *Proc Natl Acad Sci U S A* 104: 4071–6.

Shan, L. 2010. Indocyanine green–enhanced, cyclic Arg-Gly-Asp–conjugated, PEGylated single-walled carbon nanotubes. 2010 Jun 2 [Updated 2010 Jul 12]. In: Molecular Imaging and Contrast Agent Database (MICAD) [Internet]. Bethesda (MD): National Center for Biotechnology Information (US); 2004–2012. Available from: http://www.ncbi.nlm.nih.gov/books/NBK45132/.

Shen, S., Khazaeli, M.B., Gillespie, G.Y. et al. 2005. Radiation dosimetry of 131I-chlorotoxin for targeted radiotherapy in glioma-bearing mice. *J Neurooncol* 71: 113–9.

Shi, J., Kim, Y.S., Zhai, S. et al. 2009. Improving tumor uptake and pharmacokinetics of (64)Cu-labeled cyclic RGD peptide dimers with Gly(3) and PEG(4) linkers. *Bioconjug Chem* 20: 750–9.

Shi, J., Wang, L., Kim, Y.S. et al. 2008. Improving tumor uptake and excretion kinetics of 99mTc-labeled cyclic arginine-glycine-aspartic (RGD) dimers with triglycine linkers. *J Med Chem* 51: 7980–90.

Smith, C.J., Volkert, W.A., and Hoffman, T.J. 2005. Radiolabeled peptide conjugates for targeting of the bombesin receptor superfamily subtypes. *Nucl Med Biol* 32: 733–40.

Smith, M.W., and Gumbleton, M. 2006. Endocytosis at the blood–brain barrier: from basic understanding to drug delivery strategies. *J Drug Target* 14: 191–214.

Sofou, S., and Sgouros, G. 2008. Antibody-targeted liposomes in cancer therapy and imaging. *Expert Opin Drug Deliv* 5: 189–204.

Souto, E.B., and Muller, R.H. 2010. Lipid nanoparticles: effect on bioavailability and pharmacokinetic changes. *Handb Exp Pharmacol* 115–41.

Spector, N.L., and Blackwell, K.L. 2009. Understanding the mechanisms behind trastuzumab therapy for human epidermal growth factor receptor 2-positive breast cancer. *J Clin Oncol* 27: 5838–47.

Stolz, B., Weckbecker, G., Smith-Jones, P.M. et al. 1998. The somatostatin receptor-targeted radiotherapeutic [90Y-DOTA-DPhe1, Tyr3]octreotide (90Y-SMT 487) eradicates experimental rat pancreatic CA 20948 tumours. *Eur J Nucl Med* 25: 668–74.

Susini, C., and Buscail, L. 2006. Rationale for the use of somatostatin analogs as antitumor agents. *Ann Oncol* 17: 1733–42.

Tatro, J.B., Wen, Z., Entwistle, M.L. et al. 1992. Interaction of an alpha-melanocyte-stimulating hormone-diphtheria toxin fusion protein with melanotropin receptors in human melanoma metastases. *Cancer Res* 52: 2545–8.

Thakur, M.L., Aruva, M.R., Gariepy, J. et al. 2004. PET imaging of oncogene overexpression using 64Cu-vasoactive intestinal peptide (VIP) analog: comparison with 99mTc-VIP analog. *J Nucl Med* 45: 1381–9.

Tian, M., Ogawa, K., Wendt, R. et al. 2011. Whole-body biodistribution kinetics, metabolism, and radiation dosimetry estimates of 18F-PEG6-IPQA in nonhuman primates. *J Nucl Med* 52: 934–41.

Ulbrich, K., Hekmatara, T., Herbert, E. et al. 2009. Transferrin- and transferrin-receptor-antibody-modified nanoparticles enable drug delivery across the blood–brain barrier (BBB). *Eur J Pharm Biopharm* 71: 251–6.

Van de Wiele, C., Dumont, F., Dierckx, R.A. et al. 2001. Biodistribution and dosimetry of (99m)Tc-RP527, a gastrin-releasing peptide (GRP) agonist for the visualization of GRP receptor-expressing malignancies. *J Nucl Med* 42: 1722–7.

Van de Wiele, C., Phonteyne, P., Pauwels, P. et al. 2008. Gastrin-releasing peptide receptor imaging in human breast carcinoma versus immunohistochemistry. *J Nucl Med* 49: 260–4.

Veiseh, O., Sun, C., Fang, C. et al. 2009. Specific targeting of brain tumors with an optical/magnetic resonance imaging nanoprobe across the blood–brain barrier. *Cancer Res* 69: 6200–7.

Visser, C.C., Stevanovic, S., Heleen Voorwinden, L. et al. 2004. Validation of the transferrin receptor for drug targeting to brain capillary endothelial cells in vitro. *J Drug Target* 12: 145–50.

Wang, J., Yong, W.H., Sun, Y. et al. 2007. Receptor-targeted quantum dots: fluorescent probes for brain tumor diagnosis. *J Biomed Opt* 12: 044021.

Wang, W., Ke, S., Wu, Q. et al. 2004. Near-infrared optical imaging of integrin alphavbeta3 in human tumor xenografts. *Mol Imaging* 3: 343–51.

Weckbecker, G., Lewis, I., Albert, R. et al. 2003. Opportunities in somatostatin research: biological, chemical and therapeutic aspects. *Nat Rev Drug Discov* 2: 999–1017.

Wu, Y., Cai, W., and Chen, X. 2006. Near-infrared fluorescence imaging of tumor integrin alpha v beta 3 expression with Cy7-labeled RGD multimers. *Mol Imaging Biol* 8: 226–36.

Wu, Z., Li, Z.B., Cai, W. et al. 2007. 18F-labeled mini-PEG spacered RGD dimer (18F-FPRGD2): synthesis and microPET imaging of alphavbeta3 integrin expression. *Eur J Nucl Med Mol Imaging* 34: 1823–31.

Yang, H. 2010. Nanoparticle-mediated brain-specific drug delivery, imaging, and diagnosis. *Pharm Res* 27: 1759–71.

Yang, L., Peng, X.H., Wang, Y.A. et al. 2009. Receptor-targeted nanoparticles for in vivo imaging of breast cancer. *Clin Cancer Res* 15: 4722–32.

Ying, X., Wen, H., Lu, W.L. et al. 2010. Dual-targeting daunorubicin liposomes improve the therapeutic efficacy of brain glioma in animals. *J Control Release* 141: 183–92.

Zensi, A., Begley, D., Pontikis, C. et al. 2009. Albumin nanoparticles targeted with Apo E enter the CNS by transcytosis and are delivered to neurones. *J Control Release* 137: 78–86.

Zhang, H. 2007. Gly-Ser-Ser-Lys-(FITC)-Gly-Gly-Gly-Cys-Arg-Gly-Asp-Cys-CLIO-Cy5.5. 2007 Dec 17 [Updated 2008 Jan 18]. In: Molecular Imaging and Contrast Agent Database (MICAD) [Internet]. Bethesda (MD): National Center for Biotechnology Information (US); 2004-2012. Available from: http://www.ncbi.nlm.nih.gov/books/NBK23450/.

Zhang, H., Schuhmacher, J., Waser, B. et al. 2007. DOTA-PESIN, a DOTA-conjugated bombesin derivative designed for the imaging and targeted radionuclide treatment of bombesin receptor-positive tumours. *Eur J Nucl Med Mol Imaging* 34: 1198–208.

Zhang, K., Aruva, M.R., Shanthly, N. et al. 2008. PET imaging of VPAC1 expression in experimental and spontaneous prostate cancer. *J Nucl Med* 49: 112–21.

Zhang, X., Xiong, Z., Wu, Y. et al. 2006. Quantitative PET imaging of tumor integrin alphavbeta3 expression with 18F-FRGD2. *J Nucl Med* 47: 113–21.

10 Immune-Based Targeting in Cancer

Dwain K. Irvin, Chirag G. Patil, and Keith L. Black

CONTENTS

INTRODUCTION

Nanoparticles (NPs) represent small molecules usually less than 1 μm. Thanks to the advances in imaging and microscopy, we can examine and manipulate systems at the nanoscale. Furthermore, biological advances in genomics and proteomics have helped us better understand signaling mechanisms involved in processes at the cellular and molecular levels during development and/or pathological conditions. Therefore, by combining nanotechnology and current biological assays from gene therapy to immunotherapy, we have great potential to modify drug delivery systems to improve targeting and efficacy. As we continue to develop this technology for this purpose, we must always be cautious and consider the potential toxicological and safety concerns in using nanotechnological platforms for drug delivery and/or therapy, including (1) the risk of small particles penetrating through biological barriers, including the blood–brain barrier (BBB), and (2) avoiding an immune response to NPs to improve their half-life and retention time in the systemic circulation. Therefore, a major goal utilizing nanoscale drug delivery systems is to improve our ability to target the tissue or cells of interest and hence improve the local concentration of compounds to be delivered to that site, while at the same time avoiding systemic toxicity.

In many cases, nanotechnology is exemplified by coupling polymers (i.e., surfactants and polyethylene glycol-like compounds) to NPs through covalent binding. Furthermore, chemistry-altering physical properties, including size and charge of NPs, allow additional options for tissue and cell targeting. These characteristics are quite helpful when trying to optimize delivery across the BBB. In this chapter, we will consider using nanotechnology in our attempts to better understand neuroimmunology and to develop novel immunotherapy for central nervous system (CNS) diseases.

Immune-Based Targeting against Cancer

Cancer is a major cause of morbidity and mortality, and the incidence of many types has been increasing. While the most common therapies include surgery followed by chemotherapeutic drugs and radiation therapy, recent studies have demonstrated some efficacy in using immunotherapy to treat many cancers (Marr et al. 2012). A common limitation with chemotherapeutic agents is the relatively narrow therapeutic window, typically limited by targeting or getting sufficiently high doses of the chemotherapeutic agent to the target site while minimizing toxicity. Another limitation in dealing with cancer of the CNS is the BBB or the blood–tumor barrier. Novel chemotherapeutic agents and new ways to deliver them to the tumor tissue are being developed to meet these challenges. Radiation therapy of primary brain tumors shares with chemotherapy the problem of a narrow therapeutic window between effective and toxic doses.

The idea and use of immunotherapeutic approaches have shown significant promise. Various methods were employed to develop novel immunotherapeutic modalities to treat cancer, including dendritic cell-based or T cell–based approaches (Alfaro et al. 2011a,b). While some of these studies show modest improvement in survival by Kaplan–Meier analysis in humans and in animal models of disease, the biological, immune, molecular, and cellular responses to immune modulation are still being elucidated (Wheeler et al. 2008).

Combining immunotherapy with various nanoplatforms may ultimately provide several novel approaches to treating diseases and potentially improve efficacy of existing immunotherapeutic approaches. To this end, several nanoscale drug delivery systems have been approved by the Food and Drug Administration (FDA) to treat various cancers, including Doxil, which is doxorubicin encapsulated in liposomes (Leonard et al. 2009), and Abraxane, which is paclitaxel (Taxol) covalently attached to NPs (Petrelli et al. 2010). Other NPs, including exosomes, inert nanobeads, and recombinant viral particles, were used to stimulate or enhance cellular and humoral immune responses (Scheerlinck and Greenwood 2008). Several recent studies demonstrated that synthetic NPs conjugated to biological molecules have some efficacy in treating various cancers (Rieger et al. 2009). For example, recent studies utilizing exosomes derived from dendritic and tumor cells represent novel nanoscale immunotherapy treatments for cancer that prime the body's immune system to recognize and kill cancer cells (Tan et al. 2010). Exosomes are in the nanoscale range (50–100 nm) and consist of membrane vesicles, suggested to be important in cell–cell communication, although their exact biological function has not been fully elucidated (Taylor and Gercel-Taylor 2011). Exosomes are of intracellular origin and display both major histocompatability complex (MHC) class I and II molecules on their surface (Taylor and Gercel-Taylor 2011). They can mediate immune responses by activating T lymphocytes, natural killer (NK) cells, and dendritic cells (Ostrowski et al. 2010; Thery et al. 2009). Additional advantages in using biological nanoscale structures like exosomes are that they typically have low immunogenicity, a favorable biosafety profile, and a noncellular delivery platform.

Another example of an FDA-approved nanoscale delivery system that enhances immune responses to cancer is Cervarix, which uses a viral liposome platform and is efficacious in treating cervical cancer. Using cervical cancer as a prime example, current (prophylactic) treatment involves vaccination with Gardasil, which is developed from and based on virus-like particle technology, and is effective against cancer-causing human papillomavirus (Garland and Smith 2010). The approach utilizes synthetic human papillomavirus (HPV) oncopeptides that promote CD8+ and CD4+ T-cell responses in vivo that promote complete remission in patients with vulvar cancer (Chen et al. 2011; Kitchin 2011). This antiviral approach also prevents cancer formation and is thus an antiviral therapy with anticancer benefits. Another example of this type of therapy is demonstrated by the hepatitis B vaccine, which protects against hepatitis B virus and is also associated with a decreased liver cancer incidence (Kitchin 2011). Nonetheless, it is important to recognize that these cancer vaccines are actually antiviral vaccines, are prophylactic because they prevent viral infections that lead to cancer, and may not lead to remission of an established cancer because they do not directly kill cancer cells.

More recently, new nanodrugs that employ viral-like particles have shown some direct effects against cancer cells, including the use of Provenge to treat advanced prostate cancer (Sonpavde et al. 2012). This involves activation of dendritic cells ex vivo exposed to a prostatic acid phosphatase-containing fusion protein, which primes the immune system to attack prostate cancer cells (Sonpavde et al. 2012). While dendritic cell-based immunotherapy is not a nanotherapy, it does support the idea that exosomes and/or viral-like particles can be used for nanoscale cancer vaccines.

Within the tumor microenvironment, cancer cells interact with immune cells but also can evade immune surveillance through a variety of mechanisms, including releasing immunosuppressive cytokines (e.g., interleukin-10 [IL-10] and transforming growth factor-β [TGF-B]) and altered antigen presentation. Nonetheless, the complete mechanism by which the immune system tolerates cancer has not been fully elucidated. Dendritic cells are antigen-presenting cells (APCs) that process antigen, release cytokines, and prime T cells to elicit antigen-specific immune responses. Dendritic cell immunotherapy is a promising alternative to monoclonal antibody cancer immunotherapy. Several clinical trials demonstrated the therapeutic efficacy of dendritic cell-based immunotherapy against glioblastoma multiforme (Finocchiaro and Pellegatta 2011; Wheeler and Black 2011; Wheeler et al. 2008; Yu et al. 2001, 2004). This approach has several advantages over antibody therapy, including fewer side effects, an ability to activate or prime an antitumor T-cell response, and limited toxicity (Finocchiaro and Pellegatta 2011). When considering the use of NPs for cancer vaccines, one should consider that the NPs should prime the immune system to recognize specific tumor antigens and then mount an appropriate immune response toward cancer cells. While a variety of nanoplatforms are beginning to be developed and tested, the field has not achieved satisfactory results from the current products that have been tested in clinical trials: MyVax, a vaccine to be used for treating non-Hodgkin's lymphoma, failed to demonstrate any significant therapeutic response, and Provenge, although demonstrating modest improvement in survival, was potentially cost prohibitive (Bitting et al. 2011; Fadel et al. 2008; Sasada et al. 2010; Tanne 2010).

Another promising nanoplatform for anticancer immunotherapy is the use of single-walled carbon nanotubes. Carbon nanotubes may potentially be used for cancer treatments, for imaging, and as activators of an immune response. Recent evidence has shown that carbon nanotubes conjugated to tumor antigens or antibodies may elicit antitumor immune responses and may also be used as efficient vaccine delivery systems to initiate innate and adaptive immune responses (Fadel et al. 2008; Meng et al. 2008). More recently, carbon nanotubes have been used as intracellular delivery vehicles for subsequent intracellular proinflammatory activation through intracellular toll-like receptor activation (Zhao et al. 2011).

Another aspect to consider when developing nano-based immunotherapy to treat cancer is T-cell anergy due to lack of sufficient antigen presentation. To mount a strong antitumor immune response, we need to ensure that there are sufficient induction and amplification of functional antigen-specific immune responses. In turn, our bodies develop long-term immunological memory against cancers (Janikashvili et al. 2011; Khong and Restifo 2002). Furthermore, current immunotherapeutic approaches are challenged to efficiently prime or activate antitumor cytotoxic T lymphocytes (CTLs) because most tumor-associated antigens are nonmutated self-antigens with poor immunogenicity (Edwards and Evavold 2011; Khong and Restifo 2002). Dendritic cell-based vaccines are promising in both preclinical animal models and human clinical studies (Irvin et al. 2010; Szabo and Carpentier 2011; Wheeler et al. 2008; Yu et al. 2001). However, this approach is expensive and requires expertise for effective delivery. Another potential option is to combine NPs with dendritic cell immunotherapy to maximize the efficacy of antitumor immunotherapy. The therapeutic efficacy of this approach has not been determined. Developing synthetic peptides or protein-based vaccines may provide a less expensive and less complex alternative to cell-based therapy. On the other hand, using peptide or protein-based vaccines may result in uptake of vaccine immunogens by inappropriate or immature immune cells and may lead to CTL anergy (Mescher et al. 2007; Prasad et al. 2011; Srinivasan and Frauwirth 2009).

Recent studies have shown that certain nanoplatforms can deliver large doses of antigens and provide prolonged antigen presentation (Cruz et al. 2011; Sexton et al. 2009; Zhang et al. 2011). Multiple studies regarding the feasibility of antigen delivery with polymeric NPs, which gradually release conjugated molecules, have shown some therapeutic efficacy (Prasad et al. 2011; Wu et al. 2009). Furthermore, to minimize degradation and increase the concentration of antigen after in vivo delivery, some polymeric NPs were designed to encapsulate antigen and deliver it to mature APCs without proteolysis (Prasad et al. 2011). Recent studies utilizing melanoma antigen peptides encapsulated in NPs in mouse models of melanoma demonstrated significant improvement in anticancer immune responses (Ge et al. 2009; Perche et al. 2011a,b). It appears that free proteins or peptides are largely internalized through macropinocytosis by APCs, which prevents them from binding to MHC class I molecules and being presented efficiently to T cells. However, encapsulated antigens in polymeric NPs are more efficiently presented via phagocytosis and provide better presentation for

T-cell priming and activation. Nonetheless, many of these approaches are still at the developmental stage, and the true feasibility of this kind of immunotherapy against cancer has not been elucidated.

Another biological aspect that may be exploited by nanotechnology is that blood vessels within tumors typically have pores that are less than a few hundred nanometers (Kievit and Zhang 2011; Le Guelte et al. 2011; Patel et al. 2012). Therefore, NPs with diameters less than 100 nm may cross these vessels to access cancer cells. Furthermore, at the tumor site, the highly metabolic cancer cells promote de novo aberrant angiogenesis. These poorly developed vessels lack appropriate fenestrations. Many nanoplatforms exploit various physical properties of tumor-associated blood vessels and the fact that the blood–tumor barrier becomes leaky in areas of disease (Palmer 2011). The development of leaky vessels in tumors combined with reduced lymphatic drainage in the CNS results in the accumulation of molecules less than 200 nm in size (Nair et al. 2011; Patel et al. 2012; Saha et al. 2010). This phenomenon is referred to as the enhanced permeability and retention effect (EPR) of NPs in tumor tissue (Saha et al. 2010). The ability of NPs to gain access to cancer cells through the tumor's aberrant leaky de novo blood vessels suggests that NPs may assume a role in treating various cancers, especially within the CNS, where the BBB normally restricts drug delivery.

CONCLUSION

Taken together, the individual use of NPs of various forms attached to molecules that promote immune activation directly, or the combined use of NPs with immunotherapeutic approaches that are currently being used or in development, opens the door for new opportunities to treat cancer. A very important question at this stage is how soon will this technology be developed and be available to cancer patients whose current treatment options are limited to surgery, radiation, and chemotherapy. Furthermore, the development of novel NPs that have diverse physical properties (e.g., molecular weight, water/lipid solubility) makes it difficult to standardize results from different laboratories using various animal models of cancer and NP platforms. What methods or mechanisms can be employed to synergize results from different laboratories utilizing a variety of nanoplatforms to improve the efficiency of translating discoveries to clinical practice? While we are inspired to develop NPs that meet the needs of specific cancers, we are also challenged to fully develop nanoplatforms that are safe and have minimal toxicity, are biodegradable, and have little or no potential for unwanted immunogenicity. These properties, to name a few, are essential in promoting this technology and translating it to clinical trials. Nonetheless, the number of clinical trials utilizing novel immunotherapeutic approaches is growing as we better understand the role of immunity in disease. While it is challenging to bring together these novel nanoplatform applications in a comprehensive way, the promise of NPs and the advantage that their size brings to overcoming certain biological barriers provide new hope in the fight against cancer.

REFERENCES

Alfaro, C., Perez-Gracia, J.L., Suarez, N. et al. 2011a. Pilot clinical trial of type 1 dendritic cells loaded with autologous tumor lysates combined with GM-CSF, pegylated IFN, and cyclophosphamide for metastatic cancer patients. *J Immunol* 187: 6130–42.

Alfaro, C., Suarez, N., Onate, C. et al. 2011b. Dendritic cells take up and present antigens from viable and apoptotic polymorphonuclear leukocytes. *PLoS One* 6: e29300.

Bitting, R.L., Armstrong, A.J., and George, D.J. 2011. Management options in advanced prostate cancer: what is the role for sipuleucel-T? *Clin Med Insights Oncol* 5: 325–32.

Chen, J., Ni, G., and Liu, X.S. 2011. Papillomavirus virus like particle-based therapeutic vaccine against human papillomavirus infection related diseases: immunological problems and future directions. *Cell Immunol* 269: 5–9.

Cruz, L.J., Tacken, P.J., Bonetto, F. et al. 2011. Multimodal imaging of nanovaccine carriers targeted to human dendritic cells. *Mol Pharm* 8: 520–31.

Edwards, L.J., and Evavold, B.D. 2011. T cell recognition of weak ligands: roles of signaling, receptor number, and affinity. *Immunol Res* 50: 39–48.

Fadel, T.R., Steenblock, E.R., Stern, E. et al. 2008. Enhanced cellular activation with single walled carbon nanotube bundles presenting antibody stimuli. *Nano Lett* 8: 2070–6.

Finocchiaro, G., and Pellegatta, S. 2011. Immunotherapy for glioma: getting closer to the clinical arena? *Curr Opin Neurol* 24: 6417.

Garland, S.M., and Smith, J.S. 2010. Human papillomavirus vaccines: current status and future prospects. *Drugs* 70: 1079–98.

Ge, W., Li, Y., Li, Z.S. et al. 2009. The antitumor immune responses induced by nanoemulsion-encapsulated MAGE1-HSP70/SEA complex protein vaccine following peroral administration route. *Cancer Immunol Immunother* 58: 201–8.

Irvin, D.K., Jouanneau, E., Duvall, G. et al. 2010. T cells enhance stem-like properties and conditional malignancy in gliomas. *PLoS One* 5: e10974.

Janikashvili, N., Bonnotte, B., Katsanis, E. et al. 2011. The dendritic cell-regulatory T lymphocyte crosstalk contributes to tumor-induced tolerance. *Clin Dev Immunol* 2011: 430394.

Khong, H.T., and Restifo, N.P. 2002. Natural selection of tumor variants in the generation of "tumor escape" phenotypes. *Nat Immunol* 3: 999–1005.

Kievit, F.M., and Zhang, M. 2011. Cancer nanotheranostics: improving imaging and therapy by targeted delivery across biological barriers. *Adv Mater* 23: H217–47.

Kitchin, N.R. 2011. Review of diphtheria, tetanus and pertussis vaccines in clinical development. *Expert Rev Vaccines* 10: 605–15.

Le Guelte, A., Dwyer, J., and Gavard, J. 2011. Jumping the barrier: VE-cadherin, VEGF and other angiogenic modifiers in cancer. *Biol Cell* 103: 593–605.

Leonard, R.C., Williams, S., Tulpule, A. et al. 2009. Improving the therapeutic index of anthracycline chemotherapy: focus on liposomal doxorubicin (Myocet). *Breast* 18: 218–24.

Marr, L.A., Gilham, D.E., Campbell, J.D. et al. 2012. Immunology in the clinic review series; focus on cancer: double trouble for tumours: bi-functional and redirected T cells as effective cancer immunotherapies. *Clin Exp Immunol* 167: 216–25.

Meng, J., Duan, J., Kong, H. et al. 2008. Carbon nanotubes conjugated to tumor lysate protein enhance the efficacy of an antitumor immunotherapy. *Small* 4: 1364–70.

Mescher, M.F., Popescu, F.E., Gerner, M. et al. 2007. Activation-induced non-responsiveness (anergy) limits CD8 T cell responses to tumors. *Semin Cancer Biol* 17: 299–308.

Nair, B.G., Varghese, S.H., Nair, R. et al. 2011. Nanotechnology platforms; an innovative approach to brain tumor therapy. *Med Chem* 7: 488–503.

Ostrowski, M., Carmo, N.B., Krumeich, S. et al. 2010. Rab27a and Rab27b control different steps of the exosome secretion pathway. *Nat Cell Biol* 12: 19–30; sup pp 1–13.

Palmer, A.M. 2011. The role of the blood brain barrier in neurodegenerative disorders and their treatment. *J Alzheimers Dis* 24: 643–56.

Patel, T., Zhou, J., Piepmeier, J.M. et al. 2012. Polymeric nanoparticles for drug delivery to the central nervous system. *Adv Drug Deliv Rev* 15;64(7): 701–5.

Perche, F., Benvegnu, T., Berchel, M. et al. 2011a. Enhancement of dendritic cells transfection in vivo and of vaccination against B16F10 melanoma with mannosylated histidylated lipopolyplexes loaded with tumor antigen messenger RNA. *Nanomedicine* 7: 445–53.

Perche, F., Gosset, D., Mevel, M. et al. 2011b. Selective gene delivery in dendritic cells with mannosylated and histidylated lipopolyplexes. *J Drug Target* 19: 315–25.

Petrelli, F., Borgonovo, K., and Barni, S. 2010. Targeted delivery for breast cancer therapy: the history of nanoparticle-albumin-bound paclitaxel. *Expert Opin Pharmacother* 11: 1413–32.

Prasad, S., Cody, V., Saucier-Sawyer, J.K. et al. 2011. Polymer nanoparticles containing tumor lysates as antigen delivery vehicles for dendritic cell-based antitumor immunotherapy. *Nanomedicine* 7: 1–10.

Rieger, J., Freichels, H., Imberty, A. et al. 2009. Polyester nanoparticles presenting mannose residues: toward the development of new vaccine delivery systems combining biodegradability and targeting properties. *Biomacromolecules* 10: 651–7.

Saha, R.N., Vasanthakumar, S., Bende, G. et al. 2010. Nanoparticulate drug delivery systems for cancer chemotherapy. *Mol Membr Biol* 27: 215–31.

Sasada, T., Komatsu, N., Suekane, S. et al. 2010. Overcoming the hurdles of randomised clinical trials of therapeutic cancer vaccines. *Eur J Cancer* 46: 1514–9.

Scheerlinck, J.P., and Greenwood, D.L. 2008. Virus-sized vaccine delivery systems. *Drug Discov Today* 13: 882–7.

Sexton, A., Whitney, P.G., Chong, S.F. et al. 2009. A protective vaccine delivery system for in vivo T cell stimulation using nanoengineered polymer hydrogel capsules. *ACS Nano* 3: 3391–400.

Sonpavde, G., Di Lorenzo, G., Higano, C.S. et al. 2012. The role of sipuleucel-T in therapy for castration-resistant prostate cancer: a critical analysis of the literature. *Eur Urol* 61(4): 639–47.
Srinivasan, M., and Frauwirth, K.A. 2009. Peripheral tolerance in CD8+ T cells. *Cytokine* 46: 147–59.
Szabo, A.T., and Carpentier, A.F. 2011. Immunotherapy in human glioblastoma. *Rev Neurol (Paris)* 167: 668–72.
Tan, A., De La Pena, H., and Seifalian, A.M. 2010. The application of exosomes as a nanoscale cancer vaccine. *Int J Nanomedicine* 5: 889–900.
Tanne, J.H. 2010. FDA approves prostate cancer "vaccine." *BMJ* 340: c2431.
Taylor, D.D., and Gercel-Taylor, C. 2011. Exosomes/microvesicles: mediators of cancer-associated immunosuppressive microenvironments. *Semin Immunopathol* 33: 441–54.
Thery, C., Ostrowski, M., and Segura, E. 2009. Membrane vesicles as conveyors of immune responses. *Nat Rev Immunol* 9: 581–93.
Wheeler, C.J., and Black, K.L. 2011. Vaccines for glioblastoma and high-grade glioma. *Expert Rev Vaccines* 10: 875–86.
Wheeler, C.J., Black, K.L., Liu, G. et al. 2008. Vaccination elicits correlated immune and clinical responses in glioblastoma multiforme patients. *Cancer Res* 68: 5955–64.
Wu, F., Wuensch, S.A., Azadniv, M. et al. 2009. Galactosylated LDL nanoparticles: a novel targeting delivery system to deliver antigen to macrophages and enhance antigen specific T cell responses. *Mol Pharm* 6: 1506–17.
Yu, J.S., Liu, G., Ying, H. et al. 2004. Vaccination with tumor lysate-pulsed dendritic cells elicits antigen-specific, cytotoxic T-cells in patients with malignant glioma. *Cancer Res* 64: 4973–9.
Yu, J.S., Wheeler, C.J., Zeltzer, P.M. et al. 2001. Vaccination of malignant glioma patients with peptide-pulsed dendritic cells elicits systemic cytotoxicity and intracranial T-cell infiltration. *Cancer Res* 61: 842–7.
Zhang, Z., Tongchusak, S., Mizukami, Y. et al. 2011. Induction of anti-tumor cytotoxic T cell responses through PLGA-nanoparticle mediated antigen delivery. *Biomaterials* 32: 3666–78.
Zhao, D., Alizadeh, D., Zhang, L. et al. 2011. Carbon nanotubes enhance CpG uptake and potentiate antiglioma immunity. *Clin Cancer Res* 17: 771–82.

11 Directed Drug Convection Using Magnetic Nanoparticles as Therapeutic Carriers

Meeting the Challenge of Specific Brain Pharmacotherapeutics, Non-Ligand-Based Central Nervous System Targeting, Including Magnetic Focusing

David F. Moore, Vinith Yathindranath, Torsten Hegmann, Donald Miller, Marc Pelletier, and Raphael Schiffmann

CONTENTS

INTRODUCTION

The burden of neurological disease in the United States is estimated to affect many millions of patients per year (Hirtz et al. 2007; NINDS 2009). Although effective reduction in this burden in some neurological disorders such as stroke has occurred through recent therapeutic advances, especially with the use of recombinant tissue plasminogen activator (rt-PA), there is an unmet need for other significant disease burdens such as Alzheimer's and Parkinson's diseases. The requirement and challenge of drug delivery to a prespecified area of the brain are unique. This is in part due to the intricate organizational structure of the brain or central nervous system (CNS) that is comparable to a multiorgan system. The increasing burden of neurological disorders with aging societal demographics therefore demands innovation in brain therapeutics and drug delivery. Interdisciplinary research in nanoscience can potentially translate and solve CNS drug delivery challenges, where unique partition coefficient and mass transport problems exist, colloquially termed the blood–brain barrier (BBB).

Penetrating the BBB and overcoming physical and metabolic barriers existing at the brain microvascular endothelial cells (BMECs) and the "glia limitans" with xenobiotics while minimizing the toxicological consequences to vulnerable brain parenchyma are a significant challenge. For a CNS drug to be therapeutically effective, it has to cross the BBB at a significant concentration.

Almost 98% of small-molecule drugs and 100% of large-molecule drugs under current pharmacological development have significant mass transport barriers at the BBB and restricted CNS entry (Pardridge 2006). As a rule of thumb, only lipophilic molecules, not exceeding molecular weights of around 500 Da, are capable of partitioning into the CNS and crossing the BBB (Miller 2002). Many of the drugs that have been developed or are under development have a molecular weight in the range of tens of kilodaltons. Penetrating the BBB and overcoming lack of targeted and directed systemic therapy are critical to enhancing the CNS therapeutic window and reducing toxicological consequences. This would potentially allow significant reductions in dosing, particularly in the case of directed delivery of high-molecular-weight drugs to specific brain areas. Current intratumoral, intrathecal, and intravenous injection methods used to deliver drugs do not address adequately the above issues.

The most commonly used translational therapy to deliver protein or gene across the BBB is the use of adeno-associated virus (AAV) and, in particular, the AAV9 serotype. Recent studies showed efficient vector delivery to glial cells throughout the brain, dorsal root ganglia neurons, and motor neurons within the spinal cord in nonhuman primates (Fu et al. 2011). Delivery via the cerebrospinal fluid (CSF), such as using intrathecal injections, achieves a more limited but specific targeting. The main limitation of this method included the need for repeated administration in the face of likely immune response with resulting diminishing efficacy. Other therapies such as bone marrow transplantation or small molecules are either inefficient or require the presence of a particular type of mutation in the case of single-gene disorders (Khanna et al. 2010; Schiffmann 2010).

A nanocarrier for CNS targeting should facilitate transport and delivery of drug molecules at the required site even across the BBB and should not cause an adverse immune response to the nanoparticles (NPs). NP size, shape, and composition determine many of their nanoscale properties and may allow property selectivity favoring CNS targeting, cell accumulation, and in vivo functional imaging. NP size may range from 10 nm to a few hundred nanometers and allow medicinal chemistry therapeutics to be absorbed, adsorbed, encapsulated, or chemically attached to the nanocarrier. Nanocarriers are easily convected through brain capillaries ~3 μm in diameter. Specificity can be achieved by modifying nanocarrier surfaces with a cell-specific ligand.

NANOPARTICLES FOR THE CNS

To address some of these issues, versatile magnetic NPs and nanocarriers were synthesized and constructed. As a non-ligand-based targeting mechanism, magnetic focusing (MF) of NPs has a huge potential for site-specific drug delivery. One of the earliest reports on magnetic targeting (MT) of tumors with cytotoxic agents was published three decades ago (Widder et al. 1981). In these studies, external magnets were used to localize magnetic carriers at the targeted site. Liu et al. (2010) used focused ultrasound and MT using external magnets (0.2, 0.4, and 0.55 T) tied to a rat's head in synergy to increase local concentration of epirubicin linked to Fe_3O_4 NPs in the rat brain (Mathieu and Martel 2010). Using external magnets for focusing might work in small animals but will be more challenging in patients as their efficiency depends on the close proximity between magnet and target tissue.

Recently, magnetic resonance navigation (MRN) was proposed as a solution to problems faced by external magnets in deep tissue MT. MRN is achieved by a conventional magnetic resonance imaging (MRI) scanner with modified gradient coils and software to facilitate 3D magnetic focusing. It is advantageous to use clinical MRI as it is widely available in hospitals across the world. A magnetic field of 1.5 T or higher in the MRI scanner overcomes the challenge of deep tissue magnetic focusing and hence can be used for MT anywhere in the human body. At present, clinical MRI produces a gradient of 40 mT m^{-1}, which can be useful to focus millimeter-size untethered ferromagnetic objects. MF of micron-sized particles or nanoaggregates will need gradients up to 400 mT m^{-1} (Pouponneau et al. 2011). There are technical constraints in designing coils with higher field gradient amplitude, and once that is achieved, a clinical MRI scanner could be used for magnetic focusing of NPs.

Mathieu and Martel (2010) used propulsion gradient coils in an MRI scanner with a 1.5-T field to steer a suspension of Fe_3O_4 aggregates of around 11 μm in a y-shaped channel. The flow rates were carefully controlled in this in vitro experiment, and steering ratios of NPs were measured for gradients of 50, 100, 200, and 400 mT m^{-1}. It was observed that the steering ratio increases with the amplitude of the magnetic gradient and that the largest magnetic aggregates require lower magnetic gradient amplitude. Pouponneau et al. (2011) used MRN to guide and image magnetic FeCo-PLGA [iron–cobalt NPs encapsulated in poly(D,L-lactic-*co*-glycolic acid)] microparticles 58 μm in diameter through a phantom mimicking hepatic arteries. For steering studies, a 1.5-T MRI was used, with a gradient magnetic field of 400 mT m^{-1}. The microparticles formed aggregates in the MRI, which facilitated steering with the applied field gradient. Steering efficiency was measured by estimating the amount of iron and cobalt from two channels, using atomic absorption spectroscopy (AAS), and this showed successful magnetic steering.

A considerable amount of work has been carried out in designing magnetic nanocarriers that can be focused magnetically. The bulk of nanotherapeutics research on drug delivery and in vivo imaging is formulated upon well-characterized systems in terms of synthesis, chemical modifications, and biocompatibility, and in some cases that are already approved by the Food and Drug Administration (FDA). Iron oxide nanoparticles (IONPs) composed of magnetite (Fe_3O_4) and maghemite (γ-Fe_2O_3) are superparamagnetic at room and body temperature and are magnetized only in the presence of an external magnetic field. These particles are biocompatible and are already used in clinical medicine as a T2 contrast agent for MRI (Bulte et al. 2001). IONPs have the potential to enable a range of drug carrier and diagnostic options for CNS disorders. IONPs with core sizes in the tens of nanometers and suitable surface modification have the potential to permeate the BBB and tissue for drug delivery (Bulte et al. 2001; Chertok et al. 2008; Mykhaylyk et al. 2001; Rousseau et al. 1997). Drug-loaded IONPs combined with convection-enhanced diffusion (CED) techniques could be a useful strategy for drug delivery into the CNS, allowing novel treatment of brain tumors and fulminant neuronopathic disorders, such as Gaucher disease (Lonser et al. 2007). In addition to in situ MRI (Corot et al. 2006; Miyawaki et al. 2006; Tiefenauer et al. 1996), the intrinsic magnetic properties of IONPs can be exploited to provide thermal mediation for tumor thermotherapy (Gordon et al. 1979).

IONPs can be prepared with relative ease and are economically viable in comparison to many other nanocarriers. Among aqueous methods, coprecipitation of Fe(II) and Fe(III) in the presence of a base, usually aqueous ammonia, yields magnetite and maghemite (Massart 1981). In another method, hydrolysis of an organometallic iron precursor using sodium borohydride in aqueous media yielded pure magnetite using nontoxic reagents and milder conditions (Yathindranath et al. 2011). These aqueous methods yield IONPs that are more suitable for biomedical applications as they do not have remnant toxic surfactants on their surfaces. In nonaqueous methods, high temperature thermal decomposition of an organometallic precursor is widely used for the narrow size distribution of resulting NPs (Sun and Zeng 2002). The final products in most cases have hydrophobic surfactants, which have to be removed before further modifications for biomedical applications. Other methods of synthesis include sol-gel, microemulsion, and sonochemical methods (Bilecka et al. 2008; Kim et al. 2005; Sun et al. 2004).

To serve as prodrugs with adequate blood circulation half-life and targetability, IONPs need to avoid immediate phagocytosis and an unwanted immune response (Bertram 2006). Hydrophobic colloids are rapidly cleared from circulation by the reticuloendothelial system, particularly in the spleen and liver. Attachment of hydrophilic biocompatible coatings on the surface of IONPs may be used to increase their colloidal stability and circulation half-life in the bloodstream (Gref et al. 1994; Sun et al. 2008; Wadghiri et al. 2003). In addition, IONP coating is one of the available means to selectively prevent IONP irreversible aggregation.

MAGNETICALLY DIRECTED DRUG CONVECTION

In our recent work, we demonstrated a proof of principle for magnetically directed drug convection (MDDC) of a model drug protein using an IONP nanocarrier in vitro in agarose gel mimicking healthy brain tissue in a 7-T MRI (Yathindranath et al. 2009, 2011). For the studies, IONPs were synthesized using Massart's method of coprecipitation. The nanocarrier design used is schematically illustrated in Figure 11.1 and has an iron oxide core, biocompatible polymer coating, and bovine serum albumin (BSA, MW ~ 65 kD) physisorbed on the particle surface. L-Arginine was used as a stabilizer to prevent irreversible aggregation and preserve protein conformation during drying and resuspension in a physiological buffer. Biocompatible surfactants polyethylene glycol (PEG), glutamic acid, poly(ethyl methacrylate) (PEMA), and poly(2-hydroxyethyl methacrylate) (PHEMA) were used to coat IONPs. Figure 11.2 shows the high resolution transmission electron microscopy (HRTEM) and transmission electron microscopy (TEM) images of synthesized PEG-coated IONPs, a precursor for protein-immobilized particles. Protein-immobilized particles demonstrated excellent colloidal stability in water and stayed well suspended without any sign of agglomeration or settling. This is illustrated in Figure 11.3, where three differently constituted nanocarriers demonstrate their colloidal stability and their accumulation under a magnetic field.

For directing NPs magnetically in vitro, different IONPs with immobilized BSA were suspended in agarose gel (0.4% to 1.0%) and egg white while positioned in a 7-T magnetic field. A gradient of the 7-T horizontal-bore MRI magnet was applied, followed by direct magnetic resonance (MR) visualization. Agarose gel 0.6% closely resembles the mass transport properties of healthy brain tissue, and egg white is a suitable model for a complex biological fluid. Figure 11.4 illustrates the effect of increasing the agarose gel concentration across the 0.4%–1.0% range. As would be expected, there is decreasing

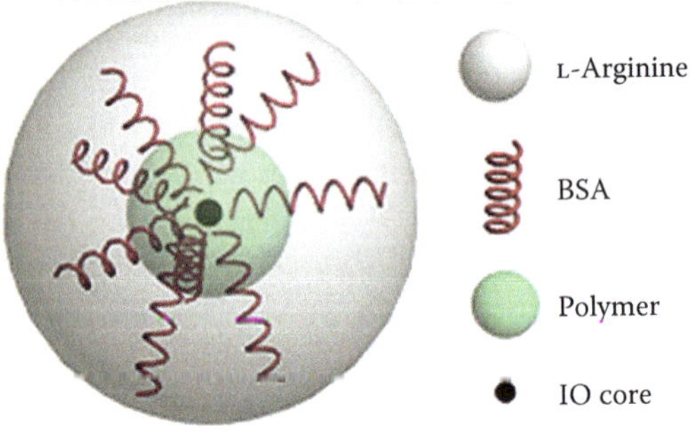

FIGURE 11.1 Schematic of the iron oxide core-shell nanoparticles (IO cs-NPs).

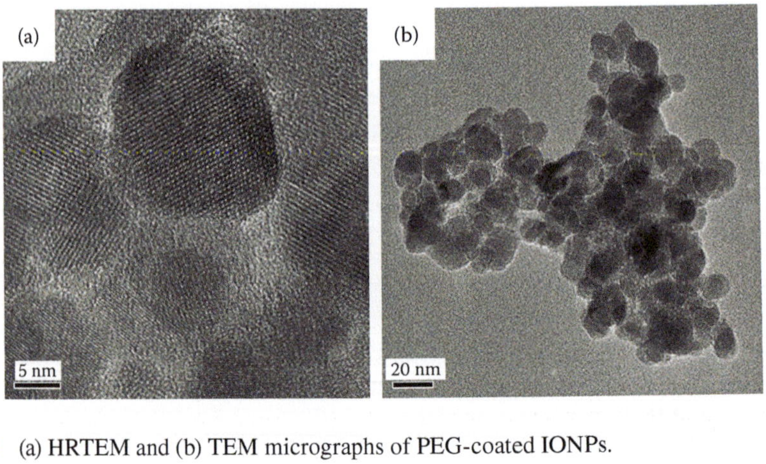

FIGURE 11.2 (a) HRTEM and (b) TEM micrographs of PEG-coated IONPs.

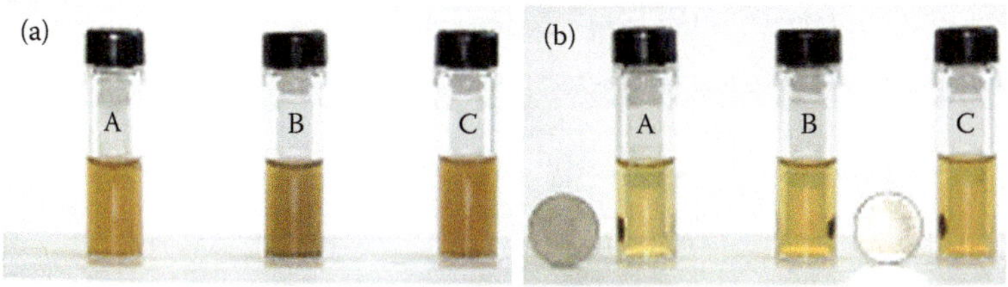

FIGURE 11.3 (a) Dispersion of BSA-immobilized on (A) PEG-, (B) dextran-, and (C) PEMA-coated IONPs in deionized (DI) water. (b) Illustration of IO cs-NP attraction to small rare earth magnet.

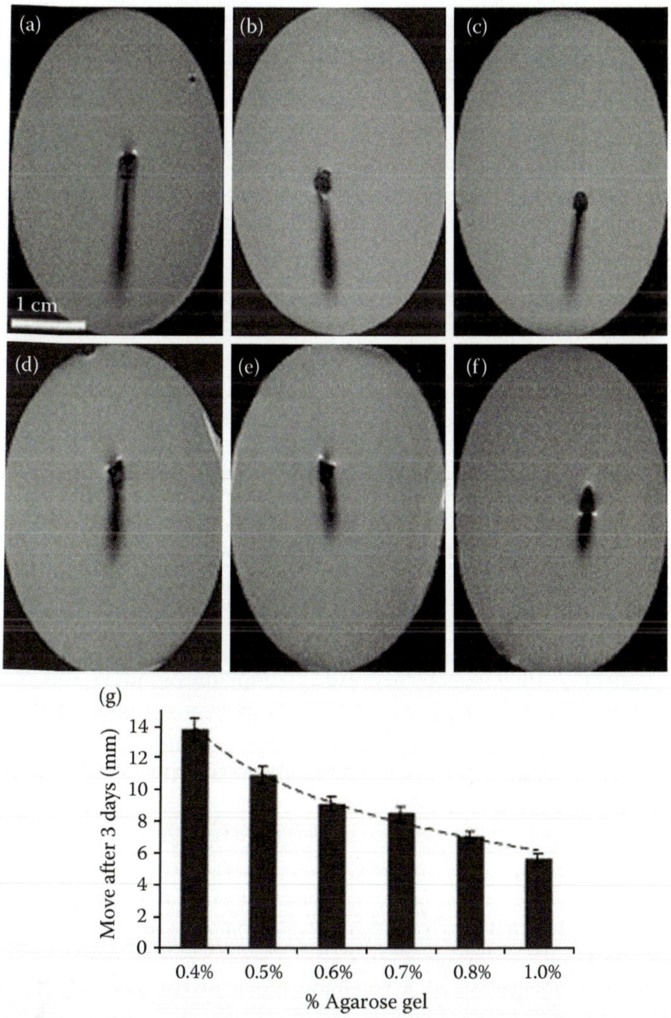

FIGURE 11.4 T2-weighted MR images of C. BSA immobilized PEMA-coated IONPs in gels prepared in DI water with different agarose concentrations (w/v): (a) 0.4%, (b) 0.5%, (c) 0.6%, (d) 0.7%, (e) 0.8%, and (f) 1.0%. (g) Plot of movement after 3 days (mm) versus % agarose gel.

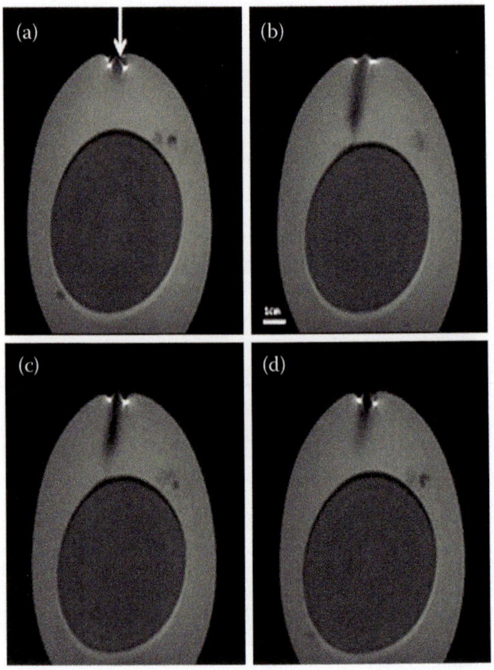

FIGURE 11.5 Representative T2-weighted MR images of an egg injected with BSA-immobilized PEMA-coated IONPs acquired (a) at time zero and (b–d) at three slice positions within the egg after 60 h of exposure to a magnetic field gradient. The magnetic field was lowest at the injection site, indicated with a white arrow, and highest at the bottom of the images shown. Scale bar represents 1 cm.

mass transport convection under the MRI gradient field. From concurrent protein staining, comigration of BSA along with a nanocarrier was observed. Figure 11.5 illustrates mass transport of PEMA-coated IONPs in egg white (Yathindranath et al. 2009). Such nanocarriers for large molecules might be able to carry and deliver large peptide molecules, such as growth factors to injured brain regions or genes for gene therapy of inherited metabolic disorders such as Gaucher disease type II and type III. It is likely that future development of therapeutic agents across the BBB will also include the combination of vascular endothelium-specific epitopes but also other means to manipulate gene expression, such as RNA interference (RNAi) in cases of dominant disorders (Chen et al. 2009; McBride et al. 2011).

As a potential treatment paradigm, MT might be helpful in achieving CNS targeting specificity of drug delivery, thereby allowing an increased therapeutic window and reducing the risk of drug toxicity while increasing drug tissue selectivity. At present, magnetic focusing of NPs is still in a development phase, and a thorough understanding of potential carrier systems coupled with advancement in technology to steer IONPs magnetically inside the body is important before MT can find its full potential, particularly for the CNS.

REFERENCES

Bertram, J. 2006. MATra—Magnet Assisted Transfection: combining nanotechnology and magnetic forces to improve intracellular delivery of nucleic acids. *Curr Pharm Biotechnol* 7: 277–85.

Bilecka, I., Djerdj, I., and Niederberger, M. 2008. One-minute synthesis of crystalline binary and ternary metal oxide nanoparticles. *Chem Commun (Camb)* 7: 886–8.

Bulte, J.W.M., Douglas, T., Witwer, B. et al. 2001. Magnetodendrimers allow endosomal magnetic labeling and in vivo tracking of stem cells. *Nat Biotechnol* 19: 1141–7.

Chen, Y.H., Chang, M., and Davidson, B.L. 2009. Molecular signatures of disease brain endothelia provide new sites for CNS-directed enzyme therapy. *Nat Med* 15: 1215–8.

Chertok, B., Moffat, B.A., David, A.E. et al. 2008. Iron oxide nanoparticles as a drug delivery vehicle for MRI monitored magnetic targeting of brain tumors. *Biomaterials* 29: 487–96.

Corot, C., Robert, P., Idee, J.M. et al. 2006. Recent advances in iron oxide nanocrystal technology for medical imaging. *Adv Drug Deliv Rev* 58: 1471–504.

Fu, H., Dirosario, J., Killedar, S. et al. 2011. Correction of neurological disease of mucopolysaccharidosis IIIB in adult mice by rAAV9 trans-blood–brain barrier gene delivery. *Mol Ther* 19: 1025–33.

Gordon, R.T., Hines, J.R., and Gordon, D. 1979. Intracellular hyperthermia—Biophysical approach to cancer treatment via intracellular temperature and biophysical alterations. *Med Hypotheses* 5: 83–102.

Gref, R., Minamitake, Y., Peracchia, M.T. et al. 1994. Biodegradable long-circulating polymeric nanospheres. *Science* 263: 1600–3.

Hirtz, D., Thurman, D.J., Gwinn-Hardy, K. et al. 2007. How common are the "common" neurologic disorders? *Neurology* 68: 326–37.

Khanna, R., Benjamin, E.R., Pellegrino, L. et al. 2010. The pharmacological chaperone isofagomine increases the activity of the Gaucher disease L444P mutant form of beta-glucosidase. *FEBS J* 277: 1618–38.

Kim, E.H., Lee, H.S., Kwak, B.K. et al. 2005. Synthesis of ferrofluid with magnetic nanoparticles by sonochemical method for MRI contrast agent. *J Magn Magn Mater* 289: 328–30.

Liu, H.L., Hua, M.Y., Yang, H.W. et al. 2010. Magnetic resonance monitoring of focused ultrasound/magnetic nanoparticle targeting delivery of therapeutic agents to the brain. *Proc Natl Acad Sci U S A* 107: 15205–10.

Lonser, R.R., Schiffman, R., Robison, R.A. et al. 2007. Image-guided, direct convective delivery of glucocerebrosidase for neuronopathic Gaucher disease. *Neurology* 68: 254–61.

Massart, R. 1981. Preparation of aqueous magnetic liquids in alkaline and acidic media. *IEEE Trans Magn* 17: 1247–8.

Mathieu, J.B., and Martel, S. 2010. Steering of aggregating magnetic microparticles using propulsion gradients coils in an MRI scanner. *Magn Reson Med* 63: 1336–45.

McBride, J.L., Pitzer, M.R., Boudreau, R.L. et al. 2011. Preclinical safety of RNAi-mediated HTT suppression in the rhesus macaque as a potential therapy for Huntington's disease. *Mol Ther* 19: 2152–62.

Miller, G. 2002. Drug targeting. Breaking down barriers. *Science* 297: 1116–8.

Miyawaki, J., Yudasaka, M., Imai, H. et al. 2006. In vivo magnetic resonance imaging of single-walled carbon nanohorns by labeling with magnetite nanoparticles. *Adv Mater* 18: 1010–4.

Mykhaylyk, O., Cherchenko, A., Ilkin, A. et al. 2001. Glial brain tumor targeting of magnetite nanoparticles in rats. *J Magn Magn Mater* 225: 241–7.

NINDS. 2009. NINDS Overview. http://www.ninds.nih.gov/about_ninds/ninds_overview.htm.

Pardridge, W.M. 2006. Molecular Trojan horses for blood–brain barrier drug delivery. *Curr Opin Pharmacol* 6: 494–500.

Pouponneau, P., Leroux, J.C., Soulez, G. et al. 2011. Co-encapsulation of magnetic nanoparticles and doxorubicin into biodegradable microcarriers for deep tissue targeting by vascular MRI navigation. *Biomaterials* 32: 3481–6.

Rousseau, V., Denizot, B., Pouliquen, D. et al. 1997. Investigation of blood–brain barrier permeability to magnetite-dextran nanoparticles (MD3) after osmotic disruption in rats. *Magn Reson Mater Phys Biol Med* 5: 213–22.

Schiffmann, R. 2010. Therapeutic approaches for neuronopathic lysosomal storage disorders. *Jo Inherit Metab Dis* 33: 373–9.

Sun, C., Veiseh, O., Gunn, J. et al. 2008. In vivo MRI detection of gliomas by chlorotoxin-conjugated superparamagnetic nanoprobes. *Small* 4: 372–9.

Sun, S., and Zeng, H. 2002. Size-controlled synthesis of magnetite nanoparticles. *J Am Chem Soc* 124: 8204–5.

Sun, S., Zeng, H., Robinson, D.B. et al. 2004. Monodisperse MFe2O4 (M = Fe, Co, Mn) nanoparticles. *J Am Chem Soc* 126: 273–9.

Tiefenauer, L.X., Tschirky, A., Kuhne, G. et al. 1996. In vivo evaluation of magnetite nanoparticles for use as a tumor contrast agent in MRI. *Magn Reson Imaging* 14: 391–402.

Wadghiri, Y.Z., Sigurdsson, E.M., Sadowski, M. et al. 2003. Detection of Alzheimer's amyloid in transgenic mice using magnetic resonance microimaging. *Magn Reson Med* 50: 293–302.

Widder, K.J., Morris, R.M., Poore, G. et al. 1981. Tumor remission in Yoshida sarcoma-bearing rats by selective targeting of magnetic albumin microspheres containing doxorubicin. *Proc Natl Acad Sci U S A* 78: 579–81.

Yathindranath, V., Hegmann, T., van Lierop, J. et al. 2009. Simultaneous magnetically directed drug convection and MR imaging. *Nanotechnology* 20: 405101.

Yathindranath, V., Rebbouh, L., Moore, D.F. et al. 2011. A versatile method for the reductive, one-pot synthesis of bare, hydrophilic and hydrophobic magnetite nanoparticles. *Adv Funct Mater* 21: 1457–64.

12 Intralesional and Intraparenchymal Convection-Enhanced Delivery of Nanoparticles

Tatsuhiro Fujii, Daniel Angeles, Li-Chen Hsu, Rahul Jandial, and Mike Y. Chen

CONTENTS

INTRODUCTION

Despite the development of promising therapeutic compounds that target neurological diseases, certain characteristics of the central nervous system (CNS) limit their penetration and efficacy. Since the first description of the blood–brain barrier (BBB) by Paul Ehrlich in 1885, investigators have found that systemic modes of delivery often fail to achieve adequate tissue concentrations of the drug in targeted areas of the brain. To circumvent the BBB, new and innovative delivery techniques were developed. One technique is convection-enhanced delivery (CED), otherwise referred to as high-flow microinfusion. A refined form of local drug delivery pioneered by Oldfield and colleagues at the National Institutes of Health (NIH), CED, at its essence, is a highly controlled form of injection. A catheter is surgically placed directly in the affected region of the brain, while the drug solution is infused through the delivery device at a predetermined rate. This technique allows delivery of highly concentrated solutions; exact placement of the catheter tip minimizes exposure of normal tissue to the drug; and the BBB restricts the infused agent from entering the systemic circulation.

The increasing number of studies that examine the nature of CED reflects the growing interest in this procedure as a viable mode of therapeutic delivery. These studies have broadened the application of CED, which has reached a level of sophistication that allows for the distribution of not only small molecules but also nanoparticles. This chapter will discuss the underlying principles that govern the effective infusion of nanoparticles using this technique and the promising findings reported in animal and human studies that lay the framework for future clinical use of CED.

PHYSICAL PRINCIPLE

CED consists of an external, automated pump that continuously drives the flow of the desired therapeutic agent through the delivery device. The delivered agent exits through a cannula or catheter that is stereotactically placed directly into the brain or tumor parenchyma. In some cases, an outer guide cannula can initially be stereotactically directed to a point that falls just above the targeted region of the brain. The inner cannula can then be guided through the outer cannula with the tip of the inner cannula reaching the desired infusion point. In doing so, this two-cannula technique maintains the integrity of the predetermined trajectory of the cannula placement (see Figure 12.1) (Lonser et al. 2007). The pressure gradient at the tip of the catheter resulting from the pump action generates bulk flow of fluid, which includes the drug solution, through the interstitial space. Although numerous factors affect permeation of the desired agent, such as hydraulic conductivity and poroelasticity of the surrounding tissue, this direct infusion of the therapeutic agent into the brain avoids many of the problems encountered by systemic modes of delivery.

However, heterogeneity of neural tissue can lead to complications due to preferential flow of the infusate along certain tracts or pathways in the brain (anisotropy); subsequent dilution of delivered fluids can occur due to diffusion along concentration gradients (Wang and Olbricht 2011). To accurately predict distribution of the infusate, various laws and mathematical expressions have been implemented to describe fluid movement during the CED process. In particular, Darcy's law, which relates fluid viscosity, discharge rate through a porous medium, and the pressure drop over a given distance, characterizes the early advective component of infusate flow (Linninger et al. 2008; Morrison et al. 1999; Neeves et al. 2006). Similarly, Fick's law of diffusion governs the latter part of infusate flow through the target tissue (Sykova and Nicholson 2008). Alongside these analytical approaches to predicting fluid transport, computational simulations were also generated to account for the simultaneous occurrence of diffusion and infusion of the delivered agent. Chen et al.'s (2004) development of an agarose gel surrogate for the in vivo brain confirmed mathematical descriptions

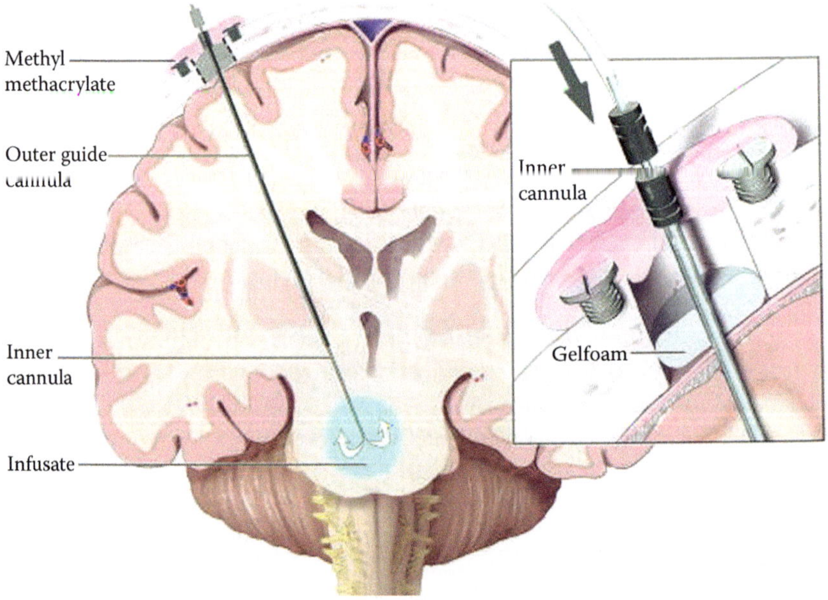

FIGURE 12.1 Illustrative coronal view of the brain showing an inner–outer two-cannula CED system. The outer guide cannula is secured to the exterior surface of the skull with anchor screws and adhering material such as methyl methacrylate. The inner infusion cannula is guided through and beyond the tip of the outer cannula to the targeted infusion site. (Reprinted from Lonser, R.R. et al., *J Neurosurg*, 107: 192, 2007. With permission.)

based on the poroelastic theory of positive-pressure infusion processes. Furthermore, their discovery of a brain tissue phantom provided a predictive and inexpensive tool for realistic strategies of intraparenchymal drug deliveries.

With the development of instruments, such as magnetic resonance imaging (MRI) and sonograms, which permit real-time tracking of infused material, more complex models of infusate distribution and flow pattern emerged (Murad et al. 2006; Saito et al. 2005; Szerlip et al. 2007). Most notably, the work of Raghavan et al. (2006) illustrated the advancement of accurate treatment planning. Their computer-generated models rely on both anatomical and diffusion tensor MRI as a source for data input. These data points are supplemented with a description of infusate movement using a stochastic differential equation that incorporates both advective and diffusive terms. Predictions of drug concentration at particular times and locations relative to the delivery site are either confirmed or manipulated using feedback from the imaging device. By following the movement of traceable material in the targeted region of the brain, these researchers collected an array of information pertaining to varying concentration gradients at neighboring points in the tissue. Not only does this model allow anisotropic variations, but it also incorporates potential sinks for the infusate such as leakage into the subarachnoid space, drainage into cavities, and fluctuations in tissue clearance rates. Subsequent human trials conducted by these researchers have shown that treatment plans can be altered while administration of the traceable material is in progress due to real-time feedback (Sampson et al. 2007). With enough time intraoperatively, any deviations from the predicted flow pattern and desired distribution can be quickly corrected by manipulating delivery parameters such as injection site and infusion rate.

Advancements in previously discussed models of infusion and real-time imaging technology enabled investigators to improve the efficacy of CED as an innovative mode of delivery. Although similar in form, the resemblance of CED to techniques of direct-injection delivery can be misleading since numerous parameters exist in the former that need to be controlled for proper infusate distribution. Particle size of the infused agent is one of the most important remaining parameters. Multiple studies have examined the feasibility of using CED to deliver nanoparticles in the CNS. This application holds particular relevance to gene therapy and the distribution of viruses and virus-sized particles, which are measured in the nanometer range. While a member of Dr. Edward Oldfield's laboratory at the NIH, the senior author performed some of the first studies examining the potential of nanosized particle distribution using CED (Chen et al. 2005). Results showed that channels existing in the extracellular space were sufficiently large to accommodate the convective distribution of polystyrene nanosphere particles ranging in size from 20 to 200 nm. Accordingly, the volume of distribution of Cy5-labeled adeno-associated virus capsids (23 nm in diameter) was similar to that of 20-nm nanospheres.

In addition to particle size, the study examined the effect of surface characteristics of nanoparticles that were convectively distributed in the brain (Chen et al. 2005). The volume of distribution of albumin-coated nanospheres was compared to that of the noncoated counterparts. Interestingly, despite the subtle increase in diameter due to the monolayer albumin coat, all coated nanospheres, with the exception of the 20-nm particles, dramatically increased in distribution volume. This finding was attributed to the increased hydrophilicity due to masking of hydrophobic polystyrene components by the albumin coat. Additionally, the albumin coat potentially masked the carboxylate groups present on the surface of the nanosphere, thereby effectively altering the surface charge that contributes to binding interactions with the extracellular matrix. Together, these findings suggest that within a given size, the surface characteristics of the infused particle substantially influence effective CED of nanoparticles.

In addition to the characteristics of the nanoparticle itself, variables of the fluid component (vehicle) may also affect the spread of the infused agent. In particular, fluid viscosity was shown to have a significant effect on the efficacy of particle distribution. Perlstein et al. (2008) used MRI to trace the convective distribution of maghemite nanoparticles (MNPs) suspended in vehicles with varying viscosity. They found greater infusate distribution in rats that had received a solution of

MNPs mixed in 3% or 6% polyethylene glycol, a highly viscous fluid, compared to animals receiving MNPs mixed in saline. Their findings were consistent with earlier works that suggested that low-viscosity fluids were more prone to leak back along the catheter, while high-viscosity infusates are more efficiently convected (Mardor et al. 2005). Mardor et al. (2005) examined the effect of viscosity by varying the amount of albumin or by adding sucrose. Both increasing the amount of albumin and adding sucrose increased the interstitial spread of the infusate. However, interpreting these two studies is not straightforward; there were other confounding factors. In Perlstein's study, addition of polyethylene glycol potentially altered surface characteristics by adding a hydrophilic coat to nanoparticles (Perlstein et al. 2008). In Mardor's experiment, the ability of albumin to act as a carrier for gadolinium and the additional diffusive component conferred by the sucrose may have influenced convective distribution as well as viscosity changes (Mardor et al. 2005).

Unrelated to the intrinsic characteristics of the infusate, there are two other parameters of CED known to influence particle distribution: rate of infusion and cannula size. Many of the first CED nanoparticle studies characterized the distribution of albumin (7.2-nm diameter) for its lack of diffusivity and inertness that prevents binding to elements of the extracellular matrix. In a rodent model, Chen et al. (1999) examined the convective distribution of C14-albumin at varying infusion rates. The experiment revealed that the distribution of the infusate was more spherical and associated with less leak back (retrograde flow of infusate around the cannula) at lower rates of infusion than at higher rates of infusion. Varying the infusate concentration as well as the preinfusion "sealing time" (time between insertion of the cannula and start of the infusion) had no observable effect on the volume of distribution, regional distribution, or infusate recovery. Szerlip et al. (2007) used real-time imaging of ferumoxtran-10, a traceable virus-sized iron oxide particle, in rats and nonhuman primates to examine the effects of the infusion rate. Significant leak back along the catheter was detected at rates above 1.5 μL/min. Although lower rates of infusion may indeed minimize any damage to brain tissue, manipulating infusion parameters must maintain a substantial pressure gradient, originating from the cannula site, to overcome interstitial pressures and intrinsic anisotropy within the targeted tissue. These researchers also observed a pattern of particle distribution that was characterized by formation of a cylindrical annulus at the distal portion of the catheter. As increasing amounts of infusate were delivered, the annulus expanded radially, adopting a more spherical distribution. Cannula size was also found to significantly affect the convection-based distribution of particles. Chen et al. (1999) reported that larger cannulae (28- and 22-gauge) caused more leak back during CED compared to the same procedure using a smaller cannula size (32-gauge).

By controlling these variables that govern the efficacy of delivery, CED allows slow, controlled infusion of the desired agent in a manner that expands the extracellular space, a condition akin to that of vasogenic edema, while keeping the integrity of the surrounding tissue. In addition to technical parameters of CED, such as the infusion rate, design of the therapeutic nanoparticles must be considered. Factors such as size and surface characteristics largely determine distribution of these nanoparticles over clinically relevant distances. With these parameters in mind, CED offers a unique and innovative method of direct intraparenchymal delivery, achieving therapeutic concentrations of antitumoral compounds in a localized area of the CNS.

ANIMAL STUDIES

Numerous experiments, using small and large animal models, were performed using CED of nanoparticles such as immunotoxins, antibodies, growth factors, and viruses to targeted tumors and regions of the brain. The intent of these studies was to investigate therapeutic applications; at the same time, however, these studies provided additional characterization of the convective delivery of nanoparticles to the CNS.

Before the procedure to surgically implant the cannula, the animal is anesthetized and secured in a stereotactic frame. The cannula is then attached to an infusion apparatus (see Figure 12.2). Sites of infusion include the striatum, brainstem, and tumors. Generally, an external pump was used to

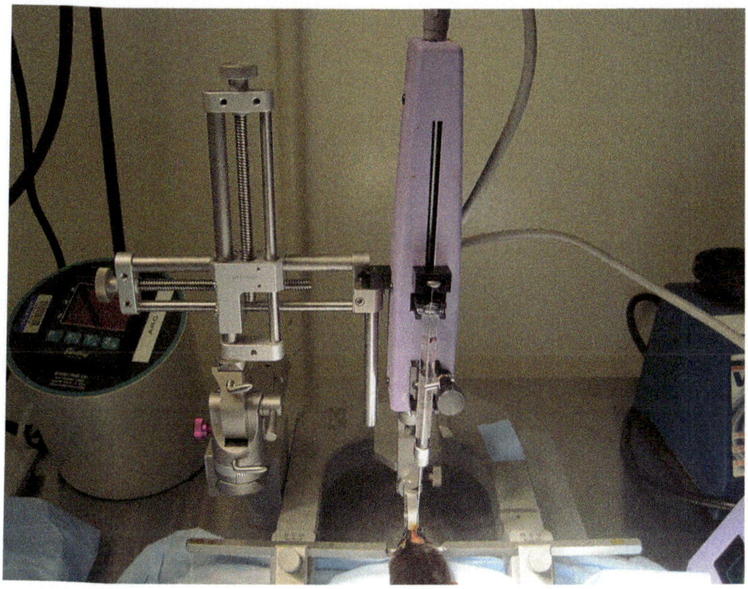

FIGURE 12.2 Photograph of the CED apparatus. The mouse is placed in a stereotactic frame, while the needle of the syringe is guided through the burr hole and directly into the cerebrum. The syringe is connected to an automated pump, allowing for a slow, controlled delivery of the infused agent.

supply the pressure gradient needed for the infusion; however, studies have occasionally used an internally implanted pump as well. The infusates in the studies are usually potential therapeutic agents, tracer solutions, or a combination of the two to allow study of flow patterns and concentration gradients. Particle size varies from small macromolecules to viruses; CED of cells (50-μm average diameter) is not feasible due to the limited size of channels within the extracellular space.

An illustrative study was performed by Murad et al. (2006) investigating the delivery of therapeutic doses of an interleukin-13 (IL13) *Pseudomonas* exotoxin conjugate (IL13-PE) to the brainstem of rats and nonhuman primates. Thirty Sprague–Dawley rats and eight adult nonhuman primates were convected with varying concentrations of IL13-PE. The animals were sacrificed at 28 days postinfusion for analysis of brain toxicity. None of the rats demonstrated any signs of toxicity in the brainstem at any of the concentrations. In the nonhuman primates, no evidence of toxicity was observed with IL13-PE at concentrations of 0.25 and 0.5 μg/mL; however, necrosis and inflammation in the infused region of the brainstem were observed at the highest concentration of IL13-PE (10 μg/mL). Despite this limitation, 0.5 μg/mL of IL13-PE was cytotoxic to malignant glioma cells. Therefore, the authors concluded that CED can be used to safely deliver glioma-toxic concentrations of IL13-PE to the primate brainstem.

Another study performed by Sampson et al. (2003) investigated the maximum tolerated dose (MTD) and dose-limiting toxicity of TP-38 in athymic rats and Rhesus macaques. Similar to IL13-PE, TP-38 is a protein nanoparticle in which transforming growth factor alpha (TGF alpha) is linked to *Pseudomonas* exotoxin. The MTD for rats was 33.3 μg/mL since higher concentrations caused demyelination and necrosis. The Rhesus macaques were treated at three different concentrations. The researchers found demyelination and necrosis in animals treated at 30 μg/mL but not at 1 nor 10 μg/mL. From these MTD observations, the authors extrapolated the MTD values for humans. Sampson et al. also tested the therapeutic efficacy of TP-38 in athymic rats. Tumor-bearing animals treated with 1 or 10 μg/mL of TP-38 had 100% and 90% survival, respectively, and evidence of tumor eradication.

Although proteins, by virtue of their size, constitute a nanoparticle, this may not fit preconceived notions of nanomedicine. Krauze et al. (2007) convected two prototypical nanoparticles, nanoliposomal (nLs) CPT-11 (irinotecan) and PEGylated liposomal doxorubicin (Doxil). Rats with intracranial malignant glioma U251MG and U87MG xenografts were infused 7 days after tumor implantation. In the U251MG model, animals receiving both the nLs-CPT-11 and Doxil survived until the end of the study (>100 days) compared to controls, which survived 36–44 days. In the U87MG model, animals receiving the combination treatment survived up to 70 days after implantation compared to the 18–21 days for rats receiving the vehicle control. Based on these results, the authors concluded that the combination treatment of nLs-CPT-11 and Doxil resulted in a significant survival advantage. However, they also note that there must be improvements in the CED technologies for successful clinical endeavors.

Another class of agents in the nanoparticle size range includes viruses. Richardson et al. (2011) investigated the feasibility of using viruses to achieve large distribution volumes in nonhuman primates for localized gene therapy. In the study, a novel MRI-guided delivery system was used to convectively distribute adeno-associated virus serotype 2 (AAV2) encoding glial-derived neurotrophic factor (GDNF) to the brains of nonhuman primates. The animals received bilateral co-infusions of the gadoteridol (Gd)/AAV2-GDNF into two sites within each putamen, and three of the subjects received larger infusion volumes into the thalamus. The GDNF expression patterns in the putamen correlated with areas of Gd distribution seen on MRI, and ratios of the volume of distribution to volume of infusion were similar in the putamen and thalamus. These findings demonstrated that large volumes of distribution can be achieved with a virus by using CED, a result that has significant positive implications for regional gene therapy of the brain.

CLINICAL TRIALS

To date, numerous human clinical trials have successfully demonstrated delivery of nanoparticles using CED. Although a variety of particles, including enzymes, antibodies, and viruses, have been convected, the vast majority of trials examined the delivery of receptor-directed toxins. Receptor-directed toxins consist of one protein that targets a particular feature on the cell surface and another protein that causes cytotoxicity (Chen et al. 2005). One previously mentioned example that has been used experimentally in patients is TP-38, which consists of TGF alpha linked to the *Pseudomonas* exotoxin (Choudhary et al. 2011).

The classic receptor-directed toxin used together with CED is Tf-CRM107, a transferrin conjugate bound to a point mutated diphtheria toxin. This therapeutic nanoparticle has received much attention as the first compound to be delivered by CED in a clinical trial. Transferrin receptors, which facilitate the uptake of iron, are overexpressed in rapidly dividing cells, such as glioblastoma, yet are much less prevalent in normal brain tissue. Transferrin-conjugated toxin therapies utilize this metabolic difference for selective targeting (Weaver and Laske 2003). In a phase I study, the delivery of Tf-CRM107 was well tolerated, reduced tumor volume by more than 50% in 9 out of 10 patients, and resulted in a median survival of 75 weeks compared to 36 weeks in nonresponders (Laske et al. 1997). In a phase II study, complete and partial tumor response without severe toxicity occurred in 35% of patients treated with Tf-CRM107 (Weaver and Laske 2003).

A further iteration of the immunotoxin approach utilizes cintredekin besudotox, a recombinant protein consisting of IL13 and a truncated form of *Pseudomonas* exotoxin, PE38QQR. The initial phase I trial showed some promising clinical results for treating recurrent glioblastoma multiforme (GBM) (Kunwar et al. 2007). However, a phase III study (PRECISE) of 296 participants who received either CED of cintredekin besudotox postoperatively or a Gliadel wafer implanted at the time of surgery demonstrated no survival benefit (Kunwar et al. 2010). Of interest, Sampson et al. (2010) reviewed the position of catheters and noted that only about 50% of catheters were placed in accordance with protocol guidelines. While it is unclear if proper placement would have changed the final outcome, Sampson et al.'s study highlights the technical challenge of using CED to perfuse the entirety of the targeted tissue.

In a particularly interesting experiment, Lonser et al. (2007) performed real-time MRI to visualize the convective delivery of a particle, the enzyme glucocerebrosidase, co-infused with a tracer agent Gd-diethylenetriamine pentaacetic acid into the brainstem of a patient with type 2 Gaucher's disease. Their results suggest that the widespread distribution of a nanoparticle in a large volume of the brainstem was precise, safe, and feasible.

Viral particles for treatment of Parkinson's disease have also been administered through CED (Richardson et al. 2011; Valles et al. 2010). Led by Bankiewicz, his team convected AAV2 encoding human aromatic L-amino acid decarboxylase (hAADC) into the putamens of 10 humans and 3 nonhuman primates (Valles et al. 2010). In both groups, MRI imaging appeared similar with T2 hyperintensity that was associated with increased uptake on positron emission tomography (PET) scans. Nonhuman primates also had immunohistochemical studies demonstrating that hAADC expression occurred in these regions. This result strongly supports the feasibility of using CED of nanoparticles such as viruses for clinical gene therapy.

In summary, CED is a technique that can be used to deliver nanoparticles regionally for "molecular neurosurgery." Although still in an experimental stage, the potential of this technique has been clearly demonstrated. Future investigations will allow further refinement of the technique and optimization of particle properties needed to traverse the interstitial space.

CONCLUSION

CED has become a viable method of targeted intracranial therapy that circumvents many of the obstacles associated with systemic drug delivery. With advancements in imaging techniques and devices, researchers can examine and fine-tune specific parameters that influence the effective delivery of therapeutic agents using CED. This chapter reviews some of these parameters that govern this effective local delivery technique. Most importantly, particle design, in terms of size and surface characteristics, can significantly influence distribution of the infused agents. Other factors such as catheter size and infusion rate have also enhanced delivery over clinically relevant distances. In addition to discussing the aforementioned parameters, we also presented some animal and human studies that illustrate the successful delivery of nanoparticles using CED. Gene therapy, in particular, is one area of application in which this method of delivery holds particular promise. Although further investigations that examine the limitations and necessary improvements of this delivery technique need to be conducted, CED remains a bright hope for therapeutic applications in the clinical setting.

REFERENCES

Chen, M.Y., Hoffer, A., Morrison, P.F. et al. 2005. Surface properties, more than size, limiting convective distribution of virus-sized particles and viruses in the central nervous system. *J Neurosurg* 103: 311–9.

Chen, M.Y., Lonser, R.R., Morrison, P.F. et al. 1999. Variables affecting convection-enhanced delivery to the striatum: a systematic examination of rate of infusion, cannula size, infusate concentration, and tissue-cannula sealing time. *J Neurosurg* 90: 315–20.

Chen, Z.J., Gillies, G.T., Broaddus, W.C. et al. 2004. A realistic brain tissue phantom for intraparenchymal infusion studies. *J Neurosurg* 101: 314–22.

Choudhary, S., Mathew, M., and Verma, R.S. 2011. Therapeutic potential of anticancer immunotoxins. *Drug Discov Today* 16: 495–503.

Krauze, M.T., Noble, C.O., Kawaguchi, T. et al. 2007. Convection-enhanced delivery of nanoliposomal CPT-11 (irinotecan) and PEGylated liposomal doxorubicin (Doxil) in rodent intracranial brain tumor xenografts. *Neuro Oncol* 9: 393–403.

Kunwar, S., Chang, S., Westphal, M. et al. 2010. Phase III randomized trial of CED of IL13-PE38QQR vs Gliadel wafers for recurrent glioblastoma. *Neuro Oncol* 12: 871–81.

Kunwar, S., Prados, M.D., Chang, S.M. et al. 2007. Direct intracerebral delivery of cintredekin besudotox (IL13-PE38QQR) in recurrent malignant glioma: a report by the Cintredekin Besudotox Intraparenchymal Study Group. *J Clin Oncol* 25: 837–44.

Laske, D.W., Youle, R.J., and Oldfield, E.H. 1997. Tumor regression with regional distribution of the targeted toxin TF-CRM107 in patients with malignant brain tumors. *Nat Med* 3: 1362–8.

Linninger, A.A., Somayaji, M.R., Mekarski, M. et al. 2008. Prediction of convection-enhanced drug delivery to the human brain. *J Theor Biol* 250: 125–38.

Lonser, R.R., Warren, K.E., Butman, J.A. et al. 2007. Real-time image-guided direct convective perfusion of intrinsic brainstem lesions. Technical note. *J Neurosurg* 107: 190–7.

Mardor, Y., Rahav, O., Zauberman, Y. et al. 2005. Convection-enhanced drug delivery: increased efficacy and magnetic resonance image monitoring. *Cancer Res* 65: 6858–63.

Morrison, P.F., Chen, M.Y., Chadwick, R.S. et al. 1999. Focal delivery during direct infusion to brain: role of flow rate, catheter diameter, and tissue mechanics. *Am J Physiol* 277: R1218–29.

Murad, G.J., Walbridge, S., Morrison, P.F. et al. 2006. Real-time, image-guided, convection-enhanced delivery of interleukin 13 bound to pseudomonas exotoxin. *Clin Cancer Res* 12: 3145–51.

Neeves, K.B., Lo, C.T., Foley, C.P. et al. 2006. Fabrication and characterization of microfluidic probes for convection enhanced drug delivery. *J Control Release* 111: 252–62.

Perlstein, B., Ram, Z., Daniels, D. et al. 2008. Convection-enhanced delivery of maghemite nanoparticles: Increased efficacy and MRI monitoring. *Neuro Oncol* 10: 153–61.

Raghavan, R., Brady, M.L., Rodriguez-Ponce, M.I. et al. 2006. Convection-enhanced delivery of therapeutics for brain disease, and its optimization. *Neurosurg Focus* 20: E12.

Richardson, R.M., Kells, A.P., Rosenbluth, K.H. et al. 2011. Interventional MRI-guided putaminal delivery of AAV2-GDNF for a planned clinical trial in Parkinson's disease. *Mol Ther* 19: 1048–57.

Saito, R., Krauze, M.T., Bringas, J.R. et al. 2005. Gadolinium-loaded liposomes allow for real-time magnetic resonance imaging of convection-enhanced delivery in the primate brain. *Exp Neurol* 196: 381–9.

Sampson, J.H., Akabani, G., Archer, G.E. et al. 2003. Progress report of a phase I study of the intracerebral microinfusion of a recombinant chimeric protein composed of transforming growth factor (TGF)-alpha and a mutated form of the Pseudomonas exotoxin termed PE-38 (TP-38) for the treatment of malignant brain tumors. *J Neuro Oncol* 65: 27–35.

Sampson, J.H., Archer, G., Pedain, C. et al. 2010. Poor drug distribution as a possible explanation for the results of the PRECISE trial. *J Neurosurg* 113: 301–9.

Sampson, J.H., Raghavan, R., Provenzale, J.M. et al. 2007. Induction of hyperintense signal on T2-weighted MR images correlates with infusion distribution from intracerebral convection-enhanced delivery of a tumor-targeted cytotoxin. *AJR Am J Roentgenol* 188: 703–9.

Sykova, E., and Nicholson, C. 2008. Diffusion in brain extracellular space. *Physiol Rev* 88: 1277–340.

Szerlip, N.J., Walbridge, S., Yang, L. et al. 2007. Real-time imaging of convection-enhanced delivery of viruses and virus-sized particles. *J Neurosurg* 107: 560–7.

Valles, F., Fiandaca, M.S., Eberling, J.L. et al. 2010. Qualitative imaging of adeno-associated virus serotype 2-human aromatic L-amino acid decarboxylase gene therapy in a phase I study for the treatment of Parkinson disease. *Neurosurgery* 67: 1377–85.

Wang, P., and Olbricht, W.L. 2011. Fluid mechanics in the perivascular space. *J Theor Biol* 274: 52–7.

Weaver, M., and Laske, D.W. 2003. Transferrin receptor ligand-targeted toxin conjugate (Tf-CRM107) for therapy of malignant gliomas. *J Neurooncol* 65; 3–13.

13 Transport of Nanoparticles across Blood–Brain Barrier

Robert T. Wicks, Zev A. Binder, and Alessandro Olivi

CONTENTS

INTRODUCTION TO BLOOD–BRAIN BARRIER

Three principle barriers regulate and limit molecular exchange between blood and neural tissue: (1) the blood–brain barrier (BBB), formed primarily by capillary endothelial cells between the blood and neuronal tissue; (2) choroid plexus epithelium, between blood and ventricular cerebral spinal fluid (CSF); and (3) arachnoid epithelium, between blood and subarachnoid CSF (Abbott 2004; Abbott et al. 2006). Due to its large surface area of approximately 20 m^2 and total blood vessel length of approximately 600 km, the most important barrier to drug delivery is the BBB (deBoer and Gaillard 2007). The distance between neurons and the nearest brain capillary is 10–15 μm, making the central nervous system (CNS) microvasculature an ideal network for global brain drug delivery (Schlageter et al. 1999) (Figure 13.1). However, almost 100% of large-molecule and >98% of small-molecule pharmaceutics cannot traverse the BBB (Pardridge 2007). Nanoparticles may offer an ideal method of transport across this barrier. A clear understanding of BBB structure and function must be attained prior to considering nanoparticle strategies.

The concept of the BBB was first proposed over 120 years ago when Dr. Paul Ehrlich described injecting acid aniline dye systemically in vivo (Figure 13.2). Aniline dye binds albumin, a ubiquitous protein. Ehrlich (1885) found that the dye stained all organs except the brain and spinal cord, hinting at a separation between blood and neural tissue (Ehrlich 1885). Ehrlich's student, Goldmann, showed that the brain and spinal cord were stained when dye was injected intrathecally but systemic organs were not, providing further evidence for a barrier between the CNS and the rest of the body (Friedmann 1942; Goldmann 1913). The majority of endothelial surfaces allow polar molecules in circulation, such as acid aniline dye, to cross paracellularly. At the BBB, however, complex tight junctions (TJs) force most molecules to take a transcellular route, strictly filtering any molecules that attempt to cross (Abbott et al. 2006; Laterra and Goldstein 2000).

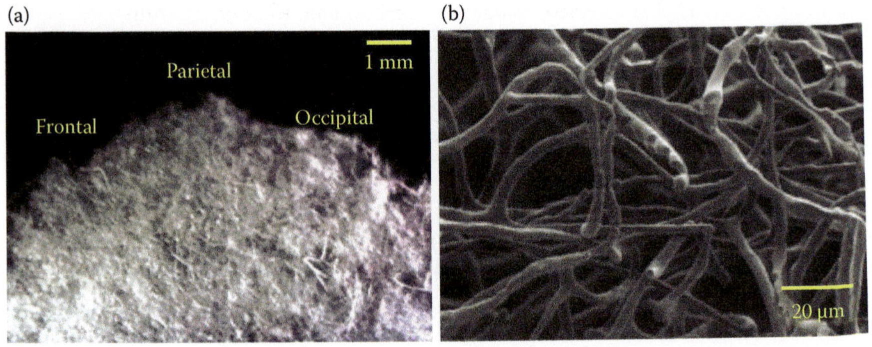

FIGURE 13.1 Complexity of the CNS microvascular network. (a) Microvascular cast of a mouse brain. The mouse was perfused with methyl methacrylate, and brain tissue was dissolved with potassium hydroxide. (b) Scanning electron micrograph of the cast. (From Zlokovic, B.V., *Neuron* 57, 178–201, 2008. With permission.)

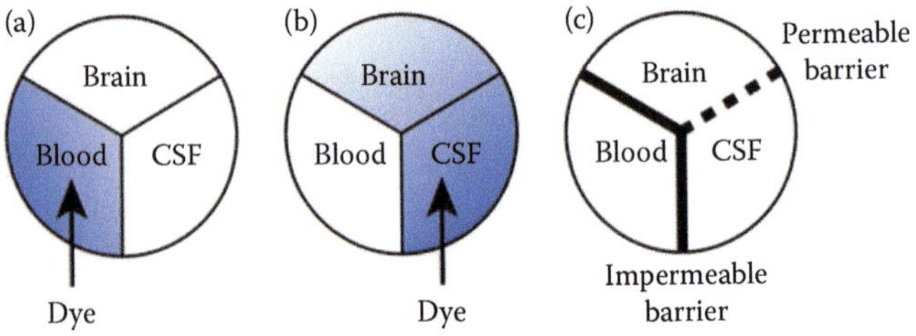

FIGURE 13.2 Evidence of a BBB from the injection of aniline dye. (a) Injection of aniline dye into systemic circulation stains all organs except the brain and spinal cord. (b) Injection of aniline dye intrathecally stains both the brain and spinal cord. (c) Combined experiments provided evidence of an impermeable barrier between the systemic circulation and the brain/spinal cord. (From Zlokovic, B.V., *Neuron* 57, 178–201, 2008. With permission.)

STRUCTURE OF BBB

Two characteristics of the CNS microvasculature make the BBB highly impermeable: (1) TJs between adjacent endothelial cells, prohibiting paracellular transport; and (2) lack of fenestrations within endothelial cells, precluding transcellular transport (Hawkins and Davis 2005). The BBB is composed of endothelial cells that line capillaries in close contact with perivascular elements: a basal lamina, astrocytic end feet, perivascular neurons, and pericytes (Cecchelli et al. 2007).

Paracellular pathways between adjacent microvascular endothelial cells are restricted by TJs and adherens junctions that are more complex than those found in other organs (Abbott et al. 2006). TJs both restrict paracellular permeability and segregate the apical and basal portions of the cell membrane so that the endothelium becomes more polarized (apical–basal) and similar to kidney epithelia than the microvasculature of other organs (Abbott et al. 2006; Hawkins and Davis 2005). The apical membrane acquires a negative surface charge that repulses negatively charged particles (deBoer and Gaillard 2007). The restricted movement of ions such as Na^+, K^+, and Cl^- leads to a transendothelial resistance of >1000 $\Omega \cdot cm^2$, far greater than the typical 2–20 $\Omega \cdot cm^2$ seen throughout peripheral capillaries (Abbott et al. 2006).

The endothelial cells lining CNS microvasculature also differ from other vascular endothelium in the uniform thickness of cell cytoplasm with a lack of fenestrae and little vesicular transport (deBoer and Gaillard 2007). CNS endothelial cells have greater mitochondrial content than other organs, which provides enhanced energy potential. This supplies the energy needed for enzymatic breakdown of compounds and selective transport mechanisms to actively transport nutrients and other compounds into and out of the brain and spinal cord (deBoer and Gaillard 2006; Persidsky et al. 2006). One endothelial cell encompasses the circumference of the capillary lumen, with the opposing ends meeting in a TJ (Hawkins and Davis 2005). The basal side of the endothelial cells is directly linked to the basal lamina through transmembrane protein focal adhesions (Cardoso et al. 2010).

The basal lamina is formed by three membranes, composed of three different extracellular matrix molecules: structural proteins (collagen and elastin), specialized proteins (fibronectin and laminin), and proteoglycans (Adibhatla and Hatcher 2008; Cardoso et al. 2010; Wolburg et al. 2009). Within the complex matrix, the basement membrane also includes receptors such as cell adhesion molecules in addition to signaling proteins (Cardoso et al. 2010).

Pericytes are integrated within the basal lamina and synthesize its many elements. Current evidence shows that these cells regulate endothelial cell proliferation, migration, and differentiation (Persidsky et al. 2006). In addition, pericytes also play a role in regulating BBB adjustments in response to stressful stimuli (Cardoso et al. 2010). Pericytes can migrate away from capillaries in response to hypoxia and traumatic brain injury (Persidsky et al. 2006). Since these are both conditions associated with increased BBB permeability, pericytes may play a critical role in maintaining TJs (Hawkins and Davis 2005).

Astrocytic end feet project and contact the basal lamina. Microglia and synaptic terminal boutons of nerve fibers are also found in the immediate perivascular space (Abbott et al. 2006). Recent evidence has shown that neurons play a critical role in regulating the BBB in response to metabolic requirements by inducing expression of enzymes that regulate vessel permeability (Cardoso et al. 2010; Persidsky et al. 2006). Astrocytes play a role in the tightness of TJs, expression of transporters such as GLUT-1 and P-glycoprotein (P-gp), and regulation of the assembly of endothelial cells and pericytes into the capillary structure (Abbott et al. 2006; Ramsauer et al. 2002). GLUT-1 is the major glucose transport for the BBB, and P-gp is an important ATP-powered transporter responsible for moving solutes out of the brain and into the vasculature. Astrocytic end feet also regulate water fluxes within the brain through expression of aquaporin 4, a membrane water channel (Koziara et al. 2006).

FUNCTIONS OF BBB AND METHODS OF CROSSING

The BBB is not a static wall but, rather, an adaptable filter able to adjust to varying circumstances. It plays physical, metabolic, and immunologic roles to allow a separate and distinct extracellular fluid compartment, the CSF (Begley 2004). The BBB maintains a medium for optimal neuronal function, protecting the brain from uncontrolled excitation due to fluctuations in ions and peripherally circulating neurotransmitters that occur with exercise or after meals (Cecchelli et al. 2007).

Molecules cross the BBB either through a paracellular route (between cells) or a transcellular route (through cells). The majority of molecules pass transcellularly due to the complexity of the TJs. Molecules most likely to cross transcellularly must be sufficiently lipophilic to pass through the cell membrane but hydrophilic enough to pass through the cytoplasm of the cell. Small gaseous molecules, such as O_2 and CO_2, and small lipophilic molecules, such as barbiturates and ethanol, can diffuse freely through the endothelial cells (Abbott et al. 2006; Hawkins and Davis 2005).

In the case of larger, essential molecules, endothelial cells contain membrane transporters that regulate transcellular trafficking between the blood and the brain. There are specific BBB transporters for amino acids, glucose, hormones, and antibodies. When the target molecule binds to its respective receptor on the endothelial luminal surface, it is transcytosed, forming a vesicle that can be transported across the cell and released on the apical side (Cecchelli et al. 2007). Carrier-mediated transport allows certain small essential molecules to cross via catalyzed transport, such as glucose entering the brain via GLUT-1 (Oldendorf 1971). Certain large molecular peptides require receptor-mediated transport mechanisms. Insulin, for example, gains access across the BBB via insulin receptors expressed on endothelial cells (Pardridge 2006). Endothelial cells also regulate the concentration of toxins and waste products through efflux pumps such as P-gp, which is thought to be one of the major causes of chemotherapy drug resistance (Cardon-Cardo et al. 1989).

The BBB also plays an important role in immune function. The CNS is considered immune privileged since it restricts the entry of lymphocytes into the brain and spinal cord parenchyma. However, cortical injury, ischemia, and disease all show that inflammatory leukocytes can accumulate within the CNS. Recent studies found that the primary form of leukocyte migration across the BBB occurs transcellularly (Wedel-Parlow et al. 2011; Weiss et al. 2009). This method diverges from the more common, paracellular, method used to cross other endothelial layers throughout the body (Springer 1994). The ability of immune cells to cross the BBB transcellularly reveals the ability of endothelial cells to accommodate the migration of relatively large objects, which offers a possible novel pathway to exploit for nanoparticle delivery.

ALTERATIONS OF BBB

Pathology can lead to increased BBB permeability, in the cases of tumor, sepsis, or traumatic brain injury (Long 1970; Morganti-Kossmann et al. 2002; Tsao et al. 2001). Tumor vessels, in particular, are unusually leaky. They often have abnormally formed and misplaced endothelial cells that lead to intercellular gaps within the vessel. These gaps allow extravasation of both proteins and erythrocytes in the form of hemorrhage (Coomber et al. 1987; Hashizume et al. 2000). Further evidence has shown that the BBB is molecularly altered within high-grade tumors, such as glioblastoma, leading to faulty TJs (Liebner et al. 2000). Other changes found with electron microscopy provide evidence for an overall disarray of barrier structure. These include increased thickness of the perivascular basal lamina, in addition to abnormal pericytes (Liebner et al. 2000). These alterations in the BBB provide a corridor for successful chemotherapy (Stewart 1994) and may yield an avenue for targeted nanoparticle delivery to the tumor bed (Juillerat-Jeanneret 2008).

Alterations of the BBB are not always pathologic. Rapoport and Thompson (1973) proposed infusing a hyperosmolar solution, such as mannitol, directly into a cerebral artery followed by administration of a pharmaceutical. The hyperosmolar solution led to increased capillary permeability, a method termed hyperosmolar BBB disruption (HBBBD) (Groothuis 2000). Zunkeler et al. (1996) studied BBB permeability using positron emission tomography (PET) scans of patients with malignant tumors. Rubidium-82 was injected into the patient's peripheral vein followed by mannitol administration. One PET scan prior to and several scans after mannitol administration looked at rubidium-82 uptake. HBBBD increased BBB permeability by 1000% in normal brain tissue but only 60% within tumors, which was concluded to be less than favorable, since the increased uptake of a therapy delivered with HBBBD primarily occurs in normal brain tissue rather than in a tumor (Groothuis 2000).

Another method to temporarily alter the BBB is to deliver chemical compounds such as bradykinin (Inamura and Black 1994), interleukin-2 (Gutman et al. 1996), and possibly phosphodiesterase type-5 inhibitors, such as vardenafil (Black et al. 2008; Hu and Gao 2010). RMP-7, a bradykinin analog, delivered intra-arterially increased the permeability of capillary TJs by ~50% in tumors without a significant increase in tumor-free brain. BBB function begins to revert to normal within 20 min of terminating the RMP-7 infusion (Black et al. 1997; Groothuis 2000; Sanovich et al. 1995).

NANOPARTICLE METHODS OF CROSSING BBB

Nanoparticles are defined as objects in the 1–100 nm range in at least one dimension. Engineered nanoparticles allow integration into one device of specific and unique chemical and physical properties that are not expressed by the individual molecules (Silva 2007). This capability makes nanoparticles an excellent platform for CNS drug development. An ideal nanoparticle drug would permit systemic delivery with minimal side effects, have the ability to localize to and cross the BBB, and diffuse to its target within the CNS to activate and carry out its purpose (Sarin et al. 2008). The specific mechanisms by which many nanoparticles cross the BBB are unknown. Two primary pathways have been described: (1) axonal translocation from peripheral sensory nerves and (2) uptake through the BBB from systemic circulation (Hu and Gao 2010).

Uptake of nanoparticles by sensory nerve endings occurs within airway epithelia or via the olfactory bulb, followed by axonal translocation to the CNS. This method has been seen largely with carbon, gold, and manganese oxide nanoparticles through inhalation studies (Oberdorster et al. 2005). Wang et al. (2007) demonstrated penetration of intranasally delivered Fe_2O_3 particles. Increased iron was localized within the olfactory bulb, olfactory nerve, and brain stem.

The second pathway, crossing the BBB from systemic circulation, has been extensively studied as a more promising route for nanoparticle drug delivery (Kreuter 2001). There are four principal methods for transporting nanoparticles from the systemic circulation across the BBB: (1) passive diffusion; (2) opening of TJs; (3) carrier/receptor-mediated transport/transcytosis; and (4) inhibition of efflux systems by coating agents (Agarwal et al. 2009; Kreuter 2001; Lockman et al. 2002) (Figure 13.3).

Passive diffusion occurs when the nanoparticle can dissolve in the lipid membrane of endothelial cells. Fenart et al. (1999) demonstrated that when a lipid bilayer containing cholesterol was applied over charged nanoparticles, a threefold to fourfold increase in brain uptake was observed. They showed that the nanoparticle remained intact through the diffusion. These nanoparticles may also have the added benefit of reducing drug uptake by the reticuloendothelial system or by macrophages in the liver and spleen, leading to greater nanoparticle concentration within the circulation. Schroeder et al. (1998) showed that when amitriptyline, which crosses the BBB, was attached to the surface of polysorbate 85-stabilized nanoparticles, serum concentration increased, in addition to brain concentration.

Local opening of CNS microvascular TJs to allow selective drug penetration has long been a proposed target of pharmaceutical development. Olivier et al. (1999) offered evidence that

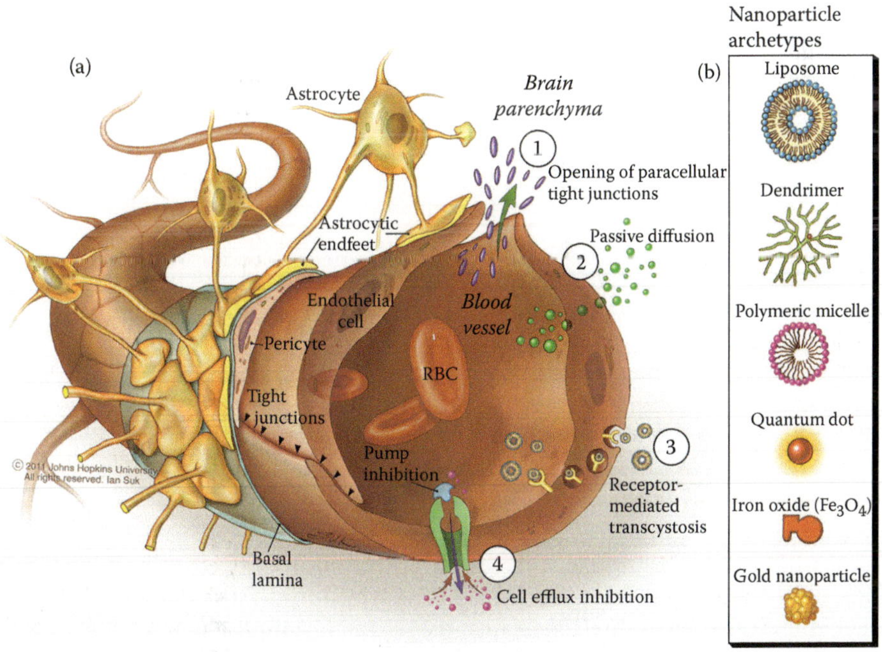

FIGURE 13.3 Nanoparticles and the BBB. (a) BBB structure with the four principle methods of nanoparticles crossing the barrier. (b) Basic structural images of six nanoparticle archetypes.

polybutylcyanoacrylate (PBCA) nanoparticles could open TJs, which led to noted drug toxicity. Kreuter et al. (2003), however, did not corroborate this evidence in their in vitro and in vivo studies. Chen et al. (2008) showed that aluminum oxide nanoparticles disrupted the BBB by decreasing expression of TJ proteins between adjacent endothelial cells. Specifically, they noted loss of occludin and claudin-5 in animal studies (Chen et al. 2008). The potential for other blood-borne substances to cross the BBB once TJs have been opened makes this strategy less than optimal for delivering nanoparticles to the CNS.

Endogenous receptor-mediated transport systems are manipulated to deliver molecular therapeutics through a method termed a "Trojan horse." Monoclonal antibodies can be used as ligands for insulin or transferrin receptors to enable transport of nanoparticles into microvascular endothelial cells (Pardridge 2006; Zlokovic 2008). The binding site of the monoclonal antibody can be separate from the endogenous binding site, so as to not interfere with normal transport (Pardridge 2006). In another method of receptor-mediated transport, PBCA nanoparticles coated with polysorbate 80 adsorb apolipoproteins B and E within the plasma. The nanoparticle–apolipoprotein fusion molecules mimic naturally occurring lipoprotein particles and are, therefore, taken up by CNS microvascular endothelial cells via receptor-mediated endocytosis (Kreuter et al. 2002; Silva 2007).

While some nanoparticles can cross the BBB, there are active transport mechanisms that return particles that cross to the systemic circulation. These transporters are found on both the luminal and abluminal surfaces and participate in the efficient efflux of potentially harmful substances (Kaur et al. 2008). P-gp, for example, transports lipophilic and cationic compounds across the cell membrane. Another transporter, multidrug resistance-associated protein (MRP), binds anionic and glutathione-conjugated compounds (Miller and Kabanov 1999). Miller et al. (1997) found that Pluronic P85 (P85) micelles inhibit P-gp function, thereby decreasing drug efflux from brain microvessel endothelial cells.

NANOPARTICLE ARCHETYPES

The many nanoparticles currently being studied for their medical application can be organized into six major categories: liposomes, micelles, dendrimers, quantum dots (QDs), iron oxide (IO), and gold nanoparticles (Mohs and Provenzale 2010). Nanoparticles have been used both in diagnostics and for treating multiple neuropathologies. Literature on the mechanisms that permit nanoparticles to cross the BBB has largely focused on therapeutic options for brain tumors. The following sections will deal primarily with nanoparticle applications in neuro-oncology.

Liposomes, micelles, and dendrimers are all structures that have been used as vehicles for crossing the BBB. While they have applications as drug delivery and imaging modalities, most publications about them have focused on drug delivery. Liposomes composed of a bilayer of phospholipids are one of the more common structures used in nanomolecular drug delivery and imaging techniques. These vehicles can carry lipophilic and hydrophilic molecules in their core for delivery. In addition, they incorporate hydrophobic elements into the bilayer itself. Most work with liposomal transport across the BBB has utilized carrier/receptor-mediated transport. Initial work with liposomes and drug delivery to tumors was hampered by rapid scavenging of the liposome vehicles by the liver and spleen. GM_1 gangliosides were first used to delay the clearance of liposomes (Allen and Chonn 1987), followed by polyethylene glycol (PEG) and modified PEG (Allen et al. 1991; Klibanov et al. 1990). Coating liposomes with PEG prolongs the half-life of the vehicles in circulation, which promotes accumulation of liposomes in solid tumors (Gabizon and Papahadjopoulos 1988).

Liposomes can also be used for targeted delivery. Antigen-specific antibodies can be incorporated into the bilayer, allowing specific targeting. By selecting a tumor-specific antigen as the target, the vehicle can be directed to the tumor where the payload is unloaded. Zhang et al. (2002) used PEGylated immunoliposomes directed against the human insulin receptor to deliver gene therapy intended to decrease EGFR expression in tumor cells.

Micelles are composed of a single layer of fatty acid, phospholipid, or similar molecules with a hydrophilic head and a hydrophobic tail. Hydrophobic molecules are carried in the hydrophobic core of the micelle. In general, micelles are smaller than liposomes and self-assemble (Prokop and Davidson 2008). Micelles can be constructed to deliver cargo on specific environmental cues. Sethuraman et al. (2008) developed micelles that respond to a lowered pH and that will release their contents in acidic conditions within tumors (Liu et al. 2008). This permits these micelles to accumulate and release their contents within the tumor despite their lack of antibodies to target tumor-specific antigens.

Micelles have other therapeutics applications involved in crossing the BBB. Liu et al. (2008) demonstrated that micelles could be used to deliver antibiotics to the brain in cases of bacterial meningitis. Drugs are encapsulated in their hydrophobic, active form within the hydrophobic core of the micelle. Early work with micelles and evaluations of their ability to cross the BBB included using antibody fragments specific to brain tissue to increase the amount of micelles localizing to the brain (Chekhonin et al. 1995). Fragments of ^{125}I-labeled antibodies against glial fibrillary acidic protein and α_2-glycoprotein were carried in micelles, and their concentrations in various rodent tissues were measured. The micelles allowed the Fab fragments to penetrate the BBB and to achieve a concentration 55% greater than in all other organs.

Polymer nanoparticles, or polymersomes, contain a nanoparticle core that has multiple layers but is otherwise closely related to liposomes and micelles (Saha et al. 2010). Carrying capacity for drug delivery is increased because polymersomes have more layers than liposomes (Tosi et al. 2008). The polymers must be biodegradable to prevent a buildup of potentially toxic breakdown products as the drug is released. Lockman et al. (2004) used the flexibility of polymersome synthesis to alter the surface charge of these nanoparticles. They found that low concentrations of anionic polymersomes disrupted the BBB and allowed polymersomes to accumulate within brain parenchyma.

Dendrimers are structures made up of highly branched polymers. The term was coined by Tomalia et al. (1985) in 1984, when they reported building "starburst polymers." The structure of dendrimers invites attachment of surface groups to each branch. The available attachment sites increase exponentially as the dendrimer gains branches and size with each generation. Similar to liposomes and micelles, dendrimers can be coated to reduce uptake by the liver and spleen, prolonging their half-life in circulation. Due to their intricate architecture, these polymers have greater versatility than linear polymers (Lee et al. 2005). A variety of different moieties can be bound to the interior and exterior of a dendrimer. Huang et al. (2008) demonstrated that dendrimers labeled with lactoferrin crossed the BBB and delivered DNA to target cells in the brain. They used a polyamidoamine (PAMAM) dendrimer as the core and added primary amino surface groups, whose positive charge induced binding to DNA. By labeling the dendrimers with lactoferrin, a transport protein for iron, they took advantage of lactoferrin receptor-mediated uptake and transcytosis in cerebral capillary endothelial cells to deliver DNA across the BBB.

The following nanoparticle cores lend themselves primarily to imaging modalities and improved delineation of CNS pathologies rather than drug delivery: QDs, IO particles, and gold nanoparticles.

QDs have been developed to improve imaging techniques. They are double-layered semiconductors that have size-dependent emission spectra (Smith et al. 2008). This allows QDs of different sizes to be detected simultaneously since they emit different wavelengths. QDs have broad excitation spectra with narrow emission spectra. This means that one excitation source can be used for more than one color of QD. In addition, there is no photobleaching of QDs, unlike traditional fluorophores, which have a limited life span. Jackson et al. (2007) demonstrated the potential use of QDs for intraoperatively visualizing a tumor, based on QD uptake by colocalized macrophages and microglial cells. However, QDs are composed of heavy metals, and they must be eliminated from the body before the metals ionize to avoid heavy metal toxicity (Hoshino et al. 2004; Lovric et al. 2005). Toxicity issues have been seen in vivo after accumulation of QDs in various tissues, including kidneys, liver, and spleen (Akerman et al. 2002; Smith et al. 2008; Yang et al. 2007). Future work with QDs is directed toward developing imaging techniques that better visualize tumor tissue. Cai et al. (2006) reported that arginine–glycine–aspartic acid tagged QDs were taken up by flank tumors in nude mice and provided visualization of tumor vasculature. If heavy metal toxicity could be overcome, QDs could be used clinically for delineating tumor margins and vascular patterns during surgery.

IO particles are used to improve imaging techniques and targeted delivery of drugs (Mohs and Provenzale 2010; Yang et al. 2007). The magnetic properties of IO particles allow them to be directed to specific locations in the body using magnetic fields, where they can deliver drugs conjugated to IO particles (Raut et al. 2010). For imaging modalities, IO is being investigated in place of gadolinium contrast. Neuwelt et al. (2004) administered dextran-coated IO particles to patients with malignant brain tumors before surgical intervention and compared their imaging properties to those of traditional gadolinium contrast in preoperative and postoperative magnetic resonance (MR) images. In a small patient sample, they observed that the IO particles remained visible by magnetic resonance imaging (MRI) in the brain for 2 to 5 days after injection, much longer than gadolinium contrast, which was beneficial for assessing tumor resection by postoperative imaging. In addition, IO signal changes were seen beyond the tumor margin outlined by gadolinium contrast. Histological specimens prepared from tissue removed at surgery showed that most IO particles were taken up by reactive cells around tumor cells rather than by tumor cells themselves. Additional sites of IO signal changes outside the gadolinium-enhancing lesions were not biopsied, so it was unknown if these sites contained tumor cells despite a lack of gadolinium enhancement there.

Gold nanoparticles exhibit specific photodynamic properties that make them useful moieties for imaging. They are biologically inert, and there are a large number of gold-based suspensions approved for clinical use, making these nanoparticles attractive for research (Mohs and Provenzale 2010). Thermal ablation therapy has taken advantage of gold, using lasers to heat gold particles and destroy tissue. Guerrero et al. (2010) conjugated amphipathic peptides to gold nanoparticles to increase the delivery of gold particles to rat brain tissue. These gold particles were previously employed to destroy β-amyloid deposits in vitro, using microwave to irradiate the particles (Araya et al. 2008).

POTENTIAL BARRIERS TO NANOPARTICLE APPLICATION

Nanotechnological approaches to brain tumor therapy and imaging have limitations. The first limitation is concern about the possible adverse consequences from exposure to nanoparticles. Nanoparticles have the potential to enter the human body via several routes: inhalation, ingestion, and dermal penetration. Once in the body, nanoparticles can readily enter the circulation, leading to local and systemic inflammation and CNS effects (Borm and Muller-Schulte 2006; Hu and Gao 2010, Silva 2007). Since some nanoparticles are not easily cleared via physiologic mechanisms, they can lead to undesirable outcomes by accumulating in organs, such as the brain (Medina et al. 2007; Sharma 2007). While few adverse effects of nanoparticle therapy have been reported, only a small number of toxicity studies have been conducted. Protocols for safely handling and administering nanoparticle therapies will have to be developed to protect patients and health care providers.

The greatest barrier to successful nanoparticle delivery to sites within the brain is lack of particle dispersion once the nanoparticle enters the brain extracellular space (ECS). Thorne and Nicholson (2006) looked at the diffusion of QDs in living tissue. Their model predicted a normal ECS width of 38–64 nm. Nanoparticles over 64 nm in diameter are therefore prohibited from moving through the ECS of normal brain tissue (Thorne and Nicholson 2006). Approaches to increasing nanoparticle distribution within the brain have been investigated. One such approach is convection-enhanced delivery, in which a solution containing a molecular drug or nanoparticle is infused directly into the brain, and the distribution of the agent depends on the hydrostatic pressure, rate, and volume of the infusion (Bobo et al. 1994; Groothuis 2000; Krauze et al. 2007; Lidar et al. 2004; Perlstein et al. 2008). Drug distribution with this method depends on whether the drug is infused into a homogeneous gray matter region or into a structure containing dense white matter tracts. When the infusion comes into contact with a white matter tract, the subsequent distribution of the agent in surrounding brain is altered accordingly by the amount and direction of bulk flow that passes along that tract

(Morrison et al. 1994). Additional methods of enhancing nanoparticle delivery within the CNS will need to be investigated as nanoparticle therapies are developed.

CONCLUSIONS

The extensive brain microvascular network makes targeted delivery across the BBB an ideal route for global brain drug delivery. Nanoparticles offer an effective vehicle for crossing this highly impermeable barrier. The two principal methods of crossing the BBB, via axonal translocation from sensory nerve endings and through transfer from the systemic circulation, have been discussed. Of the two methods, transport of nanoparticles across the BBB from systemic circulation holds the greatest promise for effective nanoparticle drug delivery.

Nanoparticles can be engineered to open paracellular TJs, diffuse passively across the BBB by becoming incorporated into the endothelial cell membrane, take advantage of endogenous receptor-mediated transport mechanisms, and inhibit endothelial cell drug reflux mechanisms such as P-gp. The major nanoparticle archetypes presented above provide a basic framework for categorizing the structures of current nanoparticles and understanding the potential of future nanoparticle development. While challenges remain concerning nanoparticle safety for patients and health care providers and nanoparticle dispersion and delivery once across the BBB, nanotechnology holds great promise for providing novel, more effective therapies for neuropathologies and more sensitive and specific imaging modalities.

REFERENCES

Abbott, N.J. 2004. Evidence for bulk flow of brain interstitial fluid: significance for physiology and pathology. *Neurochem Int* 45: 545–52.

Abbott, N., Ronnback, L., and Hansson, E. 2006. Astrocyte-endothelial interactions at the blood–brain barrier. *Nat Rev Neurosci* 7: 41–53.

Adibhatla, R.M., and Hatcher, J.F. 2008. Tissue plasminogen activator (tPA) and matrix metalloproteinases in the pathogenesis of stroke: therapeutic strategies. *CNS Neurol Disord Drug Targets* 7: 243–53.

Agarwal, A., Lariya, N., Saraogi, G. et al. 2009. Nanoparticles as novel carrier for brain delivery: a review. *Curr Pharm Des* 15: 917–25.

Akerman, M.E., Chan, W.C., Laakkonen, P. et al. 2002. Nanocrystal targeting in vivo. *Proc Natl Acad Sci U S A* 99: 12617–21.

Allen, T.M., and Chonn, A. 1987. Large unilamellar liposomes with low uptake into the reticuloendothelial system. *FEBS Lett* 223: 42–6.

Allen, T.M., Hansen, C., Martin, F. et al. 1991. Liposomes containing synthetic lipid derivatives of poly(ethylene glycol) show prolonged circulation half-lives in vivo. *Biochim Biophys Acta* 1066: 29–36.

Araya, E., Olmedo, I., Bastus, N.G. et al. 2008. Gold nanoparticles and microwave irradiation inhibit beta-amyloid amyloidogenesis. *Nanoscale Res Lett* 3: 435–43.

Begley, D.J. 2004. Delivery of therapeutic agents to the central nervous system: the problems and the possibilities. *Pharmacol Ther* 104: 29–45.

Black, K.L., Cloughesy, T., Huang, S.C. et al. 1997. Intracarotid infusion of RMP-7, a bradykinin analog, and transport of gallium-68 ethylenediamine tetraacetic acid into human gliomas. *J Neurosurg* 86: 603–9.

Black, K.L., Yin, D., Ong, J.M. et al. 2008. PDE5 inhibitors enhance tumor permeability and efficacy of chemotherapy in a rat brain tumor model. *Brain Res* 1230: 290–302.

Bobo, R.H., Laske, D.W., Akbasak, A. et al. 1994. Convection-enhanced delivery of macromolecules in the brain. *Proc Natl Acad Sci U S A* 91: 2076–80.

Borm, P.J., and Muller-Schulte, D. 2006. Nanoparticles in drug delivery and environmental exposure: same size, same risks? *Nanomedicine* 1: 235–49.

Cai, W., Shin, D.W., Chen, K. et al. 2006. Peptide-labeled near-infrared quantum dots for imaging tumor vasculature in living subjects. *Nano Lett* 6: 669–76.

Cardon-Cardo, C., O'Brien, J.P., Casals, D. et al. 1989. Multidrug-resistance gene (P-glycoprotein) is expressed by endothelial cells at the blood–brain barrier sites. *Proc Natl Acad Sci U S A* 86: 695–8.

Cardoso, F.L., Brites, D., and Brito, M.A. 2010. Looking at the blood–brain barrier: molecular anatomy and possible investigation approaches. *Brain Res Rev* 64: 328–63.

Cecchelli, R., Berezowski, V., Lundquist, S. et al. 2007. Modelling of the blood–brain barrier in drug discovery and development. *Nat Rev Neurosci* 6: 650–61.
Chekhonin, V.P., Ryabukhin, I.A., Zhirkov, Y.A. et al. 1995. Transport of hydrophobized fragments of antibodies through the blood–brain barrier. *Neuroreport* 7: 129–32.
Chen, L., Yokel, R.A., Hennig, B. et al. 2008. Manufactured aluminum oxide nanoparticles decrease expression of tight junction proteins in brain vasculature. *J Neuroimmune Pharmacol* 3: 286–95.
Coomber, B., Stewart, P., Hayakawa, K. et al. 1987. Quantitative morphology of human glioblastoma multiforme microvessels: structural basis of blood–brain barrier defect. *J Neurooncol* 5: 299–307.
deBoer, A., and Gaillard, P. 2006. Blood–brain barrier dysfunction and recovery. *J Neural Transm* 113: 455–62.
deBoer, A., and Gaillard, P. 2007. Drug targeting to the brain. *Annu Rev Pharmacol Toxicol* 47: 323–55.
Ehrlich, P. 1885. *Das Sauerstoff-Bedurfnis des Organismus: eine farbenanalytische Studie*. Berlin: Hirschward.
Fenart, L., Casanova, A., Dehouck, B. et al. 1999. Evaluation of effect of charge and lipid coating on ability of 60-nm nanoparticles to cross an in vitro model of the blood–brain barrier. *J Pharmacol Exp Ther* 291: 1017–22.
Friedmann, U. 1942. Blood–brain barrier. *Physiol Rev* 22: 125–45.
Gabizon, A., and Papahadjopoulos, D. 1988. Liposome formulations with prolonged circulation time in blood and enhanced uptake by tumors. *Proc Natl Acad Sci U S A* 85: 6949–53.
Goldmann, E. 1913. Abh. D. preuss. Akad. Wissensch. Physikal. Mathem. Klasse. 1: 1.
Groothuis, D.R. 2000. The blood–brain and blood–tumor barriers: a review of strategies for increasing drug delivery. *Neuro-oncology* 2: 45–59.
Guerrero, S., Araya, E., Fiedler, J.L. et al. 2010. Improving the brain delivery of gold nanoparticles by conjugation with an amphipathic peptide. *Nanomedicine (Lond)* 5: 897–913.
Gutman, M., Laufer, R., Eisenthal, A. et al. 1996. Increased microvascular permeability induced by prolonged interleukin-2 administration is attenuated by the oxygen-free-radical scavenger dimethylthiourea. *Cancer Immunol Immunother* 43: 240–4.
Hashizume, H., Baluk, P., Morikawa, S. et al. 2000. Openings between defective endothelial cells explain tumor vessel leakiness. *Am J Pathol* 156: 1363–80.
Hawkins, B.T., and Davis, T.P. 2005. The blood–brain barrier/neurovascular unit in health and disease. *Pharmacol Rev* 57: 173–85.
Hoshino, A., Fujioka, K., Oku, T. et al. 2004. Physicochemical properties and cellular toxicity of nanocrystal quantum dots depend on their surface modification. *Nano Lett* 4: 2163–69.
Hu, Y., and Gao, J. 2010. Potential neurotoxicity of nanoparticles. *Int J Pharm.* 394: 115–21.
Huang, R., Ke, W., Liu, Y. et al. 2008. The use of lactoferrin as a ligand for targeting the polyamidoamine-based gene delivery system to the brain. *Biomaterials* 29: 238–46.
Inamura, T., and Black, K. 1994. Bradykinin selectively opens blood tumor barrier in experimental brain tumors. *J Cereb Blood Flow Metab* 14: 862–70.
Jackson, H., Muhammad, O., Daneshvar, H. et al. 2007. Quantum dots are phagocytized by macrophages and colocalize with experimental gliomas. *Neurosurgery* 60: 524–9; discussion 529–30.
Juillerat-Jeanneret, L. 2008. The targeted delivery of cancer drugs across the blood–brain barrier: chemical modifications of drugs or drug-nanoparticles? *Drug Discov Today* 13: 1099–106.
Kaur, I.P., Bhandari, R., Bhandari, S. et al. 2008. Potential of solid lipid nanoparticles in brain targeting. *J Control Release* 127: 97–109.
Klibanov, A.L., Maruyama, K., Torchilin, V.P. et al. 1990. Amphipathic polyethyleneglycols effectively prolong the circulation time of liposomes. *FEBS Lett* 268: 235–7.
Koziara, J., Lockman, P., Allen, D. et al. 2006. The blood–brain barrier and brain drug delivery. *J Nanosci Nanotech* 6: 2712–35.
Krauze, M.T., Noble, C.O., Kawaguchi, T. et al. 2007. Convection-enhanced delivery of nanoliposomal CPT-11 (irinotecan) and PEGylated liposomal doxorubicin (Doxil) in rodent intracranial brain tumor xenografts. *Neuro Oncol* 9: 393–403.
Kreuter, J. 2001. Nanoparticulate systems for brain delivery of drugs. *Adv Drug Deliv Rev* 47: 65–81.
Kreuter, J., Ramge, P., Petrov, V. et al. 2003. Direct evidence that polysorbate-80-coated poly(butylcyanoacrylate) nanoparticles deliver drugs to the CNS via specific mechanisms requiring prior binding of drug to the nanoparticles. *Pharm Res* 20: 409–16.
Kreuter, J., Shamenkov, D., Petrov, V. et al. 2002. Apolipoprotein-mediated transport of nanoparticle-bound drugs across the blood–brain barrier. *J Drug Target* 10: 317–25.
Laterra, J., and Goldstein, G. 2000. Ventricular organization of cerebrospinal fluid: blood–brain barrier, brain edema, and hydrocephalus. In *Principles of Neural Science*, ed. Kandel, E.R., Schwartz, J.H., and Jessel, T.M., 1288–301. New York: McGraw-Hill.

Lee, C.C., MacKay, J.A., Frechet, J.M. et al. 2005. Designing dendrimers for biological applications. *Nat Biotechnol* 23: 1517–26.
Lidar, Z., Mardor, Y., Jonas, T. et al. 2004. Convection-enhanced delivery of paclitaxel for the treatment of recurrent malignant glioma: a phase I/II clinical study. *J Neurosurg* 100: 472–9.
Liebner, S., Fischmann, A., Rascher, G. et al. 2000. Claudin-1 and claudin-5 expression and tight junction morphology are altered in blood vessels of human glioblastoma multiforme. *Acta Neuropathol* 100: 323–31.
Liu, L., Venkatraman, S.S., Yang, Y.Y. et al. 2008. Polymeric micelles anchored with TAT for delivery of antibiotics across the blood–brain barrier. *Biopolymers* 90: 617–23.
Lockman, P.R., Koziara, J.M., Mumper, R.J. et al. 2004. Nanoparticle surface charges alter blood–brain barrier integrity and permeability. *J Drug Target* 12: 635–41.
Lockman, P.R., Mumper, R.J., Khan, M.A. et al. 2002. Nanoparticle technology for drug delivery across the blood–brain barrier. *Drug Dev Ind Pharm* 28: 1–13.
Long, D.M. 1970. Capillary ultrastructure and the blood–brain barrier in human malignant brain tumors. *J. Neurosurg.* 32: 127–44.
Lovric, J., Bazzi, H.S., Cuie, Y. et al. 2005. Differences in subcellular distribution and toxicity of green and red emitting CdTe quantum dots. *J Mol Med* 83: 377–85.
Medina, C., Santos-Martinez, M., Radomski, A. et al. 2007. Nanoparticles: pharmacological and toxicological significance. *Br J Pharmacol* 150: 552–8.
Miller, D.W., Batrakova, E.V., Waltner, T.O. et al. 1997. Interactions of pluronic block copolymers with brain microvessel endothelial cells: evidence of two potential pathways for drug absorption. *Biconjug Chem* 8: 649–57.
Miller, D.W., and Kabanov, A.V. 1999. Potential applications of polymers in the delivery of drugs to the central nervous system. *Colloids Surf B Biointerfaces* 16: 321–30.
Mohs, A.M., and Provenzale, J.M. 2010. Applications of nanotechnology to imaging and therapy of brain tumors. *Neuroimaging Clin N Am* 20: 283–92.
Morganti-Kossmann, M.C., Rancan, M., Stahel, P.F. et al. 2002. Inflammatory response in acute traumatic brain injury: a double-edged sword. *Curr Opin Crit Care* 8: 101–5.
Morrison, P.F., Laske, D.W., Bobo, H. et al. 1994. High-flow microinfusion: tissue penetration and pharmacodynamics. *Am J Physiol* 266: R292–305.
Neuwelt, E.A., Varallyay, P., Bago, A.G. et al. 2004. Imaging of iron oxide nanoparticles by MR and light microscopy in patients with malignant brain tumours. *Neuropathol Appl Neurobiol* 30: 456–71.
Oberdorster, G., Oberdorster, E., and Oberdorster, J. 2005. Nanotoxicology: an emerging discipline evolving from studies of ultrafine particles. *Environ Health Perspect* 113: 823–39.
Oldendorf, W.H. 1971. Brain uptake of radiolabeled amino acids, amines, and hexoses after arterial injection. *Am J Physiol* 221: 1629–39.
Olivier, J.C., Fenart, L., Chauvet, R. et al. 1999. Indirect evidence that drug brain targeting using polysorbate 80-coated polybutylcyanoacrylate nanoparticles is related to toxicity. *Pharm Res* 16: 1836–42.
Pardridge, W.M. 2006. Molecular Trojan horses for blood–brain barrier drug delivery. *Discov Med* 6: 139–43.
Pardridge, W. 2007. Blood–brain barrier delivery. *Drug Discov Today* 12: 54–61.
Perlstein, B., Ram, Z., Daniels, D. et al. 2008. Convection-enhanced delivery of maghemite nanoparticles: increased efficacy and MRI monitoring. *Neuro Oncol* 10: 153–61.
Persidsky, Y., Ramirez, S.H., Haorah, J. et al. 2006. Blood–brain barrier: structural components and function under physiologic and pathologic conditions. *J Neuroimmune Pharmacol* 1: 223–36.
Prokop, A., and Davidson, J.M. 2008. Nanovehicular intracellular delivery systems. *J Pharm Sci* 97: 3518–90.
Ramsauer, M., Krause, D., and Dermietzel, R. 2002. Angiogenesis of the blood–brain barrier in vitro and the function of cerebral pericytes. *FASEB J.* 16: 1274–6.
Rapoport, S., and Thompson, H. 1973. Osmotic opening of the blood–brain barrier in the monkey without associated neurologic deficits. *Science* 180: 971.
Raut, S.L., Kirthivasan, B., Bommana, M.M. et al. 2010. The formulation, characterization and in vivo evaluation of a magnetic carrier for brain delivery of NIR dye. *Nanotechnology* 21: 395102.
Saha, R.N., Vasanthakumar, S., Bende, G. et al. 2010. Nanoparticulate drug delivery systems for cancer chemotherapy. *Mol Membr Biol* 27: 215–31.
Sanovich, E., Bartus, R.T., Friden, P.M. et al. 1995. Pathway across blood–brain barrier opened by the bradykinin agonist, RMP-7. *Brain Res* 705: 125–35.
Sarin, H., Kanevsky, A.S., Wu, H. et al. 2008. Effective transvascular delivery of nanoparticles across the blood–brain tumor barrier into malignant glioma cells. *J Trans Med* 6: 80.
Schlageter, K.E., Molnar, P., Lapin, G.D. et al. 1999. Microvessel organization and structure in experimental brain tumors: microvessel populations with distinctive structural and functional properties. *Microvasc Res* 58: 312–28.

Schroeder, U., Sommerfeld, P., Ulrich, S. et al. 1998. Nanoparticle technology for delivery of drugs across the blood-brain barrier. *J Pharm Sci* 87: 1305–7.

Sethuraman, V.A., Lee, M.C., and Bae, Y.H. 2008. A biodegradable pH-sensitive micelle system for targeting acidic solid tumors. *Pharm Res* 25: 657–66.

Sharma, H.S. 2007. Nanoneuroscience: emerging concepts on nanoneurotoxicity and nanoneuroprotection. *Nanomedicine* 2: 753–8.

Silva, G.A. 2007. Nanotechnology approaches for drug and small molecule delivery across the blood–brain barrier. *Surg Neurol* 67: 113–6.

Smith, A.M., Duan, H., Mohs, A.M. et al. 2008. Bioconjugated quantum dots for in vivo molecular and cellular imaging. *Adv Drug Deliv Rev* 60: 1226–40.

Springer, T.A. 1994. Traffic signals for lymphocyte recirculation and leukocyte emigration: the multistep paradigm. *Cell* 76: 301–14.

Stewart, D.J. 1994. A critique of the role of the blood–brain barrier in the chemotherapy of human brain tumors. *J Neurooncol* 20: 121–39.

Thorne, R., and Nicholson, C. 2006. In vivo diffusion analysis with quantum dots and dextrans predicts the width of brain extracellular space. *Proc Natl Acad Sci* 103: 5567–72.

Tomalia, D.A., Baker, H., Dewald, J. et al. 1985. A new class of polymers—starburst-dendritic macromolecules. *Polym J* 17: 117–32.

Tosi, G., Costantino, L., Ruozi, B. et al. 2008. Polymeric nanoparticles for the drug delivery to the central nervous system. *Expert Opin Drug Deliv* 5: 155–74.

Tsao, N., Hsu, H.P., Wu, C.M. et al. 2001. Tumour necrosis factor-a causes an increase in blood–brain barrier permeability during sepsis. *J Med Microbiol* 50: 812–21.

Wang, B., Feng, W.Y., Wang, M. et al. 2007. Transport of intranasally instilled fine Fe_2O_3 particles into the brain: micro-distribution, chemical states, and histopathological observation. *Biol Trace Elem Res* 118: 233–43.

Wedel-Parlow, M.v., Schrot, S., Lemmen, J. et al. 2011. Neutrophils cross the BBB primarily on transcellular pathways: an in vitro study. *Brain Res* 1367: 67–76.

Weiss, N., Miller, F., Cazaubon, S. et al. 2009. The blood–brain barrier in brain homeostasis and neurological diseases. *Biochim Biophys Acta* 1788: 842–57.

Wolburg, H., Noell, S., Mack, A. et al. 2009. Brain endothelial cells and the glio-vascular complex. *Cell Tissue Res* 335: 75–96.

Yang, R.S., Chang, L.W., Wu, J.P. et al. 2007. Persistent tissue kinetics and redistribution of nanoparticles, quantum dot 705, in mice: ICP-MS quantitative assessment. *Environ Health Perspect* 115: 1339–43.

Zhang, Y., Jeong Lee, H., Boado, R.J. et al. 2002. Receptor-mediated delivery of an antisense gene to human brain cancer cells. *J Gene Med* 4: 183–94.

Zlokovic, B.V. 2008. The blood–brain barrier in health and chronic neurodegenerative disorders. *Neuron* 57: 178–201.

Zunkeler, B., Carson, R., Olson, J. et al. 1996. Quantification and pharmacokinetics of blood–brain barrier disruption in humans. *J Neurosurg* 85: 1056–65.

14 Nanomedicine Treatment Strategies That Exploit Unique Characteristics of Tumor Vasculature and Microenvironment

Jun Wu, Tatsuhiro Fujii, and Mike Y. Chen

CONTENTS

INTRODUCTION

Nanomedicine is a novel methodology and technology for cancer therapy that uses nanoscale molecules or particles to deliver anticancer compounds to targeted cancer cells or tissues. Investigation of tumor vasculature and the surrounding microenvironment has revealed evidence and insight into the architecture of tumor vasculature, formation and molecular regulation of tumor angiogenesis, characteristics and regulation of tumor blood flow, and the sometimes acidic and hypoxic tumor core.

Folkman (1971) introduced the concept that solid cancer tumors become angiogenesis dependent after they reach a few millimeters in diameter. Solid tumor growth relies highly on angiogenesis, the vigorous growth and formation of new tumor-related blood vessels, to respond to the heavy demand for nutrients. Tumor cells enter the new vessels and are carried through the systemic circulation to form metastatic tumors at distant sites. Despite the development of numerous anticancer compounds over the past several decades, cancer still remains a formidable lethal disease. In-depth understanding of the characteristics of tumor vascular and its microenvironment is critical to devise strategies for drug design, delivery, and targeted effectiveness that exceed those of conventional anticancer drugs.

ABERRANT TUMOR VASCULAR ARCHITECTURE

Tumor blood vessels are derived mostly from sprouting of preexisting normal blood vessels, which is induced by several growth factors, such as vascular endothelial growth factor (VEGF) and basic fibroblast growth factor (bFGF), that are secreted by neoplastic or stromal cells. Expression of these growth factors is driven by other growth factors such as hypoxic-inducing factor-1a (HIF1a), which is expressed in response to the tumor's hypoxic environment (Ferrara 2010).

In normal blood vessels, the vasculature consists of a hierarchy of arterioles, capillaries, and venules. However, tumor vasculature lacks this organization and consists of irregular vessels having enlarged diameters and containing aberrant branching and bending. On closer observation, it is apparent that in normal blood vessels, the endothelial cells, mural cells (pericytes/smooth muscle cells), and basement membrane are well organized and tightly integrated, but in tumor blood vessels, pericytes are loosely attached to endothelial cells (McDonald and Choyke 2003) or even absent (Benjamin et al. 1999). In tumor vessels, the basement membrane is impaired and of irregular thickness. The immature anatomic development of tumor vasculature results in vessels that are much more permeable (leaky) than normal vessels. There is some evidence to suggest that the tumor cells can form vessel-like structures, although the vessels may not conduct blood flow (Figure 14.1) (Folberg et al. 2000).

Ruoslahti and his colleagues demonstrated that blood vessels have organ-specific binding signatures for certain peptides (vascular zip codes). By using phage display libraries, tumor vascular homing peptides can be identified by a repeated inject–enrich–harvest–inject process (Ruoslahti 2004). These peptides can be used to guide the targeted delivery of nanomedicine (Ruoslahti et al. 2010).

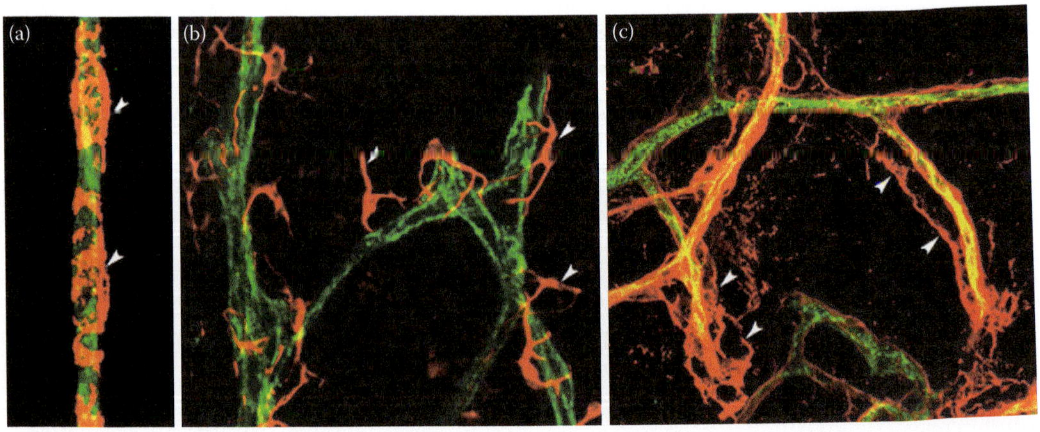

FIGURE 14.1 Imaging of tumor vessel pericytes, basement membrane, and perivascular sleeves. (a,b) Confocal microscopic images showing α-SMA (smooth muscle actin)-immunoreactive pericytes (white, arrowheads) that tightly envelop endothelial cells (green, CD31 immunoreactivity) of a normal vessel (a) but are loosely attached to blood vessels in Lewis lung carcinoma cells (b, green, CD31 immunoreactivity). (c) Confocal micrograph showing coverage of blood vessels in Lewis lung carcinoma by an abnormally loose, fragmented, and multilayered basement membrane (white, arrowheads) stained for type IV collagen immunoreactivity. Endothelial cells are marked by CD31 immunofluorescence (green). (From McDonald, D.M., and Choyke, P.L., *Nat Med*, 9, 713–25, 2003. With permission.)

ALTERED OXYGENATION LEVEL, METABOLISM, AND HOMEOSTASIS IN TUMOR MICROENVIRONMENT

The oxygen level within tumor tissue decreases with increasing distance from tumor vessels, resulting in unique hypoxic stress, which disturbs the tumor's metabolic microenvironment. Tissue hypoxia alters the glycolytic and oxidative processes within the tumor, producing an acidic environment and, often, necrosis in the center of the tumor (Figure 14.2) (Helmlinger et al. 1997; Jain 1999). The hypoxic environment greatly compromises the efficacy of chemotherapy and radiation therapy, which rely on DNA damage in actively cycling cells and on the generation of free radicals to exert cytotoxic effects. In addition, tumor cells counter oxidative stress by increasing the expression and activity of heme oxygenase-1, which protects tumor cells from attack by reactive oxygen species (Doi et al. 1999; Fang et al. 2004). Tumor tissues also have increased expression of nitric oxide synthase and elevated production of nitric oxide and its derivatives. Nitric oxide dilates blood vessels feeding and draining the tumor, increases blood flow (Fukumura et al. 1997; Jordan et al. 2003), increases blood vessel permeability (Wu et al. 1998), and facilitates tumor angiogenesis (Montrucchio et al. 1997; Wu et al. 1998).

Interstitial fluid pressure (IFP) is increased from the edge to the center of a tumor mass (Abe et al. 1985; Hori et al. 1986; Wiig et al. 1982). Interstitial fluid pressure is induced by vigorous proliferation of tumor cells, which compresses tumor blood and lymphatic vessels, impairing blood flow and lymphatic drainage. These processes in turn exacerbate interstitial fluid pressure and induce tumor necrosis (Padera et al. 2004). Elevated IFP within a tumor is considered to be a major obstacle to entry and penetration of anticancer drugs into tumor tissue (Figure 14.3) (Abe et al. 1985; Jain 1987).

Anti-VEGF therapy was considered to be a promising approach for stopping tumor angiogenesis as a strategy for cancer therapy. However, the results from clinical trials showed that antiangiogenic drugs alone could obtain only modest positive response but no benefits of long-term survival (Kerbel 2009; Narayana et al. 2009; Roodink and Leenders 2010). High doses of antiangiogenic agents administered systemically also cause severe off-target effects such as hypertension, proteinuria, and extensive alteration in normal tissues and organs. However, low doses of VEGF-blocking agents may improve survival of treated objects (Cao 2010). The idea that normalizing tumor vasculature

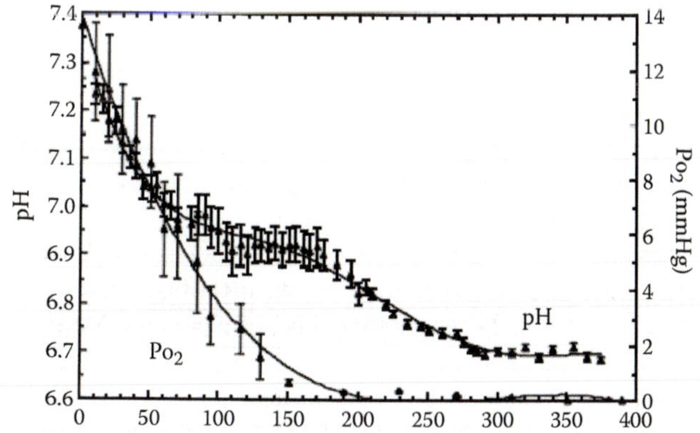

FIGURE 14.2 Spatial gradients of metabolites in tumors. pH gradients measured using fluorescence ratio imaging microscopy. Po_2 gradients measured using phosphorescence quenching. Distance (in microns) from the vessel wall is shown on the x-axis, with zero being the vessel wall. (From Helmlinger, G. et al., *Nat Med*, 3, 177–82, 1997. With permission.)

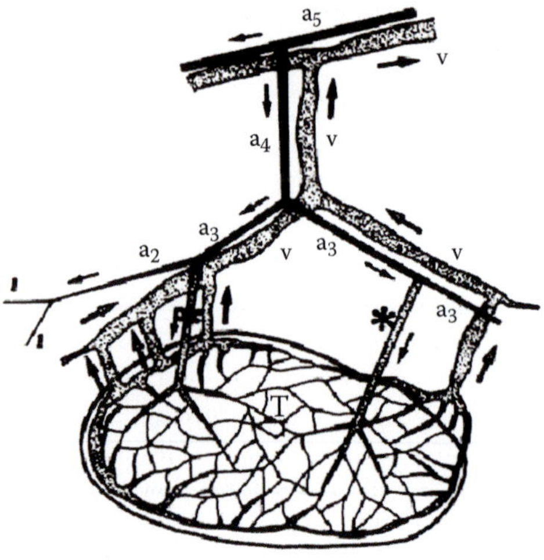

FIGURE 14.3 Schematic representation of the connection made between the preexisting vascular bed and capillaries and the tumor vascular bed. a_2, terminal arterioles; a_1–a_5, arterioles; v, venules; *, Starling vessel; arrows, direction of flow; T, tumor vascular bed. (From Hori, K. et al., *Cancer Res*, 53, 5528–34, 1993. With permission.)

by fortifying tumor vascular structure to improve tumor blood flow for better delivery of oxygen and drugs to tumor tissue eventually facilitates radiation therapy and chemotherapy has been raised (Fukumura and Jain 2007; Jain 2005).

IRREGULAR PROTEOLYTIC ACTIVITIES IN TUMOR MICROENVIRONMENT

Tumor metastasis signals increased malignant potential of the tumor, decreased cure rate with treatment, and reduced length of survival. Tumor cells begin their migration by entering the circulation through a process of invasion that includes focal, proteolytic degradation of the extracellular matrix (ECM) and basement membrane surrounding the tumor's cells and blood vessels. Proteases involved in this process include matrix metalloproteases (MMPs); cysteine proteases such as cathepsin K and cathepsin B; and serine proteases such as urokinase-type plasminogen activator (uPA) and tissue-type plasminogen activator (tPA) (Kessenbrock et al. 2010; Woodward et al. 2007). In normal physiological conditions, the proteolytic process is finely regulated by the balance of protease activators and inhibitors, but in the tumor microenvironment, the balance is tipped, and proteolytic activities predominate. Persistent proteolytic activities not only contribute to tumor cell migration and metastasis but also promote tumor angiogenesis, activate tumor growth factors and related signal transduction pathways, and enhance vascular permeability (Gialeli et al. 2011; Huang et al. 2009; Takigawa et al. 1990; Wu et al. 2001). A tumor microenvironment with increased proteolytic activity is a barrier to antibody-based therapy because antibodies can be cleaved by the tumor proteases (Brezski et al. 2009). To reduce proteolysis within tumors and enhance cancer therapy, more than 50 MMP inhibitors have been developed (Overall and Kleifeld 2006; Zucker and Cao 2009). Clinical trials of the MMP inhibitors have failed, presumably because their actions on tumor cells and MMPs were nonspecific (Drag and Salvesen 2010).

Nanomedical approaches are being developed that detect tumors by virtue of their increased proteolytic activity; this may increase the delivery of MMP inhibitors to tumor tissue. Proteolytic-activated imaging agents or prodrugs use proteolysis in tumors to expedite diagnosis and therapy (Darragh et al. 2010). Initial studies have included polymer-conjugated methotrexate (Chau et al. 2004) and nanoparticles for tumor imaging (McIntyre et al. 2004; Schellenberger et al. 2008; Scherer et al. 2008; Yao et al. 2007) and tumor therapy (Bisht et al. 2010; Veiseh et al. 2009). Recently, a multistage approach to deliver gelatin nanoparticles throughout tumor tissue has been reported (Wong et al. 2011). The size of the primary gelatin nanoparticles was about 100 nm. Initially, the particles accumulate in tumor tissue by a passive delivery mechanism described below. The MMPs in the tumor microenvironment subsequently degrade the 100-nm nanoparticles by shrinking their diameter to 10 nm, which allows the particles to penetrate further into the tumor and deliver drug more widely throughout the tumor.

HETEROGENEOUS BLOOD FLOW IN TUMOR TISSUE AND CIRCADIAN FLUCTUATION MECHANISM FOR TUMOR GROWTH

Tumor blood flow is heterogeneous within the tumor and varies between different areas of the tumor (Peterson 1991; Suzuki et al. 1984). Hori and his colleagues have investigated tumor microcirculation for more than 30 years. In 1981, they introduced a method to observe and measure microvascular pressure and tumor microcirculation within a transparent chamber in rats (Hori et al. 1981). They found that administering angiotensin II substantially increased blood flow in tumor tissue and opened the circulation in once-clogged tumor blood vessels, an event not seen in normal tissues (Hori et al. 1985, 1994; Sato et al. 2000). Angiotensin II normally induces vasoconstriction of a2 terminal arterioles within the arteriolar network and increases arterial resistance and systemic blood pressure (Hori et al. 1993). However, tumor blood vessels do not respond to angiotensin II because they are devoid of functional constriction capability as a result of aberrant mural cell integration with endothelial cells (discussed above) and reduced number of angiotensin II receptors (Kohzuki et al. 1998). Thus, angiotensin II can greatly increase tumor blood flow because it raises systemic blood pressure, even to the point of systemic hypertension, and has little effect on vascular resistance in the tumor because tumor arterioles are angiotensin II unresponsive. In this setting, tumor blood flow can increase further because increased blood flow reopens once-clogged tumor blood vessels (Figure 14.3).

The effect of angiotensin II of enhancing blood perfusion in tumor tissue provides an excellent opportunity and strategy to enhance delivery and penetration of drugs into tumor tissue to augment therapeutic effects. In two randomized controlled trials of angiotensin II-induced hypertension chemotherapy (IHC) in patients with advanced gastric carcinoma, the response rates were 42.9% compared with 10.5% in one trial and 31.3% compared with 6.7% in another trial, while toxicity frequencies did not differ statistically from those using conventional chemotherapy (Sato et al. 2000). This strategy has been applied clinically in tumor imaging and chemotherapy by other groups in pancreatic cancer (Ishikawa et al. 2007), colorectal liver metastasis and liver cancer (Burke et al. 2001; Burton et al. 1989), and breast cancer (Noguchi et al. 1988).

Hori and his colleagues reported in 1999 that a novel combretastatin A-4 derivative AC7700 (AVE8062), a microtubule-destabilizing agent, selectively induced vascular constriction at a3 and a4 arterioles (Hori et al. 1993), irreversibly shut down tumor tissue blood flow, and produced tumor necrosis (Hori et al. 1999). Moreover, AC7700 increased vascular resistance of tumor arterioles for 2–3 h after a single administration, which is sufficient to elicit vascular stasis, hemolysis of red blood cells trapped in vessel lumina, irreversible vascular thrombosis and occlusion, and tumor necrosis (Hori and Saito 2003). When administered 48 h after irradiation, the suppressive effect on tumor growth was greater than that achieved using AC7700 alone (Hori et al. 2008). Vascular disruptive agents (VDAs) such as AC7700 have attracted increased attention for preclinical testing and clinical trials because of their unique ability to target established tumor vasculature and cause vascular stasis and tumor vasculature collapse (Daenen et al. 2010).

Another interesting and crucial observation from Hori et al. is the fluctuation of blood flow in tumor tissue regulated by a circadian mechanism. They found that during the day (7 am–7 pm), blood flow is significantly lower than at night (7 pm–7 am) (20.3 mL/min/100 g in the daytime vs. 46.6 mL/min/100 g at night) (Hori et al. 1992). The tumor doubling time is also consistently shorter at night than during the day (Hori et al. 1995). Angiotensin II induced blood flow enhancement in tumor tissue during both periods, but daytime perfusion remained less than nighttime tumor perfusion (daytime increase: from 20.3 to 62.3 mL/min/100 g; nighttime increase: from 46.6 to 103.4 mL/min/100 g) (Hori et al. 1992). This finding suggests that chemotherapy administered at night and in

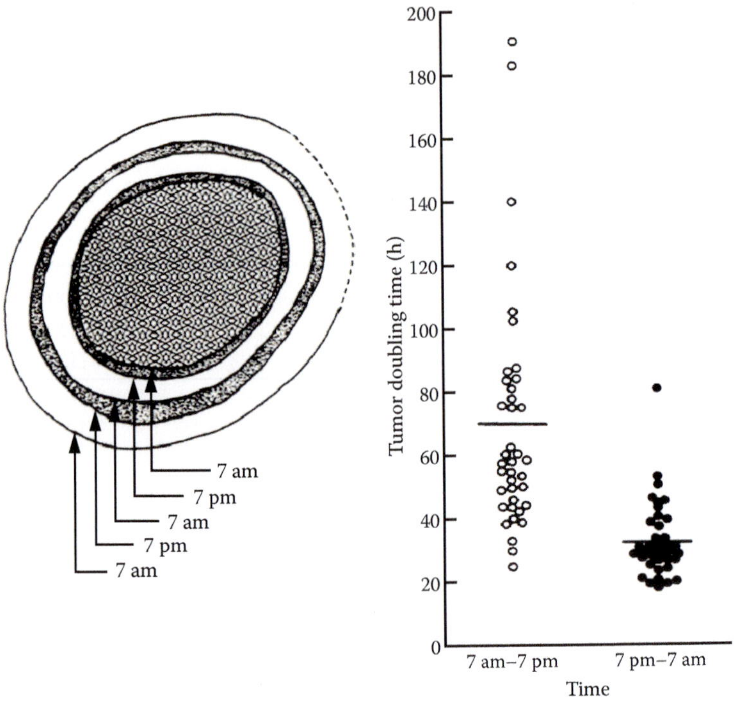

FIGURE 14.4 Circadian fluctuation growth pattern and doubling time of a rat tumor. (From Hori, K. et al., *Br J Cancer*, 71, 1163–8, 1995. With permission.)

conjunction with angiotensin II infusion will enhance anticancer drug penetration into tumor tissue, improve interaction between drug and tumor cells, and consequently lead to a better therapeutic outcome (Figure 14.4).

ENHANCED PERMEABILITY AND RETENTION EFFECT FOR MACROMOLECULAR AND LIPIDIC AGENTS

In 1983, Harold Dvorak and colleagues found that tumor cells secreted a growth factor called vascular permeability factor (VPF), which promotes fluid accumulation in tumor tissue (Senger et al. 1983). VPF was later confirmed to be identical to VEGF, previously discovered by Ferrara's lab (Keck et al. 1989; Leung et al. 1989). VPF/VEGF is a heparin-binding, mitogenic growth factor specific for vascular endothelial cells that induces vascular permeability and stimulates endothelial proliferation and new blood vessel formation (angiogenesis). Maeda et al. independently observed that lipids and macromolecules readily leak out/extravasate from tumor blood vessels and accumulate in the interstitial area surrounding tumor blood vessels because of poor lymphatic drainage. They termed this phenomenon the enhanced vascular permeability and retention effect (EPR effect) (Maeda et al. 2000; Matsumura and Maeda 1986). Macromolecules larger than 40 kDa passively accumulated in tumor tissue, while smaller-sized molecules were quickly washed out of the tumor (Noguchi et al. 1998). The EPR effect is attributed not only to defective tumor vasculature and lack of lymphatic drainage as discussed above but also to factors such as prostaglandins, bradykinin (Wu et al. 1998), nitric oxide and its derivatives, elevated protease activity (Wu et al. 2001), and VPF/VEGF. Macromolecular agents such as polymeric micelles are readily extravasated from vessels at the interface between normal tissue and the necrotic area of tumors (Hori et al. 2010) despite concerns that increased IFP in tumor tissue would hinder drug penetration there. Polymer-conjugated neocarzinostatin (SMANCS) dissolved in lipiodol has been administered into a tumor-feeding artery since 1982 and has been an excellent therapeutic agent for liver cancer (Konno et al. 1983; Maeda et al. 2001). During the past two decades, numerous polymer-conjugated drugs and liposomal, micellar, and nanoparticle-formulated drugs were developed to target tumors passively via the EPR effect (Fang et al. 2011; Maeda 2010b; Maeda and Matsumura 2011). Passive tumor targeting is a useful characteristic of nanoparticles. Low-molecular-weight compounds that are prospective conventional drugs often may have very promising in vitro effects that cannot be reproduced *in vivo* because of the short half-lives of these compounds in the systemic circulation and tumor tissues, poor solubility, poor bioavailability to tumor cells, and severe side effects because vital normal organs/tissues receive the same exposure to the agent as does the tumor tissue. By enhancing the EPR effect with a nitric oxide-releasing agent, drug delivery into tumor tissue can be augmented (Maeda 2010a). On the other hand, VEGF inhibitors, although able to reduce cerebral edema and improve quality of life in brain tumor patients, can suppress the EPR effect by decreasing tumor vascular permeability and normalizing tumor vascular structure (Figures 14.5 and 14.6) (Gerstner et al. 2009).

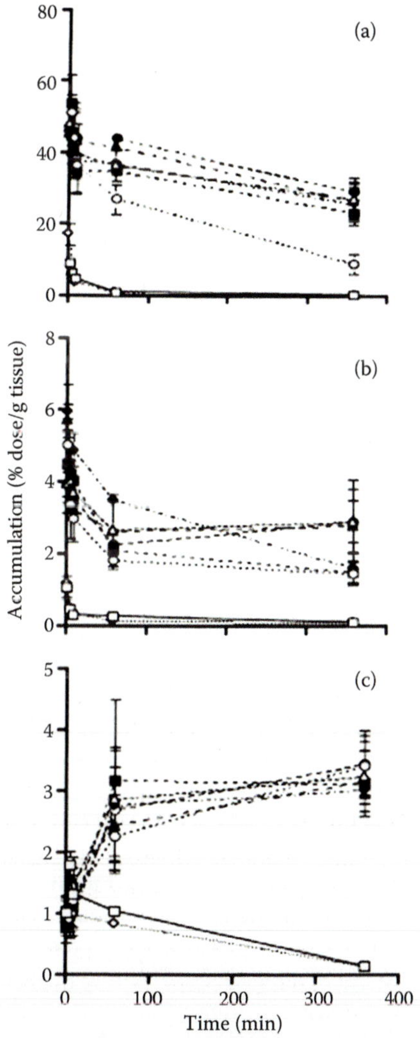

FIGURE 14.5 Distribution of 125I-labeled N-(2-Hydroxypropyl)methacrylamide (HPMA) copolymers in mice bearing S180 solid tumors. (a) plasma; (b) liver; (c) tumor. ▲, 700 kDa; ●, 500 kDa; ◆, 300 kDa; ■, 160 kDa; Δ, 85 kDa; ○, 40 kDa; ◇, 16.2 kDa; □, 4.5 kDa. (From Noguchi, Y. et al. *Jpn J Cancer Res*, 89, 307–14, 1998. With permission.)

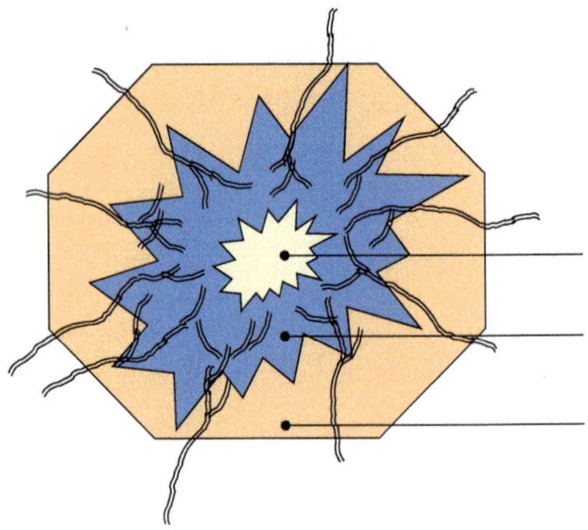

FIGURE 14.6 Scheme of the necrotic area, highly vascularized, and enhanced vascular permeability and retention effect (EPR) effect zone, and surrounding normal tissue.

CANINE CANCER MODELS FOR NANOMEDICINE PRECLINICAL TESTING

Development of nanodrugs for use in clinical trials and eventually for human clinical therapeutic application requires the selection of appropriate animal tumor models that accurately predict the *in vivo* behavior and pharmacological properties of each agent. Murine tumor models are most commonly used for preclinical testing. Although subcutaneous xenograft models in immune-incompetent animal strains are convenient for establishing and monitoring tumor growth and response to therapy, therapeutic efficacy in this context may differ greatly from the results in human trials with respect to drug sensitivity and drug availability.

Canine cancers, especially those discovered by veterinarians in pet dogs, have become better models for developing novel human cancer therapies. Canine cancers arise spontaneously and share many other similar characteristics observed in human cancers, such as (1) a long period of latency for cancer initiation, progress, and development; (2) heterogeneity of cancer cells; (3) metastasis; (4) acquired drug resistance; and (5) recurrence after incomplete therapy. Unlike murine xenograft cancer models, canine cancers develop in an intact immunocompetent and syngeneic environment. The complex interaction among cancer cells, anticancer drugs, and the immune system in a dog better replicates the human situation than do murine models. In addition, canines exhibit strong similarities to humans in terms of anatomical, physiological, genomic, metabolic, and pharmacokinetic profiles. Canine cancers also have similar cancer signal transduction pathways. Most tumor suppressor genes and oncogenes that have been identified in human cancers have been shown to contribute to the initiation and development of canine cancers. Canine cancer models provide more informative data than murine trials for predicting the subsequent outcome of human clinical trials (Paoloni and Khanna 2008; Rivera and von Euler 2011).

Each year, it is estimated that more than 1 million pet dogs in the United States are diagnosed with cancer and about 15% to 30% of them die from the disease. The spectrum of cancers seen in pet dogs is as diverse as the cancers seen in human cancer patients. According to epidemiologic studies in Denmark from 2005 to 2008, about 38% of canine neoplasms are malignant. Common cancers are adenocarcinoma (21%), mast cell tumor (20%), and lymphoma (17%) (Bronden et al. 2010). In Italy, mammary cancer is the most frequently observed cancer in female dogs (70%). Melanoma, liver cancer, osteosarcoma, prostate carcinoma, and bladder carcinoma are also quite frequently diagnosed in pet dogs (Merlo et al. 2008).

In 2003, the National Cancer Institute's Center for Cancer Research (Burke et al. 2001) launched the Comparative Oncology Program (COP) to improve the assessment of novel treatments for humans by treating pet animals—primarily cats and dogs—that have naturally occurring cancer. In 2008, the Comparative Oncology Trials Consortium (COTC; http://ccr.cancer.gov/resources/cop/COTC.asp) was established, and it functions to design and execute clinical trials in dogs with cancer to assess novel therapies (Gordon et al. 2009).

Cancer treatment methods in dogs are quite similar to those in humans and include surgery, radiation, immunotherapy, and chemotherapy. Minimization of pain and suffering is essential in treating dogs with cancer, as it is for humans. In some cases, clinical trials performed in dogs can obtain crucial answers to research questions that are difficult or impossible to obtain from human cancer patients, by using serial biopsies of tumor and surrounding normal tissues and sampling body fluids from the same cancer patient (animal) before, during, and after treatment with a novel drug. Pet owners are highly motivated to seek novel therapeutic methods for their pets, as well as to take the opportunity to contribute to the development of a cancer treatment that may eventually lead to a therapy for human cancer. From the technical point of view, the canine model provides a larger animal model than mouse models for unique drug administration methods such as intra-arterial infusion of nanomedicine. Testing of nanodrugs in large animals with spontaneous tumors is critical for optimizing the administration route and regimen of a given drug.

CONCLUSION

Nanodrug development is a process that proceeds from design and construction of a nanoparticle in the laboratory, to in vitro testing, to *in vivo* testing in a relevant animal model, and finally to clinical trials. Understanding the unique characteristics of the tumor microenvironment is essential for formulating nanodrugs that will target tumor vasculature and tumor cells, spare normal tissues, and potentially lead to improved clinical outcomes in patients with cancer.

REFERENCES

Abe, I., Suzuki, M., Hori, K. et al. 1985. Some aspects of size-dependent differential drug response in primary and metastatic tumors. *Cancer Metastasis Rev* 4: 27–39.

Benjamin, L.E., Golijanin, D., Itin, A. et al. 1999. Selective ablation of immature blood vessels in established human tumors follows vascular endothelial growth factor withdrawal. *J Clin Invest* 103: 159–65.

Bisht, S., Mizuma, M., Feldmann, G. et al. 2010. Systemic administration of polymeric nanoparticle-encapsulated curcumin (nanocurc) blocks tumor growth and metastases in preclinical models of pancreatic cancer. *Mol Cancer Ther* **9**(8): 2255–64.

Brezski, R.J., Vafa, O., Petrone, D. et al. 2009. Tumor-associated and microbial proteases compromise host igg effector functions by a single cleavage proximal to the hinge. *Proc Natl Acad Sci U S A* 106: 17864–9.

Bronden, L.B., Nielsen, S.S., Toft, N. et al. 2010. Data from the Danish veterinary cancer registry on the occurrence and distribution of neoplasms in dogs in Denmark. *Vet Rec* **166**(19): 586–90.

Burke, D., Davies, M.M., Zweit, J. et al. 2001. Continuous angiotensin ii infusion increases tumour: Normal blood flow ratio in colo-rectal liver metastases. *Br J Cancer* 85: 1640–5.

Burton, M.A., Gray, B.N., Klemp, P.F. et al. 1989. Selective internal radiation therapy: Distribution of radiation in the liver. *Eur J Cancer Clin Oncol* 25: 1487–91.

Cao, Y. 2010. Off-tumor target—beneficial site for antiangiogenic cancer therapy? *Nat Rev Clin Oncol* 7: 604–8.

Chau, Y., Tan, F.E., and Langer, R. 2004. Synthesis and characterization of dextran–peptide–methotrexate conjugates for tumor targeting via mediation by matrix metalloproteinase ii and matrix metalloproteinase ix. *Bioconjug Chem* 15: 931–41.

Daenen, L.G., Roodhart, J.M., Shaked, Y. et al. 2010. Vascular disrupting agents (VDAs) in anticancer therapy. *Curr Clin Pharmacol* **5**(3): 178–85.

Darragh, M.R., Schneider, E.L., Lou, J. et al. 2010. Tumor detection by imaging proteolytic activity. *Cancer Res* 70: 1505–12.

Doi, K., Akaike, T., Fujii, S. et al. 1999. Induction of heme oxygenase-1 nitric oxide and ischaemia in experimental solid tumours and implications for tumour growth. *Br J Cancer* 80: 1945–54.

Drag, M., and Salvesen, G.S. 2010. Emerging principles in protease-based drug discovery. *Nat Rev Drug Discov* 9: 690–701.

Fang, J., Akaike, T., and Maeda, H. 2004. Antiapoptotic role of heme oxygenase (ho) and the potential of ho as a target in anticancer treatment. *Apoptosis* 9: 27–35.

Fang, J., Nakamura, H., and Maeda, H. 2010. The EPR effect: Unique features of tumor blood vessels for drug delivery, factors involved, and limitations and augmentation of the effect. *Adv Drug Deliv Rev* 63: 136–51.

Ferrara, N. 2010. Pathways mediating vegf-independent tumor angiogenesis. *Cytokine Growth Factor Rev* 21: 21–6.

Folberg, R., Hendrix, M.J., and Maniotis, A.J. 2000. Vasculogenic mimicry and tumor angiogenesis. *Am J Pathol* 156: 361–81.

Folkman, J. 1971. Tumor angiogenesis: Therapeutic implications. *N Engl J Med* 285: 1182–6.

Fukumura, D., and Jain, R.K. 2007. Tumor microenvironment abnormalities: Causes, consequences, and strategies to normalize. *J Cell Biochem* 101: 937–49.

Fukumura, D., Yuan, F., Endo, M. et al. 1997. Role of nitric oxide in tumor microcirculation. Blood flow, vascular permeability, and leukocyte-endothelial interactions. *Am J Pathol* 150: 713–25.

Gerstner, E.R., Duda, D.G., di Tomaso, E. et al. 2009. VEGF inhibitors in the treatment of cerebral edema in patients with brain cancer. *Nat Rev Clin Oncol* 6: 229–36.

Gialeli, C., Theocharis, A.D., and Karamanos, N.K. 2011. Roles of matrix metalloproteinases in cancer progression and their pharmacological targeting. *FEBS J* 278: 16–27.

Gordon, I., Paoloni, M., Mazcko, C. et al. 2009. The Comparative Oncology Trials Consortium: Using spontaneously occurring cancers in dogs to inform the cancer drug development pathway. *PLoS Med* 6: e1000161.

Helmlinger, G., Yuan, F., Dellian, M. et al. 1997. Interstitial pH and Po_2 gradients in solid tumors *in vivo*: High-resolution measurements reveal a lack of correlation. *Nat Med* 3: 177–82.

Hori, K., Furumoto, S., and Kubota, K. 2008. Tumor blood flow interruption after radiotherapy strongly inhibits tumor regrowth. *Cancer Sci* 99: 1485–91.

Hori, K., Nishihara, M., and Yokoyama, M. 2010. Vital microscopic analysis of polymeric micelle extravasation from tumor vessels: Macromolecular delivery according to tumor vascular growth stage. *J Pharm Sci* 99: 549–62.

Hori, K., and Saito, S. 2003. Microvascular mechanisms by which the combretastatin a-4 derivative AC7700 (AVE8062) induces tumour blood flow stasis. *Br J Cancer* 89: 1334–44.

Hori, K., Saito, S., Nihei, Y. et al. 1999. Antitumor effects due to irreversible stoppage of tumor tissue blood flow: Evaluation of a novel combretastatin a-4 derivative, AC7700. *Jpn J Cancer Res* 90: 1026–38.

Hori, K., Suzuki, M., Abe, I. et al. 1981. New technique for measurement of microvascular pressure in normal and tumor vessels of rats. *Invasion Metastasis* 1: 248–60.

Hori, K., Suzuki, M., Abe, I. et al. 1985. [Changes in tumor transvascular pressure difference associated with tumor growth: Implications for angiotensin-induced hypertension chemotherapy]. *Gan To Kagaku Ryoho* 12: 1630–7.

Hori, K., Suzuki, M., Abe, I. et al. 1986. Increased tumor tissue pressure in association with the growth of rat tumors. *Jpn J Cancer Res* 77: 65–73.

Hori, K., Suzuki, M., Saito, S. et al. 1994. Changes in vessel pressure and interstitial fluid pressure of normal subcutis and subcutaneous tumor in rats due to angiotensin ii. *Microvasc Res* 48: 246–56.

Hori, K., Suzuki, M., Tanda, S. et al. 1992. Circadian variation of tumor blood flow in rat subcutaneous tumors and its alteration by angiotensin ii-induced hypertension. *Cancer Res* 52: 912–6.

Hori, K., Zhang, Q.H., Li, H.C. et al. 1995. Variation of growth rate of a rat tumour during a light–dark cycle: Correlation with circadian fluctuations in tumour blood flow. *Br J Cancer* 71: 1163–8.

Hori, K., Zhang, Q.H., Saito, S. et al. 1993. Microvascular mechanisms of change in tumor blood flow due to angiotensin II, epinephrine, and methoxamine: A functional morphometric study. *Cancer Res* 53: 5528–34.

Huang, Y., Song, N., Ding, Y. et al. 2009. Pulmonary vascular destabilization in the premetastatic phase facilitates lung metastasis. *Cancer Res* 69: 7529–37.

Ishikawa, T., Ushiki, T., Kamimura, H. et al. 2007. Angiotensin-II administration is useful for the detection of liver metastasis from pancreatic cancer during pharmacoangiographic computed tomography. *World J Gastroenterol* 13: 3080–3.

Jain, R.K. 1987. Transport of molecules in the tumor interstitium: A review. *Cancer Res* 47: 3039–51.

Jain, R.K. 1999. Transport of molecules, particles, and cells in solid tumors. *Annu Rev Biomed Eng* 1: 241–63.

Jain, R.K. 2005. Normalization of tumor vasculature: An emerging concept in antiangiogenic therapy. *Science* 307: 58–62.

Jordan, B.F., Sonveaux, P., Feron, O. et al. 2003. Nitric oxide-mediated increase in tumor blood flow and oxygenation of tumors implanted in muscles stimulated by electric pulses. *Int J Radiat Oncol Biol Phys* 55: 1066–73.
Keck, P.J., Hauser, S.D., Krivi, G. et al. 1989. Vascular permeability factor, an endothelial cell mitogen related to PDGF. *Science* 246: 1309–12.
Kerbel, R.S. 2009. Issues regarding improving the impact of antiangiogenic drugs for the treatment of breast cancer. *Breast* 18 Suppl 3: S41–7.
Kessenbrock, K., Plaks, V., and Werb, Z. 2010. Matrix metalloproteinases: Regulators of the tumor microenvironment. *Cell* 141: 52–67.
Kohzuki, M., Tanda, S., Hori, K. et al. 1998. Endothelin receptors and angiotensin II receptors in tumor tissue. *J Cardiovasc Pharmacol* 31 Suppl 1: S531–3.
Konno, T., Maeda, H., Iwai, K. et al. 1983. Effect of arterial administration of high-molecular-weight anticancer agent SMANCS with lipid lymphographic agent on hepatoma: A preliminary report. *Eur J Cancer Clin Oncol* 19: 1053–65.
Leung, D.W., Cachianes, G., Kuang, W.J. et al. 1989. Vascular endothelial growth factor is a secreted angiogenic mitogen. *Science* 246: 1306–9.
Maeda, H. 2010a. Nitroglycerin enhances vascular blood flow and drug delivery in hypoxic tumor tissues: Analogy between angina pectoris and solid tumors and enhancement of the EPR effect. *J Control Release* 142: 296–8.
Maeda, H. 2010b. Tumor-selective delivery of macromolecular drugs via the EPR effect: Background and future prospects. *Bioconjug Chem* 21: 797–802.
Maeda, H., and Matsumura, Y. 2011. EPR effect based drug design and clinical outlook for enhanced cancer chemotherapy. *Adv Drug Deliv Rev* **63**(3): 129–30.
Maeda, H., Sawa, T., and Konno, T. 2001. Mechanism of tumor-targeted delivery of macromolecular drugs, including the EPR effect in solid tumor and clinical overview of the prototype polymeric drug SMANCS. *J Control Release* 74: 47–61.
Maeda, H., Wu, J., Sawa, T. et al. 2000. Tumor vascular permeability and the EPR effect in macromolecular therapeutics: A review. *J Control Release* 65: 271–84.
Matsumura, Y., and Maeda, H. 1986. A new concept for macromolecular therapeutics in cancer chemotherapy: Mechanism of tumoritropic accumulation of proteins and the antitumor agent SMANCS. *Cancer Res* 46: 6387–92.
McDonald, D.M., and Choyke, P.L. 2003. Imaging of angiogenesis: From microscope to clinic. *Nat Med* 9: 713–25.
McIntyre, J.O., Fingleton, B., Wells, K.S. et al. 2004. Development of a novel fluorogenic proteolytic beacon for *in vivo* detection and imaging of tumour-associated matrix metalloproteinase-7 activity. *Biochem J* 377: 617–28.
Merlo, D.F., Rossi, L., Pellegrino, C. et al. 2008. Cancer incidence in pet dogs: Findings of the animal tumor registry of Genoa, Italy. *J Vet Intern Med* 22: 976–84.
Montrucchio, G., Lupia, E., de Martino, A. et al. 1997. Nitric oxide mediates angiogenesis induced *in vivo* by platelet-activating factor and tumor necrosis factor-alpha. *Am J Pathol* 151: 557–63.
Narayana, A., Kelly, P., Golfinos, J. et al. 2009. Antiangiogenic therapy using bevacizumab in recurrent high-grade glioma: Impact on local control and patient survival. *J Neurosurg* 110: 173–80.
Noguchi, S., Miyauchi, K., Nishizawa, Y. et al. 1988. Augmentation of anticancer effect with angiotensin ii in intraarterial infusion chemotherapy for breast carcinoma. *Cancer* 62: 467–73.
Noguchi, Y., Wu, J., Duncan, R. et al. 1998. Early phase tumor accumulation of macromolecules: A great difference in clearance rate between tumor and normal tissues. *Jpn J Cancer Res* 89: 307–14.
Overall, C.M., and Kleifeld, O. 2006. Towards third generation matrix metalloproteinase inhibitors for cancer therapy. *Br J Cancer* 94: 941–6.
Padera, T.P., Stoll, B.R., Tooredman, J.B. et al. 2004. Pathology: Cancer cells compress intratumour vessels. *Nature* 427: 695.
Paoloni, M., and Khanna, C. 2008. Translation of new cancer treatments from pet dogs to humans. *Nat Rev Cancer* 8: 147–56.
Peterson, H.I. 1991. Modification of tumour blood flow—a review. *Int J Radiat Biol* 60: 201–10.
Rivera, P., and von Euler, H. 2011. Molecular biological aspects on canine and human mammary tumors. *Vet Pathol* **48**(1): 132–46.
Roodink, I., and Leenders, W.P. 2010. Targeted therapies of cancer: Angiogenesis inhibition seems not enough. *Cancer Lett* **299**(1): 1–10.
Ruoslahti, E. 2004. Vascular zip codes in angiogenesis and metastasis. *Biochem Soc Trans* 32: 397–402.

Ruoslahti, E., Bhatia, S.N., and Sailor, M.J. 2010. Targeting of drugs and nanoparticles to tumors. *J Cell Biol* 188: 759–68.

Sato, H., Hori, K., Sugiyama, K. et al. 2000. [Tumor microcirculation and selective enhancement of drug delivery—clinical applications based on pathophysiological experiments]. *Gan To Kagaku Ryoho* 27: 1191–200.

Schellenberger, E., Rudloff, F., Warmuth, C. et al. 2008. Protease-specific nanosensors for magnetic resonance imaging. *Bioconjug Chem* 19: 2440–5.

Scherer, R.L., McIntyre, J.O., and Matrisian, L.M. 2008. Imaging matrix metalloproteinases in cancer. *Cancer Metastasis Rev* 27: 679–90.

Senger, D.R., Galli, S.J., Dvorak, A.M. et al. 1983. Tumor cells secrete a vascular permeability factor that promotes accumulation of ascites fluid. *Science* 219: 983–5.

Suzuki, M., Hori, K., Abe, I. et al. 1984. Functional characterization of the microcirculation in tumors. *Cancer Metastasis Rev* 3: 115–26.

Takigawa, M., Nishida, Y., Suzuki, F. et al. 1990. Induction of angiogenesis in chick yolk-sac membrane by polyamines and its inhibition by tissue inhibitors of metalloproteinases (TIMP and TIMP-2). *Biochem Biophys Res Commun* 171: 1264–71.

Veiseh, O., Gunn, J.W., Kievit, F.M. et al. 2009. Inhibition of tumor-cell invasion with chlorotoxin-bound superparamagnetic nanoparticles. *Small* 5: 256–64.

Wiig, H., Tveit, E., Hultborn, R. et al. 1982. Interstitial fluid pressure in DMBA-induced rat mammary tumours. *Scand J Clin Lab Invest* 42: 159–64.

Wong, C., Stylianopoulos, T. et al. 2011. Multistage nanoparticle delivery system for deep penetration into tumor tissue. *Proc Natl Acad Sci U S A* **108**(6): 2426–31.

Woodward, J.K., Holen, I., Coleman, R.E. et al. 2007. The roles of proteolytic enzymes in the development of tumour-induced bone disease in breast and prostate cancer. *Bone* 41: 912–27.

Wu, J., Akaike, T., Hayashida, K. et al. 2001. Enhanced vascular permeability in solid tumor involving peroxynitrite and matrix metalloproteinases. *Jpn J Cancer Res* 92: 439–51.

Wu, J., Akaike, T., and Maeda, H. 1998. Modulation of enhanced vascular permeability in tumors by a bradykinin antagonist, a cyclooxygenase inhibitor, and a nitric oxide scavenger. *Cancer Res* 58: 159–65.

Yao, H., Zhang, Y., Xiao, F. et al. 2007. Quantum dot/bioluminescence resonance energy transfer based highly sensitive detection of proteases. *Angew Chem Int Ed Engl* 46: 4346–9.

Zucker, S., and Cao, J. 2009. Selective matrix metalloproteinase (MMP) inhibitors in cancer therapy: Ready for prime time? *Cancer Biol Ther* 8: 2371–3.

15 Focused Ultrasound for Noninvasive Neuromodulation

Martin O. Culjat, Richard E. Fan, Meghedi Babakhanian, Amit P. Mulgaonkar, Rahul S. Singh, Warren S. Grundfest, and William P. Melega

CONTENTS

The development of effective therapies to treat central nervous system (CNS) disorders and diseases has historically been a challenging problem due, in part, to the invasiveness of interventional surgical procedures and to the presence of the blood–brain barrier (BBB) that restricts passage of most therapeutic drugs from the circulation into the brain. Furthermore, existing CNS drug therapies often have undesirable systemic side effects, while effective neurosurgical approaches are not applicable to large patient populations. To circumvent these limitations, there is great research interest in designing noninvasive CNS therapies. It has been hypothesized that ultrasound, with proper tuning and focusing of ultrasonic energies, can be delivered from outside the head to targeted regions in the brain to effect alterations in neuronal conduction properties and neurotransmitter release, and to induce transient opening of the BBB to facilitate drug delivery. These potential uses of ultrasound to modulate cellular and molecular functions at the micro- and nano-scale in the CNS have provided the stimulus for collaborative research among biomedical engineers, neuroscientists, and neurosurgeons that is necessary to advance this novel therapy to the clinic. The objective of this chapter is to provide an introduction to ultrasound interactions with tissue and bone and to present examples of potential CNS applications of focused ultrasound (FUS).

INTRODUCTION

Ultrasound has been in active clinical use since the 1960s, most commonly as a modality for diagnostic imaging in medicine. During this time, research into the bioeffects and therapeutic applications of ultrasound was also conducted. In fact, the first ultrasonic bioeffect was recognized in 1927, when a crude early ultrasound system was tested against a frog model (Wood and Loomis 1927). It was later observed in the 1950s that by achieving a focus, a concentration of ultrasound energy could be formed in the brain with minimal effects to surrounding tissue (Fry and Fry 1958). This discovery inspired a field of research that concentrated on noninvasive delivery of FUS and techniques to induce biological effects in the body (Figure 15.1).

The pioneers in this field observed that effects of ultrasound in the body were highly variable, depending on the targeted region and method of delivery. These effects include acoustic streaming of fluid around cells, molecular rearrangement of lipid bilayers, mechanical oscillation of cellular and molecular processes, conformational changes in enzyme and pore systems, acoustic cavitation of microbubbles and shockwave formation, and intense thermal heating. The adverse tissue effects were generated predominantly by relatively high-intensity ultrasound, while other micro- and nanoscale mechanical effects were more frequently associated with lower intensities. The high-intensity approach and its effects have already been demonstrated in clinical settings through procedures such as lithotripsy and targeted ablation of tumors or fibroids. Applications of lower-intensity therapeutic ultrasound are used clinically in physical therapy and fracture repair. There are several promising indications of low-intensity therapeutic ultrasound with reversible effects at the cellular level, such as gene therapy, sonoporation, and sonothrombolysis. Additionally, low-intensity therapies are being investigated as a way to mediate drug delivery to targeted tissues or through the blood–brain barrier (BBB). Current research is also focused on using low-intensity ultrasound to directly stimulate neural structures. A more detailed discussion of the current applications to the CNS, together with an introduction to the physical principles of ultrasound and clinical motivations for interventional strategies, follows.

Critical Barriers to Progress in the Field of Treatments for CNS Disorders

Therapies for brain diseases are presently limited to pharmacologic or invasive neurosurgical strategies. Each approach has inherent limitations. Drugs that have been designed to target a particular neurotransmitter system function or parameter generally lack regional selectivity. For example, selective serotonin reuptake inhibitors (SSRIs) affect all serotoninergic synapses in the brain as well as in the periphery. Furthermore, with SSRIs, there can be adverse effects, such as nausea, sexual dysfunction, and weight gain, as well as concerns about metabolism/metabolites and clearance in individuals with compromised hepatic and renal function. Neurosurgical interventions, such as deep brain stimulation (DBS) and resection, are invasive and have an associated morbidity risk. Nonetheless, novel CNS therapies such as viral vector transfer and stem cell transplantation have

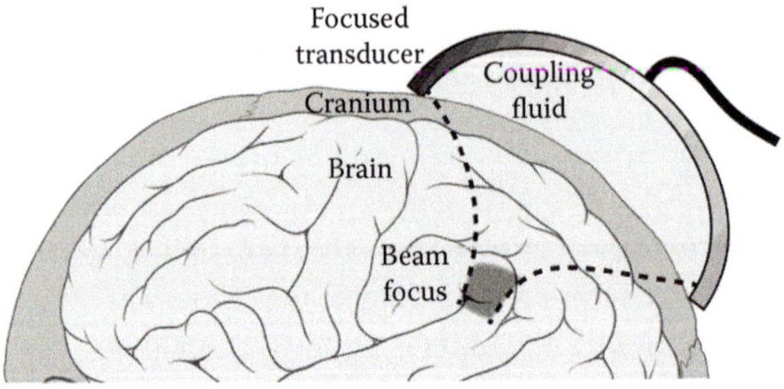

FIGURE 15.1 Illustration of focused ultrasound transducer transmitting ultrasound energy through the cranium and into the brain.

been advanced to clinical trials, while DBS is now an established clinical procedure. These invasive approaches, however, require extensive neurosurgical expertise and, due to their cost and invasive nature, are not envisioned to have wide applicability to large patient populations. Consequently, there is a pressing need for more effective treatments for brain pathologies such as Alzheimer's disease, Parkinson's disease, epilepsy, and neuropathic pain management. In particular, methods that can provide noninvasive interventions to treat diseases in the brain are of great interest, including CNS applications of ultrasound.

INTRODUCTION TO ULTRASOUND

Ultrasound waves are pressure waves that propagate through a medium with a frequency greater than 20 kHz. The audible range for humans is 20 Hz to 20 kHz; therefore, ultrasound falls immediately above the range of human hearing. The first major application of ultrasound came in the early 20th century as a tool for detecting the range (distance) of icebergs and submarines, motivated by the Titanic disaster and the advent of submarine warfare in World War I. Diagnostic medical ultrasound, or medical sonography, was first researched following World War II, using surplus naval sonar equipment (Dubose 2000).

Diagnostic Ultrasound

In the mid-1950s, simple ultrasound systems were used to take measurements of the eyes and brain (Hedrick et al. 1995). Research and development work in the 1970s led to more advanced imaging systems, in which sector scanners and arrays were constructed that could display static two-dimensional grayscale images on a screen for viewing by radiologists and ultrasound technicians. These static images were soon replaced by real-time images, and ultimately, color Doppler images and measurements became commonplace in the 1980s, allowing tissues and fluids to be characterized and viewed in motion.

Intraoperative ultrasound was first described for image guidance during neurosurgical operations in 1978 (Reid 1978). As ultrasound and neurosurgical techniques improved in the 1980s and 1990s, the two-dimensional ultrasound gained acceptance as a real time tool for locating and characterizing brain tumors and tumor margins, for image guidance during biopsy and resection, and for detecting brain shifts (Sosna et al. 2005). Use of three-dimensional ultrasound probes was more recently adopted, as was integration of ultrasound probes into neuronavigation systems for fusion with magnetic resonance (MR) and computed tomographic (CT) imagery. Intraoperative ultrasound probes are coupled directly to the dura of the brain through a craniotomy or burr hole, typically using degassed saline to improve acoustic coupling. Ultrasound imaging is infrequently performed transcranially due to the distortion and attenuation of the beam through the skull and to the subsequent requirement to use lower-frequency (~2 MHz) probes, which at that range provide poor resolution. The most common transcranial use of ultrasound today is transcranial Doppler, which is used to measure and monitor flow velocity in the arteries of the brain to diagnose stenosis, subarachnoid hemorrhage, anastomoses, and brain death. Because high-resolution imagery is not required for transcranial Doppler measurements, the effects of the cranium on the ultrasound beam are not as critical.

Therapeutic Ultrasound

Although the imaging and monitoring applications are best known, there are also many therapeutic applications of ultrasound. Therapeutic ultrasound is used to treat a specific disease or condition in the body, most commonly by heating tissues via absorption of focused acoustic waves. Because tissues in the body have different absorption coefficients, it is possible to selectively heat certain tissues while leaving others unaffected. Many procedures also depend on the nonthermal effects of therapeutic ultrasound, such as cavitation, acoustic streaming, or low-intensity mechanical oscillation. Cavitation is the expansion, contraction, and bursting of microbubbles in tissues, often causing acoustic shock waves that disrupt or damage cells. Acoustic streaming is a localized flow of liquid around the bubbles, causing shear stresses at the cell surfaces and varying the transport of ions and molecules into cells. Both thermal and nonthermal effects are often desired, although in many cases, they cannot be differentiated. Frequencies between 700 kHz and 3.3 MHz are most common in therapeutic ultrasound and with maximum energy absorption occurring at depths between 2 and 5 cm (Martinez 2010).

Therapeutic ultrasound can generally be categorized as either low or high intensity. Low-intensity therapeutic ultrasound is performed using either unfocused or focused transducers, most frequently for physical therapy of conditions such as osteoarthritis, carpal tunnel syndrome, and tennis elbow. Tissues are usually warmed to ~38°C for 10 min at intensities in the 1 W/cm^2 range (Bailey et al. 2003). Fracture healing can also be accelerated using low-intensity therapeutic ultrasound, typically in the 50 mW/cm^2 range (Warden et al. 2000). Another low-intensity application is hyperthermia, in which tumors are heated during radiation therapy to ~43°C for 1–2 h at roughly 10 W/cm^2 (Bailey et al. 2003). High-intensity applications use focused transducers or arrays and include high-intensity focused ultrasound (HIFU), shockwave lithotripsy of kidney stones, and ultrasound-enhanced drug delivery into tissues. During HIFU, tissues are typically heated to temperatures higher than 70°C for 1–3 s at intensities in the 1 kW/cm^2 range (Bailey et al. 2003). Some additional therapeutic uses of ultrasound are more vibratory in nature. These include the harmonic scalpel, for cutting and coagulation of tissues during surgery; tools used by dental hygienists for teeth cleaning; and devices for emulsification of the lens in the eye during cataract surgery.

The most advanced therapeutic ultrasound techniques rely on FUS transducers. These transducers are designed to increase energy density within a localized region, while minimizing heating of adjacent structures. Focusing allows ultrasound to travel harmlessly through tissues outside the focal zone of the transducer due to the low energy density in these regions, while heating and sometimes ablating tissues such as tumors within the focal zone (Figure 15.2). Focusing can be performed either mechanically or electronically. Mechanically focusing an ultrasound transducer is achieved either by using a focused piezoelectric material or by using an acoustic lens (Figure 15.3). The focal depth depends on the curvature and dimensions of the transducer or lens, the frequency, and the acoustic properties of the transducer and tissues. Beam position can be varied by moving the transducer by hand or by using a mechanical translation apparatus. Electronic focusing is possible when a multielement array transducer is used. In this case, the timing of the electrical input impulses into each element of the array can be varied, consequently modulating the timing of the resulting ultrasonic wave fronts. With proper time delay patterns, the ultrasonic wave fronts combine to focus the ultrasound beam in a specific region (Figure 15.4). The beam position can also be varied using a similar technique, called beam steering. Mechanical translation of an array transducer may also be used. Ultrasound transducers and focusing are described in more detail elsewhere (Singh and Culjat 2013).

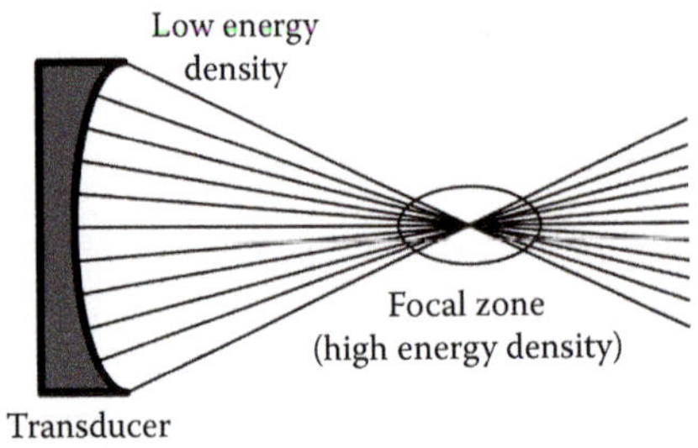

FIGURE 15.2 Focused ultrasound transducer, with low-intensity regions outside the focal zone.

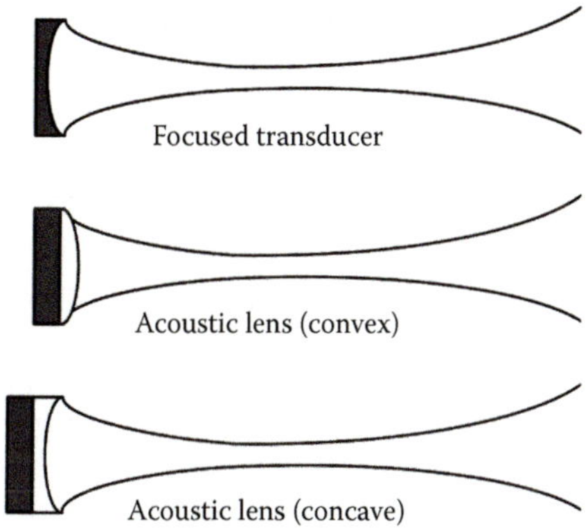

FIGURE 15.3 Focusing using a spherically focused transducer element (top) or an acoustic lens. Acoustic lenses can be either convex or concave, depending on the acoustic properties of the lens material.

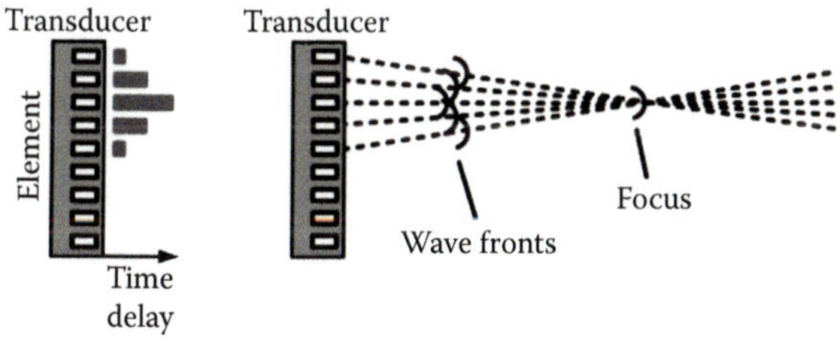

FIGURE 15.4 Electronic focusing can be performed by electronically time-delaying the electrical impulse into the elements of an array (left), consequently modulating the timing of the resulting ultrasonic wave fronts from the same elements (right).

Ultrasound Parameters and Bioeffects

Many additional techniques are used to control the amount of energy delivered to the desired location in the body. For example, heating at the target location largely depends on the electrical signal applied to the ultrasound transducer. For most ultrasound applications, the electrical signal can be fully defined by the following parameters: frequency (f), pulse repetition frequency (PRF), pulse duration, and amplitude of the signal. Waveforms illustrating these parameters are given in Figure 15.5.

The signal waveform frequency is the number of cycles that occur in 1 s, measured in hertz (1 Hz = 1 cycle/s). The time elapsed for each cycle is defined as the period, also defined as the inverse of the frequency. The wavelength is defined as the length of the space over which one cycle occurs and is the ratio of the velocity of the wave and the frequency of the wave. In the continuous wave (CW) case, these parameters, together with the amplitude of the signal, fully describe the energy. However, in imaging and therapy, the acoustic energy is commonly pulsed, with zero acoustical energy being generated between the pulses. The rate at which the pulses occur is the PRF and the length of the pulse in time is the pulse width. The time between the start of two successive pulses is the pulse repetition period (PRT), which is the inverse of the PRF.

Acoustic intensity is defined as the rate at which energy passes through a unit area and depends on the above parameters as well as the properties of the medium or material of propagation. The peak intensity is proportional to the square of the signal amplitude; therefore, if the amplitude is doubled, the total intensity is quadrupled. In medical ultrasound, intensity is commonly measured at the spatial point of the highest intensity along the beam axis and is reported as either I_{SPPA} (intensity,

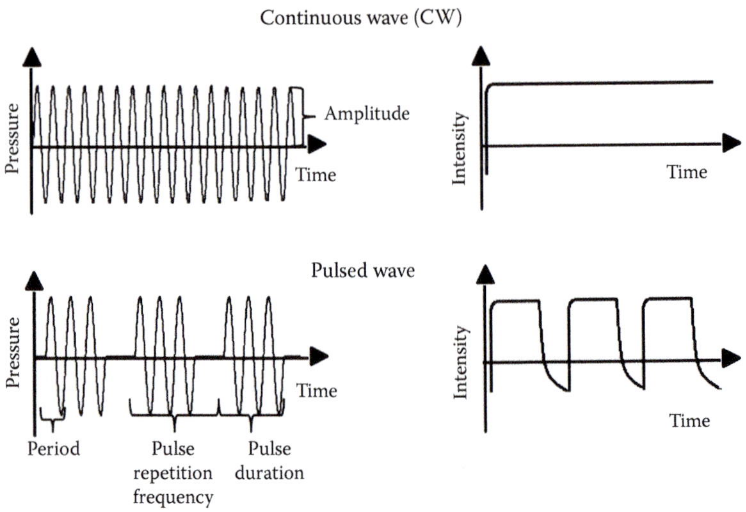

FIGURE 15.5 Graphical illustration of some of the critical parameters to vary the intensity and delivery of ultrasound energy, with a CW input (top) and a pulsed wave input (bottom) and corresponding intensity output profiles at the right.

spatial peak, pulse average) or I_{SPTA} (intensity, spatial peak, time average). I_{SPPA} measures the average intensity of each pulse and is a good indicator of mechanical bioeffects and cavitation, while I_{SPTA} provides a time-averaged intensity of all the delivered pulses and is generally an indicator of ultrasound-induced thermal effects (Fish 1990).

Biomedical Engineering Challenges to Transcranial Ultrasound

The primary impediment to using ultrasound in the brain, whether for therapeutic or imaging purposes, has been the cranium. As mentioned above, the cranium significantly impacts the transmission of ultrasound between the transducer and the desired brain region. This is due to several reasons. First, the higher density and speed of sound of the cranium results in large acoustic impedance mismatches both at the skin/cranium and cranium/dura interfaces. One consequence is that these mismatches result in refraction of the beam that modifies the position and shape of the beam focus (Figure 15.6). The mismatches also cause large reflections of the acoustic beam (~40% of the energy at each interface) that both reduce the energy transmitted into the brain and cause reverberations that clutter the acoustic signal (Figure 15.7). When using CW energy, reverberations are

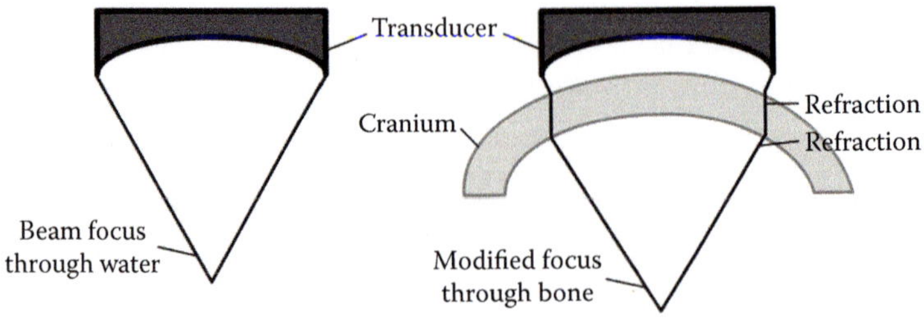

FIGURE 15.6 Beam focus from focused ultrasound transducer (left) and refracted beam through the cranium (right).

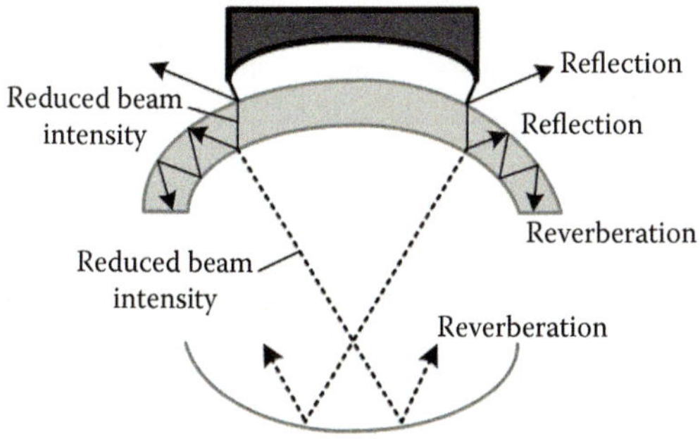

FIGURE 15.7 Reflection, reverberation, and reduced beam intensity of the acoustic beam as it passes through the cranium.

especially problematic, since the energy is not given time to dissipate. Reverberations are problematic not only at the beam's entry point into the brain but also by all internal surfaces of the cranium, which effectively form an echo chamber. As the ultrasound beam passes through the focal zone, the beam continues and reflects off the far side of the cranium, causing further interference at the desired focal zone (Figure 15.7).

Second, the heterogeneity, variable topography, and nonuniform thickness of the cranium refract the acoustic beam as it passes through the skull, making it difficult to precisely target specific regions while also broadening the beam focus (Figure 15.8). Third, the cranium can cause significant attenuation of the acoustic beam (Figure 15.9). Attenuation in cortical bone is approximately 6.9 dB/cm MHz and results in further reduction of energy traveling into the brain through both absorption and scattering (Culjat et al. 2010). However, of greater concern than the loss of energy is the absorptive conversion of acoustic energy into thermal energy, which can lead to heating of the skull and underlying tissues in the brain, especially at high intensities. Finally, acoustic coupling of the ultrasound transducer using gels or liquids is difficult due to both gravity and the presence of hair on the head.

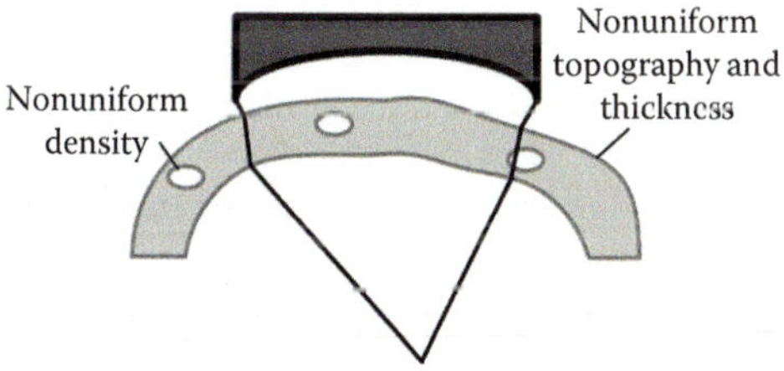

FIGURE 15.8 Modified focus through nonuniform topography, thickness, and density of the cranium. The beam focus is also broadened (not shown).

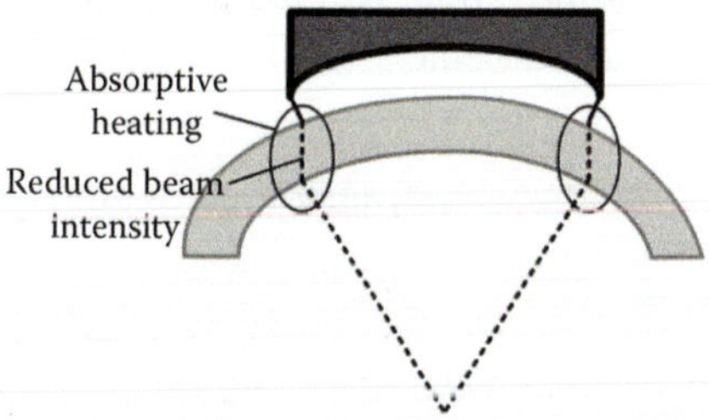

FIGURE 15.9 Reduced beam intensity and heating of the skull due to attenuation and absorptive heating.

Together, these challenges have created significant barriers for adopting ultrasound to noninvasive transcranial applications. In diagnostic transcranial ultrasound, including imaging and transcranial Doppler, the beam must travel both into the brain and back to the transducer, so the effects of the cranium are encountered twice. These applications use low intensities (<720 mW/cm^2 I_{SPTA}) (FDA 2008), high frequencies (>1 MHz), and pulsed operation. While the high frequencies used can result in relatively high acoustic attenuation, absorptive heating is not a concern because the low intensities and reverberations are mitigated by the pulsed operation. The effects of refraction can be mitigated during diagnostic ultrasound by using one of several insonation windows, or regions of the skull with thinner or flatter walls, such as the temporal region, the back of the head, below the jaw, or the eyes.

Many noninvasive therapeutic procedures have been proposed, dating back to the 1940s (Lynn and Putnam 1944; Lynn et al. 1942). However, most procedures have had limited success due to the challenges related to penetrating the cranium. High-intensity procedures are particularly troublesome in the brain, especially HIFU, which relies on high intensities and CW operation for generating maximum energy delivery and high frequencies for producing a tight beam focus that minimizes destruction of adjacent tissues.

Several innovative solutions have been recently developed that have made therapeutic procedures more practical for noninvasive clinical use. Lower-intensity techniques have been proposed for neuromodulation or drug delivery in the brain, which are not subject to the same safety concerns as HIFU. Lower frequencies (<1 MHz) can also be used, particularly in applications (aside from ablation) that do not require a tight beam focus. Pulsed operation can be used to reduce reverberations, especially for applications that do not require high energy delivery.

Many of the challenges related to heating and topography can potentially be addressed using recent advances in transducer, amplifier, modeling, and image-guidance techniques (Aubry et al. 2003; Clement et al. 2000). FUS systems have been proposed or developed by several groups using transducer arrays with dynamic focusing that can more accurately steer and focus through bone (Daum et al. 1998; Daum and Hynynen 1999). Using large arrays provides a geometric gain that can resolve the energy loss due to attenuation (Hynynen and Jolesz 1998; Thomas and Fink 1996). Hemispherical array transducers covering the surface of the scalp can also reduce energy density at the cranial surface by distributing energy across the head (Hynynen et al. 2004a). The effects of topography, heterogeneity, and thickness variations can be mitigated by using geometric and density information from preoperative CT scans of the skull and correcting the phase of the input signals into each array element to account for these nonuniformities (Clement and Hynynen 2002). Many FUS techniques are expected to benefit from MRI, which can be used to guide, monitor, and control the focus of the ultrasound beam (Hynynen et al. 2001b), and also provide temperature imaging to quantify and visualize heating resulting from the focused beam (Salomir et al. 2000). MR imaging (MRI) guidance in the context of FUS is now commonly referred to as MR-guided focused ultrasound (MRgFUS).

With all these recent advances in ultrasound transducer technologies, electronic focusing, and improved preoperative and intraoperative visualization, the prospect of noninvasive ultrasound brain therapies is possible. However, because these are only recent advances, most prior therapeutic ultrasound research efforts were performed invasively through a window in the cranium following craniotomy. These efforts, as well as transcranial efforts, are described below.

NEURAL APPLICATIONS OF FOCUSED ULTRASOUND

Focused Ultrasound Ablation

The field of FUS ablation is the best known application of therapeutic ultrasound in the brain. Researchers in this field have led the development of noninvasive transcranial FUS technologies. FUS ablation is generally associated with HIFU due to the high intensities required to ablate tissues.

HIFU was suggested many years ago as a tool to target and ablate subcutaneous tissue volumes by localized heating, specifically as a potential therapy tool for deep-seated soft tissue tumors. The application of HIFU technology was first described for various applications in the 1950s and 1960s (Fry 1954, 1958; Fry and Dunn 1962; Fry and Meyers 1962; Fry et al. 1954) through work by Fry et al. to develop four-transducer systems that produced lesions in animal models. HIFU is now approved by the Food and Drug Administration (FDA) for heating and ablating uterine fibroids in the United States; in other countries, it is used to treat prostate and bone cancers. After decades of research, the technology has evolved from a topic in research laboratories to one that is undergoing clinical trials (Hynynen and Clement 2007).

The mechanism by which HIFU systems can ablate tissues is most commonly attributed to thermal effects (ter Haar 2007). The absorption of FUS energies is used to raise the temperature of a selected tissue volume to over 55°C, a temperature threshold identified as leading to coagulative necrosis and immediate cell death (Dewhirst et al. 2003; Sapareto and Dewey 1984; ter Haar 2007).

As described in the section, "Biomedical Engineering Challenges to Transcranial Ultrasound," the primary impediment to using ultrasound in the brain, whether for therapeutic or imaging purposes, has been the cranium. Early attempts at producing lesions in the brain through the intact skull bone were unsuccessful (Lynn and Putnam 1944; Lynn et al. 1942). Propagation through the cranium results in significant distortion and attenuation of the acoustic beam, thereby limiting accuracy of the procedures and also increasing potential health risks to the patient. Due to complications of FUS transmission through the skull and the lack of adequate visualization technology in the brain, many researchers discontinued their research on HIFU in favor of radiosurgery, such as neurosurgeon Lars Leksell, who developed the gamma knife in 1967 (Jagannathan et al. 2009).

A majority of earlier ultrasound research efforts were therefore performed invasively through a window in the skull following craniotomy. Patients with Parkinson's disease were among the first to be treated with HIFU performed through a craniotomy (Fry and Fry 1958). Although the treatment helped ameliorate some Parkinson's symptoms, this research was not further pursued because of the requirement to remove a section of the skull and difficulties in obtaining accurate targeting. Many of the recent advances described previously were applied to HIFU systems to address these limitations. Commercial neurosurgical HIFU systems now exist, such as the ExAblate (Insightec, Haifa, Israel), that combine visualization of soft tissues, temperature sensitivity, and guidance of MRI with using phased-array focused transducers to provide compensation for ultrasonic distortion through the skull (Jagannathan et al. 2009).

Recent research has demonstrated the clinical feasibility of transcranial MRI-guided focused ultrasound surgery (TcMRgFUS) using the ExAblate system. In 2010, it was reported that therapeutic ultrasound could be noninvasively focused effectively through an intact skull to target a glioblastoma. The feasibility of transcranial targeting with noninvasive FUS ablation was also demonstrated, although clinical outcomes were mixed due to insufficient power (McDannold et al. 2010). Much work is still required, however, both in further developing high-power systems for effective ablation at different tissue depths and also in defining safety margins to deliver energy selectively without overheating the skull.

Ultrasound-Mediated Bioeffects

Recently, the nonthermal therapeutic effect of ultrasound on drug delivery applications has generated excitement in various medical fields. Ultrasound energy can enhance effects of thrombolytic agents, transdermal drug delivery, and anticancer drugs. In addition to the pressure oscillation experienced by cells and tissue caused by ultrasound, secondary effects such as heat and cavitation can induce strong physical and biological effects in tissues as well (Suslick and Nyborg 1990). Further investigation is required to fully understand how this phenomenon can cause biological interactions, although several potential neurosurgical applications have already been identified.

Ultrasound Facilitation of Drug Transfer through BBB

The BBB is the brain capillary barrier that separates circulating blood from cells in the CNS and plays a vital role in keeping blood-borne pathogens and toxins out of the brain. The BBB is formed primarily by capillary endothelial cells fused by tight junctions and other neural cells that further decrease capillary permeability (Bergers and Song 2005; Misra et al. 2003). The factors that determine the BBB permeability of endogenous compounds and xenobiotics in the circulation are their molecular size, charge, and lipid solubility; transporter systems can also facilitate transfer. However, these permeability factors result in the exclusion of ~98% of all small-molecule drugs (less than 400–500 Da) and approximately 100% of larger-molecule neurotherapeutics (Pardridge 2005).

Several approaches have been undertaken to facilitate drug delivery through the BBB, including the following: (1) intra-arterial injection with hyperosmotic solutions, solvents, and other chemicals (Meairs and Alonso 2007); (2) localized drug delivery into the brain parenchyma by injection into the targeted brain area; (3) direct infusion of agents via a catheter (Bobo et al. 1994) or by implanted drug-reservoir devices (Guerin et al. 2004) that slowly release drugs directly into the brain tissue; or (4) introducing a catheter into the brain vasculature and infusing a solution such as mannitol to produce diffuse disruption of the BBB downstream (Doolittle et al. 2000; Neuwelt et al. 1979). Such procedures are invasive and require skull penetration, which can potentially cause damage to nontargeted brain tissue, bleeding, and infection (Meairs and Alonso 2007).

Noninvasive disruption of the BBB using ultrasound has been studied for decades as a technique to deliver therapeutic agents into the brain (Bakay 1956; Bakay et al. 1956; Hynynen 2007; Hynynen and Clement 2007). However, it was only recently discovered that FUS could be used to produce temporary and reversible BBB disruption for drug delivery (Hynynen et al. 2001a, 2004b, 2005; McDannold et al. 2005; Mesiwala et al. 2002; Sheikov et al. 2004). Initial BBB disruption studies that used ultrasound without any additional components required significant acoustic energy to produce the effect. This increased the chance of damage to other brain structures. Vykhodtseva et al. (1995) suggested using FUS in combination with encapsulated gas-filled microbubbles (<8 μm diameter), thereby reducing the high acoustic intensities needed to produce the opening (Vykhodtseva et al. 1995). In addition, the lower acoustic energy would decrease the chances of overheating the skull and be more practical for transcranial applications. The microbubbles induced cavitation nuclei into the bloodstream and confined the ultrasound effects to the vasculature, which reduced the intensity needed to produce BBB opening and limited the effect on the surrounding parenchyma (Hynynen et al. 2004b; McDannold et al. 2005).

Extensive work has been performed to find the optimal frequency for transcranial BBB disruption with ultrasound. A frequency of 1.63 MHz combined with ultrasound contrast agents was initially demonstrated to be suboptimal due to the distorting effects and high absorption of this high-frequency ultrasound by the cranium (Hynynen et al. 2001a). Lower frequencies were subsequently adapted for this application, and successful BBB disruption has been produced with no ischemia, apoptosis, or any long-term damage detected in histology (Hynynen et al. 2005, 2006).

Sonoporation

Ultrasound may be used to enhance the uptake of molecules with varying high molecular weights into cells and tissues through sonoporation. This term refers to inducing temporary nanopores formed in the membrane of target cells by the formation and oscillation of small gas-filled bubbles. The ability of ultrasound to locally open the BBB suggests the prospect of using ultrasound to assist in delivering drugs to targeted regions in the brain (Kinoshita et al. 2006). This process can be particularly useful to facilitate introduction of a high drug concentration at a specific site. Several studies demonstrated that sonoporation can increase the uptake of specific drugs in cells, tissues, and tumors. An in vitro study showed enhanced uptake of radionuclides (^{111}In-DOTA-Tyr(3)-octreotate) in pancreatic tumor cells (van Wamel et al. 2004). Other *in vivo* studies used ultrasound for the uptake of chemotherapeutic drugs, resulting in inhibiting leukemia cell growth (Tachibana et al. 2000).

Ultrasonic Neuromodulation

In 1950, Fry et al. reported the first effects of ultrasound targeted directly on neural tissues, demonstrating the temporary suppression of nervous conduction through the ventral nerve cords of a crayfish (Fry et al. 1950). Later, demonstrating neuromodulation in the intact brain of a cat (Fry et al. 1958), Fry's early work demonstrated the ability of targeted FUS to reversibly modulate both the peripheral nervous system and CNS. These initial efforts have been expanded with a variety of studies on different neural structures conducted at varying acoustic intensities and stimulation durations. Initial trials were generally conducted at higher acoustic intensity levels. However, current efforts are focused on stimulating using low-intensity, nonthermal pulses.

While the specific mechanism underlying the neuromodulatory effects of FUS has not yet been fully elucidated, the current body of evidence suggests that the application of ultrasound, especially in the case of nonthermal low-intensity stimulation, achieves its effects by mechanically interacting with neural tissues (Tyler 2011). Prior studies indicated that many voltage-gated ion channels show mechanosensitive properties rendering their gating kinetics sensitive to transient changes in lipid bilayer tension (Morris and Juranka 2007). Systematic studies of the underlying molecular mechanisms and the role that specific acoustic pulse parameters play in effecting neuromodulation are currently underway (Babakhanian et al. 2011).

Ultrasonic Neuromodulation in the Periphery

The potential for inhibiting peripheral nerve conduction with a therapeutic goal of pain management (Ballantine et al. 1960) was recognized early on, with Young and Henneman demonstrating selective effects on cat saphenous nerve bundles thereafter (Young and Henneman 1961). Some of the earliest studies on FUS stimulation were conducted by Russian researchers in the 1970s (Gavrilov et al. 1976, 1977, 1996; Vykhodtseva et al. 1976, 1985, 1995) who observed modulation of neuronal firing in mammals and human subjects.

More recently, a permanent nerve block of a rabbit sciatic nerve *in vivo* was demonstrated using high-intensity (>1.5 kW/cm^2), 30 s continuous sonication (Foley et al. 2004). These studies, however, were confounded by temperature elevations that are accompanied by the delivery of high-intensity ultrasound. Since excessive heat and pressure associated with high-intensity ultrasound may damage neural tissue, lower-energy pulsed application of FUS sonication has been suggested as an alternative. Lower energy modulation (both up and down regulation) of the bullfrog sciatic nerve has also been demonstrated (Mihran et al. 1990; Tsui et al. 2005).

Ultrasound-Induced CNS Neuromodulation

Studies with *ex vivo* experimental setups provided supporting evidence that FUS may activate and modulate CNS neuronal activity (Bachtold et al. 1998; Rinaldi et al. 1991). Tyler et al. (2008) reported that low-intensity (pulse average intensity <3 W/cm^2) low-frequency (440 and 670 kHz) ultrasound pulses can induce reproducible excitation of neuronal circuits in *ex vivo* mouse hippocampal neurons (Tyler et al. 2008). In their experiment, short bursts (25 μs–75 ms) of low-intensity ultrasound were delivered every 100 ms for 5 min, inducing activated voltage-gated sodium and calcium channels of the hippocampal neurons. In addition to modifying calcium signaling and neuronal electrical activity, sonication altered central synaptic transmission mediated by synaptic vesicle exocytosis.

A variety of recent efforts have further demonstrated the ability of low-intensity ultrasound to stimulate *in vivo* brain circuits. Tufail et al. (2010, 2011) demonstrated general motoric activity with motor cortex stimulation as well as evoked synchronized oscillations by deep targeting the hippo campus in an intact mouse. In both cases, stimulation was achieved transcranially and demonstrated no measurable BBB disruption.

Small animal *in vivo* low-intensity FUS was also demonstrated by Yoo et al. (2009), who reported the ability to both activate and suppress changes in the blood oxygenation level-dependent (BOLD) functional magnetic resonance imaging (fMRI) signal in rabbits. The same group reversibly suppressed the response of the visual cortex to light stimulation without damaging either the BBB or brain tissue (Yoo et al. 2011).

The first potential application of low-intensity FUS to treat neural disorders has recently been demonstrated by Yoo et al. An *in vivo* study in a rodent model of pentylenetetrazol (PTZ)-induced epileptic seizures demonstrated selective modulation of regional neural activity in brain by suppressing ictal activity (Min et al. 2011; Yoo et al. 2010). The PTZ-induced epileptic rats treated with low-intensity FUS demonstrated less severe epileptic behavior based on a Racine score.

Transcranial neuromodulation in the intact brain of an *in vivo* Göttingen minipig model has also recently been demonstrated. Consistent and reversible changes in heart rate and blood pressure were recorded coincident with targeted hypothalamic stimulation with low-intensity ultrasound (Mulgaonkar et al. 2011). This effort, the first known targeted transcranial low-intensity focused ultrasonic procedure in a large animal model, was intended to demonstrate the applicability of transcranial ultrasonic stimulation and surgical targeting methods in a cranial and neural model more applicable to humans than rodents and small animals.

CONCLUSION

Until recently, FUS was not envisioned to be a viable method for delivering noninvasive neuromodulation therapies into the brain. However, human clinical studies demonstrated that tumor ablation is possible using HIFU. Moreover, recent technical advances in FUS technology and its successful applications in preclinical *in vivo* and in vitro studies suggest that low-intensity FUS may represent a new way to treat a variety of neurological disorders. Significant progress has been made in designing FUS systems and ultrasound delivery to allow for safe, noninvasive penetration of the cranium in both large animal models and in humans. Further study must be performed to better understand the underlying molecular mechanisms behind ultrasound stimuli. This may require the development of new tools that can analyze these nanoscale interactions, such as emerging X-ray imaging techniques, nano-imaging agents, and femtosecond spectroscopy systems. Collectively, the results obtained to date provide the rationale and justification for continuing research in FUS applications, with the ultimate goal of providing significant improvements in the quality of life for individuals with neurological disorders.

REFERENCES

Aubry, J., Tanter, M., Pernot, M. et al. 2003. Experimental demonstration of noninvasive transskull adaptive focusing based on prior computed tomography scans. *The Journal of the Acoustical Society of America* 113: 84.

Babakhanian, M., Fan, R.E., Mulgaonkar, A.P. et al. 2011. An in-vitro cell system for studying molecular mechanisms of action associated with low intensity focused ultrasound. Proceedings of SPIE Photonics West. San Francisco, CA.

Bachtold, M.R., Rinaldi, P.C., Jones, J.P. et al. 1998. Focused ultrasound modifications of neural circuit activity in a mammalian brain. *Ultrasound in Medicine & Biology* 24: 557–65.

Bailey, M., Khokhlova, V., Sapozhnikov, O. et al. 2003. Physical mechanisms of the therapeutic effect of ultrasound (a review). *Acoustical Physics* 49: 369–88.

Bakay, L. 1956. *The Blood–Brain Barrier*. Springfield, IL: Charles C. Thomas.

Bakay, L., Hueter, T., Ballantine, H., Jr. et al. 1956. Ultrasonically produced changes in the blood–brain barrier. *Archives of Neurology & Psychiatry* 76: 457.

Ballantine, H., Jr., H.T., Bell, E., and Manlapaz, J. 1960. Progress and problems in the neurological applications of focused ultrasound. *Journal of Neurosurgery: Pediatrics* 17.

Bergers, G., and Song, S. 2005. The role of pericytes in blood-vessel formation and maintenance. *Neuro-oncology* 7: 452.

Bobo, R.H., Laske, D.W., Akbasak, A. et al. 1994. Convection-enhanced delivery of macromolecules in the brain. *Proceedings of the National Academy of Sciences* 91: 2076.

Clement, G., and Hynynen, K. 2002. A non-invasive method for focusing ultrasound through the human skull. *Physics in Medicine and Biology* 47: 1219.

Clement, G., White, J., and Hynynen, K. 2000. Investigation of a large-area phased array for focused ultrasound surgery through the skull. *Physics in Medicine and Biology* 45: 1071–83.

Culjat, M.O., Goldenberg, D., Tewari, P. et al. 2010. A review of tissue substitutes for ultrasound imaging. *Ultrasound in Medicine & Biology* 36: 861–73.

Daum, D., Buchanan, M., Fjield, T. et al. 1998. Design and evaluation of a feedback based phased array system for ultrasound surgery. *IEEE Transactions on Ultrasonics Ferroelectrics and Frequency Control* 45: 431–8.

Daum, D., and Hynynen, K. 1999. A 256-element ultrasonic phased array system for the treatment of large volumes of deep seated tissue. *IEEE Transactions on Ultrasonics, Ferroelectrics and Frequency Control* 46: 1254–68.

Dewhirst, M., Viglianti, B., Lora-Michiels, M. et al. 2003. Basic principles of thermal dosimetry and thermal thresholds for tissue damage from hyperthermia. *International Journal of Hyperthermia: The Official Journal of European Society for Hyperthermic Oncology, North American Hyperthermia Group* 19: 267.

Doolittle, N.D., Miner, M.E., Hall, W.A. et al. 2000. Safety and efficacy of a multicenter study using intraarterial chemotherapy in conjunction with osmotic opening of the blood–brain barrier for the treatment of patients with malignant brain tumors. *Cancer* 88: 637–47.

Dubose, T. 2000. Poly-autobiography of diagnostic medical sonography. Proceedings of the American Association of History and Computing. Waco, TX.

FDA. 2008. Guidance for Industry and FDA Staff: Information for Manufacturers Seeking Marketing Clearance of Diagnostic Ultrasound Systems and Transducers. http://www.fda.gov/medicaldevices/deviceregulationandguidance/guidancedocuments/ucm070856.htm. Last accessed November, 2012.

Fish, P. 1990. *Physics and Instrumentation of Diagnostic Medical Ultrasound.* West Sussex, England: Wiley.

Foley, J.L., Little, J.W., Starr, F.L., 3rd et al. 2004. Image-guided HIFU neurolysis of peripheral nerves to treat spasticity and pain. *Ultrasound in Medicine & Biology* 30: 1199–207.

Fry, F.J., Ades, H.W., and Fry, W.J. 1958. Production of reversible changes in the central nervous system by ultrasound. *Science* 127: 83–4.

Fry, W.J. 1954. Intense ultrasound; a new tool for neurological research. *The Journal of Mental Science* 100: 85–96.

Fry, W.J. 1958. Intense ultrasound in investigations of the central nervous system. *Advances in Biological and Medical Physics* 6: 281.

Fry, W.J., and Dunn, F. 1962. Ultrasound: analysis and experimental methods in biological research. *Physical Techniques in Biological Research* 4: 261–394.

Fry, W., and Fry, F. 1958. Fundamental neurological research and human neurosurgery using intense ultrasound. *Physica Medica: PM: An International Journal Devoted to the Applications of Physics to Medicine and Biology: Official Journal of the Italian Association of Biomedical Physics (AIFB)* 37: 148.

Fry, W., and Meyers, R. 1962. Ultrasonic method of modifying brain structures. *Confinia Neurologica* 22: 315–27.

Fry, W.J., Mosberg, W.H., Jr., Barnard, J.W. et al. 1954. Production of focal destructive lesions in the central nervous system with ultrasound. *Journal of Neurosurgery* 11: 471–8.

Fry, W., Wulff, V., Tucker, D. et al. 1950. Physical factors involved in ultrasonically induced changes in living systems: I. Identification of non-temperature effects. *The Journal of the Acoustical Society of America* 22: 867.

Gavrilov, L.R., Gersuni, G.V., Ilyinsky, O.B. et al. 1976. The effect of focused ultrasound on the skin and deep nerve structures of man and animal. *Progress in Brain Research* 43: 279–92.

Gavrilov, L.R., Gersuni, G.V., Ilyinski, O.B. et al. 1977. A study of reception with the use of focused ultrasound. I. Effects on the skin and deep receptor structures in man. *Brain Research* 135: 265–77.

Gavrilov, L.R., Tsirulnikov, E.M., and Davies, I.A. 1996. Application of focused ultrasound for the stimulation of neural structures. *Ultrasound in Medicine & Biology* 22: 179–92.

Guerin, C., Olivi, A., Weingart, J.D. et al. 2004. Recent advances in brain tumor therapy: local intracerebral drug delivery by polymers. *Investigational New Drugs* 22: 27–37.

Hedrick, W.R., Hykes, D.L., Starchman, D.E. et al. 1995. *Ultrasound Physics and Instrumentation.* St. Louis, MI: Mosby.

Hynynen, K. 2007. Focused ultrasound for blood–brain disruption and delivery of therapeutic molecules into the brain. *Expert Opinion on Drug Delivery* 4: 27–35.

Hynynen, K., and Clement, G. 2007. Clinical applications of focused ultrasound-the brain. *International Journal of Hyperthermia* 23: 193–202.

Hynynen, K., Clement, G., McDannold, N. et al. 2004a. 500-element ultrasound phased array system for non-invasive focal surgery of the brain: a preliminary rabbit study with *ex vivo* human skulls. *Magnetic Resonance in Medicine* 52: 100–7.

Hynynen, K., and Jolesz, F. 1998. Demonstration of potential noninvasive ultrasound brain therapy through an intact skull. *Ultrasound in Medicine & Biology* 24: 275–83.

Hynynen, K., McDannold, N., Sheikov, N. et al. 2005. Local and reversible blood–brain barrier disruption by noninvasive focused ultrasound at frequencies suitable for trans-skull sonications. *Neuroimage* 24: 12–20.

Hynynen, K., McDannold, N., Vykhodtseva, N. et al. 2001a. Noninvasive MR imaging-guided focal opening of the blood–brain barrier in rabbits. *Radiology* 220: 640.

Hynynen, K., McDannold, N., Vykhodtseva, N. et al. 2004b. Large volume blood–brain barrier disruption by low frequency focused ultrasound: a method for targeted drug delivery and molecular imaging. *Ultrasonics Symposium, 2004 IEEE* 2: 994–7.

Hynynen, K., McDannold, N., Vykhodtseva, N. et al. 2006. Focal disruption of the blood–brain barrier due to 260-kHz ultrasound bursts: a method for molecular imaging and targeted drug delivery. *Journal of Neurosurgery* 105: 445–54.

Hynynen, K., Pomeroy, O., Smith, D. et al. 2001b. MR imaging-guided focused ultrasound surgery of fibroadenomas in the breast: a feasibility study. *Radiology* 219: 176.

Jagannathan, J., Sanghvi, N.T., Crum, L.A. et al. 2009. High-intensity focused ultrasound surgery of the brain: part 1—A historical perspective with modern applications. *Neurosurgery* 64: 201.

Kinoshita, M., McDannold, N., Jolesz, F.A. et al. 2006. Noninvasive localized delivery of Herceptin to the mouse brain by MRI-guided focused ultrasound-induced blood–brain barrier disruption. *Proceedings of the National Academy of Sciences of the United States of America* 103: 11719–23.

Lynn, J., and Putnam, T. 1944. Histology of cerebral lesions produced by focused ultrasound. *The American Journal of Pathology* 20: 637.

Lynn, J., Zwemer, R., Chick, A. et al. 1942. A new method for the generation and use of focused ultrasound in experimental biology. *The Journal of General Physiology* 26: 179.

Martinez, D.A. 2010. Therapeutic Ultrasound: A Review of the Literature. ChiroAccess. http://www.chiroaccess.com/Articles/Therapeutic-Ultrasound-A-Review-of-the-Literature.aspx?id=0000210.

McDannold, N., Clement, G., Black, P. et al. 2010. Transcranial MRI-guided focused ultrasound surgery of brain tumors: initial findings in three patients. *Neurosurgery* 66: 323.

McDannold, N., Vykhodtseva, N., Raymond, S. et al. 2005. MRI-guided targeted blood–brain barrier disruption with focused ultrasound: histological findings in rabbits. *Ultrasound in Medicine & Biology* 31: 1527–37.

Meairs, S., and Alonso, A. 2007. Ultrasound, microbubbles and the blood–brain barrier. *Progress in Biophysics and Molecular Biology* 93: 354–62.

Mesiwala, A., Farrell, L., Wenzel, H. et al. 2002. High-intensity focused ultrasound selectively disrupts the blood–brain barrier *in vivo*. *Ultrasound in Medicine & Biology* 28: 389–400.

Mihran, R.T., Barnes, F.S., and Wachtel, H. 1990. Transient modification of nerve excitability in vitro by single ultrasound pulses. *Biomedical Sciences Instrumentation* 26: 235–46.

Min, B.-K., Bystritsky, A., Jung, K.-I. et al. 2011. Focused ultrasound-mediated suppression of chemically-induced acute epileptic EEG activity. *BMC Neuroscience* 12: 23.

Misra, A., Ganesh, S., Shahiwala, A. et al. 2003. Drug delivery to the central nervous system: a review. *Journal of Pharmacy & Pharmaceutical Sciences* 6: 252–73.

Morris, C.E., and Juranka, P.F. 2007. Nav channel mechanosensitivity: activation and inactivation accelerate reversibly with stretch. *Biophysics Journal* 93: 822–33.

Mulgaonkar, A.P., Singh, R., Babakhanian, M. et al. 2011. A prototype stimulator system for noninvasive low intensity focused ultrasound delivery. *Medicine Meets Virtual Reality 19*. Newport Beach, CA: NextMed.

Neuwelt, E., Maravilla, K., Frenkel, E. et al. 1979. Osmotic blood–brain barrier disruption. Computerized tomographic monitoring of chemotherapeutic agent delivery. *Journal of Clinical Investigation* 64: 684.

Pardridge, W.M. 2005. The blood–brain barrier: bottleneck in brain drug development. *NeuroRx* 2: 3–14.

Reid, M.H. 1978. Ultrasonic visualization of a cervical cord cystic astrocytoma. *American Journal of Roentgenology* 131: 907–8.

Rinaldi, P.C., Jones, J.P., Reines, F. et al. 1991. Modification by focused ultrasound pulses of electrically evoked responses from an in vitro hippocampal preparation. *Brain Research* 558: 36–42.

Salomir, R., Vimeux, F., de Zwart, J. et al. 2000. Hyperthermia by MR-guided focused ultrasound: accurate temperature control based on fast MRI and a physical model of local energy deposition and heat conduction. *Magnetic Resonance in Medicine* 43: 342–7.

Sapareto, S.A., and Dewey, W.C. 1984. Thermal dose determination in cancer therapy. *International Journal of Radiation Oncology Biology Physics* 10: 787–800.

Sheikov, N., McDannold, N., Vykhodtseva, N. et al. 2004. Cellular mechanisms of the blood–brain barrier opening induced by ultrasound in presence of microbubbles. *Ultrasound in Medicine & Biology* 30: 979–89.

Singh, R., and Culjat, M. 2013. Medical ultrasound devices. In Culjat, M., Singh, R., and Lee, H. *Medical Devices: Surgical and Image-Guided Technologies*. Hoboken: Wiley. 303–39.

Sosna, J., Barth, M.M., Kruskal, J.B. et al. 2005. Intraoperative sonography for neurosurgery. *Journal of Ultrasound in Medicine* 24: 1671–82.

Suslick, K.S., and Nyborg, W.L. 1990. Ultrasound: its chemical, physical and biological effects. *The Journal of the Acoustical Society of America* 87: 919.

Tachibana, K., Uchida, T., Tamura, K. et al. 2000. Enhanced cytotoxic effect of Ara-C by low intensity ultrasound to HL-60 cells. *Cancer Letters* 149: 189–94.

ter Haar, G. 2007. Therapeutic applications of ultrasound. *Progress in Biophysics and Molecular Biology* 93: 111–29.

Thomas, J., and Fink, M. 1996. Ultrasonic beam focusing through tissue inhomogeneities with a time reversal mirror: application to transskull therapy. *IEEE Transactions on Ultrasonics, Ferroelectrics and Frequency Control* 43: 1122–9.

Tsui, P.H., Wang, S.H., and Huang, C.C. 2005. In vitro effects of ultrasound with different energies on the conduction properties of neural tissue. *Ultrasonics* 43: 560–5.

Tufail, Y., Matyushov, A., Baldwin, N. et al. 2010. Transcranial pulsed ultrasound stimulates intact brain circuits. *Neuron* 66: 681–94.

Tufail, Y., Yoshihiro, A., Pati, S. et al. 2011. Ultrasonic neuromodulation by brain stimulation with transcranial ultrasound. *Nature Protocols* 6: 1453–70.

Tyler, W.J. 2011. Noninvasive neuromodulation with ultrasound? A continuum mechanics hypothesis. *Neuroscientist* 17: 25–36.

Tyler, W.J., Tufail, Y., Finsterwald, M. et al. 2008. Remote excitation of neuronal circuits using low-intensity, low-frequency ultrasound. *PLoS One* 3: e3511.

van Wamel, A., Bouakaz, A., Bernard, B. et al. 2004. Radionuclide tumour therapy with ultrasound contrast microbubbles. *Ultrasonics* 42: 903–6.

Vykhodtseva, N.I., Adrianov, O.S., Konopatskaia, II et al. 1985. [Characteristics of the formation of local destructive foci in brain tissues exposed to focused ultrasound]. *Biulleten' Eksperimental'noi Biologii i Meditsiny* 99: 689–93.

Vykhodtseva, N.I., Gavrilov, L.R., Mering, T.A. et al. 1976. [Use of focused ultrasound for local destruction of different brain structures]. *Zh Nevropatol Psikhiatr Im S S Korsakova* 76: 1810–6.

Vykhodtseva, N., Hynynen, K., and Damianou, C. 1995. Histologic effects of high intensity pulsed ultrasound exposure with subharmonic emission in rabbit brain *in vivo*. *Ultrasound in Medicine & Biology* 21: 969–79.

Warden, S., Bennell, K., McMeeken, J. et al. 2000. Acceleration of fresh fracture repair using the sonic accelerated fracture healing system (SAFHS): a review. *CALCIFIED Tissue International* 66: 157–63.

Wood, E.W., and Loomis, A.L. 1927. XXXVIII. The physical and biological effects of high-frequency soundwaves of great intensity. *The London, Edinburgh, and Dublin Philosophical Magazine and Journal of Science* 4: 417–36.

Yoo, S.S., Bystritsky, A., Lee, J.H. et al. 2011. Focused ultrasound modulates region-specific brain activity. *Neuroimage* 56: 1267–75.

Yoo, S.S., Lee, J.H., Fischer, K. et al. 2009. Non-invasive regional modulation of brain function by focused ultrasound. 105.11. Proceedings of Society for Neuroscience. Chicago, IL.

Yoo, S.S., Min, B.K., Zhang, Y. et al. 2010. Non-invasive suppression of animal-model chronic epilepsy using image-guided focused ultrasound. Proceedings of SmRM–ESMRMB Joint Annual Meeting. Stockholm, Sweden.

Young, R.R., and Henneman, E. 1961. Functional effects of focused ultrasound on mammalian nerves. *Science* 134: 1521.

16 Nanosensors

J. Matthew Dubach and Heather A. Clark

CONTENTS

INTRODUCTION

The creation and synthesis of fluorescent indicators, such as fura-2, has enabled the field of intracellular analysis to expand into new frontiers and enable applications in a myriad of fields (Grynkiewicz et al. 1985; Lechleiter et al. 1991). These indicators provide a convenient method for researchers to perform calcium studies that incorporate both temporal and spatial information and are available in several wavelengths and dynamic ranges. The number of indicators available for other physiological analytes, such as potassium, sodium, and glucose, is more limited than that for calcium, and their utility for intracellular studies has proven to be narrower in some applications. A burgeoning field, the development of nanosensors for intracellular studies, has the potential to expand the number of analytes that could be measured. Nanosensors are commonly defined as particles that have one dimension on the order of a few hundred nanometers or below and measurable properties dependent on the environment. The measurable property is often electrical or optical as these two methods of measurement are commonly used in laboratory research. This review will focus on free particles with electromagnetic properties, either light or magnetic, due to their potential impact for intracellular studies. Optical sensors have good temporal and spatial resolution that is limited only by the physics of light. The decreasing cost and increasing precision of optical hardware make optical sensors ideal for cellular studies, including studies of neurons. Nanowire (Patolsky et al. 2006) and tethered sensors that can be used for neurological measurements are beyond the scope of this review. Additionally, optical sensors based on fluorescent proteins expressed by the cell itself (Fehr et al. 2003) will not be covered, though this method does have tremendous potential for neuronal studies.

The majority of nanosensors produced to date detect ions and small molecules, such as glucose. These nanosensors have the potential to extend the range of detection of ions and small molecules beyond the currently available molecular recognition dyes. Particle-based nanosensors could be developed to detect other cellular relevant species such as DNA (Chen and Leong 2006) or proteins; however, these molecules as targets are less common than ion detecting nanosensors in the literature. Most cellular studies with nanosensors have been limited to common cultured cell lines, such as HEK-293 cells. Recently, nanosensors have been used in cardiac myocytes to detect sodium signaling during action potentials (Dubach et al. 2009), which was the first use of nanosensors to measure signaling in excitable cells. Though nanosensors have not been used in neuronal cells, to our knowledge, they have unique properties that will be advantageous for use in neuroscience.

Like molecular dyes, such as fura-2, nanosensors provide spatial information of cellular events that cannot be determined with electrical techniques such as whole-cell patch clamp. However, unlike molecular dyes, nanosensors can be optimized by selecting the ion or small molecule to be measured and by adjusting the response of the nanosensors so they are most active in the concentration range of interest. Additionally, some nanosensors protect the sensing mechanism from degradation inside of cells.

There are two types of nanosensors: those with an internalized sensing mechanism and those with an external sensing mechanism. Both types have advantages depending on the type of sensing required. External recognition mechanisms are capable of sensing analytes that either are too large to enter a nanosensor or take too long to diffuse inside of the nanosensor. Internal recognition mechanisms protect the sensing mechanism from nonspecific binding and biofouling that can degrade the nanosensor response or alter the way in which the nanosensor responds to the target.

There is a vast array of sensing mechanisms that can be used to create nanosensors. Much of the development is based upon recent advances in nanoparticle synthesis that improve control of size and size distributions. By definition, nanoparticles have unique optical or magnetic properties that are imparted by virtue of their size. For example, gold nanoparticles have both fluorescent and surface-enhanced Raman spectroscopy (SERS) properties, and iron oxide nanoparticles have magnetic properties useful in magnetic resonance imaging (MRI). Additionally, the development of quantum dots, semiconductor nanoparticles that have fluorescent properties, has led to several quantum dot-based sensors that exploit the ideal quantum dot properties—a large Stokes shift and no photobleaching (Amelia et al. 2011; Chen and Leong 2006; Ruedas-Rama and Hall 2009; Zhang et al. 2005). Because nanoparticles have an extremely high surface area-to-volume ratio, the surface plays a large role in the optical and magnetic properties of the nanoparticle. This has been exploited to create sensing nanoparticles by attaching molecules or proteins, such as antibodies, to the surface of the nanoparticle. Binding of cell surface ligands to these recognition molecules can cause changes in SERS or fluorescence through, for example, Förster resonance energy transfer (FRET). Furthermore, these nanoparticles can be coated with recognition elements as well as targeting elements that may help the nanosensors cross the blood–brain barrier or the plasma membrane. Nanosensors have been developed to measure an entire range of analytes, including pH (Clark et al. 1999c; Doussineau et al. 2009; Hornig et al. 2008, 2009; Kneipp et al. 2007, 2010; Lei et al. 2010; McNamara et al. 2001; Nowak-Lovato and Rector 2009; Nowak-Lovato et al. 2010; Reibetanz et al. 2007; Schulz et al. 2009, 2010; Shen et al. 2008; Xue et al. 2010), oxygen (Amelia et al. 2011, McNamara and Rosenzweig 1998; Wu et al. 2009), potassium (Harjes et al. 2010; Ruedas-Rama and Hall 2006), sodium (Dubach et al. 2007a,b, 2009), chloride (Brasuel et al. 2003; Graefe et al. 2008), glucose (Aslan et al. 2004a,b; Billingsley et al. 2010; Xu et al. 2002), zinc (He et al. 2010; Ruedas-Rama and Hall 2009; Sumner et al. 2002), calcium (Atanasijevic et al. 2006; Clark et al. 1999c; Nguyen and Rosenzweig 2002), phosphate (Sun et al. 2008), reactive oxygen species (Baker et al. 2007; Hammond et al. 2008; He et al. 2010; Henderson et al. 2009; Kim et al. 2005, 2010; Poulsen et al. 2007), magnesium (Park et al. 2003), iron (Sumner and Kopelman 2005), copper (Arduini et al. 2007; Sumner et al. 2006), and temperature (Peng et al. 2010).

Recent research from the Jasanoff laboratory used iron oxide nanoparticles for MRI measurements (Atanasijevic et al. 2006). Iron oxide nanoparticles are commonly used as an MRI contrast agent; however, these nanosensors have a T2 signal relaxation time dependent on calcium concentration. By coating the surface with calmodulin, which is a calcium-binding protein, the nanoparticles bind to calcium in a manner that causes aggregation. This clustering of iron oxide nanoparticles in the presence of analyte alters the T2 relation properties. This development is an excellent example of a nanosensor that is dependent on surface recognition. Additionally, the nanosensor is unique in that interaction with other nanosensors causes the signal change. These nanosensors may provide a method to study calcium signaling in neurons in vivo at depths that limit the use of optical techniques.

The Kneipp laboratory has created elegant pH-sensitive SERS nanoparticles that can be used in cells to measure endosome dynamics (Kneipp et al. 2007). The basis of these sensors are gold nanoparticles that have functional groups attached to the surface (Kumar et al. 2007). The choice of gold is important for its optical and SERS properties as well as for biocompatibility. SERS nanoparticles produce a spectrum that is dependent on the environment and can be used to sense multiple chemical species at once (Kneipp et al. 2006). The limitation, to date, has been the overlap of chemical species in the spectrum, making the distinction of each species difficult. The pH-sensitive particles developed in the Kneipp laboratory incorporate an attached benzoic acid to the surface of gold nanoparticles (Kneipp et al. 2010), which has a charge that is dependent on the acidity of the environment that alters the SERS signal of the gold nanoparticles. The benzoic acid protonates as it becomes acidified in the endosomal environment, which affects the SERS signal of the gold nanoparticles. The signal can be calibrated and used inside of cells to track endosome acidity, with quantitative results.

The Kopelman laboratory was the first to develop nanosensors that entrapped the sensing mechanism within the particle itself (Clark et al. 1999a). These sensors were called PEBBLE nanosensors, an acronym for probes encapsulated by biologically localized embedding (Clark et al. 1999b,c). They were between 20 and 200 nm in diameter and were introduced into cells using a gene gun, among other techniques. The nanosensors were formed by emulsion polymerization of a hydrogel and were designed to sense calcium, oxygen, and pH. The nanosensors functioned by using a trapped fluorescent molecule inside a complex matrix that was sensitive to the ion of interest. The polymeric matrix allowed ions to freely diffuse to the fluorescent molecule but excluded interaction of larger molecules, such as proteins. PEBBLE sensors responded in the physiological range and could be imaged using standard fluorescent visualization techniques, such as an imaging microscope, and were shown to have a completely reversible response (Figure 16.1a). The PEBBLEs also created other properties useful in biological sensing. By preventing diffusion of proteins into the particle, the PEBBLE matrix completely protected the sensing mechanism from biofouling (Figure 16.1b). With a simple sensing mechanism of a pH-sensitive dye, the fluorescent signal was unaffected by bovine serum albumin in the PEBBLE matrix but significantly quenched when the dye was free in solution. This property is necessary for intracellular measurement, which occurs in an environment full of cellular machinery and mechanisms.

Heinze and Mohr have developed fluorescent polysaccharide nanoparticles that are pH sensitive at physiological pH levels (Schulz et al. 2010). These nanosensors are very similar to the PEBBLE nanosensors in that they entrap a pH-sensitive dye in a cross-linked core (Hornig et al. 2008, 2009; Schulz et al. 2009). The particles, therefore, have the advantages of PEBBLEs for intracellular measurement but explore the use of novel matrices to enhance biocompatibility. Some of the more recent work on these particles has covalently attached the sensing fluorophore to the polymer core to prevent any sensing mechanism leaching into solution and improve the robustness of the sensor

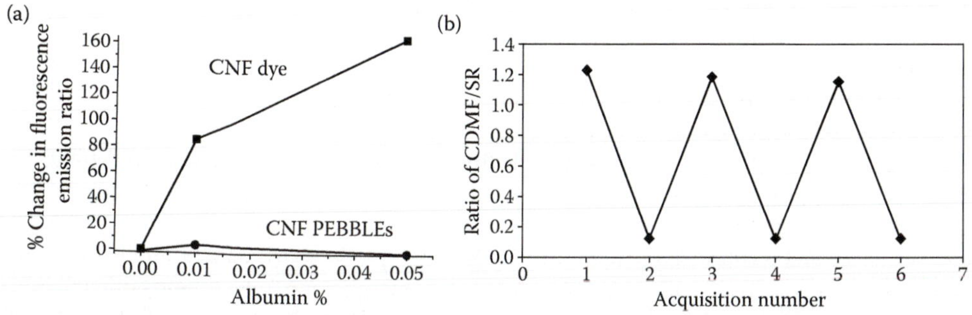

FIGURE 16.1 Using dye-loaded nanoparticles for fluorescent measurements. (a) Incorporating carbon nanofiber (CNF) dye into PEBBLEs protects the fluorescence of the dye from interactions with albumin. (From Clark, H.A. et al., *Analytical Chemistry*, 71, 4831–6, 1999. With permission.) (b) PEBBLEs show reversibility that is consistent over a number of cycles. (From Clark, H.A. et al., *Analytical Chemistry*, 71: 4837–43, 1999. With permission.)

(Schulz et al. 2010). Additionally, as with the PEBBLE sensors developed by Kopelman, a reference dye was added to enable ratiometric fluorescent measurements. Reference dyes are an internal standard that typically do not respond to the nanosensor sensing target and provide a constant fluorescence intensity that is dependent only on the concentration of the dye. Ratiometric measurements are advantageous in cellular studies because the signal ratio is independent of the concentration of nanosensors. If one cell had more nanosensors than another, a higher fluorescent signal would be measured, but it would not be from a different pH value. The ratio removes this uncertainty, and it is crucial for quantitative measurements. The authors also add a shell to the nanosensors that has the potential to increase cellular loading. This research is a good example of the additional properties that can be added or changed in nanosensors by modifications of the shell. There are limitless possibilities to change nanosensors in these ways to enhance them for the intended application.

There has been a range of nanosensors for ions and small molecules in recent years that use an optode-based mechanism. These nanosensors have fluorescent properties and behave similarly to the original PEBBLEs by extracting the target molecule or ion into the nanosensor core. The optodes, however, exchange this extracted ion for one already residing inside the sensor, a mechanism that keeps the optode in a permanent state of charge. Therefore, optode nanosensors do not rely on external recognition elements but have a more complex sensing mechanism than simply incorporating fluorescent dyes alone.

BACKGROUND ON OPTODES

Optodes were first described in the late 1980s, and they are the optical equivalents to ion-selective electrodes, such as those typically used in pH meters. This molecule, with no optical properties of its own, provides specificity over other interfering ions that exist in the biological environment. For example, sodium-responsive optode nanosensors can have selectivities of greater than three orders of magnitude. The sodium sensor will have the same response to 100-μM sodium as it does to 100-mM potassium, meaning the sensor will be able to detect small changes in sodium concentration even in the presence of much greater intracellular concentrations of potassium. Because ion-selective electrodes work by sensing electrical potential at the surface, optodes were first used as surface optodes that did not extract the ion into the matrix. However, subsequent optodes were developed as bulk optodes in which the optode was in equilibrium with the aqueous environment and the entire optode material was in the same state.

Optodes consist of a hydrophobic matrix made from a polymer and a plasticizer that creates a fluid environment in which lipophilic molecules can freely diffuse. Traditionally, the most commonly used polymer is poly(vinyl chloride) (PVC); however, multiple polymers have been used successfully. The plasticizer is typically an oil-based molecule that maintains the hydrophobicity of the optode matrix but creates fluidity that is necessary for the optode to function. There are typically three mechanistic components in an optode: chromoionophore, ionophore, and ionic additive. The chromoionophore is a hydrophobic molecule that has optical properties, absorption and, in some cases, fluorescence, that are dependent on the protonation state of the molecule. Thus, the chromoionophore will optically respond to the pH of the optode matrix. The ionophore is a hydrophobic molecule with no optical properties that specifically binds to an ion or molecule and imparts the specificity of the optode by extracting the ion or molecule that it recognizes into the optode matrix. The third component is the ionic additive that is a hydrophobic molecule that has either negative or positive charge and creates charge neutrality within the optode matrix. The main differentiating functional component between all of these sensors is the ionophore that designates the molecule or ion to which the nanosensors respond (Buhlmann et al. 1998). The affinity of the chromoionophore for hydrogen ions will balance with the affinity of the ionophore for the ion or molecule and will dictate how the optode responds.

One example of the mechanism of the optode controlled by the charge-neutral, hydrophobic matrix is illustrated in Figure 16.2. The example shown in Figure 16.2 is for a monovalent cation

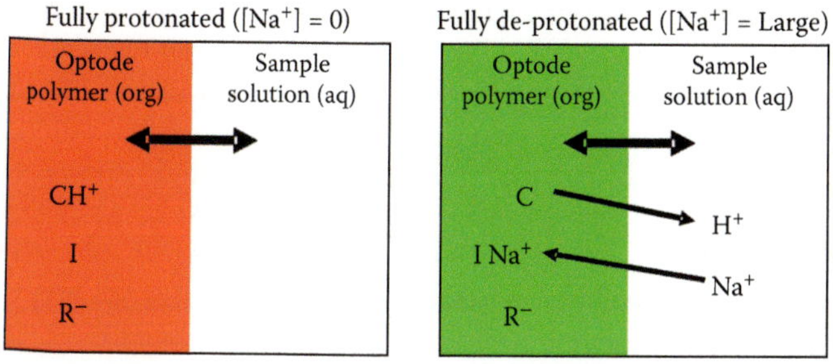

FIGURE 16.2 Optode mechanism. Depicted is an optode (red on the left and green on the right) in contact with aqueous solution. In the schematic, the optode contains chromoionophore (C), ionophore (I), and additive (R). On the left, there is no sodium present in the aqueous solution, and the optode remains fully protonated, indicated by the red color and hydrogen attached to the chromoionophore. On the right, sodium is present in the aqueous solution, and the optode is deprotonated because sodium has entered the optode by interacting with the ionophore and hydrogen has dissociated from the chromoionophore and left the optode to maintain the charge balance.

such as sodium. The bulk charge of the optode will always be neutral in physiological pH environments. The chromoionophore will have a positive charge when protonated and a neutral charge when deprotonated. The ionophore will have a positive charge when coupled with sodium and a neutral charge when uncoupled. And the additive will always have a negative charge. When there are no sodium ions present in the aqueous environment, the chromoionophore will be protonated to balance the negative charge of the ionophore. However, when high sodium concentrations are present in the surrounding solution, sodium will be extracted into the optode by the ionophore and create a transient positively charged environment. This will be corrected by optode loss of a hydrogen ion from the chromoionophore and concomitant change in the optical signal. The response of the optode to sodium, including both the dynamic range and sensitivity, can be altered by adjusting the amount of ionophore. This response can be predicted by the following equation (Bakker et al. 1997):

$$\frac{[I]}{[H]} = K_o^i \left(\frac{\alpha}{1-\alpha}\right) \frac{[R]-(1-\alpha)[C]}{[L]-\{[R]-(1-\alpha)[C]\}}, \tag{16.1}$$

where $[I]$ is the concentration of the ion of interest, $[H]$ is the concentration of hydrogen, K_o^i is the equilibrium constant dependent on the properties to the optode and mechanism, α is used in the response and indicates the degree of protonation at the ion of interest concentration, $[R]$ is the concentration of additive, $[C]$ is the concentration of chromoionophore, and $[L]$ is the concentration of ionophore.

Optodes are bulk sensors and require equilibration with the surrounding environment to produce accurate measurements, which could delay sensing of concentration changes in the ion of interest. The response time of the sensor is a bulk response and is dependent on diffusion inside the optode, which is one reason why the polymeric matrix is highly plasticized. The response time of the sensor can be modeled as follows (Bakker et al. 1997):

$$t_{95\%} = 1.13\frac{d^2}{D_\mathrm{m}}, \tag{16.2}$$

where $t_{95\%}$ is the time it takes to reach 95% response, d is the diameter of the optode, and D_m is the diffusion coefficient within the optode, which is typically on the order of 10^{-8} cm^2/s. Therefore, as the optode sensor is reduced in size, the response time decreases by the square of the size. For an optode with a thickness of 100 nm, the response time is on the order of 10 ms. This time is significantly reduced to the microsecond level when optodes are converted to nanosensors, making them practical for monitoring rapid dynamics in the cellular environment.

NANOSENSOR OPTODES

Optodes were traditionally designed to be used as environmental sensors or in vitro medical monitors. Development of optodes for cellular studies was started around the time when fluorescent molecular dyes began to show promise for intracellular imaging. The first demonstration of optode-based nanosensors came from the Kopelman laboratory (Brasuel et al. 2001, 2002). These sensors were designed to detect intracellular potassium concentrations in rat C6 glioma cells. The results showed that the nanosensors could be made to respond to potassium at different concentrations and to have a maximal response at intracellular potassium concentrations. The nanosensors averaged around 500 nm in diameter and were introduced into the cells using a gene gun. The utility of this class of nanosensors was demonstrated by the group later developing a chloride-sensing optode-based nanosensor using a similar method to the potassium nanosensor (Brasuel et al. 2003).

The Clark laboratory has recently created a new method to fabricate optode-based nanosensors (Dubach et al. 2007a). This method is performed by solvent extraction precipitation, is reproducible, takes about 10 min, and produces on the order of 10^{13} nanosensors per batch. This technique produced nanosensors that were about 100 nm in diameter and had a poly(ethylene glycol) (PEG) coating to impart nanosensor stability and biocompatibility in the cellular environment. These nanosensors were designed for intracellular sodium detection by adjusting the concentration of the sensing components until the ideal response was observed. This is important because molecular dye sensors that are commercially available have a Kd higher than that of intracellular resting sodium and are partially responsive to potassium at concentrations typically found inside cells. The optode nanosensors are completely selective for sodium over potassium in the physiologically relevant concentration range and have a Kd of around 10 mM, making the nanosensors most sensitive to sodium at the intracellular resting concentration of sodium (Figure 16.3a). The response to sodium was plotted as $1 - \alpha$ versus the activity of sodium ions divided by the activity of hydrogen ions, which is the standard way to report optode sensor response because the fluorescent component of the sensing

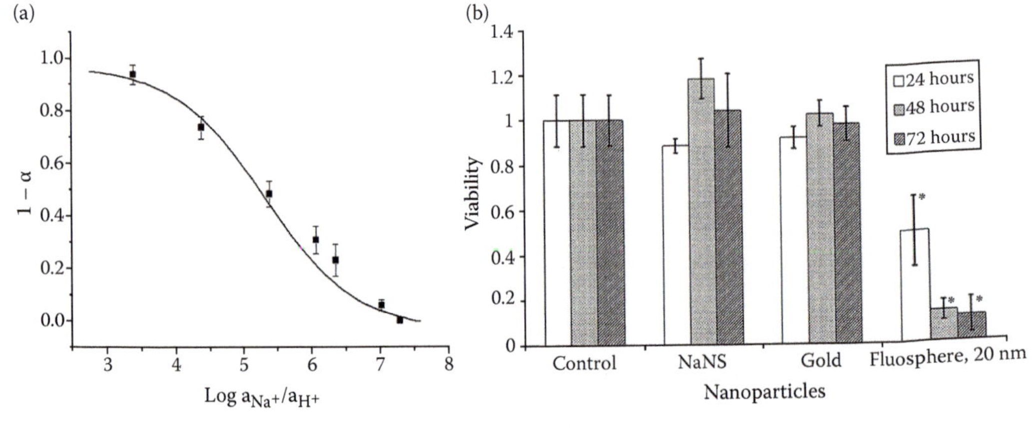

FIGURE 16.3 (a) Response of sodium optode nanosensors (NaNS) to sodium. The nanosensors have been tuned to respond best at intracellular sodium concentrations (~10 mM). (b) Cytotoxicity study of NaNS on cell culture health using an MTT assay. NaNS, gold nanoparticles, and fluospheres were tested for cytotoxicity at 24, 48, and 72 h after introduction into culture. (From Dubach, J.M. et al., *Nano Letters* 7: 1827–31, 2007.)

mechanism responds to hydrogen ion activity (alpha). Using a PEG coating that inserts into the nanosensor through a covalently attached lipid region by hydrophobic interaction allows for changes on the surface of the nanosensor. This was explored by using different surface coating molecules and measuring the size and zeta potential of the nanosensor over time. The size and response to sodium were stable for at least a week. The zeta potential is a representation of the surface charge the nanosensor will have when in solution and can affect cellular interaction, response dynamics, and lifespan. Nanosensors with large zeta potentials, either positive or negative, will not aggregate in solution.

Perhaps the most intriguing aspect of designing optode-based nanosensors is that any hydrophobic compound can easily be incorporated into the nanosensor. This was shown when quantum dots were added to the optode nanosensor to serve as a fluorescence source (Dubach et al. 2007b). In this case, the chromoionophore was used as an absorbing molecule and absorbed quantum dot fluorescence via an inner filter effect. The absorbance of the chromoionophore is dependent on the concentration of the nanosensor target in the same manner as the fluorescence is dependent. It is nearly impossible to measure absorbance of nanosensors, and, therefore, fluorescence is always used as the detection method. This nanosensor takes advantage of the brightness, large stokes shift, and photostability of quantum dots. The mechanism of sensing is the same as above; however, in this case, the fluorescence is measured by exciting the quantum dots directly and measuring quantum dot emission (Figure 16.4a). Additionally, because the chromoionophore being used has two absorption peaks and an isosbestic point, two quantum dots at wavelengths on each side of the isosbestic point were used to create a ratiometric fluorescence signal. This method produced nanosensors that reported sodium concentrations with a Kd of 17 mM, increased sensitivity, and were independent of the quantity of nanosensors being used to make the measurement (Figure 16.4b).

The Clark group also created glucose responsive nanosensors that were based on optodes (Billingsley et al. 2010). Although the mechanism was slightly different than the typical ion-sensitive optode nanosensors, the glucose nanosensors utilized a plasticized polymer matrix that selectively extracted glucose into the nanosensor. The mechanism relied on competitive binding between glucose and a dye molecule, which also contains a diol, to a boronic acid. Because glucose does not have a charge in solution, competitive binding with a dye molecule, rather than a charge balance mechanism used for ions, was needed. Glucose molecules can displace dye molecules from the boronic acid, which results in a change in fluorescence from a dye molecule. The use of an optode matrix keeps the sensing molecules within proximity of one another for continuous monitoring applications. This was the first demonstration of a small-molecule nanosensor that used an internal mechanism rather than a surface recognition technique. These glucose-sensitive nanosensors were designed for extracellular, or interstitial, detection under the skin of mice, but, because the

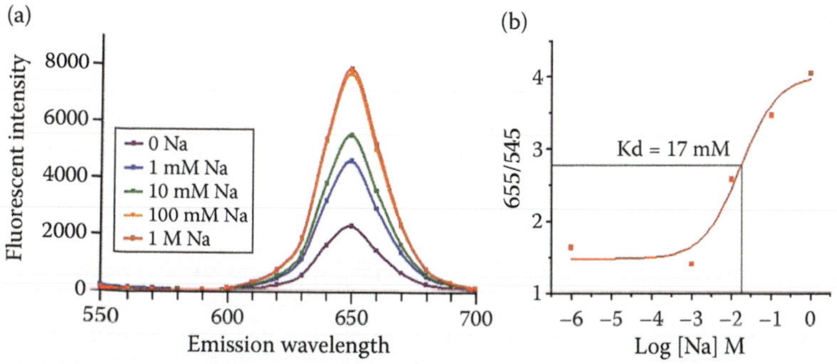

FIGURE 16.4 Using quantum dots as optical reports in optode nanosensors (Dubach et al. 2007b). (a) Spectrum of quantum dot emission at different sodium concentrations. (b) Response of the sodium nanosensors to sodium using a ratio of the intensities from two quantum dots in the nanosensors.

nanosensors use a competitive binding mechanism, the nanosensors can be easily designed to detect glucose in the concentration range found intracellularly. Glucose nanosensors were used to monitor analyte fluctuations in vivo and compared favorably to blood measurements. In addition, glucose nanosensors for intracellular measurement could greatly impact other applications, including in neuroscience, where cellular utilization of glucose is so high.

Optode-based sensors have several advantages over other nanosensors when optimized for cellular use. As mentioned above, the optode-based sensors work by extracting the target molecule or ion from the biological environment into the sensor and do not rely on surface recognition elements. This minimizes biofouling caused by interaction of the surface recognition molecules with other molecules in the biological environment. Additionally, optode-based nanosensors can be surface functionalized with various molecules that will not interfere with the recognition elements and will allow the target molecules to simply diffuse past the surface components of the sensor. The optode-based sensors can also be tuned to respond to different ion concentrations, which allows the sensors to be completely customizable for the biological target. For example, the same recognition mechanism and components were used to create nanosensors for intracellular measures (typically sodium concentration of about 10 mM) and extracellular measurements (typical sodium levels of about 140 mM) by simply adjusting the amount of each mechanism component. There are very good fluorescent calcium molecular dyes for intracellular calcium studies that have revolutionized cellular biology. However, the chemical structures of these dyes dictate the calcium concentrations to which they respond. Optode-based nanosensors can be targeted to multiple biologically relevant concentrations. Also, nanosensors are less likely to be sequestered into cellular compartments once introduced into the cell, which allows for cytosolic measurements. To date, optode-based nanosensors have been created for sodium (Dubach et al. 2007a,b, 2009), potassium (Harjes et al. 2010; Ruedas-Rama and Hall 2006), chloride (Brasuel et al. 2003), and glucose (Billingsley et al. 2010).

USE OF OPTODE NANOSENSORS IN CELLS

To realize the true utility of nanoparticle sensors for cellular studies, it is necessary to determine if there is an impact on the viability of cells when sensors are introduced. For use in biological monitoring of cellular processes, it is important that fluctuations that are measured are being caused by the biological phenomenon being studied and not the sensors themselves. Optode-based nanosensors were shown to be biocompatible when introduced to HEK-293 cells in culture (Figure 16.3b) (Dubach et al. 2007a). The nanosensors did not incur cell death, as measured by an MTT assay up to 72 h after introduction to the cells. However, the sensors did not reach the cytosol because they were endocytosed into the cells. Therefore, the nanosensors interacted with the cells only through the plasma membrane and endocytotic process. Though more detailed studies on cellular response to the introduction of nanosensors have not been performed, these preliminary studies indicate that cells can tolerate nanosensors in culture.

There have been many studies showing nanosensor uptake into cells and fluorescent tracking of pH. However, much of this work is limited to endosomal studies because, unlike molecular dyes with acetoxylmethyl ester tags, nanosensors do not freely diffuse across the plasma membrane. To overcome this limitation, a more active method of getting the nanosensors into the cell needs to be used. The Kopelman laboratory regularly used a gene gun to load nanosensors into cells (Brasuel et al. 2002). This method uses ballistics to break the nanosensors through the plasma membrane. Other methods such as electroporation or endosomal escape have been proposed and utilized to transport nanosensors and nanoparticles into cells (Webster et al. 2007). However, there remains to be a definitive method.

Recently, the Clark laboratory started to introduce optode nanosensors into adherent mammalian cells using microinjection (Dubach et al. 2009). This technique allows nanosensor loading to bypass the endocytotic pathway even if loading is only one cell at a time. Once introduced into cells, the nanosensors disperse throughout the cytosol but do not enter the nucleus (Figure 16.5a). The nanosensors were tested to determine if the response to sodium remained the same inside

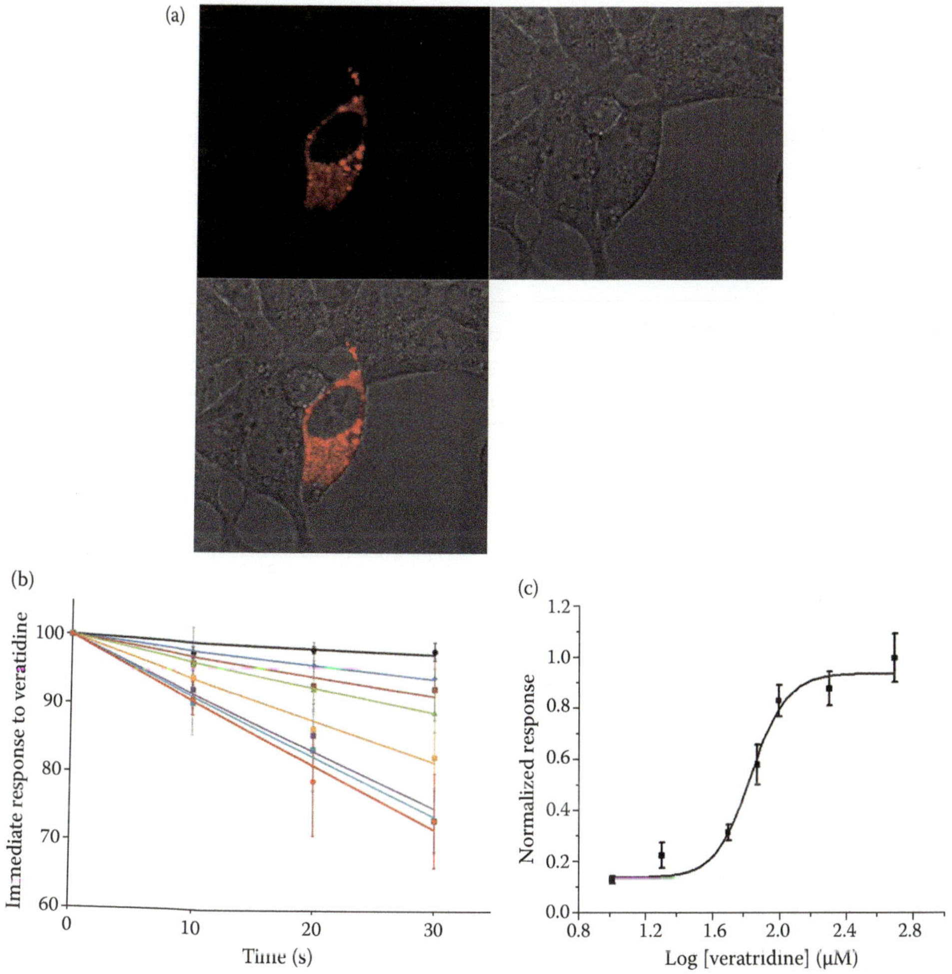

FIGURE 16.5 Use of sodium optode nanosensors in cells (Dubach et al. 2009). (a) Confocal image of an HEK-297 cell injected with sodium optode nanosensors (fluorescence, bright field, and overlay). (b) Sodium optode nanosensor response to veratridine-stimulated $Na_V1.7$ channel opening inside HEK-297 PN1 cells using various veratridine concentrations. (c) Dose–response curve of channel activation by veratridine as determined by sodium optode nanosensor fluorescence measurements.

the cell as in vitro, as is the case with molecular fluorescent dyes for calcium. This intracellular response was the same, which indicates that the nanosensors are not altered or fouled inside the cell and that calibration on the bench top will calibrate the sensors for in vivo measurements regardless of the cellular environment into which the sensors are placed. Additionally, nanosensors were injected into HEK-293 cells that were engineered to express the sodium channel $Na_V1.7$. The sodium response to channel activation by veratridine was measured, and a dose response to veratridine was determined through changes in nanosensor intensity (Figure 16.5b). This dose response was found to be similar to a dose response using patch clamp to measure channel activation (Figure 16.5c). This demonstrates that optical techniques have the potential to be used in place of electrophysiology for studying ion channel dynamics.

Sodium-responsive optode-based nanosensors were also used to measure action potentials in cardiac myocytes (Dubach et al. 2009). This was the first use of nanosensors within cells to measure action potentials. Nanosensors were microinjected into the cells that were paralyzed with a molecule that decouples the electrical and mechanical properties of the myocyte, which enables

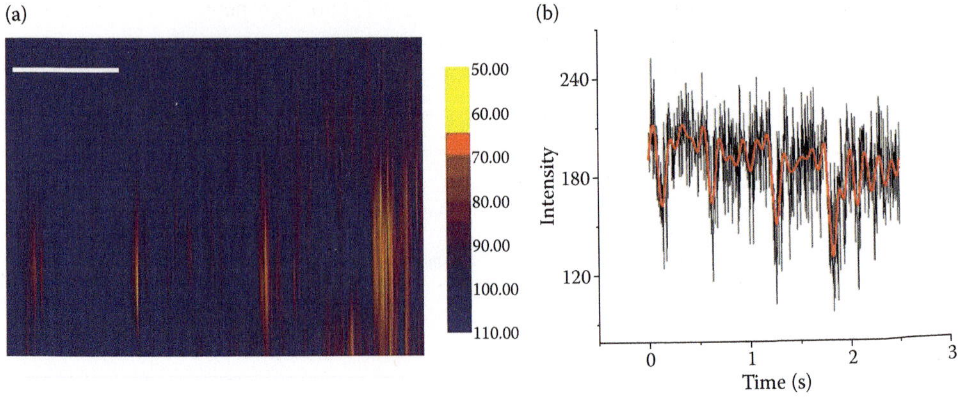

FIGURE 16.6 Use of sodium optode nanosensors in excitable cells (Dubach et al. 2009). (a) Line scan data showing sodium optode nanosensor fluorescence inside isolated cardiac myocytes. The x-axis is the time, and the y-axis is the distance along the line scan. Decreases in intensity, red and yellow coloring, indicate sodium flux into the cell during myocyte electrical stimulation. (b) Single spot of the line scan in A with raw intensity (black) and fast Fourier-filtered low-pass intensity (red). Decreases in intensity correspond to increases in intracellular sodium.

fluorescent imaging without motion artifacts. The cardiac myocytes were electrically paced with field stimulation electrodes while simultaneous confocal line scans of the cells were taken. A scan length in the scan direction versus time plot was generated to display the data (Figure 16.6a). Also, a single point from the line scan was plotted along with a low-pass fast Fourier-filtered tracing (Figure 16.6b). The image in Figure 16.6a shows that the nanosensors measure action potentials with spatial differences determined by the location of ion channel clusters in the cell membrane. This research can easily be translated into neurons to measure action potentials in these cells. Loading nanosensors through whole-cell patch clamp in neuronal tissue samples would allow both optical spatial measurements and electrical measurements.

DISADVANTAGES OF NANOSENSORS FOR APPLICATIONS IN NEURONAL CELLS

The main limitation of using nanoparticle sensors for cellular experiments is the delivery of the nanoparticle into the cytosol of the cell. Nanosensors enter the cell quite readily as they can easily undergo endocytosis. Escaping out of the endosome, however, remains elusive not only with nanosensors but also with nanoparticles used for drug or nucleic acid delivery. There have been many attempts to reliably escape the endosome by exploiting the pH change that occurs as the endosome ages. This is why there are numerous pH nanosensors that can report the pH of the endosome and do not require release out into the cytosol. To gain cellular data, though, the nanosensors need to be in the cytosol. Currently, this is accomplished by microinjection of the nanosensors, which consumes considerable time and damages cells in the process. The structure of nanosensors allows attachment of a variety of surface molecules that may help shepherd the sensors across the plasma membrane and into the cytosol. Groups have explored cell-penetrating peptides as well as other passive loading techniques. However, there has yet to be a proven, ubiquitous method to introduce nanosensors into cells. This will only be confounded by the unique properties of neurons.

SUMMARY

Nanosensors have potential to use many different sensing mechanisms, which can be tailored to respond ideally in the application of interest. This will be beneficial for neuroscience research where electrical and chemical signaling plays such a big role. The discovery of how to best load nanosensors into the cytosol would be a major breakthrough for not only neuroscience but also biology as a whole. For now, however, significant research can already be performed using nanosensors that cannot be done using any other research tool.

REFERENCES

Amelia, M., Lavie-Cambot, A., McClenaghan, N.D. et al. 2011. A ratiometric luminescent oxygen sensor based on a chemically functionalized quantum dot. *Chemical Communications* 47: 325–7.

Arduini, M., Armelao, L., Gross, S. et al. 2007. Chemical optimisation of a sol-gel procedure for the development of fluorescence Cu(II) nanosensors. *Applied Surface Science* 253: 7178–87.

Aslan, K., Lakowicz, J.R., and Geddes, C.D. 2004a. Nanogold-plasmon-resonance-based glucose sensing. *Analytical Biochemistry* 330: 145–55.

Aslan, K., Zhang, J., Lakowicz, J.R. et al. 2004b. Saccharide sensing using gold and silver nanoparticles—a review. *Journal of Fluorescence* 14: 391–400.

Atanasijevic, T., Shusteff, M., Fam, P. et al. 2006. Calcium-sensitive MRI contrast agents based on superparamagnetic iron oxide nanoparticles and calmodulin. *Proceedings of the National Academy of Sciences of the United States of America* 103: 14707–12.

Baker, N., Greenway, G.M., Wheatley, R.A. et al. 2007. A chemiluminescence nanosensor to monitor lipid peroxidation. *Analyst* 132: 104–6.

Bakker, E., Buhlmann, P., and Pretsch, E. 1997. Carrier-based ion-selective electrodes and bulk optodes. 1. General characteristics. *Chemical Reviews* 97: 3083–132.

Billingsley, K., Balaconis, M.K., Dubach, J.M. et al. 2010. Fluorescent nano-optodes for glucose detection. *Analytical Chemistry* 82: 3707–13.

Brasuel, M., Kopelman, R., Aylott, J.W. et al. 2002. Production, characteristics and applications of fluorescent pebble nanosensors: Potassium, oxygen, calcium and pH imaging inside live cells. *Sensors and Materials* 14: 309–38.

Brasuel, M., Kopelman, R., Miller, T.J. et al. 2001. Fluorescent nanosensors for intracellular chemical analysis: Decyl methacrylate liquid polymer matrix and ion exchange-based potassium pebble sensors with real-time application to viable rat c6 glioma cells. *Analytical Chemistry* 73: 2221–8.

Brasuel, M.G., Miller, T.J., Kopelman, R. et al. 2003. Liquid polymer nano-pebbles for Cl- analysis and biological applications. *Analyst* 128: 1262–7.

Buhlmann, P., Pretsch, E., and Bakker, E. 1998. Carrier-based ion-selective electrodes and bulk optodes. 2. Ionophores for potentiometric and optical sensors. *Chemistry Reviews* 98: 1593–688.

Chen, H.H., and Leong, K.W. 2006. Quantum-dots-fret nanosensors for detecting unamplified nucleic acids by single molecule detection. *Nanomedicine* 1: 119–22.

Clark, H.A., Hoyer, M., Parus, S. et al. 1999a. Optochemical nanosensors and subcellular applications in living cells. *Mikrochimica Acta* 131: 121–8.

Clark, H.A., Hoyer, M., Philbert, M.A. et al. 1999b. Optical nanosensors for chemical analysis inside single living cells. 1. Fabrication, characterization, and methods for intracellular delivery of pebble sensors. *Analytical Chemistry* 71: 4831–6.

Clark, H.A., Kopelman, R., Tjalkens, R. et al. 1999c. Optical nanosensors for chemical analysis inside single living cells. 2. Sensors for pH and calcium and the intracellular application of pebble sensors. *Analytical Chemistry* 71: 4837–43.

Doussineau, T., Trupp, S., and Mohr, G.J. 2009. Ratiometric pH-nanosensors based on rhodamine-doped silica nanoparticles functionalized with a naphthalimide derivative. *Journal of Colloid and Interface Science* 339: 266–70.

Dubach, J.M., Das, S., Rosenzweig, A. et al. 2009. Visualizing sodium dynamics in isolated cardiomyocytes using fluorescent nanosensors. *Proceedings of the National Academy of Sciences of the United States of America* 106: 16145–50.

Dubach, J.M., Harjes, D.I., and Clark, H.A. 2007a. Fluorescent ion-selective nanosensors for intracellular analysis with improved lifetime and size. *Nano Letters* 7: 1827–31.

Dubach, J.M., Harjes, D.I., and Clark, H.A. 2007b. Ion-selective nano-optodes incorporating quantum dots. *Journal of the American Chemical Society* 129: 8418–9.

Fehr, M., Lalonde, S., Lager, I. et al. 2003. In vivo imaging of the dynamics of glucose uptake in the cytosol of cos-7 cells by fluorescent nanosensors. *The Journal of Biological Chemistry* 278: 19127–33.

Graefe, A., Stanca, S.E., Nietzsche, S. et al. 2008. Development and critical evaluation of fluorescent chloride nanosensors. *Analytical Chemistry* 80: 6526–31.

Grynkiewicz, G., Poenie, M., and Tsien, R.Y. 1985. A new generation of Ca^{2+} indicators with greatly improved fluorescence properties. *The Journal of Biological Chemistry* 260: 3440–50.

Hammond, V.J., Aylott, J.W., Greenway, G.M. et al. 2008. An optical sensor for reactive oxygen species: Encapsulation of functionalised silica nanoparticles into silicate nanoprobes to reduce fluorophore leaching. *Analyst* 133: 71–5.

Harjes, D.I., Dubach, J.M., Rosenzweig, A. et al. 2010. Ion-selective optodes measure extracellular potassium flux in excitable cells. *Macromolecular Rapid Communications* 31: 217–21.

He, C.S., Zhu, W.P., Xu, Y.F. et al. 2010. Ratiometric and reusable fluorescent nanoparticles for Zn^{2+} and H_2PO_4-detection in aqueous solution and living cells. *Journal of Materials Chemistry* 20: 10755–64.

Henderson, J.R., Fulton, D.A., McNeil, C.J. et al. 2009. The development and in vitro characterisation of an intracellular nanosensor responsive to reactive oxygen species. *Biosensors & Bioelectronics* 24: 3608–14.

Hornig, S., Biskup, C., Grafe, A. et al. 2008. Biocompatible fluorescent nanoparticles for pH-sensoring. *Soft Matter* 4: 1169–72.

Hornig, S., Schulz, A., Mohr, G.J. et al. 2009. Fluorescent polysaccharide nanoparticles for pH-sensing. *Journal of Photopolymer Science and Technology* 22: 671–3.

Kim, G., Lee, Y.E.K., Xu, H. et al. 2010. Nanoencapsulation method for high selectivity sensing of hydrogen peroxide inside live cells. *Analytical Chemistry* 82: 2165–9.

Kim, S.H., Kim, B., Yadavalli, V.K. et al. 2005. Encapsulation of enzymes within polymer spheres to create optical nanosensors for oxidative stress. *Analytical Chemistry* 77: 6828–33.

Kneipp, J., Kneipp, H., McLaughlin, M. et al. 2006. In vivo molecular probing of cellular compartments with gold nanoparticles and nanoaggregates. *Nano Letters* 6: 2225–31.

Kneipp, J., Kneipp, H., Wittig, B. et al. 2007. One- and two-photon excited optical pH probing for cells using surface-enhanced Raman and hyper-Raman nanosensors. *Nano Letters* 7: 2819–23.

Kneipp, J., Kneipp, H., Wittig, B. et al. 2010. Following the dynamics of pH in endosomes of live cells with SERS nanosensors. *Journal of Physical Chemistry C* 114: 7421–6.

Kumar, S., Harrison, N., Richards-Kortum, R. et al. 2007. Plasmonic nanosensors for imaging intracellular biomarkers in live cells. *Nano Letters* 7: 1338–43.

Lechleiter, J., Girard, S., Peralta, E. et al. 1991. Spiral calcium wave-propagation and annihilation in *Xenopus laevis* oocytes. *Science* 252: 123–6.

Lei, J.Y., Wang, L.Z., and Zhang, J.L. 2010. Ratiometric pH sensor based on mesoporous silica nanoparticles and Forster resonance energy transfer. *Chemical Communications* 46: 8445–7.

McNamara, K.P., Nguyen, T., Dumitrascu, G. et al. 2001. Synthesis, characterization, and application of fluorescence sensing lipobeads for intracellular pH measurements. *Analytical Chemistry* 73: 3240–6.

McNamara, K.P., and Rosenzweig, Z. 1998. Dye-encapsulating liposomes as fluorescence-based oxygen nanosensors. *Analytical Chemistry* 70: 4853–9.

Nguyen, T., and Rosenzweig, Z. 2002. Calcium ion fluorescence detection using liposomes containing alexa-labeled calmodulin. *Analytical and Bioanalytical Chemistry* 374: 69–74.

Nowak-Lovato, K.L., and Rector, K.D. 2009. Targeted surface-enhanced Raman scattering nanosensors for whole-cell pH imagery. *Applied Spectroscopy* 63: 387–95.

Nowak-Lovato, K.L., Wilson, B.S., and Rector, K.D. 2010. SERS nanosensors that report pH of endocytic compartments during FcεRI transit. *Anal Bioanal Chem* 398: 2019–29.

Park, E.J., Brasuel, M., Behrend, C. et al. 2003. Ratiometric optical PEBBLE nanosensors for real-time magnesium ion concentrations inside viable cells. *Analytical Chemistry* 75: 3784–91.

Patolsky, F., Timko, B.P., Yu, G. et al. 2006. Detection, stimulation, and inhibition of neuronal signals with high-density nanowire transistor arrays. *Science* 313: 1100–4.

Peng, H.S., Huang, S.H., and Wolfbeis, O.S. 2010. Ratiometric fluorescent nanoparticles for sensing temperature. *Journal of Nanoparticle Research* 12: 2729–33.

Poulsen, A.K., Scharff-Poulsen, A.M., and Olsen, L.F. 2007. Horseradish peroxidase embedded in polyacrylamide nanoparticles enables optical detection of reactive oxygen species. *Analytical Biochemistry* 366: 29–36.

Reibetanz, U., Halozan, D., Brumen, M. et al. 2007. Flow cytometry of HEK 293t cells interacting with polyelectrolyte multilayer capsules containing fluorescein-labeled poly(acrylic acid) as a pH sensor. *Biomacromolecules* 8: 1927–33.

Ruedas-Rama, M.J., and Hall, E.A.H. 2006. K^+-selective nanospheres: Maximising response range and minimising response time. *Analyst* 131: 1282–91.

Ruedas-Rama, M.J., and Hall, E.A. 2009. Multiplexed energy transfer mechanisms in a dual-function quantum dot for zinc and manganese. *Analyst* 134: 159–69.

Schulz, A., Hornig, S., Liebert, T. et al. 2009. Evaluation of fluorescent polysaccharide nanoparticles for pH-sensing. *Organic & Biomolecular Chemistry* 7: 1884–9.

Schulz, A., Wotschadlo, J., Heinze, T. et al. 2010. Fluorescent nanoparticles for ratiometric pH-monitoring in the neutral range. *Journal of Materials Chemistry* 20: 1475–82.

Shen, Y., Kuang, M., Shen, Z. et al. 2008. Gold nanoparticles coated with a thermosensitive hyperbranched polyelectrolyte: Towards smart temperature and pH nanosensors. *Angewandte Chemie (International ed. in English)* 47: 2227–30.

Sumner, J.P., Aylott, J.W., Monson, E. et al. 2002. A fluorescent PEBBLE nanosensor for intracellular free zinc. *Analyst* 127: 11–6.

Sumner, J.P., and Kopelman, R. 2005. Alexa fluor 488 as an iron sensing molecule and its application in PEBBLE nanosensors. *Analyst* 130: 528–33.

Sumner, J.P., Westerberg, N.M., Stoddard, A.K. et al. 2006. Cu^+- and Cu^{2+}-sensitive PEBBLE fluorescent nanosensors using dsred as the recognition element. *Sensors and Actuators. B, Chemical* 113: 760–7.

Sun, H.H., Scharff-Poulsen, A.M., Gu, H. et al. 2008. Phosphate sensing by fluorescent reporter proteins embedded in polyacrylamide nanoparticles. *ACS Nano* 2: 19–24.

Webster, A., Coupland, P., Houghton, F.D. et al. 2007. The delivery of PEBBLE nanosensors to measure the intracellular environment. *Biochemical Society Transactions* 35: 538–43.

Wu, C., Bull, B., Christensen, K. et al. 2009. Ratiometric single-nanoparticle oxygen sensors for biological imaging. *Angewandte Chemie (International ed. in English)* 48: 2741–5.

Xu, H., Aylott, J.W., and Kopelman, R. 2002. Fluorescent nano-PEBBLE sensors designed for intracellular glucose imaging. *Analyst* 127: 1471–7.

Xue, L., Li, B., Fei, Q. et al. 2010. Carboxylate-modified squaraine dye doped silica fluorescent pH nanosensors. *Nanotechnology* 21: 215502.

Zhang, C.Y., Yeh, H.C., Kuroki, M.T. et al. 2005. Single-quantum-dot-based DNA nanosensor. *Nature Materials* 4: 826–31.

17 Implantable Infusion Pumps Incorporating Nanotechnology

Ellis Meng, Tuan Hoang, and Uttam Sinha

CONTENTS

INTRODUCTION

Advancements in drug discovery and development have led to novel nanoparticles incorporating biologics, biosimilars, and small molecules. Some of these nanoparticles require special dosing schemes (constant, pulsatile, or metronomic) to maintain sufficient therapeutic concentration for the required drug residence time to achieve optimal drug effectiveness. Certain nanopharmaceuticals cannot be administered orally since they are often digested before they can reach their target site. Others cannot be applied topically because drug size, charge, and chemistry hinder skin penetration at a rate sufficient to enable any therapeutic effect for deeper tissue targets. For these and other agents, the blood-brain barrier (BBB) is a formidable defensive barrier that prevents entry into targeted neural tissue. Delivery of nanomedicines directly to the targeted treatment site at the desired therapeutic concentration can overcome these physical, anatomic, and chemical obstacles regarding targeted drug delivery. Implantable drug delivery technologies have steadily advanced, so they can be used clinically to restrict drug delivery to the infused tissue, thereby avoiding systemic drug side effects. This chapter focuses on microtechnologies and nanotechnologies that enable implantable infusion pumps with advanced drug administration capabilities to treat chronic conditions for which localized drug infusion is required.

CURRENT STATE OF IMPLANTABLE INFUSION PUMP TECHNOLOGY

Current infusion pumps deliver drugs to tissue and compartments throughout the body and address a wide range of clinical conditions such as cancer, chronic pain, and spasticity. Chronic infusion of drug provides significant advantages over oral and parenteral dosing in terms of increased treatment duration and maintenance of optimal therapeutic concentration. Typically, with oral and parenteral dosing, a peak and valley phenomenon is encountered in which a high blood level is reached shortly after drug administration and is followed by a gradual decline until the drug is once again administered (Figure 17.1). Chronic infusion can prevent the overdosing or underdosing that can occur with these methods. Implantable infusion pumps (IIPs) were developed to provide patients with more independence during chronic infusion therapy than is possible with external infusion pumps. In particular, IIPs fulfill the need for site-specific therapy (to a target organ or compartment) and chronic infusion. Direct administration of repeated drug doses at or close to their site of action in the central nervous system (CNS) is a treatment strategy for neurological diseases that requires knowledge of tissue-specific receptors, an understanding of the mechanism of action of exogenous and endogenous ligands, which may include neuroactive peptides, and the development of highly active drugs (Lazorthes et al. 1991). Intractable cancer pain was first treated in 1978 using implanted pumps to deliver a drug, in this case morphine, into the intrathecal or intraventricular cerebrospinal fluid (CSF). More recently, intrathecal administration of Baclofen, using an implanted pump, has been described to treat severe spasticity. These two applications are now standard medical practice and are a prototype for expanding the use of IIPs for other chronic conditions that may respond to the continuous delivery of a receptor-agonist drug to a specific CNS region.

The two categories of IIPs have different pumping mechanisms. One category is the nonelectrically controlled or powered devices that utilize either passive osmotic pressure or dissolution rates. An example of the osmotic pressure-driven IIP is the Duros delivery system (Durect Corporation, Cupertino, CA), which is utilized in the VIADUR product (leuprolide acetate implant), a once-yearly implant for the palliative treatment of prostate cancer. Meanwhile, a pump based on drug dissolution is Retisert (Bausch and Lomb, Inc., Waterford, Ireland), which consists of a silicone elastomer cup containing a tablet of fluocinolone acetonide, a polyvinyl alcohol membrane, and a release orifice. This device is implanted by ophthalmologists to treat noninfectious uveitis, a chronic inflammatory disease of the eye. The performance of these passive IIPs is dictated by drug formulation, drug stability, drug payload, matrix chemistry, pore sizes, and diffusion/dissolution rates. The focus of the chapter is on the second category of IIPs, the actively driven implants requiring an intrinsic pumping force.

State-of-the art active IIPs currently in clinical use can be generally classified into two types based on their pumping mechanism. One type utilizes a fluorocarbon propellant (vapor pressure mechanism), while the other employs a battery-operated stepper motor (peristaltic mechanism). Fluorocarbon-based pumps contain a liquid that evaporates and produces a constant pressure (450 mm Hg at 37°C) at a given temperature that drives the infusion of a drug at a fixed rate. Flow rate (between 1 and 6 mL/24 h) is factory set and cannot be altered after implantation; dose can be altered only by changing the drug concentration. These pumps contain two chambers separated by metal bellows; the pumping, or charging, fluid is sealed in an outer chamber, and the drug is held in the inner chamber. Vapor pressure from the charging fluid compresses the bellows and expels

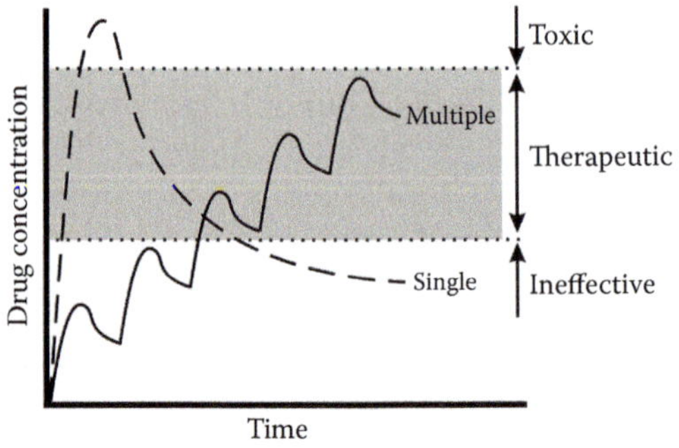

FIGURE 17.1 Therapeutic window for drug administration expressed in terms of drug concentration over time. Single and multiple bolus dosing are depicted. The advantage of multiple boluses over single dosing to maintain drug concentration in the therapeutic window is evident.

the drug through a filter, flow restrictor, and cannula into a catheter leading to the infusion site. The drug reservoir is refilled by percutaneous needle injections through a self-sealing septum. Refilling the drug chamber provides pressure against the bellows compressing and condensing the charging fluid vapor to reset the pump. Changes in ambient temperature and pressure result in flow rate variation, which can be compensated for by using a flow regulator (Dorman et al. 1981).

Active IIPs that are of the peristaltic pump type may be programmed externally to provide different patterns (e.g., bolus, multistep bolus, continuous) and rates (0.004–0.9 mL/h) of drug delivery. A stepper motor controlled by a microprocessor rotates rollers in contact with flexible tubing housed in a U-shaped track and connected to the collapsible drug reservoir. Compression occludes the tubing lumen, and rotation forces the drug to move toward the exit. Using multiple rollers positioned appropriately with respect to the housing ensures that one roller is engaged with the tubing at all times to prevent any backflow. A catheter connected to the exit port on the drug reservoir housing is directed to the site of treatment. The flow rate accuracy at the catheter tip can vary with ±15% error under ideal conditions (35°C–42°C, 0–300 mm Hg backpressure, and 0–1230 m elevation) (Decq and Keravel 1991). Under low-volume conditions (~2 cc), a Freon-like gas assists in emptying the drug reservoir, which reduces the delivery accuracy. The infusion may be stopped immediately by using the external controller; the motor idles in "stop" mode (0.00018 mL/h) (Lazorthes et al. 1991). The reservoir is periodically refilled by subcutaneous injections (Table 17.1).

Regardless of operating mechanism, current active IIPs are large, about the size of hockey pucks (about 100–180 cm^3). Thus, these pumps are typically implanted subcutaneously in a body region that is usually covered by clothing, such as the lower abdomen. IIPs operating by different mechanisms are under development but have not yet been approved by the Food and Drug Administration (FDA) for marketing. These include miniaturized pumps, which are less invasive, provide more accurate drug delivery, and can be programmed to provide more flexible drug regimens than active IIPs currently on the market.

The first IIPs delivered medications locally through a catheter at a fixed, continuous rate; they were introduced in the 1960s for insulin delivery. Vapor pressure-driven pumps were first approved by the Food and Drug Administration (FDA) in 1981 for the intravenous delivery of heparin to treat recurrent venous thromboembolism and intra-arterial infusion of 5-fluoroxyuridine (5-FUDR) into the hepatic artery for organ specific chemotherapy (Buchwald and Rohde 1984). These pumps were originally developed at the University of Minnesota, the technology licensed to Metal Bellows Corporation, and later manufactured by Infusaid Corporation, a subsidiary of Metal Bellows Corporation. Pumps were periodically refilled; each refilling procedure provided an opportunity to change the dosing regimen by changing drug concentration (Prager 2002; Wallace and Yaksh 2000).

TABLE 17.1
Comparison of Current Implantable Pumps

	SynchroMed II	SynchroMed EL	Codman 3000	IsoMed
Manufacturer	Medtronic	Medtronic	Codman & Shurtleff	Medtronic
Reservoir size (mL)	20, 40	10, 18	16, 30, 50	20, 35, 60
Flow rate (mL/day)	0.048–20	0.5–20	0.5, 1.0, 3.4[a]	0.3, 0.5, 1.0, 4.0[b]
Dosing scheme	Programmable	Programmable	Constant flow	Constant flow
Battery life	7 years[c]	6.5 years[d]	Battery free	Battery free
Weight (empty; g)	165, 175	185, 205	98, 137, 173	80
Volume (cm^3)	91, 121	123, 156.7	N/A	111.7, 135, 172.2

[a] Flow rate range for a Codman 3000 pump with a 50-mL reservoir size.
[b] Flow rate range for an IsoMed pump with a 60-mL reservoir size.
[c] Based on 0.5 mL/day flow rate for a SynchroMed II pump with a 20 mL reservoir size.
[d] Based on 0.5 mL/day flow rate for a SynchroMed EL pump with an 18-mL reservoir size.

Later, an electrically powered and programmable pump in which the flow rate could be adjusted was introduced by Medtronic, Inc. (SynchroMed) for cancer (1988) and chronic pain (1991) (Bedder 2009). The pump was battery powered, which necessitates removing and replacing the pump every 3–7 years as dictated by the battery's lifetime.

An implant such as the Ommaya reservoir, which provides long-term access to ventricular CSF and is usually used for intermittent drug administration, may be attached to an implanted pump to enable chronic continuous drug administration.

MICROTECHNOLOGIES AND NANOTECHNOLOGIES ENABLING DRUG DELIVERY

Microelectromechanical systems (MEMSs) are dramatically advancing the state of the art in infusion pumps for conventional drugs and emerging nanomedicine. MEMSs are miniature devices with characteristic sizes spanning the nanometer to millimeter range. Such tiny feature sizes are enabled by a shared lineage with the transistor. MEMSs borrow and adapt microfabrication techniques that originated from the integrated circuit industry to produce devices having more than just electronic capabilities, but also mechanical, chemical, and biological ones. Thus, many functions may be integrated into a compact form.

MEMS technology combined with biocompatible materials can be maximized to advance drug delivery. Microfabrication allows devices to be batch-fabricated with precise, reproducible features, economically scaled, thereby paving the way for inexpensively produced devices. Miniaturization facilitates devices that can be surgically placed using minimally invasive techniques. In contrast to current drug depot devices, MEMS drug delivery systems feature precise and on-demand dosing following complex regimens. Integration of multiple functions allows interesting possibilities in next-generation drug delivery systems in which drug pumps can be controlled telemetrically or can integrate sensors that provide physiological feedback, which leads to doses being delivered according to an evidence-based regimen. Such systems also improve compliance with treatment regimens and therapeutic outcomes.

Microfabricated devices for drug delivery include microparticles, microreservoirs, needles, and implantable pumps. While the focus here is on implantable pumps enabled by MEMS technology, the other relevant devices will be briefly reviewed, with the exception of needle-based approaches that are being explored for transdermal drug delivery (Donnelly et al. 2010; Henry et al. 1998; McAllister et al. 2000; Prausnitz 2004; Prausnitz et al. 2009; Sivamani et al. 2007; Vandervoort and Ludwig 2008). Three categories of microtechnology-enabled drug delivery devices are described: (1) microparticles and microreservoirs, (2) neural probes with integrated microfluidics for drug delivery, and (3) implantable pumps.

MICROPARTICLES AND MICRORESERVOIRS

The simplest form of drug delivery enabled by microtechnology involves packaging a drug into a tiny capsule that can be implanted. The drug is administered into the body either by diffusion through a microporous or nanoporous membrane or by electronically controlled on-demand selective release. Membrane-based technologies control drug release from microparticles within porous membrane-covered capsules and microreservoirs (Shawgo et al. 2002).

Drugs in microparticle format have been stored in silicon reservoirs with an integrated slow-dissolving cap. The cap dissolution rate is related to its surface chemistry and determines the period of delay between microparticle administration and drug delivery (Ahmed et al. 2002). Membrane-based technologies have also been demonstrated to physically isolate cells from the body and the immune system (i.e., immunoisolation) while still allowing therapeutic compounds produced by the cells to diffuse across the membrane into the body. Such systems typically combine a biocapsule for storing cells and a nanoporous membrane for immunoisolation and containment of cells. Immunoisolation is achieved by size exclusion. Unlike traditional polymer-based capsules, which have broad pore size distribution and low mechanical strength, robust micromachined membranes have fine and uniform pore sizes. Microfabricated membrane-based systems were demonstrated using transplanted islet cells that inhibit passage of immunologic cells (i.e., macrophages), antibodies, and large proteins (i.e., immunoglobulin G) while being permeable to small molecules such as oxygen, glucose, and insulin (Desai et al. 1998). Recently, a novel method of creating containers for particles and cells for drug delivery has been described in which polymeric polyhedral containers were assembled by self-folding upon heating (Azam et al. 2011).

Drugs can be released upon demand by electronically controlling the removal of a nonpermeable membrane. Multiple removal mechanisms have been reported. The first to be described is electrochemically dissolving a gold membrane covering a drug reservoir etched into a silicon wafer. Membranes were removed by applying anodic voltage in the presence of chloride ions. Arrays of such wells were produced for digital release of drugs in small pulses that could be combined to achieve the desired release profile (Santini et al. 1999). Using such devices, in vitro and in vivo pulsatile release of model and therapeutic molecules was achieved (Li et al. 2004, 2005; Santini et al. 1999). Li et al. (2005) demonstrated the dose-dependent inhibition of tumor growth in rats by local pulsatile delivery of 1,3-bis (2-chloroethyl)-1-nitrosourea (BCNU), a chemotherapeutic agent, and achieved results comparable to subcutaneous injections. The electrochemical dissolution mechanism was later replaced by the electrothermal removal of composite platinum and titanium membranes (Maloney et al. 2005). Electrothermally activated wells were used in the chronic programmed delivery of polypeptide leuprolide in canines for over 6 months (Prescott et al. 2006). However, removal of the entire metallic membrane electrothermally required significant electrical power. By selectively placing fuses in areas of maximum mechanical stress, membranes were removed more efficiently (Elman et al. 2010).

Additionally, a biodegradable version was reported that replaces the electronically triggered metallic membrane with a resorbable polymer-based one [poly(L-lactic acid) or poly(lactide-*co*-glycolide)], which avoids the need for electronics. The molecular masses of the membranes were varied to control the release time of the reservoir contents. Delivery of dextran, heparin, human growth hormone, glucose, and glycerol pulses was demonstrated (Richards Grayson et al. 2003, 2004a).

Responsive membranes utilize physiological or other external stimuli to regulate (gate) drug release from drug delivery chambers (Ziaie et al. 2004). The variation in pH levels within organs and tissues has been investigated as a means of controlling the swelling and shrinking behavior of hydrogel membranes. Devices responded by providing higher drug release rates at lower pH values for potentially treating tumors (Chen et al. 2009). Thermoresponsive hydrogels that respond by shrinking upon heat application were used as microvalves to provide temporal control of drug release. The hydrogels were heated using fast-responding and wirelessly activated electric heaters (Rahimi et al. 2011).

Drug delivery from microparticle and microreservoir systems is inherently limited by diffusion. To reduce the time required for drug delivery and the influence of external physiological conditions, Chung et al. (2008) integrated electrokinetic transport with a microwell and demonstrated controlled delivery rate and reduced drug delivery time from hours to seconds. Higher flow rates were later demonstrated using an electrochemically based transport mechanism for controlled vasopressin release (Chung et al. 2009). In the latter, the electrochemical dissolution of the capping metallic membrane results in gas formation, which forces the drug out of the well. A similar device used thermally generated bubbles to dose vasopressin (Elman et al. 2009). Bubbles, generated by resistive heating, result in a sudden pressure increase that bursts the sealing membrane. However, slight degradation of vasopressin was detected by High Performance Liquid Chromatography (HPLC). Instead of removing the membrane, a nanofluidic membrane consisting of etched channels can be combined with electrokinetic transport to control the magnitude, duration, and timing of drug release (Fine et al. 2011).

Neural Probes with Integrated Drug Delivery

Implantable electrodes have been employed in neural prosthetics for either recording from or stimulating neural tissue for decades. More recently, the neural prosthetics community has been interested in the potential of neural probes with integrated drug delivery for delivering chemical stimuli, neurotrophic factors, drugs, and toxins. Early devices consisted of small-diameter stainless steel or glass capillaries inserted into the brain with separate wire electrodes for recording. The electrode and fluid conduit may not be accurately colocalized if the electronics and fluidic conduit are placed sequentially. To avoid this uncertainty in targeting the desired brain structures, hybrid probes constructed by attaching the wire to the conduit or mounting the wire within a conduit were devised. Implementation of such devices was hampered by their tedious manual assembly process and large size.

Using MEMS technology, batch-fabricated silicon probes emerged that could selectively deliver chemicals in vivo and electrical recording and stimulation of neurons. The first such devices consisted of electrodes supported by a hollow-core probe with multiple fluidic orifices. Chemical stimuli (kainic acid and γ-aminobutyric acid [GABA]) were applied to the superior and inferior colliculi of guinea pigs, and the corresponding electrical responses were measured (Chen et al. 1997). These probes had open orifices, and as such, control of chemical delivery was limited. Later, fluidic probes were improved by integrating shutter valves and flow meters to control and monitor drug dosing; further schemes for integrating electronic pumping and valving were also explored for providing multiplexed chemical stimuli (Papageorgiou et al. 2001). A formal characterization of the fluidic performance of a single orifice probe was carried out followed by recording the electrical response to chemical stimulation with AMPA (α-amino-3-hydroxy-5-methyl-4-isoxalone), which mimics the effects of the neurotransmitter glutamate (Rathnasingham et al. 2004). A similar study was performed using a simplified device in which a microelectrode neural probe was affixed to a standard fused silica catheter that provided drug delivery capability. Tetrodotoxin in artificial CSF was injected into rat brain, resulting in suppression of all spiking activity near the catheter outlet (Rohatgi et al. 2009).

It has been suggested that infusions of specific chemical agents at the interface of a neural prosthesis and host tissue could reduce the reactive tissue responses that limit the chronic use of neural prostheses. In comparison with diffusion-mediated delivery (Retterer et al. 2004), active infusion under constant pressure provided drug distribution in rat brains over a greater volume than that corresponding to the reactive responses and in a more controllable and predictable manner (Retterer et al. 2008).

A disposable drug delivery system that can be attached to silicon fluidic probes for neural research was developed; most fluidic probes still utilize off-chip pumps to drive chemical delivery. Using electrically controlled heaters, the irreversible thermal expansion of microspheres was induced to infuse discrete 0.25-μL boluses into neural tissue (Spieth et al. 2010). The same group also developed a silicon probe for both neural recording and drug delivery, which has not been demonstrated in vivo (Seidl et al. 2010).

In addition to silicon-based probes, flexible and rigid polymer-based probes were also introduced using either Parylene C (Takeuchi et al. 2005; Ziegler et al. 2006) or polyimide (Metz et al. 2004), respectively, as structural materials to support both the microfluidic channels and microelectrodes. More recently, SU-8 epoxy-based probes with a single microfluidic channel and multiple electrode sites have been demonstrated. Preliminary studies indicated that less mechanical damage during insertion was observed using the softer SU-8 polymer probe compared to a standard stainless-steel stereotaxic needle in rats; Nissl staining was used to assess neuronal injury and IB4 to assess inflammation (Fernandez et al. 2009).

Similar catheter-only probe structures were developed specifically for convection-enhanced delivery. A simple probe with a single polymer microfluidic channel and orifice supported on a silicon substrate was used to infuse Evans blue and albumin into hydrogel brain phantoms and to infuse albumin into the gray matter of rats. The microfabricated probe exhibited reduced backflow along the probe–tissue interface in comparison to conventional convection-enhanced delivery (CED) needles (Neeves et al. 2006). A flexible version of this probe was explored to reduce foreign body response and tissue microtearing around the probe. The polymer-based probe was supported by a rigid biodegradable backbone (poly(DL-lactide-*co*-glycolide), [PLGA]), which supported the probe during implantation and then dissolved away into tissue. Acute injections of dye through the probe and into hydrogel brain phantoms and rat striatum were demonstrated (Foley et al. 2009).

Implantable Infusion Pumps

The development of microtechnology-based implantable pumps began in the 1990s and was facilitated by the earlier development of nonimplantable micropumps in the 1980s. The pursuit of such pumps was motivated by several factors. Compared to intermittent dosing, sustained local administration usually requires a lower total dose of drug over time to achieve the desired effect and obviates the need for frequent injections. In addition, local drug delivery limits undesirable systemic side effects because nontargeted organs are exposed to much lower drug levels than with systemic drug delivery. Long-term pulsatile dosing through an implantable pump is an alternative to continuous dosing and can provide drug levels within the therapeutic range if a dosing frequency is selected that takes into account the drug's half-life. However, the small drug volume contained in a MEMS form factor drug delivery pump is practical only to deliver high-potency drugs that are available in solution. Some MEMS pumps cannot be refilled, which limits drug delivery duration. These pumps must be surgically replaced periodically if they are to be used to treat chronic conditions.

Microfluidic pump technology may utilize many actuation mechanisms to displace fluids. These mechanisms are broadly categorized as either mechanical or nonmechanical (predominantly of the osmotic type). A brief review of implantable MEMS pumps is provided here. The reader is directed to previous reviews for further information on implantable MEMS pumps (Amirouche et al. 2009; Grayson et al. 2004b,c; Liepmann et al. 1999; Martin and Grove 2001; Polla et al. 2000; Shawgo et al. 2002; Staples et al. 2006; Tsai and Sue 2007).

Microtechnology-Enabled Implantable Pumps

Multiple pumps have utilized actuators based on a shape memory alloy (SMA) such as NiTi (trade name Nitinol). An SMA exhibits a thermally induced phase change in the solid state (austenite and martensite phases) that allows the material to be bent into various shapes. This process is reversible, and the material can be cycled repeatedly. In 1997, Reynaerts and colleagues developed SMA-based drug delivery devices suitable for both solid and liquid delivery. Solid delivery was accomplished using a clamshell-like device containing a fixed drug payload. An SMA actuator controlled the opening of the clamshell and thus release of the drug. The liquid delivery system was similar to conventional Freon-powered pumps and utilized an SMA-actuated valve to regulate drug release from a pressurized reservoir (Reynaerts et al. 1997). The high force and strain generated by SMAs was harnessed in a reciprocating diaphragm micropump (Benard et al. 1998). Pumping was achieved using SMA to actuate a flexible diaphragm. Fluid was directed using two one-way check valves to ensure unidirectional flow through the pumping chamber. Alternating pump chamber contraction and expansion forced fluid out or into the chamber, respectively. The potential of such pumps for drug delivery was described.

Piezoelectric thin films (most often lead zirconate titanate [PZT]) have interesting properties suitable for pumping applications. The crystalline structure of these materials predictably deforms in response to an applied electrical field. Maillefer and colleagues presented a PZT-driven reciprocating diaphragm pump with a limiter mechanism to accurately control the volume of each drop of fluid ejected by the pump. Thus, a highly accurate flow rate was established that bore a linear relationship to the frequency of electrically induced mechanical deformation of PZT (actuation frequency). This technology is currently marketed by Debiotech S.A. and is also featured in their nonimplantable pumps for insulin delivery applications (Maillefer et al. 1999). Cao and colleagues proposed and analyzed a potential design for a PZT-based pump featuring three separate electronically controlled actuators. Each PZT disk actuator is coupled to the flow path, and when actuated in a prescribed pattern and sequence, unidirectional peristaltic fluid pumping could be achieved

(Cao et al. 2001a). Recently, a valveless micropump driven by PZT actuation has been designed specifically to deliver drug formulations containing aggregates or particles (Su and Pidaparti 2010a,b).

The electrostatic attraction (and thus force) between oppositely charged metallic plates is utilized in electrostatic actuators. A peristaltic pump with electrostatic actuators was designed and simulated (Teymoori and Abbaspour-Sani 2005). Likewise, a reciprocating diaphragm pump using electrostatic actuation was also suggested (Bourouina et al. 1997). The linearity of such reciprocating pumps can be improved by using a circular disk attached to the pumping diaphragm (bossed membrane) (Yih et al. 2005).

A diffusion-based pump regulated by a chemomechanical valve was described for intelligent drug delivery (Cao et al. 2001b). Drug release from a reservoir was regulated by a glucose- and pH-sensitive hydrogel. The hydrogel expands in response to high pH, which presses a coupled flexible membrane against the release orifice of the drug reservoir, sealing the orifice, thus preventing drug release. Adding glucose oxidase and catalase to the hydrogel enabled the conversion of glucose concentration changes to pH changes in the hydrogel. Release of a hydrophilic dye (acid orange 8 [AO8]) as a model drug was demonstrated.

A reciprocating-type pump is being developed for intracochlear drug delivery that features alternating pumping and withdrawal stages to eliminate the possibility of net fluid volume change in the ear and thus reduce the risk of hearing damage (Sewell et al. 2009). The system is microfluidic based, and proof-of-concept prototypes were built using conventional pumps. Further miniaturization is planned to develop a fully implantable pump.

An intrathecal drug delivery system able to deliver multiple drugs was developed to treat chronic pain. To simplify the drug delivery system architecture and eliminate the need for an active pumping mechanism, mechanically pressurized reservoirs were combined with smart valves to control drug delivery. Paired reservoirs contained by polyethylene bags were pressurized using external springs. The drugs were metered using piezoelectrically actuated valves and delivered through a single delivery catheter. Pressure sensors integrated with the valves are used to determine the actual drug delivery rate. In benchtop tests, the magnetic resonance imaging (MRI)-compatible system could compensate for changes in spinal fluid pressure and maintain a steady flow rate (Evans et al. 2007, 2009, 2010a,b, 2011).

A microbolus infusion pump was developed for neuroimaging applications in freely behaving rodents. Initiation of a specific behavior (i.e., conditioned fear) wirelessly triggered the rapid release of a type of radiotracer that is freely diffusible through the BBB followed by a euthanasia agent. More radiotracer is deposited in the activated region than in surrounding regions because the brain region activated by the behavior has higher blood flow than nonactivated regions. The activated region can be detected autoradiographically on postmortem tissue sections. The pump components included elastomeric reservoirs constructed from silicone rubber and electrothermal MEMS valves containing thin metal film (platinum or titanium) supported on a Parylene polymer substrate. The valves were activated by passing a current through the resistive metal element, which melted the polymer valve membrane and allowed drug release (Givrad et al. 2008; Li et al. 2008a,b,c, 2009a, 2010a).

Osmotic pumps have been used for implantable drug delivery for several decades. These pumps harness the osmotic pressure generated from the net water flow across a salt-impermeable membrane that seeks to equilibrate an osmotic concentration gradient. No electrical power is required for pumping. Conventional systems are commercially available for both human and laboratory animal use (Durect Corporation). Such pumps have a chamber containing water separated by a salt-impermeable membrane from a chamber containing a concentrated salt solution. The water crosses the salt-impermeable membrane, which expands the chamber containing the salt, displaces an expansion membrane into a drug reservoir, and forces the drug into the infusion tubing. A microfabricated osmotic drug pump was developed but has not yet been demonstrated in vivo (Su and Lin 2004; Su et al. 2002).

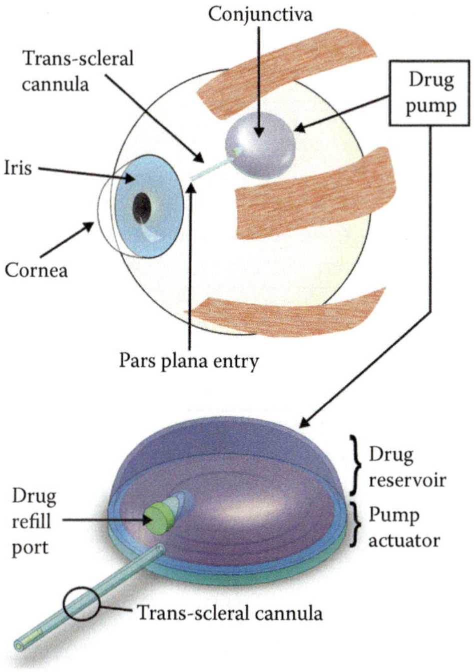

FIGURE 17.2 MEMS drug delivery device implanted in the subconjunctival space with a cannula entering at the limbus into the anterior or posterior segment. Top: Overview of implant location. Bottom: Three-dimensional illustration of pump showing major components. The valve (not shown) is located near the tip of the trans-scleral cannula.

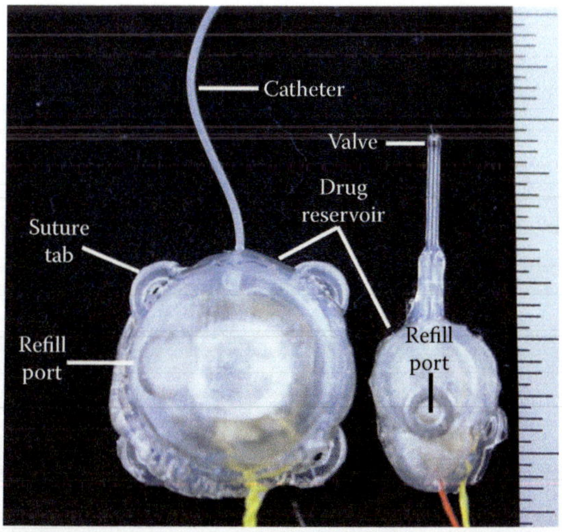

FIGURE 17.3 Photographs of two MEMS electrolysis-based pumps. The left pump infused siRNA as a radiosensitizer in an animal study; the right pump is intended for intraocular drug delivery. The major pump features are indicated. Both pumps are encapsulated in biocompatible silicone rubber.

Electrochemically driven pumps have been explored for direct intraocular delivery of therapeutics to treat chronic ocular diseases such as uveitis, glaucoma, and macular degeneration that are leading causes of blindness. The low-profile and low-power mechanism is ideal for pumps requiring placement within the eye wall (subconjunctival space) (Figure 17.2). Both the reservoir and pump, housed in a single package, were implanted in the eye wall, and a single cannula was placed in the anterior chamber. Current applied to the pump electrodes (Figure 17.3, right) initiated electrolysis, liberation of gas, and enough pressure to force the drug out of the reservoir, through the cannula, and into the eye. Such direct delivery bypasses the blood–retinal barrier, which is a major obstacle in treating retinal disease using systemic therapy (Li et al. 2007, 2008d, 2009b, 2010b; Meng et al. 2006; Saati et al. 2009, 2010).

Gene Therapy Enabled by Nanotechnology and Implantable Pumps

The aforementioned electrolysis-based pumps were adapted for use in cancer treatment involving gene delivery to reduce the radiation dosage (Gensler et al. 2010). Radiation therapy for cancer may produce collateral damage to normal tissue, resulting in severe acute and/or chronic side effects. Small interfering RNAs (siRNAs) were developed to block production of sphingosine kinase (SphK-1), which is implicated in the proliferation and antiapoptotic activity of a variety of human cancers (Spiegel and Milstien 2003). Blocking SphK-1 expression may sensitize cells to radiation therapy, permitting lower radiation doses than normally used to kill the cancer cells. To efficiently transport the siRNA across the cell membrane, a nanotechnology-mediated gene silencing approach combined with localized delivery of siRNA–nanoparticle complexes was developed. Such an approach protects the vulnerable free siRNAs from degradation and utilizes much smaller amounts of the expensive molecule than would be necessary with systemic delivery.

Surface-modified gold nanorods were used to form stable electrostatic complexes with siRNA. These "nanoplexes" were loaded into electrolysis-based pumps implanted in a mouse xenograft model of HT29 colon cancer. The animals had two tumor sites, one receiving "nanoplexes" and the other serving as a control (Figure 17.4). Comparison groups of animals received "nanoplexes" by diffusion from passive pumps or by periodic direct injection. Tumors were also treated with low doses of radiation (100 rad). In preliminary studies, active pumping of nanoplexes produced about 50% tumor shrinkage, which was significantly more than that seen with delivery of nanoplexes by diffusion or needle injection (Gensler et al. 2010).

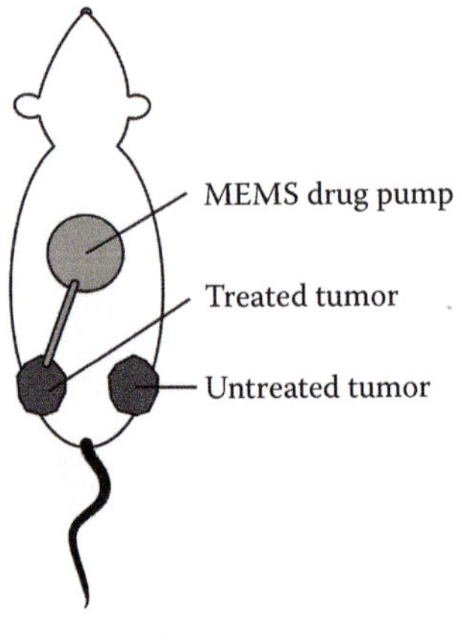

FIGURE 17.4 Illustration of MEMS electrolysis pump placement in a radiation dose reduction study. The outlet catheter of the pump was directed toward the tumor in the left flank.

FUTURE PROSPECTS

MEMS technology has enabled several advances in IIPs. The next generation of IIP technology will move beyond simply integrating electronic sensors to detect flow or pump failure and progress toward closed-loop systems that acquire real-time physiological feedback, which actively modulates the drug regimen to provide highly personalized drug therapy. Currently, pumps modulated by closed-loop systems await the development of biosensors that are sufficiently stable for clinical use. Progress toward such biosensors will enable drug delivery systems to determine when a dose is needed and automatically deliver it according to programmed instructions. Fully autonomous drug delivery systems will efficiently and effectively deliver the drug(s), minimize side effects, and reduce clinical visits. Both microtechnologies and nanotechnologies offer new ways to deliver drugs, which hope to improve patient outcome and quality of life.

DISCLOSURE STATEMENT

Dr. Ellis Meng and Dr. Tuan Hoang have a significant financial interest in Fluid Synchrony LLC.

ACKNOWLEDGMENTS

The authors acknowledge funding from the L.K. Whittier Foundation and Wallace H. Coulter Foundation Early Career Translational Research Award. The authors would like to thank members of the Biomedical Microsystems Laboratory at the University of Southern California for their assistance with this work.

REFERENCES

Ahmed, A., Bonner, C., and Desai, T.A. 2002. Bioadhesive microdevices with multiple reservoirs: a new platform for oral drug delivery. *Journal of Controlled Release* 81: 291–306.

Amirouche, F., Zhou, Y., and Johnson, T. 2009. Current micropump technologies and their biomedical applications. *Microsystem Technologies—Micro- and Nanosystems Information Storage and Processing Systems* 15: 647–66.

Azam, A., Laflin, K.E., Jamal, M. et al. 2011. Self-folding micropatterned polymeric containers. *Biomedical Microdevices* 13: 51–8.

Bedder, M.D. 2009. Intrathecal analgesics, choice of system. In *Neuromodulation*, ed. E.S. Krames, P.H. Peckham, and A.R. Rezai. St. Louis, MO: Elsevier 457–66.

Benard, W.L., Kahn, H., Heuer, A.H. et al. 1998. Thin-film shape-memory alloy actuated micropumps. *Journal of Microelectromechanical Systems* 7: 245–51.

Bourouina, T., Bosseboeuf, A., and Grandchamp, J.P. 1997. Design and simulation of an electrostatic micropump for drug-delivery applications. *Journal of Micromechanics and Microengineering* 7: 186–8.

Buchwald, H., and Rohde, T.D. 1984. Implantable infusion pumps. *Advances in Surgery* 18: 177–221.

Cao, L., Mantell, S., and Polla, D. 2001a. Design and simulation of an implantable medical drug delivery system using microelectromechanical systems technology. *Sensors and Actuators. A, Physical* 94: 117–25.

Cao, X., Lai, S., and Lee, L.J. 2001b. Design of a self-regulated drug delivery device. *Biomedical Microdevices* 3: 109–18.

Chen, J., Chu, M., Koulajian, K. et al. 2009. A monolithic polymeric microdevice for pH-responsive drug delivery. *Biomedical Microdevices* 11: 1251–7.

Chen, J.K., Wise, K.D., Hetke, J.F. et al. 1997. A multichannel neural probe for selective chemical delivery at the cellular level. *IEEE Transactions on Biomedical Engineering* 44: 760–9.

Chung, A.J., Huh, Y.S., and Erickson, D. 2009. A robust, electrochemically driven microwell drug delivery system for controlled vasopressin release. *Biomedical Microdevices* 11: 861–7.

Chung, A.J., Kim, D., and Erickson, D. 2008. Electrokinetic microfluidic devices for rapid, low power drug delivery in autonomous microsystems. *Lab on a Chip* 8: 330–8.

Decq, P., and Keravel, Y. 1991. Drug pump technical description, operative procedure and post-operative management. In *Neurosurgery for Spasticity: A Multidisciplinary Approach*, ed. M. Sindou, R. Abbot, and Y. Keravel, 97–101. New York: Springer-Verlag.

Desai, T.A., Chu, W.H., Tu, J.K. et al. 1998. Microfabricated immunoisolating biocapsules. *Biotechnology and Bioengineering* 57: 118–20.

Donnelly, R.F., Singh, T.R.R., and Woolfson, A.D. 2010. Microneedle-based drug delivery systems: microfabrication, drug delivery, and safety. *Drug Delivery* 17: 187–207.

Dorman, F.D., Rohde, T.D., Arlt, G.J. et al. 1981. A two-phase fluid powered insulin infusion pump with basal-bolus capability which compensates for pressure and temperature variability. *Transactions—American Society for Artificial Internal Organs* 27: 236–40.

Elman, N.M., Ho Duc, H.L., and Cima, M.J. 2009. An implantable MEMS drug delivery device for rapid delivery in ambulatory emergency care. *Biomedical Microdevices* 11: 625–31.

Elman, N.M., Masi, B.C., Cima, M.J. et al. 2010. Electro-thermally induced structural failure actuator (ETISFA) for implantable controlled drug delivery devices based on Micro-Electro-Mechanical-Systems. *Lab on a Chip* 10: 2796–804.

Evans, A.T., Chiravuri, S., and Gianchandani, Y.B. 2009. Transdermal power transfer for implanted drug delivery devices using a smart needle and refill port. IEEE 22nd International Conference on Micro Electro Mechanical Systems 252–5. *Piscataway*, NJ: IEEE.

Evans, A.T., Chiravuri, S., and Gianchandani, Y.B. 2010a. Transdermal power transfer for recharging implanted drug delivery devices via the refill port. *Biomedical Microdevices* 12: 179–85.

Evans, A.T., Chiravuri, S., and Gianchandani, Y.B. 2011. A multidrug delivery system using a piezoelectrically actuated silicon valve manifold with embedded sensors. *Journal of Microelectromechanical Systems* 20: 231–8.

Evans, A.T., Park, J.M., Chiravuri, S. et al. 2007. Dual drug delivery device for chronic pain management using micromachined elastic metal structures and silicon microvalves. 2008 21st IEEE International Conference on Micro Electro Mechanical Systems 252–5. *Piscataway*, NJ: IEEE.

Evans, A.T., Park, J.M., Chiravuri, S. et al. 2010b. A low power, microvalve regulated architecture for drug delivery systems. *Biomedical Microdevices* 12: 159–68.

Fernandez, L.J., Altuna, A., Tijero, M. et al. 2009. Study of functional viability of SU-8-based microneedles for neural applications. *Journal of Micromechanics and Microengineering* 19: 025007.

Fine, D., Grattoni, A., Zabre, E. et al. 2011. A low-voltage electrokinetic nanochannel drug delivery system. *Lab on a Chip* 11: 2526–34.

Foley, C.P., Nishimura, N., Neeves, K.B. et al. 2009. Flexible microfluidic devices supported by biodegradable insertion scaffolds for convection-enhanced neural drug delivery. *Biomedical Microdevices* 11: 915–24.

Gensler, H., Sheybani, R., Li, P.Y. et al. 2010. Implantable MEMS drug delivery devices for cancer radiation reduction. *MEMS 2010*, 23–6. Hong Kong, China.

Givrad, T.K., Maarek, J.-M., Li, P.-Y. et al. 2008. Implantable minipump with Parylene electrothermal valve for bolus injection in mice. *ASME Frontiers in Biomedical Devices Conference*. Irvine, CA, USA.

Henry, S., McAllister, D.V., Allen, M.G. et al. 1998. Microfabricated microneedles: a novel approach to transdermal drug delivery. *Journal of Pharmaceutical Sciences* 87: 922–5.

Lazorthes, Y., Sallerin-Caute, B., Verdie, J.C. et al. 1991. Advances in drug delivery systems and applications in neurosurgery. *Advances and Technical Standards in Neurosurgery* 18: 143–92.

Li, P.Y., Givrad, T.K., Holschneider, D.P. et al. 2008a. Mechanical and thermal modeling of a parylene electrothermal valve for mapping brain function in freely moving subjects. *12th International Conference on Miniaturized Systems for Chemistry and Life Sciences*. San Diego, CA.

Li, P.Y., Givrad, T.K., Holschneider, D.P. et al. 2008b. A wirelessly-activated Parylene electrothermal valve for mapping brain function in freely moving subjects. *Hilton Head 2008*, 32–5. Hilton Head Island, South Carolina.

Li, P.Y., Givrad, T.K., Holschneider, D.P. et al. 2009a. A Parylene MEMS electrothermal valve. *Journal of Microelectromechanical Systems* 18: 1184–97.

Li, P.Y., Givrad, T.K., Sheybani, R. et al. 2010a. A low power, on demand electrothermal valve for wireless drug delivery applications. *Lab on a Chip* 10: 101–10.

Li, P.Y., Maarek, J.M., Holschneider, D.P. et al. 2008c. Parylene electrothermal MEMS drug delivery valve. *Spring Annual Meeting American Chemical Society*, 941–2. New Orleans, Louisiana.

Li, P.Y., Sheybani, R., Gutierrez, C. et al. 2010b. A Parylene bellows electrochemical actuator. *IEEE/ASME Journal of Microelectromechanical Systems* 19: 215–28.

Li, P.Y., Sheybani, R., Kuo, J.T.W. et al. 2009b. A Parylene Bellows ELECTROCHEMICAL actuator for Intraocular Drug Delivery. *Transducers 2009*, 1461–4. Denver, CO.

Li, P.Y., Shih, J., Lo, R. et al. 2007. An electrochemical intraocular drug delivery device. *MEMS 2007*, 15–18. Kobe, Japan.

Li, P.Y., Shih, J., Lo, R. et al. 2008d. An electrochemical intraocular drug delivery device. *Sensors and Actuators. A, Physical* 143: 41–8.

Li, Y., Ho Duc, H.L., Tyler, B. et al. 2005. In vivo delivery of BCNU from a MEMS device to a tumor model. *Journal of Controlled Release* 106: 138–45.

Li, Y., Shawgo, R.S., Tyler, B. et al. 2004. In vivo release from a drug delivery MEMS device. *Journal of Controlled Release* 100: 211–9.

Liepmann, D., Pisano, A.P., and Sage, B. 1999. Microelectromechanical systems technology to deliver insulin. *Diabetes Technology & Therapeutics* 1: 469–76.

Maillefer, D., van Lintel, H., Rey-Mermet, G. et al. 1999. A high-performance silicon micropump for an implantable drug delivery system. *MEMS '99*, 541–6. Orlando, Florida, USA.

Maloney, J.M., Uhland, S.A., Polito, B.F. et al. 2005. Electrothermally activated microchips for implantable drug delivery and biosensing. *Journal of Controlled Release* 109: 244–55.

Martin, F.J., and Grove, C. 2001. Microfabricated drug delivery systems: concepts to improve clinical benefit. *Biomedical Microdevices* 3: 97–108.

McAllister, D.V., Allen, M.G., and Prausnitz, M.R. 2000. Microfabricated microneedles for gene and drug delivery. *Annual Review of Biomedical Engineering* 2: 289–313.

Meng, E., Shih, J., Li, P.-Y. et al. 2006. Electrolysis-driven drug delivery for treatment of ocular disease. *Micro Total Analysis Systems 2006*, 633–5. Tokyo, Japan.

Metz, S., Bertsch, A., Bertrand, D. et al. 2004. Flexible polyimide probes with microelectrodes and embedded microfluidic channels for simultaneous drug delivery and multi-channel monitoring of bioelectric activity. *Biosensors & Bioelectronics* 19: 1309–18.

Neeves, K.B., Lo, C.T., Foley, C.P. et al. 2006. Fabrication and characterization of microfluidic probes for convection enhanced drug delivery. *Journal of Controlled Release* 111: 252–62.

Papageorgiou, D., Bledsoe, S.C., Gulari, M. et al. 2001. A shuttered probe with in-line flowmeters for chronic in-vivo drug delivery. 2001 IEEE 14th International Conference on Micro Electro Mechanical Systems 212–5. *Piscataway*, NJ: IEEE.

Polla, D.L., Erdman, A.G., Robbins, W.P. et al. 2000. Microdevices in medicine. *Annual Review of Biomedical Engineering* 2: 551–76.

Prager, J.P. 2002. Neuraxial medication delivery: the development and maturity of a concept for treating chronic pain of spinal origin. *Spine* 27: 2593–605; discussion 2606.

Prausnitz, M.R. 2004. Microneedles for transdermal drug delivery. *Advanced Drug Delivery Reviews* 56: 581–7.

Prausnitz, M.R., Mikszta, J.A., Cormier, M. et al. 2009. Microneedle-based vaccines. *Current Topics in Microbiology and Immunology* 333: 369–93.

Prescott, J.H., Lipka, S., Baldwin, S. et al. 2006. Chronic, programmed polypeptide delivery from an implanted, multireservoir microchip device. *Nature Biotechnology* 24: 437–8.

Rahimi, S., Sarraf, E.H., Wong, G.K. et al. 2011. Implantable drug delivery device using frequency controlled wireless hydrogel microvalves. *Biomedical Microdevices* 13: 267–77.

Rathnasingham, R., Kipke, D.R., Bledsoe, S.C. et al. 2004. Characterization of implantable microfabricated fluid delivery devices. *IEEE Transactions on Biomedical Engineering* 51: 138–45.

Retterer, S.T., Smith, K.L., Bjornsson, C.S. et al. 2004. Model neural prostheses with integrated microfluidics: a potential intervention strategy for controlling reactive cell and tissue responses. *IEEE Transactions on Biomedical Engineering* 51: 2063–73.

Retterer, S.T., Smith, K.L., Bjornsson, C.S. et al. 2008. Constant pressure fluid infusion into rat neocortex from implantable microfluidic devices. *Journal of Neural Engineering* 5: 385–91.

Reynaerts, D., Peirs, J., and VanBrussel, H. 1997. An implantable drug-delivery system based on shape memory alloy micro-actuation. *Sensors and Actuators. A, Physical* 61: 455–62.

Richards Grayson, A.C., Choi, I.S., Tyler, B.M. et al. 2003. Multi-pulse drug delivery from a resorbable polymeric microchip device. *Nature Materials* 2: 767–72.

Richards Grayson, A.C., Cima, M.J., and Langer, R. 2004a. Molecular release from a polymeric microreservoir device: Influence of chemistry, polymer swelling, and loading on device performance. *Journal of Biomedical Materials Research A* 69: 502–12.

Rohatgi, P., Langhals, N.B., Kipke, D.R. et al. 2009. In vivo performance of a microelectrode neural probe with integrated drug delivery. *Neurosurgical Focus* 27: 1–11.

Saati, S., Lo, R., Li, P.Y. et al. 2009. Mini drug pump for ophthalmic use. *Transactions of the American Ophthalmological Society* 107: 60–71.

Saati, S., Lo, R., Li, P.Y. et al. 2010. Mini drug pump for ophthalmic use. *Current Eye Research* 35: 192–201.

Santini, J.T., Cima, M.J., and Langer, R. 1999. A controlled-release microchip. *Nature* 397: 335–8.

Seidl, K., Spieth, S., Herwik, S. et al. 2010. In-plane silicon probes for simultaneous neural recording and drug delivery. *Journal of Micromechanics and Microengineering* 20: 11.

Sewell, W.F., Borenstein, J.T., Chen, Z. et al. 2009. Development of a microfluidics-based intracochlear drug delivery device. *Audiology & Neuro-otology* 14: 411–22.

Shawgo, R.S., Richards Grayson, A.C., Li, Y.W. et al. 2002. BioMEMS for drug delivery. *Current Opinion in Solid State & Materials Science* 6: 329–34.

Sivamani, R.K., Liepmann, D., and Maibach, H.I. 2007. Microneedles and transdermal applications. *Expert Opinion on Drug Delivery* 4: 19–25.

Spiegel, S., and Milstien, S. 2003. Sphingosine-1-phosphate: an enigmatic signalling lipid. *Nature Reviews. Molecular Cell Biology* 4: 397–407.

Spieth, S., Schumacher, A., Kallenbach, C. et al. 2010. Neuromedicator—a disposable drug delivery system with silicon microprobes for neural research. *23rd IEEE International Conference on Micro Electro Mechanical Systems (MEMS 2010)*, 983–6. Piscataway, NJ: IEEE.

Staples, M., Daniel, K., Cima, M.J. et al. 2006. Application of micro- and nano-electromechanical devices to drug delivery. *Pharmaceutical Research* 23: 847–63.

Su, G.G., and Pidaparti, R.M. 2010a. Drug particle delivery investigation through a valveless micropump. *Journal of Microelectromechanical Systems* 19: 1390–9.

Su, G.G., and Pidaparti, R.M. 2010b. Transport of drug particles in micropumps through novel actuation. *Microsystem Technologies—Micro- and Nanosystems Information Storage and Processing Systems* 16: 595–606.

Su, Y.C., and Lin, L.W. 2004. A water-powered micro drug delivery system. *Journal of Microelectromechanical Systems* 13: 75–82.

Su, Y.C., Lin, L.W., and Pisano, A.P. 2002. A water-powered osmotic microactuator. *Journal of Microelectromechanical Systems* 11: 736–42.

Takeuchi, S., Ziegler, D., Yoshida, Y. et al. 2005. Parylene flexible neural probes integrated with microfluidic channels. *Lab on a Chip* 5: 519–23.

Teymoori, M.M., and Abbaspour-Sani, E. 2005. Design and simulation of a novel electrostatic peristaltic micromachined pump for drug delivery applications. *Sensors and Actuators, A, Physical* 117: 222–9.

Tsai, N.C., and Sue, C.Y. 2007. Review of MEMS-based drug delivery and dosing systems. *Sensors and Actuators. A, Physical* 134: 555–64.

Vandervoort, J., and Ludwig, A. 2008. Microneedles for transdermal drug delivery: a minireview. *Frontiers in Bioscience* 13: 1711–5.

Wallace, M., and Yaksh, T.L. 2000. Long-term spinal analgesic delivery: a review of the preclinical and clinical literature. *Regional Anesthesia and Pain Medicine* 25: 117–57.

Yih, T.C., Wei, C., and Hammad, B. 2005. Modeling and characterization of a nanoliter drug-delivery MEMS micropump with circular bossed membrane. *Nanomedicine* 1: 164–75.

Ziaie, B., Baldi, A., Lei, M. et al. 2004. Hard and soft micromachining for BioMEMS: review of techniques and examples of applications in microfluidics and drug delivery. *Advanced Drug Delivery Reviews* 56: 145–72.

Ziegler, D., Suzuki, T., and Takeuchi, S. 2006. Fabrication of flexible neural probes with built-in microfluidic channels by thermal bonding of Parylene. *Journal of Microelectromechanical Systems* 15: 1477–82.

18 Nanostructured Surfaces for Intracerebral Neurotransmitter Recording

David A. Price, Artin Petrossians, Pooja M. Talauliker, Francois Pomerleau, Peter Huettl, Florian Mansfeld, James Weiland, J. Todd Hastings, and Greg A. Gerhardt

CONTENTS

INTRODUCTION

Neurotransmitters are the molecular intermediaries of synaptic transmission, a process that enables interneuronal communication in the central nervous system (CNS). Neurochemical changes underlie complex behaviors, are modulated by pharmacological agents, and contribute to or coincide with neuropathologies. As such, neurotransmitter systems represent a cornerstone in diagnosing and treating addictive disorders, epilepsy, mental health disorders, metabolic disturbances, and neurodegenerative diseases (Siegel 2006). Extracellular neurotransmitters are present in both synaptic and extrasynaptic pools (Kullmann 2000). While the synaptic pool contains neurotransmitters of neuronal origin that are necessary and sufficient to provide rapid neurotransmitter transients (i.e., phasic release), spillover from the synaptic cleft and nonneuronal release (i.e., tonic release) contribute to an extrasynaptic pool that modulates neural activity. Thus, phasic and tonic neurotransmitter release represents different facets of synaptic transmission and, thereby, influences neuronal signaling to different extents (Barbour 2001; Farrant and Nusser 2005). Neurochemical changes in the extracellular space are shaped in time and space by the intrinsic complexity and plasticity of neural processes. Diseases that disrupt these processes have an additional impact on spatial–temporal aspects of extracellular neurotransmitter dynamics. The rising prevalence of neuropathologies together with the need for early diagnosis and novel treatments on the basis of fundamental neurochemistry reiterates the continued need to develop and use technologies for rapid, reliable, and spatially resolved measures of CNS neurotransmitters in real time for basic science and preclinical and clinical applications.

Analytical approaches have been essential to identify and quantify neurotransmitter levels in the CNS. The abundance and varying concentrations of chemical substances in the CNS require that these approaches have high selectivity and sensitivity for the analyte of interest. Microdialysis has been the conventional method for measuring neurotransmitters in vivo (Watson et al. 2006) and has been utilized for investigational and neurochemical monitoring in the clinical setting (Goodman and Robertson 2009). Microdialysis samples the extracellular space through the exchange of compounds along a concentration gradient within the dialysis probe, which collects the analyte(s) of interest for subsequent analysis. The technique is advantageous because analyte(s) collected in the dialysate can be quantified with a high degree of selectivity and sensitivity through the tandem use of powerful analytical techniques such as high performance liquid chromatography. However, the poor spatial (i.e., 1–4 mm probe size) and temporal (i.e., 1–20 min sampling rate) resolutions of microdialysis limit sampling to the extrasynaptic pool (Del Arco et al. 2003; Drew et al. 2004; van der Zeyden et al. 2008). The large probe size further constrains the spatial–temporal resolution of sampling through secondary effects caused by edema, extensive tissue damage, and neuroinflammation (Clapp-Lilly et al. 1999).

The laboratory of R. N. Adams was the first to apply electrochemical methods coupled to miniature electrodes for intracerebral detection of neurotransmitters (Kissinger et al. 1973). At the most basic level, in vivo electrochemical methods with microelectrodes record when electroactive molecules come into direct contact with a polarized electrode-recording surface being directly oxidized or reduced (Justice 1987). Although microdialysis measures have superior selectivity, electrochemical techniques are advantageous due to their high spatial (i.e., micron-sized sampling areas) and temporal (i.e., sub-second) resolutions and can, therefore, be viewed as a complementary approach alongside microdialysis. Indeed, electrochemical recordings enable the monitoring of physiological events that are inaccessible by other analytical approaches (Borland and Michael 2007). To capitalize on the power and utility of in vivo electrochemical technologies (including recording devices and instrumentation), our laboratory (Stephens et al. 2008, 2010) and others (Bledsoe et al. 2009; Sandberg et al. 2011) initiated translational aspects by adapting this technology for use in preclinical (e.g., nonhuman primates) and clinical models.

Our laboratory has pioneered the development, design, and implementation of ceramic-based microelectrode arrays (MEAs) for in vivo monitoring of neuromolecules (Burmeister and Gerhardt 2001; Burmeister et al. 2000; Hascup et al. 2007; Stephens et al. 2008). An emerging area of investigation has been to further understand the enhanced performance of our MEA relative to other recording surfaces by characterizing the intrinsic physical properties of the MEA device (Talauliker et al., 2011). Here, we provide an account of our MEA technology for intracerebral monitoring of neurotransmitters and discuss the nanostructured surfaces of our platinum recording sites. In addition, we expand on a new arena that is aimed at enhancing the nanostructure of the recording surface by electrodepositing platinum–iridium alloy film on top of the native thin-film MEA platinum recording sites. The presence of nanofeatures on the MEA recording surface is important to the function of the device because it increases the electroactive surface area relative to the planar geometric surface area. Ultimately, these features enhance the performance of the MEA and continue to be an important consideration for improving MEA recording lifetime and performance postimplantation.

MICROELECTRODE ARRAYS FOR INTRACEREBRAL NEUROTRANSMITTER MEASUREMENTS

MEAs containing 4–16 thin-film platinum recording sites are fabricated on 125-μm-thick inert ceramic alumina (Al_2O_3) substrates (Coors Ceramic Company, Golden, CO) using photolithography (Hybrid Circuits, Inc., Sunnyvale, CA) as previously described (Burmeister and Gerhardt 2001, 2006; Burmeister et al. 2000; Hascup et al. 2007; Stephens et al. 2008). Briefly, microelectrode recording sites, connecting lines, and bonding pad images are patterned onto the ceramic substrate through a series of custom-designed masks. Thin-film recording sites, connecting lines, and bonding pads are defined by sputtering the ceramic substrate with a 500-Å adhesion layer of titanium and

2300-Å layer of elemental platinum in a vacuum-sealed sputter chamber. Importantly, all circuitry (except for the platinum recording sites) is insulated with a layer of polyimide (1.2 μm) that protects against aqueous environments to minimize cross-talk.

Ceramic is an ideal substrate for use as a neural device since it is biocompatible, durable, and rigid, with prior use in medical applications (Kohal et al. 2008; Navarro et al. 2008). The strength of ceramic permits ease of machining (e.g., sawing individual MEAs with a computer-controlled diamond saw) and is an important consideration as we begin to design and fabricate new MEA devices that are, overall, smaller (e.g., columnar design: width, 175 μm; depth, 125 μm; Figure 18.1c). In addition, the insulating property of ceramic is useful for electrochemical measurements because it helps to reduce cross-talk (Burmeister et al. 2000). Neurological stimulating and recording procedures need to be conducted through electrochemically stable electrodes that meet specific criteria: good biocompatibility, corrosion resistance, low impedance, and high charge injection capability (Margalit et al. 2002). Platinum has widespread use as an electrochemical recording surface, which can be attributed to its biocompatibility, corrosion resistance, electrocatalytic properties, and ability to transfer charge through reversible reactions without adverse consequences on living tissue (Merrill et al. 2005).

Representative MEA designs are shown in Figure 18.1. In addition to platinum recording sites with different geometric surface areas (e.g., 50 × 50 μm, 2500 μm^2; 333 × 15 μm, 4995 μm^2; 250 × 20 μm, 5000 μm^2), recording sites can be arranged in spatially defined configurations. Figure 18.1a illustrates the small size of the implanted portion of the MEA by comparing the distal end (including $n = 5$ platinum recording sites, each 250 × 20 μm) of an R8e MEA to a single strand of human hair. New, cutting-edge designs allow for double-sided recording capabilities through mirrored platinum recording sites on both the front and back surfaces of the MEA (Figure 18.1b). These features improve the spatial resolution of the device, especially for accomplishing multiple tasks simultaneously (see below). In addition, MEA technology continues to evolve as we design smaller devices such as the new ultra columnar C-series of MEAs (Figure 18.1c). A backscattered electron micrograph of the front and side surfaces of an S2 MEA is shown in Figure 18.2a. The ceramic substrate, insulating layer of polyimide, and platinum recording sites are clearly distinguishable based on their gray intensity levels. Higher-resolution secondary electron micrographs detail the textured

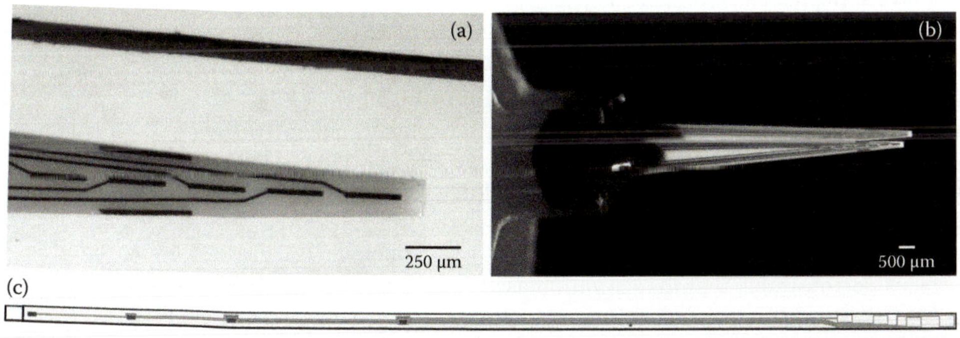

FIGURE 18.1 Ceramic-based MEA designs. (a) Portion of an R8e MEA is shown versus a single strand of human hair. (b) Double-sided R8e MEA containing $n = 16$ thin-film platinum recording sites (each 250 × 20 μm) arranged in a row (100 μm between sites) with $n = 8$ recording sites on the front and $n = 8$ sites on the back is shown. The lower image corresponds to the front of the MEA, and the upper is a mirror image reflecting the recording sites on the back of the MEA. This particular electrode permits electrochemical recording from multiple brain structures or depths within a single region. (c) Schematic of our new Ultra C-series MEA design is shown. The ceramic substrate of the Ultra C-series is a rectangular design that is 125 μm thick and 175 μm wide. Although a variety of electrode configurations are being fabricated, the featured design contains $n = 4$ thin-film platinum recording sites (50 × 100 μM) arranged in a row on the front and back (not shown) of the MEA. The spacing between electrodes is 1–2 mm to allow recording from multiple brain structures or depths.

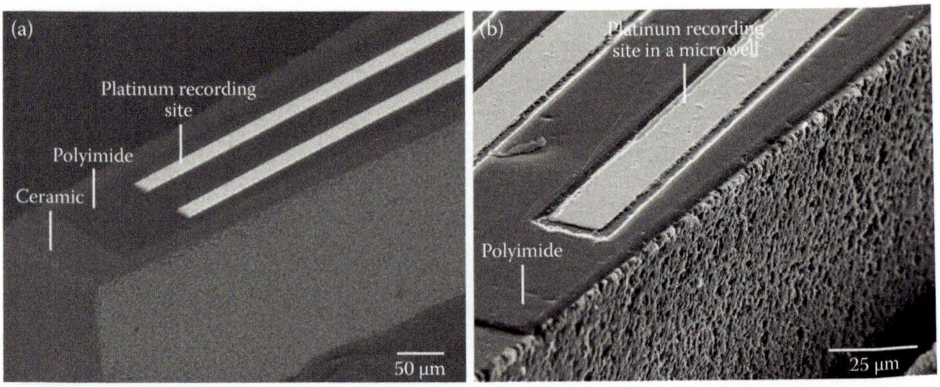

FIGURE 18.2 Electron micrographs highlight the basic aspects of MEAs. (a) Backscattered electron micrograph of the front and side surfaces at the distal end of an S2 MEA is shown. The polyimide coating, ceramic substrate, and two thin-film platinum recording sites are labeled. Note that the polyimide layer is homogenous and does not cover the platinum recording sites. (b) Higher-magnification secondary electron micrograph of the front and side surface of an S2 MEA is shown. A continuous layer of polyimide encloses the entire front surface, except for the platinum recording sites. The side surface of the MEA assumes a rough appearance following diamond sawing postfabrication. Platinum recording sites reside in a recessed microwell structure.

ceramic substrate on the side of the MEA that is a direct result of sawing out individual MEAs with a diamond saw following fabrication with photolithography (Figure 18.2b).

The availability of multiple, spatially defined recording sites makes the MEA a versatile neural device that can do the following: (1) concurrently measure multiple neuromolecules; (2) record from multiple brain regions or layered structures (e.g., hippocampus or prefrontal cortex); (3) monitor electrophysiological (e.g., single-unit or multiunit activity and field potentials) and chemical activity at the same recording site; and (4) employ self-referencing techniques to quantify resting levels of neurotransmitters. Most importantly, these measurements have sub-second resolution, and each task can be accomplished simultaneously in time and space. In addition to coating with exclusion layers (e.g., Nafion) to enhance electrode selectivity (Gerhardt et al. 1984), enzymes can be immobilized onto individual recording sites to convert a molecule that is not inherently electroactive (e.g., glutamate) into a reporter molecule (e.g., H_2O_2) that is oxidized upon contact with an electrode surface where an appropriate potential is applied (Burmeister and Gerhardt 2001; Burmeister et al. 2000, 2008; Hascup et al. 2007). The presence of the platinum recording sites in a microwell structure (Figure 18.2b) is important for using enzymes because microwells aid in protein immobilization (Jung et al. 2008).

A comparison of sensitivity per unit area for H_2O_2, glutamate, and glucose on different recording surfaces demonstrates that the ceramic-based MEAs have a highly active recording surface that exceeds the sensitivity previously reported by many groups (Talauliker et al. 2011). While the performance of the MEA is crucial to the acute recording capabilities of the device, improved performance is also essential for enhancing the recording lifetime of the MEA for both reusability and use in chronic recordings. In this regard, we have shown that MEA platinum recording sites maintain a high signal-to-noise ratio up to 17 days in vivo for glutamate (Rutherford et al. 2007) and 30 days in vitro for H_2O_2 (unpublished data). On the other hand, MEA-based single-unit neuronal activity measurements have been achieved up to 24 weeks postimplantation (Moxon et al. 2004) (unpublished data). These findings support the possibility that factors other than the physical properties of the MEA device contribute to the loss of the signal-to-noise ratio (e.g., extracellular environment, exclusion layers, and/or immobilized enzyme). Although ongoing efforts are addressing each of these concerns, it remains important to further advance our understanding of the physical properties of the MEA as we continue to develop a neural device that will be most suitable for both acute and long-term electrochemical recordings.

NANOSTRUCTURE OF MICROELECTRODE ARRAY RECORDING SITES

Smaller electrodes and, thereby, devices are highly desirable for in vivo recordings. Such improvements are important as they enhance the selectivity of neural stimulation and recording by increasing the spatial resolution of the implantable system. However, decreasing the geometric surface area of the electrode increases impedance and, subsequently, reduces implant efficiency through loss of sensitivity (Negi et al. 2010). Since (1) MEA sensitivity per unit area is better than that achieved using other platinum recording surfaces and (2) the photolithographic fabrication of thin-film platinum on ceramic-based MEAs is unique within the field, we are interested in further understanding the physical properties of the MEA device. As much as these efforts will help to understand the unique performance capabilities of the MEA, they are also instrumental for expanding on them.

To address this, we recently used complimentary investigative techniques to quantitatively characterize the surface topography of the platinum recording sites. MEA recording sites, which were composed of elemental platinum, were marked by nonuniform, nanoscale irregularities in the form of elevations and depressions (absolute feature height <5 nm) in the recording surface (Talauliker et al. 2011). Typical microscale and nanoscale topographies of the platinum recording surfaces are shown for 225 μm² (Figure 18.3a) and 1 μm² (Figure 18.3b) sampled areas, respectively. Using a gradient shows variations in surface height, with lighter and darker shading corresponding to regions with raised topography and indentations in the recording surface, respectively. Thus, native MEA platinum recording surfaces retain unique physical qualities from the photolithographic and thin-film processing. Because the nanostructure of the recording surface impacts the total surface area that is accessible to electroactive molecules, these nanoscale irregularities contribute to the enhanced performance of the MEA device and may also explain subtle differences between individual recording sites on the same MEA (Talauliker et al. 2011).

To improve the charge transfer properties at the electrode/tissue interface and, thereby, enhance MEA performance, our laboratory has recently modified the native MEA platinum recording surface. Electrodes with enhanced surface areas, much greater than their geometrical area, can be formed by electrodeposition. This has commonly been used in preparing platinum black electrodes, which can have surface areas hundreds of times larger than their geometric area. Because

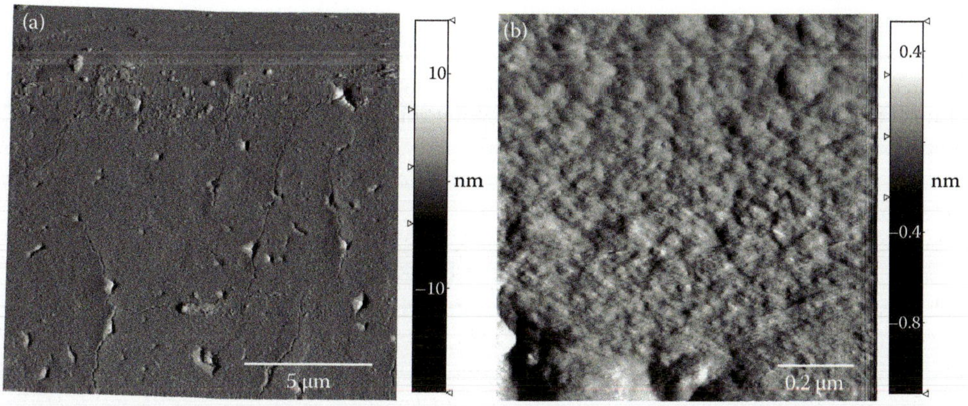

FIGURE 18.3 Atomic force microscopy reveals the typical microscale and nanoscale topographies of native MEA thin-film platinum recording sites. Topographical representations of the native platinum recording surface for (a) 225 μm² (deflection range = 34.77 nm) and (b) 1 μm² (deflection range = 1.65) sampled regions. Variations in surface height are shown in the form of a gradient, with lighter shading corresponding to regions with raised topography and darker coloring corresponding to indentations or cavities in the recording surface.

platinum black electrodes are mechanically weak and require using a lead plating bath, this particular surface modification has been of limited utility for implantable devices (Schuettler et al. 2005). Although lead-free plating solutions have recently been developed (Whalen et al. 2006), platinum electrodes with enhanced surface areas that contain structures with high aspect ratios may be subject to mechanical damage due to the softness of platinum. Platinum–iridium alloy electrodes have been used in biomedical and industrial applications. Importantly, iridium adds mechanical stiffness to the platinum (Petrossians et al. 2011). However, only a limited number of methods are available for producing platinum–iridium films, including methods that are inefficient (e.g., chemical vapor deposition and plasma sputtering) and electrodeposition methods that have not been studied comprehensively.

Electrodeposition of high-surface-area platinum–iridium films has many advantages over other methods for the surface modification of electrochemical recording surfaces. The method can be used on many different electrically conductive surfaces and is an alternative approach to more conventional techniques for thin-film deposition and metal plating. Platinum–iridium alloy electrodes also have much larger charge storage capacity and lower electrochemical impedance compared to pure platinum that is commonly used in biomedical applications (Piersma and Greatbatch 1987; Robblee and Rose 1990). We used a reproducible method for the deposition of platinum–iridium coatings with a composition of approximately 60% platinum and 40% iridium on gold or platinum base electrodes. This process produces recording sites with low impedance and improved mechanical properties that can transfer large amounts of electrical charge (Petrossians et al. 2011).

Platinum–iridium alloy films were formed on native MEA platinum recording sites using a potential sweep technique (scan rate = 0.2 mV/s for two to three potential sweeps) in a three-electrode system. The potential range of $E = +0.1$ to -0.1 V versus Ag/AgCl reference was selected for electrodeposition since this range includes potentials at which both platinum and iridium would be deposited while staying below the equilibrium potentials of the two metals. This approach resulted in adherent high-surface-area films on top of the native MEA platinum recording surfaces. The electrodeposition solution was constituted by adding K_2SO_4 (0.3 M) and agitating with an ultrasonic homogenizer (Misonix, Inc., Newtown, CT) at a frequency = 20 kHz and power = 8 W to maintain constant mass transfer while keeping the temperature of the plating bath between 56°C and 65°C during the electrodeposition process. Quantitative compositional analysis was performed for the electrodeposited thin films using energy dispersive spectroscopy (Oxford INCA Energy 300 X-Ray EDS System, Bucks, UK). In contrast to the native MEA platinum recording surface that was composed primarily of elemental platinum (Talauliker et al. 2011), the average chemical composition for all electrodeposited samples was 57.86 ± 0.70% platinum and 42.14 ± 0.70% iridium. Historically, platinum pacemaker electrodes are composed of 5%–40% iridium (Piersma and Greatbatch 1987). Such electrodes are fabricated through more sophisticated processes (e.g., high temperature casting, extrusion or sintering of particles) that create high surface areas with low impedance. Here, we utilized a novel electrochemical synthesis method to fabricate high-surface-area platinum–iridium alloy films where nearly 40% iridium composition was achieved. This process may be valuable for low-cost coating of biomedical electrodes.

A field emission scanning electron microscope (Zeiss LEO 1550VP, Carl Zeiss Inc., Germany) was used to evaluate the surface topography of electrodeposited platinum–iridium films. Micrographs of an S2 ceramic-based MEA are shown before (Figure 18.4a) and after (Figure 18.4b) electrodeposition with platinum–iridium films that were grown using potential sweeps as previously described. The native thin-film platinum recording sites on the S2 MEA are shown in Figure 18.4a. Electrodeposition of platinum–iridium created a surface coating that was granular, consisting of numerous nodules ranging in size from 50 to 500 nm (Figure 18.4b and c). The electrodeposits tended to be dendritic, with large nodules that contributed to the increased surface-area platinum–iridium alloy film. Electroplated platinum–iridium films initially formed a thin, dense layer (data not shown) but, with increasing electrodeposition time, developed the rough, nodular topography (Figure 18.4c). Comparison of the native MEA thin-film platinum recording site (Figure 18.4a) with

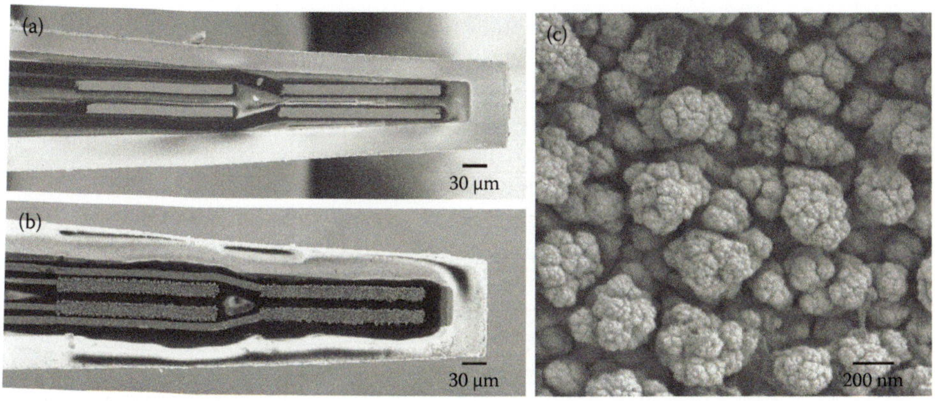

FIGURE 18.4 Electrodeposition of platinum–iridium alloy film on top of the native MEA thin-film platinum recording sites enhances the nanoscale topography of the recording surface. Scanning electron micrographs of a MEA device before and after electrodeposition with platinum–iridium alloy are shown. The electrode shown here is an S2 MEA consisting of an array of four platinum recording sites (333 × 15 μm) that are geometrically arranged as two side-by-side electrode pairs (100 μm between pairs, 30 μm between sites within a pair). (a) S2 MEA containing native thin-film platinum recording surfaces. (b) S2 MEA following electrodeposition of platinum–iridium on top of the native platinum recording surface. (c) Higher-magnification micrograph reveals the high-surface-area platinum–iridium films, which consist of numerous nanosized nodules.

that of the enhanced platinum–iridium film (Figure 18.4b) suggested that the electrodeposited films were rough and possessed a nanocrystalline structure with an average grain size <50 nm.

Cyclic voltammograms were collected for the enhanced platinum–iridium films to evaluate their electrochemical properties and provide a qualitative assessment regarding the increased surface area. Electrodeposited platinum–iridium films on MEAs were electrochemically cleaned in 0.05 M H_2SO_4 by cycling the potential over the range $E = -0.3$ to +1.2 V versus Ag/AgCl reference at a scan rate = 200 mV/s until the cyclic voltammograms reached steady state. The electrochemical properties of native MEA platinum recording sites (Figure 18.4a) and those with enhanced platinum–iridium films (Figure 18.4b and c) were characterized using cyclic voltammetry in 0.05 M H_2SO_4 over the potential range $E = -0.3$ to +1.2 V versus Ag/AgCl reference at a scan rate = 50 mV/s (Figure 18.5). The cyclic voltammogram of MEAs electrodeposited with platinum–iridium film exhibited

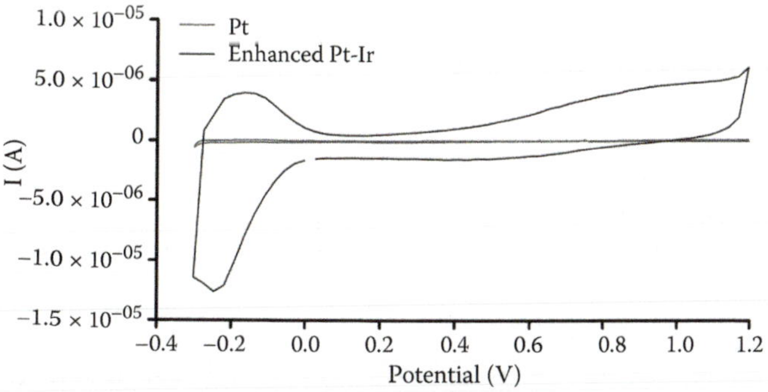

FIGURE 18.5 MEA recording surfaces with electrodeposited platinum–iridium films have a larger charge storage capacity than native MEA thin-film platinum recording surfaces. The performance of native MEA thin-film platinum recording sites is compared to that of an enhanced MEA with electrodeposited platinum–iridium alloy. The MEA recording sites were immersed in a solution of 0.05 M H_2SO_4, and cyclic voltammograms were obtained.

some characteristics of a "pure" platinum electrode in H_2SO_4 but lacked distinct platinum oxidation/reduction peaks and hydrogen adsorption/desorption peaks. The enhanced platinum–iridium MEAs showed a broad platinum oxidation shoulder in the anodic sweep without distinct peaks associated with specific oxide transition states (Farebrother et al. 1991; Pell et al. 2002). For neurostimulation applications, the area enclosed by the cyclic voltammogram curve corresponds to the charge capacity of the electrode. This measure is an indicator of the performance of the neurostimulating electrode material and is often used to compare performance with other well-studied electrodes (Robblee and Rose 1990). The cyclic voltammogram for the enhanced platinum–iridium MEA was significantly larger than that of the native MEA thin-film platinum (Figure 18.5), indicating a higher charge storage capacity for electrodeposited platinum–iridium films (Robblee and Rose 1990).

Electrochemical impedance spectroscopy (EIS) was utilized as a nondestructive technique to determine any differences in the electrochemical properties of native thin-film platinum versus enhanced platinum–iridium recording surfaces. EIS was performed in 0.05 M H_2SO_4 at the open-circuit potential with an applied ±50 mV amplitude alternating current (AC) signal over a frequency range of 100 kHz to 5 MHz. Experimental data were fitted to a one-time constant equivalent circuit, and the values of the solution resistance (R_s), polarization resistance (R_p), and capacitance (C) were determined using the ANALEIS software (Mansfeld 2006; Mansfeld et al. 1992, 1993). EIS data for the native platinum (geometric surface area for 333 × 15 μm recording sites = 4.995×10^{-5} cm^2) and electrodeposited platinum–iridium recording sites are shown as Bode plots (Figure 18.6a). The impedance for electrodeposited platinum–iridium surfaces was significantly reduced relative to the native platinum surfaces. The impedance measured at 100 Hz was drastically decreased from 2.7×10^6 to 2.3×10^3 Ω after electroplating the native platinum surface with the high-surface-area platinum–iridium film. Similar impedance results have been reported for electroplating platinum

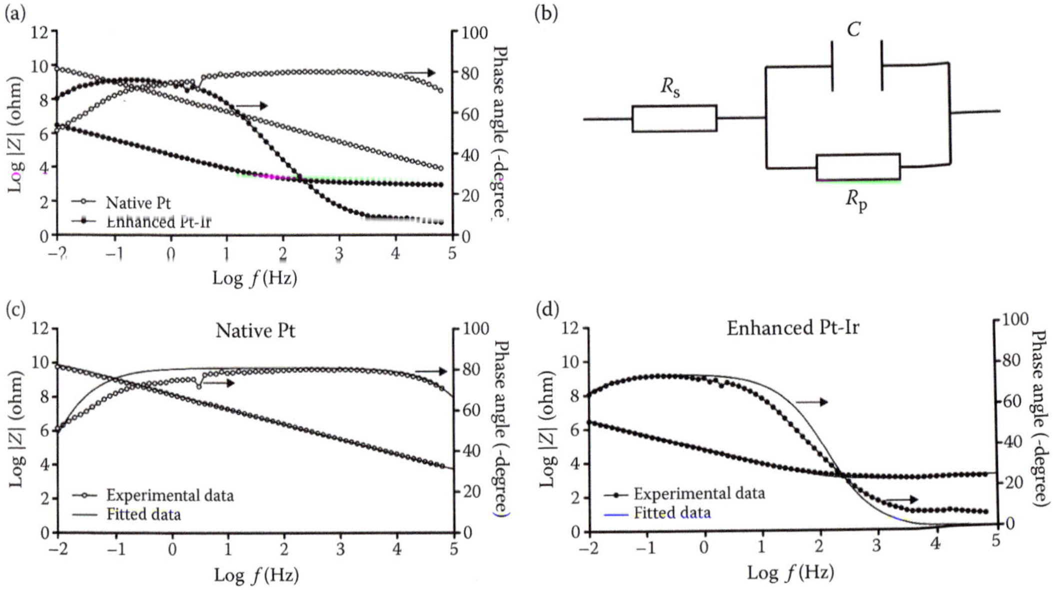

FIGURE 18.6 Electroactive surface area of electrodeposited platinum–iridium MEAs is greatly increased compared to native thin-film platinum MEA recording sites. (a) Bode plots for native platinum and electrodeposited platinum–iridium MEA recording surfaces. The impedance for each device was measured using electrochemical impedance spectroscopy in 0.05 M H_2SO_4. (b) Equivalent circuit used to analyze the impedance spectra data. Bode plots of experimental (circles) and fitted (solid line) data are shown for (c) native MEA thin-film platinum recording sites and (d) enhanced, electrodeposited platinum–iridium MEA recording sites. Arrow indicates curves that correspond to the right y-axis.

TABLE 18.1
Fit Parameters for Electrochemical Impedance Spectra

Sampled Surface	C (F)	R_p (Ω)	R_s (Ω)
Native platinum	3.31×10^{-5}	1.40×10^{10}	1.40×10^{3}
Enhanced platinum–iridium	1.37×10^{-1}	3.92×10^{7}	1.70×10^{3}

black (Franks et al. 2005) and poly(3,4-ethylenedioxythiophene) (PEDOT) (Richardson-Burns et al. 2007) on platinum electrode surfaces. Since the impedance modulus |Z| is inversely proportional to the electrode capacitance, C ($|Z| = 1/2\pi fC$, where f is the frequency of the applied AC signal) in the capacitive region, the observed decrease in |Z| is due to an increased C that is attributed to the increase in the real surface area.

The impedance spectra for the native platinum and electrodeposited platinum–iridium films are shown in Bode-plot format in Figure 18.6a. These spectra were analyzed using the equivalent circuit shown in Figure 18.6b. Figure 18.6c and d demonstrates good agreement between the experimental and fitted data for native platinum and enhanced platinum–iridium films, respectively. The fit parameters C, R_p, and R_s are given in Table 18.1. Comparison of the C values for native platinum and electrodeposited platinum–iridium demonstrates that the real surface area of the platinum–iridium films was greatly increased.

CONCLUSIONS

Synaptic neurotransmission represents the primary form of interneuronal communication in the mammalian CNS. Because neurotransmitter dynamics vary in time and space and most often occur on a sub-second time scale, electrochemical techniques coupled with microelectrodes have started to move toward the forefront of intracerebral neurotransmitter monitoring. The sub-second sampling rates offered by electrochemical approaches have provided insight into neurochemical events that are missed by microdialysis sampling and, thus, have been a complementary approach (Borland and Michael 2007). The versatility found in the MEA device exemplifies this. The abilities to simultaneously measure multiple analytes, record from different structures, and quantify resting levels of neurotransmitters are a few important hallmarks of MEA technology. The MEA implant is biocompatible, minimally invasive, and functionally superior to other platinum-based electrodes (Hascup et al. 2009; Talauliker et al. 2011). Many of these features are a direct result of our unique fabrication process, which yields spatially defined, reproducible recordings sites with nanoscale roughness that has not been observed with other designs (Talauliker et al. 2011). Ultimately, this translates to an increased electroactive area and is an important consideration as we design future generations of MEAs that are further miniaturized. The ability to enhance the nanofeatures of the platinum recording surface through the electrodeposition of platinum–iridium provides further evidence that the nanoscale topography of the recording sites can greatly impact their performance. Future studies aimed at assessing the in vivo recording performance, lifetime, and stability of the enhanced platinum–iridium recording surfaces will be essential for exploiting these surface modifications. These studies represent an ongoing effort by our laboratory to advance our understanding of the intrinsic physical properties and chemistry of MEAs to improve the in vivo electrochemical performance of the device.

ACKNOWLEDGMENTS

We thank Larry Rice and George Turpin (University of Kentucky Electron Microscopy Center, Lexington, KY) for their assistance with data processing. This work was supported by grants from United States Public Health Service (NS39787, DA017186, AG13494, T32 AG000242) and National Science Foundation (EEC-0310723).

DISCLOSURE STATEMENT

Greg A. Gerhardt is the owner of Quanteon, LLC (Nicholasville, KY), which makes the FAST-16 recording system and MEAs.

REFERENCES

Barbour, B. 2001. An evaluation of synapse independence. *J Neurosci* 21: 7969–84.

Bledsoe, J.M., Kimble, C.J., Covey, D.P. et al. 2009. Development of the wireless instantaneous neurotransmitter concentration system for intraoperative neurochemical monitoring using fast-scan cyclic voltammetry. *J Neurosurg* 111: 712–23.

Borland, L.M., and Michael, A.C. 2007. An introduction to electrochemical methods in neuroscience. In *Electrochemical Methods for Neuroscience*, ed. A.C. Michael and L.M. Borland, 1–15. Boca Raton, FL: CRC Press.

Burmeister, J.J., and Gerhardt, G.A. 2001. Self-referencing ceramic-based multisite microelectrodes for the detection and elimination of interferences from the measurement of L-glutamate and other analytes. *Anal Chem* 73: 1037–42.

Burmeister, J.J., and Gerhardt, G.A. 2006. Neurochemical arrays. In *Encyclopedia of Sensors*, ed. C. Grimes, E. Dickey, and M.V. Pishko, 525–37. Stevenson Ranch, CA: American Scientific Publishers.

Burmeister, J.J., Moxon, K., and Gerhardt, G.A. 2000. Ceramic-based multisite microelectrodes for electrochemical recordings. *Anal Chem* 72: 187–92.

Burmeister, J.J., Pomerleau, F., Huettl, P. et al. 2008. Ceramic-based multisite microelectrode arrays for simultaneous measures of choline and acetylcholine in CNS. *Biosens Bioelectron* 23: 1382–9.

Clapp-Lilly, K.L., Roberts, R.C., Duffy, L.K. et al. 1999. An ultrastructural analysis of tissue surrounding a microdialysis probe. *J Neurosci Methods* 90: 129–42.

Del Arco, A., Segovia, G., Fuxe, K. et al. 2003. Changes in dialysate concentrations of glutamate and GABA in the brain: an index of volume transmission mediated actions? *J Neurochem* 85: 23–33.

Drew, K.L., Pehek, E.A., Rasley, B.T. et al. 2004. Sampling glutamate and GABA with microdialysis: suggestions on how to get the dialysis membrane closer to the synapse. *J Neurosci Methods* 140: 127–31.

Farebrother, M., Goledzinowski, M., Thomas, G. et al. 1991. Early stages of growth of hydrous platinum oxide films. *J Electroanal Chem (Lausanne Switz)* 297: 469–88.

Farrant, M., and Nusser, Z. 2005. Variations on an inhibitory theme: phasic and tonic activation of GABA(A) receptors. *Nat Rev Neurosci* 6: 215–29.

Franks, W., Schenker, I., Schmutz, P. et al. 2005. Impedance characterization and modeling of electrodes for biomedical applications. *IEEE Trans Biomed Eng* 52. 1295–302.

Gerhardt, G.A., Oke, A.F., Nagy, G. et al. 1984. Nafion-coated electrodes with high selectivity for CNS electrochemistry. *Brain Res* 290: 390–5.

Goodman, J.C., and Robertson, C.S. 2009. Microdialysis. Is it ready for prime time? *Curr Opin Crit Care* 15: 110–7.

Hascup, E.R., af Bjerken, S., Hascup, K.N. et al. 2009. Histological studies of the effects of chronic implantation of ceramic-based microelectrode arrays and microdialysis probes in rat prefrontal cortex. *Brain Res* 1291: 12–20.

Hascup, K.N., Rutherford, E.C., Quintero, J.E. et al. 2007. Second-by-second measures of L-glutamate and other neurotransmitters using enzyme-based microelectrode arrays. In *Electrochemical Methods for Neuroscience*, ed. A.C. Michael and L.M. Borland, 407–50. Boca Raton, FL: CRC Press.

Jung, E., Manessis, D., Neumann, A. et al. 2008. Lamination and laser structuring for a microwell array. *Microsyst Technol* 14: 931–6.

Justice, J.B. 1987. Introduction to in vivo voltammetry. In *Voltammetry in the Neurosciences: Principles, Methods, and Applications*, ed. J.B. Justice, 3–101. Clifton, NJ: The Humana Press, Inc.

Kissinger, P.T., Hart, J.B., and Adams, R.N. 1973. Voltammetry in brain tissue—a new neurophysiological measurement. *Brain Res* 55: 209–13.

Kohal, R.J., Att, W., Bachle, M. et al. 2008. Ceramic abutments and ceramic oral implants. An update. *Periodontol 2000* 47: 224–43.

Kullmann, D.M. 2000. Spillover and synaptic cross talk mediated by glutamate and GABA in the mammalian brain. *Prog Brain Res* 125: 339–51.

Mansfeld, F. 2006. Electrochemical impedance spectroscopy. In *Analytical Methods in Corrosion Science and Engineering*, ed. P. Marcus, and F. Mansfeld, 463–505. Boca Raton, FL: CRC Press.

Mansfeld, F., Shih, H., Greene, H. and Tsai, C.H. 1993. Analysis of EIS data for common corrosion processes. In *Electrochemical Impedance: Analysis and Interpretation*, ed. J.R. Scully, M.W. Kendig, and D.C. Silverman, 37–53. Philadelphia, PA: American Standard for Testing and Materials.

Mansfeld, F., Tsai, C.H. and Shih, H. 1992. Software for simulation and analysis of electrochemical impedance spectroscopy (EIS) data. In *Computer Modeling in Corrosion*, ed. R.S. Munn, 186–96. Philadelphia, PA: American Standard for Testing and Materials.

Margalit, E., Maia, M., Weiland, J.D. et al. 2002. Retinal prosthesis for the blind. *Surv Ophthalmol* 47: 335–56.

Merrill, D.R., Bikson, M., and Jefferys, J.G. 2005. Electrical stimulation of excitable tissue: design of efficacious and safe protocols. *J Neurosci Methods* 141: 171–98.

Moxon, K.A., Leiser, S.C., Gerhardt, G.A. et al. 2004. Ceramic-based multisite electrode arrays for chronic single-neuron recording. *IEEE Trans Biomed Eng* 51: 647–56.

Navarro, M., Michiardi, A., Castano, O. et al. 2008. Biomaterials in orthopaedics. *J R Soc Interface* 5: 1137–58.

Negi, S., Bhandari, R., Rieth, L. et al. 2010. In vitro comparison of sputtered iridium oxide and platinum-coated neural implantable microelectrode arrays. *Biomed Mater* 5: 15007.

Pell, W.G., Zolfaghari, A., and Conway, B.E. 2002. Capacitance of the double-layer at polycrystalline Pt electrodes bearing a surface-oxide film. *J Electroanal Chem* 532: 13–23.

Petrossians, A., Whalen, J.J., Weiland, J.D. et al. 2011. Electrodeposition and characterization of thin-film platinum–iridium alloys for biological interfaces *J Electrochem Soc* 158: D269.

Piersma, B.J., and Greatbatch, W. 1987. Coupling reactions at the metal–tissue interface in electrical stimulation with cardiac pacemaker electrodes. *J Electrochem Soc* 134: 2458–64.

Richardson-Burns, S.M., Hendricks, J.L., Foster, B. et al. 2007. Polymerization of the conducting polymer poly(3,4-ethylenedioxythiophene) (PEDOT) around living neural cells. *Biomaterials* 28: 1539–52.

Robblee, L.S., and Rose, T.L. 1990. Electrochemical guidelines for selection of protocols and electrode materials for neural stimulation. In *Neural Prostheses*, ed. W.F. Agnew and D.B. McCreery, 25–66. Englewood Cliffs, NJ: Prentice Hall.

Rutherford, E.C., Pomerleau, F., Huettl, P. et al. 2007. Chronic second-by-second measures of L-glutamate in the central nervous system of freely moving rats. *J Neurochem* 102: 712–22.

Sandberg, S., Kishida, K., Comair, Y. et al. 2011. In vivo voltammetric measurement of dopamine release in human striatum during a behavioral task. *44th Annual Winter Conference on Brain Research*, P60. Keystone, CO.

Schuettler, M., Doerge, T., Wien, S.L. et al. 2005. Cytotoxicity of platinum black. *10th Annual Conference of the International Functional Electrical Stimulation Society*, 343–5. Montreal, Canada.

Siegel, G.J. 2006. *Basic Neurochemistry: Molecular, Cellular, and Medical Aspects*. Burlington, MA: Elsevier Academic Press.

Stephens, M.L., Pomerleau, F., Huettl, P. et al. 2010. Real-time glutamate measurements in the putamen of awake rhesus monkeys using an enzyme-based human microelectrode array prototype. *J Neurosci Methods* 185: 264–72.

Stephens, M.L., Spencer, D.D., Cavus, I. et al. 2008. Microelectrode-based epilepsy therapy: a hybrid neural prosthesis incorporating seizure prediction and intervention with biomimetic maintenance of normal hippocampal function. In *Computational Neuroscience in Epilepsy*, ed. I. Soltesz and K. Staley, 540–63. Amsterdam, Netherlands: Elsevier Academic Press.

Talauliker, P.M., Price, D.A., Burmeister, J.J. et al. 2011. Ceramic-based microelectrode arrays: recording surface characteristics and topographical analysis. *J Neurosci Methods* 198: 222–9.

van der Zeyden, M., Oldenziel, W.H., Rea, K. et al. 2008. Microdialysis of GABA and glutamate: analysis, interpretation and comparison with microsensors. *Pharmacol Biochem Behav* 90: 135–47.

Watson, C.J., Venton, B.J., and Kennedy, R.T. 2006. In vivo measurements of neurotransmitters by microdialysis sampling. *Anal Chem* 78: 1391–9.

Whalen, J.J.I., Young, J., Weiland, J.D. et al. 2006. Electrochemical characterization of charge injection at electrodeposited platinum electrodes in phosphate buffered saline. *J Electrochem Soc* 153: C834–9.

19 Therapy for Neuronopathic Lysosomal Storage Diseases

Ayaz M. Khawaja, John D. Heiss, and Roscoe O. Brady

CONTENTS

Lysosomal storage diseases (LSDs) have phenotypic manifestations in multiple organ systems, most notably the central nervous system (CNS). They are classically characterized as "storage disorders" resulting from enzymatic abnormalities, either actual or functional, which lead to accumulation of substrates. As many as 50–60 acid hydrolases are centered within the lysosomes (Journet et al. 2002)—organelles that degrade various metabolites in the extracellular and intracellular environments and participate in endocytosis and autophagy (Harada 2011; Rautou et al. 2010). Theoretically, any of the enzymes may become dysfunctional, leading to phenotypic expression of an LSD. This was originally purported to be due to accumulation of a single substrate that is normally part of the degradation chain managed by the respective enzyme. This was believed to result in various ultrastructural modifications, such as enlargement of the lysosome and cell due to mass effect, deranged cellular functions, and consequent apoptosis (Desnick et al. 1976; Hers 1965). However, lysosomes interact with various other degradation mechanisms including endosomal, autophagosomal, and ubiquitin–proteasomal systems, and are best recognized as being part of the "greater lysosomal system" due to their complex effects on genetic expression and signal transduction (Walkley 2007). Indeed, transcriptional factor DNA section that binds this complementary transcription factor to initiate gene transcription, also known as E-box (EB) can coordinate transcription of genes responsible for lysosomal biogenesis and function and to increase its expression under conditions of lysosomal stress or overload (Sardiello et al. 2009). Enzyme deficiency within the neuron results in abnormal storage of substrate that affects many signaling pathways, disrupts numerous cellular functions, increases synthetic activity, and enhances autophagy, leading to the development of diverse and complex clinical manifestations (Walkley 2009).

LSDs can be classified broadly as sphingolipidosis (SL), mucopolysaccharidosis (MPS), mucolipidosis (ML), multiple enzyme deficiencies (MED), oligosaccharidoses (OS), and neuronal ceroid lipofuscinosis (NCF) (Jardim et al. 2010). Briefly, SL results in accumulation of lipids derived from the aliphatic amino-alcohol sphingosine. Gaucher, Niemann–Pick (NPD), Krabbe, and Farber diseases and metachromatic leukodystrophy (MLD) are characterized within this category. Hurler, Hunter, and Sanfilippo syndromes are typical examples of MPS, which result from the accumulation of glycosaminoglycans (GAGs). Sialidosis type II, inclusion cell (I-cell) disease, and pseudo-Hurler polydystrophy are examples of ML, which manifests symptoms of SL and MPS (Jardim et al. 2010). MED is due to either multiple sulfatase deficiency (Austin disease) or galactosialidosis (deficient ß-galactosidase and neuraminidase). OS is due to deficient removal of glycoprotein carbohydrate chains. Examples include alpha- and beta-mannosidosis. NCF results in lysosomal accumulation of autofluorescent lipofuscin and ceroid.

Another classification system for LSDs is concerned with the actual disease mechanism. Deficient enzyme function is required in all cases; for example, NPD type C is due to a defective cholesterol transporter (Bellettato and Scarpa 2010). Theoretically, a defect anywhere along a biochemical synthetic pathway can cause LSD, including synthesis in the endoplasmic reticulum, modification in the Golgi apparatus, subsequent targeting to lysosomes, or dysfunctional enzyme activators. Some lysosomal enzymes may not become bound to mannose-6-phosphate (M6P) receptors in the trans-Golgi network (TGN) and are instead released into circulation, where they are scavenged by cells and transferred to lysosomes through M6P-modulated endocytosis. This mechanism has become the basis for stem cell transplantation therapy and enzyme replacement therapy (ERT) (Beck 2010).

Two-thirds of all LSDs involve the CNS and display typical clinical features (Meikle et al. 1999). Neurons are especially affected due to their limited regenerative capacity. Neurons develop well-defined structural alterations due to the LSD pathology. For example, in Tay–Sachs disease, GM2 ganglioside accumulation results in swelling of the axon hillock, of the compartment between the axon hillock and the first axonal segment enlargement, known as the meganeurite, and of storage bodies (Bellettato and Scarpa 2010). Discrete enlargements of axonal segments called axonal spheroids interfere with action potential conduction and bidirectional axonal transport (Bellettato and Scarpa 2010). Ectopic dendrites may also arise from the axon hillock due to gangliosides' permissive effect on neurite growth. Eventually a combination of many factors results in manifestation of cellular dysfunction and disease progression, including the following: (1) impaired autophagy, (2) mitochondrial dysfunction, (3) altered calcium homeostasis, (4) altered Golgi and endoplasmic reticulum function, (5) neuroinflammation, (6) altered axon transport mechanisms, (7) oxidative stress, (8) altered iron storage, and (9) oxidative stress (Ballabio and Gieselmann 2009; Bellettato and Scarpa 2010).

CURRENT STANDARD OF CARE AND RESEARCH FOR LSDs

As mentioned, the secretion of lysosomal enzymes into the systemic circulation by cells and eventual capture by the same and other cells form the basis for treatment of LSDs using ERT. However, ERT has proven ineffective for alleviating neuropathic features of LSDs primarily because of inadequate transport of enzyme across the blood–brain barrier (BBB). Various strategies have been formulated to address CNS involvement in LSDs and are discussed below, with particular reference to Beck's excellent review of therapy for LSDs (Beck 2010) (Table 19.1).

Enzyme Replacement Therapy

1. Gaucher disease

   For Gaucher disease, ERT was first proposed in 1966 and was evaluated by Barton and Brady et al. in a clinical trial, where 12 patients with the non-neuronopathic form were

TABLE 19.1
Lysosomal Storage Diseases and Their Treatment

LSD	Enzyme Deficient	Accumulated Substance	Clinical Features	Therapy
Tay–Sachs disease	Hexosaminidase A	GM2 gangliosides	Dementia, symptoms of brain stem dysfunction, macular cherry-red spot, motor dysfunction.	No approved treatment. Options are ERT, gene therapy, and substrate reduction therapy.
Krabbe disease	Galactocerebroside beta-galactosidase	Galactosyl ceramide, psychosine	Early onset causes nonspecific symptoms such as seizures, failure to thrive, motor dysfunction. Late onset causes visual and gait difficulties.	No approved treatment. Bone marrow transplantation is a viable option.
Gaucher disease	Glucocerebrosidase	Glucosyl ceramide	Hepatosplenomegaly, pancytopenia, bone pain and osteoporosis, motor dysfunction, seizures, and dementia.	ERT is FDA approved (alglucerase, imiglucerase, velaglucerase alfa, teleglucerase alfa). Substrate reductive therapy (miglustat) is FDA approved.
Niemann–Pick disease	Sphingomyelinase	Sphingomyelin	Hepatosplenomegaly, gait, gaze and motor dysfunction, speech difficulty, seizures. Cherry-red macular spot and nonspecific symptoms in infantile form.	No approved therapy in the United States. Miglustat approved in Europe for type C.
Pompe disease	Acid alpha-glucosidase	Glycogen	Early onset results in death from CVS and respiratory complications. Late onset results in muscle weakness, followed by respiratory compromise and infection.	Myozyme is approved for early onset and Lumizyme for late-onset.
Fabry disease	Alpha-galactosidase A	Glycolipid (globotriaosylceramide)	Cardiomyopathy, renal insufficiency, acroparesthesia, angiokeratomas, ocular and neurological symptoms.	Agalsidase beta FDA approved.
Neuronal ceroid lipofuscinosis	Various enzymes in different subtypes	Lipofuscin	All subtypes exhibit evidence of neurodegeneration with nonspecific symptoms.	No approved treatment.

(continued)

TABLE 19.1 (Continued)
Lysosomal Storage Diseases and Their Treatment

LSD	Enzyme Deficient	Accumulated Substance	Clinical Features	Therapy
Metachromatic leukodystrophy	Arylsulfatase A	Various sulfatides	Early onset with nonspecific symptoms of developmental delay, motor dysfunction, seizures, paralysis, dysphagia. Late onset with dementia.	No approved treatment.
Mucolipidosis	*N*-acetylglucosamine 1 phosphotransferase (types 2 and 3). Sialidase (type 1).	Glycoproteins in types 1–3	Various subtypes with features similar to mucopolysaccharidosis and sphingolipidosis.	No approved treatment.

infused intravenously with placenta-derived ß-glucocerebrosidase (alglucerase) (Barton et al. 1991). The enzyme had exposed mannose residues to target the mannose receptors on macrophages. Hemoglobin concentration increased in all patients, and platelet count increased in seven patients. Acid phosphatase activity decreased in 10 patients. Splenic volume decreased in all patients and hepatic volume in five. Alglucerase (Ceredase, Genzyme Therapeutics, Cambridge, MA) was subsequently approved for Gaucher disease. Later, imiglucerase (Cerezyme, Genzyme Therapeutics) was tested for efficacy versus alglucerase in a group of 30 patients with 15 assigned to each drug group; imiglucerase was found to generate fewer antibodies (Grabowski et al. 1995). Imiglucerase was also approved to treat Gaucher disease and had a favorable safety profile but was not useful against the neurological deficits in Gaucher types 2 and 3 (Germain 2004). In the type 2 variety, administration of ERT in early infancy in two patients did not alter disease course (Prows et al. 1997). Two new forms of the enzyme were developed, one derived from plant cells that is currently in phase III clinical trials (Aviezer et al. 2009) and another that is gene derived and has twofold better internalization into macrophages than imiglucerase (Brumshtein et al. 2010).

2. Fabry disease

Two types of alpha-galactosidase (agalsidase) are available, agalsidase alfa produced from the Chinese hamster ovary cells and agalsidase beta from human cell lines. Agalsidase alfa (Replagal, Shire, Lexington, MA) was evaluated in a clinical trial with 15 heterozygous female patients receiving the enzyme weekly for 55 weeks (Baehner et al. 2003). None of the patients developed antibodies or experienced infusion reactions. Decrease in left ventricular size and QRS duration, increase in quality of life, stable renal function, and a favorable safety profile were noted. Agalsidase beta (Fabrazyme, Genzyme) was evaluated in a clinical trial in which 82 patients with adult Fabry disease with mild-to-moderate kidney dysfunction were randomly assigned to receive either agalsidase beta infusion (51) or placebo (31) bimonthly for 35 months (Banikazemi et al. 2007). A greater proportion of patients in the placebo group, compared to the enzyme group, experienced a clinical episode, defined as a renal/cardiac/CVS (cerebrovascular system) event, or death. Only infusion-related events were greater in the enzyme group compared to the placebo group, with greater clinical benefit for ERT noted if the glomerular filtration rate was greater than 55 mL/min. A literature review by Beck showed agalsidase to be very efficient in improving and stabilizing organ function and anatomy in Fabry patients (Beck 2009). Fabrazyme is the only version of agalsidase approved by the Food and Drug Administration (FDA).

3. Pompe disease

   Alglucosidase alfa was evaluated in a clinical trial of 21 infants with advanced Pompe disease (negligible enzyme activity and abnormal left ventricular mass index), with bimonthly infusions for up to 168 weeks (Nicolino et al. 2009). Fifteen patients survived to the end of the study. The enzyme preparation significantly reduced the risk of death (by 79%) and of invasive ventilation (by 58%), improved motor and functional parameters, and decreased the left ventricular (LV) mass index. Another trial from Germany noted improvement in various neuromuscular parameters in 38 patients, albeit with a variable course, and stabilization of the rampant progression of the disease (Regnery et al. 2012). Myozyme (Genzyme) is the form of alglucosidase alfa that is FDA approved for use in infants and children.

4. MPS

   Clinical trials of ERT in several subtypes of MPS resulted in some improvement in non-CNS disease. Laronidase (alpha I iduronidase), idursulfase (iduronate 2 sulfatase), and arylsulfatase B were shown to be beneficial for MPS type 1 (Clarke et al. 2009), MPS type 2 (Wraith et al. 2008), and MPS type 4 (Harmatz et al. 2008). Improved lung and joint function and organomegaly were noted. The CNS manifestations of these diseases did not improve, however.

Research into Effective Use of ERT

Replacement enzymes must enter cells to have their desired effect. The use of ERT requires the "capture" of enzymes located in the extracellular milieu by the reticuloendothelial system (RES) through mannose receptors (Figure 19.1) or by other cells through M6P receptors (Figure 19.2). These enzymes are then internalized through a clathrin-mediated endocytic mechanism and subsequently transferred to the lysosomes through endosome–lysosome fusion (Neufeld 1980). Enzymes developed for clinical use are modified to enhance cellular uptake. For example, exposing mannose links on enzymes after production increases cellular binding (Furbish et al. 1981). Specific targeting can be achieved due to exclusive mannose receptor expression in RES cells (Achord et al. 1978), which can be induced by dexamethasone to improve delivery (Zhu et al. 2004). More generalized targeting is accomplished due to the presence of Mannose-6-phosphate receptors (M6P-R) on cell surfaces (Sands et al. 2001). Other targeting strategies concentrate on either binding the enzyme to a different domain of the M6P-R, a different receptor complex altogether, or non-receptor-mediated uptake. Conjugating the enzyme with insulin-like growth factor II (IGF-II) enables binding to a different domain of M6P-R and internalization (LeBowitz et al. 2004). Tat (trans-activator of transcription) protein modification of an enzyme promoted internalization of the enzyme through electrostatic interactions with cell membrane components (Brooks et al. 2005) and transfer to lysosomes through M6P-independent and -dependent pathways (Lee et al. 2005). A similar strategy was employed to investigate transport across the BBB through transcytosis. Three receptors were studied in animal models: M6P-R, insulin, and transferrin. M6P-R-mediated transport was efficient only in newborn mice (Urayama et al. 2008). Enzymes conjugated to anti-insulin receptor peptide and transferrin suppressed GAG accumulation in adult animal models, indicating successful transport of the enzymes across the BBB (Boado et al. 2008; Osborn et al. 2008).

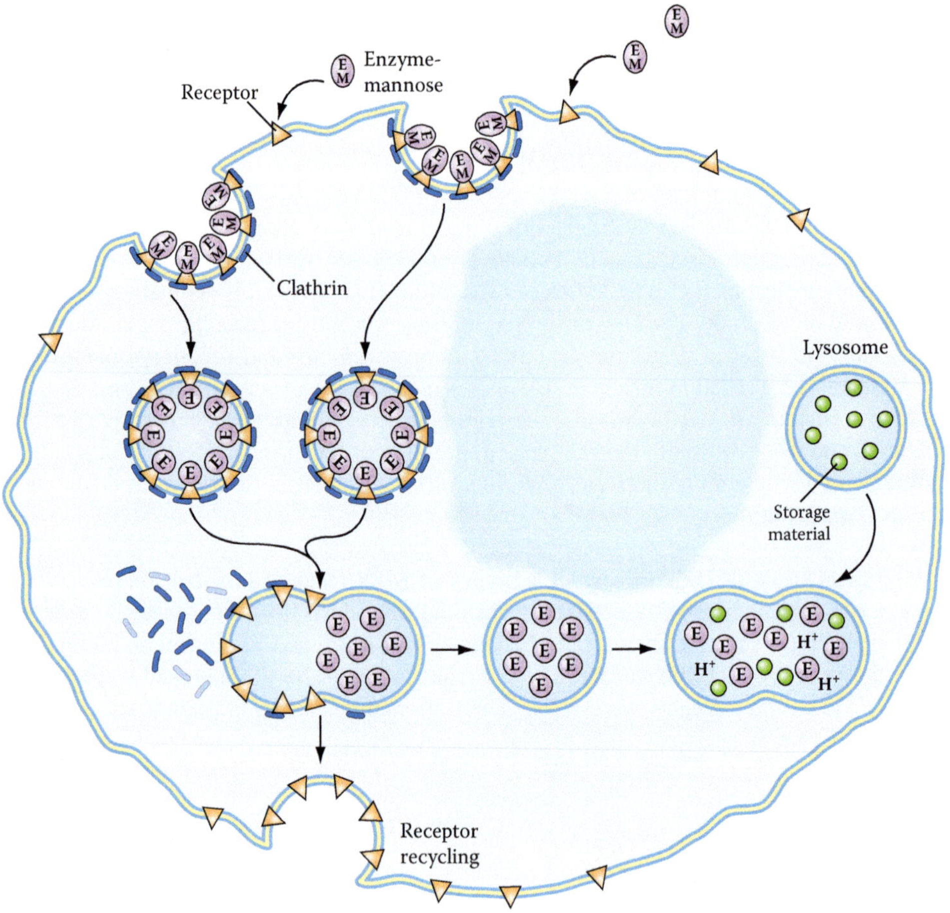

FIGURE 19.1 Posttranslational modification of the enzyme adds mannose groups. Macrophage cell surface harbors the mannose receptor and directs these enzymes toward macrophages. The enzyme–mannose receptor complex is internalized via clathrin mediated endocytosis. In a subsequent series of steps outlined here, the enzymes are directed toward the diseased lysosomes containing the storage material, where they exert their therapeutic action.

Stem Cell Transplantation

Stem cell transplantation uses stem cells from a normal donor to replace stem cells in the host bone marrow (BM) that cannot produce the deficient enzyme. For Hurler syndrome, an outcomes study was performed in 27 patients who received either human leukocyte antigen (HLA) matched or HLA-mismatched BM transplants. Although 4 subjects had graft failures and 4 died, 21 had very low graft-versus-host disease (GVHD) incidence and improved cognitive function (Souillet et al. 2003). A significant improvement in organ size, cognitive function, airway obstruction, and other characteristics such as coarse facial features was noted (Guffon et al. 1998). BM transplantation demonstrated efficacy for other LSD syndromes as well, including MLD, mannosidosis, globoid cell leukodystrophy, Krabbe disease, and Gaucher disease (Krivit et al. 1999). Infection, graft rejection, and GVHD are the major obstacles to the successful BM transplantation, and strategies to avoid these factors as well as increase the pool of donors would support BM transplantation as a strong treatment option. To be effective, BM transplant must be done before there is evidence of CNS involvement (Cartier

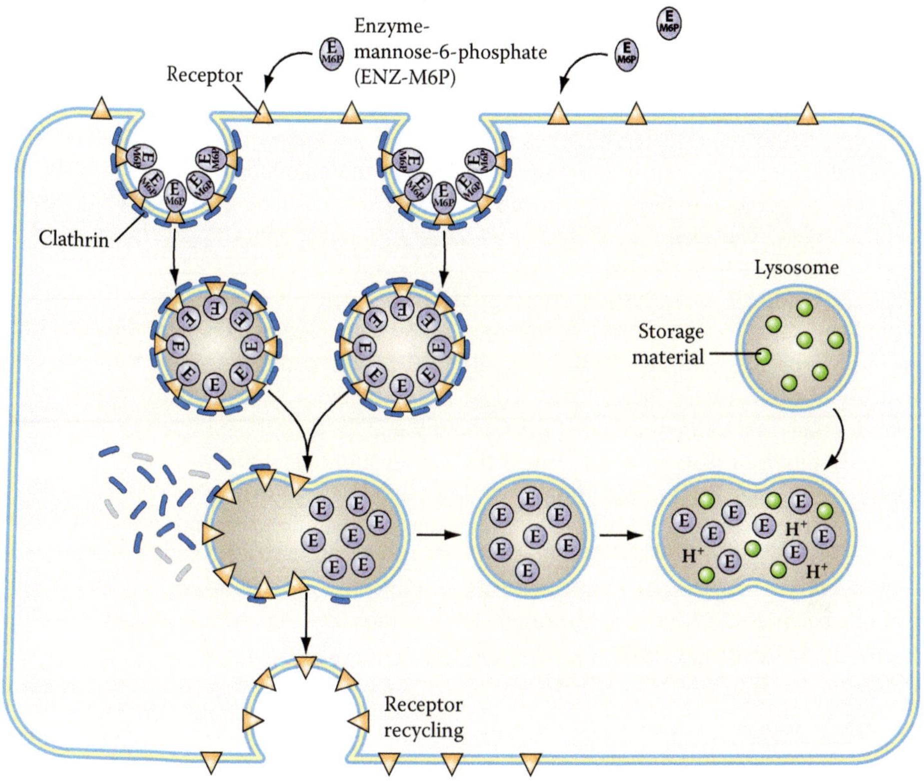

FIGURE 19.2 Posttranslational modification of the enzyme adds mannose-6-phosphate groups. All cells except macrophages include the mannose-6-phosphate receptor on their surface. After administering the enzyme, the mannose-6-phosphate groups on the enzyme bind to the mannose-6-phosphate receptors on the cell surface. The enzyme–receptor complexes are internalized via clathrin-mediated endocytosis. The enzyme dissociates from the receptors and is directed toward the diseased lysosomes containing the storage material, where it exerts its therapeutic action.

and Aubourg 2008). Graft failure and GVHD can be avoided by using autologous stem cells that have been genetically modified to express the missing enzyme (Cartier and Aubourg 2008).

Gene Therapy

Gene therapy appears to be most applicable for single-gene disorders such as LSDs because it has the potential to provide a constant source of the missing enzyme; only a fraction of normal enzyme activity (15%–20%) is required to achieve therapeutic efficacy (Beck 2010; Sands and Davidson 2006). Gene therapy can be provided using in vivo and ex vivo methods. The in vivo method normally delivers adenovirus, retrovirus, or lentivirus using various methods to establish "depots" in the liver or lung, which then continually secrete the enzyme into the circulation. Ex vivo methods involve (1) the previously described genetically modified allogenic BM transplant constituting CD 34+ cells with direct access to the circulation and the capability to transport enzyme to all organ systems (Sands and Davidson 2006), including the CNS (Ashcuer et al. 2004), and (2) the use of artificial organoids, which are composed of a genetically modified, nonmigratory cell line that is transplanted into the host to express the deficient enzyme (Sands and Davidson 2006). These cells can be safely modified with viral or nonviral vectors and explanted from the site if necessary (Sands and Davidson 2006).

Animal and human studies have shown the potential of in vivo gene therapy (Li et al. 2002) of LSDs. CNS involvement cannot be treated using the method in which peripheral organ systems serve as depots because the secreted enzymes are not transported across the BBB (Beck 2010). In an animal model of late infantile neuronal cerold lipofuscinosis (NCL), direct intracranial injection of adeno-associated viral vectors containing CLN2 gene that encodes the enzyme tripeptidyl peptidase I led to marked reduction, compared to controls, in producing the autofluorescent storage material that leads to neurodegeneration in NCL (Passini et al. 2006). A clinical trial in 10 patients was conducted (Passini et al. 2006). The adeno-associated virus serotype 2 vector with CLN2 complementary DNA (cDNA) was intracranially implanted in the patients, after which they were assessed using a neurological rating scale and magnetic resonance imaging (MRI) parameters. Over the course of the trial, there was a significant reduction in neurological decline, slower MRI parameter decline rate, and less NCL progression compared to controls (Passini et al. 2006). One patient died following status epilepticus.

The utility of ex vivo gene therapy was shown in animal studies. Genetically modified hematopoietic stem progenitor cells distributed throughout the CNS and arylsulfatase A expression prevented manifestation of disease and reversed the neuropathological and neurological deficits in a mouse model of MLD (Biffi et al. 2006). In comparison, a peripheral source of enzyme did not result in any improvement. The artificial organoid method was investigated in a mouse model of Sly syndrome. Beta-glucuronidase-deficient fibroblasts were transfected with a retroviral vector, embedded in collagen lattices, and intraperitoneally implanted into the mice (Moullier et al. 1993). The implants became vascularized, all mice had detectable enzyme levels, and storage bodies in RES organs thereafter disappeared.

Substrate Reduction

This strategy of substrate reduction therapy (SRT) works by reducing the concentration of substrates for the step normally catalyzed by the defective enzyme, rather than replacing the defective enzyme itself. Several clinical trials and drug approvals documented the utility and clinical efficacy of this strategy. *N*-butyldeoxynojirimycin (miglustat), a glucose analog, inhibits glucosidases and glucosyltransferase and the formation of glucosylceramide-based glycosphingolipids (Platt et al. 1994). Miglustat was investigated in a clinical trial of 28 patients with the non-neuronopathic form of Gaucher disease who could not be given ERT. Patients received 100 mg of miglustat three times daily over a period of 6–12 months and experienced a significant decrease in RES organ volume and improved hematological parameters (Cox et al. 2000). No serious side effects were seen. Another clinical trial of 10 patients confirmed the hematological and organ size findings in the previous study and additionally reported improvement in osteoporosis and bone pain in most patients (Pastores et al. 2005). Another inhibitor of glucosylceramide synthase, Genz-112638, was effective in reducing levels of glucosylceramide and Gaucher cells in RES organs in murine models of Gaucher disease (McEachern et al. 2007). Miglustat is marketed under the trade name Zavesca by Actelion Pharmaceuticals and is approved for treatment of Gaucher disease not amenable to ERT.

Other clinical trials focused on the utility of Miglustat in NPD type C. In one trial, 20 patients over 12 years of age received miglustat, and 9 received standard treatment (Patterson et al. 2007). Horizontal saccadic eye movement (HSEM) was set as the primary end point due to its correlation with disease progression. HSEM velocity improved in all patients who were given miglustat compared to those who received standard care. The miglustat group also had better swallowing capacity, auditory acuity, and ambulatory index scores compared to the control group (Patterson et al. 2007). Miglustat is approved in Europe to treat NPD type C (Beck 2010).

For MPS type 3A, weekly administration of rhodamine B (inhibitor of synthesis of GAG) in a mouse model of MPS3A significantly improved CNS function measured by escape latency and correct/incorrect entries, compared to untreated mice (Roberts et al. 2007). Genistein (another inhibitor of GAG formation) was shown in a pilot study of 10 patients with MPS type 3 (Sanfilippo syndrome) (5 with MPS3A, 5 with MPS3B) to significantly decrease urinary GAG excretion in all

MPS3A and 2 MPS3B patients and to improve hair morphology and cognitive function (Piotrowska et al. 2008). The authors call for additional clinical trials to further evaluate the effects of genistein.

Chaperones

The function of chaperones is primarily to aid in forming more complex structural configurations of a protein, failure of which would result in degradation by the ubiquitin–proteasomal system. Missense mutations result in an amino acid substitution that changes enzyme structure, which can prevent the enzyme from assuming its normal folded structure and inactivate its catalytic function; aiding the enzyme folding process with chaperones may allow the abnormal enzyme to provide at least partial enzyme activity (Beck 2010). Chaperones are normally distributed throughout the CNS. 1-Deoxy-galactonojirimycin, an inhibitor of alpha-galactosidase A, increased maturation and transport of mutant enzyme in Fabry lymphoblasts and raised enzyme activity in vivo in transgenic mice (Fan et al. 1999). *N*-butyldeoxynojirimycin, another chaperone, enhanced delivery of alpha-glucosidase to lysosomes and increased maturation and stability in Pompe fibroblasts (Porto et al. 2009). This synergistic effect of chaperones with ERT was also shown in a mouse model of Pompe disease by the same authors (Porto et al. 2009).

Reducing Degradation of Deficient Enzyme

Missense mutations in the glucocerebrosidase gene may reduce glucocerebrosidase catalytic activity in Gaucher disease primarily because the mutated enzyme has reduced localization to the lysosome and increased proteasomal degradation rather than decreased intrinsic catalytic function (Lu et al. 2011). Improving lysosomal localization and reducing proteasomal degradation of the mutated enzyme could restore cellular glucocerebrosidase catalytic activity in these cases. Histone deacetylase inhibitors (HDACis) were tested in vitro to assess their effect on glucocerebrosidase catalytic activity and half-life in cultured fibroblasts derived from Gaucher disease patients and HeLa cells transfected with mutations seen in non-neuronopathic (N370S) and neuronopathic (L444P) Gaucher disease. The HDACis decreased binding of the mutated enzyme to heat shock protein (HSP) 90, which normally directs the protein toward the degradation pathway, and increased binding to HSP70, which is involved in proper protein folding (Lu et al. 2011). This resulted in both increased half-life of glucocerebrosidase and cellular glucocerebrosidase catalytic activity (Lu et al. 2011). HDACis require further study in clinical trials to establish if they can play a role in treating neuronopathic and non-neuronopathic Gaucher disease.

USING NANOTECHNOLOGY TO DEVELOP NOVEL THERAPEUTICS FOR LSDs

Nanotechnology could potentially improve the treatment of CNS manifestations of LSDs by improving delivery to the CNS of enzymes, drugs for substrate reduction, or genes. LSDs are multisystem disorders involving anatomically distinct organs systems, such as the skeletal system and CNS, and, therefore, the targets of the nanoparticle system are diverse and wide ranging. Consequently, the targeting functionalization of the nanoparticle carrying the payload needs to be sufficiently specific to be internalized by affected cells while avoiding other cells that would adversely affect nanoparticle biodistribution. Cells of the immune system primarily limit the effectiveness of ERT. Chronic administration of a foreign molecule into the bloodstream induces generating of specific antibodies and rapid clearing of the molecule from the circulation (Wang et al. 2008). Recognition of foreign substances by macrophages was exploited to promote enzyme uptake by macrophages by exposing specific mannose residues through posttranslational modification of the enzyme. Macrophages constitute a very mobile component of the immune system, can traverse the BBB and penetrate extracellular spaces, and originate form monocytes in RES organs such as spleen, liver, and other immunologic aggregates. This strategy of macrophage enzyme transport may allow stem cell transplantation to have an effect on CNS disease because macrophages can traverse the BBB, differentiate into microglia, and produce enzyme within the CNS.

Hypersensitivity reactions, including life-threatening anaphylaxis and serum sickness, can occur with intravenous administration of a foreign molecule. In practice, ERT for LSDs is relatively well tolerated, with gradual development of immune tolerance and decline in antibody titer (Beck 2010). Adverse effects are uncommon. In a study of 16 male Fabry patients, 11 developed Immunoglobulin G (IgG) antibodies against the administered alpha-galactosidase A and had evidence of significantly reduced enzyme activity compared to patients who did not develop antibodies, but there were no adverse immunologic events (Linthorst et al. 2004). Antibody production was also observed in a study of ERT involving 1122 Gaucher patients, where seroconversion was observed in 12.8% of the total number of patients (Rosenberg et al. 1999). The vast majority of the seroconverted patients had only transient production of inhibitory antibody and eventually became tolerant to enzyme administration, with negligible effect on enzyme efficacy (Rosenberg et al. 1999).

Structural Modifications of Nanoparticles

Pegylation of nanoparticles is a common strategy to enhance the hydrophilicity of the nanoparticle, evade immune recognition and clearance from the circulation, enhance the half-life in circulation, and avoid rapid renal elimination. Pegylated versions of degrading enzymes enabled clearance of accumulated material and prevented its reaccumulation in an in vivo animal model of LSD in which the liver accumulated "storage materials" (Mumtaz and Bachhawat 1994). Liposomal encapsulation was also evaluated in the same study and enhanced enzyme delivery, stability, and activity. Liposomal encapsulation enabled a degree of control over distribution of the enzyme, as positively charged liposomal ß-glucuronidase was retained much longer in murine liver compared to the negatively charged liposomal construct, although both had efficient localization in the liver (Steger and Desnick 1977). While pegylated and liposomal formulations evade immune recognition and uptake, such constructs can be targeted specifically for phagocytic recognition through immunoglobulin functionalization and consequent accumulation in lysosomes (Weissmann et al. 1975). Another formulation, entrapment of enzyme within calcium alginate microspheres, was also investigated to enable a sustained release and supply of the enzyme in an in vitro fibroblast culture model of Gaucher disease (Barrias et al. 2005). The cells internalized the glucocerebrosidase construct efficiently, and activity levels were achieved that were comparable to internalization of the free enzyme, using only half the dose of glucocerebrosidase. The authors argue for the potential of this construct to target bone disease in Gaucher disease through local implantation of the sustained-release enzyme version (Barrias et al. 2005). Enzyme delivery to lysosomes can be modulated using a pH-delivery mechanism; this was accomplished with trimethyl chitosan polyelectric complexes of alpha-galactosidase that exhibit selective accumulation within lysosomes and release the enzyme with an acidic pH trigger (Giannotti et al. 2011). Aptamer targeting was also employed to evaluate the usefulness of exploiting an endocytic mechanism that does not require uptake of the enzyme through M6P and M6P-R interactions but instead targets the transferrin receptor (Chen et al. 2008). In an in vitro culture of mouse fibroblasts that did not express alpha-L-iduronidase, an aptamer conjugate of alpha-L-iduronidase that had its M6P residues removed could correct GAG accumulation within the cells; uptake kinetics demonstrated saturation and inhibition of this effect by excess free aptamers. Lysosomal accumulation was demonstrated by using aptamer–streptavidin conjugates and characterizing location with confocal microscopy (Chen et al. 2008).

Modification of Nanoparticles for CNS Targeting Enzymes

The BBB presents a considerable obstacle to developing and global marketing of drugs with indications for CNS disorders that is not encountered in developing drugs targeted to other organ systems (Khawaja 2011). Several modifications are necessary to ensure passage of a drug through the BBB and to avoid drug efflux. A nanoparticle can be functionalized to enhance its transport across the BBB, or "transcytosis." In animal models, transcytosis was increased by targeting M6P-R, transferrin-R, and insulin-R, as discussed previously (Boado et al. 2008; Osborn et al. 2008; Urayama et al. 2008). Transferrin-R is a potentially useful target to utilize for nanoparticles because it mediates endothelial uptake and transport across the BBB through clathrin- and caveolin-mediated endocytic pathways (Huang et al. 2007) and is expressed by most neuronal and glial cells in the CNS (Moos and Morgan 2000). Lactoferrin-R targeting also aids in transport of nanoparticles across the BBB through similar endocytic mechanisms (Huang et al. 2008). Such specific additional modulations are required to traverse the BBB only for CNS delivery, as the nonmodulated version with only the M6P residues on the enzyme is sufficient to target most diseased areas outside the CNS. Targeting multiple displayed antigens on the cell surface increases the affinity of the construct through a multivalency effect. This was shown in an in vivo murine model using a nanoparticle targeted with monoclonal antibodies that exhibited affinity against different antigens, with better accumulation of the nanoparticle at the target site compared to using a single monoclonal antibody (Fujita et al. 2007).

Nanoparticles may be particularly suited to deliver enzymes because of their high surface-to-volume ratio, an inert internal environment that does not interfere with protein configuration, enhanced thermal stability, and broad working pH range (Ansari and Husain 2012). Cross-linked enzyme aggregates were used with various nanostructures (Kim et al. 2006a, 2007). These add stability to the enzyme structure. Cross-linked enzymes were investigated using lysozyme (muramidase) linked by glutaraldehyde and immobilized onto polystyrene/poly(styrene-*co*-maleic anhydride) (PS/PSMA) nanofibers (Kim et al. 2011). Such a construct exhibited optimal activity at higher pH and temperature and retained enzymatic potential for a much longer time compared to the free enzyme (Kim et al. 2011). Other characterized structures include single-enzyme nanoparticles that are only a few nanometers in size and consist of a thin and porous network shell around a single enzyme molecule (Kim and Grate 2003; Kim et al. 2006b). The nanoparticle shell adds exceptional stability to the enzyme system even under adverse conditions yet provides a large surface area by virtue of its nanoporous covering. Cross-linked enzyme aggregates and single enzyme nanoparticles could prove effective if they can be functionalized to cross the BBB and enter affected neurons.

CNS targeting can be potentially enhanced using convection-enhanced delivery (CED) methods. CED provided adequate biodistribution of glucocerebrosidase at target sites within the brain in a large animal model. Gadolinium-DTPA diethylenetriaminepentacetate (Gd-DTPA), an MRI contrast agent, was coadministered by CED with glucocerebrosidase and was distributed to nearly the same volume as the enzyme, providing a method for following enzyme distribution using real-time MRI (Lonser et al. 2007). This method was used in a patient with neuronopathic Gaucher disease, and CED resulted in excellent distribution of the enzyme–Gd-DTPA solution to targeted brain regions (frontal lobe and brainstem) with no evidence of toxicity. CED could also distribute adeno-associated viral capsids and ultrasmall supraparamagnetic iron oxide (USPIO) nanoparticles safely and effectively throughout the frontal lobe and striatum in small and large animal models (Szerlip et al. 2007). CED could be employed in LSDs to deliver viral capsids carrying a normal replacement gene that codes for the deficient enzyme (Passini et al. 2006).

Intercellular Adhesion Molecule 1-Mediated Targeting and Endocytosis

Intercellular adhesion molecule 1 (ICAM-1), also known as CD54, belongs to the immunoglobulin superfamily. It is expressed by many types of cells and is an attractive target for carrier systems. ICAM-1 expression is significantly increased with inflammation and a prothrombotic state in patients with Fabry disease (DeGraba et al. 2000). Although the ICAM-1 system does not normally participate in endocytosis, nanoparticles carrying human acid sphingomyelinase (ASM) conjugated to anti-ICAM antibody were shown to be efficiently taken up via ICAM-mediated endocytosis in vitro by fibroblasts cultured from a patient with NPD (Muro et al. 2006). This type of endocytosis is independent of clathrin- and mannose-mediated internalization. Inflammation upregulates ICAM-1 expression, and ICAM targeting may be helpful in increasing enzyme efficacy in LSDs since these disorders exhibit local inflammation (Muro et al. 2006). ICAM-mediated endocytosis depends upon "multivalent engagement" of the extracellular domains of the receptor and activation of discrete cytosolic signaling pathways such as tyrosine phosphorylation-mediated activation of GTPases, which ultimately lead to actin polymerization and cell membrane invagination to orchestrate endocytosis (Muro 2010). This receptor-mediated process can accommodate structures that vary in size from 100 to 5000 nm with very diverse configurations, in stark contrast to clathrin- and caveolin-mediated endocytic pathways that can accommodate only smaller nanoparticles (Muro 2010). The nanocarrier is directed to the lysosome, where it releases its payload. ICAM-1 later recycles back to the cell membrane, where it is available to bind more ligands (Figure 19.3).

ICAM-mediated endocytosis represents an ideal target for ERT because of its widespread expression at CNS and non-CNS disease sites. Such exceptional proclivity for inflamed areas was demonstrated in both in vitro and in vivo studies, where accumulation in inflamed endothelium is manyfold higher than in noninflamed endothelium and in untargeted structures (Sakhalkar et al. 2003). Nanocarriers made of biodegradable polystyrene and poly(lactic-*co*-glycolic acid) were conjugated with ASM and anti-ICAM antibody and investigated for their ability to accumulate at target sites in vitro and in animal models of NPD types A and B (Garnacho et al. 2008). Encapsulation

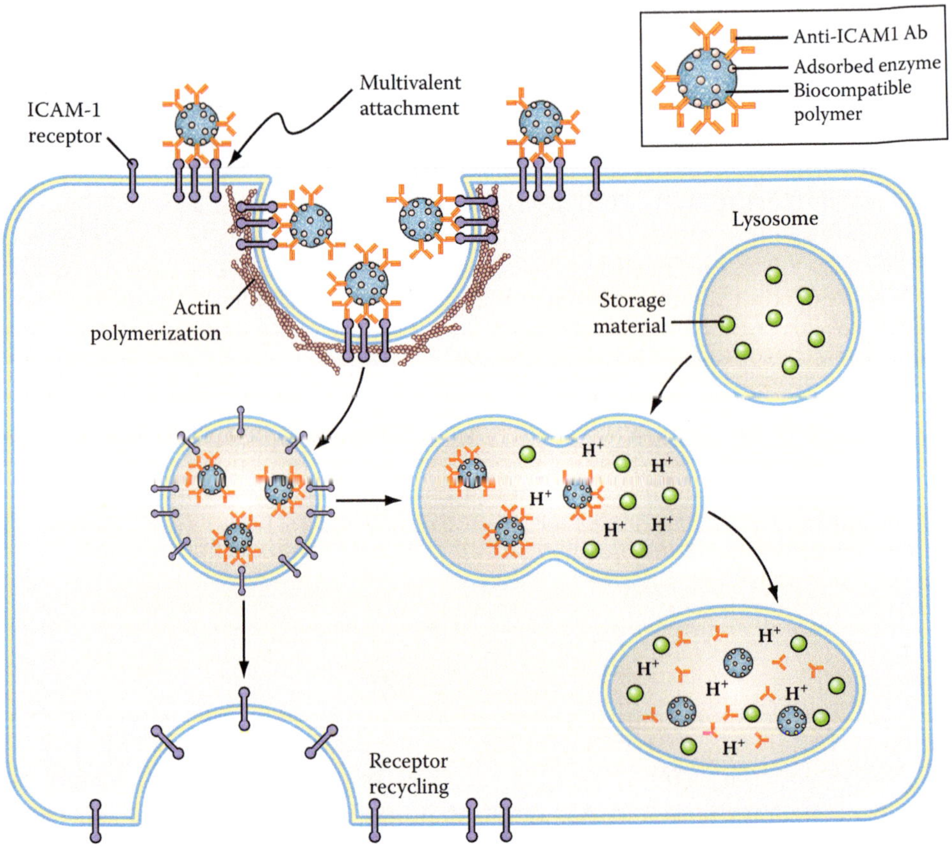

FIGURE 19.3 Inset shows a nanoparticle with enzyme adsorbed onto the surface of a biocompatible polymer (polystyrene and poly(lactic-*co*-glycolic acid)). The anti-ICAM-1 antibody enables engagement of multiple ICAM-1 receptors (multivalent engagement), and internalization occurs via a pathway independent of clathrin and caveolin. The receptors are recycled, and the nanoparticles are directed to the diseased lysosomes containing the storage material. The acidic milieu of the lysosome releases the antibody and enzyme, which exerts the therapeutic function.

of the enzyme was not necessary because the enzymes were prodrugs, able to activate in the acidic environment of the lysosomes. Such targeting increased accumulation of the nanocarriers as much as 100-fold in multiple organs after intravenous administration, as demonstrated by postmortem analysis of the vascular system and endothelia of organs using fluorescence microscopy. Additionally, no evidence of organ damage or raised vascular permeability was observed (Garnacho et al. 2008). Such nanocarrier constructs bypass the conventional clathrin and caveolin pathways and are delivered to lysosomes where anti-ICAM is degraded and ASM is processed to a configuration that remains stable for long periods of time, restoring normal levels of sphingomyelinase in the lysosome (Muro et al. 2006). These nanocarriers can also potentially be delivered directly to the CNS by CED.

CONCLUSIONS

Many treatments for LSDs already completed clinical trials and were adopted as standard care. Despite the success of ERT in ameliorating non-CNS manifestations of LSDs, CNS manifestations have resisted ERT primarily because the BBB blocks delivery of the replacement enzyme to the CNS. SRT using miglustat is a way to treat type 1 Gaucher disease in patients who cannot be treated with ERT. At this point, the use of nanotechnology is primarily being evaluated to improve existing therapeutic methods. Nanotechnology may be able to improve immune tolerance to the foreign enzyme and the present inability to target the neuronopathic manifestations of LSDs. Nanotechnology may offer a solution for CNS involvement by delivering the therapeutic payload, for enzymatic, genetic, or substrate reductive therapies, across the BBB and into CNS. Additionally, novel methods, such as cross-linked enzyme aggregates and single enzyme nanoparticles, can potentially improve the stability of the administered enzymes in ERT and provide a way to evade the immune system recognition and premature clearance from the circulation. Current research into using ICAM targeting for delivering a large therapeutic payload to the CNS and other sites is ongoing. CED of replacement enzymes was reported in animal and human studies. Using gene therapy delivered by CED of viral or nanoparticle vectors is also an area of research interest. Better treatments are required for the CNS involvement that accompanies many of these disorders and that often results in severe morbidity and early mortality.

REFERENCES

Achord, D.T., Brot, F.E., Bell, C.E. et al. 1978. Human beta-glucuronidase: in vivo clearance and in vitro uptake by a glycoprotein recognition system on reticuloendothelial cells. *Cell* 15: 269–78.

Ansari, S.A., and Husain, Q. 2012. Potential applications of enzymes immobilized on/in nano materials: a review. *Biotechnol Adv* 30: 512–23.

Asheuer, M., Pflumio, F.o., Benhamida, S. et al. 2004. Human CD34+ cells differentiate into microglia and express recombinant therapeutic protein. *Proc Natl Acad Sci U S A* 101: 3557–62.

Aviezer, D., Brill-Almon, E., Shaaltiel, Y. et al. 2009. A plant-derived recombinant human glucocerebrosidase enzyme—a preclinical and phase I investigation. *PLoS One* 4: e4792.

Baehner, F., Kampmann, C., Whybra, C. et al. 2003. Enzyme replacement therapy in heterozygous females with Fabry disease: results of a phase IIIB study. *J Inherit Metab Dis* 26: 617–27.

Ballabio, A., and Gieselmann, V. 2009. Lysosomal disorders: from storage to cellular damage. *Biochim Biophys Acta* 1793: 684–96.

Banikazemi, M., Bultas, J., Waldek, S. et al. 2007. Agalsidase-beta therapy for advanced Fabry disease: a randomized trial. *Ann Intern Med* 146: 77–86.

Barrias, C.C., Lamghari, M., Granja, P.L. et al. 2005. Biological evaluation of calcium alginate microspheres as a vehicle for the localized delivery of a therapeutic enzyme. *J Biomed Mater Res A* 74: 545–52.

Barton, N.W., Brady, R.O., Dambrosia, J.M. et al. 1991. Replacement therapy for inherited enzyme deficiency—macrophage-targeted glucocerebrosidase for Gaucher's disease. *N Engl J Med* 324: 1464–70.

Beck, M. 2009. Agalsidase alfa for the treatment of Fabry disease: new data on clinical efficacy and safety. *Expert Opin Biol Ther* 9: 255–61.

Beck, M. 2010. Therapy for lysosomal storage disorders. *IUBMB Life* 62: 33–40.

Bellettato, C.M., and Scarpa, M. 2010. Pathophysiology of neuropathic lysosomal storage disorders. *J Inherit Metab Dis* 33: 347–62.

Biffi, A., Capotondo, A., Fasano, S. et al. 2006. Gene therapy of metachromatic leukodystrophy reverses neurological damage and deficits in mice. *J Clin Invest* 116: 3070–82.

Boado, R.J., Zhang, Y., Xia, C.F. et al. 2008. Genetic engineering of a lysosomal enzyme fusion protein for targeted delivery across the human blood–brain barrier. *Biotechnol Bioeng* 99: 475–84.

Brooks, H., Lebleu, B., and Vives, E. 2005. Tat peptide-mediated cellular delivery: back to basics. *Adv Drug Deliv Rev* 57: 559–77.

Brumshtein, B., Salinas, P., Peterson, B. et al. 2010. Characterization of gene-activated human acid-β-glucosidase: crystal structure, glycan composition, and internalization into macrophages. *Glycobiology* 20: 24–32.

Cartier, N., and Aubourg, P. 2008. Hematopoietic stem cell gene therapy in Hurler syndrome, globoid cell leukodystrophy, metachromatic leukodystrophy and X-adrenoleukodystrophy. *Curr Opin Mol Ther* 10: 471–8.

Chen, C.H., Dellamaggiore, K.R., Ouellette, C.P. et al. 2008. Aptamer-based endocytosis of a lysosomal enzyme. *Proc Natl Acad Sci U S A* 105: 15908–13.

Clarke, L.A., Wraith, J.E., Beck, M. et al. 2009. Long-term efficacy and safety of laronidase in the treatment of mucopolysaccharidosis I. *Pediatrics* 123: 229–40.

Cox, T., Lachmann, R., Hollak, C. et al. 2000. Novel oral treatment of Gaucher's disease with *N*-butyldeoxynojirimycin (OGT 918) to decrease substrate biosynthesis. *Lancet* 355: 1481–5.

DeGraba, T., Azhar, S., Dignat-George, F. et al. 2000. Profile of endothelial and leukocyte activation in Fabry patients. *Ann Neurol* 47: 229–33.

Desnick, R.J., Thorpe, S.R., and Fiddler, M.B. 1976. Toward enzyme therapy for lysosomal storage diseases. *Physiol Rev* 56: 57–99.

Fan, J.Q., Ishii, S., Asano, N. et al. 1999. Accelerated transport and maturation of lysosomal alpha-galactosidase A in Fabry lymphoblasts by an enzyme inhibitor. *Nat Med* 5: 112–15.

Fujita, M., Lee, B.S., Khazenzon, N.M. et al. 2007. Brain tumor tandem targeting using a combination of monoclonal antibodies attached to biopoly(beta-L-malic acid). *J Control Release* 122: 356–63.

Furbish, F.S., Steer, C.J., Krett, N.L. et al. 1981. Uptake and distribution of placental glucocerebrosidase in rat hepatic cells and effects of sequential deglycosylation. *Biochim Biophys Acta* 673: 425–34.

Garnacho, C., Dhami, R., Simone, E. et al. 2008. Delivery of acid sphingomyelinase in normal and Niemann–Pick disease mice using intercellular adhesion molecule-1-targeted polymer nanocarriers. *J Pharmacol Exp Ther* 325: 400–8.

Germain, D.P. 2004. [Gaucher disease: clinical, genetic and therapeutic aspects]. *Pathol Biol (Paris)* 52: 343–50.

Giannotti, M.I., Esteban, O., Oliva, M. et al. 2011. pH-responsive polysaccharide-based polyelectrolyte complexes as nanocarriers for lysosomal delivery of therapeutic proteins. *Biomacromolecules* 12: 2524–33.

Grabowski, G.A., Barton, N.W., Pastores, G. et al. 1995. Enzyme therapy in type 1 Gaucher disease: comparative efficacy of mannose-terminated glucocerebrosidase from natural and recombinant sources. *Ann Intern Med* 122: 33–9.

Guffon, N., Souillet, G., Maire, I. et al. 1998. Follow-up of nine patients with Hurler syndrome after bone marrow transplantation. *J Pediatr* 133: 119–25.

Harada, M. 2011. Autophagy in liver diseases. *J UOEH* 33: 337–44.

Harmatz, P., Giugliani, R., Schwartz, I.V. et al. 2008. Long-term follow-up of endurance and safety outcomes during enzyme replacement therapy for mucopolysaccharidosis VI: final results of three clinical studies of recombinant human *N*-acetylgalactosamine 4-sulfatase. *Mol Genet Metab* 94: 469–75.

Hers, H.G. 1965. Inborn lysosomal diseases. *Gastroenterology* 48: 625–33.

Huang, R., Ke, W., Liu, Y. et al. 2008. The use of lactoferrin as a ligand for targeting the polyamidoamine-based gene delivery system to the brain. *Biomaterials* 29: 238–46.

Huang, R.Q., Qu, Y.H., Ke, W.L. et al. 2007. Efficient gene delivery targeted to the brain using a transferrin-conjugated polyethyleneglycol-modified polyamidoamine dendrimer. *FASEB J* 21: 1117–25.

Jardim, L.B., Villanueva, M.M., de Souza, C.F. et al. 2010. Clinical aspects of neuropathic lysosomal storage disorders. *J Inherit Metab Dis* 33: 315–29.

Journet, A., Chapel, A., Kieffer, S. et al. 2002. Proteomic analysis of human lysosomes: application to monocytic and breast cancer cells. *Proteomics* 2: 1026–40.

Khawaja, A.M. 2011. The legacy of nanotechnology: revolution and prospects in neurosurgery. *Int J Surg* 9: 608–14.

Kim, J., and Grate, J.W. 2003. Single-enzyme nanoparticles armored by a nanometer-scale organic/inorganic network. *Nano Lett* 3: 1219–22.

Kim, J., Grate, J.W., and Wang, P. 2006a. Nanostructures for enzyme stabilization. *Chem Eng Sci* 61: 1017–26.

Kim, J., Jia, H.F., Lee, C.W. et al. 2006b. Single enzyme nanoparticles in nanoporous silica: a hierarchical approach to enzyme stabilization and immobilization. *Enzym Microb Technol* 39: 474–80.

Kim, M., Park, J.M., Yoon, J. et al. 2011. Synthesis and characterization of CLEA-lysozyme immobilized PS/PSMA nanofiber. *J Nanosci Nanotechnol* 11: 7894–900.

Kim, M.W., Lee, S.G., Kim, T.T. et al. 2007. Experimental demonstration of bending and splitting of self-collimated beams in two-dimensional photonic crystals. *Appl Phys Lett* 90: 113121–3.

Krivit, W., Peters, C., and Shapiro, E.G. 1999. Bone marrow transplantation as effective treatment of central nervous system disease in globoid cell leukodystrophy, metachromatic leukodystrophy, adrenoleukodystrophy, mannosidosis, fucosidosis, aspartylglucosaminuria, Hurler, Maroteaux-Lamy, and Sly syndromes, and Gaucher disease type III. *Curr Opin Neurol* 12: 167–76.

LeBowitz, J.H., Grubb, J.H., Maga, J.A. et al. 2004. Glycosylation-independent targeting enhances enzyme delivery to lysosomes and decreases storage in mucopolysaccharidosis type VII mice. *Proc Natl Acad Sci U S A* 101: 3083–8.

Lee, K.O., Luu, N., Kaneski, C.R. et al. 2005. Improved intracellular delivery of glucocerebrosidase mediated by the HIV-1 TAT protein transduction domain. *Biochem Biophys Res Commun* 337: 701–7.

Li, C., Ziegler, R.J., Cherry, M. et al. 2002. Adenovirus-transduced lung as a portal for delivering alpha-galactosidase A into systemic circulation for Fabry disease. *Mol Ther* 5: 745–54.

Linthorst, G.E., Hollak, C.E., Donker-Koopman, W.E. et al. 2004. Enzyme therapy for Fabry disease: neutralizing antibodies toward agalsidase alpha and beta. *Kidney Int* 66: 1589–95.

Lonser, R.R., Schiffman, R., Robison, R.A. et al. 2007. Image-guided, direct convective delivery of glucocerebrosidase for neuronopathic Gaucher disease. *Neurology* 68: 254–61.

Lu, J., Yang, C., Chen, M. et al. 2011. Histone deacetylase inhibitors prevent the degradation and restore the activity of glucocerebrosidase in Gaucher disease. *Proc Natl Acad Sci U S A* 108: 21200–5.

McEachern, K.A., Fung, J., Komarnitsky, S. et al. 2007. A specific and potent inhibitor of glucosylceramide synthase for substrate inhibition therapy of Gaucher disease. *Mol Genet Metab* 91: 259–67.

Meikle, P.J., Hopwood, J.J., Clague, A.E. et al. 1999. Prevalence of lysosomal storage disorders. *JAMA* 281: 249–54.

Moos, T., and Morgan, E.H. 2000. Transferrin and transferrin receptor function in brain barrier systems. *Cell Mol Neurobiol* 20: 77–95.

Moullier, P., Bohl, D., Heard, J.M. et al. 1993. Correction of lysosomal storage in the liver and spleen of MPS VII mice by implantation of genetically modified skin fibroblasts. *Nat Genet* 4: 154–9.

Mumtaz, S., and Bachhawat, B.K. 1994. Enhanced intracellular stability and efficacy of PEG modified dextranase in the treatment of a model storage disorder. *Biochim Biophys Acta* 1199: 175–82.

Muro, S. 2010. New biotechnological and nanomedicine strategies for treatment of lysosomal storage disorders. *Wiley Interdiscip Rev Nanomed Nanobiotechnol* 2: 189–204.

Muro, S., Schuchman, E.H., and Muzykantov, V.R. 2006. Lysosomal enzyme delivery by ICAM-1-targeted nanocarriers bypassing glycosylation- and clathrin-dependent endocytosis. *Mol Ther* 13: 135–41.

Neufeld, E.F. 1980. The uptake of enzymes into lysosomes: an overview. *Birth Defects Orig Artic Ser* 16: 77–84.

Nicolino, M., Byrne, B., Wraith, J.E. et al. 2009. Clinical outcomes after long-term treatment with alglucosidase alfa in infants and children with advanced Pompe disease. *Genet Med* 11: 210–19.

Osborn, M.J., McElmurry, R.T., Peacock, B. et al. 2008. Targeting of the CNS in MPS-IH using a nonviral transferrin-[alpha]-L-iduronidase fusion gene product. *Mol Ther* 16: 1459–66.

Passini, M.A., Dodge, J.C., Bu, J. et al. 2006. Intracranial delivery of CLN2 reduces brain pathology in a mouse model of classical late infantile neuronal ceroid lipofuscinosis. *J Neurosci* 26: 1334–42.

Pastores, G.M., Barnett, N.L., and Kolodny, E.H. 2005. An open-label, noncomparative study of miglustat in type I Gaucher disease: efficacy and tolerability over 24 months of treatment. *Clin Ther* 27: 1215–27.

Patterson, M.C., Vecchio, D., Prady, H. et al. 2007. Miglustat for treatment of Niemann–Pick C disease: a randomised controlled study. *Lancet Neurol* 6: 765–72.

Piotrowska, E., Jakobkiewicz-Banecka, J., Tylki-Szymanska, A. et al. 2008. Genistin-rich soy isoflavone extract in substrate reduction therapy for Sanfilippo syndrome: an open-label, pilot study in 10 pediatric patients. *Curr Ther Res Clin Exp* 69: 166–79.

Platt, F.M., Neises, G.R., Dwek, R.A. et al. 1994. *N*-butyldeoxynojirimycin is a novel inhibitor of glycolipid biosynthesis. *J Biol Chem* 269: 8362–5.

Porto, C., Cardone, M., Fontana, F. et al. 2009. The pharmacological chaperone *N*-butyldeoxynojirimycin enhances enzyme replacement therapy in Pompe disease fibroblasts. *Mol Ther* 17: 964–71.

Prows, C.A., Sanchez, N., Daugherty, C. et al. 1997. Gaucher disease: enzyme therapy in the acute neuronopathic variant. *Am J Med Genet* 71: 16–21.

Rautou, P.E., Mansouri, A., Lebrec, D. et al. 2010. Autophagy in liver diseases. *J Hepatol* 53: 1123–34.

Regnery, C., Kornblum, C., Hanisch, F. et al. 2012. 36 months observational clinical study of 38 adult Pompe disease patients under alglucosidase alfa enzyme replacement therapy. *J Inherit Metab Dis* 35: 837–45.

Roberts, A.L., Rees, M.H., Klebe, S. et al. 2007. Improvement in behaviour after substrate deprivation therapy with rhodamine B in a mouse model of MPS IIIA. *Mol Genet Metab* 92: 115–21.

Rosenberg, M., Kingma, W., Fitzpatrick, M.A. et al. 1999. Immunosurveillance of alglucerase enzyme therapy for Gaucher patients: induction of humoral tolerance in seroconverted patients after repeat administration. *Blood* 93: 2081–8.

Sakhalkar, H.S., Dalal, M.K., Salem, A.K. et al. 2003. Leukocyte-inspired biodegradable particles that selectively and avidly adhere to inflamed endothelium in vitro and in vivo. *Proc Natl Acad Sci U S A* 100: 15895–900.

Sands, M.S., and Davidson, B.L. 2006. Gene therapy for lysosomal storage diseases. *Mol Ther* 13: 839–49.

Sands, M.S., Vogler, C.A., Ohlemiller, K.K. et al. 2001. Biodistribution, kinetics, and efficacy of highly phosphorylated and non-phosphorylated beta-glucuronidase in the murine model of mucopolysaccharidosis VII. *J Biol Chem* 276: 43160–5.

Sardiello, M., Palmieri, M., di Ronza, A. et al. 2009. A gene network regulating lysosomal biogenesis and function. *Science* 325: 473–7.

Souillet, G., Guffon, N., Maire, I. et al. 2003. Outcome of 27 patients with Hurler's syndrome transplanted from either related or unrelated haematopoietic stem cell sources. *Bone Marrow Transplant* 31: 1105–17.

Steger, L.D., and Desnick, R.J. 1977. Enzyme therapy. VI: Comparative in vivo fates and effects on lysosomal integrity of enzyme entrapped in negatively and positively charged liposomes. *Biochim Biophys Acta* 464: 530–46.

Szerlip, N.J., Walbridge, S., Yang, L. et al. 2007. Real-time imaging of convection-enhanced delivery of viruses and virus-sized particles. *J Neurosurg* 107: 560–7.

Urayama, A., Grubb, J.H., Sly, W.S. et al. 2008. Mannose 6-phosphate receptor-mediated transport of sulfamidase across the blood–brain barrier in the newborn mouse. *Mol Ther* 16: 1261–6.

Walkley, S.U. 2007. Pathogenic mechanisms in lysosomal disease: a reappraisal of the role of the lysosome. *Acta Paediatr Suppl* 96: 26–32.

Walkley, S.U. 2009. Pathogenic cascades in lysosomal disease—why so complex? *J Inherit Metab Dis* 32: 181–9.

Wang, J., Lozier, J., Johnson, G. et al. 2008. Neutralizing antibodies to therapeutic enzymes: considerations for testing, prevention and treatment. *Nat Biotechnol* 26: 901–8.

Weissmann, G., Bloomgarden, D., Kaplan, R. et al. 1975. A general method for the introduction of enzymes, by means of immunoglobulin-coated liposomes, into lysosomes of deficient cells. *Proc Natl Acad Sci U S A* 72: 88–92.

Wraith, J.E., Scarpa, M., Beck, M. et al. 2008. Mucopolysaccharidosis type II (Hunter syndrome): a clinical review and recommendations for treatment in the era of enzyme replacement therapy. *Eur J Pediatr* 167: 267–77.

Zhu, Y., Li, X., Schuchman, E.H. et al. 2004. Dexamethasone-mediated up-regulation of the mannose receptor improves the delivery of recombinant glucocerebrosidase to Gaucher macrophages. *J Pharmacol Exp Ther* 308: 705–11.

20 Nanotechnology for Cerebral Aneurysm Treatment

Jean Paul Allain, Teodoro Tigno, and Rocco Armonda

CONTENTS

INTRODUCTION

The sudden onset of a sentinel headache (Beck et al. 2006) in an otherwise healthy individual typically signals an incipient spontaneous subarachnoid hemorrhage (SAH). The catastrophic event of aneurysmal subarachnoid hemorrhage (aSAH) carries a mortality rate of 25%–50% (ACROSS 2000; Anderson 2009), with 10%–15% of patients dying before reaching a hospital. aSAHs from the 30,000 cerebral aneurysms that rupture each year in the United States account for 2%–5% of all strokes. With some geographical exceptions (al-Rajeh et al. 1998; Awada and al Rajeh 1999), aSAH exhibits a female preponderance. In more than 90% of the patients, SAH is sporadic. This treacherous disease entity debilitates as many as 46% of SAH survivors with long-term cognitive and other neurologic impairments, which significantly impact their functional status and quality of life (Audibert et al. 2007).

Treatment strategies offered for cerebral aneurysms have evolved over time. In the 18th century, several investigators (Phillips 1832; Ransohoff 1886; Velpeau 1831) experimented with inserting needles into thoracic, aortal, or limb aneurysms to induce thrombosis, a feat that was abandoned due to unpredictable and often unsatisfactory results. The deliberate Hunterian ligation of the ipsilateral cervical carotid artery followed as a treatment to reduce blood flow to the aneurysm despite great risks of an accompanying cerebral infarction (Davis and Wetzel 1956; Fager 1958; Hardy et al. 1958; Shenkin et al. 1958). The era of improved dissection skills and surgical proficiency dawned with the reported successful cases of Drake (Batjer and Duckworth 2009), Suzuki (Suzuki and Miura 1967), and others using open techniques to occlude the aneurysm neck with adjustable aneurysm clips. Yasargil (Yasargil 1969; Yasargil and Fox 1975; Yasargil et al. 1970) and Rhoton (Rhoton 2007; Rhoton and Natori 1996) ushered in the era of the surgical microscope and intraoperative appreciation of the detailed anatomy of the neurovascular micro-environment. Amidst earlier independent initiatives with equivocal results (Fogarty et al. 1963; Hunner 1900; Luessenhop and Velasquez 1964; Werner et al. 1941), an article published by Serbinenko (1974) reporting successful balloon catheterization and cerebral vessel occlusion heralded the further development of the neuroendovascular alternative for cerebral aneurysm treatment, including the subsequent advent of the Guglielmi (Guglielmi et al. 1991a,b) detachable coil, the current neuroendovascular first-line therapy for cerebral aneurysms.

The landmark International Subarachnoid Trial (ISAT) (Molyneux et al. 2002a,b) study was a multicenter randomized control trial comparing aneurysm clipping through an open craniotomy with neuroendovascular coiling. The results of neuroendovascular coiling compared favorably to those of aneurysm clipping. Koebbe et al. (2006) depicted the evolving practice paradigm to treat cerebral aneurysms shifting from the open craniotomy approach toward endovascular treatment since 2000.

However, endovascular coiling, the current first-line treatment for intracerebral aneurysms since 2002, has often led to complications, making revisions of aneurysmal coiling an accepted part of neuroendovascular practice (Kole et al. 2005; Mathis et al. 1997; Raymond et al. 2003; Rordorf et al. 2001; Shinokuma et al. 2000; Soeda et al. 2004; Stiebel-Kalish et al. 2003; Thornton et al. 2000; Uda et al. 1998; Wang et al. 2004; Workman et al. 2002). We present the case in Figure 20.1 demonstrating the failure of a conventional stent/coil combination to treat a complex giant middle cerebral artery (MCA) aneurysm. This patient did not present with a delayed MCA thrombosis but, rather, a significant short-term enlargement of the aneurysm. The limitations of current endovascular protocols are displayed showing a significant recurrence of this aneurysm despite advances in coils, self-expanding nitinol stents, and microcatheters. In treating this case, the hazards of microsurgery are

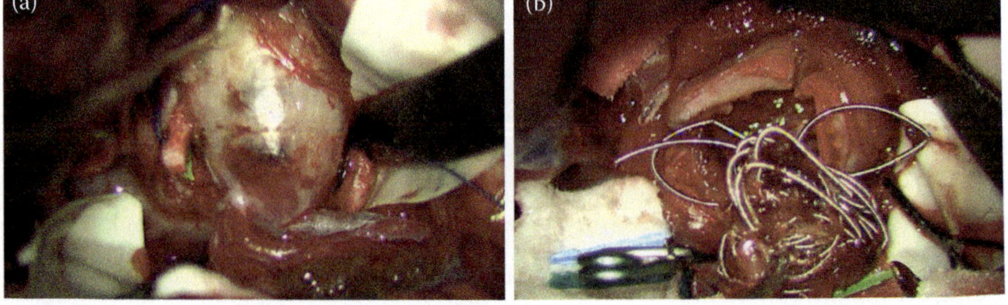

FIGURE 20.1 (a) Intraoperative view of failure of a conventional stent/coil combination to treat a complex giant MCA aneurysm that led to significant short-term aneurysmal enlargement. (b) Intraoperative dissection of coil from aneurysm sac to enable clipping.

amplified since there is a large thrombogenic coil mass in the aneurysm, preventing clipping at the aneurysm neck. Additionally, the presence of the endoluminal stent in the MCA increases the risk of parent artery thrombosis during surgery due to the necessary temporary clipping to open the aneurysm, remove the coil mass, and apply a clip at the neck. It was in this setting of uncertain durability of current first-line treatment of cerebral aneurysms that the authors, Tigno and Armonda, developed a neuroendovascular treatment alternative in 2006 requiring five strategic elements: (1) use of stents as bioactive scaffolds; (2) the need of tissue nidus to repair a vascular defect across a scaffold; (3) using the arterial wall as a source of stem cell derivatives for tunica media reconstruction; and (4) confining induced thrombogenic activity extraluminally, with recognition of the role of a focal magnetic field in this process and (5) orienting smooth muscle regrowth using a pulsatile environment. Based on these strategic elements author Allain visualized the emergent role of nanotechnology in neurovascular and cerebrovascular disease and treatment. In particular, he proposed how scaffold surfaces could be functionalized to manipulate and regulate cellular behavior at subcellular scales and thus direct tissue proliferation along vascular defects.

This chapter is divided into two parts. The first is a discussion of the five strategic elements that comprise the novel neuroendovascular treatment alternative devised by Tigno and Armonda. The second part explains the role nanotechnology plays in such treatment and in the broader context of cerebrovascular disease and endovascular treatment.

FIVE STRATEGIES TO A NEUROENDOVASCULAR TREATMENT ALTERNATIVE

The idea that stents can be used as a primary treatment for cerebral aneurysms had few followers (Benitez et al. 2004; Hanel et al. 2005). Reported cases of angiographically proven occlusion of difficult-to-clip or coil aneurysms after stenting (Lanzino et al. 1999) demonstrated a potential to induce closure of the aneurysmal sac through flow diversion (FD) techniques and progressive thrombosis, which led to development of the pipeline embolization device (PED) (ev3, Plymouth, MN) and the Silk FD (Montmorency, France). However, the unabated buildup of intra-aneurysmal thrombosis in these FD devices had been considered a positive factor for delayed aneurysm rupture after FD treatment (Kulcsar et al. 2011). Wehman et al. (2006) illustrated a variable-porosity stent in 2006 but could not follow up with a fabricated prototype, much less its clinical application. In 2009, Ionita and colleagues, in applying variable-porosity stents (asymmetric vascular stents [AVSs]) in rabbits, showed complete occlusion in all nine AVS cases but had three deaths during AVS treatment (Ionita et al. 2009). Clearly, for the endovascular strategy of stenting to be viable, the stents must lessen blood entry into the aneurysm and induce aneurysm thrombosis but not lead to thrombogenic occlusion of the parent artery or stent thrombosis (ST).

Although clinically used mainly for coronary stenosis/occlusion treatment, drug-eluting stents (DESs) were previously offered to counter ST, but recent studies (Ferreira-Gonzalez et al. 2010; Park et al. 2008) showed DES to have a higher late-onset ST risk (Ong and Serruys 2005). Inversely but accurately stated, ST is an inherent consequence of stent placement and endoluminal manipulation. To use it for its original intention (to dilate a stenosed duct or vessel) is to accept varying degrees of ST, which can only be minimized but not totally removed as a complication. The major role of minimizing endoluminal manipulation in preventing intimal hyperplasia has been demonstrated by Christen et al. (2001) in porcine coronary arteries. But what would be the effect if stents were used to induce closure of aneurysms endoluminally? Would inherent ST still be seen solely as a "complication"? Is not one aim of the current first-line treatment for cerebral aneurysms to induce endothelial proliferation through introduction of thrombogenic coils within the sac? Thus begins the first portion of the endovascular treatment strategy developed to treat cerebral aneurysms: to redirect the *stent's inherent thrombogenicity* to induce hyperproliferative tissue closure solely at the aneurysm orifice.

Thinking of Stents as More Than Scaffolds

Aneurysms are defects within the arterial walls with failed attempts at vascular remodeling. Understanding the hemodynamics on shear stress and arterial wall tension underlying aneurysm formation led to the first successful demonstration of cerebral aneurysm formation in primates in 1987 (Hazama and Hashimoto 1987), an advancement from the previous glass models.

By thinking of stents as passive scaffolds maintaining the patency of the parent artery to include the area of failed vascular remodeling (aneurysm) instead of utilizing it solely as a tool for endoluminal manipulation to coil the aneurysmal sac, one could readily appreciate its potential to repair the arterial wall defect. Repair means not only the thrombogenic induction of a weak neointimal film across the aneurysmal orifice, as has been repeatedly shown with coiling (Abrahams et al. 2001) or placement of variable-porosity stents, but also the actual ingrowth of the missing arterial wall across the stent scaffold. Of special mention is the tunica media or vascular smooth muscle layer, which will provide the resilience to counter the wall tension and shear stress created by hemodynamic forces and contribute to durability of treatment. Will the provision of a scaffold across the area of failed vascular remodeling (aneurysm orifice) be sufficient to induce repair across this defect? We postulate that a targeted biological response at the area of the orifice is required for growth to develop beyond the neointimal layer allowing for arterial medial ingrowth across the aneurysm neck and its eventual endoluminal closure.

The Need for Tissue Nidus to Repair Vascular Defects across a Scaffold

Since a critical limit for repair exists for bone beyond which an intervening homo/allograft is needed to induce osteogenesis across a defect (Jager et al. 2005), the same critical limit exists for vasculogenesis (Keefer et al. 2010), as can be seen in the spontaneous healing of a tiny puncture wound in arterial walls as compared to the failed vascular remodeling seen in aneurysms. To really induce repair (reestablishment of the tunica media), an intervening homo/allograft would be needed over the defect, which would serve as the nidus to which the peripheral rim of the defect (aneurysm neck orifice) could direct its growth.

Porcine coronary smooth muscle (PCSM) was selected as the nidus of choice for the approach proposed by Tigno and Armonda since PCSM cells retain their phenotypic plasticity in culture and, thus, in vitro mimic their in vivo differentiation states as shown in different vascular studies (Christen et al. 1999; Poliseno et al. 2006). Currently, a Part I study by Allain and coworkers is being conducted on fabricating and developing a stent prototype that would induce aneurysm closure endoluminally. Should the preliminary findings of this Part I Pseudoaneurysm Study yield results favorable for a Part II animal study, the animal of choice would be the porcine model based on previous animal studies performed to test endovascular procedures (Murayama et al. 1999, 2003), making the animal model histocompatible with the nidus of choice. A *PCSM nidus* is the second strategic element in the conceptual development of Tigno and Armonda's neuroendovascular treatment strategy for cerebral aneurysms.

Using the Arterial Wall as a Source of Stem Cell Derivatives for Tunica Media Reconstruction

In undertaking reconstitutive repair for various pathologies in both research and clinical areas of medicine, stem cells are usually provided from an external source. For example, they may originate from umbilical cord blood, discarded eggs from in vitro fertilization (IVF) trials, or cultured pluripotent stem cells. An internal source for stem cells residing in the organ needing repair has not received the equivalent attention it deserves, although there is sufficient evidence in the literature for this. Hu et al. (2002) demonstrated that smooth muscle cells (SMCs) of neointimal and atherosclerotic lesions found in allografts were derived from the recipient and that non-bone marrow–derived progenitor cells are a possible source of SMCs in atherosclerotic lesions. Further vascular studies (Abedin 2004; Tintut et al. 2003) show that the arterial wall contains cells that have the potential for multiple lineages similar to mesenchymal stem cells but with a unique differentiation repertoire. In carrying out experiments of magnetically targeted endothelial cells (ECs) in stented vessels, Pislaru et al. (2006) noted that the labeled cultured porcine ECs found their way not only to the targeted magnetized stent but also to the adjacent (injured) denuded vessel wall. The fact that *resident stem cells* within the arterial wall will respond to injury deserves to be included in the development of the endovascular treatment strategy proposed by Tigno and Armonda for cerebral aneurysms.

Confining Induced Thrombogenic Activity Extraluminally and the Role of Focal Magnetic Field

There are multiple choices for investigators to prevent the spread of ST throughout the parent artery (see Figure 20.2) in light of its radial placement within the blood vessel wall. As stated before, the ST consequence can only be minimized and not totally removed. A one-time placement of the stent is, of course, the least endoluminal manipulation possible. In fact, patients who have undergone and continue to undergo stent placement harbor a certain degree of ST, but it does not reach clinical significance usually because of ongoing antiplatelet therapy. Other pharmaceutical options (Jeger et al. 2009; Koo et al. 2008) could be considered, but pharmaceuticals may produce unpredictable effects on the complex biochemical interactions in an arterial wall that is radially stressed by the stent and an aneurysmal orifice requiring closure. A bioabsorbable stent (Krucoff et al. 2008) was thought to basically provide no wall stress. However, the period of biodegradability had to be synchronized to the needs for adequate aneurysm closure; lack of overlap of stent biodegradability and tunica media formation could open the aneurysm and reintroduce the risk of premature rupture. With the stent in place as a scaffold and the PCSM nidus impregnated within the stent area covering the aneurysm orifice, the more predictable remaining factor in redirecting thrombogenic endothelial proliferation extraluminally (into the aneurysm) would be a physical factor rather than a biochemical one.

Among the physical factors, magnetic field activity is one that has been studied in depth within biological systems. Magnetic fields were shown to affect cell cleavage in plants (Rochalska 2009). The field of hematology had its beginnings with the blood banking system, which began when Waller et al. (1985) noted that older red blood cells had a different hemoglobin composition (methemoglobin). Molday et al. (1977) and Owen (1978) later induced this ferric state (methemoglobin) by

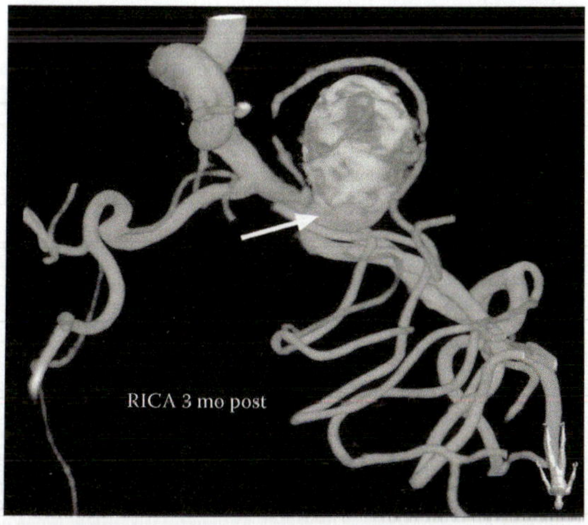

FIGURE 20.2 3D cerebral angiography of right middle cerebral aneurysm presenting as recurrence (white arrow) 3 months postcoiling.

oxidizing the iron atoms in hemoglobin and rendering them paramagnetic, leading to high-gradient magnetic separation of erythrocytes from whole blood. Berry et al. (2003) demonstrated the ability of magnetic nanoparticles (MNPs) to induce fibroblastic growth and activity as seen both in comparative growth media and under the scanning electron microscope. *Retention of cells in the presence of blood flow using a magnetic field* is a requirement for intravascular tissue repair in the fourth part of Tigno and Armonda's neuroendovascular treatment concept. To ensure a directed path of growth extraluminally but not intraluminally within the parent artery and stent, the magnetic field should emanate solely from the side of the stent facing the aneurysm. Allain and coworkers introduced nanotechnology as a viable design strategy to effectively direct cell growth, as will be discussed later in the chapter.

Pulsatile Environment Orients Smooth Muscle Growth

A fifth element of the strategy is based on a study (Kanda and Matsuda 1994) that showed the role of pulsatile stress in creating hybrid muscle tissue, which mimics the molecular orientation and physiologic function of native tunica media within a radial blood vessel wall. This study proved that *pulsatile stress is needed for an oriented growth of smooth muscle tissue*. Since all phases of our proposed study will be carried out in a simulated and actual hemodynamic milieu, this fifth factor is already taken into account. The future outcomes of this study will provide Tigno, Armonda, and Allain with clues as to whether the placement of a rigid stent in a selected section of an elastic blood vessel would counter the pulsatile stress needed for oriented smooth muscle growth. Realizing the scale and size of the stent sections involved is the daunting challenge faced in fabricating and developing a prototype. Others have speculated on the impact of nanotechnology on the future treatment of cerebral aneurysms stating that, "The potential for nanotechnology in this field and the future endovascular suite are envisioned" (Hopkins and Ecker 2008) p. 1483.

NANOTECHNOLOGY AND CEREBROVASCULAR DISEASE

Introduction

Nanotechnology can dramatically impact development of minimally invasive protocols for treating neurovascular and cerebrovascular disease. For example, nanotechnology is already influencing clinical neurology and neurosurgery, including the following: medical imaging; therapies for neurologic disease at the nanoscale; nerve tissue engineering; and nanoscale sensors in the brain (Bremner and Pautler 2010; Collins et al. 2010; Gonsalves et al. 2008; Merkoçi 2009; Zhavnerko and Marletta 2006). Nanotechnology is also playing a more prominent role in neurovascular intervention such as treatment of pseudoaneurysms and intracranial aneurysms. Figure 20.3 demonstrates the multiscale nature of endovascular neurosurgery, down to the nanoscale. Advances include optimizing embolization techniques and nanoscale aneurysm positioning sensors with nanorobots (Cavalcanti et al. 2007). In 2006, Tigno postulated that a stent used as tissue scaffold with a magnetic face directed at the aneurysmal neck orifice to induce closure would be the most efficient strategy to treat pseudoaneurysms based on human fibroblast interaction with MNPs previously shown by Berry et al. (2003) and the known nondurability of pseudoaneurysm cases treated with coiling (Kole et al. 2005). This approach ultimately allowed a fast endovascular treatment minimizing cerebral trauma and promptly reconstructing the aneurysmal neck defect.

Hench and Polak (2002) described third-generation biomaterials that could control cell function at the molecular level. Herein lies the important role that nanotechnology potentially can play in endovascular neurosurgery by introducing new advanced, multifunctional materials. For example, copolymer micelles constructed by incubating spinal cord tissue with 60-nm-diameter micelles were used to repair spinal cord injuries (Shi et al. 2010). In the context of endovascular treatment of a cerebral aneurysm, a biocompatible multifunctional stent with a nanostructured scaffold surface

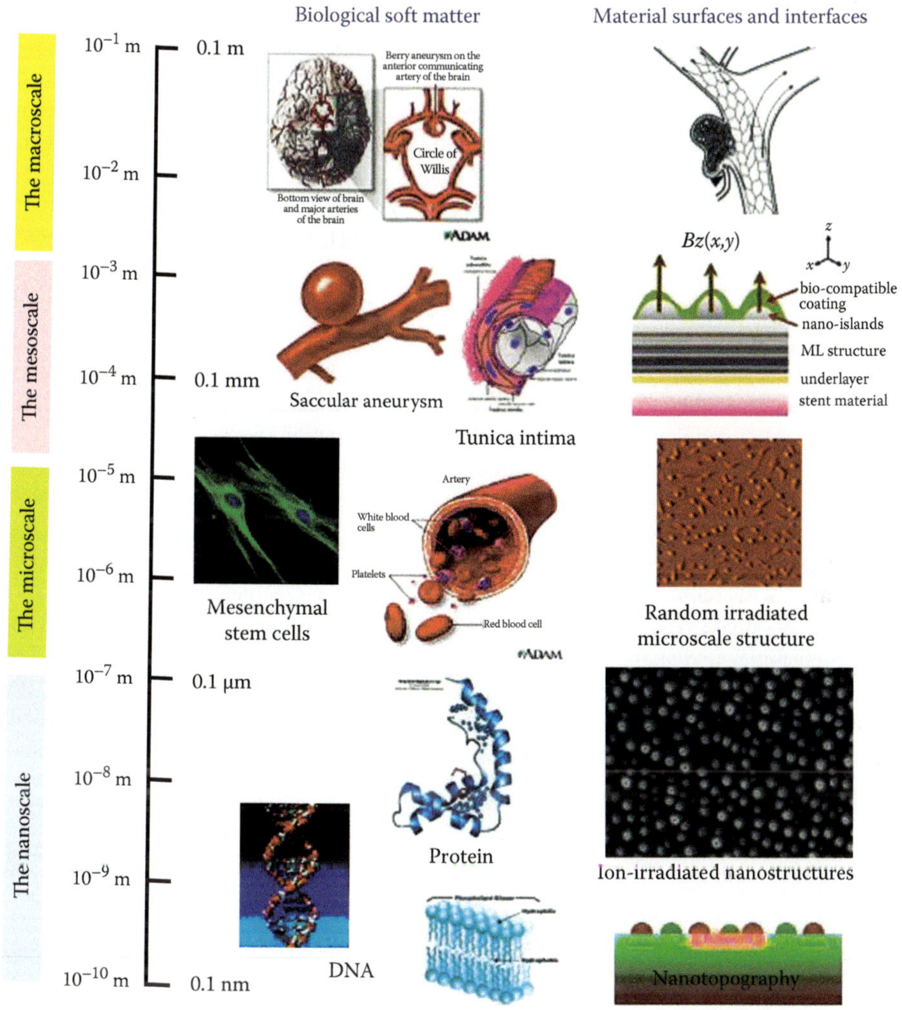

FIGURE 20.3 Organic/inorganic multiscale interface in endovascular nanoneurosurgery.

coating facing the aneurysmal neck defect can be used. Attraction of magnetic or magnetized cells can be locally manipulated with the nanostructured surface promoting tissue growth and facilitating reconstruction of the absent tunica media. Numerous studies provide increasing evidence that nanotopography significantly influences cell behavior. Mesenchymal stem cells respond to cues provided by surface chemistry, stress-induced effects, and the nature of the topography (Aguirre et al. 2010; Dalby et al. 2007; Huang et al. 1999; McNamara et al. 2010). This approach to directed cell growth forms the basis to provide protocols using functionalized stents as scaffolds that can induce in situ reconstruction of the absent tunica media across a pseudoaneurysm or aneurysm neck orifice while preventing stenosis of the parent artery. The proposed endovascular procedure for aneurysm treatment requires stent placement alone, which can be accomplished more rapidly than aneurysm coiling, the current initial endovascular treatment for cerebral aneurysms. Furthermore, by carefully designing multifunctional properties on the scaffold patch, the opportunity to use the arterial wall as a source of stem cell derivatives as mentioned earlier provides for in situ tunica media reconstruction.

Nanotechnology is playing an important role in developing new endovascular treatments. Cell targeting mediated by magnetic attraction methods has recently garnered much attention. In particular, endothelial-related cell therapies have been used as a strategy to treat restenosis within injured arteries (Forbes et al. 2008; Pislaru et al. 2006; Polyak et al. 2008). Conventional strategies have used MNPs or MNP-impregnated ECs and external magnetic fields that attract them to regions around stents. However, these methods use complete symmetric magnetizing stents and, thus, are not very efficient at delivering MNPs to local *asymmetric* regions around the stent and, hence, the injured arterial wall. Recently, work by Polyak et al. (2008) has used a dual magnetic attraction mechanism where not only MNPs but also magnetic stents are used that introduce high field gradients that attract MNPs to the stent scaffold. However, these methods treat the attraction of MNPs and/or magnetized endothelial cells only at a macroscopic level and target the whole stent structure. Since MNPs are attracted to all parts of the scaffold without tunable control, these methods are unsuitable for applications requiring localized delivery to a region of the stent (i.e., in aneurysmal neck defect region). Control and manipulation of particles to the cerebral aneurysmal neck orifice had its early beginnings with the pioneering work of John Alksne (Alksne 1971; Alksne and Fingerhut 1965; Alksne et al. 1966; Fingerhut and Alksne 1966; JAMA 1966). He directed magnetized particles to the region of the cerebral aneurysmal neck and fundus to thrombose intracranial aneurysms.

Rendering a specific region of the stent material magnetic and with nanoscale topography to control cell behavior is part of a cutting-edge neurovascular intervention strategy (Patent Pending, Allain et al.). This, in part, can be accomplished by a novel technique known as *directed irradiation synthesis (DIS)* where materials are designed from "the bottom up" under conditions "far from equilibrium." This approach facilitates athermal synthesis mechanisms of surface nanostructures and metastable phase transformations at the nanoscale, thus providing a means to design and implement a robust nanoscale heterostructure with multiple functionalities tailored to influence cell behavior at the submicron and nanoscale level. DIS, which can be seen in Figure 20.4, provides a complementary approach toward inorganic nanostructure and nanopatterning growth driven by self-organization mechanisms in contrast to self-assembly synthesis approaches (Barth et al. 2005). One example, attracting and manipulating magnetized endothelial cells at submicron levels, makes

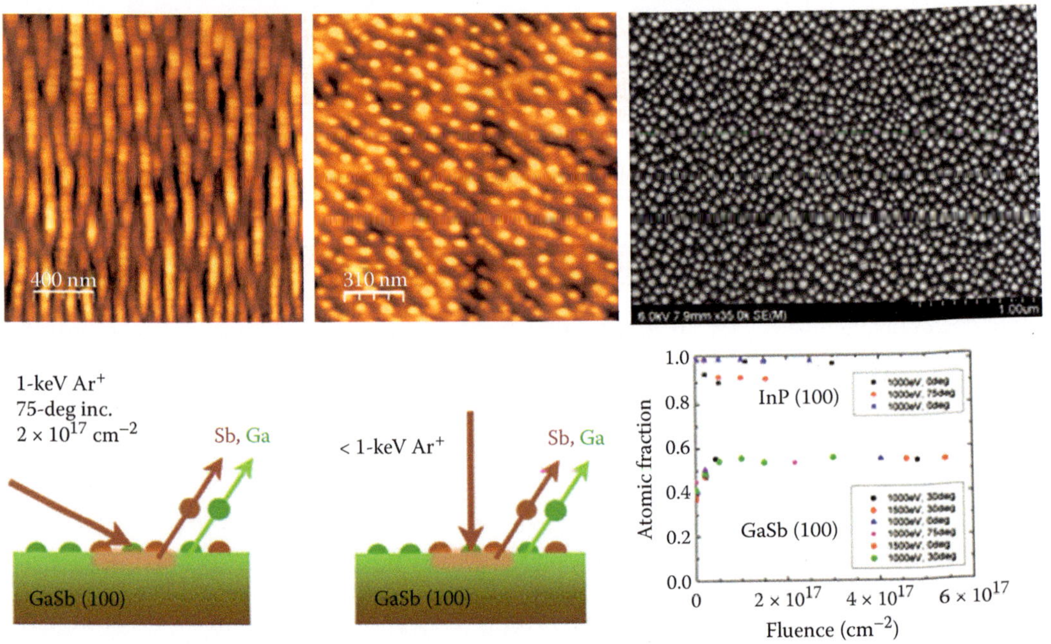

FIGURE 20.4 Nanostructured GaSb (100) interfaces created using directed irradiation synthesis. Tailoring surface structure and in situ measurements of electronic properties and performance using methods including x-ray photoelectron spectroscopy (XPS), angle-resolved photoemission spectroscopy (ARPES), ultraviolet photoemission spectroscopy (UPS), and spectroscopic ellipsometry.

the control and manipulation of tissue engineering intrinsically a "bottom-up" process. The concept of inducing stem cell growth around the orifice of cerebral aneurysms to seed reconstruction of the tunica intima and tunica media is supported by clinical studies of growth along stents (Wehman et al. 2006). The key challenge is devising how to promote growth along the stent surface regions while maintaining the material's structural properties.

Endovascular Techniques and Nanotechnology

Koebbe et al. (2006) described the evolving practice since 2000 of the treatment of cerebral aneurysms shifting from the open surgical clipping approach toward endovascular treatment. Despite endovascular treatment gaining progressive support in the 15-year period after introduction of the Guglielmi detachable coil (Guglielmi et al. 1991a,b), various authors documented suboptimal aneurysmal sac occlusions and postcoiling recurrences. Inadequate occlusion rates were reported to range from 21.3% for small aneurysms to 50.6% for large aneurysms, and from 52.3% of wide-neck aneurysms to 23.7% for aneurysms with small neck size (Kole et al. 2005; Raymond et al. 2003). The same authors found large aneurysmal remnants in 27% of cases examined immediately after postcoiling. Similarly, increased remnant size by angiography was found in 19.1% of 131 postcoiled cases, 14.5% of which required aneurysm recoiling. Thromboembolic events and vessel stenosis postcoiling were also reported (Wang et al. 2004; Workman et al. 2002). Despite these reports, impetus on the further clinical use of endovascular procedures for aneurysms persisted. This was due, in part, to the International Subarachnoid Aneurysm Trial results, which established Level 1 evidence showing significantly reduced disability or death with endovascular treatment compared with surgical clipping (Molyneux et al. 2002a,b).

Broader support for the endovascular alternative continues to be driven by the successive availability of newer stents and stent techniques, coils (bioactive, three-dimensional, ultrasoft, PED), and coil placement techniques (Doerfler et al. 2005; Fiorella et al. 2009; Liebig et al. 2006) as well as thrombogenic materials (Kailasnath and Chaloupka 2002). Despite these new endovascular alternatives, intraprocedural complications continue to be documented. A review of cases involving particular subsets of aneurysms (e.g., MCA, broad-necked, fusiform) (Gorji et al. 2001; Kole et al. 2005; Raymond et al. 2003) reiterates inherent limitations of current endovascular approaches to aneurysm treatment. Endovascular treatment of aneurysms remains focused on toughening the aneurysmal sac rather than rebuilding the aneurysmal neck defect, with the intention of excluding the aneurysm from parent artery perfusion and hemodynamic consequences. Although endovascular delivery of growth factors (e.g., vascular endothelial growth factor, transforming growth factor, fibroblast growth factor) and gene therapy have been attempted in animal models to induce neointimal wall regeneration across the neck (Abrahams et al. 2001; Matsumoto et al. 2003; Poliseno et al. 2006), an effective strategy toward reconstructing the tunica media that is absent at the aneurysmal neck, which is the basic pathology underlying aneurysm formation, remains elusive. Herein lie both the challenge and the opportunity for nanotechnology-based strategies that can impact a localized endovascular approach for treating cerebrovascular disease. Stent and stent-based device surface coatings can be tested as scaffolds for tissue proliferation and reconstruction at the aneurysmal neck region.

Scaffolds, Cerebrovascular Disease, and Nanotechnology

The concept of scaffolds to organize tissue repair is not new and may be applied to vascular tissue repair. Nanotechnology has had a critical impact in the repair and reconstruction of tissue using tailored (nanostructured) scaffolding to influence cell surface response and regulation (Fletcher and Mullins 2010; Liu and Webster 2007; Stevens and George 2005). Figure 20.5 demonstrates how cells respond to scaffolded surfaces. In the context of cell regulation, differentiation, and tissue proliferation, nanotechnology plays an important role in the context of scaffold surface engineering to

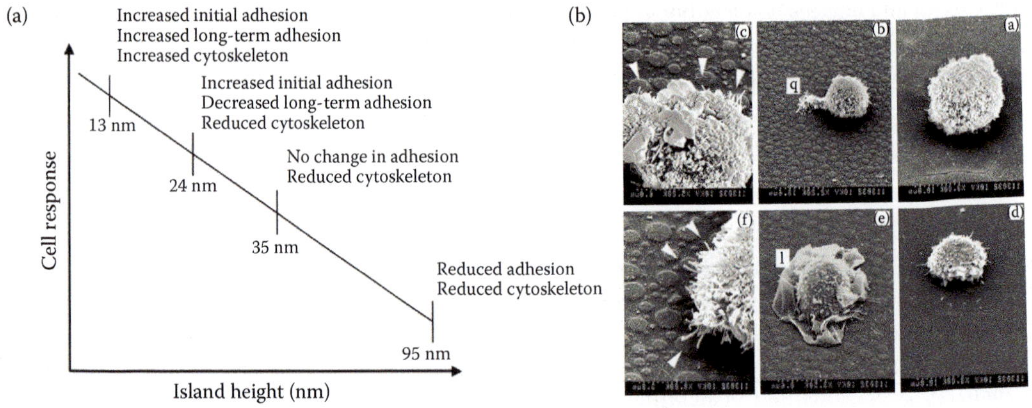

FIGURE 20.5 (a) Plot showing cell response with island height in the nanoscale. Evidence shows correlation of nanoscale height with increase in adhesion and sustainable viability (Gonsalves 2008). (Reprinted with permission of John Wiley & Sons Inc.) (b) Scanning electron micrographs of human fibroblasts on 13 nm and higher nanostructures. Evidence for strong interaction between filopodia and nanoislands is shown (Dalby 2002a).

influence tissue, cellular, and subcellular response (Aguirre et al. 2010; Dalby et al. 2002a,b, 2007; McNamara et al. 2010). The capability of using advanced materials that are both biocompatible (e.g., polymer-based materials) and bioactive (e.g., influencing the cellular biochemical response) drives the influence of nanotechnology on cell and tissue proliferation. In the context of cerebrovascular disease, nanotechnology can play a pivotal role by influencing cell regulation and tissue proliferation on scaffold surfaces placed at the cerebral aneurysmal neck defect.

Synthesizing Nanopatterned Scaffold Surfaces

Traditionally, patterning of material surfaces is conducted with top-down processes such as lithography. Today's advanced lithography uses 193 nm light to expose feature sizes of about 45–70 nm. These structures are routinely achieved at high-volume manufacturing levels. However, in applications related to nanopatterning of scaffold surfaces, the challenge becomes one of conformal deposition on three-dimensional surfaces of stents used as scaffolds to manipulate tissue proliferation. This approach is limited by the method of nanopatterning and the conditions of the stent surface (e.g., roughness, etc.), since they can dramatically alter the surface morphology of the deposited coatings. For example, a stent surface roughness in the order of 10–100 nm is similar to what can be deposited on a 1–5 μm thin film with conventional techniques such as magnetron sputtering at room temperature. Dalby et al. (2007) found control of cell regulation optimized for surface nanostructures between 10 and 100 nm, therefore placing stringent requirements on the substrate interface roughness (Dalby et al. 2007). Electron-beam lithography and other nanoimprint, maskless lithography techniques can produce feature sizes of less than 10 nm. However, it is more challenging to pattern the surfaces of predominantly three-dimensional structures than of two-dimensional structures. Ion-beam irradiation synthesis is becoming an attractive alternative due to its ability to both smoothen surfaces and induce nanopatterning, depending on how one tunes incident particle energy, angle, and fluence (number of particles) (Rokusek et al. 2009).

Directed Irradiation Synthesis: Using Ion-Irradiation Protocol

Tailoring nanopatterned surfaces either with long-range order or with randomized nanostructures is possible using irradiation-driven processes. Irradiation-driven systems have been limited to the study of dissipative systems leading to permanent damage by operating in energy and flux regimes that limit self-organization of low-dimensional state structures. However, energetic radiation such as low-energy heavy-ion beams can be used to induce patterned structures and unique topography at the nanoscale by means of sputtering and other surface-related processes (Chan and Chason 2007; Erlebacher et al. 1999; Facsko et al. 1999; Frost and Rauschenbach 2003; Frost et al. 2000; Makeev and Barabasi 1997; Wei et al. 2009; Ziberi et al. 2005). This approach can have significant implications in designing nanostructured surfaces used for cell engineering and cell biology (Curtis et al. 2006; Dalby 2009; Dalby et al. 2007; Stevens and George 2005). Irradiation-driven systems have been explored in similar moderate energy regimes dominated by knock-on atom displacement regimes (Sigmund 1969) for semiconductor metallization microstructure control (Harper and Rodbell 1997), artificial texturing of ceramics (Yu et al. 1985), engineering nanostructured carbon (Krasheninnikov and Banhart 2007), and compositional patterning of immiscible alloys (Enrique and Bellon 2001). Such findings suggest that irradiation-driven self-organized structures can have critical effects on surface properties of low-dimensional state structures. Despite numerous experiments and models that have investigated ion-induced nanostructure self-organization on surfaces, the application for biomaterial functionalization remains in its infancy. Ion-induced strategies are used only for biocompatibility enhancement. One important limitation in current nanostructure synthesis is its dependence on naturally self-ordered processes that balance kinetic and thermodynamic dissipative forces in the absence of irradiation techniques (Barth et al. 2005). One example is the use of electron nanolithography.

Current cutting-edge work in the area of DIS will transform the synthesis and design of nanostructured systems by identifying the composition-dependent mechanisms that drive self-organization on irradiation-driven nanostructures, enabling tunability and control of their chemical and electronic properties. This is critical to the introduction of novel process design and synthesis pathways tuning bioactive properties of thin-film coatings used in the stents used as scaffolds for neuroendovascular reconstruction. Likely what is learned will be equally applicable for any irradiation-driven synthesis strategy of nanostructures and will have significant impact on future developments in nanoelectronics, nuclear materials, hydrogen storage, and biomaterials, including applications in angiogenesis, orthopedics, and vascular healing.

DIS of Nanostructured Bioactive Stent-Coating Materials

Several techniques can be devised to provide simultaneous irradiation and thermal deposition of irradiation-driven nanostructures on various stent model systems that include thin films of Au, Pd, and Pt. Stainless-steel 316 (SS 316) and nitinol (nickel–titanium alloy) are the conventional materials used for intracranial stents. Both stents are inserted via catheter, SS is expanded using a balloon ("balloon-expandable stent"), and nitinol is expanded by its shape memory effect ("self-expandable stent"). Nitinol is a shape memory alloy with superelastic properties that allow it to be deformed up to 6% without exceeding its elastic limit or causing plastic deformation of the metal (Cragg et al. 1993). The shape memory property of nitinol allows it to expand to its original shape once inserted and heated to body temperature. After implantation, nitinol is known to exhibit significant resistance to deformation, damage, or kink (Gupta et al. 2003). Stainless steel is frequently used in the medical industry, and it is already known to be biocompatible except for some thrombosis issues; nitinol likewise has demonstrated good biocompatibility (Bertrand et al. 1998) although limited by Ni surface enrichment.

Irradiation Modification Improving Biocompatibility of Inorganic Interfaces

Ion irradiation has also been demonstrated to improve biocompatibility of material interfaces (Cui and Luo 1999; Sioshansi and Tobin 1996) and is another example where nanotechnology is influencing control on the multifunctional properties of biomaterials. As the ions are implanted into the target material, they deposit their energies, displace the target atoms, and break the initial bonds. Therefore, the chemical properties as well as the mechanical properties of a material surface are changed when ion implantation is performed (Cui and Luo 1999; Jagielski et al. 2006; Suzuki 2003). Ion implantation has been applied to the modification of metals for improving their wear and corrosion resistance. It has been demonstrated that an order-of-magnitude improvement in wear resistance can be achieved for nitinol modified by oxygen ion implantation. This is because a beneficial precipitate layer develops in the subsurface. The biocompatibility of titanium can be improved by fluoride-ion implantation (Liu et al. 2008). A change in the surface chemical composition, specifically the creation of TiF_3 and TiO_2, accompanied this process. Cells attach more rapidly and strongly to fluoride ion-implanted titanium than to nonirradiated titanium.

Ion implantation has also been applied to the surface modification of polymers to improve compatibility with blood (Pignataro et al. 1997). It improves cell adhesion to polymer surfaces. Complete spreading of the endothelial cells was observed on the ion-implanted domain of polystyrene, while cell adhesion and spreading were reduced or even absent on the nonimplanted region (Suzuki 2003). Significant improvement in cell adhesion, spreading, and proliferation was observed after ion implantation of polyethersulfones and polyurethane (Pignataro et al. 1997). The fundamental reason for such improvement is that ion implantation alters the surface layer to make it more conducive to cellular attachment. Ion implantation cracks the original chemical bonds to form new phases and expose specific functional chemical groups on the polymer surface. It has numerous advantages for modifying biomaterial surfaces, including selectivity, controllability, and reproducibility. Another aspect of ion irradiation is improving a material's wettability and tuning either hydrophilicity or hydrophobicity. Ion-beam irradiation techniques used to improve biocompatibility can be coupled to DIS strategies to synthesize nanostructured and patterned surfaces. This approach can be integrated with a single modification source.

Cell Mediation and Control with Nanostructured Surfaces

A strong correlation, although not completely understood, exists between surface topography and cell function (Curtis et al. 2006; Dalby 2009; Dalby et al. 2007; Stevens and George 2005). Patterned nanostructured and microstructured surfaces can increase cell adhesion because they have increased surface wettability and surface area, and influence cell shape (Ito 1999; Kim et al. 2005; Norman and Desai 2006). Both the topography of the nanopatterned surface and its surface chemistry play a critical role in cell function, including regulating growth factor and intracellular signaling, which are important for proliferating and reconstructing endovascular tissue (Stevens and George 2005). Surface wettability is determined by the contact angle, with stronger cell adhesion on a hydrophilic surface being encouraged by an angle being between 0° and less than 90° (Grinnell and Feld 1982). Nanostructures also improve cell growth by acting as directional cues. Cells are often directed by the ridges/grooves of the nanostructures and will align themselves with the direction of these ridges. Multiple ridges/grooves increase the cell contact inhibition effect as well as make it difficult for cells to grow over the ridges. This "contact guidance" phenomenon generates cell orientation and an orderly cellular pattern over a large surface area (Dalby et al. 2002b,c; Grinnell and Feld 1982; Zhu et al. 2004a,b). Nanostructures, such as the naturally occurring extracellular matrix (ECM), provide support and anchorage for cells as well as directional cues.

Cell shape is a major determinant of cell growth because as cells spread, they elongate. In response to nanostructures, endothelial cells exhibited the same arcuate, or curved, morphology that is typical for vascular endothelium (Dalby et al. 2002a). This well-spread morphology and a well-developed cytoskeleton have been observed for a nanoisland height of 13 nm (Dalby et al. 2002b; Gonsalves et al. 2008). Once the cells have adhered and formed a monolayer, a second specifically aligned monolayer can form on top of it without even having direct contact with the nanostructure substrate (Zhu et al. 2004a,b). This could allow multiple aligned layers of cells in a desired region of a scaffold, for example. The goal of nanostructures is to mimic the in vivo environment for cell growth and proliferation.

Use of Copolymer Micelles (20 nm in Size) and Nidus Cells

Several studies demonstrated enhancement of growth of porcine coronary smooth muscle (PCSM) cells by copolymer micelles and nanotexturing of surfaces of a scaffold. Other cell lines showed similar effects, including human aortic endothelial cells (HAEC) and human umbilical vein endothelial cells (HUVEC) (Invitrogen, Carlsbad, CA). Iron oxide (IO) nanoparticles are well suited for biological testing of cells since they have a long blood retention time, provide magnetic resonance imaging (MRI) contrast, are biodegradable, and exhibit low toxicity (Harisinghani et al. 2003; Peng et al. 2008). Briefly, one can label cells with IO nanocrystals (Ocean NanoTech, Springdale, AK) by incubating the nanocrystals with the cells for 2–3 h at 37°C (Yang et al. 2009) and allowing the cells to internalize the particles. Once the cells take up the IO particles, becoming magnetic, they can be washed of any free nanoparticles and can be observed by transmission electron microscopy (TEM) to verify particle internalization. Once labeled, cells are ready to use immediately. Furthermore, the IO nanoparticles are terminated with a streptavidin end that easily binds to a biotin-conjugated fluorescent antibody or another protein of interest that can be used for further detection, such as by conventional flow-activated cell sorting (FACS) and analysis.

Copolymer micelles enhance cell proliferation on interfaces. Copolymer micelles have been used at the 60-nm scale to treat traumatic spinal cord injuries by manipulating early stages of spinal cord injury (Shi et al. 2010). Copolymer micelles also support cell proliferation on nanopatterned substrates. Micellization of quantum dots (QDs) in copolymer amphiphilic ligands is achieved by incorporation in a nonpolar organic solvent, which is then hydrated in aqueous solution (saline or pure water) at 50°C for 5 min. The nonpolar "heads" of the copolymer repel the hydration, forming micelles and encapsulating the QDs. The crude suspension is centrifuged (20,000*g*, 10 min) to break up aggregates and consequently purified by ultrafiltration (Vivaspin 500, 10,000 MWCO) and ultracentrifugation with a pH 8.5 borate buffer (120,000*g*, 90 min) (Roullier et al. 2008). Monolayers of the micelles can be formed over nanostructured substrates using the dip-coating method (reasonable pulling velocity of 5–20 mm/min). Tapping mode atomic force microscopy (AFM) can be used to characterize their morphology, as shown in Figure 20.6 (Aytun et al. 2008). The same sample can

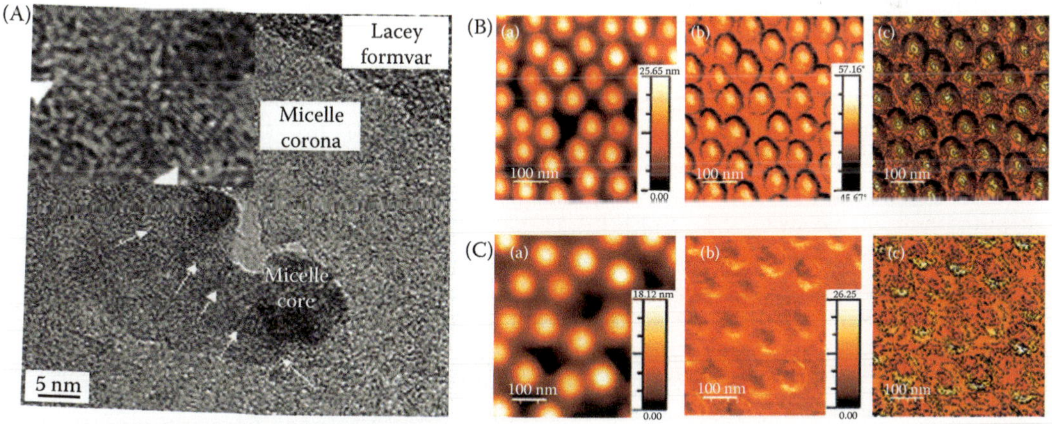

FIGURE 20.6 (A) Bright-field TEM image of the ZnO particles (indicated by the white arrows) produced inside a micelle reactor. The micron bar corresponds to 5 nm. The inset shows a few of the nanocrystals enlarged to facilitate inspection. (B) AFM height (a) and phase (b) images of zinc acetate loaded PS-b-P2VP diblock copolymer micelles with A0 = 40.5 nm, rsp = 0.2. The image in (c) is a contour plot representation of the phase image, which facilitates visualization of the phase contrast of the micelle feature. (C) AFM topo (a) and phase (b) images of zinc acetate loaded PS-b-P2VP diblock copolymer micelles with A0 = 20.25 nm and rsp = 0.6. (c) Contour plot representation of the phase image in part (b) facilitates visualizing the compositional variation in the micelle monolayers.

then be characterized using AFM in liquid that mimics the real environment in which the stent covered with the micellar monolayers will be used. The physical properties of the QDs can also be characterized by TEM and their magnetic fields measured using a superconducting quantum interference device (SQUID) (Roullier et al. 2008).

Magnetic Attraction Protocols for Tissue Engineering and Cell Manipulation

Cell targeting mediated by magnetic attraction methods has garnered much attention due to its high efficiency and its potential use for endothelial-related cell therapies of restenosis of injured arteries (Forbes et al. 2008; Pislaru et al. 2006; Polyak et al. 2008). Conventional strategies localize cells labeled with MNPs to the entire stent region using external magnetic fields. Adequate localization of cells using these conventional magnetic attraction methods requires either an external magnetic field of several kilogauss (Forbes et al. 2008; Polyak et al. 2008) or partial occlusion of blood flow (Pislaru et al. 2006). Figure 20.7 shows a computational model of the attraction of MNPs to a Ni thin film.

Cell attraction by a magnetic protocol is therefore an interesting application of nanoscale functional properties (e.g., magnetic attraction, magnetized nanoparticles, etc.) that can be used in cerebral endovascular aneurysm treatments. The proposed treatment protocol involves (1) use of a stent with a several-micrometer-thick magnetic coating functionalized at the nanoscale that is sufficient to attract cells to the stent surface. Placement of the stent provides an environment more conducive for cell growth within the aneurysm because it reduces the blood flow velocity and vortex within the aneurysm and the drag force acting on the endothelial cells. After placing the stent, (2) catheterization can be used to deliver nidus cells directly to the aneurysm neck orifice, with magnetized cells attracted to the magnetic stent region. Magnetization of cells can be easily achieved using magnetized nanoparticles (Krishnan 2010); superparamagnetic IO nanoparticles (MNP) for cell labeling are commercially available. The velocity profile and flow domain of the blood flow change significantly around the aneurysm neck orifice due to the aneurysm and implantation of the stent. This modified flow pattern has been well studied, and Tateshima et al. (2003, 2009) concluded that implanting the stents reduces the intra-aneurysmal flow velocity of MCA aneurysms by 43%–64% and changes flow patterns in their domes (Aenis et al. 1997; Lieber et al. 1997; Stuhne and Steinman 2004). The primary cause for these changes increased resistance to flow between the parent vessel and the aneurysmal sac as a result of the stent pores.

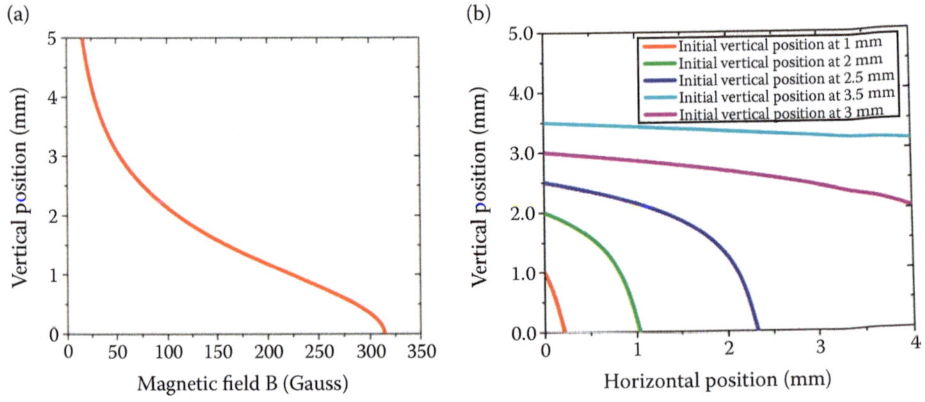

FIGURE 20.7 (a) Model result of magnetic field strength as a function of vertical position of MNP. (b) Model results of vertical position MNP movement toward surface against drag force.

An alternative magnetic protocol to attract cells on an asymmetric region of the stent surface is by perpendicular magnetic anisotropy (dipole ([field]) direction). Perpendicular magnetic anisotropy is currently of great interest and often is attributed to interfacial contribution (interactions at an interface). Sputtered Pd/Co superlattices that are ultrathin display perpendicular anisotropy (Carcia 1988); the dependence on Co thickness is still not understood. Annealing the superlattices at high temperatures causes interdiffusion between layers and perpendicular magnetic anisotropy correlated with interface sharpness. When the Pd/Co was heated above 400°C, the magnetic easy axis (direction of spontaneous magnetization) changed from perpendicular to in plane with a magnetization increase of ~15%. Much of this has been attributed to interface anisotropy because the Co atoms at Pd interfaces are more easily polarized perpendicular to the film surface. Changes in coercivity (resistance to demagnetization) can be attributed to the face-centered cubic (fcc [lattice structure]) to face-centered tetragonal (fct) structural transition. This transition produces high crystal magnetic anisotropy energy (up to 7×10^7 erg/cm^3) and large coercivity values. The coercivities of as-deposited FePt films were small, with the magnetic easy axis lying in the film plane; however, square loops and large coercivities were found after annealing at the transformation temperature (Luo and Sellmyer 1995). The thickness of the soft layer's effect on the coercivity value remains controversial among studies. FePt multilayers of different thicknesses that were annealed at their transition temperature were found to have large coercivities at thicknesses ranging from 300 to 600 Å (Luo and Sellmyer 1995). Yet with thickness greater than 600 Å or less than 300 Å, the coercivity decreased sharply. In another study, FePt/Fe bilayers showed an enhanced remnant magnetization (M_r = 750 against 599 emu/cm^3) over a single hard layer (Casoli et al. 2005). The perpendicular magnetic anisotropy and coercivity seem influenced greatly by the annealing process, the growing method, as well as interface interactions. As for materials, Pt has been widely cited in many papers, with Co/Pt multilayers seemingly promising. $L1_0$ compounds, which are composed of 3d ferromagnetic elements Fe, Co, and Ni, often alloyed with 5d transition elements (i.e., Pd, Rh, Pt, and Ir), are also of interest. "A large interface anisotropy may occur not only if magnetic and nonmagnetic elements are combined in a ML (monolayer), but also if magnetic elements *only* are used. This leads to the prospect of new multilayered films combining a perpendicular magnetization with a high saturation magnetization" (Johnson et al. 1996).

Techniques to Study Nanoscale Bioactive and Biocompatible Scaffold Coatings

In situ characterization of surface chemical properties and nanostructure topography during ion irradiation is critical to the design and manufacture of bioactive stents. As mentioned earlier, cells cue to their environment through subcellular and molecular-level signaling, which depend strongly on both surface chemistry and nanoscale topography. Therefore, when using DIS to induce nanopatterning on scaffold surfaces, the objective is to correlate surface composition and nanotopography evolution during irradiation by heavy-ion sputtering (e.g., Ar, Xe, etc.) with formation of patterns at the nanoscale as shown in Figure 20.8. Figure 20.9 shows a novel methodology integrating complementary surface-sensitive techniques to correlate surface topography evaluated by scanning electron microscopy with surface chemical composition evaluated by surface chemistry techniques (including x-ray photoelectron spectroscopy, low-energy ion spectroscopy, ultraviolet photoelectron spectroscopy, Raman spectroscopy) in one system *during* irradiation of the patterned surface at the nanoscale. Allain and colleagues are performing experiments in the Particle and Radiation Interaction with Hard and Soft Matter (PRIHSM) facility at Purdue University to pioneer the use of postionization secondary neutral mass spectrometry to measure the relative amount of ion-induced desorbed species and to correlate them to changes in surface morphology, cell behavior response, and cellular proliferation (Figure 20.9). As with all surface-sensitive techniques, the requirement of an ultrahigh vacuum environment limits their use for studying cell behavior. However, various synthesis-driven material protocols can be readily studied and then correlated with in vitro

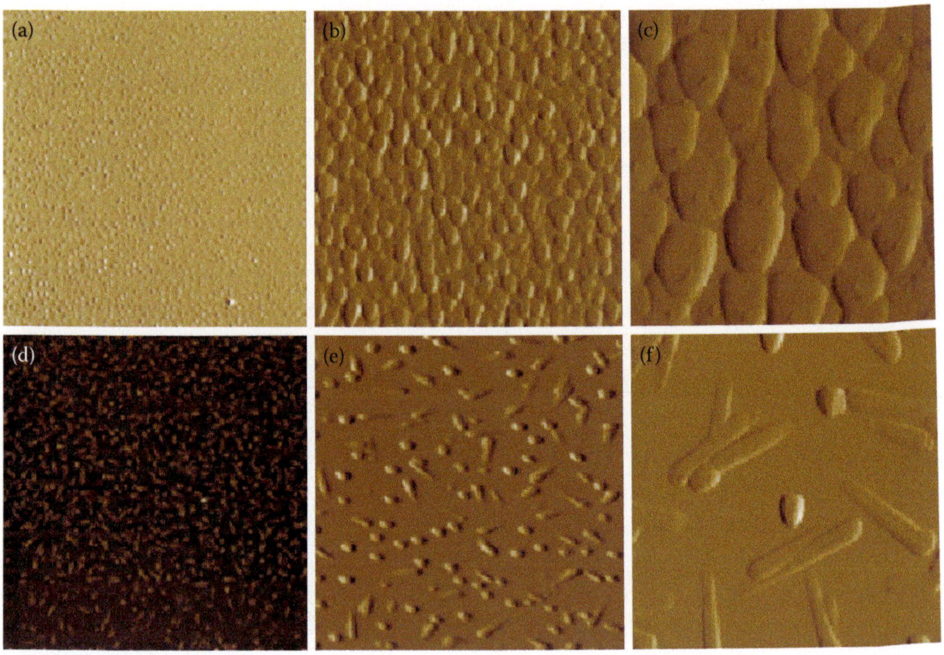

FIGURE 20.8 (a) Deposition of Sn particles imaged with low AFM magnification. (b) Deposition of Sn particles imaged with medium AFM magnification. (c) Deposition of Sn particles imaged with high AFM magnification. (d) AFM low-magnification image of Xe-ion dominant flux. (e) AFM medium-magnification image of Xe-ion dominant flux. (f) AFM high-magnification image of Xe-ion dominant flux. Particle and radiation interactions with hard and soft matter (PRIHSM).

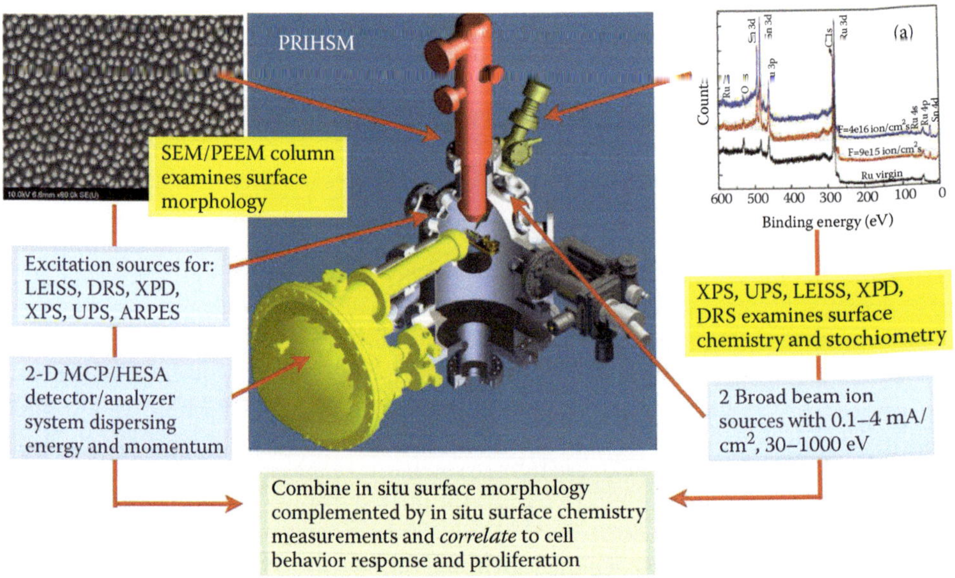

FIGURE 20.9 PRIHSM facility at Purdue University.

cell regulation studies. Furthermore DIS can be integrated with in situ surface characterization as demonstrated in Figures 20.10 and 20.11. Here, the scanning electron microscopy (SEM) technique is used to observe the unique synthesis control of DIS to alter surface morphology by combining ion-beam irradiation and thermal deposition. The materials used are thermal Sn atoms on Ru substrates. Ion-beam irradiation allows control of morphology as a function of ion fluence (number of ion particles per unit area, in this case using Xe low-energy ions). In Figure 20.10, one can observe the ability to control and vary atom density distribution over a surface measuring 10–50 μm.

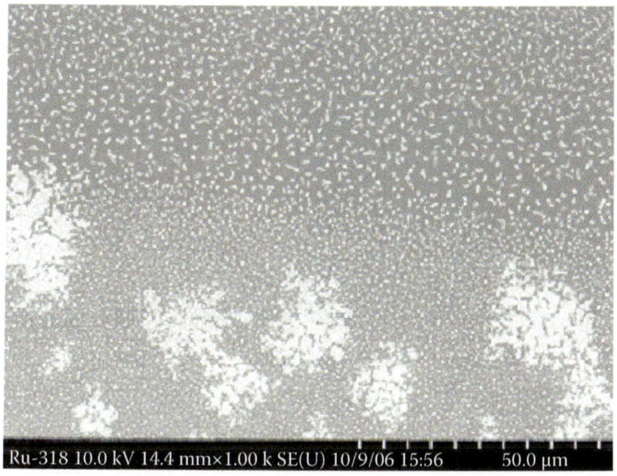

FIGURE 20.10 Scanning electron micrograph with 1000× magnification showing two separate regions exposed to high and low Xe-ion irradiation compared to Sn thermal deposition on a Ru substrate.

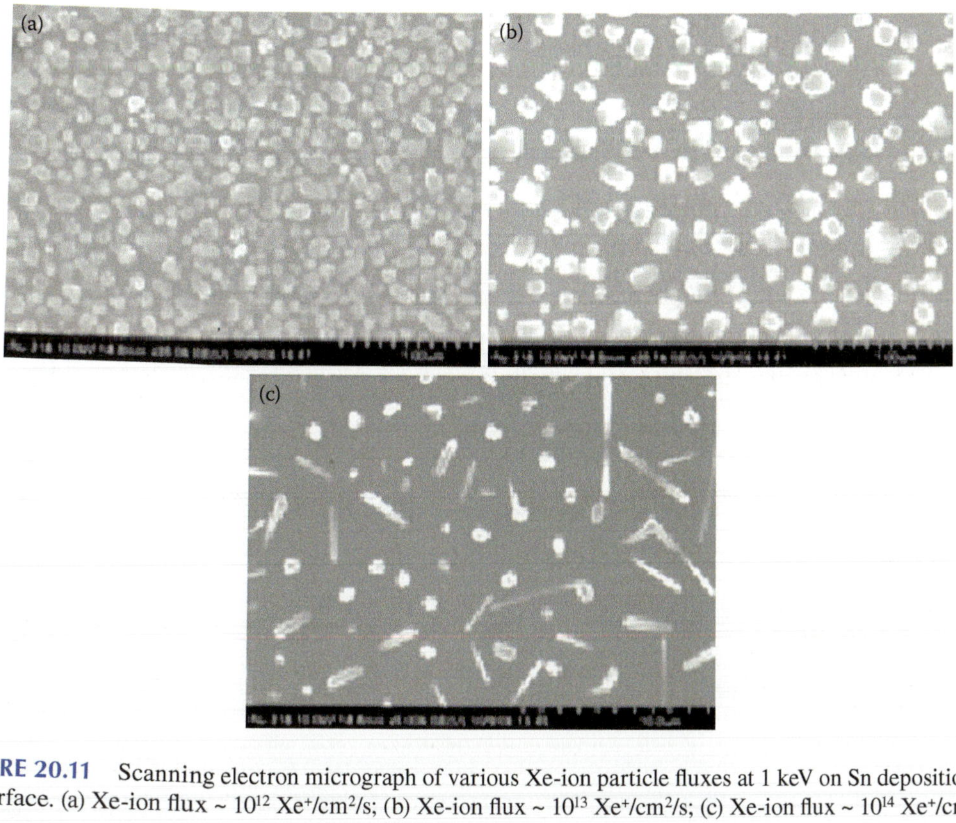

FIGURE 20.11 Scanning electron micrograph of various Xe-ion particle fluxes at 1 keV on Sn deposition of Ru surface. (a) Xe-ion flux ~ 10^{12} $Xe^+/cm^2/s$; (b) Xe-ion flux ~ 10^{13} $Xe^+/cm^2/s$; (c) Xe-ion flux ~ 10^{14} $Xe^+/cm^2/s$.

CONCLUSIONS AND OUTLOOK

The scope and breadth of nanotechnology are vast, and its application to treatment of cerebrovascular disease imminent. There are emerging methods to manipulate cells and tissue near and around the neck orifice of cerebral aneurysms. The challenge is to control cells by giving them cues that lead to their proliferation and ultimately to tissue reconstruction that closes the aneurysmal neck orifice and strengthens the arterial wall defect. Nanotechnology can be envisioned as providing engineering design pathways not only to control thrombosis with endovascular protocols but also to provide a multiscale model for molecular, subcellular, cellular, and tissue manipulation.

ACKNOWLEDGMENTS

A special acknowledgement is given to Eric Yang, Emily Walker, and Alexandra Kacic for assistance in typesetting, scholarly research, and illustration preparation for this chapter. We thank Osman El-Atwani for section on copolymer micelles. Similarly, Lisa Reece of the Birck Nanotechnology Center is acknowledged for lending her perspectives on magnetic nanoparticle applications. We thank our sponsors Nuclear Regulatory Commission (NRC), Purdue Research Foundation, Paul C. Zmola Scholar Award, and the Birck Nanotechnology Center. We acknowledge the support of the Henry M. Jackson Foundation for the Advancement of Military Medicine, Inc. and the Defense Medical Research and Development Program (DMRDP) Grant No. W81XWH-11-2-0067 for realizing the continuing studies that had made this contribution possible.

REFERENCES

Abedin, M., Tintut, Y., and Demer, L.L. 2004. Mesenchymal stem cells and the artery wall. *Circ. Res* 95: 671–6.

Abrahams, J.M., Forman, M.S., Grady, M.S. et al. 2001. Delivery of human vascular endothelial growth factor with platinum coils enhances wall thickening and coil impregnation in a rat aneurysm model. *AJNR Am J Neuroradiol* 22: 1410–7.

ACROSS. 2000. Epidemiology of aneurysmal subarachnoid hemorrhage in Australia and New Zealand: incidence and case fatality from the Australasian Cooperative Research on Subarachnoid Hemorrhage Study (ACROSS). *Stroke* 31: 1843–50.

Aenis, M., Stancampiano, A.P., Wakhloo, A.K. et al. 1997. Modeling of flow in a straight stented and nonstented side wall aneurysm model. *J Biomech Eng* 119: 206–12.

Aguirre, A., Planell, J.A., and Engel, E. 2010. Dynamics of bone marrow-derived endothelial progenitor cell/mesenchymal stem cell interaction in co-culture and its implications in angiogenesis. *Biochem Biophys Res Commun* 400: 284–91.

al-Rajeh, S., Larbi, E.B., Bademosi, O. et al. 1998. Stroke register: experience from the eastern province of Saudi Arabia. *Cerebrovasc Dis* 8: 86–9.

Alksne, J.F. 1971. Stereotactic thrombosis of intracranial aneurysms. *N Engl J Med* 284: 171–4.

Alksne, J.F., and Fingerhut, A.G. 1965. Magnetically controlled metallic thrombosis of intracranial aneurysms. A preliminary report. *Bull Los Angeles Neurol Soc* 30: 153–5.

Alksne, J.F., Fingerhut, A.G., and Rand, R.W. 1966. Magnetically controlled focal intravascular thrombosis in dogs. *J Neurosurg* 25: 516–25.

Anderson, T. 2009. Current and evolving management of subarachnoid hemorrhage. *Crit Care Nurs Clin North Am* 21: 529–39.

Audibert, G., Bousquet, S., Charpentier, C. et al. 2007. [Subarachnoid haemorrhage: epidemiology, genomic, clinical presentation]. *Ann Fr Anesth Reanim* 26: 943–7.

Awada, A., and al Rajeh, S. 1999. The Saudi Stroke Data Bank. Analysis of the first 1000 cases. *Acta Neurol Scand* 100: 265–9.

Aytun, T., Mutaf, O.F., el-Atwani, O.J. et al. 2008. Nanoscale composition mapping of segregation in micelles with tapping-mode atomic force microscopy. *Langmuir* 24: 14183–7.

Barth, J.V., Costantini, G., and Kern, K. 2005. Engineering atomic and molecular nanostructures at surfaces. *Nature* 437: 671–9.

Batjer, H.H., and Duckworth, E.A. 2009. Selected Drake teachings: an affectionate look back and a look forward—the Charles G. Drake lecture: 2006. *Neurosurgery* 65: 360–9; discussion 370–1.

Beck, J., Raabe, A., Szelenyi, A. et al. 2006. Sentinel headache and the risk of rebleeding after aneurysmal subarachnoid hemorrhage. *Stroke* 37: 2733–7.

Benitez, R.P., Silva, M.T., Klem, J. et al. 2004. Endovascular occlusion of wide-necked aneurysms with a new intracranial microstent (Neuroform) and detachable coils. *Neurosurgery* 54: 1359–67; discussion 1368.

Berry, C.C., Wells, S., Charles, S. et al. 2003. Dextran and albumin derivatised iron oxide nanoparticles: influence on fibroblasts in vitro. *Biomaterials* 24: 4551–7.

Bertrand, O.F., Sipehia, R., Mongrain, R. et al. 1998. Biocompatibility aspects of new stent technology. *J Am Coll Cardiol* 32: 562–71.

Brenner, S.A., and Pautler, M. 2010. Nanotechnology applications in vascular disease. *J Nanotechnol Eng Med* 1: 034001.

Carcia, P.F. 1988. Perpendicular magnetic anisotropy in Pd/Co and Pt/Co thin-film layered structures. *J Appl Phys* 63: 5066–73.

Casoli, F., Albertini, F., Fabbrici, S. et al. 2005. Exchange-coupled FePt/Fe bilayers with perpendicular magnetization *IEEE Trans Magn* 41: 3877–9.

Cavalcanti, A., Shirinzadeh, B., Fukuda, T. et al. 2007. Hardware architecture for nanorobot application in cerebral aneurysm. *7th IEEE Conference on Nanotechnology (IEEE-NANO)*, 237–42. Hong Kong.

Chan, W.L., and Chason, E. 2007. Making waves: kinetic processes controlling surface evolution during low energy ion sputtering. *J Appl Phys* 101: 121301.

Christen, T., Bochaton-Piallat, M.L., Neuville, P. et al. 1999. Cultured porcine coronary artery smooth muscle cells. A new model with advanced differentiation. *Circ Res* 85: 99–107.

Christen, T., Verin, V., Bochaton-Piallat, M. et al. 2001. Mechanisms of neointima formation and remodeling in the porcine coronary artery. *Circulation* 103: 882–8.

Collins, K.L., Orringer, D.A., and Patil, P.G. 2010. Nanotechnology in neurosurgery. *J Nanotechnol Eng Med* 1: 034501.

Cragg, A.H., De Jong, S.C., Barnhart, W.H. et al. 1993. Nitinol intravascular stent: results of preclinical evaluation. *Radiology* 189: 775–8.

Cui, F.Z., and Luo, Z.S. 1999. Biomaterials modification by ion-beam processing. *Surf Coat Technol* 112: 278–85.

Curtis, A.S., Dalby, M., and Gadegaard, N. 2006. Cell signaling arising from nanotopography: implications for nanomedical devices. *Nanomedicine (London)* 1: 67–72.

Dalby, M.J. 2009. Nanostructured surfaces: cell engineering and cell biology. *Nanomedicine (London)* 4: 247–8.

Dalby, M.J., Gadegaard, N., Tare, R. et al. 2007. The control of human mesenchymal cell differentiation using nanoscale symmetry and disorder. *Nat Mater* 6: 997–1003.

Dalby, M.J., Marshall, G.E., Johnstone, H.J. et al. 2002a. Interactions of human blood and tissue cell types with 95-nm-high nanotopography. *IEEE Trans Nanobioscience* 1: 18–23.

Dalby, M.J., Riehle, M.O., Johnstone, H. et al. 2002b. In vitro reaction of endothelial cells to polymer demixed nanotopography. *Biomaterials* 23: 2945–54.

Dalby, M.J., Yarwood, S.J., Riehle, M.O. et al. 2002c. Increasing fibroblast response to materials using nanotopography: morphological and genetic measurements of cell response to 13-nm-high polymer demixed islands. *Exp Cell Res* 276: 1–9.

Davis, R.A., and Wetzel, N. 1956. An analysis of the results of treatment of intracranial aneurysms by common carotid artery ligation. *Surg Forum* 6: 510–4.

Doerfler, A., Becker, W.P., Wanke, I. et al. 2005. A novel flexible, retrievable endovascular stent system for small-vessel anatomy: preliminary in vivo data. *AJNR Am J Neuroradiol* 26: 862–8.

Enrique, R., and Bellon, P. 2001. Compositional patterning in immiscible alloys driven by irradiation. *Phys Rev B* 63: 134111.

Erlebacher, J., Aziz, M.J., Chason, E. et al. 1999. Spontaneous pattern formation on ion bombarded Si(001) *Phys Rev Lett* 82: 2330.

Facsko, S., Dekorsy, T., Koerdt, C. et al. 1999. Formation of ordered nanoscale semiconductor dots by ion sputtering. *Science* 285: 1551–3.

Fager, C.A. 1958. The effectiveness and risk of carotid ligation in the treatment of intracranial aneurysms. *Lahey Clin Bull* 10: 226–33.

Ferreira-Gonzalez, I., Marsal, J.R., Ribera, A. et al. 2010. Background, incidence, and predictors of antiplatelet therapy discontinuation during the first year after drug-eluting stent implantation. *Circulation* 122: 1017–25.

Fingerhut, A.G., and Alksne, J.F. 1966. Thrombosis of intracranial aneurysms. An experimental approach utilizing magnetically controlled iron particles. *Radiology* 86: 342–3.

Fiorella, D., Lylyk, P., Szikora, I. et al. 2009. Curative cerebrovascular reconstruction with the Pipeline embolization device: the emergence of definitive endovascular therapy for intracranial aneurysms. *J NeuroIntervent Surg* 1: 56–65.

Fletcher, D.A., and Mullins, R.D. 2010. Cell mechanics and the cytoskeleton. *Nature* 463: 485–92.

Fogarty, T.J., Cranley, J.J., Krause, R.J. et al. 1963. A method for extraction of arterial emboli and thrombi. *Surg Gynecol Obstet* 116: 241–4.

Forbes, Z.G., Yellen, B.B., Halverson, D.S. et al. 2008. Validation of high gradient magnetic field based drug delivery to magnetizable implants under flow. *IEEE Trans Biomed Eng* 55: 643–49.

Frost, F., and Rauschenbach, B. 2003. Nanostructuring of solid surfaces by ion-beam erosion. *Appl Phys A* 77: 1–9.

Frost, F., Schindler, A., and Bigl, F. 2000. Roughness evolution of ion sputtered rotating InP surfaces: pattern formation and scaling laws. *Phys Rev Lett* 85: 4116–9.

Gonsalves, K.E., Halberstadt, C.R., Laurencin, C.T. et al. 2008. *Biomedical Nanostructures*. Hoboken, NJ: John Wiley & Sons, Inc.

Gorji, R., Willoughby, P.H., Cozza, C. et al. 2001. Guglielmi detachable coil: report of intraoperative complications. *J Neurosurg Anesthesiol* 13: 40–2.

Grinnell, F., and Feld, M.K. 1982. Fibronectin adsorption on hydrophilic and hydrophobic surfaces detected by antibody binding and analyzed during cell adhesion in serum-containing medium. *J Biol Chem* 257: 4888–93.

Guglielmi, G., Vinuela, F., Dion, J. et al. 1991a. Electrothrombosis of saccular aneurysms via endovascular approach. Part 2: preliminary clinical experience. *J Neurosurg* 75: 8–14.

Guglielmi, G., Vinuela, F., Sepetka, I. et al. 1991b. Electrothrombosis of saccular aneurysms via endovascular approach. Part 1: electrochemical basis, technique, and experimental results. *J Neurosurg* 75: 1–7.

Gupta, V., Johnson, A.D., Martynov, V. et al. 2003. Nitinol thin film three-dimensional devices: fabrication and applications. In *International Conference on Shape Memory and Superelastic Technologies*, ed. A.R. Pelton and T. Duerig, 639–50. Pacific Grove, CA: SMST Society, Inc.

Hanel, R.A., Boulos, A.S., Sauvageau, E.G. et al. 2005. Stent placement for the treatment of nonsaccular aneurysms of the vertebrobasilar system. *Neurosurg Focus* 18: E8.

Hardy, W.G., Thomas, L.M., Webster, J.E. et al. 1958. Carotid ligation for intracranial aneurysm; a follow-up study of 54 patients. *J Neurosurg* 15: 281–9.

Harisinghani, M.G., Barentsz, J., Hahn, P.F. et al. 2003. Noninvasive detection of clinically occult lymph-node metastases in prostate cancer. *N Engl J Med* 348: 2491–9.

Harper, J.M.E., and Rodbell, K.P. 1997. Microstructure control in semiconductor metallization. *J Vac Sci Technol B* 15: 763.

Hazama, F., and Hashimoto, N. 1987. An animal model of cerebral aneurysms. *Neuropathol Appl Neurobiol* 13: 77–90.

Hench, L.L., and Polak, J.M. 2002. Third-generation biomedical materials. *Science* 295: 1014–7.

Hopkins, L.N., and Ecker, R.D. 2008. Cerebral endovascular neurosurgery. *Neurosurgery* 62: 1483–501; discussion 1501–2.

Hu, Y., Davison, F., Ludewig, B. et al. 2002. Smooth muscle cells in transplant atherosclerotic lesions are originated from recipients, but not bone marrow progenitor cells. *Circulation* 106: 1834–9.

Huang, X.L., Takakura, N., and Suda, T. 1999. In vitro effects of angiopoietins and VEGF on hematopoietic and endothelial cells. *Biochem Biophys Res Commun* 264: 133–8.

Hunner, G.L. 1900. Aneurysm of the aorta treated by the insertion of a permanent wire and galvanism (Moore-Corradi method). *Bull Johns Hopkins Hosp* 11: 263–79.

Ionita, C.N., Paciorek, A.M., Dohatcu, A. et al. 2009. The asymmetric vascular stent: efficacy in a rabbit aneurysm model. *Stroke* 40: 959–65.

Ito, Y. 1999. Surface micropatterning to regulate cell functions. *Biomaterials* 20: 2333–42.

Jager, M., Sager, M., Lensing-Hohn, S. et al. 2005. The critical size bony defect in a small animal for bone healing studies (II): implant evolution and surgical technique on a rat's femur. *Biomed Tech (Berl)* 50: 137–42.

Jagielski, J., Piatkowska, A., Aubert, P. et al. 2006. Ion implantation for surface modification of biomaterials. *Surf Coat Technol* 200: 6355–61.

JAMA. 1966. Medical news: magnet attracts iron to thrombose aneurysms. *JAMA* 195: 28–9.

Jeger, R.V., Brunner-La Rocca, H.P., Hunziker, P.R. et al. 2009. Drug-eluting stents and glycoprotein IIb/IIIa inhibitors in vessels at low anatomic risk: a retrospective analysis of previously published data from the Basel Stent Kosten Effektivitats Trial. *Clin Ther* 31: 2886–93.

Johnson, M.T., Bloemen, P.J.H., den Broeder, F.J.A. et al. 1996. Magnetic anisotropy in metallic multilayers. *Rep Prog Phys* 59: 1409–58.

Kailasnath, P., and Chaloupka, J.C. 2002. Quantitative assessment of polymerization-binding mechanics of cyanoacrylates: model development and validation. *AJNR Am J Neuroradiol* 23: 772–8.

Kanda, K., and Matsuda, T. 1994. In vitro reconstruction of hybrid arterial media with molecular and cellular orientations. *Cell Transplant* 3: 537–45.

Keefer, A., Hislop, S., Singh, M.J. et al. 2010. The influence of aneurysm size on anatomic suitability for endovascular repair. *J Vasc Surg* 52: 873–7.

Kim, P., Kim, D.H., Kim, B. et al. 2005. Fabrication of nanostructures of polyethylene glycol for applications to protein adsorption and cell adhesion. *Nanotechnology* 16: 2420–6.

Koebbe, C.J., Veznedaroglu, E., Jabbour, P. et al. 2006. Endovascular management of intracranial aneurysms: current experience and future advances. *Neurosurgery* 59: S93–S102; discussion S3–S13.

Kole, M.K., Pelz, D.M., Kalapos, P. et al. 2005. Endovascular coil embolization of intracranial aneurysms: important factors related to rates and outcomes of incomplete occlusion. *J Neurosurg* 102: 607–15.

Koo, M.H., Nawarskas, J.J., and Frishman, W.H. 2008. Prasugrel: a new antiplatelet drug for the prevention and treatment of cardiovascular disease. *Cardiol Rev* 16: 314–8.

Krasheninnikov, A.V., and Banhart, F. 2007. Engineering of nanostructured carbon materials with electron or ion beams. *Nat Mater* 6: 723–33.

Krishnan, K.M. 2010. Biomedical nanomagnetics: a spin through possibilities in imaging, diagnostics, and therapy. *IEEE Trans Magn* 46: 2523–58.

Krucoff, M.W., Kereiakes, D.J., Petersen, J.L. et al. 2008. A novel bioresorbable polymer paclitaxel-eluting stent for the treatment of single and multivessel coronary disease: primary results of the COSTAR (Cobalt Chromium Stent With Antiproliferative for Restenosis) II study. *J Am Coll Cardiol* 51: 1543–52.

Kulcsar, Z., Houdart, E., Bonafe, A. et al. 2011. Intra-aneurysmal thrombosis as a possible cause of delayed aneurysm rupture after flow-diversion treatment. *AJNR Am J Neuroradiol* 32: 20–5.

Lanzino, G., Wakhloo, A.K., Fessler, R.D. et al. 1999. Efficacy and current limitations of intravascular stents for intracranial internal carotid, vertebral, and basilar artery aneurysms. *J Neurosurg* 91: 538–46.

Lieber, B.B., Stancampiano, A.P., and Wakhloo, A.K. 1997. Alteration of hemodynamics in aneurysm models by stenting: influence of stent porosity. *Ann Biomed Eng* 25: 460–9.

Liebig, T., Henkes, H., Reinartz, J. et al. 2006. A novel self-expanding fully retrievable intracranial stent (SOLO): experience in nine procedures of stent-assisted aneurysm coil occlusion. *Neuroradiology* 48: 471–8.

Liu, H.Y., Wang, X.J., Wang, L.P. et al. 2008. Effect of fluoride-ion implantation on the biocompatibility of titanium for dental applications. *Appl Surf Sci* 254: 6305–12.

Liu, H., and Webster, T.J. 2007. Nanomedicine for implants: a review of studies and necessary experimental tools. *Biomaterials* 28: 354–69.

Luessenhop, A.J., and Velasquez, A.C. 1964. Observations on the tolerance of the intracranial arteries to catheterization. *J Neurosurg* 21: 85–91.

Luo, C.P., and Sellmyer, D.J. 1995. Magnetic properties and structure of Fe/Pt thin films. *IEEE Trans Magn* 31: 2764–6.

Makeev, M.A., and Barabasi, A.-L. 1997. Ion-induced effective surface diffusion ion sputtering. *Appl Phys Lett* 71: 2800.

Mathis, J.M., Evans, A.J., DeNardo, A.J. et al. 1997. Hydrophilic coatings diminish adhesion of glue to catheter: an in vitro simulation of NBCA embolization. *AJNR Am J Neuroradiol* 18: 1087–91.

Matsumoto, H., Terada, T., Tsuura, M. et al. 2003. Basic fibroblast growth factor released from a platinum coil with a polyvinyl alcohol core enhances cellular proliferation and vascular wall thickness: an in vitro and in vivo study. *Neurosurgery* 53: 402–7; discussion 407–8.

McNamara, L.E., McMurray, R.J., Biggs, M.J. et al. 2010. Nanotopographical control of stem cell differentiation. *J Tissue Eng* 2010: 120623.

Merkoçi, A. 2009. *Biosensing Using Nanomaterials*. Hoboken, NJ: John Wiley & Sons.

Molday, R.S., Yen, S.P., and Rembaum, A. 1977. Application of magnetic microspheres in labelling and separation of cells. *Nature* 268: 437–8.

Molyneux, A., Kerr, R., Stratton, I. et al. 2002a. International Subarachnoid Aneurysm Trial (ISAT) of neurosurgical clipping versus endovascular coiling in 2143 patients with ruptured intracranial aneurysms: a randomised trial. *Lancet* 360: 1267–74.

Molyneux, A., Kerr, R., Stratton, I. et al. 2002b. International Subarachnoid Aneurysm Trial (ISAT) of neurosurgical clipping versus endovascular coiling in 2143 patients with ruptured intracranial aneurysms: a randomized trial. *J Stroke Cerebrovasc Dis* 11: 304–14.

Murayama, Y., Tateshima, S., Gonzalez, N.R. et al. 2003. Matrix and bioabsorbable polymeric coils accelerate healing of intracranial aneurysms: long-term experimental study. *Stroke* 34: 2031–7.

Murayama, Y., Vinuela, F., Suzuki, Y. et al. 1999. Development of the biologically active Guglielmi detachable coil for the treatment of cerebral aneurysms. Part II: an experimental study in a swine aneurysm model. *AJNR Am J Neuroradiol* 20: 1992–9.

Norman, J.J., and Desai, T.A. 2006. Methods for fabrication of nanoscale topography for tissue engineering scaffolds. *Ann Biomed Eng* 34: 89–101.

Ong, A.T., and Serruys, P.W. 2005. Drug-eluting stents: current issues. *Tex Heart Inst J* 32: 372–7.

Owen, C.S. 1978. High gradient magnetic separation of erythrocytes. *Biophys J* 22: 171–8.

Park, D.W., Yun, S.C., Lee, S.W. et al. 2008. Stent thrombosis, clinical events, and influence of prolonged clopidogrel use after placement of drug-eluting stent data from an observational cohort study of drug-eluting versus bare-metal stents. *JACC Cardiovasc Interv* 1: 494–503.

Peng, X.-H., Qian, X., Mao, H. et al. 2008. Targeted magnetic iron oxide nanoparticles for tumor imaging and therapy. *Int J Nanomed* 3: 311–21.

Phillips, B. 1832. *A Series of Experiments Performed for the Purpose of Shewing that Arteries may be Obliterated without Ligature, Compression or the Knife.* London, UK: Longman, Rees, Orme, Brown, and Green.

Pignataro, B., Conte, E., Scandurra, A. et al. 1997. Improved cell adhesion to ion beam-irradiated polymer surfaces. *Biomaterials* 18: 1461–70.

Pislaru, S.V., Harbuzariu, A., Gulati, R. et al. 2006. Magnetically targeted endothelial cell localization in stented vessels. *J Am Coll Cardiol* 48: 1839–45.

Poliseno, L., Cecchettini, A., Mariani, L. et al. 2006. Resting smooth muscle cells as a model for studying vascular cell activation. *Tissue Cell* 38: 111–20.

Polyak, B., Fishbein, I., Chorny, M. et al. 2008. High field gradient targeting of magnetic nanoparticle-loaded endothelial cells to the surfaces of steel stents. *Proc Natl Acad Sci U S A* 105: 698–703.

Ransohoff, J. 1886. A case of aortic aneurism treated by the insertion of wire. *JAMA* 7: 481–5.

Raymond, J., Guilbert, F., Weill, A. et al. 2003. Long-term angiographic recurrences after selective endovascular treatment of aneurysms with detachable coils. *Stroke* 34: 1398–403.

Rhoton, A.L.J. 2007. *Rhoton's Cranial Anatomy and Surgical Approaches.* Philadelphia, PA: Lippincott Williams & Wilkins.

Rhoton, A.L.J., and Natori, Y. 1996. *The Orbit and Sellar Region: Microsurgical Anatomy and Operative Approaches.* New York: Thieme Medical Publishers.

Rochalska, M. 2009. The influence of magnetic fields on flora and fauna. *Med Pr* 60: 43–50.

Rokusek, D.L., Wagener, C.R., Nieto-Perez, M. et al. 2009. In-situ surface characterization of mechanical and electronic behavior during ion-induced nanostructure synthesis. *Mater Res Soc Symp*, EE05-16. Warrendale, PA: Materials Research Society.

Rordorf, G., Bellon, R.J., Budzik, R.F., Jr. et al. 2001. Silent thromboembolic events associated with the treatment of unruptured cerebral aneurysms by use of Guglielmi detachable coils: prospective study applying diffusion-weighted imaging. *AJNR Am J Neuroradiol* 22: 5–10.

Roullier, V., Grasset, F.F.A., Boulmedais, F. et al. 2008. Small bioactivated magnetic quantum dot micelles. *Chem Mater* 20: 6657–65.

Serbinenko, F.A. 1974. Balloon catheterization and occlusion of major cerebral vessels. *J Neurosurg* 41: 125–45.

Shenkin, H.A., Polakoff, P., and Finneson, B.E. 1958. Intracranial internal carotid artery aneurysms; results of treatment by cervical carotid artery ligation. *J Neurosurg* 15: 183–91.

Shi, Y., Kim, S., Huff, T.B. et al. 2010. Effective repair of traumatically injured spinal cord by nanoscale block copolymer micelles. *Nat Nanotechnol* 5: 80–7.

Shinokuma, T., Wada, A., Morita, M. et al. 2000. Right middle cerebral artery occlusion during endovascular embolisation of cerebral aneurysm by Guglielmi detachable coil: a case report. *Masui* 49: 1024–6.

Sigmund, P. 1969. Theory of sputtering, I. Sputtering yield of amorphous and polycrystalline targets. *Phys Rev* 184: 383–416.

Sioshansi, P., and Tobin, E.J. 1996. Surface treatment of biomaterials by ion beam processes. *Surf Coat Technol* 83: 175–82.

Soeda, A., Sakai, N., Sakai, H. et al. 2004. Endovascular treatment of asymptomatic cerebral aneurysms: anatomic and technical factors related to ischemic events and coil stabilization. *Neurol Med Chir (Tokyo)* 44: 456–65; discussion 466.

Stevens, M.M., and George, J.H. 2005. Exploring and engineering the cell surface interface. *Science* 310: 1135–8.

Stiebel-Kalish, H., Maimon, S., Amsalem, J. et al. 2003. Evolution of oculomotor nerve paresis after endovascular coiling of posterior communicating artery aneurysms: a neuro-ophthalmological perspective. *Neurosurgery* 53: 1268–73; discussion 1273–4.

Stuhne, G.R., and Steinman, D.A. 2004. Finite-element modeling of the hemodynamics of stented aneurysms. *J Biomech Eng* 126: 382–7.

Suzuki, J., and Miura, Y. 1967. Surgery of multiple cerebral aneurysm. *Shujutsu* 21: 67–76.

Suzuki, Y. 2003. Ion beam modification of polymers for the application of medical devices. *Nucl Instrum Methods Phys Res Sect B* 206: 501–6.

Tan, L., Dodd, R.A., and Crone, W.C. 2003. Corrosion and wear-corrosion behavior of NiTi modified by plasma source ion implantation. *Biomaterials* 24: 3931–9.

Tateshima, S., Tanishita, K., Hakata, Y. et al. 2009. Alteration of intraaneurysmal hemodynamics by placement of a self-expandable stent. Laboratory investigation. *J Neurosurg* 111: 22–7.

Tateshima, S., Vinuela, F., Villablanca, J.P. et al. 2003. Three-dimensional blood flow analysis in a wide-necked internal carotid artery-ophthalmic artery aneurysm. *J Neurosurg* 99: 526–33.

Thornton, J., Dovey, Z., Alazzaz, A. et al. 2000. Surgery following endovascular coiling of intracranial aneurysms. *Surg Neurol* 54: 352–60.

Tintut, Y., Alfonso, Z., Saini, T. et al. 2003. Multilineage potential of cells from the artery wall. *Circulation* 108: 2505–10.

Uda, K., Goto, K., Ogata, N. et al. 1998. Embolization of cerebral aneurysms using Guglielmi detachable coils—problems and treatment plans in the acute stage after subarachnoid hemorrhage and long-term efficiency. *Neurol Med Chir (Tokyo)* 38: 143–52; discussion 152–4.

Velpeau, A. 1831. Memoir on the pique or puncture of arteries in the treatment of aneurysms. *Gaz Med Paris* 2: 1–4.

Waller, H.D., Schlegel, B., Muller, A.A. et al. 1985. Der Hemoglobininhalt in Alternden Erythrocyten. *Klin Wochenschr* 42: 898.

Wang, D., Ling, F., and Wang, A. 2004. Complication analysis of intracranial aneurysm embolization with controllable coils. *Chin Med Sci J* 19: 51–5.

Wehman, J.C., Hanel, R.A., Levy, E.I. et al. 2006. Giant cerebral aneurysms: endovascular challenges. *Neurosurgery* 59: S125–S38; discussion S3–S13.

Wei, Q.M., Zhou, X.L., Joshi, B. et al. 2009. Self-assembly of ordered semiconductor nanoholes by ion beam sputtering. *Adv Mat* 21: 2865.

Werner, S.C., Blakemore, S.H., and King, B.G. 1941. Aneurysm of the internal carotid artery within the skull: wiring and electrothermic coagulation. *JAMA* 116: 578–82.

Workman, M.J., Cloft, H.J., Tong, F.C. et al. 2002. Thrombus formation at the neck of cerebral aneurysms during treatment with Guglielmi detachable coils. *AJNR Am J Neuroradiol* 23: 1568–76.

Yang, L., Peng, X.H., Wang, Y.A. et al. 2009. Receptor-targeted nanoparticles for in vivo imaging of breast cancer. *Clin Cancer Res* 15: 4722–32.

Yasargil, M.G. 1969. Possibilities and limitations of vascular surgery of the central nervous system. *Bull Schweiz Akad Med Wiss* 24: 487–93.

Yasargil, M.G., and Fox, J.L. 1975. The microsurgical approach to intracranial aneurysms. *Surg Neurol* 3: 7–14.

Yasargil, M.G., Krayenbuhl, H.A., and Jacobson, J.H., 2nd. 1970. Microneurosurgical arterial reconstruction. *Surgery* 67: 221–33.

Yu, L.S., Harper, J.M.E., Cuomo, J.J. et al. 1985. Alignment of thin films by glancing angle ion bombardment during deposition. *Appl Phys Lett* 47: 932.

Zhavnerko, G., and Marletta, G. 2006. Self-organization at the molecular level and surface patterning. In *Encyclopedia of Surface and Colloid Science*, ed. P. Somasundaran, 5544–57. Boca Raton, FL: Taylor & Francis.

Zhu, B., Lu, Q., Yin, J. et al. 2004a. Effects of laser-modified polystyrene substrate on CHO cell growth and alignment. *J Biomed Mater Res B Appl Biomat* 70B: 43–8.

Zhu, B., Zhang, Q., Lu, Q. et al. 2004b. Nanotopographical guidance of C6 glioma cell alignment and oriented growth. *Biomaterials* 25: 4215–23.

Ziberi, B., Frost, F., Höche, T. et al. 2005. Ripple pattern formation on silicon surfaces by low-energy ion-beam erosion: experiment and theory. *Phys Rev B* 72: 235310.

21 Improving Treatment of Stroke through Nanotechnology

Francesca Nicholls, Natalia Gorenkova, Ellen Bible, Christy Franco, David Chau, Christos Tolias, Kevin Shakesheff, Jennifer West, and Michel Modo

CONTENTS

INTRODUCTION

Stroke is the main cause of adult disability in the Western world, with an annual incidence of 94 per 100,000 (Feigin et al. 2009). Ischemic stroke occurs when blood flow to an area of the brain is disrupted, normally caused by occlusion of a cerebral artery by thrombi or atherosclerotic plaques. Reduced blood flow (Figure 21.1) initially causes loss of electrical activity in neurons, with other cellular functions unaffected (Astrup et al. 1977). As ischemia persists beyond approximately 30 min (Lipton 1999), vital cell functions, such as membrane ion pumps, start to fail, leading to morphological changes and, eventually, cell death (Bell et al. 1985). This stepwise process yields a core of dead tissue, temporarily surrounded by the penumbra, consisting of cells that have lost their electrical activity but not yet their other functions. Cells within the penumbra are potentially salvageable with return of blood flow. Current treatments aim to dissolve thrombi and to restore blood flow to rescue cells in the penumbral region. From more than 1026 experimental treatments tested (e.g., neuroprotective factors), no efficacious approach has emerged to save more dying cells within the penumbra in stroke patients than return of blood flow alone (O'Collins et al. 2006). There are also currently no therapies available to replace tissue that has already been lost.

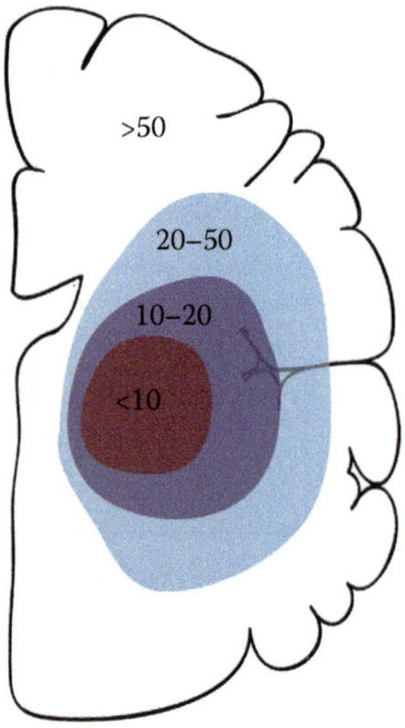

FIGURE 21.1 Damage gradient after ischemic stroke. Numbers represent approximate cerebral blood flow (in mL/100 g/min) based on Moustafa and Baron (2008). The blue area represents tissue that would be salvageable by reperfusion alone. Tissue within the purple area would additionally require neuroprotective treatment. The red area represents the ischemic core, where no protective treatment can improve outcome. Only de novo tissue formation within the core could lead to possible restoration of function within this area.

Stem cell therapy is proving to be a promising area of research for stroke. These cells not only can potentially provide a localized delivery of prosurvival factors, such as glial cell-derived neurotrophic factor (GDNF), and modulate inflammatory response, but also can potentially repair damaged tissue. Although proof-of-concept clinical trials (Bang et al. 2005; Kondziolka et al. 2000, 2005) aim to implement efficacy previously demonstrated in animal studies (Borlongan et al. 1998; Stroemer et al. 2009; Wakabayashi et al. 2010), no robust clinical efficacy has so far been demonstrated. At present, it is unlikely that efficacy of cell transplants is caused by replacement of lost cells or tissues; instead, behavioral improvements are generally ascribed to beneficial effects on host cells (Jiang et al. 2006). Putative mediators of recovery include regional neuroprotection, neuronal plasticity, neurogenesis, and angiogenesis (Shen et al. 2010). Beneficial effects mediated through the vasculature are also thought to underpin improvements associated with intravascular infusion or intraparenchymal injection to deliver non-neuronal cells (e.g., bone marrow stem cells, umbilical cord cells). These cells typically exhibit low brain penetration and survival. However, neural stem cells (NSCs) vary widely in their survival, with functional recovery being reported even in the absence of significant transplant survival (Saporta et al. 1999). Importantly, NSCs implanted into the lesion cavity exhibit poor survival and migrate into existing tissue, rather than regenerating lost tissue (Buhnemann et al. 2006). This is significantly different from fetal tissue grafts implanted into the lesion cavity. In addition to NSCs, fetal tissue grafts contain endothelial cells, microglia, extracellular matrix, and partially differentiated cells. Fetal tissue transplants therefore form a "tissue clump" within the cavity that poorly integrates with the host brain. Still, this tissue clump survives with the formation of some axonal connections and the integration of blood vessels (Grabowski et al. 1995). Survival of fetal grafts, in contrast to NSCs, after transplantation may be attributed to the presence of a complex microenvironment composed of the extracellular matrix, stromal cells, and soluble factors that support fetal progenitors after delivery. This, quite literally, leaves a niche that could be filled by nanotechnology (Orive et al. 2009). Integration of nanotechnology to deliver, support, and improve therapeutic interventions is essential to limit the impact of stroke or to replace lost tissue through tissue engineering.

IMPROVING THERAPEUTIC SUCCESS THROUGH NANOTECHNOLOGY

Limiting the Vascular Impact of Stroke

The first aim for stroke is to limit the duration of this ischemic event. Nanocarriers, such as liposomes, injected intravenously could be functionalized to respond to proteins expressed on thrombi. Nanoparticles attaching to fibrin contained within the thrombus could anchor drug-delivering carriers and ensure a high dose of antithrombotic factor release within the occlusion, hence avoiding side effects in other unaffected areas (Marsh et al. 2007). Moreover, antithrombotic therapies could be maintained after a stroke within the lesion area to ensure that the disintegrated fibrin loosened from the clot does not reaggregate to re-form a new clot or to limit blood flow (Shrivastava et al. 2011).

Once the clot has been cleared, the next therapeutic challenge lies in preventing endothelial cell death. Compromised viability of blood vessels and the blood–brain barrier (BBB) can lead to small hemorrhages in the core or in the peri-infarct area, causing infiltration of blood cells into the brain, which is deleterious to recovery (Belayev et al. 1996). Being able to intravenously administer protective antiapoptotic agents or factors that stabilize unstable reperfused arteries could limit a second wave of neuronal damage associated with ischemic stroke. Ideally, nanoparticles would attach to damaged endothelial cells (Winer et al. 2010). Surface characteristics of nanoparticles are known to determine this interaction, but cellular uptake and transcytosis should also be considered (Georgieva et al. 2010; Taite and West 2006). Damage to the BBB typically results in increased infiltration of lymphocytes, which are generally thought to hamper recovery and cause additional damage (Arumugam et al. 2005; Nilupul Perera et al. 2006; Wang et al. 2007). Anti-inflammatory agents can be effectively delivered using nanoparticles that target areas affected by stroke while maintaining a functional immune system to fight off infections. However, at present, there are no reports of studies using nanoparticles to target blood vessel survival or inflammation after stroke, although an extensive effort in ophthalmology is underway (Souto et al. 2010).

Additional blood supply to the stroke area can be achieved by promoting angiogenesis, a process characteristic of the perilesional area (Arai et al. 2009). Release of proangiogenic factors, such as vascular endothelial growth factor (VEGF) or platelet-derived growth factor (PDGF), will increase the formation of new blood vessels (Moon et al. 2010; Saif et al. 2010; Saik et al. 2011; Bible 2012a). Formation of new blood vessels is likely to be a major factor in maintaining and rescuing brain cells that were affected by ischemia but remain viable. Indeed, transplantation of endothelial cells into the stroke-damaged brain reduces infarct size and promotes behavioral recovery (Fan et al. 2010). These endothelial cells not only improve vascular supply, but also deliver growth factors that are beneficial to the survival of neuronal cells (Moubarik et al. 2011).

Preserving Dying Neurons

Decreased blood supply will inevitably lead to brain cell death. Engineering of nanoparticles or nanofibers to secrete neuroprotective factors (Bertram et al. 2010; Gouhier et al. 2000; Luk et al. 2004; Ward et al. 2007) or more conventional pharmaceutical approaches, such as the *N*-methyl-D-aspartate antagonist memantine (Joshi et al. 2010; Liu et al. 2009), could dramatically reduce the impact of stroke. Current strategies suffer from poor BBB penetration, quick metabolization of compounds, and potential significant side effects due to drugs affecting other organ systems (Ciofani et al. 2008; Dirnagl 2006). Administration of low doses in clinical trials has, for instance, been identified as one of the key factors that stymie the translation of efficacy from animals to humans (Lees 1997). A targeted intralesional approach using nanocarriers would enable effective doses to be administered directly to the affected area without systemic effects, such as hypertension. For instance, superoxide dismutase, a free-radical scavenger, can be incorporated into biodegradable poly(D,L-lactide-*co*-glycolide) nanoparticles and achieve a 65% reduction in lesion volume, whereas the same dose in solution only achieved a 25% decrease (Reddy and Labhasetwar 2009). Therefore, localized delivery can dramatically improve therapeutic intervention compared to conventional drug delivery.

Importantly, the release of factors over protracted periods, from minutes to weeks, could essentially be used to nurse dying cells back to life and functionality. However, little work has been conducted in this potentially promising area. One challenge is controlling the release of chemical factors in a poorly characterized environment. Although in vitro efforts are underway to establish how various factors within an ischemic environment influence the release of compounds at physiologically relevant doses (Kurtoglu et al. 2009), the complexity of this environment is easily underestimated. Not only are cytokines released in response to stroke, but biomaterials themselves might engender a response. The environment also changes with time. For instance, initially, the tissue mostly retains its extracellular matrix and integrity, possibly limiting the spread of nanoparticles (Klose et al. 2009). As cells die, macrophages gradually clear the debris, and the ensuing cavity is filled with either cerebrospinal fluid or extracellular fluid. It is likely that nanoparticles will degenerate differently within this evolving environment. Time-consuming in vivo studies might therefore be required to establish a biodegradation profile and its appropriate factor release. There are currently no reports indicating how different biomaterials degrade within the stroke environment. Degradation characteristics and drug release could therefore be difficult to estimate.

Therapeutic Targeting

Although intralesional administration is likely to be the most effective means of achieving a large number of nanoparticles at the site of stroke, it is an invasive procedure that could itself lead to brain damage. Therefore, an important aspect of intralesional delivery is the use of imaging to guide the injection site and volume of particles within the lesion environment (see below for a detailed discussion) (Bible et al. 2009a). Alternatively, with magnetic nanoparticles, it is possible to localize therapy, such as thrombolysis, to a specific area (Torno et al. 2008). The use of neodymium magnets to direct therapy is fairly crude and would not allow the targeting of a stroke area seated deep inside the brain. However, one study involving the intravenous delivery of magnetically labeled stem cells to areas of stroke damage proves that magnetic targeting can improve therapeutic efficacy (Song et al. 2010). An alternative method for achieving brain penetration is administering therapeutic microparticles intranasally (Dalpiaz et al. 2008), potentially circumventing the BBB. Application of microbubble ultrasound in addition to an antithrombolytic agent, termed sonothrombolysis, also improves clot breakup and decreases the risk of hemorrhage (Brown et al. 2010; Hitchcock and Holland 2010). These approaches, however, are fairly unspecific and do not discriminate between "diseased" and healthy tissues.

A more specific approach is developing functionalized nanoparticles that incorporate a targeting moiety. This will enable drug delivery directly to the site of highest efficacy—whether the delivery of a neuroprotective factor to penumbral cells or the delivery of anticoagulants to a thrombus. A sophisticated multifaceted targeting would be desirable. Ideally, nanoparticles would first target the damaged and inflamed vascular endothelium present within the ischemic territory. This would prevent particles in nondiseased areas from being taken up. Targeting of overexpressed VEGF receptors in the ischemic microvasculature, for instance, is one approach that can be exploited for this (Lee et al. 2010a). In a second step, these particles access the brain. Various methods to improve the uptake of nanoparticles into the central nervous system have already been described (Tosi et al. 2010). Although most therapeutic agents will diffuse once they reach the interstitial space, a potential final step would require uptake into ischemia-damaged cells (Kubo et al. 2009) to deliver payloads that aim to alter cellular functions, such as small interfering RNA (Kim et al. 2010a).

Functionalization of nanoparticles also affords control over biodegradation, which can be used to release therapeutic factors (e.g., anti-inflammatory versus antiapoptotic agents) in response to a particular environment. For example, various hydrogels are responsive to inflammatory environments owing to degradation by hydroxyl radicals (Yui et al. 1992) or human neutrophil elastase (Aimetti et al. 2009). These could consequently be engineered to only release anti-inflammatory agents in this environment. Incorporating responsiveness to another therapeutic target, such as apoptosis, could produce a sophisticated treatment paradigm that specifically targets particular pathological mechanisms. Multicomponent microparticles that could disintegrate into small nanoparticles in response to a particular pathology would be useful not only for the interstitial delivery of, for instance, angiogenic factors, but also for the intracellular delivery (Chen et al. 2010a) of therapeutic genes (Kim et al. 2010a).

OPPORTUNITIES FOR COMBINATORIAL TREATMENTS

Gene and Cell Therapy

Gene delivery could also be an important avenue of treatment for stroke because gene expression provides a steady release of protein directly to the site without having to transport large amounts across the BBB. Although a collagen-coated endovascular microcoil (normally used to treat cerebral aneurysms) with immobilized antiadenovirus antibody can be used for vector tethering (Abrahams et al. 2002), most gene therapy studies for stroke use intracerebral injection of a viral vector (Badin et al. 2009). However, these injections are distributed differentially across the brain parenchyma and are known to result in variable expression. Virus-sized particles can improve the biodistribution of therapeutic genes in the central nervous system (Chen et al. 2005). Disulfide bond-based polypeptides are being used more commonly as vectors because their cytoplasm-induced reductive response facilitates the release of nucleic acids, thereby increasing transfection efficiency (Kim et al. 2009, 2010b). Gene therapy can be used to target not only diseased cells with survival factors but also remote regions to support diaschisis (i.e., neuronal plasticity that facilitates recovery) (Lim et al. 2010). Biodegradable polyamidoamine (PAMAM) ester can, for instance, deliver small interfering RNA into neurons (Kim et al. 2010a), whereas reducible poly(oligo-D-arginines) (rPOA) can act as carriers of the heme oxygenase-1 (HO-1) gene to decrease apoptosis (Hyun et al. 2010). Both strategies (i.e., decreasing apoptosis and inducing plasticity) can reduce the impact of stroke. One of the most exciting developments for gene therapy, however, would be the conversion of astrocytes in the glial scar into functional neurons (Heinrich et al. 2010).

Most commonly, a combination of gene-based and cell-based therapies is envisaged to improve outcome after a stroke. Because viral vectors generally indiscriminately transfect host cells where injected, transplanted cells can reach areas of damage and provide a flexible gene therapy approach. Cells overexpressing various growth factors can enhance recovery from stroke compared to unmodified cells (Kim 2004; Kurozumi et al. 2005; Lee et al. 2010b; Zhu et al. 2004).

Protecting Grafts from Host Environment

To enhance the efficiency of this approach, we can improve graft survival using biomaterials. For instance, cells transplanted with Matrigel were shown to improve outcome compared to cells alone (Jin et al. 2010; Yu et al. 2010). Transplant survival can also be improved with nanoparticles secreting survival factors (e.g., superoxide dismutase) that protect grafted cells from oxidative stress present within the stroke environment (Reddy et al. 2008) (Figure 21.2). In addition to supporting graft survival, shielding cells from a hostile host environment (immune rejection, poor nutrient supply) also depends on the use of biomaterials.

Microencapsulation of transplanted cells shields them from hostile environments where an ongoing immune response, for instance, would be deleterious to graft survival. In addition, xenogenic grafts, such as fetal porcine-derived cells, can be shielded from the immune system without the need for additional immunosuppression (Aebischer et al. 1991). This strategy is especially relevant if the aim of transplanted cells is to administer growth factors, such as GDNF, brain-derived neurotrophic factor (BDNF), or nerve growth factor (NGF), using xenogenic cells. This approach has shown some benefit in animals with stroke damage, where the porcine choroid plexus secreting GDNF, BDNF, and NGF improved behavioral recovery and reduced infarct volume (Borlongan et al. 2004).

Encapsulation also ensures a certain amount of safety for the host (e.g., removal of encapsulated cells) in cases where cells are genetically engineered to deliver particular growth factors, such as GDNF, or neurotransmitters, such as dopamine, to the host (Sautter et al. 1998). Current clinical trials demonstrate that this approach can provide continued local delivery of growth factors in neurological conditions

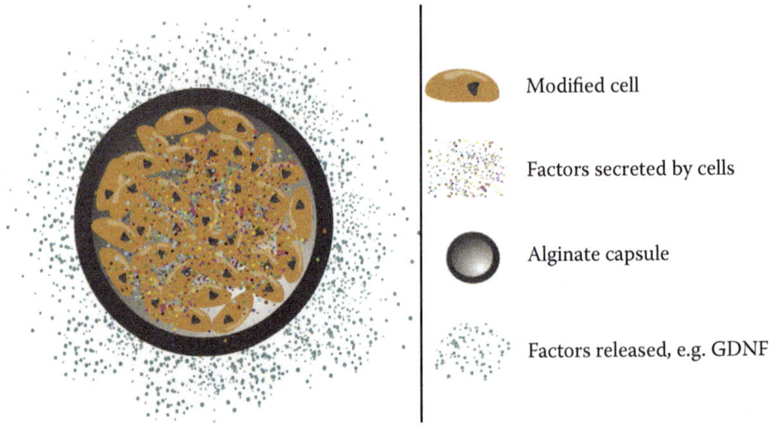

FIGURE 21.2 Encapsulated cells. Cells modified to overexpress particular factors (e.g., GDNF) are encapsulated within, for instance, alginate for protection from the host environment. Because alginate is a porous material through which chemical factors can permeate, cells will remain within the capsule while delivering growth factors to areas of damage. Novel engineering of encapsulation technology at the nanoscale will provide functional pores that, for instance, will only allow particular chemical factors, such as GDNF, to transfuse into the brain while retaining within the capsule all other factors secreted from the cells (potentially deleterious). Essentially, this approach would provide a local "bioreactor" to produce compounds that are needed for continued therapeutic effect. This approach is only relevant if the aim is to merely replace or deliver specific chemical factors, rather than to promote the integration of cells with the brain.

without any serious side effects of systemic infusion or minipumps (Aebischer et al. 1996; Bloch et al. 2004). Cotransplantation of encapsulated supporting cells that secrete growth factors, such as NGF, to enhance graft survival of NSCs is another means that could improve existing cell therapies that generally suffer from poor graft survival (Date et al. 1996). Combining the scaffolding properties of biomaterials with the secretion of particular chemical factors that influence cell fate (survival, differentiation, and migration) can be exploited to develop tissue engineering in the brain. The aim of such an approach would be to replace tissue that was lost because of stroke.

Replacing Lost Tissue through Tissue Engineering

The simplest approach to enhancing the survival of transplanted cells within the lesion cavity is to provide some structural support. A wide variety of biomaterials appear to be compatible with NSCs (i.e., allow survival and differentiation) (Thonhoff et al. 2008), but long-term testing under biodegradable conditions is required to establish the influence of degraded scaffold components on cellular fate. Intralesional transplantation of NSCs with biomaterials can lead to the formation of primitive neural tissue in the cavity as support materials degrade (Bible et al. 2009b; Park et al. 2002). Hydrogel support significantly increases cell survival of neural progenitors within the lesion cavity (Zhong et al. 2010). It remains unclear if these approaches will translate into improved efficacy; however, transplanting NSCs onto polymer scaffolds can improve recovery after a traumatic injury of the spinal cord (Chen et al. 2010b; Olson et al. 2009; Teng et al. 2002).

Although structural support appears essential to retain NSCs within a lesion cavity, it is unlikely that this will be sufficient to produce behavioral recovery or tissue regeneration. Demonstrating functional de novo tissue is much more challenging because a myriad of other processes might require tweaking and controlling. For instance, even if NSCs establish connections between themselves, they require the supply of nutrients through blood vessels. Unless an efficient and sufficient vascular supply can be established, it is unlikely that NSCs will prosper in this environment. Cotransplantation of endothelial cells and NSCs, as well as in vitro preformed neurovascular units, could overcome this issue (Figure 21.3). However, controlled factor release from biomaterials is

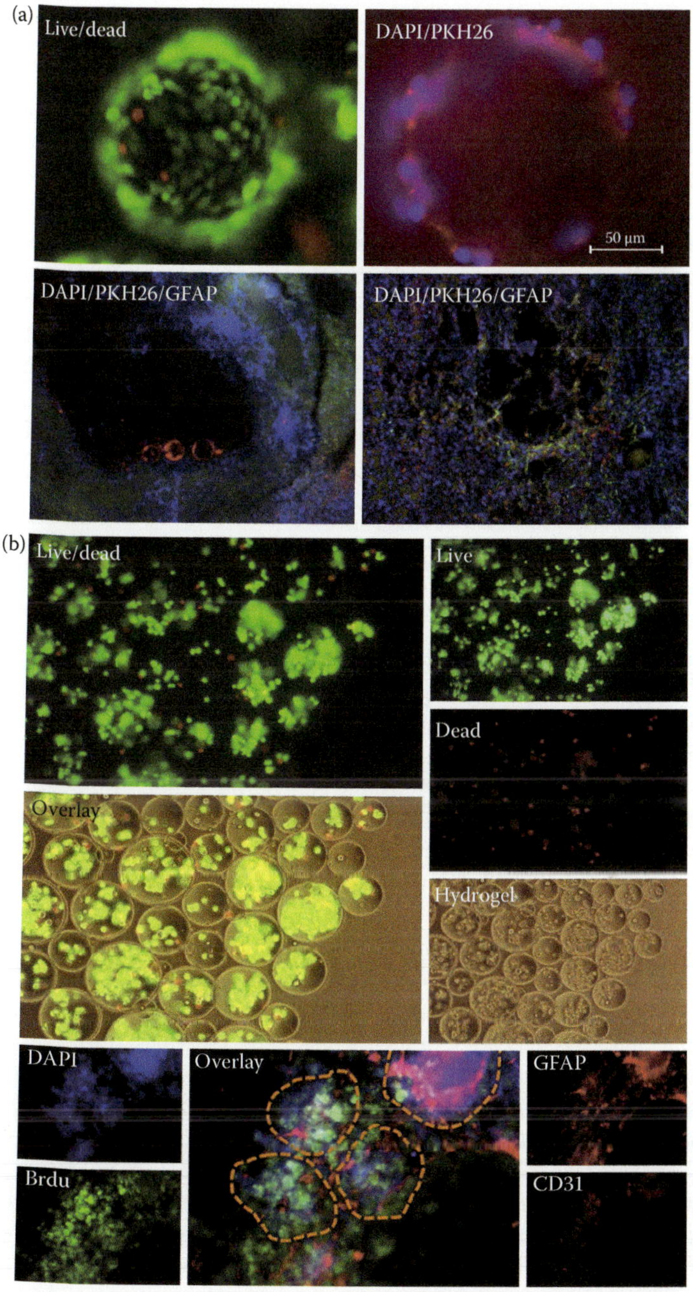

FIGURE 21.3 Tissue engineering. (a) NSCs need structural support to promote reconstruction of damaged tissue within the stroke cavity. Poly(D,L-lactide-*co*-glycolic) acid (PLGA) particles allow the attachment of NSCs without affecting their viability (green indicates live cells). The particles can then be injected into the brain through a small-bore needle, with cells being labeled with PKH26 for postmortem detection. Once injected, particles and cells attach to the host tissue. If sufficient cells and particles are present within the damaged area, they can create a primitive de novo tissue that still contains PLGA particles with transplanted cells connecting between this scaffold. (b) An alternative approach is to contain multiple cell types, such as endothelial cells and NSCs, within a polyethylene-glycol hydrogel that also does not significantly affect cell viability. Upon transplantation, individual hydrogel particles can be detected with astrocytes (glial fibrillary acidic protein [GFAP]$^{+}$ cells), anchoring these in place within the stroke lesion. Both approaches (PLGA and hydrogel particles) provide potential means to support and promote tissue regeneration after stroke. DAPI, 4′,6-diamidino-2-phenylindole.

likely to be required to ensure a concerted response and integration with the host brain. For example, Saif et al. (2010) showed that a scaffold-producing VEGF, hepatocyte growth factor (HGF), and angiopoietin-1 enhanced neovascularization in hindlimb ischemia compared to cell therapy alone. Another factor to consider is the site-appropriate differentiation of transplanted cells. This would be particularly important when transplanting directly into the lesion. Although relevant differentiation cues may be present at the periphery of the lesion, they are unlikely to be present at the core of neuroscaffolding. Small synthetic molecules can, however, be used to promote the differentiation of human striatal NSCs (El Akabawy et al. 2011). An in vivo-directed differentiation is therefore conceivable. Nevertheless, the logistics required in combining NSCs with biomaterials are considerable and require detailed protocols that can be easily transferred between different sites. Success will also heavily depend on novel noninvasive imaging approaches that allow transplantation into a suitable site at an appropriate time point (Bible et al. 2009a). Multiple combinations of the therapies described above can potentially be used to target particular processes involved in stroke. However, their application in patients will depend on the information derived from noninvasive imaging such as magnetic resonance imaging (MRI) and positron emission tomography (PET).

NONINVASIVE IMAGING OF THERAPEUTIC SUCCESS

Image-Guided Therapy Selection

One of the challenges in stroke is administering the appropriate treatment at the appropriate moment in the evolution of damage. Although time per se is a factor, the degree and site of occlusion, as well as reperfusion, vary widely and affect the physiological state of the ischemic area dramatically. Presently, noninvasive whole-brain imaging techniques are the only means to assess the emerging pathology caused by vessel occlusion (Moustafa and Baron 2008). Although, typically, the first images acquired provide a differential diagnosis with intracerebral hemorrhage, imaging of the occlusion and, more specifically, the thrombus is essential to improve thrombolysis. Fibrin-specific MRI contrast agents can provide a direct visualization of the thrombus location (Uppal et al. 2010), hence delivering a positive diagnostic marker to select thrombolysis as a needed therapeutic intervention. This molecular MRI can also be applied to visualize the immediate after effects of successful thrombolysis and reperfusion. Notably, endothelial activation leading to inflammation can be visualized using contrast agents targeted to P-selectin (Jacobin-Valat et al. 2010; van Kasteren et al. 2009) or damage to the basement membrane (Bennett et al. 2008). Based on the presence of the P-selectin marker, it is possible to administer anti-inflammatory agents. Intracerebral inflammation that can be detected using [^{11}C]PK11195 (Schroeter et al. 2009)—a PET radio-ligand that binds with high affinity to peripheral benzodiazepine receptors, which are highly expressed in activated microglia—can provide further indication to apply anti-inflammatory agents. If there is no indication of inflammation, it is unlikely that anti-inflammatory intervention will be effective. In clinical trials aiming to establish the efficacy of new medication, patients who do not present positive selection criteria should be excluded.

Further engineering of nanoparticles is required to develop more specific and multifunctional contrast agents. One of the challenges of these molecular imaging agents is their clearance from the blood stream, but additionally their crossing of a fairly intact BBB, and selective uptake into target cells remains an issue (Karatas et al. 2009; Tosi et al. 2011; Vernon et al. 2010; Yang 2010). Surface modification with *o*-carboxymethyl chitosans, for instance, can allow iron oxide nanoparticles to evade the reticuloendothelial system, increasing circulation time and hence favoring target-specific uptake (Amstad et al. 2009; Fan et al. 2011). In contrast, improving intracellular uptake requires adding molecules that are actively taken up into cells. For a selective imaging of only cells of interest, it is also necessary to prevent the agent from entering nontargeted cells. This approach can yield very specific imaging that only targets cells of interest (Nguyen et al. 2010) (Figure 21.4). In stroke, selective cellular imaging would potentially provide information about apoptosis in penumbral

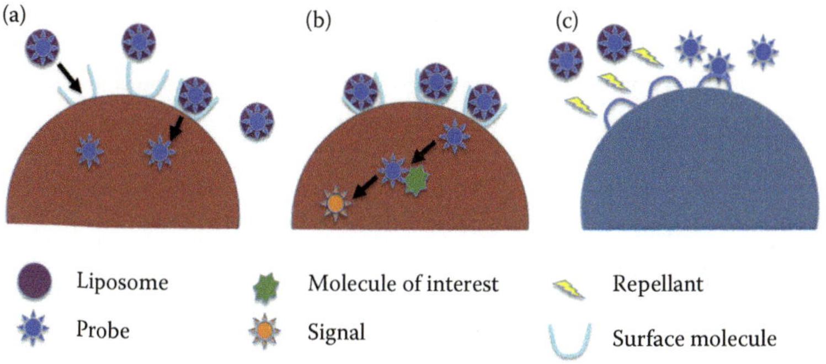

FIGURE 21.4 Molecular and cellular imaging in stroke. Selective imaging of apoptotic cells in the penumbral region would allow monitoring of the effectiveness of neuroprotective therapies. For selective molecular imaging, probes (blue stars) need to be specifically engineered to target the cell of interest through a surface molecule (a) that leads to the internalization of the probe if the target is located within the cell. In some cases, it is desirable to lose the targeting moiety (e.g., liposome used in a Trojan horse strategy to deliver an imaging probe). However, the targeted cells (for instance, neurons) might not express the molecules of interest (a), such as cleaved caspases, and hence should not produce a signal. Only when the probe (blue star) binds to the molecule of interest (green star) will it produce a signal (orange star) that can be detected using noninvasive imaging (b). Ideally, engineering of the imaging probe not only targets the cells of interest but also is designed to be repelled (yellow lightning) from cells that do not contain the molecule of interest (e.g., astrocytes) (c). Without this repellant property, probes might stick to nontargeted cells or might be endocytosed.

regions or monitor glial scar formation. Molecular imaging could provide more specific information about the physiological condition of the lesion environment (Gao et al. 2010). It could also potentially provide an indication of secondary Wallerian degeneration (i.e., death of cells or receding of axons in regions removed from the area of ischemia). Secondary damage is typically associated with poor behavioral recovery.

Monitoring Therapeutic Delivery and Success

Establishing the correct diagnosis and selecting a therapeutic agent are essential to patient management. But as therapies become more complex and specific, it is increasingly important to monitor their administration. For this purpose, MRI contrast agents can, for instance, be incorporated into cells in vitro prior to being administered for cell therapy. This approach affords visualization of the intracerebral location and migration of cells over time (Modo et al. 2009; Bible 2012b). It is important to understand how these cells exert a beneficial impact on outcome because the location of cells is indicative of their site of action (Modo et al. 2005). Intracellular magnetic labeling can also aid the targeting of cells to a specific location by applying magnetic field gradients (Riegler et al. 2010; Song et al. 2010). However, this approach needs to be further refined to avoid potential adverse effects that may only emerge over protracted periods (Modo et al. 2009). An alternative approach is incorporating therapeutic cells into perfluorooctyl bromide-modified alginate beads (Barnett et al. 2011). This system would potentially avoid the possible adverse effects of cellular incorporation and allow detection by MRI, ultrasonography (US), and computed tomography (CT) imaging.

Utilizing biomaterials to deliver or improve a therapeutic agent also requires possible use of noninvasive imaging for patient management. Many biomaterials used in neurosurgery produce susceptibility artifacts on magnetic resonance (MR) images (Matsuura et al. 2002). Although, in some cases, this can help identify the presence of biomaterial within the area of stroke (Bible et al. 2009b), it will interfere with patient management because of its effects on pathology detection. Biomaterials, however, can be engineered to provide a specific MR signature that will potentially fade gradually over time to provide an imaging marker of their degree of degradation (Colomb et al. 2010). Integration of biomaterials as a scaffold for tissue regeneration will eventually require distinctive imaging of the pathology, transplanted cells, and biomaterials. If growth factors are incorporated into the biomaterial, selective detection of this release would also be desirable. Smart polymer matrices encapsulating magnetic nanoparticles have been described for this purpose (Gupta and Bajpai 2011) and could be integrated with large scaffolding biomaterials.

Theranostics

The concept whereby nanotechnology is used to simultaneously image, monitor, and treat disease by carrying multiple molecules is known as thera(g)nostics, a portmanteau of therapeutics and diagnostics (Pene et al. 2009). In principle, theranostics will provide greater control over the diagnosis and faster treatment. One of the key challenges in stroke medicine is diagnosing patients quickly using noninvasive imaging. Ideally, within an hour of arrival at a hospital, a neurologist could diagnose stroke and deliver therapy. The advantage here of a theranostic agent would be that, as the diagnosis is being established, treatment is already being started, and there is no further loss of time to initiate therapy. The sooner therapy can be administered to the patient, the better is the prognosis.

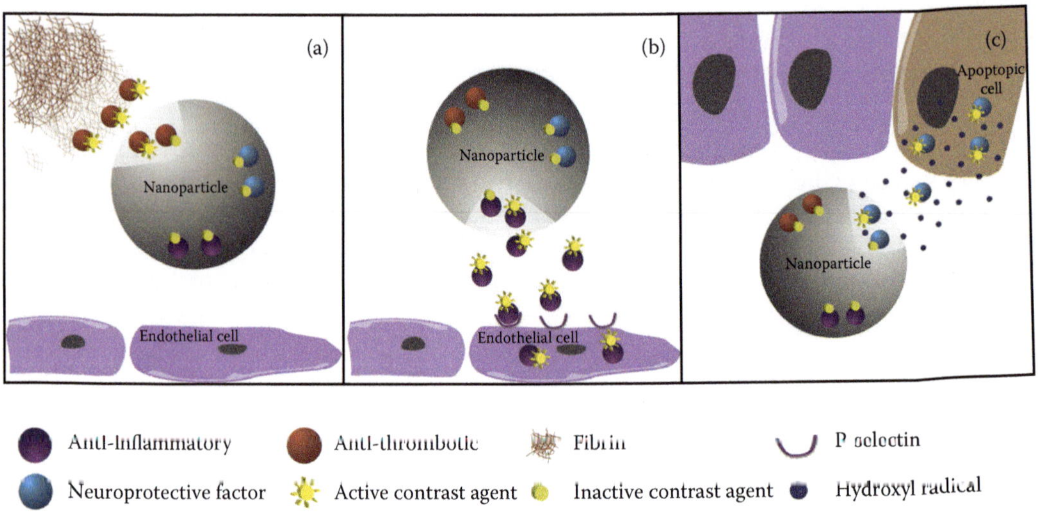

FIGURE 21.5 Hypothetical theranostic nanoparticle. This nanoparticle is multifunctional and adapts to various environments. If the nanoparticle encounters fibrin, fibrin will interact with its molecular structure and lead to the degradation of the part of the nanoparticle that in turn frees an antithrombotic agent. The release of an antithrombotic agent can hence specifically target a fibrin clot at its location (a). This would allow a much higher concentration of antithrombotic agent to be delivered because it will only have a local effect. Because this theranostic nanoparticle may have many components, it is possible to engineer multiple therapeutics within a single particle. For instance, a clotting agent could be part of the same agent, allowing safe administration of the same theranostic agent irrespective of the stroke being ischemic or hemorrhagic. An anti-inflammatory agent could also be part of the same agent not only to affect the cause of stroke, but also to prevent the spread of the ischemic cascade. Selective release of anti-inflammatory agents in response to, for instance, the expression of P-selectin could reduce endothelial inflammation (b). This inflammation induces an immune response that will allow macrophages to infiltrate the area of stroke and potentially worsen outcome. If the nanoparticle can cross the BBB, the presence of hydroxyl radicals could lead to the release of neuroprotective factors, hence providing a localized and specific treatment only within the area of stroke (c). In lieu of these factors (or in conjunction with these), contrast agents could be released to indicate whether the patient had suffered an ischemic stroke or a hemorrhagic stroke or whether, for instance, neuroprotective agents are released within the area of infarction. Chemical exchange saturation transfer (CEST) agents would be especially appropriate for this application because multiple agents can be detected with specific radio-frequency pulses within the same milieu.

The evolution of ever-smarter theranostics is promising. Multilayered nanoparticles could contain multiple potential diagnostic and therapeutic agents (Haglund et al. 2009; Yang 2010). Although there have been no reported studies on the use of theranostic agents in stroke, preliminary studies on brain tumors are starting to generate important parameters that will define appropriate physiochemical characteristics for these agents. For instance, systemically administered theranostic agents should be sized between 7 and 10 nm so that they would accumulate specifically within the tumor tissue rather than within the vasculature (Sarin 2010). Nevertheless, although this knowledge will be essential to designing environment-responsive agents, very little is known about how these agents interact with molecules within diseased tissues. In stroke, activating a specific diagnostic moiety that is responsive to the presence of fibrin, for example, could lead to the release of an antithrombotic agent. This would control the administration of therapy to only conditions where fibrin clots are present, but keep the theranostic agent inactive in all other conditions, such as intracranial bleeding. Because multiple elements can be bound within one particle, it would also be possible to release anti-inflammatory agents in the same location if inflammatory molecules are present, besides dissolving the clot with an antithrombotic agent. Therefore, theranostic agents potentially afford the opportunity to engineer decision trees (algorithms) based on physiological and pathological markers typically used in patient management. This approach will eventually lead to faster therapy and improved outcome for stroke patients (Figure 21.5).

CONCLUSION

It is well known that the ideal therapy for stroke would target several aspects at once, with early intervention remaining the key factor to therapeutic success. Novel approaches that bundle diagnosis and therapeutics into a theranostic agent are likely to have a major impact on stroke management, but, currently, little research effort is dedicated to this approach. However, even delayed therapeutic applications, such as cell therapy, will benefit from developments in nanotechnology. Imaging magnetic particles offers novel views of developing methods to monitor cells and drug delivery. Multifunctional biomaterials are key to providing greater control over therapeutic interventions, but they will require integration with the complex in vivo biology of stroke. The concerted effort of bioengineers and neuroscientists, as well as neurologists, neurosurgeons, and neuroimagers, is required to usher in this new era of treatment for stroke.

ACKNOWLEDGMENTS

We gratefully acknowledge the Biotechnology and Biological Sciences Research Council (grant BB/D014808/1 to K.S. and M.M.), the Medical Research Council (grant G0802552 to M.M.), and the National Institute of Biomedical Imaging and Bioengineering (grant 1 P20 EB007076-01 to J.W. and M.M.) for generously supporting our work on the transplantation of cells and biomaterials. F.N. was supported by a PhD studentship from the Medical Research Council.

REFERENCES

Abrahams, J.M., Song, C., DeFelice, S. et al. 2002. Endovascular microcoil gene delivery using immobilized anti-adenovirus antibody for vector tethering. *Stroke* 33: 1376–82.

Aebischer, P., Schluep, M., Deglon, N. et al. 1996. Intrathecal delivery of CNTF using encapsulated genetically modified xenogeneic cells in amyotrophic lateral sclerosis patients. *Nat Med* 2: 696–9.

Aebischer, P., Tresco, P.A., Winn, S.R. et al. 1991. Long-term cross-species brain transplantation of a polymer-encapsulated dopamine-secreting cell line. *Exp Neurol* 111: 269–75.

Aimetti, A.A., Machen, A.J., and Anseth, K.S. 2009. Poly(ethylene glycol) hydrogels formed by thiol-ene photopolymerization for enzyme-responsive protein delivery. *Biomaterials* 30: 6048–54.

Amstad, E., Zurcher, S., Mashaghi, A. et al. 2009. Surface functionalization of single superparamagnetic iron oxide nanoparticles for targeted magnetic resonance imaging. *Small* 5: 1334–42.

Arai, K., Jin, G., Navaratna, D. et al. 2009. Brain angiogenesis in developmental and pathological processes: neurovascular injury and angiogenic recovery after stroke. *FEBS J* 276: 4644–52.

Arumugam, T.V., Granger, D.N., and Mattson, M.P. 2005. Stroke and T-cells. *Neuromol Med* 7: 229–42.

Astrup, J., Symon, L., Branston, N.M. et al. 1977. Cortical evoked potential and extracellular K^+ and H^+ at critical levels of brain ischemia. *Stroke* 8: 51–7.

Badin, R.A., Modo, M., Cheetham, M. et al. 2009. Protective effect of post-ischaemic viral delivery of heat shock proteins in vivo. *J Cereb Blood Flow Metab* 29: 254–63.

Bang, O.Y., Lee, J.S., Lee, P.H. et al. 2005. Autologous mesenchymal stem cell transplantation in stroke patients. *Ann Neurol* 57: 874–82.

Barnett, B.P., Ruiz-Cabello, J., Hota, P. et al. 2011. Fluorocapsules for improved function, immunoprotection, and visualization of cellular therapeutics with MR, US, and CT imaging. *Radiology* 258: 182–91.

Belayev, L., Busto, R., Zhao, W. et al. 1996. Quantitative evaluation of blood–brain barrier permeability following middle cerebral artery occlusion in rats. *Brain Res* 739: 88–96.

Bell, B.A., Symon, L., and Branston, N.M. 1985. CBF and time thresholds for the formation of ischemic cerebral edema, and effect of reperfusion in baboons. *J Neurosurg* 62: 31–41.

Bennett, K.M., Zhou, H., Sumner, J.P. et al. 2008. MRI of the basement membrane using charged nanoparticles as contrast agents. *Magn Reson Med* 60: 564–74.

Bertram, J.P., Rauch, M.F., Chang, K. et al. 2010. Using polymer chemistry to modulate the delivery of neurotrophic factors from degradable microspheres: delivery of BDNF. *Pharm Res* 27: 82–91.

Bible, E., Chau, D.Y., Alexander, M.R. et al. 2009a. Attachment of stem cells to scaffold particles for intracerebral transplantation. *Nat Protoc* 4: 1440–53.

Bible, E., Chau, D.Y., Alexander, M.R. et al. 2009b. The support of neural stem cells transplanted into stroke-induced brain cavities by PLGA particles. *Biomaterials* 30: 2985–94.

Bloch, J., Bachoud-Levi, A.C., Deglon, N. et al. 2004. Neuroprotective gene therapy for Huntington's disease, using polymer-encapsulated cells engineered to secrete human ciliary neurotrophic factor: results of a phase I study. *Hum Gene Ther* 15: 968–75.

Borlongan, C.V., Skinner, S.J., Geaney, M. et al. 2004. Intracerebral transplantation of porcine choroid plexus provides structural and functional neuroprotection in a rodent model of stroke. *Stroke* 35: 2206–10.

Borlongan, C.V., Tajima, Y., Trojanowski, J.Q. et al. 1998. Transplantation of cryopreserved human embryonal carcinoma-derived neurons (NT2N cells) promotes functional recovery in ischemic rats. *Exp Neurol* 149: 310–21.

Brown, A.T., Flores, R., Hamilton, E. et al. 2010. Microbubbles improve sonothrombolysis in vitro and decrease hemorrhage in vivo in a rabbit stroke model. *Invest Radiol* 46: 202–7.

Buhnemann, C., Scholz, A., Bernreuther, C. et al. 2006. Neuronal differentiation of transplanted embryonic stem cell derived precursors in stroke lesions of adult rats. *Brain* 129: 3238–48.

Chen, C., Geng, J., Pu, F. et al. 2010a. Polyvalent nucleic acid/mesoporous silica nanoparticle conjugates: dual stimuli-responsive vehicles for intracellular drug delivery. *Angew Chem Int Ed Engl* 50: 882–6.

Chen, C., Hu, Y.R., Wan, H. et al. 2010b. Functional recovery following traumatic spinal cord injury mediated by a unique polymer scaffold seeded with neural stem cells and Schwann cells. *Chin Med J (Engl)* 123: 2424–31.

Chen, M.Y., Hoffer, A., Morrison, P.F. et al. 2005. Surface properties, more than size, limiting convective distribution of virus-sized particles and viruses in the central nervous system. *J Neurosurg* 103: 311–9.

Ciofani, G., Raffa, V., Pizzorusso, T. et al. 2008. Characterization of an alginate-based drug delivery system for neurological applications. *Med Eng Phys* 30: 848–55.

Colomb, J., Louie, K., Massia, S.P. et al. 2010. Self-degrading, MRI-detectable hydrogel sensors with picomolar target sensitivity. *Magn Reson Med* 64: 1792–9.

Dalpiaz, A., Gavini, E., Colombo, G. et al. 2008. Brain uptake of an anti-ischemic agent by nasal administration of microparticles. *J Pharm Sci* 97: 4889–903.

Date, I., Ohmoto, T., Imaoka, T. et al. 1996. Cografting with polymer-encapsulated human nerve growth factor-secreting cells and chromaffin cell survival and behavioral recovery in hemiparkinsonian rats. *J Neurosurg* 84: 1006–12.

Dirnagl, U. 2006. Bench to bedside: the quest for quality in experimental stroke research. *J Cereb Blood Flow Metab* 26: 1465–78.

El Akabawy, G., Martinez-Medina, L., Jeffries, A.R. et al. 2011. Purmorphamine increases DARPP-32 differentiation in human striatal neural stem cells through the hedgehog pathway. *Stem Cells Dev* 20: 1873–87.

Fan, C., Gao, W., Chen, Z. et al. 2011. Tumor selectivity of stealth multi-functionalized superparamagnetic iron oxide nanoparticles. *Int J Pharm* 404: 180–90.

Fan, Y., Shen, F., Frenzel, T. et al. 2010. Endothelial progenitor cell transplantation improves long-term stroke outcome in mice. *Ann Neurol* 67: 488–97.

Feigin, V.L., Lawes, C.M., Bennett, D.A. et al. 2009. Worldwide stroke incidence and early case fatality reported in 56 population-based studies: a systematic review. *Lancet Neurol* 8: 355–69.

Gao, G.H., Lee, J.W., Nguyen, M.K. et al. 2010. pH-responsive polymeric micelle based on PEG-poly(beta-amino ester)/(amido amine) as intelligent vehicle for magnetic resonance imaging in detection of cerebral ischemic area. *J Controlled Release* 155: 11–7.

Georgieva, J.V., Kalicharan, D., Couraud, P.O. et al. 2010. Surface characteristics of nanoparticles determine their intracellular fate in and processing by human blood–brain barrier endothelial cells in vitro. *Mol Ther* 19: 318–25.

Gouhier, C., Chalon, S., Venier-Julienne, M.C. et al. 2000. Neuroprotection of nerve growth factor-loaded microspheres on the D2 dopaminergic receptor–positive striatal neurones in quinolinic acid-lesioned rats: a quantitative autoradiographic assessment with iodobenzamide. *Neurosci Lett* 288: 71–5.

Grabowski, M., Johansson, B.B., and Brundin, P. 1995. Neocortical grafts placed in the infarcted brain of adult rats: few or no efferent fibers grow from transplant to host. *Exp Neurol* 134: 273–6.

Gupta, R., and Bajpai, A.K. 2011. Magnetically guided release of ciprofloxacin from superparamagnetic polymer nanocomposites. *J Biomater Sci Polym Ed* 22: 893–918.

Haglund, E., Seale-Goldsmith, M.M., and Leary, J.F. 2009. Design of multifunctional nanomedical systems. *Ann Biomed Eng* 37: 2048–63.

Heinrich, C., Blum, R., Gascon, S. et al. 2010. Directing astroglia from the cerebral cortex into subtype specific functional neurons. *PLoS Biol* 8: e1000373.

Hitchcock, K.E., and Holland, C.K. 2010. Ultrasound-assisted thrombolysis for stroke therapy: better thrombus break-up with bubbles. *Stroke* 41: S50–3.

Hyun, H., Won, Y.W., Kim, K.M. et al. 2010. Therapeutic effects of a reducible poly (oligo-D-arginine) carrier with the heme oxygenase-1 gene in the treatment of hypoxic–ischemic brain injury. *Biomaterials* 31: 9128–34.

Jacobin-Valat, M.J., Deramchia, K., Mornet, S. et al. 2010. MRI of inducible P-selectin expression in human activated platelets involved in the early stages of atherosclerosis. *NMR Biomed* 24: 413–24.

Jiang, Q., Zhang, Z.G., Ding, G.L. et al. 2006. MRI detects white matter reorganization after neural progenitor cell treatment of stroke. *Neuroimage* 32: 1080–9.

Jin, K., Mao, X., Xie, L. et al. 2010. Transplantation of human neural precursor cells in Matrigel scaffolding improves outcome from focal cerebral ischemia after delayed postischemic treatment in rats. *J Cereb Blood Flow Metab* 30: 534–44.

Joshi, S.A., Chavhan, S.S., and Sawant, K.K. 2010. Rivastigmine-loaded PLGA and PBCA nanoparticles: preparation, optimization, characterization, in vitro and pharmacodynamic studies. *Eur J Pharm Biopharm* 76: 189–99.

Karatas, H., Aktas, Y., Gursoy-Ozdemir, Y. et al. 2009. A nanomedicine transports a peptide caspase-3 inhibitor across the blood–brain barrier and provides neuroprotection. *J Neurosci* 29: 13761–9.

Kim, I.D., Lim, C.M., Kim, J.B. et al. 2010a. Neuroprotection by biodegradable PAMAM ester (e-PAM-R)-mediated HMGB1 siRNA delivery in primary cortical cultures and in the postischemic brain. *J Controlled Release* 142: 422–30.

Kim, S.H., Jeong, J.H., Kim, T.I. et al. 2009. VEGF siRNA delivery system using arginine-grafted bioreducible poly(disulfide amine). *Mol Pharmacol* 6: 718–26.

Kim, S.U. 2004. Human neural stem cells genetically modified for brain repair in neurological disorders. *Neuropathology* 24: 159–71.

Kim, T.I., Lee, M., and Kim, S.W. 2010b. A guanidinylated bioreducible polymer with high nuclear localization ability for gene delivery systems. *Biomaterials* 31: 1798–804.

Klose, D., Laprais, M., Leroux, V. et al. 2009. Fenofibrate-loaded PLGA microparticles: effects on ischemic stroke. *Eur J Pharm Sci* 37: 43–52.

Kondziolka, D., Steinberg, G.K., Wechsler, L. et al. 2005. Neurotransplantation for patients with subcortical motor stroke: a phase 2 randomized trial. *J Neurosurg* 103: 38–45.

Kondziolka, D., Wechsler, L., Goldstein, S. et al. 2000. Transplantation of cultured human neuronal cells for patients with stroke. *Neurology* 55: 565–9.

Kubo, M., Egashira, K., Inoue, T. et al. 2009. Therapeutic neovascularization by nanotechnology-mediated cell-selective delivery of pitavastatin into the vascular endothelium. *Arterioscler Thromb Vasc Biol* 29: 796–801.

Kurozumi, K., Nakamura, K., Tamiya, T. et al. 2005. Mesenchymal stem cells that produce neurotrophic factors reduce ischemic damage in the rat middle cerebral artery occlusion model. *Mol Ther* 11: 96–104.

Kurtoglu, Y.E., Navath, R.S., Wang, B. et al. 2009. Poly(amidoamine) dendrimer–drug conjugates with disulfide linkages for intracellular drug delivery. *Biomaterials* 30: 2112–21.
Lee, C.M., Kim, E.M., Cheong, S.J. et al. 2010a. Targeted molecular imaging of VEGF receptors overexpressed in ischemic microvasculature using chitosan–DC101 conjugates. *J Biomed Mater Res A* 92: 1510–7.
Lee, H.J., Lim, I.J., Lee, M.C. et al. 2010b. Human neural stem cells genetically modified to overexpress brain-derived neurotrophic factor promote functional recovery and neuroprotection in a mouse stroke model. *J Neurosci Res* 88: 3282–94.
Lees, K.R. 1997. Cerestat and other NMDA antagonists in ischemic stroke. *Neurology* 49: S66–9.
Lim, S.T., Airavaara, M., and Harvey, B.K. 2010. Viral vectors for neurotrophic factor delivery: a gene therapy approach for neurodegenerative diseases of the CNS. *Pharmacol Res* 61: 14–26.
Lipton, P. 1999. Ischemic cell death in brain neurons. *Physiol Rev* 79: 1431–568.
Liu, C., Lin, N., Wu, B. et al. 2009. Neuroprotective effect of memantine combined with topiramate in hypoxic–ischemic brain injury. *Brain Res* 1282: 173–82.
Luk, Y.O., Chen, W.Y., Wong, W.J. et al. 2004. Treatment of focal cerebral ischemia with liposomal nerve growth factor. *Drug Delivery* 11: 319–24.
Marsh, J.N., Senpan, A., Hu, G. et al. 2007. Fibrin-targeted perfluorocarbon nanoparticles for targeted thrombolysis. *Nanomedicine* 2: 533–43.
Matsuura, H., Inoue, T., Konno, H. et al. 2002. Quantification of susceptibility artifacts produced on high-field magnetic resonance images by various biomaterials used for neurosurgical implants. Technical note. *J Neurosurg* 97: 1472–5.
Modo, M., Beech, J.S., Meade, T.J. et al. 2009. A chronic 1 year assessment of MRI contrast agent-labelled neural stem cell transplants in stroke. *Neuroimage* 47 (Suppl 2): T133–42.
Modo, M., Hoehn, M., and Bulte, J.W. 2005. Cellular MR imaging. *Mol Imaging* 4: 143–64.
Moon, J.J., Saik, J.E., Poche, R.A. et al. 2010. Biomimetic hydrogels with pro-angiogenic properties. *Biomaterials* 31: 3840–7.
Moubarik, C., Guillet, B., Youssef, B. et al. 2011. Transplanted late outgrowth endothelial progenitor cells as cell therapy product for stroke. *Stem Cell Rev* 7: 208–20.
Moustafa, R.R., and Baron, J.C. 2008. Pathophysiology of ischaemic stroke: insights from imaging, and implications for therapy and drug discovery. *Br J Pharmacol* 153 (Suppl 1): S44–54.
Nguyen, Q.T., Olson, E.S., Aguilera, T.A. et al. 2010. Surgery with molecular fluorescence imaging using activatable cell-penetrating peptides decreases residual cancer and improves survival. *Proc Natl Acad Sci U S A* 107: 4317–22.
Nilupul Perera, M., Ma, H.K., Arakawa, S. et al. 2006. Inflammation following stroke. *J Clin Neurosci* 13: 1–8.
O'Collins, V.E., Macleod, M.R., Donnan, G.A. et al. 2006. 1,026 experimental treatments in acute stroke. *Ann Neurol* 59: 467–77.
Olson, H.E., Rooney, G.E., Gross, L. et al. 2009. Neural stem cell- and Schwann cell-loaded biodegradable polymer scaffolds support axonal regeneration in the transected spinal cord. *Tissue Eng* 15: 1797–805.
Orive, G., Anitua, E., Pedraz, J.L. et al. 2009. Biomaterials for promoting brain protection, repair and regeneration. *Nat Rev Neurosci* 10: 682–92.
Park, K.I., Teng, Y.D., and Snyder, E.Y. 2002. The injured brain interacts reciprocally with neural stem cells supported by scaffolds to reconstitute lost tissue. *Nat Biotechnol* 20: 1111–7.
Pene, F., Courtine, E., Cariou, A. et al. 2009. Toward theragnostics. *Crit Care Med* 37: S50–8.
Reddy, M.K., and Labhasetwar, V. 2009. Nanoparticle-mediated delivery of superoxide dismutase to the brain: an effective strategy to reduce ischemia–reperfusion injury. *FASEB J* 23: 1384–95.
Reddy, M.K., Wu, L., Kou, W. et al. 2008. Superoxide dismutase-loaded PLGA nanoparticles protect cultured human neurons under oxidative stress. *Appl Biochem Biotechnol* 151: 565–77.
Riegler, J., Wells, J.A., Kyrtatos, P.G. et al. 2010. Targeted magnetic delivery and tracking of cells using a magnetic resonance imaging system. *Biomaterials* 31: 5366–71.
Saif, J., Schwarz, T.M., Chau, D.Y. et al. 2010. Combination of injectable multiple growth factor-releasing scaffolds and cell therapy as an advanced modality to enhance tissue neovascularization. *Arterioscler Thromb Vasc Biol* 30: 1897–904.
Saik, J.E., Gould, D.J., Watkins, E.M. et al. 2011. Covalently immobilized platelet-derived growth factor-BB promotes angiogenesis in biomimetic poly(ethylene glycol) hydrogels. *Acta Biomater* 7: 133–43.
Saporta, S., Borlongan, C.V., and Sanberg, P.R. 1999. Neural transplantation of human neuroteratocarcinoma (hNT) neurons into ischemic rats. A quantitative dose-response analysis of cell survival and behavioral recovery. *Neuroscience* 91: 519–25.
Sarin, H. 2010. On the future development of optimally-sized lipid-insoluble systemic therapies for CNS solid tumors and other neuropathologies. *Recent Pat CNS Drug Discovery* 5: 239–52.

Sautter, J., Tseng, J.L., Braguglia, D. et al. 1998. Implants of polymer-encapsulated genetically modified cells releasing glial cell line-derived neurotrophic factor improve survival, growth, and function of fetal dopaminergic grafts. *Exp Neurol* 149: 230–6.

Schroeter, M., Dennin, M.A., Walberer, M. et al. 2009. Neuroinflammation extends brain tissue at risk to vital peri-infarct tissue: a double tracer [^{11}C]PK11195- and [^{18}F]FDG-PET study. *J Cereb Blood Flow Metab* 29: 1216–25.

Shen, L.H., Li, Y., and Chopp, M. 2010. Astrocytic endogenous glial cell derived neurotrophic factor production is enhanced by bone marrow stromal cell transplantation in the ischemic boundary zone after stroke in adult rats. *Glia* 58: 1074–81.

Shrivastava, S., Singh, S.K., Mukhopadhyay, A. et al. 2011. Negative regulation of fibrin polymerization and clot formation by nanoparticles of silver. *Colloids Surf B* 82: 241–6.

Song, M., Kim, Y.J., Kim, Y.H. et al. 2010. Using a neodymium magnet to target delivery of ferumoxide-labeled human neural stem cells in a rat model of focal cerebral ischemia. *Hum Gene Ther* 21: 603–10.

Souto, E.B., Doktorovova, S., Gonzalez-Mira, E. et al. 2010. Feasibility of lipid nanoparticles for ocular delivery of anti-inflammatory drugs. *Curr Eye Res* 35: 537–52.

Stroemer, P., Patel, S., Hope, A. et al. 2009. The neural stem cell line CTX0E03 promotes behavioral recovery and endogenous neurogenesis after experimental stroke in a dose-dependent fashion. *Neurorehabil Neural Repair* 23: 895–909.

Taite, L.J., and West, J.L. 2006. Poly(ethylene glycol)-lysine dendrimers for targeted delivery of nitric oxide. *J Biomater Sci Polym Ed* 17: 1159–72.

Teng, Y.D., Lavik, E.B., Qu, X. et al. 2002. Functional recovery following traumatic spinal cord injury mediated by a unique polymer scaffold seeded with neural stem cells. *Proc Natl Acad Sci U S A* 99: 3024–9.

Thonhoff, J.R., Lou, D.I., Jordan, P.M. et al. 2008. Compatibility of human fetal neural stem cells with hydrogel biomaterials in vitro. *Brain Res* 1187: 42–51.

Torno, M.D., Kaminski, M.D., Xie, Y. et al. 2008. Improvement of in vitro thrombolysis employing magnetically-guided microspheres. *Thromb Res* 121: 799–811.

Tosi, G., Bondioli, L., Ruozi, B. et al. 2011. NIR-labeled nanoparticles engineered for brain targeting: in vivo optical imaging application and fluorescent microscopy evidences. *J Neural Transm* 118: 145–53.

Tosi, G., Vergoni, A.V., Ruozi, B. et al. 2010. Sialic acid and glycopeptides conjugated PLGA nanoparticles for central nervous system targeting: in vivo pharmacological evidence and biodistribution. *J Controlled Release* 145: 49–57.

Uppal, R., Ay, I., Dai, G. et al. 2010. Molecular MRI of intracranial thrombus in a rat ischemic stroke model. *Stroke* 41: 1271–7.

van Kasteren, S.I., Campbell, S.J., Serres, S. et al. 2009. Glyconanoparticles allow pre symptomatic in vivo imaging of brain disease. *Proc Natl Acad Sci U S A* 106: 18–23.

Vernon, A.C., Ballard, C., and Modo, M. 2010. Neuroimaging for Lewy body disease: is the in vivo molecular imaging of alpha-synuclein neuropathology required and feasible? *Brain Res Rev* 65: 28–55.

Wakabayashi, K., Nagai, A., Sheikh, A.M. et al. 2010. Transplantation of human mesenchymal stem cells promotes functional improvement and increased expression of neurotrophic factors in a rat focal cerebral ischemia model. *J Neurosci Res* 88: 1017–25.

Wang, Q., Tang, X.N., and Yenari, M.A. 2007. The inflammatory response in stroke. *J Neuroimmunol* 184: 53–68.

Ward, M.S., Khoobehi, A., Lavik, E.B. et al. 2007. Neuroprotection of retinal ganglion cells in DBA/2J mice with GDNF-loaded biodegradable microspheres. *J Pharm Sci* 96: 558–68.

Winer, I., Wang, S., Lee, Y.E. et al. 2010. F3-targeted cisplatin–hydrogel nanoparticles as an effective therapeutic that targets both murine and human ovarian tumor endothelial cells in vivo. *Cancer Res* 70: 8674–83.

Yang, H. 2010. Nanoparticle-mediated brain-specific drug delivery, imaging, and diagnosis. *Pharm Res* 27: 1759–71.

Yu, H., Cao, B., Feng, M. et al. 2010. Combinated transplantation of neural stem cells and collagen type I promote functional recovery after cerebral ischemia in rats. *Anat Rec* 293: 911–7.

Yui, N., Okano, T., and Sakurai, Y. 1992. Inflammation responsive degradation of crosslinked hyaluronic acid gels. *J Controlled Release* 22: 105–16.

Zhong, J., Chan, A., Morad, L. et al. 2010. Hydrogel matrix to support stem cell survival after brain transplantation in stroke. *Neurorehabil Neural Repair* 24: 636–44.

Zhu, W., Zhou, L.F., Wang, Y. et al. 2004. Transplantation of gene-transfected neural stem cells for transient cerebral ischemia in rats. *Zhonghua Yi Xue Za Zhi* 84: 1029–34.

22 Molecular Medical Devices for Nanoneurosurgery

Rutledge Ellis-Behnke

CONTENTS

INTRODUCTION

Surgery brings to mind visions of operating rooms with patients—almost completely covered with drapes—lying beneath doctors and machines. The multitude of instruments surrounding the surgical field are mostly manipulated by humans. However, with the introduction of operating robots, doctors' arms and hands are being replaced with mechanical devices that are manipulated remotely by a human. Surgery is defined as a branch of medicine that is concerned with the treatment of body injuries or disorders by incision or manipulation, especially with instruments. The concept of treatment and manipulation with instruments needs to be clarified; although surgical instruments are commonly associated with stainless steel or titanium, they can be manufactured using different molecules and need not be sized to fit into a hand.

This chapter is about nanoneurosurgery, so what exactly does nano have to do with neurosurgery? Nano refers to size. At first glance, it might seem rather silly to use size to modify a field such as neurosurgery; however, nano creates many new possibilities and allows for many new types of interventions that other larger sizes simply cannot provide. As has been seen before, when resolution is increased by a power of 10, most of the lines of many fields become blurred and are then redrawn as the new field develops. The field of neurosurgery is no different. What is currently thought to be the realm of highly skilled surgeons—making very small cuts with great care—could someday be ruled by a skilled chemist who can create a new tool or instrument that enables the treatment of an injury or disorder just by making a chemical incision in the exact area where it is needed (e.g., resecting an unwanted brain tumor that is later excreted). This would be performed without the need to physically open the skull.

The brain and its surrounding structures, such as the skull, are manipulated many times during a neurosurgical procedure. There is an endless quest for ever smaller and more targeted interventions that leave either no footprints (or at least footprints that cannot be seen by the immune system) or no disruptions after neurosurgery—from minimally invasive pituitary tumor resection to removal of clots in clogged brain arteries after a stroke using tissue plasminogen activator (tPA). With these minimally invasive procedures, collateral damage is also drastically reduced.

A perfect surgery would allow an army of small objects to seek out what is wrong and repair it while refraining from creating more damage. Although this sounds like a fantastic concept, there is a distinct advantage with that idea. In that idea, everything is under direct human control. The ability of some of these materials and procedures to act autonomously either by design or by reaction to the local environment will be addressed here.

This chapter focuses on the new and exciting world of nanoneurosurgery in the laboratory that hopefully will soon be available in the clinic—how recent developments in nanotechnology have enabled the probing, testing, monitoring, and repair of the central nervous system (CNS). It is the beginning of nanoneurosurgery. In addition, there will be speculations as to what can be performed with a little more work using what is already known about the ability of the body to repair itself. Each of these technologies and the experiments described have already been performed and are in the process of being clinically tested. The Conclusion section of this chapter will speculate on the future: With the advances and added functionality that nanotechnology affords in the realm of nanoneurosurgery, where is humankind going?

Before stepping into the future, let me start at the beginning. Humans are self-assembled and self-organized beings, so it is important to show how human assembly allows for the repair and mitigation of damage in the CNS.

Origins of Neurosurgery

In the early days of neurosurgery, tools were self-assembled at the molecular scale; humans fashioned these tools from obsidian and other hard, sharp stones. There was evidence supporting the claim that, some 13,000 years ago, pre-Columbian Incas performed trephination, an early form of neurosurgery that involved opening the skull to let out everything from evil spirits (probably epilepsy) to headaches.

Organization of Tissue

Neurulation

Neurulation is the formation of the neural tube from the blastocyst's ectoderm. The egg changes from a single cell to a blastocyst, which separates the three domains of cells: (1) endoderm, which forms the lining of the gut and the esophagus; (2) mesoderm, which forms the organs and the circulatory system; and (3) ectoderm, which becomes the skin and the CNS (Figure 22.1) (Purves and

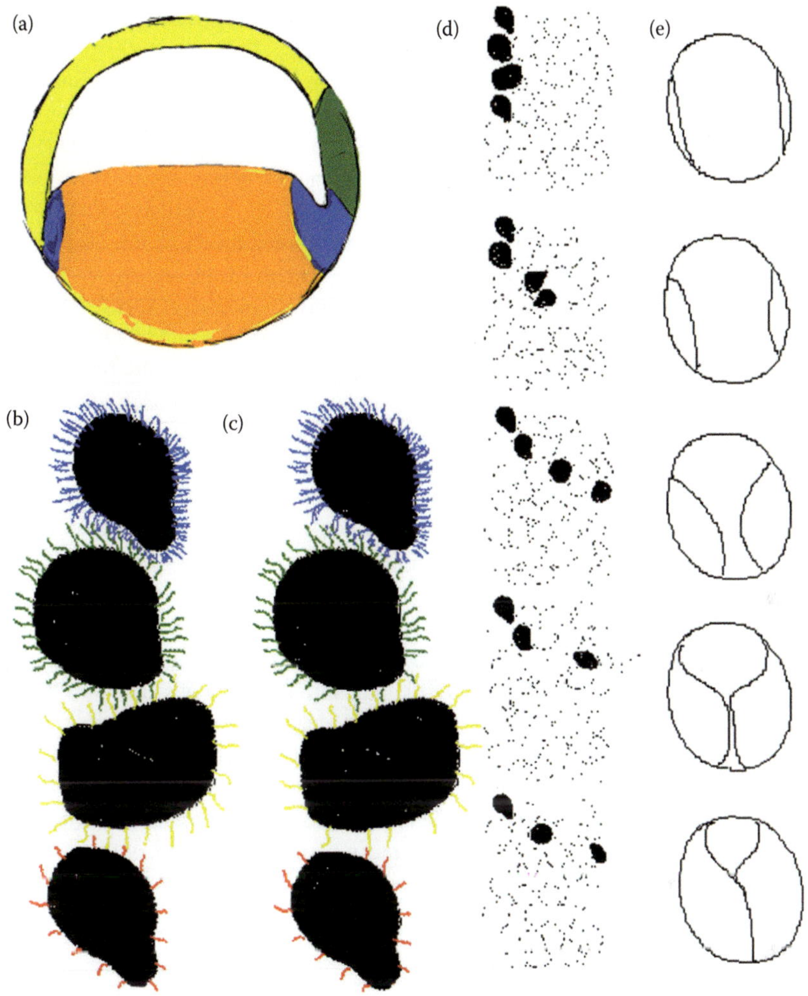

FIGURE 22.1 Neurulation. (a) Embryo showing the ectoderm (yellow) and the start of the neural plate precursor (green). Also shown are the mesoderm and the endoderm. (b and c) Cells in four different colors to illustrate their differential movement during the formation of the neural tube. (b) If two sides of the cell encounter low density, then the movement (or the on–off rate) will be high on/high off. This will make it appear that the cells are moving past each other at high rates of speed. In this first case, ECM density is uniform for each color. However, the interaction of each color represents differential complementarity: the red cell would move faster when passing the yellow cell; the yellow cell would move faster than the green cell; and the green cell would move faster than the blue cell. This is one scenario. In the second case, the cells are polarized and have different ECM densities on each of the faces. (c) Depending on the orientation of the cells, the green or red cell may move at the same rate past the yellow cell if the yellow cell changes its orientation, presenting the face with the lower density toward the green cell and not the red cell. This type of ECM interaction is an example of how some self-assembling nanomaterials work—allowing certain cells to move faster while other cells move slower to repair the damaged area. The ECM will become entangled and interact, thus slowing or even stopping cell migration. Although this example uses only two cells, the formation of the CNS has three dimensions that will interact in a graded fashion based on the orientation of each cell and the corresponding ECM density. (d) How cells move during the formation of the neural tube. Note that the red cell moves fastest and is first to reach the midline (arrow indicates the direction of midline), where the neural tube starts to zip up. The rest of the cells follow at a slower rate. The center of the spinal cord joins first and then proceeds in both directions toward the head and the tail. (e) Neural tube zipping up in relation to cell migration.

Lichtman 1985). The brain and skin come from the same precursor tissue. During neurulation, the external surfaces of the cells have different densities of extracellular matrix (ECM) (Schneider et al. 2005). Figure 22.1 ignores signaling for the sake of simplicity; however, signaling plays a major role in cellular organization and movement.

An additional parameter that is independent of ECM density is complementarity. If the ECM sequence is complementary, then the likelihood of the interaction being longer is much higher owing to the molecular interaction of the ECM. Likewise, if the ECM is repulsive, then the interaction will be much shorter and the cells will move past one another much faster, despite the absence of additional signaling molecules used to cause changes in the ECM and in receptor expression. Interactions in a molecular environment will change the cell's movement, speed, and direction.

The formation of the neural tube illustrates what can be accomplished when nanoscale interactions are manipulated. Although this is made clearer later in the chapter, the fundamental point is that molecular complementarity is a cornerstone of the creation of an environment that fundamentally changes how the CNS heals and how it reacts to outside stimuli after trauma or during tumor formation. These mechanical and electrical interactions will either allow a tumor to implant and slowly transform the local environment to make it more hospitable to the tumor, or stop the tumor from spreading and allow the immune system to remove it before the tumor creates an environment that resident defenses can no longer attack and eliminate.

Although this is not surgery, it is a dynamic reorganization and organization of the body and the CNS. This process uses many of the same techniques used to create a new environment by tricking the local environment into thinking that this is natural and that the clock needs to be reset for the cell to regrow as it did when it formed. Turning back the clock or playing with the clock has been performed by several people, but the best example is the work of Salomoni and Calegari (2010). They changed the duration of the G1 phase and were able to reset the cell, tricking the cell into thinking that it was actually younger and that it was time to grow again. Such work needs a lot more experimentation before it can be applied to humans, but one could imagine that it could attempt to change the G1 phase perhaps in a kidney cell, in a cell at the margin of a missing limb, or even in brain tissue lost because of trauma or stroke. This would then allow the cell to regrow back into the area to fill in the missing area.

Central Nervous System

There is one last issue that needs to be addressed before delving into the nanorealm. The eye is also part of the CNS. However, depending on whose perspective—neurosurgeon or ophthalmologist—the eye is an outcropping of the brain (the view of most neurosurgeons), or the brain is an outcropping of the eye (the view of ophthalmic surgeons). The point here is that much of nanotechnology will impact the eye before it impacts the brain in any procedure from drug delivery to surgical intervention.

The goal of all work discussed from this point forward is to reconnect disconnected parts of the CNS. Not only are axons regenerated but they are also functional, as demonstrated by their behavior. Many experiments were performed on the visual system because of its isolated group of accessible neurons that can be clearly and concisely manipulated. There was also a clear behavioral readout when the disconnection happened, and the reconnection was just as clear, with evidence of behavior and not only of histology. Although spinal cord studies showed return of behavior, it was very difficult to clearly differentiate regeneration from unmasking of neurons in a diaschetic state (spinal shock). Along with the restoration of the environment in the spinal cord are many reports of miraculous recoveries in very short periods (i.e., injection of stem cells and recovery of some function in 1 week). In these types of treatments, it is nearly impossible to differentiate among true regeneration, restoration of disconnected neurons, or simply restoration of function in neurons that are connected but nonfunctional.

Distinction must be made between the three different populations of neurons: (1) connected and functional; (2) disconnected and nonfunctional; and (3) connected but nonfunctional, which, to date, is impossible to distinguish. This begs the question: What does a neuronal injury look like? In many cases, there is no complete disconnection but a partial disconnection with secondary damage, caused first by inflammation and then by ischemia. In many cases, this is managed with steroids or nonsteroidal anti-inflammatory drugs (NSAIDs) to mitigate secondary damage. It can also be managed with hypothermia of the local or global system. Although there is still much controversy attached to the use of hypothermia to mitigate inflammation and to preserve neurons, it is very clear that there is a benefit. However, objective measures are still trying to attain the level of acuity necessary to truly measure what is happening to the neuronal population.

Four Ps of CNS Regeneration

To better understand the regenerative environment, I developed a framework to address the issues and to measure each stage of the regeneration process. Each of the "four Ps" addresses one of the main problems encountered and provides an objective measure of overcoming each of these issues. The four Ps are preserve, permit, promote, and plasticity (Figures 22.2 and 22.3) (Ellis-Behnke 2007).

Preserve

When any disconnection is encountered, the first and most important step is to preserve the cells or to keep them alive—not only neurons but also glia and microglia. This can be performed by limiting inflammation and antiapoptotic factors, stabilizing the wound environment (which can include hemostasis), or creating an environment that keeps bacteria and other foreign invaders out of the wound, either from the outside or from within the body. Unlike other types of regeneration, neurons in the brain are very similar to trees; they regrow from the point from which they are disconnected, much the same as an oak tree regrows from the stump when cut down. The second part of

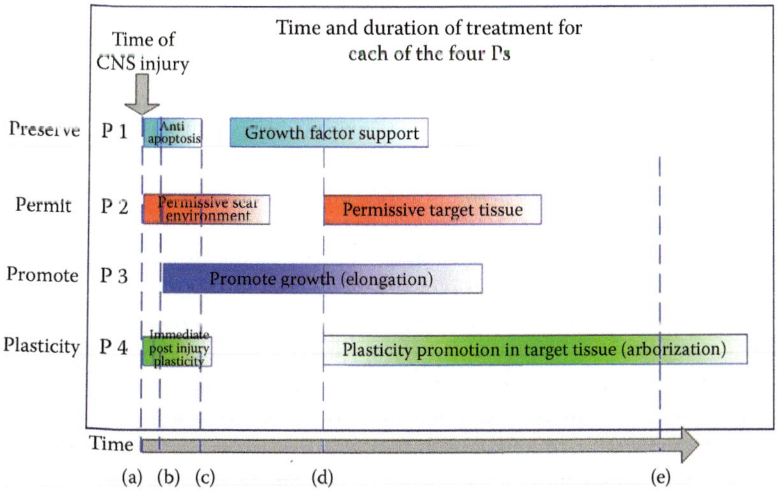

FIGURE 22.2 Four Ps of CNS regeneration. Each of the four Ps is represented by a horizontal bar depicting the time of treatment, as well as the duration needed for the treatment to remain active. P1: *preserve* cells. This is broken down into two parts: antiapoptosis and antinecrosis treatment, and growth factor support during elongation. P2: create a *permissive* environment. This is also broken down into two parts: creating a permissive environment around the site of injury and making the target tissue permissive to reinnervating axons. P3: *promote* the growth of axons in an elongation mode. P4: *plasticity*. This has two parts: reducing immediate postinjury plasticity so that filling-in is kept to a minimum and dysregulating plasticity during the time that axons are reconnecting in the target tissue. This can be accomplished with factors or training. The timeline runs from left to right and starts at the time of injury: (a) time of injury; (b) 24 h after injury; (c) 72–96 h after injury; (d) 2 weeks to 4 months or longer; (e) years.

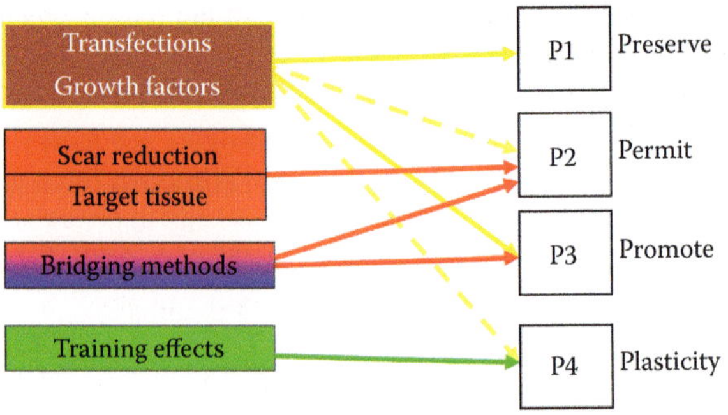

FIGURE 22.3 Mapping experiments to the four Ps of CNS regeneration. There can be interplay with each of the four Ps, or one therapy could impact all of the Ps at once. Pay close attention to the correct time to pull each of the levers.

preservation is of longer duration—keeping the cell alive during the regeneration and reconnection phases of growth into the target tissue.

Permit

The second P refers to the creation of a permissive environment. A permissive environment has two parts. The first part is creating a local environment that the cells perceive as being part of the normal tissue but only at a lower density, which allows the cells to migrate in and start reclaiming the lost area. It also signals to the neurons that the environment will enable growth through the local area. The second part is making the neurons unable to perceive inhibitory signals during the regrowth phase back to the target tissue.

Promote

The third P refers to promotion of growth. If successful (the cells remain alive and the environment is permissive to growth), the goal of this P will be to speed regeneration to the target tissue.

Plasticity

Finally, the fourth P is plasticity, which has two parts as well. The first part is stopping immediate postinjury plasticity. Once there is damage, there is a filling-in of denervated structures. This can cause prolonged rehabilitation because the connections not only need to reach the target but also need to unwind the filled-in connection and to reinnervate the target. The second part of plasticity is the target tissue, which is reconnected and reorganized, driven by rehabilitation. As rehabilitation proceeds, the connection can be strengthened if the behavior is rewarded. This reward refers to the typical Hebbian synapse formation and strengthening caused by a correct temporal feedback based on actions. In animals, it is usually associated with food reward, but it originates from increased signal strength and feedback caused by directed behavioral recovery (Hebb 1967). Although this is primarily used for CNS regeneration, it can be useful for other types of injury with minor modifications, the first being tissue stabilization after injury.

NANOTECHNOLOGY

As it is already known that nano is one billionth of a meter or 10^{-9} m, the erection of scaffold using amino acids will be presented here. This is quite unique owing to the fact that a scaffold is built and used to create an environment that allows reconnection. After its use, the scaffold is broken down. The components of the scaffold are a series of amino acids that are, on average, 16 amino acids (AA) long, with alternating hydrophilic and hydrophobic components that enable assembly due to the complementarity of their charges. The materials can be injected. They form a structure, and when the structure is no longer needed, the body disassembles it and then excretes it through the urine. The structure formed is not permanent, and there is no covalent bonding that permits dynamic assembly and disassembly as cells migrate through the area or as axons grow through the environment. Although much of this chapter focuses on one or two materials, there are well over 600. This tool kit shows promising potential for the reconstruction of the human body.

Nanomedicine is the use of nanotechnology to create a new ability in an existing treatment, or the creation of new treatments and classes of therapies and devices based on their size. Nanomedicine is different from small molecules in that most small molecules, although appearing to be nano, are in fact aggregations of 100,000 or more molecules that are delivered in a clump. Some are broken down by the body; however, many never reach the target because of solubility, distribution, and metabolic pathways. The promise of true nanomedicine is the presentation of one or several molecules to a specific site, followed by a careful destruction or excretion of these materials without side effects.

NANOMATERIALS

Materials and Self-Assembly

To understand how these materials work, one has to first understand what they are as much as what they are not. Many materials are used for biomedical applications. While some of these materials are hard and used as implants, such as hip replacements, others are soft, break down, and are excreted. There are also a plethora of materials in-between, such as materials used as drug delivery receivers, which release therapeutic agents in the local environment during breakdown. Other materials, such as poly(lactide-*co*-glycolide) (PLGA) and polymethyl methacrylate (PMMA), are directly implanted into the injury site, allowing some regeneration. However, many of these materials, upon breakdown, create an acidic environment and can cause a secondary immune reaction. The materials are also covalently bonded to one another, creating a stable structure that must be chemically degraded. Each of these materials has a use and is ideally structured for specific needs. There are other classes of materials (liposomes and micelles) used for loading therapeutic agents; these can have a targeting moiety placed on the outer shell, enabling either active or passive targeting to a tumor or a specific tissue. Much of the work for these types of systems has focused on cancer treatment.

The self-assembly process is important because it is bidirectional. Imagine that a scaffold has been built but part of the structure needs to be changed. This scaffold will rearrange itself based on new cues from the nanodomain in that environment. This is what happens in the blood–brain barrier (BBB), the luminal wall of the intestine, and other examples throughout the body.

The materials discussed in this chapter are those that self-assemble. This self-assembly is caused by both the structure of the material and the sequences used. The dominant forces are hydrophobic and hydrophilic interactions that allow complementary assembly: from small 16 AA sequences to larger secondary structures such as sheets and fibers. Other forces that cause these to assemble are van der Waals forces and weak hydrogen bonds, which can be initiated by salt or charge. The assembly is very similar to that in the ECM. Repeated patterns are very similar to those seen in the ECM in early development and in the temporary structures used during organogenesis, such as the tight junction between the endothelial tissues in blood vessels and the dynamic barriers in the kidney separating urine from blood.

A sample sequence is the (RADA)4 sequence, which is 16 AA long. It first forms a β-sheet structure then forms fibers. The fibers increase in size and begin to organize in water, creating a highly hydrated material with a water content of 99.9%. With the addition of salt or any ionic solution, water condenses as it is replaced by structure. The size change is imperceptible until very high concentrations of salt are reached. This is similar to DNA condensation. In some of the first experiments with DNA, high salt levels were used to cause DNA to condense and precipitate at the bottom of the test tube (Altundag et al. 2007). This work was started in 1869 by Johannes Freidrich Miescher, who was the first to isolate nuclein—now known as nucleic acid—from white blood cells.

This type of chemistry causes DNA to condense and precipitate. Condensation adds a great deal of flexibility to some of these materials because by changing both the concentration and the type of salt, it is possible to slightly change the structure. This type of manipulation adds a new dimension to the fine-tuning of the environment.

Self-Assembling Peptides

(RADA)4 self-assembling peptide (SAP) is a synthetic biological material formed through the assembly of ionic self-complementary peptides and is designed by using alternating positive and negative L-amino acids that form highly hydrated scaffolds in the presence of physiological-concentration salts (i.e., saline, tissue culture media, physiological solutions, or human body fluids such as cerebrospinal fluid). The scaffold consists of β-sheet ionic peptides containing 50% charged residues. (RADA)4 consists of interwoven nanofibers; individual fibers are ~10 nm in diameter (Figure 22.4) (Ellis-Behnke et al. 2006a).

Nanofiber density correlates with the concentration of peptide solution. This designed peptide nanofiber scaffold provides several benefits over currently available polymer biomaterials. (1) The peptide scaffold forms a network of nanofibers that are similar in scale to the native ECM and therefore provides an in vivo environment for cell growth, migration, and differentiation. (2) It can be broken down into natural L-amino acids and can potentially be used by the surrounding tissue because much of the material is excreted in the urine. (3) It is synthetic and free of chemical and biological contaminants that are typically present in animal-derived biomaterials such as many collagens. (4) It appears to be immunologically inert, having no apparent immune response or tissue inflammation when introduced into animals, thus avoiding the problem of neural tissue rejection. (5) It has a true three-dimensional (3D) environment for cell growth and migration, similar to the native ECM. (6) It has excellent physiological compatibility and minimal cytotoxicity because it is composed of naturally occurring amino acids (Ellis-Behnke et al. 2006a; Gao et al. 2007). All these attributes make it very attractive for use as a peptide nanofiber scaffold in both in vitro and in vivo studies.

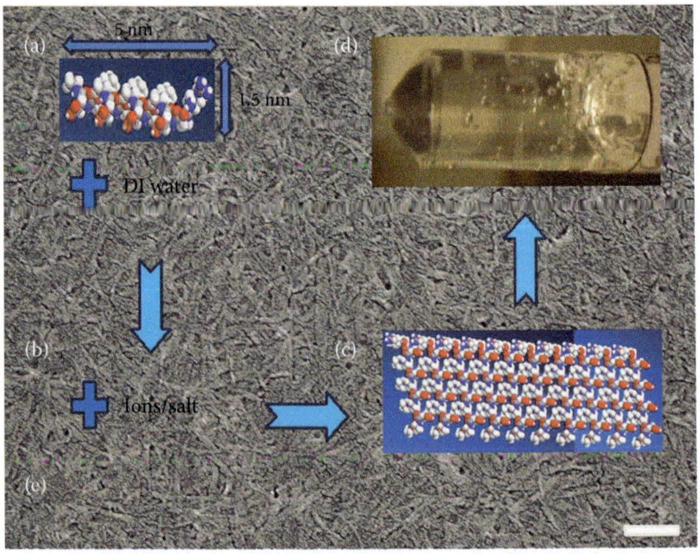

FIGURE 22.4 How self-assembled materials work, what they look like, and what the resultant fiber formation looks like under a scanning electron microscope (SEM). (a) (RADA)4 molecule with dimensions. This is mixed in deionized water at various concentrations. (b) (RADA)4 mixture added to the body, which is water with high salt content. The material assembles into sheets as a result of the aqueous environment due to the hydrophobic part trying to hide from water. (c) Sheets assembling into fibers and forming a 3D structure in the tube. (d) High hydration with 99% water. (e) The 20-nm fiber structure is visualized using an SEM. Scale bar represents 200 nm.

The fact that self-assembly is a key component that occurs during introduction to a charged environment changes how biomaterials are viewed. Rather than manufacturing and milling something to the size needed, liquid can be injected into the cavity formed by the damage, and it will self-organize to interact with the smallest parts of the wound environment.

When an amino acid sequence and water are injected into the site, the nucleation process starts at the interface of the tissue. As the materials are exposed to the charge, they will orient to interact with the damaged areas of the ECM and with the components of the cells that have spilled out into the wound. This assembly proceeds very quickly. Soon, there is a clear structure that appears to match the modulus of the local tissue. If the color were the same, it would be indistinguishable. In this case, the (RADA)4 sequence assembles into a β-sheet with a slight twist: pull on the tissue and it expands and contracts much the same way as undamaged neural tissues surrounding the wound.

The 3D mesh created can be tuned to match the modulus of the local tissue by one of several means. The first is through structure: a peptide sequence such as EAKA is much stiffer and is like putting a nail into a pudding (Holmes et al. 2000). As the brain moves during pulsations of the heart and the circulatory system, the hard material will shear from the interface and possibly create more damage, whereas the (RADA)4 sequence will move with the tissue at the same concentration. The variations in concentration and salt content are far beyond the scope of this chapter, but each of these axes can be manipulated to elicit additional functionality. This is in homogeneous repeats. When a fourth dimension, such as heterogeneous sequences, is added by combining different sequences or by mixing a solution of two different sequences, the combinations and diversity are astounding.

Nanomedicine

When one thinks about nanomedicine, it seems absurd to refer to medicine based on size. However, there is a fundamental break from the norm when descending from the microscale to the nanoscale. Many of the properties change, and there are additional benefits with the use of nanomaterials in manipulating the environment. For example, making a hole in a cell would cause instant cell death or would at least lead the cell to succumb eventually to trauma. However, in the nanorealm, holes can be made in cells, and proteins and pH can be monitored without detrimental effects on the cell's normal functions. This brings back readers to the age of cellular surgery, which had its beginnings in in vitro fertilization (IVF) clinics. In 1997, with assisted reproductive technology (ART), the successful fertilization and production of an embryo, leading to the delivery of a baby, used not two partners but three: DNA from the nucleus of an egg, DNA from the cytoplasm of another egg, and DNA from a sperm, each from a different donor (Cohen et al. 1997). This seemingly small start of nanosurgery of the cell is now routine in IVF clinics.

NANONEUROSURGERY

The surgical field is often obscured with blood and debris during an operation. In addition, clearing debris from the surgical field usually requires irrigating the site with saline. Saline is only a temporary solution and needs to be continuously applied to maintain a clear surgical field. Note that (RADA)4 is clear and allows for clear viewing and easier manipulation of the surgical field.

CNS Regeneration

Reconnecting Disconnected Parts of the Brain

Reconnecting Severed Optic Nerve with Fibrin Glue

Before the discovery of self-assembled materials, I was looking for a natural material that would (1) not be rejected by the body, (2) be able to easily flow into a wound, (3) fill a damaged area, and (4) form a seal that would enable tissue to reconnect. In the late 1950s, fibrin glue was used to coat

Dacron arterial grafts to make them more biologically compatible (Cooley et al. 1981; Debakey et al. 1964).

Since preliminary studies have shown that the optic nerve (ON) could be regenerated using a sciatic nerve bridge that was connected from the back of the eye to the superior colliculus (SC), reconnection, regeneration of the ON, and return of function were all possible (Ellis-Behnke et al. 2000).

The components of fibrin glue were used to make reattachment easier. In the peripheral nerve (PN) bridge work, care was taken not to disrupt blood flow to the eye by opening the nerve sheath and by carefully inserting the sciatic nerve into the sheath after transecting the ON inside the sheath. The sciatic nerve was then inserted over the ON that was cut and then sewn into the back of the eye with one or two sutures. This did not disrupt blood supply to the eye (Ellis-Behnke et al. 2001).

Taking it one step further, I created a new paradigm to cut the ON and to disconnect the ON sheath from the back of the eye. This was designed not only to test the ability of retinal ganglion cells (RGCs) to regrow but also to determine the amount of RGC death when the entire nerve sheath and the nerve itself are disconnected.

This study of a self-assembled material and a self-polymerizable natural material was performed prior to my work with (RADA)4, about which the rest of the chapter is devoted. There were three important lessons derived from this seemingly simple experiment. First, employing a material that could flow into a wound, create a seal, and be used for reattachment could be very useful. Second, sealing tissues with a somewhat biologically inert material could help keep a viable cell population alive longer than previously reported for a similar injury because the reestablishment of membrane integrity may play an important role in preserving these cells and tissues. Third, a completely disrupted main blood supply to the eye could be reopened with a correct material that would permit the breakdown and reconnection of the vasculature.

The third lesson is most interesting because the fact that the blood supply was reestablished very quickly completely changed the perception that the retina is as fragile as once thought. It must be kept in mind that the subjects were rodents and that the blood supply was smaller and may have had a different ability to reconnect than in humans. Also, the mechanism that dissolves the artificial clot formed may be different in humans than in rodents. Finally, the mixtures of the different components gave different results. Since this was only a pilot study, it was hard to draw concrete conclusions about the mixtures; however, a big difference from one mixture to the next was observed.

Some of the difficulties encountered were related to the storage of components. They had to be refrigerated and mixed within the week of usage or they would not polymerize and hold the tissue together. Also, in humans, there was always the possibility of transmitting hepatitis.

The use of natural components that will form structures, be broken down by the body, and facilitate tissue reconstruction may play an important role in a self-assembling nanomaterial. This polymerized material forms covalent bonds that are broken down, while the other SAPs discussed in this chapter dynamically reconfigure and allow cells to migrate into the material without breaking it down. The intersection of the paradigms self-assembly and dynamic reconfiguration, either by change in charge or by dissolution and breakdown, may have a synergistic effect that can be applied to the CNS in a more robust way. The fact that the fibrin mixture allowed survival may be an advantage for some other materials, as this ability to preserve keeps cells alive longer. The mechanism for this is unknown but deserves further investigation.

Surgical Method

After the animal was completely anesthesized, the superior rectus and lateral rectus muscles of the eye were protruded and the ON was exposed. In some cases, the superior oblique muscle also needed to be cut. Care was taken to allow suturing of the muscles when the eye was placed back into the socket. After the eye was protruded, both the ON and the nerve sheath were severed posterior to the lamina cribosa at the point where the ON sheath joined the globe. This gave the maximum surface area for the adherence of fibrin mixture. The mixture was composed of thrombin, fibrinogen, and calcium chloride, which were obtained from Sigma and placed in separate containers. Each of these was placed in a syringe, and the syringes were taped together so they could be applied simultaneously and in correct proportion to achieve maximum binding and shortened polymerization time. Since this was a pilot study, several different concentrations were tried to achieve the best results based on the shortest bonding time and the least amount of material. This would ensure the best possibility for the healing process and would create the least obstacle to the restoration of blood flow and the reinnervation of the ON. The animals were allowed to survive for 8 weeks, and histology was performed to determine RGC survival, tissue integrity, and regeneration.

Results

The fibrin glue formed by the mixture allowed revascularization of the globe in less than 24 h. Although direct measurement of the blood flow was not accomplished, RGC survival was used as a metric for revascularization. In the past, the retina of rats and hamsters would quickly die if there was no perfusion for 48–72 h. Since less than 20% of the RGCs died in the best-case scenario, it was surmised that blood flow to the ophthalmic artery was restored. In the worst-case scenario, 50% of the RGCs survived. This was in sharp contrast to others' work showing that 80% of the RGCs died when the ON was severed behind the globe. Blood will dissolve fibrin and allow reestablishment of the connection. The ON sheath healed. However, there was no reinnervation of the RGCs past the transection. This was assessed with the injection of cholera toxin B (CTB)–fluorescein isothiocyanate (FITC) into the eye, the use of antibody to CTB-FITC, and visualization with a fluorescent dye (Figure 22.5).

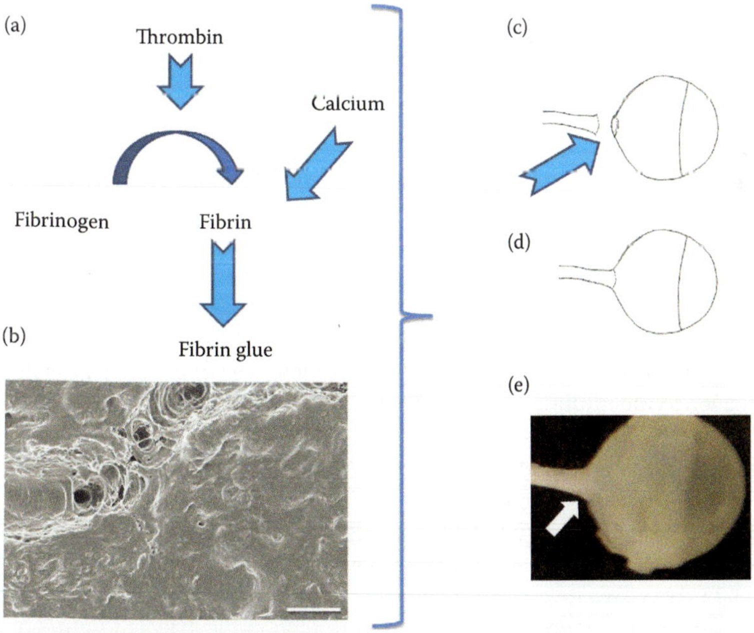

FIGURE 22.5 Repairing the brain with fibrin glue. (a) Components of fibrin glue and cascade of how thrombin and fibrinogen, along with calcium, are added to form fibrin glue. (b) Scanning electron microscopy of fibrin glue and the clot that forms as a result of the clotting cascade. (c) Locations of the disconnection and treatment of the mixture. (d) How ON is reconnected at the same location. (e) An eye showing how reconnection happens and how the resulting tissue looks like after being reconnected. This was performed in hamsters, rats, and mice. Picture shown is that of a fixed rat eye. The tissue appears normal as the fibrin dissolves and as the tissue heals. Scale bar represents 10 μm.

CNS Regeneration after Acute Injury

(RADA)4 was used in the "nano neuro knitting" study (Ellis-Behnke et al. 2006a) because it spontaneously formed nanofibers, creating a scaffold-like tissue-bridging structure that provided a framework for partial reinnervation by axons with regenerative potential in young and adult animals. Because peptide fibers are in nanoscale, it was hypothesized that there would be a direct interaction between the peptide scaffold, the ECM, and the neural tissue on both sides of the lesion. By knitting tissue together, the (RADA)4 scaffold created a permissive environment that enabled reconnection of brain tissue and allowed axonal growth while discouraging or preventing scar formation that normally occurs at an early stage. The result was a complete reversal of blindness, as evidenced by functional behavior.

Prior to (RADA)4, no treatment has created a permissive environment for axonal regeneration that allowed growth through the center of a lesion. This study demonstrated that, with a single (RADA)4 solution treatment at the site of injury, it was possible to overcome a major barrier to CNS regeneration in the optic tract (OT) of hamsters.

The use of nanobiomedical technology and molecular self-assembly to repair brain structures opened up a new field and a new source of hope for efficacious treatment of CNS trauma. This successful outcome gives a glimpse of what reconstructive brain surgery may hold in the future.

Surgical and Behavioral Methods

Young Animals In vivo applications to brain wounds were carried out by using 53 P2 Syrian hamster pups anesthesized with whole-body cooling. The scalp was opened, and the OT within the SC was completely severed with a deep knife wound through a slot cut into the cartilaginous skull, extending 1.5 mm below the surface from the midline to a point beyond the lateral margin of the SC. Animals were kept alive for 24 h, 72 h, 30 days, and 60 days after surgery. During surgery, 10 animals were treated with a 10-μL injection of 1% (RADA)4 into the brain wound. Control animals with the same lesions included 6 animals injected with isotonic saline (10 μL), 27 earlier cases with knife cuts and no injection that were allowed to live for 6–9 days, and 10 animals that were lesioned negative controls and allowed to survive for 24 h, 72 h, 30 days, and 60 days after the lesion.

Adult Animals Sixteen adult Syrian hamsters were anesthesized with an intraperitoneal injection of sodium pentobarbital (50 mg/kg) and then fitted in a head holder. The cortex was exposed and aspirated to reveal the rostral edge of the SC and the brachium of the superior colliculus (BSC) on the left side. Each animal received a complete transection of the BSC. Lesions are shallow in many CNS regeneration models. To ensure that there would be no regeneration around the cut, the model used a 2-mm-deep cut. The increased depth prevented axons from growing around the bottom of the cut, ensuring that they grew through the center of the lesion site rather than around it. In addition, special attention was paid to ensuring that there was complete transection of the BSC from the lateral edge to the midline. In controls, no axons grew through or around the 2-mm-deep lesion site.

Animal Behavior Assessment All adult animals treated with (RADA)4 were tested for visually elicited orienting movements (Darragh et al. 2010). Briefly, for orienting movements to be elicited, sunflower seeds were presented first by hand and later with the aid of a white metal wire, at the end of which was a small black rubber ball slotted for holding a seed. Animals were tested two to three times weekly for 15 days. Two investigators independently tested each animal for 10–20 min every other day, or less frequently, before any responses were seen in the affected field. On each test day, a minimum of 10 presentations were made on each side, with a random choice of side. Visually elicited orienting movements were tested repeatedly by presenting a sunflower seed to a part of the hamster's visual field. Left and right visual fields were tested; all trials were videotaped with an overhead camera. A trial was counted if a response occurred within 2 s of stimulus presentation. Completion of a turn was signaled by the animal's head coming to a stationary position within 5° of the stimulus for at least one-third of a second.

Results

In the first experiment using young animals, 10 μL of 1% (RADA)4 solution was applied within the knife wound (1.5 mm deep and 2.0 mm wide) inflicted on the midbrain (in the rostral part of the SC) of P2 hamsters (Figure 22.6). A schematic illustration of a parasagittal section of the dorsal midbrain of a hamster shows the injury's location. In all (RADA)4-treated cases, the gap created in the tissue during injury was reduced in the first 24 h after surgery and was reduced or completely eliminated at all later time points (72 h, 30 days, and 60 days after surgery).

In the second experiment, 8-week-old adult animals were used. The OT was transected at the brachium of the SC, and 30 μL of 1% (RADA)4 was injected into the injury site of treated cases. In all cases examined, there were no gaps between the injured tissue and the appearance of "healed" tissue at 30, 45, or 90 days after injury. These adult cases showed results similar to P2-operated animals at 30- and 60-day time points. The lesions were healed in both young and adult animals. The two sides of the lesion came back together, and the tissue appeared to be knitted together, with little sign of the original lesion.

For anatomical evidence of axons regenerating through the lesion site, a fluorescently tagged CTB fragment, which is taken up in retinal ganglion cells and transported to axon terminals, was used. This tracer served as a marker of regeneration across wounds in treated animals or as a marker of regeneration failure in lesioned controls. In preparation for axon tracing, the animals with post-operative survival times of 30, 45, 60, and 90 days received 1-μL injections of 1% CTB conjugated with FITC into the vitreous humor of the eye, contralateral to the SC surgery site, 4 days before death. Axon tracing results for the young and adult treated animals in both experiments showed evidence of axonal growth into the SC caudal to the lesion site. Figure 22.7 shows the results for adult animals. Histological results revealed that 92% of peptide scaffold-treated cases showed labeled regenerated axons in the SC. There were no labeled axons in the SC in lesioned control animals.

Behavioral testing of the adult treated animals in experiment 2 was performed to evaluate functional vision (indicated by the ability to orient toward a small object), and this visual ability was found to have returned in 75% of these animals (Figure 22.8). The controls remained blind. The amount of axon reinnervation correlated with the return of functional vision in (RADA)4-treated adult animals (Spearman rank correlation = 0.91, $n = 8$, $P = 0.01$), with 75% of the animals demonstrating visually guided behavior, although the animals' turning responses to stimulus were 29% slower than the responses of normal animals. The graph illustrates two very important points.

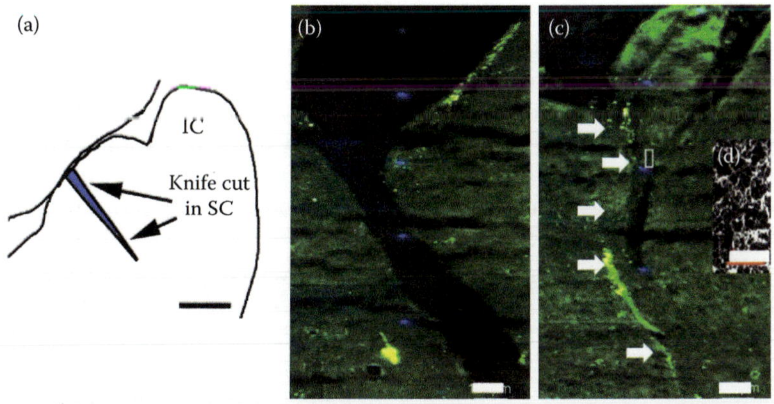

FIGURE 22.6 Repairing the brain of young animals. (a) Parasagittal view of an adult hamster brain showing a schematic representation of the lesion (blue) that transects the OT in the middle of the SC in young animals. (b) Control at 1 month after lesion showing the gap formed without regeneration. (c) Treated case at 1 month after lesion and treatment showing repair of the brain (arrows). (d) Inset: Scanning electron microscopy of material as it self-assembles in the brain. Scale bar represents (a) 500 μm, (b and c) 100 μm, and (d) 1 μm.

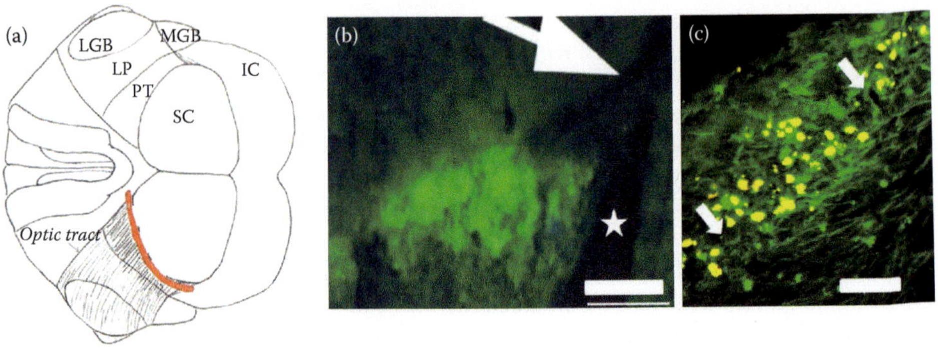

FIGURE 22.7 Repairing the brain of adult animals. (a) Dorsal view reconstruction of the hamster brain with the cortex removed. Rostral to the left, and caudal to the right. Anatomical structures include the SC, pretectal area (PT), lateral posterior nucleus (LP), medial geniculate body (MGB), lateral geniculate body (LGB), and inferior colliculus (IC). The red line depicts the location of the transection of the OT (brachium of the SC) made at 3 months. (b) Enlarged view of the area rostral to the lesion site showing CTB-FITC-labeled axons stopped before the gap (large arrow) created by the injury in the saline-treated control animal. The gap was formed with saline treatment (star). (c) Labeled regenerating axons crossing the injured region and growing toward the target in the (RADA)4-treated animal. The location of the healed lesion and the approximate location of the injury (arrows) are shown. Scale bars represent (b) 100 μm and (c) 25 μm.

(1) Seventy-five percent of the treated animals responded well to the stimulus (in blue), whereas the lesioned controls (in yellow) spontaneously turned at a rate below 30%, consistent with the spontaneous response rate in blind animals. Data points represent the average of each week's behavioral testing on the animals. Two treated animals did not reach the minimum reinnervation density and responded the same as blind controls; this was caused by vascular injury during surgery, and 95% of the SC was gone. (2) As testing progressed, the treated group's vision improved, and the responses became more robust, suggesting that this effect can be amplified with sufficient reinnervation and long-term rehabilitation. Tests of pupillary response to light showed positive responses in all animals that also showed visual orienting and in only one animal without such orienting but with a small amount of regeneration.

Addition of Agents

Another way to add functionality is by adding agents to these materials. One approach is the direct addition of growth factors and therapeutic agents to the sequence. Although this appears to be a good idea, the regulatory environment needed for this type of combination system can be very complex. The breakdown of these materials and the subsequent signaling events can become unmanageable in the current regulatory framework. Second, because of the small amounts of some of these materials, many breakdown products may be below detectable limits in the blood. This is the primary safety determinant for many therapeutic agents today.

In another investigation of OT brain lesion repair, combinations of (RADA)4 and chondroitinase ABC (Chon-ABC) appeared to create a more permissive environment, resulting in earlier return of vision (Ellis-Behnke 2007).

Surgical Method

Seventy-nine adult golden hamsters were anesthesized with an intraperitoneal injection of sodium pentobarbital (50 mg/kg) and then fitted in a head holder. The cortex was exposed and aspirated to reveal the rostral edge of the SC and the BSC on the left side. Each animal received a complete transection of the BSC, and 1% (RADA)4, 2.5 U/mL Chon-ABC, or both were injected into the injury site. Reinnervation of the SC was monitored at 4, 6, and 12 weeks after the lesion using antero-retrograde labeling with an intravitreal injection of CTB-FITC-conjugated CTB.

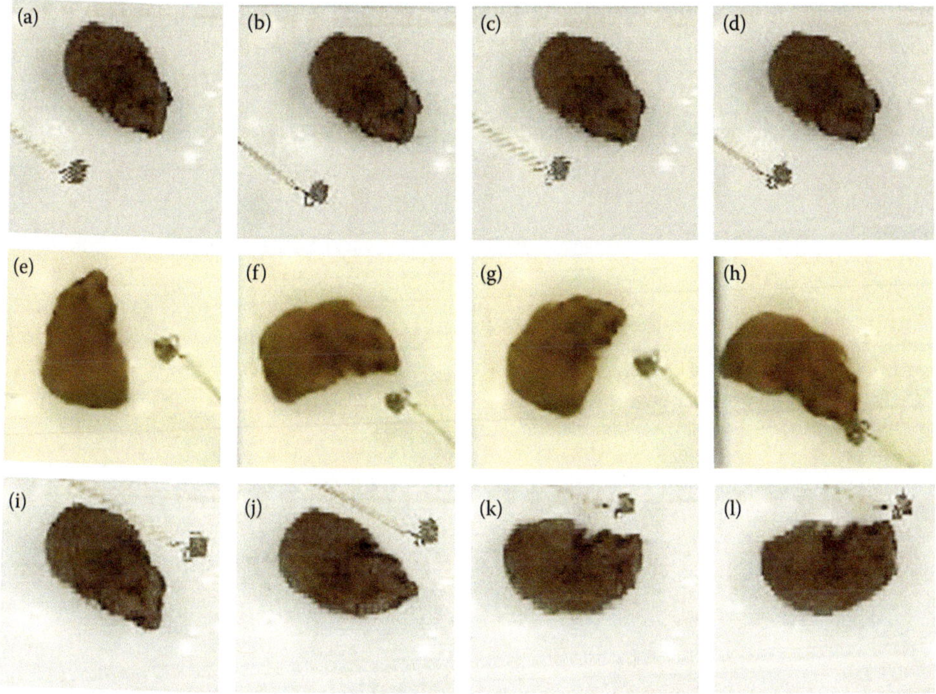

FIGURE 22.8 Series of animal behavior responses when a seed is presented to the visual field at the end of a wire. Stimulus presentations were random and made up of a series of frames captured from behavioral analysis videos. (a–d) Animal whose eye had been disconnected from the SC (saline-injected control at the location of the lesion). Each step lasted 0.75 s. The duration of the stimulus presentation was longer here than in any of the other presentations, but the animal still did not make any orienting movements toward the stimulus. This also served as an olfactory control. If the animal could have smelled the seed, it would have turned toward it. However, there was no indication of an olfactory cue in any of the tests. (e–h) Regenerated animal turning in small steps toward the stimulus at the affected left visual field and prolonged here by movements of the stimulus away from it. Each frame was taken from a single turning movement at 0.00, 0.27, 0.53, and 0.80 s from movement initiation. The animal reached the stimulus just after the last frame. This was about 0.20 s slower than most turns by a normal animal. (i–l) Control animal. The series was 0.20 s apart. The animal reached the stimulus in approximately 0.8 s from the initial turn.

Results

(RADA)4 and Chon-ABC, when used together, helped retinal ganglion cell axons to regenerate across the lesion site, with a synergistic effect evidenced by the larger area of reinnervation compared with either (RADA)4 or Chon-ABC alone. This effect was observed in all treatment groups as early as 4 weeks after the transection, while saline controls showed no reinnervation at any time point (Figure 22.9). In a behavioral study (Schneider et al. 2006), adult hamsters showed a functional return of vision in (RADA)4–Chon-ABC-treated cases beginning at 6 weeks after surgery (Figure 22.10).

CNS Regeneration after Chronic Injury

Real-time measurement of axon regeneration in the CNS in vivo is essential to CNS injury assessment to speed the recovery process and to determine the optimal timing and delivery of therapeutic agents and rehabilitation. Axonal regeneration accompanied by behavioral recovery had been previously shown in hamsters after acute injury and treatment with (RADA)4 (Ellis-Behnke et al. 2006a). However, for enhanced CNS recovery after chronic injury, real-time measurement of axon

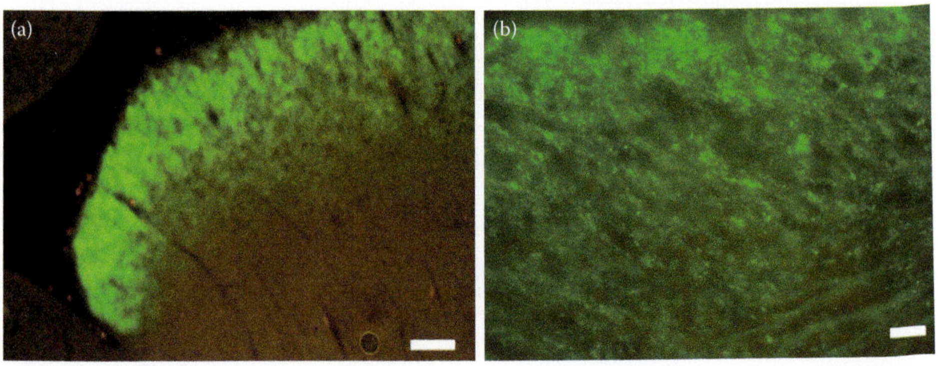

FIGURE 22.9 Combination treatment Chon-ABC. (a) Series dark-field photo, at 12 weeks after surgery, of a parasagittal section of the brain of a hamster treated with (RADA)4 and Chon-ABC at the time of surgery at the lesion site. This was released over time as the (RADA)4 material disassembled, prolonging the window for regeneration at the injury site. (b) Density, axon terminals, and fiber bundles looking almost normal. Scale bars represent (a) 100 μm and (b) 25 μm.

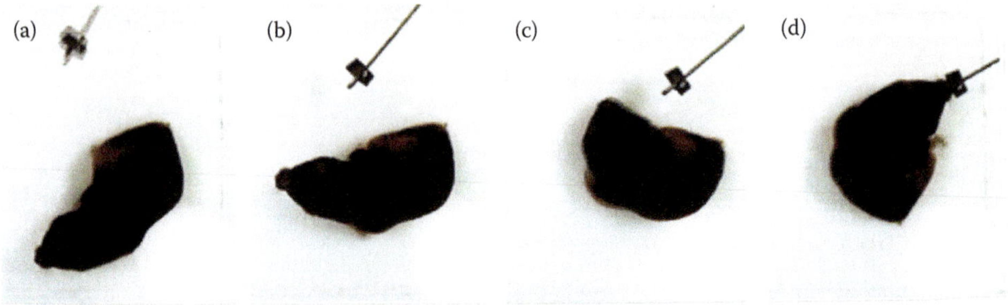

FIGURE 22.10 Behavior with combination treatment Chon-ABC. This is a series of frames taken from the behavior recordings of hamsters during trials. This adult animal had a combination treatment of (RADA)4 and Chon-ABC coinjected at the injury site. The animal turned in small steps toward the stimulus at the affected left visual field and was prolonged here by movements of the stimulus away from it. Each frame was taken from a single turning movement at (a) 0.00 s, (b) 0.32 s, (c) 0.6 s, and (d) 1.1 s from movement initiation. The animal reached the stimulus at the last frame. This was about 0.20 s slower than most turns by a normal animal.

regeneration in vivo was essential to assess the extent of injury and the optimal timing and delivery of therapeutic agents and rehabilitation.

For this work to be clinically relevant, it was necessary to develop a chronic animal model with an in vivo measurement technique to provide real-time monitoring and feedback. It was also important to precisely pinpoint the time when regeneration of axons attained a suitable stage of reinnervation at the target tissue to determine the correct timing for rehabilitation.

With the framework of the four Ps (preserve, promote, permit, and plasticity) of CNS regeneration as guide (Ellis-Behnke 2007), the next steps were (1) assessing the degree of injury to determine the optimal extent and duration of a permissive environment for axonal regeneration to occur, and (2) measuring real-time axonal growth in acute and chronic CNS injury to determine the optimal time for the delivery of therapeutic agents and rehabilitation. In this study, it was hypothesized that magnetic resonance imaging (MRI) and a manganese chloride ($MnCl_2$) tracer nano contrast agent (NCA) could be used to show axon regeneration in hamsters in a chronic CNS injury model (Liang et al. 2011) because MRI is a noninvasive approach to studying details of the body's inner depth in vivo and $MnCl_2$ can label the nervous system intra-axonally.

This work created a chronic model of CNS injury that has multiple readouts—one that is reproducible and employs three separate modes for measuring CNS regeneration. The first mode provides real-time feedback before, during, and after OT transection through MRI. The second mode, behavioral testing, uses visually guided orienting behavior. The third mode is histological assessment. The treatment method created a channel at the site of injury; the material was deposited into the channel to give axons a growth path to follow during the regeneration process, as opposed to attempting to apply treatment at the original wound site. With the use of MRI, the continuity of both connection and subsequent disconnection was tested. The development of the injury site and the subsequent reconnection in some animals were followed.

This new paradigm used MRI to detect axons in the OT of hamsters before, during, and after regeneration in a chronic injury treatment model. With combined chronic injury and (RADA)4 treatment, as well as the use of an NCA to monitor the progress of axon regeneration in mammals, the model appears to be robust and can be used to carefully monitor the progress of axon regeneration and to establish a time table for implementing each of the four Ps.

The ability to follow the progress of the healing and reinnervation of the vacated target area within the CNS opens up another venue for studying the time window within which promotion of wound healing and axonal regeneration can be enhanced under this noninvasive approach.

It may be possible to use this in humans to provide real-time feedback on regenerating tracts, as opposed to current methods that can take months or years to show behavioral return. Measurement is the key to enhancing feedback so that patients do not lose hope. MRI appears to be more sensitive to future return of function than other traditional behavioral methods of measurement. This is a way to change and tailor treatments and to obtain real-time feedback without having to wait for function as a sign of recovery.

Surgical, Imaging, and Behavioral Methods

Baseline Imaging Young adult Syrian golden hamsters (*Mesocricetus auratus*) aged 6–8 weeks and weighing 90–100 g were used. The eyes of anesthesized hamsters in group 1 were first injected with an NCA. The animals were then scanned with a 7-T MRI. $MnCl_2$-enhanced MRI (MEMRI) was performed on normal young adult hamsters to determine the optimal window for maximum contrast in the intact hamster visual system and to establish a comparative baseline for experimental results.

Transection of BSC and (RADA)4 Treatment The left OT in a group of experimental young adult hamsters was completely transected with a scalpel incision at the BSC, as previously described (Ellis-Behnke et al. 2006a). Briefly, an incision extending 1–2 mm below the surface of the brainstem anterior to the SC was made from the midline of the exposed BSC to a point beyond the lateral margin of the SC.

After the normal visual pathway was established with 7-T MRI scanning, a group of experimental animals was prepared. On day 1, the BSC was exposed and transected. On day 45, visually guided behavioral assessment began. The 45-day delay was necessary to mimic the recovery period used in the acute transection and treatment model. On day 90, the contralateral eye was injected and scanned to verify the extent and completeness of the surgery, as well as to assess regeneration prior to treatment. On day 105, the transected brachium was re-exposed and treated with (RADA)4. On day 152, visually guided behavioral assessment was conducted before the contralateral eye was injected and scanned to assess the regeneration and reinnervation of the SC after (RADA)4 treatment in the experimental group of animals. Concurrently, the untreated animal group was injected and scanned for use as a control group to compare the effectiveness of (RADA)4 treatment over the same period. Finally, after the last scan, the animals were injected with CTB-FITC to trace RGC axons and then sacrificed 4 days later. Brains were dissected, sectioned, processed for histology, and examined under a fluorescent microscope.

Animal Behavioral Assessment There were two rounds of visually guided behavioral assessment: (1) 45 days after transection of the BSC and (2) 45 days after treatment. All adult animals were tested for visually elicited orienting movements, as previously described (Schneider et al. 2006).

Treatment Surgery Two-thirds of the animals were randomly selected as the experimental group (n = 10) and treated with (RADA)4, which was applied to new cuts. The other third of the animals were cut (n = 5), but not treated, and used as sham controls. The BSC was re-exposed 105 days after the first surgery, and three perpendicular cuts were made across the old transected region; 30 μL of 1% (RADA)4 solution was injected into the channels.

Intravitreal $MnCl_2$ Injections Approximately 24 h prior to MRI, the hamsters were anesthesized with an intraperitoneal injection of sodium pentobarbital 50 mg/kg. Two microliters of 0.2 M $MnCl_2$ dissolved in Milli-Q water mixture was injected into the vitreous chamber of the eye through a glass micropipette inserted into the right eye contralateral to the inflicted OT lesion at the BSC at the temporal corneal–scleral junction. The micropipette was held inside the eyeball for ~20 s before removing it to prevent leakage of the injected solution. The experimental animals received one $MnCl_2$ injection 24 h before scanning at each of the three time points: days 0, 90, and 152.

Horseradish Peroxidase OT Tracing Five microliters of 25% wheat germ agglutinin (WGA) conjugated to horseradish peroxidase (HRP) solution was injected into the posterior chamber of a normal animal's left eye to trace retinofugal projections. Two days later, the animal was perfused with Karnosky's fixative, and the brain was dissected out from the skull and postfixed in the same fixative for another 4 h at 4°C. The isolated brain was rinsed, and the meninges covering the brain were removed with a soft brush (to avoid damaging the surface) until the brain was clean to improve the penetration of chromogen into the brain tissue. The whole brain was reacted for HRP histochemistry using tetramethyl-benzidine (TMB) as chromogen. The reacted brain was rinsed several times before photographs were taken to reveal the HRP-filled retinofugal tract extending toward, and covering, the entire SC.

Three-Dimensional MRI In a group of control animals, the eyes were injected with NCA and imaged in a 7-T MRI to determine the optimal postinjection scan time and the intactness of the visual system. All MRI measurements were acquired using a 7-T Bruker scanner. Under inhaled isoflurane anesthesia (3% induction and 1.5% maintenance), animals were kept warm on a ~35°C heating pad and placed in prone position during scanning. Images were acquired using an isotropic T_1-weighted 3D FLASH sequence with the following settings: T_R/T_E = 21.7/3.1 ms, flip angle = 30°, acquisition matrix = 168 × 168 × 128, and voxel resolution = 226 × 226 × 226 μm³. Four averages were used; the total acquisition time was 23.5 min for each animal. Three months after (RADA)4, or 6 months after the initial BSC transaction, MEMRI was performed again to assess the effect of (RADA)4 treatment on the regeneration and reinnervation of axotomized retinal fibers.

Results

Three different measures show regeneration and return of function: (1) the dose of $MnCl_2$ (400 nmol) was sufficient to visualize any sparse connections that may have been spared during surgery; (2) eye injections could show the intact visual system, the disconnected visual system with contrast agent in axons next to the SC, and the reconnection of axons from the eye after treatment but before behavioral recovery; and (3) this finding was confirmed with CTB to trace the regenerated axons of the retinal ganglion cells that had regenerated across the treated and healed area. All controls showed a large gap with no healing or reconnection.

Posttreatment MEMRI

Eight of 10 (RADA)4-treated animals showed that (RADA)4-treated channels were healed, confirming previously reported cases that (RADA)4 enables wound healing in the brain. The signal intensity, compared before and after treatment, showed a significant difference, indicating that some regeneration had occurred. The ratio of the left SC to the right SC showed a significant signal increase after treatment ($P < 0.01$ for the eight animals that showed reconnection). Moreover, some fibers containing NCA regenerated and extended their axonal processes into the vacated SC through newly created channels across the original BSC-transected site (Figure 22.11).

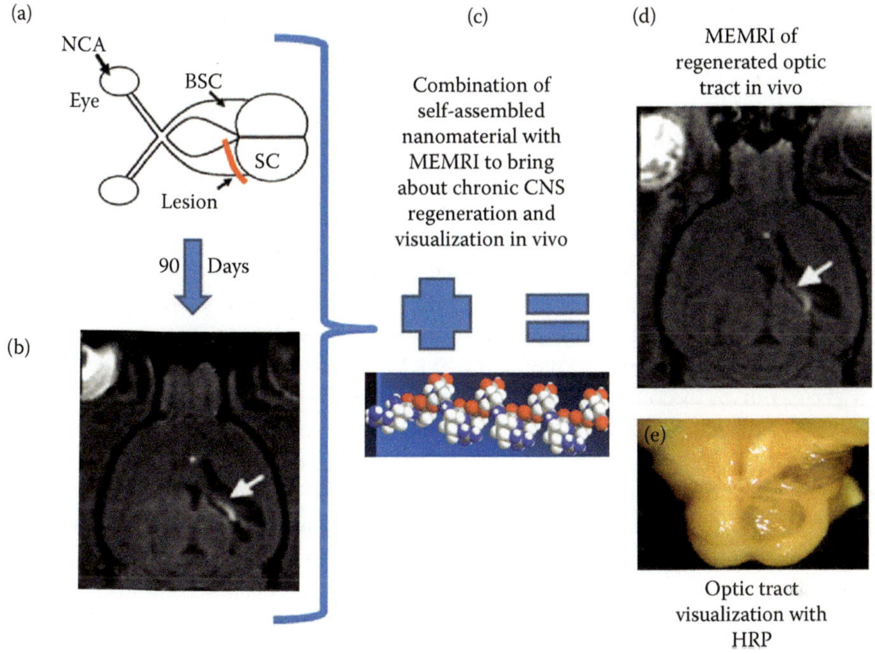

FIGURE 22.11 Chronic CNS injury and subsequent reconnection after treatment upon confirmation of complete disconnection using both behavior and MEMRI scans. (a) Paradigm and location of $MnCl_2$ injection (NCA) in the right eye and of the lesion in the left SC. (b) MEMRI scan showing complete disconnection of the SC from the OT (arrow). Note that the axons were still alive and able to transport the contrast agent, demonstrating that direct projections of retinal ganglion cells were still alive. (c) Area treated with (RADA)4 injection at the location of disconnection. (d) Scan at 170 days after lesion and 65 days after treatment. Reconnection was seen at the area where the contrast agent (arrow) was in both the SC and the OT. (e) Brainstem after injection of HRP. The OT and the SC appeared black owing to the reaction product of HRP injected into the eye. This matched the basic signal of $MnCl_2$ in MRI. Chronic CNS regeneration and visualization were demonstrated using a combination of self-assembled materials and MEMRI in vivo.

Regenerating the Spinal Cord

Traumatic spinal cord injury results in a loss of nervous tissue characterized by formation of large cystic cavities and partial or complete loss of neurological functions. For restoration of function after spinal cord injury, damaged axons must regenerate across and beyond the site of injury. Injecting cells into the host can create problems owing to the large cavities that form, necessitating a scaffold to contain these cells and to bridge the gap.

The successful use of (RADA)4 to repair the brain (Ellis-Behnke et al. 2006a) might also be effective in treating spinal cord injury. This study (Gao et al. 2007) was designed to determine if (RADA)4 (1) could provide a friendly 3D environment for neural progenitor cell (NPC) survival, migration, and differentiation in vitro and in vivo; (2) could integrate into and bridge the injured spinal cord; and (3) would elicit axon regeneration alone or with the addition of Schwann cells (SchCs) or NPCs.

Many types of cells, tissues, or biomaterials were used for transplantation in an attempt to repair the spinal cord's conductive function. To date, most have not integrated with surrounding tissues as a result of the difficulty of filling and holding injected cells in the lesion cavity. These cavities, or cysts, formed in the injured spinal cord are one of the most important physical obstacles to axon regeneration, structure rebuilding, and return of function. It is important and necessary to pretreat (RADA)4 in culture medium before transplantation to buffer pH in order to fill the cavity and to build a bridge. Untreated (RADA)4 elicited serious inflammation in the host tissue, creating distinct gaps and cysts surrounding the implants, and the primary reason for this is believed to be low pH. After (RADA)4 was pretreated in culture medium, boosting pH to a level similar to that of blood as in a previous study of hemostasis (Ellis-Behnke et al. 2006a), it was able to acclimatize to the host's microenvironment.

(RADA)4 appeared to be completely biocompatible both in vivo and in vitro, integrating very well with the host spinal cord in vivo and reducing inflammation caused by trauma. In vitro, merged stacks of multiple confocal optical sections taken by a two-photon microscope 4 weeks after the cells were cultured within (RADA)4 show images of live SchCs and NPCs that survived and migrated three-dimensionally 4 weeks after the cells were cultured within (RADA)4.

The therapeutic potential of NPCs and SchCs is well established: NPCs promote axonal regeneration either by reconstituting a "bridge" through a lesion site that is able to support axonal attachment and growth or by secreting diffusible molecules, such as growth factors, to attract injured axons. SchCs are successful in spinal cord repair, expressing a variety of neurotrophic factors that support the growth of axons, axonal growth-supporting cell adhesion molecules, and axonal growth-promoting substrates such as laminin and fibronectin. In addition, SchCs were also shown to remyelinate regenerated axons. Pretreated (RADA)4 transplants can fill and bridge the cavities caused by acute trauma, suggesting that grafted biomaterials combined with SchCs or NPCs may be the best strategy to bridge the injured spinal cord and is a potential solution for repair.

The cells migrated in both directions. First, 4′,6-diamidino-2-phenylindole (DAPI) staining showed large amounts of host cells that had migrated into the implant of pretreated (RADA)4. Second, where pretreated (RADA)4 was transplanted into the injured spinal cord of green fluorescent protein (GFP)–transgenic rats, GFP-positive cells also migrated from host to implant. Finally, grafted cells of (RADA)4 with GFP-positive SchCs or NPCs transplanted into the spinal cord of wild-type rats survived very well in the implants, with some of them even migrating into the host tissue. In another important finding, numerous blood vessels were observed in the implants, indicating the presence of blood supply to allow healing and regeneration.

Artificial 3D scaffolds that store or attract cells are critically important in regenerative medicine. The (RADA)4 scaffold demonstrated its ability to support the survival and migration of cells after transplantation into the spinal cord. Both NPCs and SchCs can survive, migrate, and differentiate within the (RADA)4 scaffold three-dimensionally, not only in vitro but also in vivo. In addition, there were both robust migration of host cells and growth of blood vessels into the scaffold. This indicates that (RADA)4 may have the potential to repair damaged tissue and to bridge injured axons across the lesion site after spinal cord injury.

Surgical and Treatment Methods

Four materials were used to test axon regeneration: (1) (RADA)4 alone, (2) (RADA)4 pretreated to adjust pH, (3) (RADA)4 cultured with SchCs, and (4) (RADA)4 cultured with NPCs.

Pretreatment of (RADA)4 Because untreated (RADA)4 has a very low pH (about 3–4) and the preliminary study showed that untreated (RADA)4 would damage the host tissue in either the dorsal column–transected spinal cord or the right-hemisected spinal cord (data not shown), (RADA)4 was neutralized in the culture medium before transplantation.

SchCs and NPCs in 3D Culture Within (RADA)4 Before transplantation, SchCs and NPCs were respectively cultured within the (RADA)4 scaffold. Briefly, SchCs and NPCs were collected and finally adjusted to 5 × 105 cells/μL. Then, 1 μL of the cell suspension was mixed with 9 μL of (RADA)4 peptide. The mixture was gently and quickly plated to dishes in Dulbecco's modified Eagle's medium (DMEM)/F12 with 10% fetal bovine serum (FBS). The medium was changed twice at 1, 10, and 30 min after plating, and once every 3 days on the following days. On day 7, the cultures were used for transplantation. Some of the cultures were maintained for 4 weeks; images of living cells were taken by a two-photon confocal microscope. Four weeks later, the cultures were fixed with 4% paraformaldehyde for immunostaining.

Surgical Procedures Thirty adult female wild-type Sprague–Dawley rats (220–250 g) and three adult female GFP-transgenic Sprague–Dawley rats (220–250 g) were used. The rats were divided into five groups: (1) saline control, (2) uncultured (RADA)4, (3) precultured (RADA)4 alone, (4) (RADA)4 seeded with NPCs, and (5) (RADA)4 seeded with SchCs. Each group had six wild-type rats; three GFP-transgenic rats were added to the precultured (RADA)4-alone group. Rats were anesthesized with ketamine (80 mg/kg) and xylazine (10 mg/kg). Dorsal laminectomy was performed on the sixth and seventh cervical vertebrae. The dura was incised longitudinally and pulled laterally. Spinal cord dorsal column transection was performed between C6 and C7, followed by the removal of 1 mm of dorsal column tissue. With the assistance of a microscope, 5 μL of precultured (RADA)4 scaffold or an equal volume of the cultured mixture of (RADA)4 and SchCs or NPCs, respectively, was transferred from the culture dish to the lesion cavity. As controls, 5 μL of uncultured (RADA)4 or saline was placed in the lesion cavity. After transplantation, the dura was closed by suturing with an 11-0 suture. Muscle layers and skin were also closed with a suture. Manual bladder expression was performed twice daily until the recovery of the bladder reflex.

Results

The spinal cord injury created a large cavity at the injury site, and multitudinous macrophages (ED1-positive cells) surrounded the cavity. (RADA)4 that was directly transplanted into the injured spinal cord without being pretreated with the culture medium did not integrate well with the host and caused serious inflammation. There were obvious gaps and cysts between the implanted tissue and the host tissue.

In contrast, when (RADA)4 was pretreated with culture medium before transplantation, it bridged the lesion area properly. The implants integrated very well with the host tissue, with no obvious cavities or gaps between the implanted material and the host. In addition, there was much less inflammation (as shown by ED1-positive cells) surrounding the lesion site. DAPI staining showed large numbers of host cells that had migrated into the implants in the case of pretreated (RADA)4, which was transplanted into injured cells that had moved from the host to the implant. After the (RADA)4 with GFP-positive SchCs or NPCs had been transplanted into the spinal cord of wild-type rats, grafted cells survived very well in the implants, with some of them migrating into the host tissue (Figure 22.12).

In the spinal cord of GFP-transgenic rats, GFP-positive hematoxylin-eosin staining showed that much of the interface between the implants [either pretreated (RADA)4 alone or pretreated (RADA)4 seeded with grafted cells] and the host integrated very well. Another important finding was that there were numerous blood vessels in the implants; some vessels were very large, especially at the edges of the implants.

Anti-neurofilament (NF) immunolabeling revealed that axons grew into the implants whether (RADA)4 was used alone or was seeded with SchCs or NPCs. However, the number and distribution of the axons in the implants varied in the different groups. Very few NF axons were observed in (RADA)4-alone transplant animals. (RADA)4 implants with either NPCs or SchCs showed more axons than the implants of (RADA)4 alone. (RADA)4 transplanted with SchCs showed many more axons than (RADA)4 with NPCs.

5-Hydroxytryptamine (5HT) and calcitonin gene-related peptide (CGRP) immunohistochemistry was performed to assess whether raphe spinal axons and primary sensory axons grew in the implants. Similar to NF-positive axons, more 5HT-positive and CGRP-positive axons were found in the transplant group (RADA)4 plus SchCs than in other groups.

Stroke

Intracerebral hemorrhage (ICH) has the highest mortality rate among all types of stroke. Hematoma growth is associated with high systolic blood pressure (SBP) and increased mortality. The effects of (RADA)4 on a renovascular hypertension (RVHT) rat model of ICH plus blood clot aspiration were tested. Administration of (RADA)4 and blood clot aspiration after ICH protected the brain against

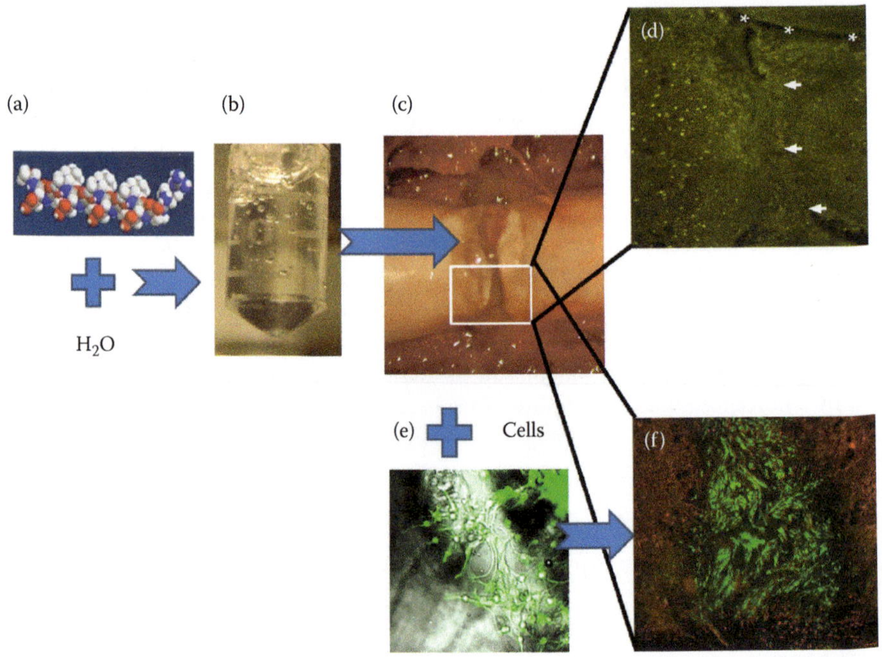

FIGURE 22.12 Treatment of spinal cord injury and restoration of tissue structure. (a) (RADA)4 used in this case. It was then mixed with deionized water. (b) The material forming a clear liquid after being mixed. (c) Rat spinal cord transected with the material expressed in the gap. (d) The material facilitating tissue restoration and blocking the formation of cyst that normally appears. (e) Addition of cells to the material in culture. The cells were GFP-expressing cells that were later implanted into the spinal cord. (f) Cells implanted into the spinal cord that filled the gap and facilitated regeneration. There was no cyst formation, and implanted cells started to invade the undamaged spinal cord.

both mechanical and chemical injury factors, resulting in reduced inducible nitric oxide synthase (iNOS) expression and cell apoptosis (Sang et al. 2009).

Surgical Method

RVHT was achieved by applying a silver clip onto the left renal artery of male Sprague–Dawley rats. At 4–6 weeks after operation, rats with SBP higher than 150 mm Hg were selected to undergo ICH. At 3.5 h after ICH onset, manual aspiration was performed to remove blood clots. After aspiration, a 1% (RADA)4 solution in saline was injected into the lesion. Hematoma volume was quantified using ImageJ software at 24 h after ICH. At 3 days after ICH, perihematomal cell death was determined with terminal deoxynucleotidyl transferase dUTP nick end labeling (TUNEL) staining. iNOS-immunoreactive cells were detected immunohistochemically.

Results

(RADA)4 treatment prevented hematoma enlargement compared with the aspiration-only and saline control group rats. The number of TUNEL-positive cells in perihematomal regions was reduced in the (RADA)4 treatment group (79.0 ± 20.4 cells/mm^2) as compared with the other groups (333.2 ± 34 cells/mm^2 for the ICH-only group, 386.7 ± 21.0 cells/mm^2 for the aspiration-only group, and 214.9 ± 54.1 cells/mm^2 for the saline control group). At 3 days after ICH, iNOS immunoreactive–positive cells were present in the ipsilateral hemisphere. (RADA)4 treatment reduced iNOS expression (43.3 ± 32.7 cells/mm^2) as compared with the saline control group (234.3 ± 133.3 cells/mm^2) and the aspiration-only group (307.5 ± 70.8 cells/mm^2).

Objective Measures

The most difficult measurements are made in the CNS because of the high resolution necessary to make accurate diagnoses and predictions of outcomes. Objective measures are needed to understand not only what has been disconnected but the extent of the disconnection and the secondary damage that ensues due to inflammation. The resolution needed for such diagnosis and outcome measure goes from the meter scale to the millimeter scale. The resolution must increase from the millimeter scale to the micron scale to see individual axons and neurons in the brain. Some single-cell measurements are being used today on cells that were labeled and injected into the subventricular zone of the brain. They are then followed to understand their fate. This was performed on rodents using a 7-T MRI machine. There are 18.8-T machines that can measure a fruit fly in 100-μm sections (Null et al. 2008). For improved visualization, measurement contrast agents are added to create a signal only in target axon bundles.

As mentioned previously, there are three types of neurons present in every trauma: neurons that are disconnected and nonfunctional, neurons that are connected and functional, and neurons that are still connected but nonfunctional.

The first type of neurons is easy to detect. The functional readout of disconnected neurons is well established using many different scales such as the American Spinal Injury Association (ASIA) scale for spinal cord injury. The second type of neurons is also easily measured using the same scale because there is function, and this is evident in patients when they can feel and move different parts of the body. However, the challenge exists in the third type. This set of neurons, in what is commonly known as spinal shock or diaschesis, comprises neurons that may start to function sometime in the future or may not, depending on possible scenarios that are too numerous to mention. They are known to exist because of reports of miraculous cures 1 week after an injection of stem cells into the spinal cord. Since neurons do not grow quickly, these must be cases of this third set of neurons regaining function once again.

Several neuroprotection clinical trials have failed because the outcome measure is so crude that the result is obscured by noise. MRI measurements are no exception. Although much progress has been made recently, some of these advances need to be considered carefully owing to technical difficulties in taking these measurements.

The need for objective measures is even more critical in stroke and traumatic brain injury. In both of these areas, patients are discharged from the hospital after a week or two, depending on severity, and are placed in a rehabilitation hospital. They spend the next 6–24 weeks working on how to become independent and then are sent home. Most are told that they will not achieve any additional return of function or that any function that they will achieve will be minimal. This is now slowly changing, with some of the new rehabilitation techniques, such as the MYOMO device (Page et al. 2011; Stein 2009), actually using a patient's own feedback system to help rebuild connections. The longest time after a stroke that a patient showed recovery was 19 years; this was attributable to complete neglect.

What is needed is the ability to take patients and follow their injury and recovery over time. It would be great to put them in a scanner, tell them what is disconnected and what is connected but not functional (this will give them a pretty good idea of what is functioning), and write a prescription based on diffuser tension imaging (DTI) and functional imaging, together with magnetic resonance spectroscopy (MRSpec), to determine what is actually still functional. The difficulty with this is that there are very few standards for this type of measurement, and that there is some variation when transiting from one machine to another. This can be remedied by using blanks and by tuning the machine much like how a gas chromatograph (GC) is tuned, with standard. But it is expensive and complicated. This is what is now being performed to synchronize measurements and to come up with a standard so that MRI can be used for clinical trial endpoint.

The disconnection and development of CNS injury and the subsequent reconnection can be followed using MEMRI (Liang et al. 2011) so that treatments can someday be adjusted in real time and not after 5 years. This type of objective measurement is critical to the success of any restoration of function in order to determine when to apply the correct amount of rehabilitation and to measure its effectiveness.

As this work moves forward, better imaging agents can be used to assess the health of neurons. Possibly, soon, there will be the ability to make changes in treatment decisions long before there is any outward physical sign of recovery. Although MRI does not have nanoscale resolution, the NCAs being tested are nano and enable amazing details.

One last point about objective measures is that they should never be performed outside of regulation. For better objective measures, facilitation of regulation needs to move forward to allow new technology to be used to its optimum potential. This is also true for nanotechnology. Regulation will always lag behind technology because it takes time to sort out these issues and to determine what is safe and effective. However, there is a need to also remember that regulation should not create a bottleneck that stops innovation and progress. Objective measures will always provide a better metric for measuring progress, or the lack of it. But not being able to use a more sensitive technique for a clinical trial endpoint keeps many new therapies from the market.

Creating a 3D Environment to Preserve Stem Cells

The ability to keep cells young during implantation is very important to the reconstitution of tissues or organs being replaced. In the emerging field of stem cells, there should be an environment that can be tuned to the needs of the cells to slow differentiation or proliferation in vitro or in vivo while remaining invisible to the immune system.

It is important to place cells in a defined system within a 3D environment to allow better control over growth and differentiation and to effectively mimic the ECM both in vitro and in vivo, particularly by (1) determining the appropriate concentration of material based on the type of cells being implanted; (2) determining if the material needs to be prebuffered or if the cells need to be precultured in material before they are injected into the lesion area; (3) controlling the physical environment to make it more permissive to cell migration while at the same time slowing or even arresting the proliferation, elongation, and differentiation of implanted cells; and (4) modulating the immune system with a unique physical environment around the implant site.

For successful reconstruction of tissues and organs, cellular therapies must integrate into the injury site. For CNS injuries, regeneration is not only replacement but also regrowth of the lost neuronal circuitry, followed by promotion of the plasticity of spared and regenerated neurons.

When cells are placed in a defined system, their proliferation, differentiation, and maturation can be delayed, depending on the density of the cell population, the density of the matrix, and the local environment. This is performed in a 3D environment that allows the benefits of cell therapy to be more fully realized (Ellis-Behnke et al. 2009).

Methods

Pheochromocytoma (PC12) Cell Culture PC12 cells were purchased, plated on 10-cm polystyrene dishes, and grown in a mixture of Ham's F-12K medium with 10% horse serum and 2.5% FBS. Penicillin and streptomycin were added along with 2 mM L-glutamine and 1.5 g/L sodium bicarbonate. Cells were used after passage 6. Cells were removed with trypsin–ethylenediaminetetraacetic acid (EDTA) for transfer to (RADA)4. (RADA)4 was prebuffered in DMEM. (RADA)4 at various concentrations was placed in separate wells, and DMEM was added and changed at 1-, 10-, and 30-min intervals. Phenol red in the culture medium was used as an indicator of balanced pH condition. The cells were transferred to 0.5%, 1%, or 2% (RADA)4 and implanted into the material. On day 10, nerve growth factor (NGF) was added to assess the viability of the cell after proliferation delay. Cells were incubated at 37°C in 5% CO_2. The medium was changed every 3 days.

Schwann Cells SchCs were isolated from rat sciatic nerves. Briefly, sciatic nerves from adult GFP-transgenic Sprague–Dawley rats ["green rat CZ-004" SD TgN (act-EGFP) OsbCZ-004] were disassociated and placed in culture dishes with DMEM/F12 supplemented with 10% FBS. When the outgrowth of migratory cells (predominantly fibroblasts) reached a near-confluent monolayer around explants (about 7 days), the explants were transferred to new culture dishes with fresh medium. After three to five such passages (3–5 weeks), the cells that emerged from the explants were primarily SchCs. The explants were then transferred to a 35-mm dish containing 1.25 μg/mL dispase, 0.05% collagenase, and 15% FBS in DMEM/F12 for incubation overnight at 37°C in 5% CO_2. On the following day, the explants were dissociated, and the cells were plated onto poly-L-lysine-coated dishes in DMEM/F12 with 10% FBS. Later, the cultures were refed with the same medium supplemented with 20 μg/mL pituitary extract and 2 μM forskolin for division. When the SchCs reached confluence, they were rinsed in Ca^{2+}-free and Mg^{2+}-free Hanks balanced salt solution (CMF-HBSS) and briefly treated with 0.05% trypsin and 0.02% EDTA in CMF-HBSS. Cells were washed twice in DMEM/F12 with 10% FBS and transferred into new dishes at a density of 2×10^6 cells per 100-mm dish. When the cells reached confluence again, they were collected for transplantation.

Neural Precursor Cells The hippocampi of embryonic day 16 embryos of GFP-transgenic Sprague–Dawley rats were dissected in cooled CMF-HBSS and dissociated mechanically. The cells were collected by centrifugation and resuspended in DMEM/F12 supplemented with B27 supplement (2%), N_2 (1%), epidermal growth factor (EGF; 20 ng/mL), basic fibroblast growth factor (bFGF; 20 ng/mL), penicillin (100 U/mL), and streptomycin (100 μg/mL). The cells were adjusted to 1×10^5 cells/mL and introduced into culture flasks. Half of the medium was replaced every 3 days. Typically, the cells were mechanically dissociated approximately once a week, reseeded at approximately 1×10^5 cells/mL, and then implanted into (RADA)4 at concentrations of 1%, 2%, and 3%. Pictures were then taken at 5, 7, 9, 12, 15, and 34 days. This was repeated in a non-GFP-expressing cell line, and pictures were taken at 3, 5, 7, and 9 days.

SchCs in 3D Culture Within (RADA)4 Before transplantation, SchCs obtained from the sciatic nerves of GFP rats were cultured within a 1% (RADA)4 scaffold. Briefly, the SchCs were collected and finally adjusted to 5×10^5 cells/μL. Then, 1 μL of the cell suspension was mixed with 9 μL of (RADA)4. The mixture was gently and quickly plated to a dish in DMEM/F12 with 10% FBS. The medium was changed at 1, 10, and 30 min after plating, and once every 3 days during the following days. Images of living cells were taken by a two-photon confocal microscope.

Young Animals In vivo applications to brain wounds were carried out by using 16 postnatal day 5 (P5) Syrian hamster pups anesthesized with whole-body cooling. The scalp was opened, and the OT within the SC was completely severed with a deep knife wound through a slot cut in the cartilaginous skull, extending 1.5 mm below the surface from the midline to a point beyond the lateral margin of the SC. Animals were kept alive 30 days after surgery. During surgery, eight animals were treated by injecting 10 μL of 1% (RADA)4 into the brain wound. Control animals with the same lesions included eight animals injected with 10 μL of isotonic saline.

Adult Animals Thirty-two adult Sprague–Dawley rats (220–250 g) were used and divided into six groups: (1) (RADA)4 only; (2) uncultured (RADA)4 with SchC at 1, 4, and 8 weeks; and (3) precultured (RADA)4 with SchC alone at 2 and 8 weeks. Group 1 had two wild-type rats; groups 2 and 3 had six wild-type rats at each time point. Rats were anesthesized with ketamine (80 mg/kg) and xylazine (10 mg/kg).

Surgery Dorsal laminectomy was performed on the sixth and seventh cervical vertebrae. The dura was incised longitudinally and pulled laterally. Spinal cord dorsal column transection was performed between C6 and C7, followed by removal of a 1-mm dorsal column tissue. Ten microliters of 1% (RADA)4 solution was injected into the lesion site, or an equal volume of the cultured mixture of (RADA)4 and SchCs was transferred from the culture dish to the lesion cavity. The dura was closed after treatment. The muscle layers and skin were also closed with a suture. Manual bladder expression was performed twice daily until the recovery of bladder reflex.

In Vitro Results

Primary Neurons

PC12 Cells The cells appeared senescent in the 3D environment of (RADA)4, whereas they did not divide for 3 days in 0.5% (RADA)4 or for 5 days in 1% (RADA)4 under serum-free (SF) conditions. When the cells reached the boundary of (RADA)4, they proliferated at a normal rate until they reached confluence and then stopped dividing. This process was repeated three times. The use of 2% (RADA)4 delayed the proliferation of PC12 cells to day 7 under SF conditions. Care was taken during cell implantation not to allow the culture medium to contact the cells. When NGF was added, the processes extended normally but with a (RADA)4 concentration-dependent delay. This was repeated under serum conditions; proliferation was only delayed by 1 day past control under the 0.5% condition. However, at the higher concentrations of 1% and 2%, division was observed on days 3 and 5, respectively.

Schwann Cells At a concentration of 5000 cells/μL, no processes were observed 3 days after implantation in 1% (RADA)4 in vitro. After 5 days, processes were seen extending from the cell body. When SchCs were implanted in 2% (RADA)4, process extension was not observed until 5 days after (RADA)4 implantation. FBS was used in both conditions. When the concentration of cells was increased to 10,000 cells/μL, process formation was delayed to 5 days in the 1% (RADA)4 concentration.

Neural Precursor Cells

Cells Implanted in (RADA)4 at 1% Concentration A maximum of 3 + 1 processes per cell were reached by day 12 of culture under SF conditions (Figure 22.13a). With serum added, a maximum of 9 + 1 processes per cell were reached by day 7. Branching of the processes showed statistically significant differences. Under SF conditions, only 28% of the processes were branched by day 15, and 45% of the total processes were branched with serum added.

Cells Implanted in (RADA)4 at 2% Concentration A maximum of 3 + 1 processes per cell were reached by day 12 of culture under SF conditions (Figure 22.13b). With serum added, a maximum of 9 + 1 processes per cell were reached by day 7. Branching of the processes showed statistically significant differences. Under SF conditions, only 35% of the processes were branched by day 15, while 58% of the total processes were branched with serum added. If cells were plated at 2%, it took 5–7 days before the processes emerge. The processes continued up to 30 days.

Cells Implanted in (RADA)4 at 3% Concentration A maximum of 3.5 + 1 processes per cell were reached by day 15 of culture under SF conditions (Figure 22.13c). With serum added, the cells reached a maximum of 11.5 + 1 processes per cell by day 12. Branching of the processes showed statistically significant differences. Under SF conditions, only 22% of the processes were branched by day 15, and 48% of the total processes branched by day 12 with serum added.

Under all SF conditions, the number of branched processes per cell remained below 1. The serum-added conditions after day 5 had more than one branched process per cell. As the concentration of (RADA)4 increased, there was an increase in the number of branches under serum conditions. The maximum number of branched processes per cell reached 5.5 + 1 under the serum-added 3% (RADA)4 condition on day 12.

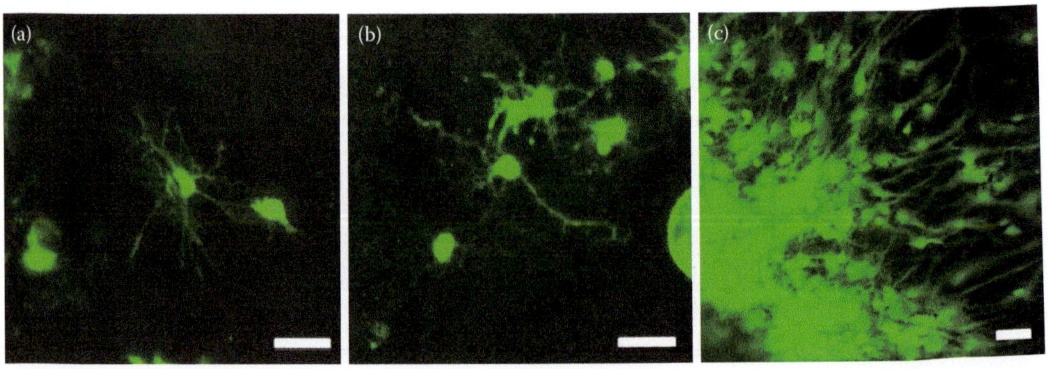

FIGURE 22.13 Neural precursor cells. Series of photographs taken 7 days after the implantation of neural precursor cells cultured in 1% (a), 2% (b), and 3% (c) (RADA)4 with serum. Note the size of the process and the number of branches from each cell. Just by changing the density of the environment, it was possible to change how the cells differentiate and migrate. The processes were less defined as the concentration of the material increased at this early time point. Scale bar represents 25 μm.

In Vivo Results

Brain

In P5 hamsters, cells migrated into the 1% (RADA)4 implantation after transection in the SC in seven of eight cases. All controls had cavities at the location of the lesion at 1-month survival.

Spinal Cord

One percent (RADA)4 implanted into the spinal cord alone eliminated the cysts when there were enough cerebrospinal fluid and other ionic materials left in the surgical site to buffer the material implanted at the lesion site. This was repeated twice.

SchC Implantation into the Spinal Cord

SchCs mixed with unbuffered 1% (RADA)4 and directly implanted showed apoptosis, necrosis, and reduced cyst formation at 4 and 8 weeks compared with untreated controls.

Prebuffered

One percent (RADA)4 with cells showed no cyst or cavity formation at 2 or 8 weeks. Cells migrated into and out of the implantation site and were evident at up to 1 cm from the lesion site.

Immediate Hemostasis during Surgery

Stabilization of the tissue environment, by stopping bleeding in less than 15 s, blocks much of the inflammatory signal, which appears to impact secondary inflammation by halting ischemia and by preserving tissue in the wound environment. With global ischemia blocked, the second wave of tissue death—along with secondary necrosis and the resulting cycle of death, inflammation, ischemia, and more death—can be blocked.

Hemostasis during Brain Surgery

Stabilizing tissue after injury is the most critical component, and this is where the journey begins. As previously mentioned, skin and brain come from the same precursor tissue. From the standpoint of cell migration into the wound site, the skin and brain heal in the same way, which can be a very powerful model for CNS healing. This also includes how tissue reacts to burns, which is very similar to the use of cauterization to stop bleeding in the brain and skin. Measures used to control bleeding, and not surgery per se, usually lead to most damage (Ellis-Behnke et al. 2006a).

Surgical Method

Part of the overlying skull was removed in an adult rat—a complete transection of a venous branch to the superior sagittal sinus in the brain of rats (n = 15) and hamsters (n = 15). The areas were treated with 20 μL of a 1% w/v solution of (RADA)4 or with iced saline (Figure 22.14).

Results

In the groups treated with (RADA)4, hemostasis was achieved in less than 10 s in both hamsters and rats. Control group hamsters (n = 5) and rats (n = 5) irrigated with saline bled for more than 3 min, completely obscuring the surgical field. Student's t test for two independent samples in both hamsters and rats showed highly significant differences ($P < 0.0001$).

Hemostasis during Spinal Cord Surgery

The goal was to determine if the material could quickly control bleeding to reduce secondary damage caused by surgery (Ellis-Behnke et al. 2006a) since blood has been shown to be toxic to neural tissue and the spinal environment has been shown to be more sensitive to this type of insult than the brain.

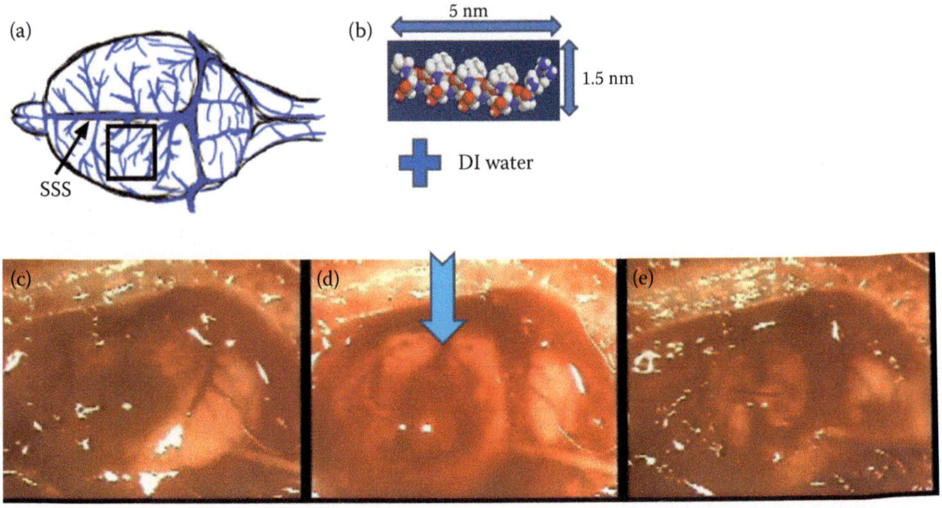

FIGURE 22.14 Hemostasis of the brain. (a) Dorsal view of the rat brain. Blue lines depict blood vessels superficial to the cortex. The boxed area corresponds to the location of surgery and treatment. The arrow points to the superior sagittal sinus (SSS). (b) (RADA)4 molecule mixed with deionized water at 1% concentration and applied to the wound to stop bleeding. Part of the overlying skull was removed in an adult rat, and one of the veins of the SSS was transected and treated with 1% self-assembling (RADA)4. (c) Exposed brain and veins of the SSS. (d) Cutting the vein (arrow). (e) The same area 5 s after the application of (RADA)4 to the cut (arrow), as seen under optically transparent (RADA)4.

Surgical Method

After laminectomy and removal of the dura, the spinal cord was hemisected at T_2 from the dorsal aspect to the ventral aspect. The injury also created a large cavity at the site, with macrophages (ED1-positive cells) surrounding the cavity (Figure 22.15).

Results

After treatment with 20 μL of 1% wt/v (RADA)4, hemostasis was achieved in just over 10 s. In saline controls, bleeding continued for as long as 5 min, and the surgical field was obscured.

Crystal-Clear Surgery

The surgical field is often obscured with blood and debris during an operation. In addition, clearing debris from the surgical field usually requires irrigating the site with saline, which is only a temporary solution and needs to be continuously applied to maintain a clear surgical field. This poses several issues: (1) any contamination in existence will easily spread; (2) a small opening will require alternating irrigation and operation; and (3) the use of saline can spread infection, leading to postoperative complications. (RADA)4 provides a transparent environment for the surgical field because it is an optically clear liquid that allows operation through the resultant liquid-gel mixture (Ellis-Behnke et al. 2008).

Surgical Method

The cavity that formed when the skull was removed was filled with (RADA)4. Forceps were used to remove the dura. Note the tip of the forceps, which were clearly visible during the entire procedure (Figure 22.16). As bleeding started, the material flowed in and stopped the bleeding. The dura was pulled to the right.

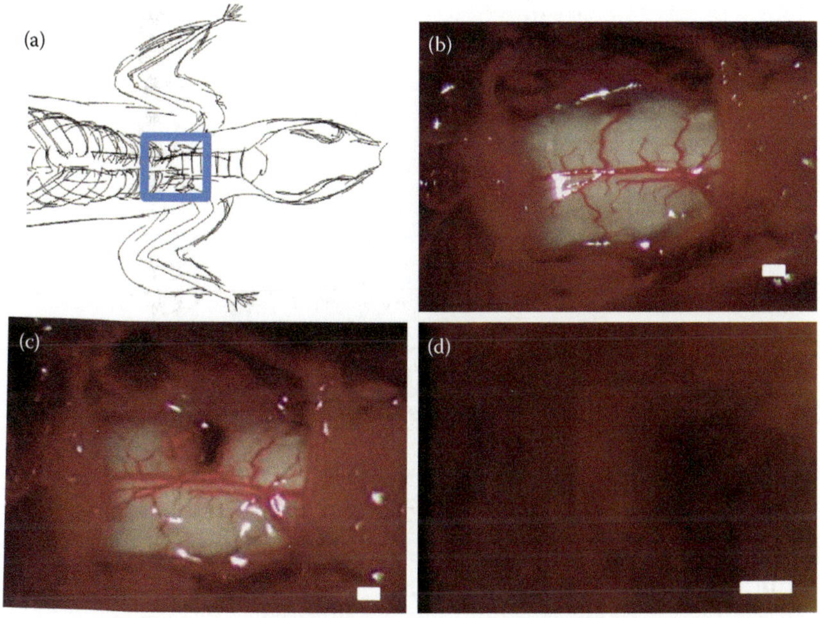

FIGURE 22.15 Hemostasis of the spinal cord. (a) Anterior portion of the rat showing the skeleton. The blue box represents the location of laminectomy performed to expose the spinal cord. (b) Exposed spinal cord of the rat. The spinal cord was cut and (RADA)4 was added to bring about hemostasis. (c) Cessation of bleeding while the material is still inside the hemisection site. (d) Control using iced saline and packing for hemostasis. The surgical field is completely occluded with blood after the transection. Scale bar represents 1 mm.

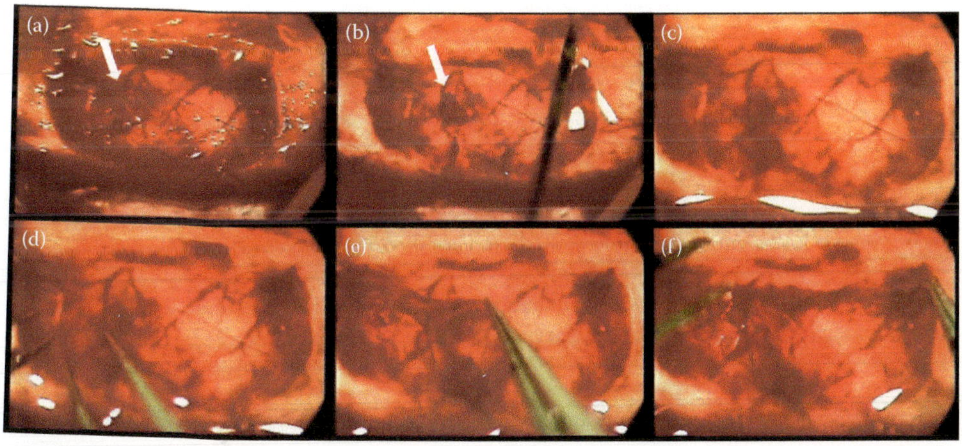

FIGURE 22.16 Crystal-clear surgery. (a) Overlying skull removed in an adult rat. The area was irrigated with (RADA)4 before removing the dura. (b) Arrow showing the area where there was some bleeding, which stopped as the material flowed into the area. (c) Cavity resulting from skull removal filled with (RADA)4. (d) Forceps starting to remove the dura. Note the tip of the forceps, which were clearly visible during the entire procedure. As bleeding started, the material flowed in and stopped the bleeding. (e) Dura being pulled to the right. Note capillary bleeding as the dura was lifted; however, the surgical field remained clear. (f) The material formed a transparent barrier during the surgical procedure while blocking infiltration of materials to the wound.

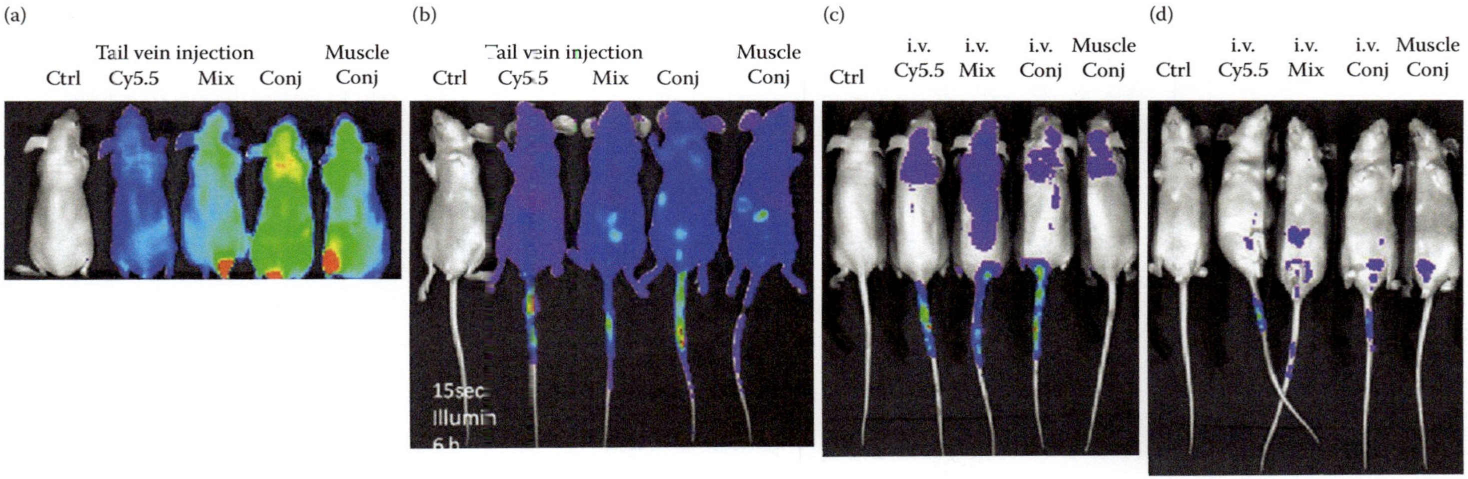

FIGURE 22.17 Elimination of material from the body (ADME). This experiment was performed in a series of mice that were injected with Cy5.5, a mixture of Cy5.5 and (RADA)4, or Cy5.5-conjugated (RADA)4 into the tail vein (intravenous) or muscle. (a) Day 1 dorsal view. (b) Day 1 ventral view. (c) Day 15 dorsal view. (d) Day 15 ventral view. The material was excreted by the mice. There were no signs of toxicity or immunogenicity.

Results

As the dura was lifted, capillary bleeding occurred; however, the surgical field remained clear. The material formed a transparent barrier during the surgical procedure while blocking infiltration of the wound.

NANOTECHNOLOGY AND SAFETY

Many popular books have proclaimed both the doom and the salvation that nanotechnology will bring to the world. Since the SAPs used are simple proteins, their toxicity is not as great a concern as with a totally new material that the body has never encountered before. It is important to remember that some of the simplest proteins have the most potent toxins. Thus, the first rule of thumb in using nanomaterials is to test for safety to ensure that there will be no breakdown products from known toxins or in case of an inflammatory response.

This study focused on simple structures; any additions were usually coinjections that allowed the agent to be released, disassembled, and excreted by the kidneys. ADME, which stands for absorption, distribution, metabolism, and excretion, is an acronym used to describe pharmacokinetics and pharmacology. This series of tests determines the distribution of the compound and its safety for human use. Previous experiments were unable to raise antibodies to (RADA)4; therefore, there was a need to understand blood/(RADA)4 interaction to determine if there were any residual effects caused by the materials. This is better illustrated in simple ADME experiments to determine where in the body the materials are located and how they move through it. The materials were injected directly into the veins of many rodents. They were found not to cause any secondary thrombi and were cleared from the blood by the kidneys.

Method

Infrared dyes were conjugated with the materials and injected into animals for ADME to show that (RADA)4 breaks down in the body (Figure 22.17). This can also be used to assess healing progress without removing the bandage, 3D localization, and concentration of the dyes in any area of the body. Through the addition of a tracer, the closure of a wound and the movement and breakdown of the materials can be monitored in real time. This can be used to monitor the rate of healing by replacing the materials at the site of injury. Since these dyes can be localized in three dimensions, this permits exact localization of the site with a low-cost detector, enabling deployment in the field. Multispectral imagers that are already being used in the field can be used to localize these materials with the correct dye.

Results

This experiment was performed in a series of mice that were injected in the tail vein or muscle with Cy5.5, a mixture of Cy5.5 and (RADA)4, or Cy5.5-conjugated (RADA)4. The material was excreted by the mice. There were no signs of toxicity or immunogenicity.

CONCLUSION

From wound stabilization to CNS regeneration, the field of nanotechnology holds the key to a vast amount of new functionalities and an exciting new frontier in medicine.

Although this may seem surprising, each time there is a power-of-10 increase in resolution, there is a breakdown of barriers between fields, where a new field slowly emerges as the old one begins to disappear. The nanomedicine revolution is no exception: with new functionalities and the ability to fine-tune "what" and "where," it will be possible to deliver many new therapies that are not yet conceivable.

The size difference and the resultant new frontier being developed can perhaps now be appreciated. The two concepts that previously had the greatest impact on medicine were anesthetics and antibiotics. Some might also say genetics, although this has not yet been realized. However, nanotechnology has enabled smaller amounts of DNA to be used such that the genome of a single cell can now be appreciated. In addition, direct reading of the genome is also closely enabled by nanotechnology.

I believe that nanotechnology and its application to medicine will be viewed as one of the greatest advances of the late 20th century or the early 21st century. But at the end of the day, nobody would care unless this technology can be used to help humans. The main goal and focus of all of the work covered in this chapter will benefit mankind in a very real way: from stabilizing tissues to reconnecting disconnected parts of the CNS.

Quality of life is the most important part of life. Everyone defines it differently, but someone who has no ability to communicate or even react to the touch or words of family members has a very real problem. Very little has been and can be performed to restore the ability to communicate. It is usually caused by a brainstem stroke that has resulted in the "locked-in syndrome." This syndrome disconnects the brain from the rest of the body. Measuring the disconnection and surgically reconnecting these parts to restore that person's life will, hopefully, soon be a reality.

By the year 2020, 80 million people will be blind worldwide as a result of glaucoma. Nearly US$3 trillion per year are being spent globally for 733 million people who have impaired vision. This cost is expected to dramatically increase by 2020. By 2030 in the United States, the number of people suffering from stroke will double; 25% of Baby Boomers (born 1949–1961) will be living with diabetes; and 60% of them will be managing more than one chronic disease (American Hospital Association and First Consulting Group 2007). This may sound like a doomsday prediction. However, note that it was only 100 years ago when catching the flu was likely a death sentence. Today, one can barely remember the last time one had a cold. I believe that nanotechnology and neurosurgery will play primary roles in restoring life to these growing populations.

REFERENCES

Altundag, K., Dede, D.S., and Purnak, T. 2007. Albumin-bound paclitaxel (ABI-007; Abraxane) in the management of basal-like breast carcinoma. *J Clin Pathol* 60: 958.

American Hospital Association and First Consulting Group. 2007. *When I'm 64: How Boomers Will Change Health Care*. Chicago, IL: American Hospital Association.

Cohen, J., Scott, R., Schimmel, T. et al. 1997. Birth of infant after transfer of anucleate donor oocyte cytoplasm into recipient eggs. *Lancet* 350: 186–7.

Cooley, D.A., Romagnoli, A., Milam, J.D. et al. 1981. A method of preparing woven Dacron aortic grafts to prevent interstitial hemorrhage. *Cardiovasc Dis* 8: 48–52.

Darragh, M.R., Schneider, E.L., Lou, J. et al. 2010. Tumor detection by imaging proteolytic activity. *Cancer Res* 70: 1505–12.

Debakey, M.E., Jordan, G.L., Jr., Abbott, J.P. et al. 1964. The fate of Dacron vascular grafts. *Arch Surg* 89: 757–82.

Ellis-Behnke, R.G. 2007. Nano neurology and the four P's of central nervous system regeneration: preserve, permit, promote, plasticity. *Med Clin North Am* 91: 937–62.

Ellis-Behnke, R.G., Carter, D.A., So, K.F. et al. 2000. *Optic-Nerve Regeneration Through a Long Peripheral Nerve Bridge to Midbrain Leads to Functional Recovery of Vision*. New Orleans, LA: Society for Neuroscience.

Ellis-Behnke, R.G., Liang, Y.X., Tay, D.K. et al. 2006a. Nano hemostat solution: immediate hemostasis at the nanoscale. *Nanomedicine* 2: 207–15.

Ellis-Behnke, R.G., Liang, Y.X., You, S.W. et al. 2006b. Nano neuro knitting: peptide nanofiber scaffold for brain repair and axon regeneration with functional return of vision. *Proc Natl Acad Sci U S A* 103: 5054–9.

Ellis-Behnke, R.G., Liang, Y.X., Tay, D.K.C. et al. 2008. *Crystal Clear Surgery Using Self-Assembling Molecules That Act as a Bio Barrier in the Brain and Spinal Cord*. Chicago, IL: Society for Neuroscience.

Ellis-Behnke, R.G., Liang, Y.X., Guo, J. et al. 2009. Forever young: how to control the elongation, differentiation, and proliferation of cells using nanotechnology. *Cell Transplant* 18: 1047–58.

Ellis-Behnke, R.G., Okobi, A., and Schneider, G. 2001. *Re-wiring the Adult Brain: Recovery of Vision in Hamsters After Severance and Re-routing of the Optic Tract Using Peripheral Nerve Bridges*. San Diego, CA: Society for Neuroscience.

Gao, W.D., Liu, X.S., Han, X. et al. 2007. [The application of artificial dermis and recombinant bFGF afer immersion bath in residual burn wound]. *Zhonghua Shao Shang Za Zhi* 23: 40–2.

Hebb, D.O. 1967. *The Organization of Behavior; A Neuropsychological Theory*. New York: Wiley.

Holmes, T.C., de Lacalle, S., Su, X. et al. 2000. Extensive neurite outgrowth and active synapse formation on self-assembling peptide scaffolds. *Proc Natl Acad Sci U S A* 97: 6728–33.

Liang, Y.X., Cheung, S.W., Chan, K.C. et al. 2011. CNS regeneration after chronic injury using a self-assembled nanomaterial and MEMRI for real-time in vivo monitoring. *Nanomedicine* 7: 351–9.

Null, B., Liu, C.W., Hedehus, M. et al. 2008. High-resolution, in vivo magnetic resonance imaging of *Drosophila* at 18.8 Tesla. *PLoS One* 3: e2817.

Page, S.J., Hermann, V.H., Levine, P.G. et al. 2011. Portable neurorobotics for the severely affected arm in chronic stroke: a case study. *J Neurol Phys Ther* 35: 41–6.

Purves, D. and Lichtman, J.W. 1985. *Principles of Neural Development*. Sunderland, MA: Sinauer Associates.

Salomoni, P. and Calegari, F. 2010. Cell cycle control of mammalian neural stem cells: putting a speed limit on G1. *Trends Cell Biol* 20: 233–43.

Sang, Y.H., Liang, Y.X., Liu, L.G. et al. 2013. A rat model of intracerebral hemorrhage permitting hematoma aspiration plus intralesional injection. *Experimental Animals* 62: 63–39.

Schneider, G.E., Ellis-Behnke, R.G., Liang, Y.X. et al. 2006. Behavioral testing and preliminary analysis of the hamster visual system. *Nat Protoc* 1: 1898–905.

Schneider, G.E., Liang, Y.-X., Tay, D.K.C. et al. 2005. *Skin and Brain: Common Origins and Common Cures After Trauma; In Vivo Experiments in Mammals*. Washington, DC: Society for Neuroscience.

Stein, J. 2009. e100 NeuroRobotic system. *Expert Rev Med Devices* 6: 15–9.

23 Applications of Nanotechnology in Epilepsy

Paul R. Carney

CONTENTS

INTRODUCTION

Epilepsy is a common chronic neurological disorder characterized by recurrent unprovoked seizures (Blume et al. 2001; Fisher et al. 2005). It is the most common neurological condition in children and the third most common in adults after Alzheimer's disease and stroke. The World Health Organization (2001) estimates that there are 40–50 million people with epilepsy worldwide. Seizures are transient epochs attributable to abnormal, excessive, or synchronous neuronal activity in the brain (Fisher et al. 2005). Epilepsy is a generic term used to define a family of seizure disorders. A person with recurring seizures is said to have epilepsy. Currently, there is no cure for epilepsy. Many patients' seizures can be controlled, but not cured, with medications. Those resistant to medications may become candidates for surgical intervention.

Roughly 70% of cases present with no known cause. Of the remaining 30%, the following are the most frequent causes: brain tumor, stroke, or both; head trauma, especially from automobile accidents, gunshot wounds, sports accidents, and falls and blows; poisoning, such as lead poisoning and substance abuse; infection, such as meningitis, viral encephalitis, lupus erythematosus, and, less frequently, mumps, measles, diphtheria, and others; and maternal injury, infection, or systemic illness that affects the developing brain of the fetus during pregnancy. All people inherit varying degrees of susceptibility to seizures. The genetic factor is assumed to be greater when no specific cause can be identified. Mutations in several genes have been linked to some types of epilepsy. Several genes that code for protein subunits of voltage-gated and ligand-gated ion channels are associated with forms of generalized epilepsy and infantile seizure syndromes (Meisler and Kearney 2005).

The first-line treatment for epilepsy are antiepileptic drugs (AEDs). In some cases, implantation of a vagus nerve stimulator, a special ketogenic diet, or a resective epilepsy surgery can be very helpful. While AED treatment is the standard therapy for epilepsy, one-third of all patients remain unresponsive to currently available medication. There is general agreement that, despite pharmacological and surgical advances in epilepsy treatment, seizures cannot be controlled in many patients, and new therapeutic approaches are needed (Hauser and Hesdorffer 1990; Jacobs et al. 2001; Theodore and Fisher 2004). Unfortunately, for those responsive to medication, many AEDs have significant side effects that negatively impact quality of life. Some side effects can be of particular concern for women, children, and the elderly. For these reasons, the need for more effective treatments for pharmacoresistant epilepsy was among the driving forces behind a recent White House-initiated conference on epilepsy, "Curing Epilepsy: Focus on the Future," which emphasized specific research directions and benchmarks for discovering effective and safe treatments for people with epilepsy (Frazin 2000). There is growing awareness that the development of new antiepileptic therapies has slowed, and that novel approaches are needed to move toward new and more effective therapies (Stables et al. 2002). In this chapter, recent advances in neurosurgery and nanotechnology that aim to prevent and treat epilepsy are discussed. Many of these advances will hopefully offer individuals with uncontrolled epilepsy freedom from seizure without side effects.

EVALUATION FOR EPILEPSY SURGERY: NANONEUROSURGERY ADVANCEMENTS

Approximately 2.5 million Americans live with epilepsy and epilepsy-related deficits today—more than those disabled by Parkinson's disease or brain tumors. The impact of seizures and epilepsy on the United States is significant, with a total cost of approximately US$12.5 billion. Epilepsy consists of more than 40 clinical syndromes affecting 40 million people worldwide. Approximately 25% of individuals receiving AEDs have inadequate seizure control; however, 80% of individuals with medication-resistant epilepsy might be cured through surgery if seizure focus could be localized precisely. Neurosurgical operations for epilepsy can be palliative, reducing the frequency or severity of seizures; for some patients, an operation can be curative. However, about 25% of people with epilepsy will continue to experience seizures even with the best available treatment (Annegers 1997). When AEDs are ineffective in controlling seizures or when anticonvulsant therapy is associated with significant side effects, excisional brain surgery can be considered. The goal of epilepsy surgery is to remove the minimal amount of brain tissue necessary to render the patient seizure-free without unintended neurological deficits.

MAGNETOENCEPHALOGRAPHY TO LOCALIZE SEIZURE ONSET

Magnetoencephalography (MEG) measures the magnetic fields produced by current flow within the brain. MEG has demonstrated its value in assessing seizure onset in patients with intractable focal epilepsy who are under evaluation for epilepsy surgery. Importantly, MEG can noninvasively assist with identifying epileptogenic foci at millimeter-level and millisecond-level resolutions (Engel et al. 2007). MEG signals arise from pyramidal neurons' intracellular tangential current flow that is detectable within a magnetic field. A commonly used source modeling in MEG is single equivalent dipole, which works best for stationary nondistributed sources such as epileptic spikes. In brief, several studies have suggested that MEG is a useful tool in epilepsy surgical workup. Importantly,

MEG, when in good hands, can often be highly predictive of a clinically good surgical outcome (Assaf et al. 2004; Barkley 2004; Knowlton et al. 2006; Mamelak et al. 2002; Manoharan et al. 2007; Stefan et al. 2003; Wheless et al. 1999).

Covalently Conjugated Magnetonanoparticles

An exciting and recent development in technology is the use of nonradioactive nanoparticles to assist with seizure localization. Specific classes of nanoparticles (NPs) that cross the blood–brain barrier (BBB) can concentrate exclusively on epileptogenic foci (Akhtari et al. 2008). Nonradioactive α-methyl tryptophan was covalently attached to magnetonanoparticles composed of iron oxide and dextran (Figure 23.1). These particles not only crossed the BBB but also were taken up intraparenchymally and were visible on magnetic resonance imaging (MRI) (Figure 23.2). The particle uptake correlated with the physiological epileptogenic onset zone. This approach has the potential to accurately localize epileptogenic brain regions and thereby assist in evaluating presurgical resective surgery assessment.

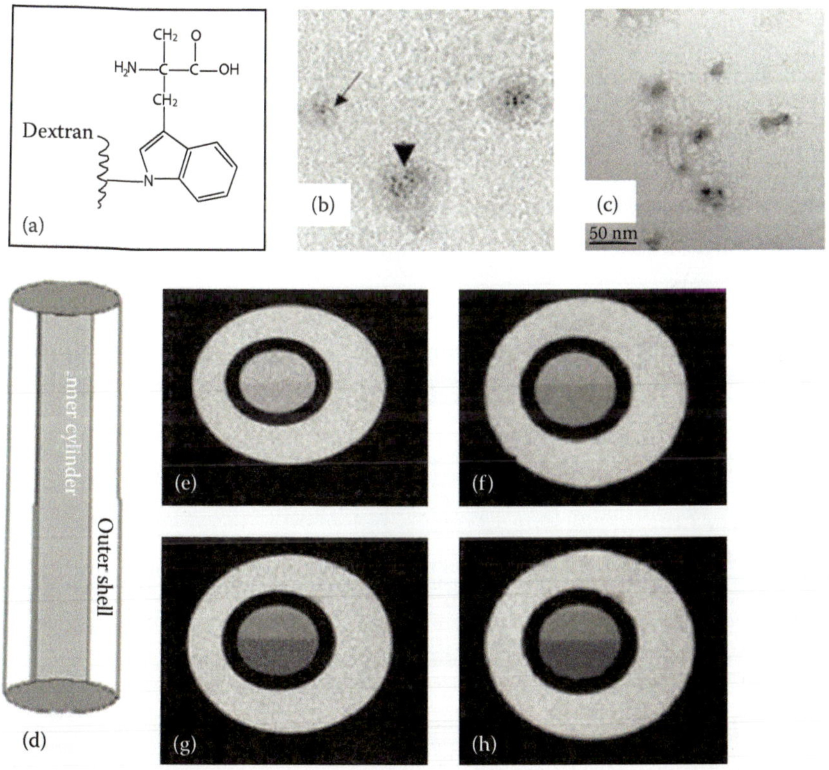

FIGURE 23.1 Magnetonanoparticle characterization. (a) Schematic diagram of α-methyl tryptophan (AMT) conjugated to dextran coating. (b) Transmission electron microscopy showing single particle core (arrow) and aggregates (arrowhead). (c) Transmission electron microscopy showing the size and morphology of individual AMT-magnetonanoparticles (MNP) particles. (d) Typical phantom studies in which the phantom serves as an artificial object of known dimensions and properties. These studies are employed to test the homogeneity, imaging performance, and orientation aspects of the nanoparticles under study. The figure shows a phantom experimental study and control study, respectively, with the inner cylinder containing magnetonanoparticles and with the outer shell containing water alone. (e–h) Signal dropout as T_E time was increased. (From Akhtari, M. et al., *Epilepsia*, 49, 1419–30, 2008. With permission.)

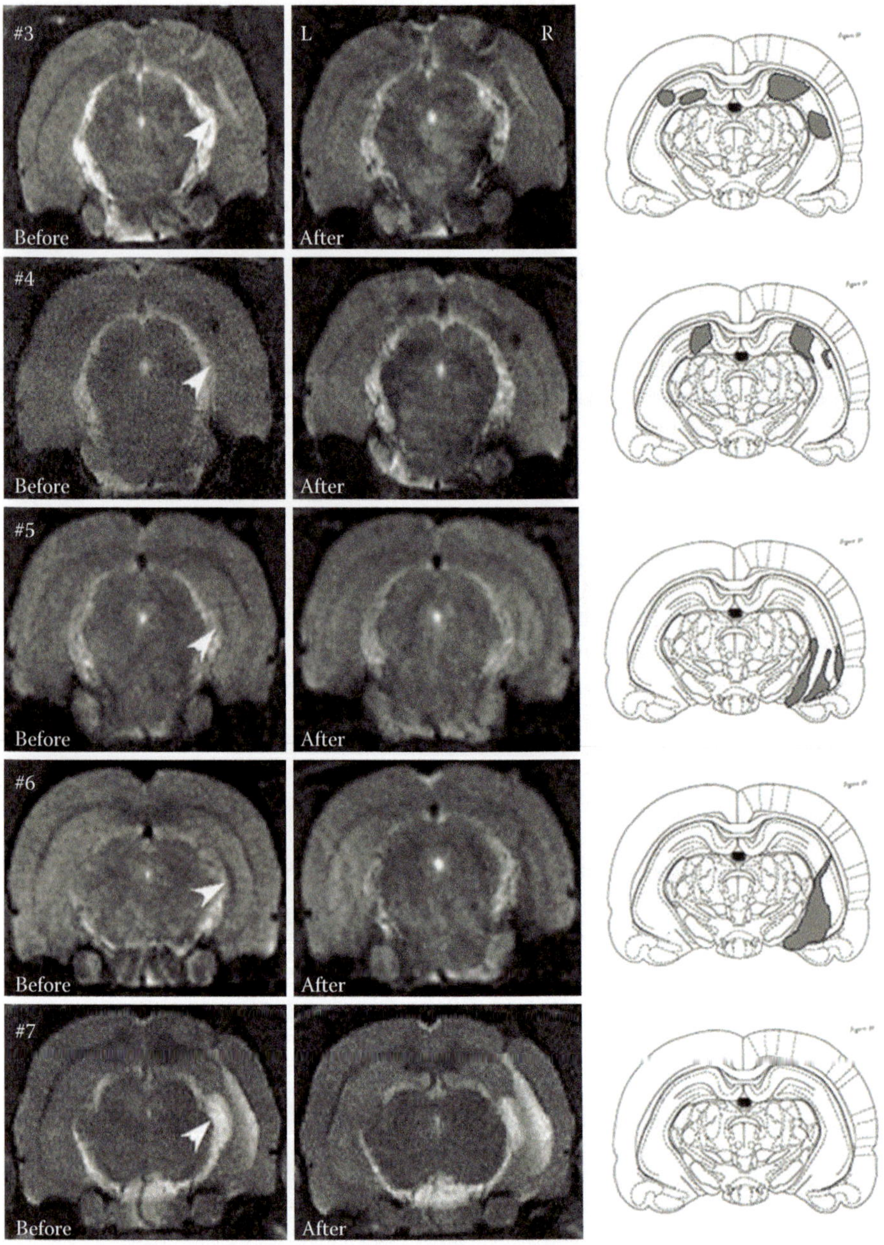

FIGURE 23.2 Magnetonanoparticles. Magnetic brain images of rats (top to bottom) are shown before (left column) and after (middle column) the tail vein injection of AMT-MNP particles. Rats 3 and 4 showed bilateral uptake of particles; rats 5 and 6 showed unilateral uptake; and rat 7 had no particle uptake. (From Akhtari, M. et al., *Epilepsia*, 49, 1419–30, 2008. With permission.)

Localizing Electroencephalographic Abnormalities Using Neuroimaging

Advanced neuroimaging techniques that combine scalp electroencephalography (EEG) and functional MRI (fMRI) in real time suggest that increased blood flow to specific brain regions may correlate with epileptogenic foci. Combining EEG and fMRI was previously hampered by technical constraints. The traditional approach to removing magnetic resonance (MR)-induced noise is replaced with increased signal-to-noise ratio of EEG acquired during MR scanning, analog noise cancellation techniques, and matched reference for each scalp electrode. The combined EEG-fMRI approach has great potential as a tool for presurgical mapping and localization of epileptogenic zones (Lemieux 2004; Zijlmans et al. 2007).

Diffusion Tensor Imaging and Tractography

Several groups are developing a set of MR acquisition and software tools for MRI data fusion, segmentation, registration, and visualization. High-resolution MRI brain images are collected together with fMRI and 128-channel scalp EEG. Anatomical images are registered using rigid transformation (Rueckert et al. 1999; Ruiz-Alzola et al. 2002; Viola and Wells 1997).

Diffusion tensor imaging (DTI) is being developed to detect abnormal white matter tracts and association fibers that connect seizure foci. If successful, DTI and tractography may provide additional anatomical correlations for epileptogenic foci. DTI is a noninvasive MR sequence that measures the structural integrity of white matter tracts and thus is an extremely useful tool for investigating brain connections. The directional features of water diffusion allow one to infer the connectivity patterns prevalent in tissue and track changes in this connectivity over time. The recent application of DTI to the study of the epileptic brain provided important information that is suggestive of underlying structural characteristics in temporal lobe epilepsy. The major drawback of DTI is that it can only reveal a single fiber orientation in each voxel and fails in voxels with orientational heterogeneity, making DTI inappropriate for examining detailed connectivity relationships in regions of the brain containing crossing or bifurcating fiber bundles (von dem Hagen and Henkelman 2002). These complex nervous tissue regions are likely to be critical to the development of abnormal connectivity relationships in epilepsy.

The limited ability of DTI to visualize regions of complex tissue structure has prompted interest in finding an improved image acquisition scheme called high-angular-resolution diffusion imaging (HARDI) MRI (Tuch et al. 2001), in which apparent diffusion coefficients are measured along a large number of directions—a practical approach to existing clinical MRI systems. By sampling diffusion signals with HARDI, the complexity of the tissue structure can be visualized in each image voxel by determining the displacement probability density function for water diffusion in the tissue. An advantage of this approach is that the acquired HARDI data can also be viewed and compared with the usual DTI model. These tools are being developed to provide accurate planning by identifying crucial regions, including eloquent cortices, white matter tracks, and epileptogenic lesions. In conclusion, these approaches may provide surgeons another tool for a safe and therapeutic removal of epileptogenic foci.

Optical Imaging

Optical recording of intrinsic signals (ORIS) is a useful tool for studying brain dynamics and functions. With ORIS, optical fibers typically deliver visible light (~570–610 nm) from a white light source to the exposed cortex of an animal or a human. A charge-coupled device (CCD) camera is used to take a fast image of the tissue area of interest from which spatial distributions of oxyhemoglobin (HbO), deoxyhemoglobin (HbR), blood volume, and oxygenation can be calculated using data analysis based on the Beer–Lambert law (Bahar et al. 2006; Sheth et al. 2004). When used intraoperatively, optical imaging can provide dynamic (surface) maps of functional and epileptiform activities in rat, monkey, and human somatosensory cortices with high spatial and temporal resolutions (Haglund and Hochman 2004; Haglund et al. 1992). Optical recording techniques have the potential to map the epileptic circuit in near-real time relative to standard electrophysiological recording methods (Shariff et al. 2006) such as EEG or subdural electrocorticography (ECoG). For instance, Bahar et al. (2006) reported that a focal and prolonged reflectance drop could be reliably seen in the ictal onset zone after 4-aminopyridine-induced seizures in an animal model. This persistent drop in light reflectance, called "epileptic dip," corresponded to decreased oxygenated hemoglobin in the seizure focus. This finding indicates that, during the course of a seizure, metabolic demands increase at the cortical level, which can be imaged using optical techniques (Bahar et al. 2006). The major drawback of ORIS, however, is that it is a superficial imaging technique that requires the exposure of the cortex for brain imaging, an invasive procedure that is associated with many risks.

Modern neuroimaging has revolutionized epilepsy surgery by increasing the visualization of abnormal anatomy and physiology associated with seizure onset. MRI has been used most frequently for epilepsy assessments because of its high anatomical resolution. However, some patients with subtle structural or "functional" abnormalities remain undetected with this imaging modality. Positron emission tomography (PET) can provide data on regional cerebral blood flow (rCBF), glucose metabolism, and binding of specific ligands to receptors. Single-photon emission computed tomography (SPECT) can obtain images representative of rCBF at the time of seizures. The principal advantage of SPECT over PET is that the former is much less expensive and the equipment is more widespread (Duncan 1997). Owing to their low spatial resolution, PET and SPECT modalities are interpreted with high-quality anatomical MRI to provide a structural–functional correlation (Duncan 1997). Although MRI and PET are excellent tools for providing a structural description of an abnormal brain, one important limitation is their inability to image the real-time dynamics of an evolving epileptic seizure.

In near-infrared spectroscopy (NIRS), light from diode lasers at multiple near-infrared (NIR) wavelengths is delivered, via optical fibers, through the scalp at a depth of approximately 3 cm below the skin. The scattered (reflected) light from the brain is typically collected by multiple optical fibers at different locations and detected by sensitive photodetectors. Near-infrared spectroscopy (NIRS) has been used to obtain cerebral blood volume and blood oxygenation in epilepsy patients (Sokol et al. 2000; Steinhoff et al. 1996; Watanabe et al. 2000). Cerebral activation induces local blood flow fluctuations, leading to changes in HbO and HbR concentrations. HbO and HbR present different absorption coefficients for NIR light. Wavelengths between 680 and 1000 nm in the NIR spectrum for HbO (830 nm) and HbR (690 nm) absorption coefficients allow simultaneous measurement of HbO and HbR concentration changes in the brain (Boas et al. 2001). Based on a modified Beer–Lambert law, the amount of absorption of HbO and HbR wavelengths can be estimated. Accordingly, changes in HbO and HbR concentrations reflect associated neural activation in specific brain areas (Gratton and Fabiani 2007; Villringer and Chance 1997). Results suggest that NIRS, when applied to epilepsy, may potentially contribute to presurgical epileptogenic zone evaluation and language lateralization (Gallagher et al. 2007, 2008a,b; Kennan et al. 2002; Noguchi et al. 2002; Watanabe et al. 1998, 2000; Watson et al. 2004) by identifying areas of seizure neural activation and language activation, respectively. However, at present, limitations related to spatial coverage and depth have slowed NIR as a widely useful clinical tool for epilepsy.

Recent studies suggest that diffusion optical tomography (DOT) may overcome some of the shortcomings of NIR since it offers real-time temporal resolution coupled with hemodynamic functions and cellular morphologic information that are complementary to those available from existing techniques. In DOT, multispectral NIR light sequentially excites the tissue at multiple locations, and the scattered light is measured at various positions along the scalp in the case of brain imaging. The measured multispectral light is then used to reconstruct the spatial distribution of tissue absorption and scattering coefficients at each wavelength through a light propagation model in tissue. Images of absorption and scattering spectra are used to derive images of functional parameters via the Beer–Lambert law, as well as images of cellular morphology via the particle scattering theory. In DOT, an effective model-based reconstruction algorithm is critical for accurate image reconstruction. To date, various linear and nonlinear algorithms have been developed (Barbour et al. 1993; Colak et al. 1997; Jiang et al. 1995; Matson and Liu 1999; Schotland et al. 1993; Walker et al. 1997). In the area of neuroimaging, DOT was applied to image intraventricular hemorrhage (IVH) in the neonatal brain (Benaron et al. 2000; Hebden et al. 2002). DOT was also used to measure hemodynamically evoked response in several cortical areas in both humans and small animals (Boas et al. 2001; Durduran et al. 2004; Hielscher et al. 2002; Hillman et al. 2007). These studies showed that DOT

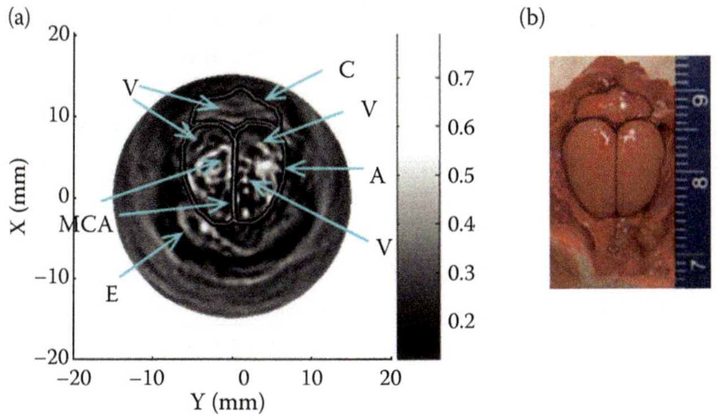

FIGURE 23.3 PAT. (a) Recovered photoacoustic image from one rat measurement. The gray scale (right) represents optical absorption of tissue (arbitrary units). (b) Photograph of the rat brain with the skull opened. A, increased absorption caused by seizure; C, cerebellum; E, eye; MCA, middle cerebral artery; V, blood vessel.

can detect hemodynamic responses spatially equivalent to fMRI. However, as with NIRS, DOT has relatively low spatial resolution (2–4 mm) attributable to strong tissue scattering.

Noninvasive laser-induced photoacoustic tomography (PAT) is an emerging imaging modality that has the potential to image the dynamic function of the brain owing to its unique ability to image biological tissues with high optical contrast and ultrasound resolution (Zhang et al. 2008). Zhang et al. (2008) reported the first application of finite element-based PAT for imaging epileptic seizures in an animal model. In vivo photoacoustic images were obtained in rats with focal seizures induced by a microinjection of bicuculline, a γ-aminobutyric acid type A antagonist, into the neocortex (Figure 23.3). The seizure focus was accurately localized by PAT, as confirmed with standard EEG. PAT is concerned with an inverse problem where a single short-pulsed light beam illuminates an object and the photoacoustic waves excited by thermoelastic expansion are measured using wideband ultrasound transducers in multiple locations around the object. The geometry of the object and the spatial distribution of the optical absorption are obtained from the measured scattered fields using a reconstruction algorithm. An effective reconstruction algorithm is critical in PAT, where the core of the reconstruction algorithm is a model describing light-induced acoustic wave propagation in tissue, which provides a tractable basis for image reconstruction. From a clinical perspective, PAT, as compared to existing neuroimaging modalities, has both high spatial and temporal resolutions in a single imaging modality and is also portable and low-cost, making it possible to bring brain imaging to the bedside.

CLOSED-LOOP SEIZURE PREVENTION NANOSYSTEMS

Over the past 25 years, numerous seizure prediction algorithms have emerged from several centers worldwide (Carney et al. 2011). Advanced seizure warning could allow patients or caregivers to take action and thereby minimize risk of injury. Depending on its configuration, an automatic seizure warning system could be made to trigger pharmacological intervention in the form of fast-acting drugs or electrical stimulation. Most seizure detection and prediction algorithms begin by extracting specific features from either subdural ECoG or scalp EEG. The appropriate neural signal scale remains a hotly debated subject. Two recent methods at the nanoscale level suggest that detection and prediction algorithms may be enhanced by recording signals at the neuronal and vascular levels of resolution (Carney et al. 2011; Zhang et al. 2008).

Microelectrode Recordings in Seizure Localization, Detection, and Prediction

For a better understanding of spatiotemporal dynamics at the action potential and neuronal ensemble levels, it is necessary to develop recordings and data acquisition that best capture the fundamental properties of the network. While recording from every neuron in the brain is an unreasonable goal, studying the mechanistic and self-organizing processes of statistically representative samples of neurons provides a direct view into systems of neurons. Many other methods, such as macroscopic recording and imaging, can aid in understanding network function; but in the end, all these indirect observations should be translated back into the fundamental elements of the nervous system—the neuron and the local neural circuitry. The parallel and distributed processing capabilities of the brain suggest that neural information is spread across populations of neurons (Chapin 2004). Therefore, compared to serial single-neuron recordings, more information can be extracted via simultaneous recordings from a large number of neurons. This basic hypothesis of systems neuroscience has sparked the development of a variety of invasive recording techniques and electrode arrays that can chronically record the activity of tens to hundreds of neurons for long periods of time. The functionally representative modulation of activity in neural assemblies is the signal of interest. Two major types of fixed probes commonly used for neural recordings are (1) conventional wire microelectrodes assembled from insulated tungsten (or platinum, nickel, stainless steel) wires and (2) micromachined electrodes fabricated using integrated circuit microfabrication technologies (Kipke et al. 2003) (Figure 23.4). Conventional wire microelectrodes are still the "work horse" of neural recordings owing to their good recording characteristics in terms of cell yields. Development of the aforementioned methods has allowed neural recordings to be analyzed at both the level of neuronal populations and the level of a single neuron. Because a large collection of neurons is recorded simultaneously, one can directly measure activity modulations that occur in neuronal populations during a variety of epilepsy states (Buzsaki 2004). The first step in developing functional relationships in neuronal groups is identifying physiological signals that modulate and correlate with the behavioral outputs or sensory inputs of the organism. A broad foundation of research supports the use of multimicroelectrode techniques in interactions with large groups of neurons, which include modulating photoreceptive cells to generate phosphenes in retinal implants (Zrenner 2002); reading pyramidal cell rate codes in motor systems (Fetz and Finocchio 1975);

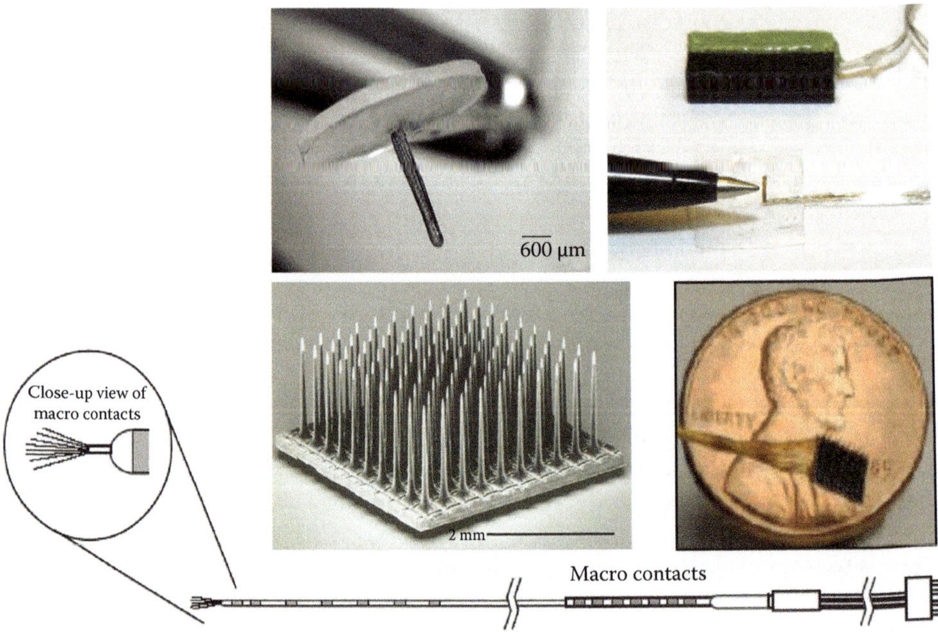

FIGURE 23.4 Microelectrode arrays. Top: Laminar microelectrode array. Middle: NeuroPort system. Bottom: Adtech microwire array. (From Schachter, S.C. et al., *Epilepsy Behav.* 16, 3–46, 2009. With permission.)

stimulating the tonotopic organization of neurons for cochlear implants (Loeb 1990); and disrupting abnormal basal ganglia neuronal firing in deep brain stimulation (DBS) for Parkinson's disease (DeLong 1990). This ability to statistically account for modulation in neuronal populations provides a more representative assessment of assembly activation compared to acute recordings of the same single neurons over many trials. For epilepsy research, one can utilize this highly resolved spatiotemporal approach to measure complex changes in neuronal population activity recorded over many time frames, ranging from fractions of seconds to several hours or days. Several studies, both in animal models (Bower and Buckmaster 2008; Sanchez et al. 2006) and in humans (Truccolo et al. 2011), suggest that neuronal firing increases in advance of the seizure, suggesting a potentially new approach to anticipating seizures.

Closed-Loop Seizure Detection

To date, seizure detection algorithms have been based solely on analyses of surface or intracranial EEGs (Carney et al. 2011). Recently, hemodynamic changes in the brain that occur prior to seizure onset were demonstrated by employing optical imaging methods such as NIR tomography. Specifically, spatially modulated NIR illumination with concurrent real-time video/EEG recording demonstrated a reduced optical scattering coefficient prior to EEG seizure onset in well-defined epilepsy mouse models (Cuccia et al. 2005). Thus, optical imaging could result in a minimally invasive optical seizure detection/prediction algorithm, which could then be coupled with a "closed-loop" seizure prevention system.

Hybrid Neuroprosthetics

Hybrid cell silicon devices integrate exogenous cells as a drug/neurotransmitter source with a closed-loop electrical device to provide controllable chemical delivery with a chronic chemical source. These devices could be coupled with prediction, detection, and treatment delivery systems for closed-loop therapies (Schachter et al. 2009). A similar device uses an implantable subdural hybrid neuroprosthesis (HNP) to treat seizures. The subdural HNP includes a subdural electrode–drug delivery strip (a signal conditioner embedded in cranial bones and connected to the rest of the apparatus via tunneled wires and tubing); a microprocessor for both seizure detection/prediction and hardware control; a two-way radiofrequency communication module for postimplantation software adjustment and emergency signaling; a dual peristaltic minipump; and a transcutaneously rechargeable power supply (Ludvig et al. 2005). As with hybrid cell silicon devices, HNP is currently undergoing preclinical testing for safety and efficacy.

DRUG DELIVERY NANOTECHNOLOGY

The current approach to drug delivery is to give a particular AED orally or intravenously at doses that have been proven to provide therapeutic blood levels. However, this approach does not ensure that the brain receives the desired active compound owing to a variety of biological factors. In both cases, the drug must enter the brain via the BBB. Even if one can get past the BBB, toxic effects are common since, in addition to the target brain area, so-called "bystander" brain areas also receive the drug. This often leads to intolerable adverse side effects. A variety of approaches to delivering AEDs directly into the central nervous system have been proposed to overcome these limitations. These approaches open up the possibility for potentially powerful AEDs such as neuropeptides, which are neither orally active nor BBB-permeable. Some direct delivery approaches involve targeted application into brain regions responsible for the initiation or spread of seizures. Other approaches include BBB modification, direct drug modification, and convection-enhanced delivery (CED) (Bennewitz and Saltzman 2009). These various approaches are summarized in the following sections.

Direct Brain Delivery Systems

Intracerebroventricular Administration

With intracerebroventricular (ICV) administration, drugs are directly infused into the cerebrospinal fluid from an implantable reservoir such as an Ommaya reservoir or a pump (Bakhshi and North 1995). The ICV mode of drug delivery mitigates many potential intravascular problems, including opsonization by serum proteins, serum drug metabolism, and systemic toxicity (Serralta et al. 2006). However, one significant problem with ICV administration relates to minimal drug penetration into the brain parenchyma due to ventricular ependymal and astrocytic cell lining barrier restrictive factors. Thus, ICV substances often concentrate within the ependymal cell lining rather than penetrating the brain parenchyma (Pardridge 1997). Another potentially significant drawback of the ICV route is that this method is invasive and may carry an increased risk of local and systemic infections, as well as unwanted elevation in intracranial pressure (Boogerd et al. 2004; Sandberg et al. 2000). Various preclinical animal studies have evaluated anticonvulsant drug administration, with ICV administration showing the highest degree of efficacy and the lowest side-effect profile for valproic acid (Serralta et al. 2006) and gabapentin (Oommen et al. 2007) relative to intraperitoneal (IP) administration.

Intracerebral Administration

The two main advantages of the intracranial administration of antiseizure drugs and implants include decreased systemic toxicity and direct targeting of the seizure focus (Yildirim and Marangoz 2007). However, although intracerebral administration is a very attractive mode of drug delivery in theory, it has many potential pitfalls in practice, including the potential for subtherapeutic drug concentrations due to pharmacodynamic factors (e.g., drug elimination, degradation, and metabolism) and brain interstitial space diffusion barrier factors (Saltzman and Olbricht 2002). One suggested approach to overcoming some of these limitations is using nanoparticles that contain properties of controlled drug release. For instance, biodegradable polymers—including polyanhydrides, such as polybis (*p*-carboxyphenoxy) propane-sebacic acid (p(CPP-SA)), and polyesters, such as PLA, PGA, and poly(lactic-*co*-glycolic acid) (PLGA)—are possible alternatives (Sawyer et al. 2006), as are nondegradable polymers such as polyethylene-*co*-vinyl acetate (EVAc) intracranial implants (Sawyer et al. 2006). For instance, copolymeric matrices containing γ-aminobutyric acid (Kokaia et al. 1994), thyrotropin-releasing hormone (Kubek et al. 1998), and phenytoin (Tamargo et al. 2002) tested in electrically kindled rats resulted in a sustained reduction of seizure thresholds. Bioceramic materials such as sol–gel titania reservoirs containing anticonvulsant drugs have been implanted into limbic structures in rats with varying results (Lopez et al. 2007a,b). These studies of direct intracerebral delivery may allow the use of new classes of drugs. Within the last few years, several groups have injected a variety of AEDs intracerebrally and observed their effects on animal models of epilepsy. More recently, newer classes of potentially efficacious antiseizure drugs that may not easily cross the BBB or are not absorbed via the gastrointestinal route are being tested via direct intracerebral administration (Mirnajafi-Zadeh et al. 2002; Tolmacheva and van Luijtelaar 2007).

Convection-Enhanced Delivery

Local drug delivery methods, such as CED, are being used to circumvent the BBB and to distribute BBB-impermeable therapeutic anticonvulsant agents over large selected volumes within the central nervous system (Heiss et al. 2005). By bypassing the BBB, local drug delivery allows the drug to be targeted to a specific location and to be delivered in high concentrations (Gasior et al. 2007; Lonser et al. 1999). Experimental studies have provided insight into the relation between CED methods and drug delivery efficacy (Figure 23.5). For instance, small molecules seem to diffuse farther than large molecules, molecules with neutral or negative charge travel greater distances, and liposomes diffuse more readily than positively charged agents of similar size (MacKay et al. 2005).

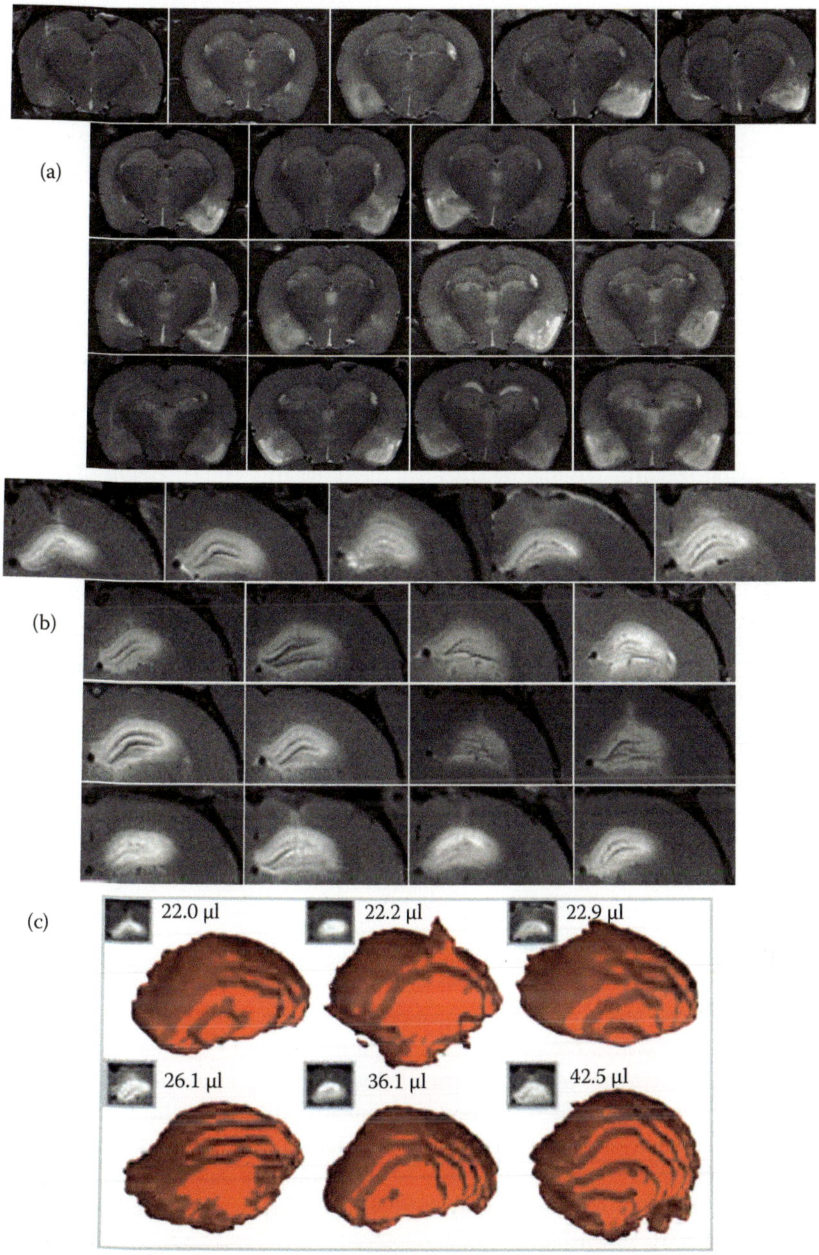

FIGURE 23.5 CED after experimental status epilepticus. (a) MRI T_2-weighted coronal images showing that an episode of status epilepticus resulted in structural changes in the limbic system within 24 h. Injury index was rated by edema in the unilateral piriform cortex and amygdalar nuclei, septal nuclei, middle thalamic nuclei, lateral thalamic nuclei, ventral subiculum, and bilateral piriform cortex and amygdala nuclei. (b) MRI T_1-weighted coronal brain images taken 24 h after status epilepticus showing the distribution of gadolinium–albumin within the dorsal hippocampus, contralateral to the site of electrode implantation. Gadolinium is a nontoxic paramagnetic contrast enhancement agent utilized in MRI. When injected during a scan, gadolinium will tend to change signal intensities by shortening T_1 in its surroundings. Similarly to previously published data from control animals, the contrast agent travels along hippocampal circuitry pathways with minimal backflow or exposure to other regions. Infusions of gadolinium–albumin into the injured hippocampus resulted in larger volume distribution profiles that correlated with increased MR-measured edema. (c) MRI hippocampal segmentations.

Limitations with CED include backflow along the catheter and tissue plugging (Chen et al. 1999; Neeves et al. 2006; Raghavan et al. 2006). In epilepsy, prolonged infusion of muscimol via CED was shown to be both safe and efficacious in rhesus monkeys (Heiss et al. 2005). A similar study involving ω-conotoxin GVIA and ω-conotoxin MVIIA in rats showed that CED can be both safe and effective for treating seizures (Gasior et al. 2007).

BBB Modification

Rather than directly trying to bypass the BBB completely, an alternate approach is to transiently modify the BBB to allow for both efficient drug delivery across the BBB and permanence in the brain parenchyma. In epilepsy, P-glycoproteins and several multidrug-resistant associate proteins are overexpressed in epileptic tissue (Aronica et al. 2003, 2004; Chengyun et al. 2006; Dombrowski et al. 2001; Lazarowski et al. 2004; Potschka et al. 2003; Sisodiya et al. 2001, 2006; Tishler et al. 1995; Vogelgesang et al. 2004). These proteins hydrolyze ATP and move molecules against their concentration gradients into the systemic circulation (Loscher and Potschka 2005). The movement of anticonvulsant drugs out of the brain may be overcome by blocking ATP-binding cassette transporters in the brain. In epilepsy, verapamil, a calcium-channel blocker, was shown to inhibit the blocking action of P-glycoproteins and to have an overall therapeutic effect on human status epilepticus (Iannetti et al. 2005; Summers et al. 2004), where pumping action was inhibited via calcium-channel blockade. Similar studies showed that probenecid, an inhibitor of multidrug-resistant associate proteins 1 and 3, resulted in increased phenytoin concentrations in rat brain (Potschka et al. 2003).

Increasing the osmolarity of the blood has also been used to increase transport across the BBB. Hyperosmolar mannitol given intravascularly temporarily widens endothelial cells' tight junctions and thus increases the permeability of the BBB to substances (Haluska and Anthony 2004). Although this method appears to improve drug delivery in cancer treatment, hyperosmolar opening of the BBB in epilepsy was associated with worsening of seizures in rats (van Vliet et al. 2005).

Drug Modification

Nanoscale anticonvulsant drug delivery systems consist of a drug encapsulated within a nanoparticle (Figure 23.6). Such systems can range in size from 10 nm to 1000 nm and consist of biodegradable polymers, with PLGA being one of the most commonly used polymers. PLGA is of particular interest because of its known safety profile, biocompatibility, and Food and Drug Administration approval status. Other commonly used platforms include classes of liposomes; less commonly used platforms include dendrimers, micelles, carbon nanotubes, emulsions, solid lipid nanoparticles, and nanostructured lipid carriers (Muller et al. 2007; Saad et al. 2008; Yang et al. 1999). Some of these platforms are particularly advantageous because of their stability and time-releasing and degradation properties (Barratt 2000; Couvreur and Vauthier 2006; Dutt and Khuller 2001; Sharma and Sharma 1997; Verdun et al. 1990).

Liposomes range in diameter from approximately 50 nm to 1 μm and essentially consist of multiple bilayers of phospholipids (Couvreur and Vauthier 2006). Liposomes deliver drugs by direct entry or drug release into extracellular space. Further drug release can be enhanced by temperature-sensitive or pH-sensitive methods (Drummond et al. 2000; Paasonen et al. 2007). Hydrophilic polymers such as poly(ethylene glycol) (PEG) have been conjugated to the surfaces of NPs and liposomes to increase circulation time (van Vlerken et al. 2007).

Another delivery system that was recently added to clinical practice are prodrugs. Fosphenytoin sodium (5,5-diphenyl-3-[(phosphonooxy)methyl]-2,4-imidazolidinedione disodium salt; Cerebyx), a prodrug intended for parenteral administration, is composed of phenytoin attached to a phosphate ester (Rautio et al. 2008). When given intramuscularly or intravenously, fosphenytoin is cleaved by alkaline phosphatases to yield the active drug phenytoin. The net result is that fosphenytoin can be administered at a higher rate than phenytoin alone and without phenytoin's drug infusion adverse

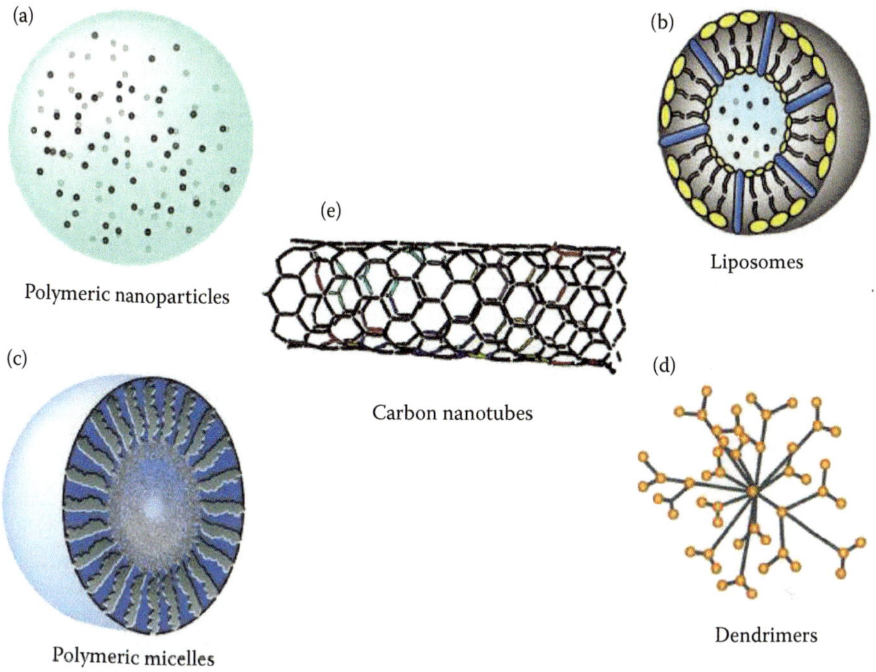

FIGURE 23.6 Nanocolloidal carriers utilized in drug delivery. (a) Polymeric nanoparticles typically made of biodegradable polymers that entrap drugs via encapsulation or polymer–drug conjugation. (b) Liposomes composed of phospholipids and containing an aqueous core surrounded by a lipid bilayer. Hydrophilic drugs can be incorporated into the core, while hydrophobic and amphiphilic drugs can be integrated into the bilayer. (c) Polymeric micelles formed in an aqueous environment from the associations of block copolymers containing both hydrophilic and hydrophobic segments. The hydrophobic core can be loaded with lipophilic drugs, while the hydrophilic surface increases the stability of micelles in water. (d) Dendrimers containing many polymeric monomers that form branched tree-like structures, allowing drugs to be attached to its many arms. (e) Carbon nanotubes, made up of benzene rings, carrying drugs inside the lumen of the tube or attached to the sides. (From Bennewitz, M.F., and Saltzman, W.M., *Neurotherapeutics*, 6, 323–36, 2009. With permission.)

effects (Knapp and Kugler 1998). Other prodrugs include valproic acid (DP-VPA DP16; Depakote), whose attached moiety makes the drug more lipophilic, thereby increasing its ability to cross the BBB (Trojnar et al. 2004). DP VPA, upon reaching the seizure focus, releases active valproic acid by phospholipase A2 cleavage of lecithin (Bialer et al. 2001).

CONCLUSIONS

A wide array of possibilities will, hopefully, soon be available for patients with medically intractable epilepsy and will eventually facilitate treatment tailored to their unique genetic makeup. Developments in the fields of neural engineering, material science, neuroinformatics and modeling, pharmacogenomics, and drug discovery will be increasingly important. Neurosurgeons and neurologists will be able to consider individualized therapies for patients and choose the safest and most effective treatment.

ACKNOWLEDGMENTS

This project was funded, in part, by the National Institutes of Health, the Epilepsy Foundation of America, the Children's Miracle Network, and the B.J. and Eve Wilder Endowment Fund for Epilepsy Research. I also thank Lucia Manubens-Carney for editing the article.

REFERENCES

Akhtari, M., Bragin, A., Cohen, M. et al. 2008. Functionalized magnetonanoparticles for MRI diagnosis and localization in epilepsy. *Epilepsia* 49: 1419–30.

Annegers, J.F. 1997. United States perspective on definitions and classifications. *Epilepsia* 38: S9–12.

Aronica, E., Gorter, J.A., Jansen, G.H. et al. 2003. Expression and cellular distribution of multidrug transporter proteins in two major causes of medically intractable epilepsy: focal cortical dysplasia and glioneuronal tumors. *Neuroscience* 118: 417–29.

Aronica, E., Gorter, J.A., Ramkema, M. et al. 2004. Expression and cellular distribution of multidrug resistance–related proteins in the hippocampus of patients with mesial temporal lobe epilepsy. *Epilepsia* 45: 441–51.

Assaf, B.A., Karkar, K.M., Laxer, K.D. et al. 2004. Magnetoencephalography source localization and surgical outcome in temporal lobe epilepsy. *Clin Neurophysiol* 115: 2066–76.

Bahar, S., Suh, M., Zhao, M. et al. 2006. Intrinsic optical signal imaging of neocortical seizures: the "epileptic dip." *NeuroReport* 17: 499–503.

Bakhshi, S., and North, R.B. 1995. Implantable pumps for drug delivery to the brain. *J Neurooncol* 26: 133–9.

Barbour, R.L., Graber, H.L., Chang, J. et al. 1993. A perturbation approach for optical diffusion tomography using continuous-wave and time-resolved data. In *Medical Optical Tomography: Functional Imaging and Monitoring*, ed. G. Muller, 87–120. Bellingham, WA: SPIE Press.

Barkley, G.L. 2004. Controversies in neurophysiology. MEG is superior to EEG in localization of interictal epileptiform activity. *Proc Clin Neurophysiol* 115: 1001–9.

Barratt, G.M. 2000. Therapeutic applications of colloidal drug carriers. *Pharm Sci Technol Today* 3: 163–71.

Benaron, D.A., Hintz, S.R., Villringer, A. et al. 2000. Noninvasive functional imaging of human brain using light. *J Cereb Blood Flow Metab* 20: 469–77.

Bennewitz, M.F., and Saltzman, W.M. 2009. Nanotechnology for delivery of drugs to the brain for epilepsy. *Neurotherapeutics* 6: 323–36.

Bialer, M., Johannessen, S.I., Kupferberg, H.J. et al. 2001. Progress report on new antiepileptic drugs: a summary of the Fifth Eilat Conference (EILAT V). *Epilepsy Res* 43: 11–58.

Blume, W.T., Luders, H.O., Mizrahi, E. et al. 2001. Glossary of descriptive terminology for ictal semiology: report of the ILAE Task Force on Classification and Terminology. *Epilepsia* 42: 1212–8.

Boas, D.A., Gaudette, T., Strangman, G. et al. 2001. The accuracy of near infrared spectroscopy and imaging during focal changes in cerebral hemodynamics. *Neuroimage* 13: 76–90.

Boogerd, W., van den Bent, M.J., Koehler, P.J. et al. 2004. The relevance of intraventricular chemotherapy for leptomeningeal metastasis in breast cancer: a randomised study. *Eur J Cancer* 40: 2726–33.

Bower, M.R., and Buckmaster, P.S. 2008. Changes in granule cell firing rates precede locally recorded spontaneous seizures by minutes in an animal model of temporal lobe epilepsy. *J Neurophysiol* 99: 2431–42.

Buzsaki, G. 2004. Large-scale recording of neuronal ensembles. *Nat Neurosci* 7: 446–51.

Carney, P.R., Myers, S., and Geyer, J.D. 2011. Seizure prediction: methods. *Epilepsy Behav* 22 (Suppl 1): S94–101

Chapin, J.K. 2004. Using multi-neuron population recordings for neural prosthetics. *Nat Neurosci* 7: 452–5.

Chen, M.Y., Lonser, R.R., Morrison, P.F. et al. 1999. Variables affecting convection-enhanced delivery to the striatum: a systematic examination of rate of infusion, cannula size, infusate concentration, and tissue–cannula sealing time. *J Neurosurg* 90: 315–20.

Chengyun, D., Guoming, L., Elia, M. et al. 2006. Expression of multidrug resistance type 1 gene (MDR1) P-glycoprotein in intractable epilepsy with different aetiologies: a double-labelling and electron microscopy study. *Neurol Sci* 27: 245–51.

Colak, S.B., Papaioannou, D.G., t Hooft, G.W. et al. 1997. Tomographic image reconstruction from optical projections in light-diffusing media. *Appl Opt* 36: 180–213.

Couvreur, P., and Vauthier, C. 2006. Nanotechnology: intelligent design to treat complex disease. *Pharm Res* 23: 1417–50.

Cuccia, D.J., Bevilacqua, F., Durkin, A.J. et al. 2005. Modulated imaging: quantitative analysis and tomography of turbid media in the spatial-frequency domain. *Opt Lett* 30: 1354–6.

DeLong, M.R. 1990. Primate models of movement disorders of basal ganglia origin. *Trends Neurosci* 13: 281–5.

Dombrowski, S.M., Desai, S.Y., Marroni, M. et al. 2001. Overexpression of multiple drug resistance genes in endothelial cells from patients with refractory epilepsy. *Epilepsia* 42: 1501–6.

Drummond, D.C., Zignani, M., and Leroux, J. 2000. Current status of pH-sensitive liposomes in drug delivery. *Prog Lipid Res* 39: 409–60.

Duncan, J.S. 1997. Imaging and epilepsy. *Brain* 120 (Pt 2): 339–77.

Durduran, T., Yu, G., Burnett, M.G. et al. 2004. Diffuse optical measurement of blood flow, blood oxygenation, and metabolism in a human brain during sensorimotor cortex activation. *Opt Lett* 29: 1766–8.

Dutt, M., and Khuller, G.K. 2001. Liposomes and PLG microparticles as sustained release antitubercular drug carriers—an in vitro–in vivo study. *Int J Antimicrob Agents* 18: 245–52.

Engel, J., Jr., Pedley, T.A., Aicardi, J. et al. 2007. *Epilepsy: A Comprehensive Textbook*. Philadelphia, PA: Lippincott-Raven.

Fetz, E.E., and Finocchio, D.V. 1975. Correlations between activity of motor cortex cells and arm muscles during operantly conditioned response patterns. *Exp Brain Res* 23: 217–40.

Fisher, R.S., van Emde Boas, W., Blume, W. et al. 2005. Epileptic seizures and epilepsy: definitions proposed by the International League Against Epilepsy (ILAE) and the International Bureau for Epilepsy (IBE). *Epilepsia* 46: 470–2.

Frazin, N. 2000. White House-Initiated conference on epilepsy. *Curing Epilepsy: Focus on the Future*. Bethesda, MD: National Institute of Neurological Disorders and Stroke, National Institutes of Health.

Gallagher, A., Bastien, D., Pelletier, I. et al. 2008a. A noninvasive, presurgical expressive and receptive language investigation in a 9-year-old epileptic boy using near-infrared spectroscopy. *Epilepsy Behav* 12: 340–6.

Gallagher, A., Lassonde, M., Bastien, D. et al. 2008b. Non-invasive pre-surgical investigation of a 10-year-old epileptic boy using simultaneous EEG-NIRS. *Seizure* 17: 576–82.

Gallagher, A., Theriault, M., Maclin, E. et al. 2007. Near-infrared spectroscopy as an alternative to the Wada test for language mapping in children, adults and special populations. *Epileptic Disord* 9: 241–55.

Gasior, M., White, N.A., and Rogawski, M.A. 2007. Prolonged attenuation of amygdala-kindled seizure measures in rats by convection-enhanced delivery of the N-type calcium channel antagonists omega-conotoxin GVIA and omega-conotoxin MVIIA. *J Pharmacol Exp Ther* 323: 458–68.

Gratton, G., and Fabiani, M. 2007. Optical imaging of brain function. In *Neuroergonomics: The Brain at Work*, eds. R. Parasuraman and M. Rizzo, 65–81. Oxford, UK: Oxford University Press.

Haglund, M.M., and Hochman, D.W. 2004. Optical imaging of epileptiform activity in human neocortex. *Epilepsia* 45 (Suppl 4): 43–7.

Haglund, M.M., Ojemann, G.A., and Hochman, D.W. 1992. Optical imaging of epileptiform and functional activity in human cerebral cortex. *Nature* 358: 668–71.

Haluska, M., and Anthony, M.L. 2004. Osmotic blood–brain barrier modification for the treatment of malignant brain tumors. *Clin J Oncol Nurs* 8: 263–7.

Hauser, W.A., and Hesdorffer, D.C. 1990. *Causes and Consequences*. New York: Demos Medical Publishing.

Hebden, J.C., Gibson, A., Yusof, R.M. et al. 2002. Three-dimensional optical tomography of the premature infant brain. *Phys Med Biol* 47: 4155–66.

Heiss, J.D., Walbridge, S., Morrison, P. et al. 2005. Local distribution and toxicity of prolonged hippocampal infusion of muscimol. *J Neurosurg* 103: 1035–45.

Hielscher, A.H., Bluestone, A.Y., Abdoulaev, G.S. et al. 2002. Near-infrared diffuse optical tomography. *Dis Markers* 18: 313–37.

Hillman, E.M., Devor, A., Bouchard, M.B. et al. 2007. Depth-resolved optical imaging and microscopy of vascular compartment dynamics during somatosensory stimulation. *Neuroimage* 35: 89–104.

Iannetti, P., Spalice, A., and Parisi, P. 2005. Calcium-channel blocker verapamil administration in prolonged and refractory status epilepticus. *Epilepsia* 46: 967–9.

Jacobs, M.P., Fischbach, G.D., Davis, M.R. et al. 2001. Future directions for epilepsy research. *Neurology* 57: 1536–42.

Jiang, H., Paulsen, K.D., Osterberg, U.L. et al. 1995. Simultaneous reconstruction of optical absorption and scattering maps in turbid media from near-infrared frequency-domain data. *Opt Lett* 20: 2128–30.

Kennan, R.P., Kim, D., Maki, A. et al. 2002. Non-invasive assessment of language lateralization by transcranial near infrared optical topography and functional MRI. *Hum Brain Mapp* 16: 183–9.

Kipke, D.R., Vetter, R.J., Williams, J.C. et al. 2003. Silicon–substrate intracortical microelectrode arrays for long-term recording of neuronal spike activity in cerebral cortex. *IEEE Trans Neural Syst Rehabil Eng* 11: 151–5.

Knapp, L.E., and Kugler, A.R. 1998. Clinical experience with fosphenytoin in adults: pharmacokinetics, safety, and efficacy. *J Child Neurol* 13 (Suppl 1): S15–8; discussion S30–2.

Knowlton, R.C., Elgavish, R., Howell, J. et al. 2006. Magnetic source imaging versus intracranial electroencephalogram in epilepsy surgery: a prospective study. *Ann Neurol* 59: 835–42.

Kokaia, M., Aebischer, P., Elmer, E. et al. 1994. Seizure suppression in kindling epilepsy by intracerebral implants of GABA- but not by noradrenaline-releasing polymer matrices. *Exp Brain Res* 100: 385–94.

Kubek, M.J., Liang, D., Byrd, K.E. et al. 1998. Prolonged seizure suppression by a single implantable polymeric-TRH microdisk preparation. *Brain Res* 809: 189–97.

Lazarowski, A., Ramos, A.J., Garcia-Rivello, H. et al. 2004. Neuronal and glial expression of the multidrug resistance gene product in an experimental epilepsy model. *Cell Mol Neurobiol* 24: 77–85.

Lemieux, L. 2004. Electroencephalography-correlated functional MR imaging studies of epileptic activity. *Neuroimaging Clin North Am* 14: 487–506.

Loeb, G.E. 1990. Cochlear prosthetics. *Annu Rev Neurosci* 13: 357–71.

Lonser, R.R., Corthesy, M.E., Morrison, P.F. et al. 1999. Convection-enhanced selective excitotoxic ablation of the neurons of the globus pallidus internus for treatment of parkinsonism in nonhuman primates. *J Neurosurg* 91: 294–302.

Lopez, T., Ortiz, E., Quintana, P. et al. 2007a. A nanostructured titania bioceramic implantable device capable of drug delivery to the temporal lobe of the brain. *Colloids Surf A* 300: 3–10.

Lopez, T., Quintana, P., Ascencio, J. et al. 2007b. The determination of dielectric constants of mixtures used in the treatment of epilepsy and the encapsulation of phenytoin in a titania matrix. *Colloids Surf A* 300: 99–105.

Loscher, W., and Potschka, H. 2005. Blood–brain barrier active efflux transporters: ATP-binding cassette gene family. *NeuroRx* 2: 86–98.

Ludvig, N., Kovacs, L., Medveczky, G. et al. 2005. Toward the development of a subdural hybrid neuroprosthesis for the treatment of intractable focal epilepsy. *Epilepsia* 46: 270–70.

MacKay, J.A., Deen, D.F., and Szoka, F.C., Jr. 2005. Distribution in brain of liposomes after convection enhanced delivery; modulation by particle charge, particle diameter, and presence of steric coating. *Brain Res* 1035: 139–53.

Mamelak, A.N., Lopez, N., Akhtari, M. et al. 2002. Magnetoencephalography-directed surgery in patients with neocortical epilepsy. *J Neurosurg* 97: 865–73.

Manoharan, A., Bowyer, S.M., Mason, K. et al. 2007. Localizing value of MEG in refractory partial epilepsy: surgical outcomes. *Int Congr Ser* 1300: 657–60.

Matson, C.L., and Liu, H. 1999. Analysis of the forward problem with diffuse photon density waves in turbid media by use of a diffraction tomography model. *J Opt Soc Am A* 16: 455–66.

Meisler, M.H., and Kearney, J.A. 2005. Sodium channel mutations in epilepsy and other neurological disorders. *J Clin Invest* 115: 2010–7.

Mirnajafi-Zadeh, J., Mortazavi, M., Fathollahi, Y. et al. 2002. Effect of transient hippocampal inhibition on amygdaloid kindled seizures and amygdaloid kindling rate. *Brain Res* 954: 220–6.

Muller, R.H., Petersen, R.D., Hommoss, A. et al. 2007. Nanostructured lipid carriers (NLC) in cosmetic dermal products. *Adv Drug Deliv Rev* 59: 522–30.

Neeves, K.B., Lo, C.T., Foley, C.P. et al. 2006. Fabrication and characterization of microfluidic probes for convection enhanced drug delivery. *J Controlled Release* 111: 252–62.

Noguchi, Y., Takcuchi, T., and Sakai, K.L. 2002. Lateralized activation in the inferior frontal cortex during syntactic processing: event-related optical topography study. *Hum Brain Mapp* 17: 89–99.

Oommen, J., Kraus, A.C., and Fisher, R.S. 2007. Intraventricular administration of gabapentin in the rat increases flurothyl seizure threshold. *Neurosci Lett* 417: 308–11.

Paasonen, L., Romberg, B., Storm, G. et al. 2007. Temperature-sensitive poly(*N*-(2-hydroxypropyl)methacrylamide mono/dilactate)-coated liposomes for triggered contents release. *Bioconjugate Chem* 18: 2131–6.

Pardridge, W.M. 1997. Drug delivery to the brain. *J Cereb Blood Flow Metab* 17: 713–31.

Potschka, H., Fedrowitz, M., and Loscher, W. 2003. Multidrug resistance protein MRP2 contributes to blood–brain barrier function and restricts antiepileptic drug activity. *J Pharmacol Exp Ther* 306: 124–31.

Raghavan, R., Brady, M.L., Rodriguez-Ponce, M.I. et al. 2006. Convection-enhanced delivery of therapeutics for brain disease, and its optimization. *Neurosurg Focus* 20: E12.

Rautio, J., Kumpulainen, H., Heimbach, T. et al. 2008. Prodrugs: design and clinical applications. *Nat Rev Drug Discov* 7: 255–70.

Rueckert, D., Sonoda, L.I., Hayes, C. et al. 1999. Nonrigid registration using free-form deformations: application to breast MR images. *IEEE Trans Med Imaging* 18: 712–21.

Ruiz-Alzola, J., Westin, C.F., Warfield, S.K. et al. 2002. Nonrigid registration of 3D tensor medical data. *Med Image Anal* 6: 143–61.

Saad, M., Garbuzenko, O.B., Ber, E. et al. 2008. Receptor targeted polymers, dendrimers, liposomes: which nanocarrier is the most efficient for tumor-specific treatment and imaging? *J Controlled Release* 130: 107–14.

Saltzman, W.M., and Olbricht, W.L. 2002. Building drug delivery into tissue engineering. *Nat Rev Drug Discov* 1: 177–86.

Sanchez, J.C., Mareci, T.H., Norman, W.M. et al. 2006. Evolving into epilepsy: multiscale electrophysiological analysis and imaging in an animal model. *Exp Neurol* 198: 31–47.

Sandberg, D.I., Bilsky, M.H., Souweidane, M.M. et al. 2000. Ommaya reservoirs for the treatment of leptomeningeal metastases. *Neurosurgery* 47: 49–54; discussion 54–5.
Sawyer, A.J., Piepmeier, J.M., and Saltzman, W.M. 2006. New methods for direct delivery of chemotherapy for treating brain tumors. *Yale J Biol Med* 79: 141–52.
Schachter, S.C., Guttag, J., Schiff, S.J. et al. 2009. Advances in the application of technology to epilepsy: the CIMIT/NIO Epilepsy Innovation Summit. *Epilepsy Behav* 16: 3–46.
Schotland, J.C., Haselgrove, J.C., and Leigh, J.S. 1993. Photon hitting density. *Appl Opt* 32: 448–53.
Serralta, A., Barcia, J.A., Ortiz, P. et al. 2006. Effect of intracerebroventricular continuous infusion of valproic acid versus single i.p. and i.c.v. injections in the amygdala kindling epilepsy model. *Epilepsy Res* 70: 15–26.
Shariff, S., Suh, M., Zhao, M.R. et al. 2006. Recent developments in oximetry and perfusion-based mapping techniques and their role in the surgical treatment of neocortical epilepsy. *Epilepsy Behav* 8: 363–75.
Sharma, A., and Sharma, U.S. 1997. Liposomes in drug delivery: progress and limitations. *Int J Pharm* 154: 123–40.
Sheth, S.A., Nemoto, M., Guiou, M. et al. 2004. Linear and nonlinear relationships between neuronal activity, oxygen metabolism, and hemodynamic responses. *Neuron* 42: 347–55.
Sisodiya, S.M., Lin, W.R., Squier, M.V. et al. 2001. Multidrug-resistance protein 1 in focal cortical dysplasia. *Lancet* 357: 42–3.
Sisodiya, S.M., Martinian, L., Scheffer, G.L. et al. 2006. Vascular colocalization of P-glycoprotein, multidrug-resistance associated protein 1, breast cancer resistance protein and major vault protein in human epileptogenic pathologies. *Neuropathol Appl Neurobiol* 32: 51–63.
Sokol, D.K., Markand, O.N., Daly, E.C. et al. 2000. Near infrared spectroscopy (NIRS) distinguishes seizure types. *Seizure* 9: 323–7.
Stables, J.P., Bertram, E.H., White, H.S. et al. 2002. Models for epilepsy and epileptogenesis: report from the NIH workshop, Bethesda, Maryland. *Epilepsia* 43: 1410–20.
Stefan, H., Hummel, C., Scheler, G. et al. 2003. Magnetic brain source imaging of focal epileptic activity: a synopsis of 455 cases. *Brain* 126: 2396–405.
Steinhoff, B.J., Herrendorf, G., and Kurth, C. 1996. Ictal near infrared spectroscopy in temporal lobe epilepsy: a pilot study. *Seizure* 5: 97–101.
Summers, M.A., Moore, J.L., and McAuley, J.W. 2004. Use of verapamil as a potential P-glycoprotein inhibitor in a patient with refractory epilepsy. *Ann Pharmacother* 38: 1631–4.
Tamargo, R.J., Rossell, L.A., Kossoff, E.H. et al. 2002. The intracerebral administration of phenytoin using controlled-release polymers reduces experimental seizures in rats. *Epilepsy Res* 48: 145–55.
Theodore, W.H., and Fisher, R.S. 2004. Brain stimulation for epilepsy. *Lancet Neurol* 3: 111–8.
Tishler, D.M., Weinberg, K.I., Hinton, D.R. et al. 1995. MDR1 gene expression in brain of patients with medically intractable epilepsy. *Epilepsia* 36: 1–6.
Tolmacheva, E.A., and van Luijtelaar, G. 2007. Absence seizures are reduced by the enhancement of GABA-ergic inhibition in the hippocampus in WAG/Rij rats. *Neurosci Lett* 416: 17–21.
Trojnar, M.K., Wierzchowska-Cioch, E., Krzyzanowski, M. et al. 2004. New generation of valproic acid. *Pol J Pharmacol* 56: 283–8.
Truccolo, W., Donoghue, J.A., Hochberg, L.R. et al. 2011. Single-neuron dynamics in human focal epilepsy. *Nat Neurosci* 14: 635–41.
Tuch, D.S., Wedeen, V.J., Dale, A.M. et al. 2001. Conductivity tensor mapping of the human brain using diffusion tensor MRI. *Proc Natl Acad Sci U S A* 98: 11697–701.
van Vlerken, L.E., Vyas, T.K., and Amiji, M.M. 2007. Poly(ethylene glycol)-modified nanocarriers for tumor-targeted and intracellular delivery. *Pharm Res* 24: 1405–14.
van Vliet, E.A., Redeker, S., Aronica, E. et al. 2005. Expression of multidrug transporters MRP1, MRP2, and BCRP shortly after status epilepticus, during the latent period, and in chronic epileptic rats. *Epilepsia* 46: 1569–80.
Verdun, C., Brasseur, F., Vranckx, H. et al. 1990. Tissue distribution of doxorubicin associated with polyisohexylcyanoacrylate nanoparticles. *Cancer Chemother Pharmacol* 26: 13–8.
Villringer, A., and Chance, B. 1997. Non-invasive optical spectroscopy and imaging of human brain function. *Trends Neurosci* 20: 435–42.
Viola, P., and Wells, W.M. 1997. Alignment by maximization of mutual information. *Int J Comput Vis* 24: 137–54.
Vogelgesang, S., Kunert-Keil, C., Cascorbi, I. et al. 2004. Expression of multidrug transporters in dysembryoplastic neuroepithelial tumors causing intractable epilepsy. *Clin Neuropathol* 23: 223–31.
von dem Hagen, E.A., and Henkelman, R.M. 2002. Orientational diffusion reflects fiber structure within a voxel. *Magn Reson Med* 48: 454–9.

Walker, S.A., Fantini, S., and Gratton, E. 1997. Image reconstruction by backprojection from frequency-domain optical measurements in highly scattering media. *Appl Opt* 36: 170–4.

Watanabe, E., Maki, A., Kawaguchi, F. et al. 1998. Non-invasive assessment of language dominance with near-infrared spectroscopic mapping. *Neurosci Lett* 256: 49–52.

Watanabe, E., Maki, A., Kawaguchi, F. et al. 2000. Noninvasive cerebral blood volume measurement during seizures using multichannel near infrared spectroscopic topography. *J Biomed Opt* 5: 287–90.

Watson, N.F., Dodrill, C., Farrell, D. et al. 2004. Determination of language dominance with near-infrared spectroscopy: comparison with the intracarotid amobarbital procedure. *Seizure* 13: 399–402.

Wheless, J.W., Willmore, L.J., Breier, J.I. et al. 1999. A comparison of magnetoencephalography, MRI, and V-EEG in patients evaluated for epilepsy surgery. *Epilepsia* 40: 931–41.

World Health Organization. 2001. Epilepsy: epidemiology, etiology and prognosis. http://www.who.int/inf-fs/en/fact165.html.

Yang, S.C., Lu, L.F., Cai, Y. et al. 1999. Body distribution in mice of intravenously injected camptothecin solid lipid nanoparticles and targeting effect on brain. *J Controlled Release* 59: 299–307.

Yildirim, M., and Marangoz, C. 2007. Anticonvulsant effects of focal and intracerebroventricular adenosine on penicillin-induced epileptiform activity in rats. *Brain Res* 1127: 193–200.

Zhang, Q., Liu, Z., Carney, P.R. et al. 2008. Non-invasive imaging of epileptic seizures in vivo using photo-acoustic tomography. *Phys Med Biol* 53: 1921–31.

Zijlmans, M., Huiskamp, G., Hersevoort, M. et al. 2007. EEG-fMRI in the preoperative work-up for epilepsy surgery. *Brain* 130: 2343–53.

Zrenner, E. 2002. Will retinal implants restore vision? *Science* 295: 1022–5.

24 Applications of Nanotechnology to Diseases of the Spine

Bryn S. Moore and Steven A. Toms

CONTENTS

INTRODUCTION

Nanotechnology involves manipulating the chemical properties of objects less than 100 nm in size. This chapter focuses on the structural spinal elements of intervertebral discs and vertebrae since central nervous system (CNS) regeneration and tumors are well covered in other chapters of this volume. Nanotechnology has led to the creation of new materials for treating structural spinal disorders, including disc replacement and bone scaffolding materials, which are currently being tested in animals. In this chapter, these and other materials arising from nanotechnology are described, including those that are designed to improve osteoblast adherence and proliferation, to increase injectability of bone scaffolding materials, and to improve nucleus pulposus replacement. Some of the major applications of nanotechnology for treating spinal diseases are reviewed, and nanotechnology that has already been implemented experimentally is described.

DISEASES OF APPLICATION

Degenerative Disc Disease

Intervertebral discs link vertebral bodies and have two main components: (1) the nucleus pulposus, consisting of type II collagen, proteoglycan, and water; and (2) the annulus fibrosus, enclosing the nucleus pulposus and consisting of type I collagen, including cartilaginous end plates (Modic and Ross 2007). As the disc ages and degenerates, there is loss of proteoglycan and other biomolecules, resulting in morphologic changes in the disc and a drop in its osmotic pressure (Urban and Roberts 2003). Disc degeneration is associated with disc herniation or prolapse, which alters disc height and spinal column mechanics. In the lumbar spine, sciatica results when a herniated disc compresses a nerve root (Urban and Roberts 2003). Degenerative disc disease (DDD) is thought to result from inadequate nutrient supply to disc cells (Choi 2009; Nachemson et al. 1970), abnormal mechanical load on the disc (Lipson and Muir 1981; Lotz et al. 1998), and genetic predisposition (Battie et al. 1995a,b; Sambrook et al. 1999). Pain relief usually occurs after the herniated or degenerated disc tissue is removed, thereby relieving nerve root compression. However, because the disc transfers load between adjacent vertebrae, its removal transfers weight bearing to facet joints at that spinal segment, increases stress on adjacent discs, and compromises spinal function (Meisel et al. 2006).

Spinal Fractures and Vertebral Augmentation

Although spinal surgery is usually performed to treat the consequences of progressive DDD, including acute disc herniation and degenerative spinal stenosis, spinal fusion may be required in cases of spinal instability and fracture. In addition, vertebral augmentation procedures may be needed to relieve pain associated with acute osteoporotic vertebral fractures. Vertebral fractures occur from acute trauma and are more common in people with osteoporosis. The progressive loss of bone mass that characterizes osteoporosis increases susceptibility to fracture in both the axial skeleton and the appendicular skeleton. Vertebral compression fractures caused by osteoporosis are a leading cause of disability and morbidity in the elderly (Phillips 2003). In the United States, there are approximately 1.5 million osteoporotic fractures each year, with nearly half affecting the spine (Riggs and Melton 1995). Vertebral compression fractures occur in 20% of people older than 70 years and in 16% of postmenopausal women (Cohen 1990). Fracture of one vertebra can lead to increased stress on adjacent vertebrae and increased risk of their fracture (Browner et al. 2009; Phillips 2003). Nonoperative treatment is the preferred initial approach for osteoporotic fractures of the spine. However, surgery can be considered in cases involving progressive neurologic deficit, intractable pain refractory to conservative care, and progressive spinal deformity (Browner et al. 2009).

CURRENT APPLICATIONS OF NANOTECHNOLOGY IN SPINAL SURGERY

Vertebroplasty and Kyphoplasty for Osteoporotic Fracture

Approximately 38,000 vertebroplasties and 16,000 kyphoplasties were performed in 2002 (Nussbaum et al. 2004) to treat vertebral compression fractures. Vertebroplasties and kyphoplasties involve placing a large-bore needle into the vertebral body through which a bone substitute, such as polymethylmethacrylate (PMMA), is injected. Kyphoplasty involves inflating a balloon tamp inside the vertebral body prior to injecting PMMA to help restore the original vertebral height (Browner et al. 2009; Phillips 2003; Rao and Singrakhia 2003). Indications for vertebroplasty or kyphoplasty include stabilization of osteoporotic vertebral fractures or damaged vertebrae caused by metastases or multiple myeloma (Phillips 2003). The ideal bone substitute should maximize the osteogenic effect of bone morphogenic proteins (BMPs) by maintaining their bioactivity, optimizing their release profile, and prolonging their duration of action at the implantation site. It should also provide an osteoconductive scaffold with a suitable pore structure for vascularization and bone formation (Wei et al. 2007). There are many exciting nanotechnology-influenced developments in the field of bone substitutes that aim to make them easier to inject, stronger, and better able to act as osteoconductive scaffolds.

PMMA as Bone Substitutes

Although PMMA is commonly utilized as a bone substitute, there are potential hazards associated with its use. These include thermal damage to adjacent structures induced by exothermic polymerization during the curing process and complications resulting from a lack of long-term biointegration of the cement (Phillips 2003).

Calcium Phosphate Cement as Bone Substitutes

An alternative to PMMA is calcium phosphate cement (CPC), which is a mixture of tetracalcium phosphate and anhydrous dicalcium phosphate. CPC can be mixed with liquid and sculpted during surgery to conform to defects in hard tissue. CPC self-hardens to form resorbable hydroxyapatite (HA); however, it is brittle and has a low strength limit (Xu et al. 2008). Nanotechnology is being employed to increase the strength of CPC and to improve its applicability as a bone substitute.

PMMA was compared to CPC combined with recombinant human bone morphogenic protein-2 (rhBMP-2) in a rhesus monkey model system of vertebroplasty. New vessels and bone were visible in CPC/rhBMP-2 after 2 months. After 6 months, the cement was almost completely replaced with mature bone tissue (Figure 24.1). In the PMMA group, new bone did not form, and material

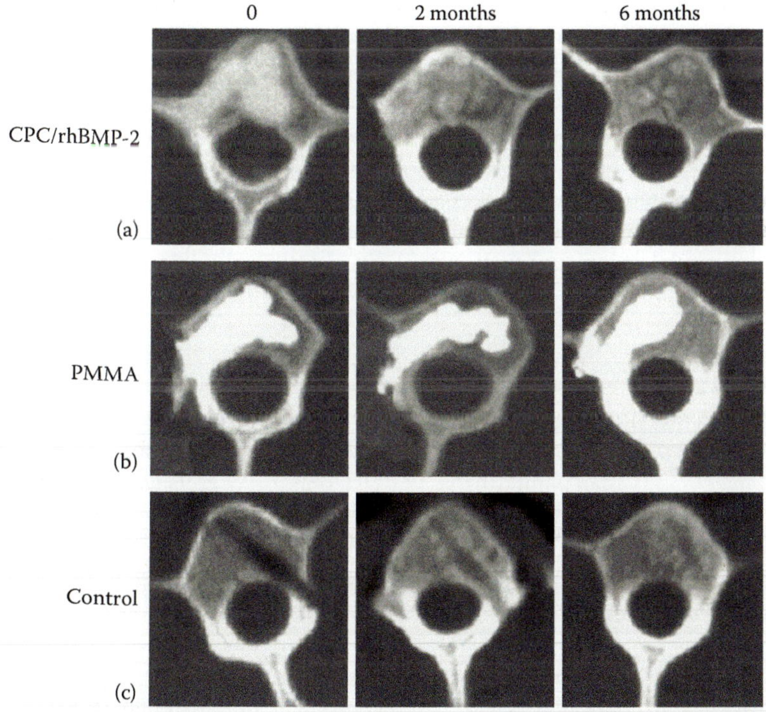

FIGURE 24.1 Computed tomography scans from a rhesus monkey after vertebroplasty filled with CPC/rhBMP-2, PMMA, or nothing (control) taken at 0, 2, and 6 months after surgery. (a) CPC/rhBMP-2 filled in a monkey's vertebrae, degraded gradually, and was replaced by new bone. (b) PMMA in a monkey's vertebrae showed no degradation at 2 or 6 months. (c) The bone canal filled in nothing during the operation; new bone deposited at 2 months and completely healed at 6 months. (From Bai, B. et al., *Spine (Phila, Pa 1976)*, 34, 1887–92, 2009. With permission.)

degradation was present (Bai et al. 2009). These results suggest that the CPC/rhBMP-2 bone substitute is potentially a better long-term bone substitute than PMMA.

Vlad et al. (2008) investigated the setting, flow, and biocompatibility properties of new iron-modified calcium phosphate bone cement. Iron oxide particles increased the initial and final setting times of the bone cement, thereby improving its workability and fluidity without disturbing mechanical properties. Decreasing the viscosity of CPC with iron oxide improves cement injectability but could potentially increase the possibility of bone cement flowing out of the vertebra and into the spinal canal. The iron oxide additive neither induced toxicity nor modified setting reactions. The evolved apatitic microstructure of the cement produced had a higher compressive strength at saturation than nonmodified cement (Vlad et al. 2008). These studies suggest that nanotechnology can improve the characteristics of current bone substitutes to make them easier to work with in a surgical setting.

Spinal Fusion for DDD or Fracture

Operative procedures for fusing segments of the cervical, thoracolumbar, and lumbar spine have greatly increased, with more than 200,000 fusions performed in 2003 (Cowan et al. 2006). This increase is especially prevalent in patients older than 40 years, in whom fusions are used to treat spinal instability resulting from DDD, scoliosis, and fracture (Cowan et al. 2006). Spinal fusion depends on a bone graft to induce bone growth between adjacent vertebrae, resulting in a large bone mass that essentially welds together the adjacent vertebrae. Spinal fusion is the standard treatment for spinal instability resulting from DDD; however, fusion eliminates motion between the fused levels and alters the biomechanics of the spine, which can accelerate degeneration of adjacent intervertebral discs. Spinal fusion procedures traditionally have required autogenous bone grafts to provide osteoinductive factors and osteoconductive scaffolds that support the formation of new bone (Liao et al. 2003). As an alternative to harvesting bone grafts from patients, tissue engineering is developing biomaterial scaffolds to support bone growth and spinal fusion (Nesti et al. 2008). Bone scaffolding materials are designed to be osteoconductive matrices that can attract, stimulate, and support the formation of new bone by host osteogenic cells. Nanotechnology is supporting new developments in bone scaffolding that encourage the growth of bone between spinal levels and may eventually eliminate the use of large autogenous bone grafts for spinal fusion.

CPC as Bone Scaffold

Xu et al. (2008) created an improved form of CPC that is strong, macroporous, and osteoconductive. The modified CPC, which consists of tetracalcium phosphate and dicalcium phosphate, together with a scaffold of absorbable fibers, biopolymer chitosan, and mannitol porogen, had better fracture resistance and injectability than unmodified CPC. Osteoblast-like cells were included in the CPC by placing them in hydrogel beads, which protected them from destruction during the cement-setting process. Growth factors can also be incorporated into this CPC to further encourage new bone growth. Although still in early development, this modified CPC may serve as a bone substitute to fill in and strengthen osteoporotic bone lesions at risk for fracture (Xu et al. 2008).

Poly(lactic-*co*-glycolide) as Bone Scaffold

A scaffold providing a controlled release of recombinant human bone morphogenic protein-7 (rhBMP-7) was developed by Wei et al. (2007). Here, rhBMP-7 was encapsulated in poly(lactic-*co*-glycolide) (PLGA) nanospheres and immobilized within three-dimensional macroporous and nanofibrous poly(L-lactic acid) (PLLA) scaffolds. The interconnected porous structure of the scaffolding material supported bone growth, and the sustained release of bioactive rhBMP-7 induced bone formation. The release kinetics of rhBMP-7 were controlled by the nanospheres. BMP's ability to stimulate bone formation depended on the properties (e.g., material type, geometry, porosity, and pore size) of the carriers (Wei et al. 2007). The properties of the nanospheres could be modified to optimize the release kinetics of each growth factor.

Deproteinized cancellous bovine bone matrices, under the name of Bio-Oss®, have been substituted for bone grafts in jaw and nose surgeries (Kim and Kim 2008). The osteogenic potential of a synthetic bone substitute, apatite-coated PLGA/nano-HA particulates, was compared to that of Bio-Oss. A mixture of fibrin gel and either apatite-coated PLGA/HA particulates or Bio-Oss was implanted into rats with parietal bone defects. Eight weeks after treatment, the new bone formations around grafting materials in PLGA/HA-filled defects and Bio-Oss-filled defects were similar. This study showed that a synthetic bone graft material consisting of apatite-coated PLGA/HA particulates had bone regeneration potential comparable to Bio-Oss (Kim and Kim 2008), which was derived from bone xenograft. Although apatite-coated PLGA/HA particulates have been tested only in animal studies and have not yet been tested in the spine, this bone substitute has the potential to fill and solidify bony defects of the spine, such as those occurring in osteoporosis.

HA as Bone Scaffold

Liao et al. (2003) combined the mineralized nano-HA collagen/polylactic acid (nHAC/PLA) matrix with or without the growth factor rhBMP-2 during spinal fusion procedures in the rabbit lumbar spine. Although autologous bone graft is the gold standard for spinal fusion, its use involves donor site morbidity, blood loss, and increased operative time; sometimes, insufficient bone graft is available at the donor site. The availability of suitable substitutes for autologous bone graft could prevent these problems. The nHAC/PLA matrix has several favorable properties for a bone substitute, including high porosity, mechanical strength, and a hierarchical microstructure that simulates cancellous bone. nHAC/PLA is an effective carrier for rhBMP-2, which promotes new bone regeneration (Figure 24.2). When rhBMP-2 is added to nHAC/PLA, the resulting spinal fusion is stronger and stiffer than that observed with autologous bone graft alone. In cases where autologous bone graft was the initial choice for fusion material but the donor site could not provide a sufficient amount of bone graft, a surgeon can supplement the available autologous bone graft with nHAC/PLA because they are biocompatible and yield a similar strength of spinal fusion (Liao et al. 2003).

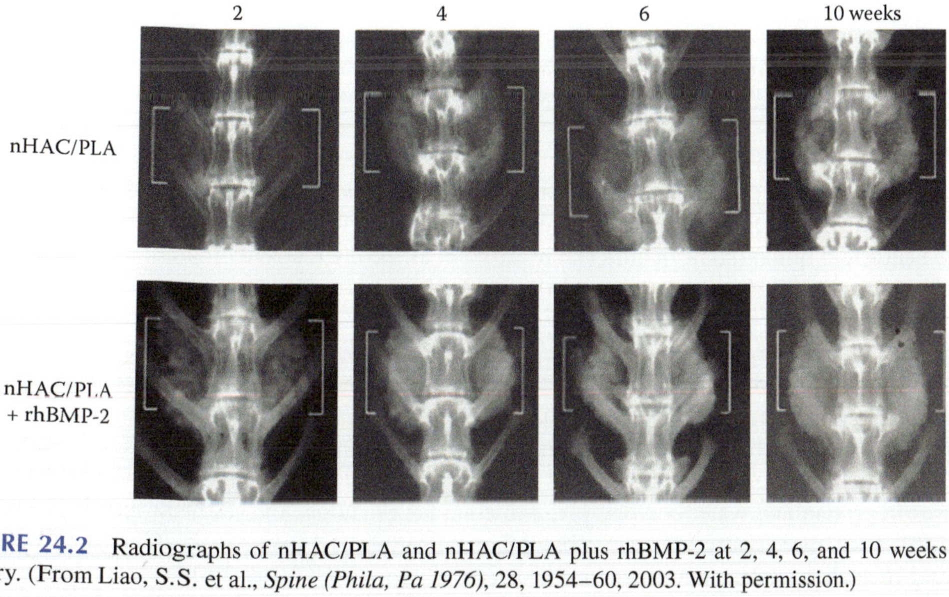

FIGURE 24.2 Radiographs of nHAC/PLA and nHAC/PLA plus rhBMP-2 at 2, 4, 6, and 10 weeks after surgery. (From Liao, S.S. et al., *Spine (Phila, Pa 1976)*, 28, 1954–60, 2003. With permission.)

Coating of Structural Implants

Orthopedic implants, traditionally, have had a high failure rate, which is partially related to the incompatibility between osteoblasts and implant materials such as titanium (Jayaraman et al. 2004). Titanium implants are frequently used in spinal fusion and other spinal surgeries. Nanotechnology can be used to design and engineer implants with surfaces that mimic the microanatomic structure of bone and to improve the integration of the implant with the surrounding endogenous bone. Current orthopedic implants are modified with micron-sized metal beads or etched to establish micron-sized surface features, encouraging new bone formation around the implant (Friedman et al. 1996). PLGA is a common coating biomaterial for devices used in orthopedics, providing a biodegradable matrix for bone cell adhesion and differentiation (Smith et al. 2007). Helical rosette nanotubes (HRNs) mimic the nanostructure of collagen and undergo self-assembly in solution. HRNs are amenable to the addition of various functional groups and have biologically inspired surface features, which may create bone-like environments and improve osteointegration with juxtaposed bone (Zhang et al. 2008). HRNs can be functionalized with a diverse range of peptides that are known to enhance osteoblast adhesion (Chun et al. 2005). HA is osteoconductive and thus can encourage bone growth that bonds the bone to the implant (Zhang et al. 2008). Coating titanium with HRN and HA/HRN increases osteoblast adhesion onto titanium. Additionally, Smith et al. (2007) showed that the osteoblast cell density of implants could be increased by adding nanotopography to titanium implants either by anodization of the titanium or by addition of nanoetched PLGA.

Intervertebral Disc Substitutes for DDD

There are presently two lumbar total disc replacement implants that have been approved for human use (i.e., Charite and Prodisc-L) and two others that are far along in the Food and Drug Administration approval process (i.e., Maverick, Flexi-core) (Nesti et al. 2008). These implants preserve flexion, extension, and lateral bending motions, and can restore disc height. However, these implants do not address compressive forces, have unknown longevity, and are unable to biointegrate and remodel (Nesti et al. 2008). Nanotechnology has been used to construct a synthetic disc in vitro that addresses some of the problems with current implants.

Whole Disc Substitutes

Nesti et al. (2008) seeded adult human mesenchymal stem cells into a construct consisting of a biodegradable nanofibrous scaffold surrounding a hyaluronic acid hydrogel center. Cell growth in vitro resulted in the formation of a cartilaginous hyaluronic acid–nanofibrous scaffold (HANFS) that resembled an intervertebral disc. The cartilaginous tissue contained components similar to the annulus fibrosus and the nucleus pulposus (Nesti et al. 2008) (Figure 24.3).

Nucleus Pulposus Substitute

Replacing the nucleus pulposus is a tempting alternative to spinal fusion procedures because it could potentially slow adjacent level degeneration, restore normal loads to the diseased level, and restore segmental spinal biomechanics (Di Martino et al. 2005). Current prosthetic nucleus pulposus substitutes are still being developed (Wilke et al. 2001).

Injectable scaffolds for disc repair must be mechanically strong, durable, biocompatible, and sterilizable. Additionally, they need to have minimal setting time, low viscosity, and the ability to promote both tissue formation and matrix formation while degrading over time. A recombinant protein copolymer, NuCore, which consists of amino acid sequence blocks derived from silk and elastin structural proteins, has been injected and has prevented disc height loss and associated biomechanical and biochemical changes due to reduced disc height and volume. NuCore mimicked the nucleus pulposus in protein and water content, pH, and complex modulus, and restored the biomechanics of the disc after microdiscectomy while being biocompatible and nontoxic. NuCore is suitable for replacing the native nucleus pulposus after discectomy (Boyd and Carter 2006).

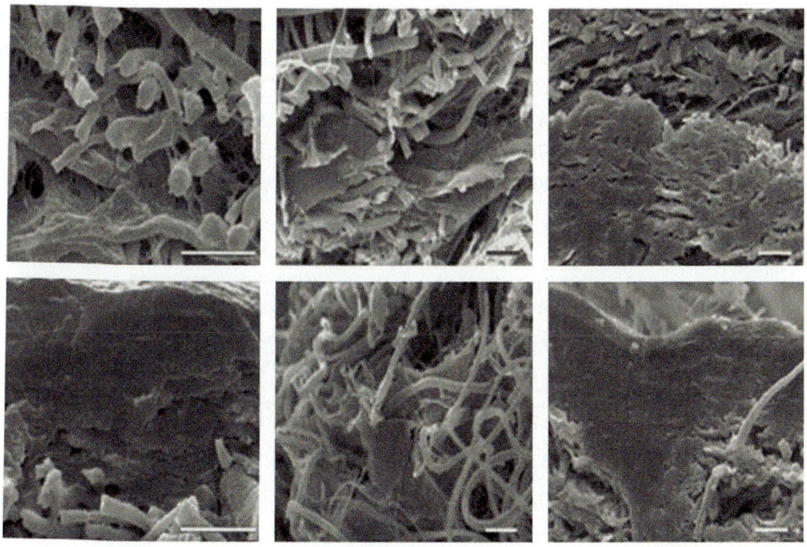

FIGURE 24.3 Scanning electron microscopy of HANFS constructs at 7, 14, and 28 days. Cells adhered to individual nanofibers in both the annulus fibrosus (AF) and the nucleus pulposus (NP) regions of the construct, and increasing amounts of extracellular matrix were deposited as a function of time. (From Nesti, L.J. et al., *Tissue Eng. Part A*, 14, 1527–37, 2008. With permission.)

Condensed collagen matrix with barium sulfate, which was used as a nucleus pulposus replacement, restored the height and improved the physiological stability of the intervertebral disc. However, the implant can extrude similar disc material through the hole in the annulus fibrosus through which it was originally implanted (Wilke et al. 2006). An alternative method of nucleus pulposus substitution that utilizes nanotechnology is injecting scaffolding material, such as the collagen matrix, seeded with cartilage-producing cells. Annulus cells attach well to randomly oriented electrospun polyamide nanofibers and can assemble type II collagen and chondroitin sulfate. The positive charge on nanofibers increases proteoglycan production (Gruber et al. 2009). Nanotechnology applications are playing a large role in developing intervertebral disc replacements.

FUTURE APPLICATIONS FOR NANOTECHNOLOGY IN SPINAL SURGERY

In this chapter, many applications of nanotechnology for treating diseases of the structural spine that affect vertebral bodies and discs have been explored. Applications such as HA injection and onlay grafting for bony augmentation are already being used clinically to improve fracture healing and bony regrowth. There are ongoing clinical trials and animal studies that are evaluating other new treatments utilizing nanotechnology. In 2002, a prospective randomized clinical trial comparing discectomy versus discectomy and autologous disc chondrocyte transplantation (ADCT) found that patients who received ADCT had greater pain reduction and demonstrated a significant increase in the fluid content of their treated and adjacent discs (Meisel et al. 2006). Injection of scaffolds to hold cells in place or to locate them centrally in the disc might enhance the effect of chondrocyte transplantation. Most of the potential applications of nanotechnology to diseases of the spine have only been tested in vitro or in animal studies. A few (Kim and Kim 2008) have been tested for repairing defects in the axial skeleton but not in the spine. The role of nanotechnology in treating spinal disorders is likely to evolve from solely providing novel methods of structural support and infection prevention to include nanotopological systems that promote cellular replacement therapy. These systems will preserve the motion of the spine, restore degenerative discs, and strengthen osteoporotic vertebrae using temporary biodegradable materials that support and direct the growth of cellular elements. The spinal column remains a dynamic structure in which normal cellular elements are reduced by diseases such as osteoporosis and DDD. Current therapy attempts to improve the dysfunction caused by these diseases but does not reverse the disease process. An alternative treatment strategy would aim to repair skeletal elements affected by osteoporosis and DDD by regenerating normal tissue using stem cells, growth factors, and biodegradable tissue scaffolds.

REFERENCES

Bai, B., Yin, Z., Xu, Q. et al. 2009. Histological changes of an injectable rhBMP-2/calcium phosphate cement in vertebroplasty of rhesus monkey. *Spine (Phila Pa 1976)* 34: 1887–92.

Battie, M.C., Haynor, D.R., Fisher, L.D. et al. 1995a. Similarities in degenerative findings on magnetic resonance images of the lumbar spines of identical twins. *J Bone Joint Surg Am* 77: 1662–70.

Battie, M.C., Videman, T., Gibbons, L.E. et al. 1995b. 1995 Volvo Award in Clinical Sciences. Determinants of lumbar disc degeneration. A study relating lifetime exposures and magnetic resonance imaging findings in identical twins. *Spine (Phila Pa 1976)* 20: 2601–12.

Boyd, L.M., and Carter, A.J. 2006. Injectable biomaterials and vertebral endplate treatment for repair and regeneration of the intervertebral disc. *Eur Spine J* 15 (Suppl 3): S414–21.

Browner, B.D., Levine, A.M., Jupiter, J.B. et al. 2009. *Skeletal Trauma: Basic Science, Management, and Reconstruction*. Philadelphia, PA: W. B. Saunders Company.

Choi, Y.S. 2009. Pathophysiology of degenerative disc disease. *Asian Spine J* 3: 39–44.

Chun, A.L., Moralez, J.G., Webster, T.J. et al. 2005. Helical rosette nanotubes: a biomimetic coating for orthopedics? *Biomaterials* 26: 7304–9.

Cohen, L.D. 1990. Fractures of the osteoporotic spine. *Orthop Clin North Am* 21: 143–50.

Cowan, J.A., Jr., Dimick, J.B., Wainess, R. et al. 2006. Changes in the utilization of spinal fusion in the United States. *Neurosurgery* 59: 15–20; discussion 15–20.

Di Martino, A., Vaccaro, A.R., Lee, J.Y. et al. 2005. Nucleus pulposus replacement: basic science and indications for clinical use. *Spine (Phila Pa 1976)* 30: S16–S22.

Friedman, R.J., An, Y.H., Ming, J. et al. 1996. Influence of biomaterial surface texture on bone ingrowth in the rabbit femur. *J Orthop Res* 14: 455–64.

Gruber, H.E., Hoelscher, G., Ingram, J.A. et al. 2009. Culture of human anulus fibrosus cells on polyamide nanofibers: extracellular matrix production. *Spine (Phila Pa 1976)* 34: 4–9.

Jayaraman, M., Meyer, U., Buhner, M. et al. 2004. Influence of titanium surfaces on attachment of osteoblast-like cells in vitro. *Biomaterials* 25: 625–31.

Kim, S.S., and Kim, B.S. 2008. Comparison of osteogenic potential between apatite-coated poly(lactide-co-glycolide)/hydroxyapatite particulates and Bio-Oss. *Dent Mater J* 27: 368–75.

Liao, S.S., Guan, K., Cui, F.Z. et al. 2003. Lumbar spinal fusion with a mineralized collagen matrix and rhBMP-2 in a rabbit model. *Spine (Phila Pa 1976)* 28: 1954–60.

Lipson, S.J., and Muir, H. 1981. Experimental intervertebral disc degeneration: morphologic and proteoglycan changes over time. *Arthritis Rheum* 24: 12–21.

Lotz, J.C., Colliou, O.K., Chin, J.R. et al. 1998. Compression-induced degeneration of the intervertebral disc: an in vivo mouse model and finite-element study. *Spine (Phila Pa 1976)* 23: 2493–506.

Meisel, H.J., Ganey, T., Hutton, W.C. et al. 2006. Clinical experience in cell-based therapeutics: intervention and outcome. *Eur Spine J* 15 (Suppl 3): S397–405.

Modic, M.T., and Ross, J.S. 2007. Lumbar degenerative disk disease. *Radiology* 245: 43–61.

Nachemson, A., Lewin, T., Maroudas, A. et al. 1970. In vitro diffusion of dye through the end-plates and the annulus fibrosus of human lumbar inter-vertebral discs. *Acta Orthop Scand* 41: 589–607.

Nesti, L.J., Li, W.J., Shanti, R.M. et al. 2008. Intervertebral disc tissue engineering using a novel hyaluronic acid–nanofibrous scaffold (HANFS) amalgam. *Tissue Eng Part A* 14: 1527–37.

Nussbaum, D.A., Gailloud, P., and Murphy, K. 2004. A review of complications associated with vertebroplasty and kyphoplasty as reported to the Food and Drug Administration medical device related web site. *J Vasc Interv Radiol* 15: 1185–92.

Phillips, F.M. 2003. Minimally invasive treatments of osteoporotic vertebral compression fractures. *Spine (Phila Pa 1976)* 28: S45–53.

Rao, R.D., and Singrakhia, M.D. 2003. Painful osteoporotic vertebral fracture. Pathogenesis, evaluation, and roles of vertebroplasty and kyphoplasty in its management. *J Bone Joint Surg Am* 85-A: 2010–22.

Riggs, B.L., and Melton, L.J., III. 1995. The worldwide problem of osteoporosis: insights afforded by epidemiology. *Bone* 17: 505S–11S.
Sambrook, P.N., MacGregor, A.J., and Spector, T.D. 1999. Genetic influences on cervical and lumbar disc degeneration: a magnetic resonance imaging study in twins. *Arthritis Rheum* 42: 366–72.
Smith, L.J., Swaim, J.S., Yao, C. et al. 2007. Increased osteoblast cell density on nanostructured PLGA-coated nanostructured titanium for orthopedic applications. *Int J Nanomed* 2: 493–9.
Urban, J.P., and Roberts, S. 2003. Degeneration of the intervertebral disc. *Arthritis Res Ther* 5: 120–30.
Vlad, M.D., del Valle, L.J., Barraco, M. et al. 2008. Iron oxide nanoparticles significantly enhances the injectability of apatitic bone cement for vertebroplasty. *Spine (Phila Pa 1976)* 33: 2290–8.
Wei, G., Jin, Q., Giannobile, W.V. et al. 2007. The enhancement of osteogenesis by nano-fibrous scaffolds incorporating rhBMP-7 nanospheres. *Biomaterials* 28: 2087–96.
Wilke, H.J., Heuer, F., Neidlinger-Wilke, C. et al. 2006. Is a collagen scaffold for a tissue engineered nucleus replacement capable of restoring disc height and stability in an animal model? *Eur Spine J* 15 (Suppl 3): S433–8.
Wilke, H.J., Kavanagh, S., Neller, S. et al. 2001. Effect of a prosthetic disc nucleus on the mobility and disc height of the L4–5 intervertebral disc postnucleotomy. *J Neurosurg* 95: 208–14.
Xu, H.H., Weir, M.D., and Simon, C.G. 2008. Injectable and strong nano-apatite scaffolds for cell/growth factor delivery and bone regeneration. *Dent Mater* 24: 1212–22.
Zhang, L., Chen, Y., Rodriguez, J. et al. 2008. Biomimetic helical rosette nanotubes and nanocrystalline hydroxyapatite coatings on titanium for improving orthopedic implants. *Int J Nanomed* 3: 323–33.

25 Nanoparticle-Based Treatment and Imaging of Brain Tumors

Potentials and Limitations

Bong-Seop Lee, Christopher J. Hanson, Rina Amatya, Esther Chung, Mansur Ghani, Allen Yen, Keith L. Black, and John S. Yu

CONTENTS

INTRODUCTION

Despite intensive therapeutic interventions, including surgery, radiation, and chemotherapy, brain tumors remain one of the most difficult cancers to treat and cure. The median overall survival of glioblastoma multiforme (GBM), the most malignant form of human astrocytoma, does not exceed 15 months after first-line therapy (Stupp et al. 2007). Gliomas infiltrate the functioning parenchyma, causing surgical resection to jeopardize healthy brain tissue and neurological function (Cha 2006). In more than 90% of GBM cases, tumors recur within 2 years of the first-line therapy, usually at the original irradiated site.

Advancements in the chemotherapy of central nervous system (CNS) tumors are much slower than those of other cancers, as even the most effective drug therapy regimens only increase the 2-year survival by 5% (Stewart 2002). Challenges arise from the inability of many compounds to permeate the specialized vascular system of the brain, called the blood–brain barrier (BBB), and to avoid detection by efflux pumps of multidrug resistance (MDR) pathways prevalent in tumor cells (Juillerat-Jeanneret 2008). Treating brain tumors via systemically administered drugs is very difficult because the BBB hinders the transport of active compounds into the brain, thus decreasing their concentration and therapeutic effectiveness. Approximately 98% of small-molecule drugs and nearly all high-molecular-weight biotechnological products, such as proteins, peptides, antisense oligonucleotides, and gene-based drugs, cannot cross the BBB (Pardridge 2005).

Significant advancements in the field of nanoparticle engineering indicate that delivery of promising chemotherapeutics drugs and contrast agents using nanoparticle may yield significant strides in brain tumor treatment. Nanoparticles are solid colloids ranging from 1 to 1000 nm that consist of a bulk matrix of material within which an active compound can be dissolved, entrapped, or externally attached. This allows the conjugation of desired therapeutics with a vast range of peptides, antibodies, polymers, and ligands that confer advantageous traits to the nanoparticles, such as an altered surface charge or a targeted delivery to cell-presenting antigens, and includes methods of permeating the BBB and avoiding the MDR pathway (Juillerat-Jeanneret 2008).

BRAIN TUMOR

Overview

There are over 120 different forms of brain cancer, including meningiomas, astrocytomas, gliomas, pituitary tumors, acoustic neuromas, and GBM (National Brain Tumor Society 2011). Tumors are classified by cell origin, rate of growth, and malignancy. In terms of origin, there are two major classifications: primary and secondary. Primary brain tumors (PBTs) are a heterogeneous group of malignancies that originate and reside within the brain. PBTs form in the brain's tissues, blood vessels, bony and fatty coverings, and ventricle linings. Sometimes cancer cells originating from elsewhere in the body travel through the bloodstream or lymphatic system, which transports them into the brain (secondary brain tumor).

For classification of PBTs, the rate of growth (and thereby malignancy) is assigned a grade level between grade I and grade IV, with grade IV representing the most aggressive tumors with the worst prognosis. Statistically, grade IV gliomas have been the most frequent primary intraparenchymal tumors in the elderly, whereas medulloblastomas have the highest incidence in children (Vescovi et al. 2006). Adding to the critical exigency for effective treatment, GBM or grade IV astrocytoma, the most aggressive malignant PBT, accounts for about 15% of all primary malignant brain tumors (National Cancer Institute 2011). Most succumb to GBM within 3–5 months of diagnosis; other forms of brain cancer display only slightly better prognoses.

Current Treatments

Earlier stages of intracranial cancer are hard to detect, allowing progression to a form that is more difficult to treat. By the time altered neurological function is noticed, the tumor has likely developed into a significant mass, with the potential to migrate into the neuraxis (Gajjar et al. 1995). The quick growth and resilience of glioblastoma cells cause the development of large tumors that endanger and affect the anatomical structures critical for basic motor, cognitive, reflexive, and other functions performed by the brain. This and the dangerous tendency for indistinct secondary tumor development entail the necessity of quick follow-up therapy.

Current treatment of brain tumors follows the conventional multimodal approach to removing and stopping tumor proliferation. The multimodal approach consists of an invasive surgery for tumor removal followed by noninvasive adjuvant (chemotherapy/radiation) therapy for suppression of possible secondary tumor growths. Complete surgical resection of the tumor, to eradicate every tumor cell, is impossible. In many cases, some of the tumor is left intact—a practice called surgical debulking—to avoid damaging critical structures in the brain (Lesniak and Brem 2004). Successful operations will resect cancerous tissue without damaging crucial structures near the tumor, but they can never be expected to remove all tumor cells.

Radiation therapy (RT) can be used to kill the remaining cancer cells and has also shown reasonable effectiveness in halting the growth of early-stage tumors (Sheline 1977). Owing to whole-brain RT-induced radiation necrosis, involved field radiation therapy (IFRT) utilizing focal radiation fields has become the standard treatment protocol (Choucair et al. 1986). Further enhanced specificity has been achieved with three-dimensional (3D) conformal RT (Thornton et al. 1991); intensity-modulated RT avoids radiation-sensitive anatomy. Stereotactic radiosurgery involves substantially higher dosages of irradiation than IFRT and is applied to smaller tumors (Nwokedi et al. 2002). Despite vast technological improvements, adult patients often experience a decline in cognitive function and encounter the risk of secondary malignancy in irradiated areas (Armstrong et al. 2002). Moreover, pediatric patients later display signs of severely impaired brain development (Taphoorn 2003).

Some treatment methods involve the combination of therapeutic methods such as chemotherapy and radiation, which is currently the standard of care. The DNA-alkylating drug temozolomide (TMZ) has proven to be the most effective and best-tolerated agent for treating malignant gliomas. The combination of TMZ and IFRT is advantageous, posing a 27.2% 2-year survival compared to 10.9% with IFRT alone (Stupp et al. 2009). Other chemotherapy treatments include carmustine (bis-chloroethylnitrosourea [BCNU]; via DNA and RNA alkylation) (Brandes et al. 2004) and bevacizumab, a monoclonal antibody that binds vascular endothelial growth factor and inhibits angiogenesis (Vredenburgh et al. 2007).

Another conventional treatment method involves local delivery of chemotherapeutic drugs by direct infusion via catheter or by implantation of gel wafers with an entrapped antineoplastic agent. These processes require surgical implantation and are therefore a risky treatment option (Valtonen et al. 1997). However, using a Gliadel wafer (a BCNU-loaded biodegradable polymer) lengthens survival time and controls symptoms of high-grade gliomas longer than surgery or radiation alone (McGirt et al. 2009).

Limitations of Conventional Drug Treatment

Multidrug Resistance

MDR is a condition in which chemotherapy fails because cancer cells are resistant to certain drugs in the treatment regimen. MDR occurs when certain integral membrane transporter proteins, which are responsible for expelling drugs from cells, are overexpressed on the surface of cancer cells (Juranka et al. 1989). The system uses P-glycoprotein (Pgp), an ATP-dependent transporter that is sometimes appropriately referred to as an MDR-associated protein. Pgp is a product of the MDR gene in humans that actively transports a wide range of lipid-soluble products before they can cross into the brain parenchyma (Loscher and Potschka 2005). The drug verapamil caused increased drug accumulation in certain patients by competitively inhibiting Pgp (Miller et al. 1991). Another Pgp blocker, valspodar, considerably increases the therapeutic effect of paclitaxel when coadministered with it (Fellner et al. 2002).

Cells develop MDR to a variety of drugs. Not all cancer cells in an individual express MDR transporters; target cells that are not resistant are killed, leaving behind drug-resistant cells to grow and proliferate. When chemotherapy fails, it is often attributable to MDR causing residual drug-resistant cells to dominate the tumor population. Fortunately, nanocarriers that are internalized by cells through specific ligand–receptor targeting interactions bypass the activity of these MDR transporters (Peer and Margalit 2006).

Blood–Brain Barrier

The BBB is a highly regulated interface between the peripheral circulation and the CNS that is formed by the unique lining of endothelial cells around cerebral capillaries. The BBB is a dynamic regulator of ion balance, a facilitator of nutrient transport, and a barrier to potentially harmful molecules (Hawkins and Davis 2005). Its ability to act as a physical barrier is attributed to tight cell–cell junctions between the microvascular endothelium, characterized by a lack of fenestrations, minimal pinocytotic activity, and increased mitochondrial content, which provide energy for specific BBB transport proteins (Figure 25.1).

The "neurovascular unit" ensuring the health and function of the CNS consists of this endothelium, along with astrocytes, pericytes, neurons, and the extracellular matrix (Figure 25.1). Astrocytes and pericytes have been shown to be critical to the induction of BBB properties. Astrocytes secrete chemical agents such as transforming growth factor, glial-derived neurotrophic factor, and basic fibroblast growth factor that induce aspects of the BBB phenotype in endothelial cells (Abbott et al. 2006), while pericytes attached to the abluminal surface of endothelial cells can induce endothelial expression of occludin. Occludin, claudins, and junctional adhesion molecules are transmembrane proteins that contribute to the structure of tight junctions, seal the paracellular pathway, and contribute to the high transendothelial electrical resistance that restricts passage of small ions such as Na^+ and Cl^-. Endothelial cells are surrounded by a 30- to 40-nm-thick basal membrane composed of collagen type IV, heparin sulfate proteoglycan, laminin, and fibronectin, providing physical support and controlling cellular migration (Hawkins and Davis 2005).

Although some solutes may successfully diffuse through the membrane, enzymes in mitochondria-dense endothelial cells rapidly degrade most peptides (Brown 1978). Furthermore, Pgp and multispecific organic anion transporters are both found in high concentrations in the

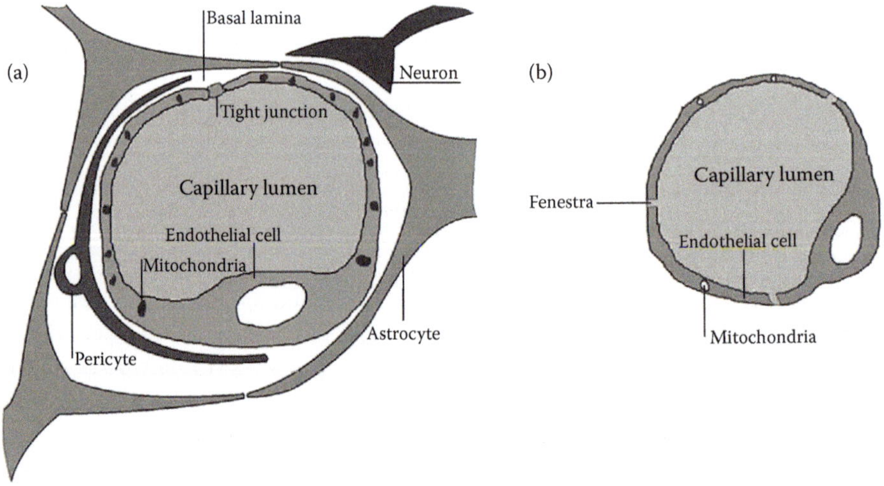

FIGURE 25.1 BBB. (a) BBB vasculature characterized by tight cellular junctions, presence of astrocytes and pericytes, endothelial cells lacking fenestrations, and higher mitochondria content. (b) Normal blood vessel.

cytoplasm of the BBB's endothelial cells (Cordon-Cardo et al. 1989). Both transporters actively expulse perceived endotoxins and exotoxins along with, disappointingly, most drug treatments for brain metastases, affective disorders, epilepsy, neurodegenerative disorders, infections, and human immunodeficiency virus (HIV) encephalopathy (Hendrikse et al. 2001).

The BBB prevents foreign molecules—except those that are small, nontoxic, uncharged, and lipid-soluble—from penetrating the barrier by passive diffusion. For a small-molecule drug to cross the BBB in a pharmacologically significant amount, it must have not only lipid solubility but also a molecular mass below the 400- to 500-Da threshold (Pardridge 2005). Moreover, permeability through the BBB exponentially decreases with every addition of a pair of hydrogen bonds to the chemical structure of a drug. Medicinal chemistry has been used to increase drug lipophilicity, which allows easier penetration of the BBB. However, this increases penetration of the drug across all membranes, thereby decreasing targeting capabilities and decreasing the effects of drug uptake on brain tumor tissue. These hindrances to drug delivery render the most potentially effective therapeutic drugs useless in treating CNS conditions, presenting one of the most significant challenges in neuroscience today. While few drugs, such as TMZ, have proven to permeate the BBB, their mechanism of accomplishment is unknown, and such success is rare.

Blood–Tumor Barrier

Under certain pathological conditions, the integrity of the BBB is disrupted, allowing transport of drugs across the BBB. However, vascular disruption in the BBB is localized to the tumor; anticancer drugs cannot penetrate the healthy BBB of the tumor's surrounding tissue and thus fail to effectively reach the proliferating mass. The blood–tumor barrier (BTB) is the barrier that forms between the circulating blood and the tumor tissue. While the BBB is impaired in most glioma cases, residual BTB function generally limits the distribution of antineoplastic agents to subtherapeutic levels. In one study, the average permeability of radioactive molecules into brain metastases was 2%–5% that of peripheral breast tumor, as detected by autoradiography (Lockman et al. 2010).

The heterogeneity of BTB blood vessels leads to spotty drug delivery. Furthermore, as a tumor grows in the area, the relative vascular surface area decreases, leading to reduced diffusion across the vessel endothelium (Jain 1989). Intracapillary distance increases, forcing the drugs to diffuse across greater lengths, making drug accumulation less likely in neoplastic target tissue (Misra et al. 2003).

Blood–Cerebrospinal Fluid Barrier

One way for systemically administered drugs to enter the brain parenchyma is to first pass the epithelium of the choroid plexus (CP) and accumulate in the cerebrospinal fluid (CSF). The CP consists of an extensive capillary network within a single layer of epithelium interconnected by tight junctions as seen in the BBB, effectively creating a blood–cerebrospinal fluid barrier (BCSFB) (Engelhardt and Sorokin 2009). The CP is a critical factor in the transport of peptide drugs because it is the site of both CSF production and exchange between the CSF and the extracellular fluid (ECF). This membrane is sealed primarily by tight junctions and is composed of ependymal cells that have microvilli on the CSF side, basolateral folding, and plentiful mitochondria. The BCSFB acts through the epithelial expression of the multidrug-resistant Pgp and the MDR-associated protein. A relative brain–CSF barrier also arises because the distance between the CSF and brain interstitial fluid is too great for any molecule to diffuse through (Pardridge 1998). Therefore, successful accumulation in the CSF does not necessarily mean that the drug will penetrate into the brain tissue.

Traditional Methods for Overcoming the BBB

Pharmaceutical researchers have attempted various tactics to overcome the BBB and to deliver drugs to the brain. Transcranial drug delivery methods and chemical and biochemical methods are conventional clinical techniques used to deliver drugs to the brain parenchyma. However, as detailed below, these methods lack the safety and specificity desired for a proper and effective drug delivery to the CNS.

Transcranial Delivery Methods

Transcranial drug delivery is a method used to circumvent the BBB and to achieve therapeutic levels of active agents in localized areas of the brain by craniotomy. This approach allows administration of drugs by intracerebroventricular (ICV) injection, intracerebral (IC) injection, and convection-enhanced delivery (CED). The effectiveness of both ICV and IC injections is drastically lessened by their reliance on free diffusion as a method of transport through the parenchyma and into the afflicted area. In one study of ICV injection, a limited amount was administered at a single-point depot site and affected less than 1% of the total brain volume, while concentrations at depths of 1–2 mm dropped to 2% of administered levels (Pardridge 2005). These deficiencies prove that this technique is clearly insufficient to treat most brain diseases.

CED utilizes convection, instead of free diffusion, by applying a pressure gradient and has been reported to show clear advantages over ICV injection (Koo et al. 2006). The process allows the discharge of high concentrations over a large distance with a more constant concentration profile and well-defined diffusion boundaries. CED enables the delivery of a variety of substances, such as conventional chemotherapeutic agents, monoclonal antibodies, viruses, and nanocarriers (Allard et al. 2009). Negative results highlighting the limitations of the technology have been reported, as the brain was not designed for significant intraparenchymal volume flow. Studies have shown the procedure to lead to diffuse white matter astrogliosis and demyelination around the catheter and along white matter tracts (Ai et al. 2003).

Each transcranial method for overcoming the BBB is limited by the need for a craniotomy to administer the drug and for a perpetually maintained wound to continually dispense it over time. While this has exhibited some success in the past, a more effective method must undoubtedly be developed.

Chemical and Biochemical Methods

A classical method for attempting to permeabilize the BBB is transient osmotic disruption, which is performed by shrinking endothelial cells with a hyperosmolar agent such as mannitol. This results in a reversible opening of the tight junctions of the BBB that allows free passage of drugs in the systemic circulation. Such a method is approved and often used to supplement chemotherapy. However, since mannitol will permeabilize the BBB indiscriminately, it enables the entry of harmful toxins and unwanted molecules that the BBB is intended to inhibit. There is increased uptake of plasma proteins (most notably albumin), which are very toxic to the cells of the disrupted areas of the brain (Pardridge 2005). One study indicated that mannitol induced electrographic and behavioral seizures in one-quarter of patients immediately after BBB disruption despite anesthetic dosages of the anti-epileptic drug thiopental (Marchi et al. 2007). Similarly, biochemical opening of the BBB can be induced by bradykinin analogs such as RMP-7 (Bartus et al. 1996) and by the selective bradykinin B1 agonist Sar-[D-Phe8]des-Arg9BK (Cardoso et al. 2004).

Like cell membranes in general, the BBB may be opened chemically using solvents such as high-dose ethanol and dimethyl sulfoxide, or detergents such as sodium dodecyl sulfate and polysorbate-80 (Borisenko 1988; Pan et al. 2008; Saija et al. 1997). Freund's adjuvant was found to recruit the immune system to the injection site and to cause an inflammatory response, opening the BBB and allowing the recruited antibodies to enter the brain. Ultrasound has been found to deliver a localized and reversible image-guided disruption of the BBB when performed in the presence of a circulating agent consisting of preformed gas bubbles, if the ultrasound wave distortion of the skull is particularly corrected (Hynynen et al. 2005). Cytokines and chemokines can attract white blood cells to disrupt the BBB (Anthony et al. 1998), and proper use of hypothermic saline infusions and high-dose free fatty acids or manipulation of intracarotid pH has each been shown to contribute to permeation (Oztas and Kucuk 1995; Sztriha and Betz 1991).

A drug's permeability across the BBB strongly correlates with its lipophilicity. A majority of large hydrophilic drugs do not cross the BBB. These include most biotechnological large-molecule products such as monoclonal antibodies, peptides, antisense oligonucleotides, or recombinant proteins. Therefore, a possible strategy to improve the BBB permeability of drugs is to make them more lipophilic by chemical derivatization. Lipophilic analogs of small hydrophilic drugs may more readily cross the BBB. However, this strategy has often been disappointing, as shown in the lipophilic analogs of nitrosoureas. In a quantitative structural activity relationship (QSAR) study, it was demonstrated that the antineoplastic activity of the lipophilic analogs of nitrosoureas was inversely proportional to their lipophilicity because the more lipophilic analogs are less soluble in the aqueous physiological milieu and bind readily to plasma proteins, lowering the concentration of available drugs.

Enhanced Permeability and Retention Effect

When tumor cells multiply into large-enough clusters to form 2- to 3-mm masses, angiogenesis is induced to meet the tumor's ever-increasing oxygen and nutrition demands (Folkman 1971). This causes new capillaries with frequent fenestrations to form in the tumor, causing loss of the permselectivity of the tumor mass due to vessel pores up to 550 nm in diameter (Jain et al. 2007). Leaky tumor microvessel walls and heterogeneous hyperpermeability are part of the enhanced permeability and retention (EPR) effect of solid tumors (Maeda et al. 1994). The creation of dilated leaky blood vessels causes extravasation into the tumor tissue and sequestration of crucial cell growth components. Tumors exhibit poor venous return; once substances leave the blood vessels and enter the tumor tissue, they have great difficulty backing out (Maeda et al. 2000). This is a novel mechanism that tumors employ when recruiting compounds for unmediated growth. However, scientists can use this to their advantage: the tissue's poor lymphatic drainage ensures the retention of macromolecules, such as nanoparticles, polymeric drugs, and lipid particles, in the solid tumor (Iyer et al. 2006). Manipulation of the EPR effect has thus become the "gold standard of anticancer drug design" in the treatment of cancerous tumors.

NANOPARTICLES FOR TREATING AND IMAGING BRAIN TUMORS

A promising recent advancement in the treatment of brain tumors is the rapid development of the field of nanotechnology, where research is directed toward an efficacious delivery of contrast or therapeutic agents precisely to the affected area. Brain tumor nanoparticle research has focused on developing materials or devices at the nanometer scale (1–300 nm) that are composed of solid colloidal particles (commonly of phospholipids or insoluble polymers) conjugated to a desired agent by adsorption, adsorption, encapsulation, or covalent linkage (Gilmore et al. 2008; Kumar 2000). Nanoparticles are largely developed to deliver a compound to the brain tumor site, such as magnetic resonance imaging (MRI) contrast agents, fluorescent dyes, chemotherapeutic agents, or photosensitizers (Orringer et al. 2009). The conventional administration is peroral or intravenous delivery of these compounds unaltered, resulting in indiscriminate exposure to both tumor and nontumor tissues and potentially causing unwanted systemic side effects (Iyer et al. 2006). Interest in this drug delivery mechanism stems from the small size of nanoparticles and its benefits to drug delivery, as well as the large surface area/volume ratio, which allows greatly increased amounts of functionalized drugs.

Encapsulating a drug within the matrix of a nanoparticle masks MDR antigens from recognition by cancer cells, greatly enhancing delivery (Koo et al. 2006). Effective imaging and therapy via systemic delivery also suffer from the impediments of the BBB, BTB, and BCSFB (Lesniak and Brem

2004). Precise nanoparticle development can lead to permeation of these barriers and increased targeting selectivity for brain tumor cells.

Methods of transport through the BBB include passively diffusing across irregular defects via EPR effect or targeting systems of carrier-mediated transport, receptor-mediated endocytosis, and adsorption-mediated endocytosis by attachment of endogenous peptides, ligands, or monoclonal antibodies (Beduneau et al. 2007). Assembling recognition molecules on the surface of nanoparticles can also result in tumor-specific delivery systems with mechanisms for locating tumor cells or associated vasculature, fibroblasts, and immune cells by targeting highly expressed cancer-specific ligands and antigens (Juillerat-Jeanneret 2008).

A huge impediment to the effective systemic delivery of nanoparticles is their appropriation by the monocytes and macrophages of the reticuloendothelial system (RES) and their uptake into the liver and spleen. Aggregation of complement proteins and immunoglobulins, which alert the RES, depends on the surface properties of the nanoparticles—largely on their level of hydrophobicity (Juillerat-Jeanneret 2008). Coating nanoparticles with poly(ethyleneglycol) (PEG) is a widely employed method for eluding the RES (Moghimi and Szebeni 2003), thereby reducing protein adsorption and surface opsonization. This creates longer systemic clearance times, so that a sudden delivery at therapeutic levels can be performed without administering the compound at a toxic concentration.

Drug loading of therapeutic agents is accomplished via absorption within the nanoparticle matrix during nanoparticle preparation, adsorption to the surface based on surface characteristics (hydrophobic/hydrophilic), encapsulation within a micelle compartment, or covalent linkage between functional groups. The consequences of the preparation methods on the physiochemical properties of nanoparticles should be well characterized because a precise understanding of nanoparticle properties is needed to comprehend biocompatibility and drug delivery efficiency.

Biodegradation of the encompassing matrix determines the nature of drug release into the body and greatly affects its therapeutic effectiveness. Drug escape mechanisms can be related to anionic–cationic surface charge and the pH of the environment; thus, many drug release approaches have been engineered based on pH effects between cellular compartments such as lysosomes and surface-functionalized drug-loaded nanoparticles (Gu et al. 2008). Other release mechanisms depend on enzymatic polymer degradation, erosion, desorption, or diffusion, as well as on nanoparticle reaction to the tumor environment.

Nanoparticle-Based Anticancer Treatments

The high incidence of brain tumors and the particular challenges inherent in treating the CNS have led to an extensive amount of laboratory research into the molecular pathogenesis of brain tumors, particularly targets for effective chemotherapy (Sathornsumetee et al. 2007). Cancer drugs can target the cell cycle, cellular DNA, cytoskeleton, vasculature, growth receptors, and cellular signaling pathways. Limitations in chemotherapy now result less from ineffectual drugs and more from lack of drug selectivity and delivery past the BBB. Nanoparticle engineering has the potential to solve the problems related to the BBB and MDR—impediment of delivery and drug selectivity. Common compositions of nanoparticle matrices include organic and inorganic polymers, phospholipids, various metals, ceramics, and carbon shaped into tubes, rods, spheres, and heterogeneous and homogeneous bulk structures. The choice depends on intended use (Ferrari 2005). The following is a discussion of commonly preferred nanoparticles for enhanced chemotherapy delivery to brain tumors.

Liposomes

Liposomes are spherical, self-enclosed nanoscale vesicles formed by one or several concentric phospholipid bilayers. Such an arrangement creates an aqueous core for sequestering and masking therapeutic drugs, which protects from nonspecific binding and early degradation, and allows increased permeation of the cerebral endothelium and delivery to the brain parenchyma (Albrecht et al. 2001). Liposomes suffer from short clearance time and low transport rate, but the addition of a PEG coating and targeting ligands can increase systemic residence time and further protect against premature degradation (Yang 2010). Novel advancements are being made with liposomes. One group designed a cysteine-cleavable phospholipid PEG to mask coupled targeting ligands from RES clearance and to enhance passive targeting of doxorubicin to the tumor. After PEG chains detach at the target site, folate ligands on the liposomes are exposed, enabling receptor-mediated uptake (McNeeley et al. 2009).

Solid Lipid Nanoparticles

Solid lipid nanoparticles (SLNs) are nontoxic solid lipid matrices prepared by the high-pressure homogenization of lipid microemulsions and stabilized by surfactants, which provide controlled drug release. The SLN design allows for increased endocytosis of nanoparticles across the BBB, likely by recruiting adsorption of circulating plasma proteins onto its surface (Nagayama et al. 2007). SLNs readily allow modification with active ligands. SLNs with incorporated doxorubicin and polysorbate coating increased drug concentrations 50-fold in glioblastoma cells compared to a free drug solution (Steiniger et al. 2004).

Polymers

Polymeric nanoparticles are colloidal particles consisting of matrices (generally from 10 to 1000 nm in diameter) made of natural or artificial polymers with desirable qualities of biocompatibility and biodegradability and within which therapeutic drugs can be contained. They are the most commonly researched materials for nanoparticle drug therapy because they can be easily chemically modified to trap drugs in a way that results in preferential drug efflux (Lockman et al. 2002). Research has shown the ability of poly(lactic acid) (PLA), poly(lactic-*co*-glycolic acid) (PLGA), poly(butyl cyanoacrylate) (PBCA), polycaprolactone (PCL), chitosan, poly(alkylcyanoacrylate) (PACA), poly(lysine), and polyaspartic acid (PAsp) to comprise the nanoparticle core matrix (Gref et al. 2003).

In 1999, doxorubicin bound to PBCA nanoparticles coated with 1% polysorbate-80 was tested in an in vivo rat model. This formulation allowed BBB permeation and indicated that the brain concentration of doxorubicin was over 60-fold greater than free drug administration (Gulyaev et al. 1999). Many polymer nanoparticles incorporate polysorbate-80 as surfactant coating because it has shown 20-fold higher amounts of endocytotic uptake by BBB endothelial cells than uncoated particles (Ramge et al. 2000).

Polymer Micelles

Polymer micelles are made from the spontaneous association of amphiphilic copolymers. This results in a hydrophobic inner core in an aqueous phase, allowing hydrophobic compounds to achieve therapeutic activity by masking against degradation and rapid clearance. The hydrophobic portion of the copolymer is a polyester like PLA, a poly(L-amino acid) like PAsp, or a phospholipid like phosphatidylethanolamine (PE). The hydrophilic portions are commonly polyethers such as poly(propylene oxide) (PPO) and PEG (Beduneau et al. 2007).

One in vivo rat study used PEG with a cholesterol-terminated end and conjugation of a trans-activating transcriptional activator (TAT) peptide for brain targeting. This TAT–PEG–cholesterol was a 200-nm micelle, with the chemotherapeutic antibiotic ciprofloxacin encapsulated in the hydrophobic cholesterol core. It effectively crossed the BBB (Liu et al. 2008). Another study compared the growth-inhibitory effects on glioblastoma of a micelle with an incorporated SN-38 chemotherapeutic agent to the effects of the unprotected prodrug CPT-11. The micelle-induced growth inhibition was 34- to 444-fold more effective than the CPT-11-induced growth inhibition, with markedly enhanced and prolonged distribution (Kuroda et al. 2009) (Table 25.1).

TABLE 25.1
Common Chemotherapeutics Effective for Use in a Nanoparticle Drug Carrier System, with Laboratory-Tested Nanoparticle Formulations and Modes of Attack of Chemotherapeutics at the Tumor Site

Chemotherapeutic Agent	Nanoparticle Carrier	Mode of Attack	Reference
Doxorubicin	PEGylated phospholipid	Inhibition of topoisomerase II and RNA polymerase II, generating reactive oxygen species	McNeeley et al. (2009)
Paclitaxel	PEGylated PTMC	Inhibition of mitotic spindle formation; stimulation of the secretion of interleukin-1b and tumor necrosis factor-α	Jiang et al. (2011)
Camptothecin	Amphiphilic cyclodextrin	Intercalation into chromosomal DNA; inhibition of topoisomerase I	Cirpanli et al. (2011)
Topotecan	Non-PEGylated liposome	Intercalation into chromosomal DNA; inhibition of topoisomerase I	Grahn et al. (2009)
Carmustine	Catanionic SLNs	DNA alkylation, forming lethal double-strand cross-links	Kuo and Liang (2011)
TMZ	PCBA	DNA methylation, causing cycling of DNA mismatch repair pathway, in turn causing DNA double-strand breaks	Tian et al. (2011)
Carboplatin	Liquid crystalline cubic phases	Inhibition of cell growth by forming interstrand DNA cross-links	Von Eckardstein et al. (2005)
5-Fluorouracil	PAMAM dendrimer	Misincorporation of fluoronucleotides into RNA and DNA; inhibition of the nucleotide synthetic enzyme thymidylate synthase	Bhadra et al. (2003)
Irinotecan	Polymer micelle	Intercalation into chromosomal DNA; inhibition of topoisomerase I	Kuroda et al. (2009)

Note: PTMC = poly(trimethylene carbonate); PCBA = poly(*n*-butyl cyanoacrylate); PAMAM = polyamidoamine.

Dendrimers

Dendrimers are macromolecules containing symmetrically arranged 3D nanoscale branches, which arise from a multifunctional core and grow by outward-controlled polymerization, to which a precise number of terminal groups are added stepwise. They have a very low polydispersity index and high functionality, allowing a high drug load bound at the surface (Esfand and Tomalia 2001). Polyamidoamine (PAMAM) dendrimers receive the most research attention owing to their easy drug-conjugating ability, high water solubility, and biocompatibility. A 2007 in vivo study used the brain-targeting ligand transferrin (Tf) conjugated to PAMAM via PEG, yielding PAMAM–PEG–Tf. Results suggested that it effectively and quickly penetrated the BBB and targeted the brain via noninvasive administration attributed to the transepithelial capability of PAMAM dendrimers and Tf-induced receptor-mediated endocytosis (Huang et al. 2007).

Other Nanoparticles for Killing Cancer Cells

Nanoparticle Photodynamic Therapy

Photodynamic therapy (PDT) was developed to overcome some fundamental issues associated with classical cancer therapy, as it imparts little systemic morbidity and is not affected by MDR systems. This method consists of the selective delivery of photosensitizers to the brain tumor site, where—when excited with the appropriate associated wavelength of visible or near-infrared (NIR) light—it produces a reactive oxygen species that will react readily with most other molecules, resulting in chemical destruction of nearby tissue (Tang et al. 2005). A common PDT method is vascular targeting to produce irreversible damage on the tumor vascular system, leading to ischemia and tumor necrosis (Reddy et al. 2006). Improved photosensitizer delivery to the tumor site is possible with nanoparticles.

In vivo tests on implanted 9L gliomas, using the photosensitizer porfimer sodium (Photofrin) in a polyacrylamide nanoparticle, showed that nanoparticle encapsulation provided a significant survival benefit (median, 33 vs. 13 days) (Reddy et al. 2006). A 2010 study reported a tumor-targeting PDT nanoparticle with methylene blue succinimidyl ester as a photosensitizer conjugated to polyacrylamide with a PEGylated surface and tumor-specific F3 peptides. It was 20–30 nm in diameter, exhibited optimal singlet oxygen production, and effectively targeted and killed a 9L tumor cell line (Hah et al. 2011).

Photothermal Ablation

Magnetic particle hyperthermia (MPH) is a method for inducing tumor necrosis that is commonly delivered in the form of nanoshells formed from a silica inner core and a metallic outer layer, often of gold. The characteristic wavelength of light that the nanoshell absorbs or scatters can be altered by changing the ratio of the inner-shell composition to the outer-shell composition, a characteristic referred to as "spectral tunability." MPH allows absorbance of light in the NIR region, which is optimal for transmitting light through tissue. A nanoshell can produce heat when it receives an appropriate wavelength of light or radiation; when targeted correctly, this heat can be used for thermal ablation of tumors (Dickerson et al. 2008). Currently in development is a nanoshell of gold silica conjugated with an antibody against interleukin-13 receptor-α2 (IL13Rα2). Glioma cells that took up the nanoshell conjugate in vitro were photothermally ablated when subsequently exposed to NIR laser light (Bernardi et al. 2008). They demonstrated selective killing of only the high-grade glioma cells expressing IL13Rα2. These are promising results, since nanoshells are capable of surface modification with targeting moieties, are small enough to accumulate via the EPR effect, and only induce cell death in regions exposed to NIR light. Also, cancer cells developing resistance to thermal ablation is very unlikely (Bernardi et al. 2008).

Nucleotide Delivery

Traditional gene therapy has relied on the viral vector as the delivery mechanism of genes to target cells. Viral vectors are effective but present the risks of reverting to wild type and inducing an immunogenic reaction. Also, viral vectors are excluded from passing through the BBB.

Nanoparticles have thus been considered as an alternative for gene delivery because a correct formulation with cationic molecules allows negatively charged DNA to bind. These nanoparticles were made of polyethyleneimine, dextran, calcium phosphate, and chitosan (Juillerat-Jeanneret 2008). In vivo tests have proven the applicability of this concept. One group formulated a cationic PEGylated nanoparticle with conjugated albumin for adsorption-mediated endocytosis across the BBB. They incorporated a plasmid encoding proapoptotic Apo2 ligand/tumor necrosis factor–related apoptosis-inducing ligand (Apo2l/TRAIL). Release of pDNA in the nucleus and induction of apoptosis occurred 48 h after injection, significantly delaying tumor growth (Lu et al. 2006).

Gene expression can also be altered by using nanoparticles as transport vehicles for antisense oligonucleotides. The principle of antisense therapy is the sequence-specific binding of an antisense oligonucleotide to target mRNA, which then prevents gene translation (Tamm et al. 2001). One group studied the delivery of antisense oligonucleotides in an anionic pH-sensitive liposome to NG108-15 brain tumor cells in vitro (Skalko-Basnet et al. 2002). pH-sensitive liposomes are stable at physiological pH (7.4) but destabilize—releasing their contents—when exposed to an acidic environment, such as the low pH found in endosomes (5.0–6.0) (Simoes et al. 2004). pH-sensitive liposomes showed significantly better cellular uptake than conventional liposomes (Table 25.2).

TABLE 25.2

Clinical Trials on the Benefits and Effects of Nanoparticle-Based Brain Tumor Treatment, with Nanoparticle Formulation and Incorporated Chemotherapeutics or Other Modes of Attack

Nanoparticle Carrier	Mode of Attack	Clinical Trial
Poly(isohexyl cyanoacrylate) nanoparticles	Doxorubicin	Phase 1 (Kattan et al. 1992)
Human serum albumin colloid	Paclitaxel	Phase 1 (Ibrahim et al. 2002)
Polymer NK105 micelle	Paclitaxel	Phase 1 (Hamaguchi et al. 2007)
Polyethylene glycol liposome	Doxorubicin	Phase 1 (Hau et al. 2004)
Cationic liposome	Interferon β gene therapy	Phase 1 (Yoshida et al. 2004)
Aminosilane SPIO	Photothermal ablation	Phase 2 (van Landeghem et al. 2009)
		Phase 2 (Maier-Hauff et al. 2011)
Polyethylene glycol liposome	Doxorubicin	Phase 2 (Fabel et al. 2001)

Nanoparticles for Brain Tumor Imaging

There is a clear and distinct correlation between the ability to visualize the CNS and the ability to effectively treat its disorders. MRI, with its ability to utilize differences in the magnetic dipole properties of tissues, is the commonest and furthest developed strategy for clearly imaging the CNS and provides the best images with the most precise contrast of any modality. The most commonly used MRI contrast agent is paramagnetic gadolinium [Gd(III)]. Gd(III) exhibits many favorable characteristics, as well as undesirable complicating factors. For example, it has an elimination half-life of only 1.6 h, does not cross the BBB, and tends to image areas of prominent BBB breakdown instead of being preferential to the tumor. Owing to a weak paramagnetic effect, it produces a hyperintense signal on T_1-weighted magnetic resonance (MR) images (Winer et al. 2011). Many of these complications can be alleviated with the use of new imaging agents and the flexibility of nanoparticles.

Superparamagnetic Iron Oxide

Nanoparticle formulations for administering contrast agents have almost exclusively incorporated iron oxide. In fact, iron oxide's popularity has become so vast that it has since been applied in nanoparticle systems for cellular therapy, drug delivery, magnetically induced hyperthermia, tissue repair, and magnetic separation, and as sensor for biomolecules (McCarthy and Weissleder 2008). However, its primary application has been as contrast agent for MRI, where it is classified either as superparamagnetic iron oxide (SPIO; 120–180 nm), such as ferumoxides, or as ultrasmall superparamagnetic iron oxide (USPIO; 20–50 nm), such as ferumoxtran-10 and ferumoxytol (Hunt and Neuwelt 2010). Such small diameters are beneficial to slowing opsonization and endocytosis. The addition of dextran coating, with its innate affinity for iron oxide, imbues the nanoparticle with a greatly increased half-life in the range of 24–30 h, which allows imaging for many hours after administration (Neuwelt et al. 2004). It also affords simple bonding of ligands or functional groups and facilitates uptake through cell transport systems (Winer et al. 2011). USPIO shows a tendency to congregate in intracranial tumors owing to uptake by phagocytic cells at tumor margins, but only in tumors with an active inflammatory component (Varallyay et al. 2002). As concentration increases, USPIO produces an increased signal intensity on T_1-weighted images but a decreased intensity on T_2-weighted images, both of which are advantageous characteristics for MR tumor delineation (Sun et al. 2008). An advantage to iron oxide contrast agents over gadolinium is that they are detectable with iron stain in biopsy specimens and with electron microscopy in pathological studies. It has been shown that ferumoxides, ferumoxtran-10, and ferumoxytol are nontoxic to the brain parenchyma, and that the iron is found only in inflammatory cells at the tumor surface (Muldoon et al. 2005). Like many nanoparticles for delivery of therapeutics, PEG attachment protects USPIO

TABLE 25.3

Clinical Trials on the Benefits and Effects of Nanoparticle-Based Brain Tumor Imaging Treatments

Nanoparticle Carrier	Imaging Agent	Clinical Trial
Dextran-SPIO	Ferumoxides and ferumoxtran	Phase 1 (Varallyay et al. 2002)
Dextran-USPIO	Ferumoxtran-10	Phase 1 (Enochs et al. 1999)
Dextran-USPIO	Ferumoxtran-10	Phase 1 (Hunt et al. 2005)
Dextran-USPIO	Ferumoxtran-10	Phase 1 (Murillo et al. 2005)
Dextran-USPIO	Ferumoxtran-10	Phase 1 (Neuwelt et al. 2004)

Note: Most employed iron oxide-based nanoparticles for MRI.

nanoparticles from uptake by the RES and increases systemic circulation. Iron oxide nanoparticles have been conjugated to tumor-specific antibodies (to produce MRI enhancement of LX-1 brain tumors in mice) (Remsen et al. 1996), tumor cell-permeating drugs such as methotrexate (Kohler et al. 2006), and targeting peptides such as chlorotoxin and F3 (Veiseh et al. 2005). Chlorotoxin is an effective tumor-targeting ligand because it binds to many cancer cells, including glioma and medulloblastoma, via matrix metalloproteinase-2 (Sun et al. 2008) (Table 25.3).

Intraoperative MRI

The long systemic half-life of USPIO allows initial delineation of tumor boundaries preoperatively, as well as MRI for surgical assistance intraoperatively. This yields updated images of the progress of operation while maintaining the patient in the operating room and provides the ability to determine the size and location of craniotomy and the extent of resection (Hunt and Neuwelt 2010). In contrast, the short half-life of gadolinium forces repeated administration throughout the operation and suffers from the pitfall of MRI enhancement within surgical disruptions in the brain instead of solely concentrating around the tumor (Knauth et al. 2001).

The use of fluorescence to define tumor boundaries during an operation may help resect the tumor completely. Iron oxide nanoparticles have been designed to incorporate the molecule Cy5.5, which will fluoresce in the NIR region, allowing real-time updates of resected tumor margins. This nanoparticle was localized within tumor and inflammatory cells, with no uptake into healthy tissue, in the 9L model (Trehin et al. 2006). Quantum dots, composed of a metalloid crystalline core and a shielding shell, are another popular fluorescence agent owing to their size-tunable narrow emission spectra in the 400- to 2000-nm range and their ability to receive pharmacokinetic/targeting surface modification (Gao et al. 2005).

Nanoparticles for Treatment and Imaging

The ability of nanoparticles to incorporate multiple pharmacological molecules has led to the discovery that a single nanoparticle can provide both imaging and treatment functions. Recent studies investigated a PEGylated liposome carrying gadolinium diethylenetriaminepentaacetic acid as an MR agent and doxorubicin as a chemotherapeutic agent. It imaged brain blood vessels and the cancerous lesion at high resolution for up to 3 h and showed localization of liposomes within the lesion 48 h after injection, which was later confirmed by histological analysis. This provides the advantage of tracking the fate of systemically administered nanoparticle therapy (Karathanasis et al. 2007). In 2006, in vivo tests were performed on a light-activated theranostic polyacrylamide nanoagent incorporating iron oxide for MRI and the photosensitizer Photofrin with a PEG coating (Reddy et al. 2006). The targeting mechanism was F3 peptide, which binds to nucleolin on the surface of tumor cells and induces receptor-mediated endocytosis (McCarthy and Weissleder 2008). MRI showed significant enhancement of the tumor 10 min after injection, and targeting provided a tumor residence half-life of 123 min. PDT was effective, killing nearly 90% of targeted cells and resulting in two of five animals becoming disease-free after 6 months.

Another compelling advantage to incorporating iron oxide into nanoparticles is its superparamagnetic property, which allows nanoparticles to be guided by an externally positioned magnet, actively increasing their deposition at the target site. An ambitious application of this trait was the study of nanoparticles for the synergistic delivery of epirubicin for therapy and of iron oxide for MRI monitoring, with induction of BBB permeability via focused ultrasound and targeting of tumor bed with magnetic targeting. Calculations showed that these passive and active targeting strategies increased delivery to the tumors by at least an order of magnitude, as determined by MRI monitoring (Liu et al. 2010).

Future of Nanoparticles

Clinical application of nanoparticles to image and treat brain tumors is still a goal, not the reality. Numerous issues must be resolved before the vast multitudes of nanoparticles are deemed safe for human use. Nanoparticles do not affect organs, tissues, and cells in the same way that larger particles of the same composition do; thus, evaluation of nanoparticle toxicity cannot be extrapolated from the toxicity of bulk material. Also, the ability of nanoparticles to be diversely composed limits the determination of toxicity because the combination of various drugs, targeting moieties, and structural molecules will affect the body differently than each individual constituent would (Juillerat-Jeanneret 2008). Many constituents are eliminated via the liver and excreted into bile, risking the possibility of long-term toxicity from extended build-up. Thus, an appropriate and universal approach to the clearance and toxicity of complex nanoparticles must be established.

Clinical approval of nanoparticles will depend on our ability to create effective, repeatable, and precise formulations. Our effort needs to be directed toward (Orive et al. 2010)

- Achieving a precise targeted delivery to the tumor bed, vasculature, or extracellular matrix while avoiding exposure to healthy cells
- Detecting newly formed neoplasms earlier with improved imaging processes, tumor identification, and tumor delineation
- Enhancing the ability to track and label cancer cells in real time during surgical resection to assist in defining the entire tumor location
- Positively supplementing and engaging the host immune system to invade tumors

These can be achieved by developing suitable screening methodologies to determine the ideal characteristics of nanocarriers to optimize their abilities. Future research needs to focus on only the most promising drug therapies and the nanoparticle carriers most appropriate for their encapsulation and delivery, in ways that are less invasive and more selective for brain tumors. Finding delivery methods that can substitute for intravenous administration should be emphasized, as this is highly demanded by patients. Simplicity in design is critical because increased complexity will increase the risk of adverse reactions, complicate the scale-up process, and increase cost.

CONCLUSIONS

Treatment of intracranial tumors is complicated by the intricate system of barriers between the systemic circulation and the brain, CSF, and tumor itself, characterized by tight cellular junctions between endothelial cells and the presence of degrading enzymes, which inhibit the majority of compounds from diffusing through. A further challenge to effectively killing proliferating tumor cells is the presence of specific integral membrane proteins overexpressed in cancer cells, which expel recognized chemotherapeutics and imbue the tumor with MDR. One method of overcoming these inhibitions and successfully delivering cancer-killing compounds to their site of function is by incorporating them into nanoparticles. A compelling aspect of nanoparticles is their ability to achieve numerous purposes by incorporating different functional compounds—those for therapy, those that target particular areas of the body, those that hide nanoparticles from macrophage uptake, and those that can send and receive signals from outside the body. They will advance the field of diagnosis, enlighten surgeons during operation, destroy neoplasms, and prevent metastasis. As our knowledge of nanoparticle engineering grows, they will play an ever expanding role in brain tumor management.

REFERENCES

Abbott, N., Ronnback, L., and Hansson, E. 2006. Astrocyte–endothelial interactions at the blood–brain barrier. *Nat Rev Neurosci* 7: 41–53.

Ai, Y., Markesbery, W., Zhang, Z. et al. 2003. Intraputamenal infusion of GDNF in aged rhesus monkeys: distribution and dopaminergic effects. *J Comp Neurol* 461: 250–61.

Albrecht, K.W., de Witt Hamer, P.C., Leenstra, S. et al. 2001. High concentration of daunorubicin and daunorubicinol in human malignant astrocytomas after systemic administration of liposomal daunorubicin. *J Neuro-Oncol* 53: 267–71.

Allard, E., Passirani, C., and Benoit, J.P. 2009. Convection-enhanced delivery of nanocarriers for the treatment of brain tumors. *Biomaterials* 30: 2302–18.

Anthony, D., Dempster, R., Fearn, S. et al. 1998. CXC chemokines generate age-related increases in neutrophil-mediated brain inflammation and blood–brain barrier breakdown. *Curr Biol* 8: 923–6.

Armstrong, C.L., Hunter, J.V., Ledakis, G.E. et al. 2002. Late cognitive and radiographic changes related to radiotherapy: initial prospective findings. *Neurology* 59: 40–8.

Bartus, R.T., Elliott, P.J., Dean, R.L. et al. 1996. Controlled modulation of BBB permeability using the bradykinin agonist, RMP-7. *Exp Neurol* 142: 14–28.

Beduneau, A., Saulnier, P., and Benoit, J.P. 2007. Active targeting of brain tumors using nanocarriers. *Biomaterials* 28: 4947–67.

Bernardi, R.J., Lowery, A.R., Thompson, P.A. et al. 2008. Immunonanoshells for targeted photothermal ablation in medulloblastoma and glioma: an in vitro evaluation using human cell lines. *J Neuro-Oncol* 86: 165–72.

Bhadra, D., Bhadra, S., Jain, S. et al. 2003. A PEGylated dendritic nanoparticulate carrier of fluorouracil. *Int J Pharm* 257: 111–24.

Borisenko, S.A. 1988. [Effect of carbidopa and dimethyl sulfoxide on the permeability of the hemato-encephalic barrier for ^{14}C-tyrosine and ^{14}C-DOPA in rats]. *Biull Eksp Biol Med* 106: 686–8.

Brandes, A.A., Tosoni, A., Amista, P. et al. 2004. How effective is BCNU in recurrent glioblastoma in the modern era? A phase II trial. *Neurology* 63: 1281–4.

Brown, W.J. 1978. The capillaries in acute and subacute multiple sclerosis plaques: a morphometric analysis. *Neurology* 28: 84–92.

Cardoso, R.C., Lobao-Soares, B., Bianchin, M.M. et al. 2004. Enhancement of blood–tumor barrier permeability by Sar-[D-Phe8]des-Arg9BK, a metabolically resistant bradykinin B1 agonist, in a rat C6 glioma model. *BMC Neurosci* 5: 38.

Cha, S. 2006. Update on brain tumor imaging: from anatomy to physiology. *AJNR Am J Neuroradiol* 27: 475–87.

Choucair, A.K., Levin, V.A., Gutin, P.H. et al. 1986. Development of multiple lesions during radiation therapy and chemotherapy in patients with gliomas. *J Neurosurg* 65: 654–8.

Cirpanli, Y., Allard, E., Passirani, C. et al. 2011. Antitumoral activity of camptothecin-loaded nanoparticles in 9L rat glioma model. *Int J Pharm* 403: 201–6.

Cordon-Cardo, C., O'Brien, J.P., Casals, D. et al. 1989. Multidrug-resistance gene (P-glycoprotein) is expressed by endothelial cells at blood–brain barrier sites. *Proc Natl Acad Sci U S A* 86: 695–8.

Dickerson, E.B., Dreaden, E.C., Huang, X. et al. 2008. Gold nanorod assisted near-infrared plasmonic photothermal therapy (PPTT) of squamous cell carcinoma in mice. *Cancer Lett* 269: 57–66.

Engelhardt, B., and Sorokin, L. 2009. The blood–brain and the blood–cerebrospinal fluid barriers: function and dysfunction. *Semin Immunopathol* 31: 497–511.

Enochs, W.S., Harsh, G., Hochberg, F. et al. 1999. Improved delineation of human brain tumors on MR images using a long-circulating, superparamagnetic iron oxide agent. *J Magn Reson Imaging* 9: 228–32.

Esfand, R., and Tomalia, D.A. 2001. Poly(amidoamine) (PAMAM) dendrimers: from biomimicry to drug delivery and biomedical applications. *Drug Discov Today* 6: 427–36.

Fabel, K., Dietrich, J., Hau, P. et al. 2001. Long-term stabilization in patients with malignant glioma after treatment with liposomal doxorubicin. *Cancer* 92: 1936–42.

Fellner, S., Bauer, B., Miller, D.S. et al. 2002. Transport of paclitaxel (Taxol) across the blood–brain barrier in vitro and in vivo. *J Clin Invest* 110: 1309–18.

Ferrari, M. 2005. Cancer nanotechnology: opportunities and challenges. *Nat Rev Cancer* 5: 161–71.

Folkman, J. 1971. Tumor angiogenesis: therapeutic implications. *N Engl J Med* 285: 1182–6.

Gajjar, A., Bhargava, R., Jenkins, J.J. et al. 1995. Low-grade astrocytoma with neuraxis dissemination at diagnosis. *J Neurosurg* 83: 67–71.

Gao, X., Yang, L., Petros, J.A. et al. 2005. In vivo molecular and cellular imaging with quantum dots. *Curr Opin Biotechnol* 16: 63–72.

Gilmore, J.L., Yi, X., Quan, L. et al. 2008. Novel nanomaterials for clinical neuroscience. *J Neuroimmune Pharmacol* 3: 83–94.

Grahn, A.Y., Bankiewicz, K.S., Dugich-Djordjevic, M. et al. 2009. Non-PEGylated liposomes for convection-enhanced delivery of topotecan and gadodiamide in malignant glioma: initial experience. *J Neuro-Oncol* 95: 185–97.

Gref, R., Couvreur, P., Barratt, G. et al. 2003. Surface-engineered nanoparticles for multiple ligand coupling. *Biomaterials* 24: 4529–37.

Gu, J., Cheng, W.P., Liu, J. et al. 2008. pH-triggered reversible "stealth" polycationic micelles. *Biomacromolecules* 9: 255–62.

Gulyaev, A.E., Gelperina, S.E., Skidan, I.N. et al. 1999. Significant transport of doxorubicin into the brain with polysorbate 80-coated nanoparticles. *Pharm Res* 16: 1564–9.

Hah, H.J., Kim, G., Lee, Y.E. et al. 2011. Methylene blue-conjugated hydrogel nanoparticles and tumor-cell targeted photodynamic therapy. *Macromol Biosci* 11: 90–9.

Hamaguchi, T., Kato, K., Yasui, H. et al. 2007. A phase I and pharmacokinetic study of NK105, a paclitaxel-incorporating micellar nanoparticle formulation. *Br J Cancer* 97: 170–6.

Hau, P., Fabel, K., Baumgart, U. et al. 2004. Pegylated liposomal doxorubicin—efficacy in patients with recurrent high-grade glioma. *Cancer* 100: 1199–207.

Hawkins, B.T., and Davis, T.P. 2005. The blood–brain barrier/neurovascular unit in health and disease. *Pharmacol Rev* 57: 173–85.

Hendrikse, N.H., Bart, J., de Vries, E.G. et al. 2001. P-glycoprotein at the blood–brain barrier and analysis of drug transport with positron-emission tomography. *J Clin Pharmacol* Suppl: 48S–54S.

Huang, R.Q., Qu, Y.H., Ke, W.L. et al. 2007. Efficient gene delivery targeted to the brain using a transferrin-conjugated polyethyleneglycol-modified polyamidoamine dendrimer. *FASEB J* 21: 1117–25.

Hunt, M.A., Bago, A.G., and Neuwelt, E.A. 2005. Single-dose contrast agent for intraoperative MR imaging of intrinsic brain tumors by using ferumoxtran-10. *AJNR Am J Neuroradiol* 26: 1084–8.

Hunt, M.A., and Neuwelt, E.A. 2010. Magnetic resonance imaging of brain tumors using iron oxide nanoparticles. In *Methods of Cancer Diagnosis, Therapy, and Prognosis, Vol. 8. Brain Tumors*, ed. M.A. Hayat, 297–304. Dordrecht: Springer Science + Business Media BV.

Hynynen, K., McDannold, N., Sheikov, N.A. et al. 2005. Local and reversible blood–brain barrier disruption by noninvasive focused ultrasound at frequencies suitable for trans-skull sonications. *Neuroimage* 24: 12–20.

Ibrahim, N.K., Desai, N., Legha, S. et al. 2002. Phase I and pharmacokinetic study of ABI-007, a Cremophor-free, protein-stabilized, nanoparticle formulation of paclitaxel. *Clin Cancer Res* 8: 1038–44.

Iyer, A.K., Khaled, G., Fang, J. et al. 2006. Exploiting the enhanced permeability and retention effect for tumor targeting. *Drug Discov Today* 11: 812–8.

Jain, R.K. 1989. Delivery of novel therapeutic agents in tumors: physiological barriers and strategies. *J Natl Cancer Inst* 81: 570–6.

Jain, R.K., di Tomaso, E., Duda, D.G. et al. 2007. Angiogenesis in brain tumours. *Nat Rev Neurosci* 8: 610–22.

Jiang, X., Xin, H., Sha, X. et al. 2011. PEGylated poly(trimethylene carbonate) nanoparticles loaded with paclitaxel for the treatment of advanced glioma: in vitro and in vivo evaluation. *Int J Pharm* 420: 385–94.

Juillerat-Jeanneret, L. 2008. The targeted delivery of cancer drugs across the blood–brain barrier: chemical modifications of drugs or drug-nanoparticles? *Drug Discov Today* 13: 1099–106.

Juranka, P.F., Zastawny, R.L., and Ling, V. 1989. P-glycoprotein: multidrug-resistance and a superfamily of membrane-associated transport proteins. *FASEB J* 3: 2583–92.

Karathanasis, E., McNeeley, K., Agarwal, A. et al. 2007. MR trackable, chemotherapeutic nanoparticles for patient specific glioma therapy. *Microsc Microanal* 13: 44–45.

Kattan, J., Droz, J.P., Couvreur, P. et al. 1992. Phase I clinical trial and pharmacokinetic evaluation of doxorubicin carried by polyisohexylcyanoacrylate nanoparticles. *Invest New Drugs* 10: 191–9.

Knauth, M., Egelhof, T., Roth, S.U. et al. 2001. Monocrystalline iron oxide nanoparticles: possible solution to the problem of surgically induced intracranial contrast enhancement in intraoperative MR imaging. *AJNR Am J Neuroradiol* 22: 99–102.

Kohler, N., Sun, C., Fichtenholtz, A. et al. 2006. Methotrexate-immobilized poly(ethylene glycol) magnetic nanoparticles for MR imaging and drug delivery. *Small* 2: 785–92.

Koo, Y.E., Reddy, G.R., Bhojani, M. et al. 2006. Brain cancer diagnosis and therapy with nanoplatforms. *Adv Drug Deliv Rev* 58: 1556–77.

Kumar, M.N.V.R. 2000. Nano and microparticles as controlled drug delivery devices. *J Pharm Pharm Sci* 3: 234–58.

Kuo, Y.C., and Liang, C.T. 2011. Inhibition of human brain malignant glioblastoma cells using carmustine-loaded catanionic solid lipid nanoparticles with surface anti-epithelial growth factor receptor. *Biomaterials* 32: 3340–50.

Kuroda, J., Kuratsu, J., Yasunaga, M. et al. 2009. Potent antitumor effect of SN-38-incorporating polymeric micelle, NK012, against malignant glioma. *Int J Cancer* 124: 2505–11.

Lesniak, M.S., and Brem, H. 2004. Targeted therapy for brain tumours. *Nat Rev Drug Discov* 3: 499–508.

Liu, H.L., Hua, M.Y., Yang, H.W. et al. 2010. Magnetic resonance monitoring of focused ultrasound/magnetic nanoparticle targeting delivery of therapeutic agents to the brain. *Proc Natl Acad Sci U S A* 107: 15205–10.

Liu, L., Venkatraman, S.S., Yang, Y.Y. et al. 2008. Polymeric micelles anchored with TAT for delivery of antibiotics across the blood–brain barrier. *Biopolymers* 90: 617–23.

Lockman, P.R., Mittapalli, R.K., Taskar, K.S. et al. 2010. Heterogeneous blood–tumor barrier permeability determines drug efficacy in experimental brain metastases of breast cancer. *Clin Cancer Res* 16: 5664–78.

Lockman, P.R., Mumper, R.J., Khan, M.A. et al. 2002. Nanoparticle technology for drug delivery across the blood–brain barrier. *Drug Dev Ind Pharm* 28: 1–13.

Loscher, W., and Potschka, H. 2005. Drug resistance in brain diseases and the role of drug efflux transporters. *Nat Rev Neurosci* 6: 591–602.

Lu, W., Sun, Q., Wan, J. et al. 2006. Cationic albumin-conjugated pegylated nanoparticles allow gene delivery into brain tumors via intravenous administration. *Cancer Res* 66: 11878–87.

Maeda, H., Noguchi, Y., Sato, K. et al. 1994. Enhanced vascular permeability in solid tumor is mediated by nitric oxide and inhibited by both new nitric oxide scavenger and nitric oxide synthase inhibitor. *Jpn J Cancer Res* 85: 331–4.

Maeda, H., Wu, J., Sawa, T. et al. 2000. Tumor vascular permeability and the EPR effect in macromolecular therapeutics: a review. *J Controlled Release* 65: 271–84.

Maier-Hauff, K., Ulrich, F., Nestler, D. et al. 2011. Efficacy and safety of intratumoral thermotherapy using magnetic iron-oxide nanoparticles combined with external beam radiotherapy on patients with recurrent glioblastoma multiforme. *J Neuro-Oncol* 103: 317–24.

Marchi, N., Angelov, L., Masaryk, T. et al. 2007. Seizure-promoting effect of blood–brain barrier disruption. *Epilepsia* 48: 732–42.

McCarthy, J.R., and Weissleder, R. 2008. Multifunctional magnetic nanoparticles for targeted imaging and therapy. *Adv Drug Deliv Rev* 60: 1241–51.

McGirt, M.J., Than, K.D., Weingart, J.D. et al. 2009. Gliadel (BCNU) wafer plus concomitant temozolomide therapy after primary resection of glioblastoma multiforme. *J Neurosurg* 110: 583–8.

McNeeley, K.M., Karathanasis, E., Annapragada, A.V. et al. 2009. Masking and triggered unmasking of targeting ligands on nanocarriers to improve drug delivery to brain tumors. *Biomaterials* 30: 3986–95.

Miller, T.P., Grogan, T.M., Dalton, W.S. et al. 1991. P-glycoprotein expression in malignant lymphoma and reversal of clinical drug resistance with chemotherapy plus high-dose verapamil. *J Clin Oncol* 9: 17–24.

Misra, A., Ganesh, S., Shahiwala, A. et al. 2003. Drug delivery to the central nervous system: a review. *J Pharm Pharm Sci* 6: 252–73.

Moghimi, S.M., and Szebeni, J. 2003. Stealth liposomes and long circulating nanoparticles: critical issues in pharmacokinetics, opsonization and protein-binding properties. *Prog Lipid Res* 42: 463–78.

Muldoon, L.L., Sandor, M., Pinkston, K.E. et al. 2005. Imaging, distribution, and toxicity of superparamagnetic iron oxide magnetic resonance nanoparticles in the rat brain and intracerebral tumor. *Neurosurgery* 57: 785–96; discussion 785–96.

Murillo, T.P., Sandquist, C., Jacobs, P.M. et al. 2005. Imaging brain tumors with ferumoxtran-10, a nanoparticle magnetic resonance contrast agent. *Therapy* 2: 871–82.

Nagayama, S., Ogawara, K., Fukuoka, Y. et al. 2007. Time-dependent changes in opsonin amount associated on nanoparticles alter their hepatic uptake characteristics. *Int J Pharm* 342: 215–21.

National Brain Tumor Society. 2011. *Tumor Types*. http://www.braintumor.org/patients-family-friends/about-brain-tumors/tumor-types/.

National Cancer Institute. 2011. *Glioblastoma Multiforme. The Cancer Genome Atlas.* http://cancergenome.nih.gov/cancersselected/glioblastomamultiforme.

Neuwelt, E.A., Varallyay, P., Bago, A.G. et al. 2004. Imaging of iron oxide nanoparticles by MR and light microscopy in patients with malignant brain tumours. *Neuropathol Appl Neurobiol* 30: 456–71.

Nwokedi, E.C., DiBiase, S.J., Jabbour, S. et al. 2002. Gamma knife stereotactic radiosurgery for patients with glioblastoma multiforme. *Neurosurgery* 50: 41–6; discussion 46–7.

Orive, G., Ali, O.A., Anitua, E. et al. 2010. Biomaterial-based technologies for brain anti-cancer therapeutics and imaging. *Biochim Biophys Acta* 1806: 96–107.

Orringer, D.A., Koo, Y.E., Chen, T. et al. 2009. Small solutions for big problems: the application of nanoparticles to brain tumor diagnosis and therapy. *Clin Pharmacol Ther* 85: 531–4.

Oztas, B., and Kucuk, M. 1995. Intracarotid hypothermic saline infusion: a new method for reversible blood–brain barrier disruption in anesthetized rats. *Neurosci Lett* 190: 203–6.

Pan, W., Barron, M., Hsuchou, H. et al. 2008. Increased leptin permeation across the blood–brain barrier after chronic alcohol ingestion. *Neuropsychopharmacology* 33: 859–66.

Pardridge, W.M. 1998. CNS drug design based on principles of blood–brain barrier transport. *J Neurochem* 70: 1781–92.

Pardridge, W.M. 2005. The blood–brain barrier: bottleneck in brain drug development. *NeuroRx* 2: 3–14.

Peer, D., and Margalit, R. 2006. Fluoxetine and reversal of multidrug resistance. *Cancer Lett* 237: 180–7.

Ramge, P., Unger, R.E., Oltrogge, J.B. et al. 2000. Polysorbate-80 coating enhances uptake of polybutylcyanoacrylate (PBCA) nanoparticles by human and bovine primary brain capillary endothelial cells. *Eur J Neurosci* 12: 1931–40.

Reddy, G.R., Bhojani, M.S., McConville, P. et al. 2006. Vascular targeted nanoparticles for imaging and treatment of brain tumors. *Clin Cancer Res* 12: 6677–86.

Remsen, L.G., McCormick, C.I., Roman-Goldstein, S. et al. 1996. MR of carcinoma-specific monoclonal antibody conjugated to monocrystalline iron oxide nanoparticles: the potential for noninvasive diagnosis. *AJNR Am J Neuroradiol* 17: 411–8.

Saija, A., Princi, P., Trombetta, D. et al. 1997. Changes in the permeability of the blood–brain barrier following sodium dodecyl sulphate administration in the rat. *Exp Brain Res* 115: 546–51.

Sathornsumetee, S., Reardon, D.A., Desjardins, A. et al. 2007. Molecularly targeted therapy for malignant glioma. *Cancer* 110: 13–24.

Sheline, G.E. 1977. Radiation therapy of brain tumors. *Cancer* 39: 873–81.

Simoes, S., Moreira, J.N., Fonseca, C. et al. 2004. On the formulation of pH-sensitive liposomes with long circulation times. *Adv Drug Deliv Rev* 56: 947–65.

Skalko-Basnet, N., Tohda, M., and Watanabe, H. 2002. Uptake of liposomally entrapped fluorescent antisense oligonucleotides in NG108 15 cells: conventional versus pH-sensitive. *Biol Pharm Bull* 25: 1583–7.

Steiniger, S.C., Kreuter, J., Khalansky, A.S. et al. 2004. Chemotherapy of glioblastoma in rats using doxorubicin-loaded nanoparticles. *Int J Cancer* 109: 759–67.

Stewart, L.A. 2002. Chemotherapy in adult high grade glioma: a systematic review and meta-analysis of individual patient data from 12 randomised trials. *Lancet* 359: 1011–8.

Stupp, R., Hegi, M.E., Gilbert, M.R. et al. 2007. Chemoradiotherapy in malignant glioma: standard of care and future directions. *J Clin Oncol* 25: 4127–36.

Stupp, R., Hegi, M.E., Mason, W.P. et al. 2009. Effects of radiotherapy with concomitant and adjuvant temozolomide versus radiotherapy alone on survival in glioblastoma in a randomised phase III study: 5-year analysis of the EORTC-NCIC trial. *Lancet Oncol* 10: 459–66.

Sun, C., Veiseh, O., Gunn, J. et al. 2008. In vivo MRI detection of gliomas by chlorotoxin-conjugated superparamagnetic nanoprobes. *Small* 4: 372–9.

Sztriha, L., and Betz, A.L. 1991. Oleic acid reversibly opens the blood–brain barrier. *Brain Res* 550: 257–62.

Tamm, I., Dorken, B., and Hartmann, G. 2001. Antisense therapy in oncology: new hope for an old idea? *Lancet* 358: 489–97.

Tang, W., Xu, H., Kopelman, R. et al. 2005. Photodynamic characterization and in vitro application of methylene blue-containing nanoparticle platforms. *Photochem Photobiol* 81: 242–9.

Taphoorn, M.J. 2003. Neurocognitive sequelae in the treatment of low-grade gliomas. *Semin Oncol* 30: 45–8.

Thornton, A.F., Jr., Hegarty, T.J., Ten Haken, R.K. et al. 1991. Three-dimensional treatment planning of astrocytomas: a dosimetric study of cerebral irradiation. *Int J Radiat Oncol Biol Phys* 20: 1309–15.

Tian, X.H., Lin, X.N., Wei, F. et al. 2011. Enhanced brain targeting of temozolomide in polysorbate-80 coated polybutylcyanoacrylate nanoparticles. *Int J Nanomed* 6: 445–52.

Trehin, R., Figueiredo, J.L., Pittet, M.J. et al. 2006. Fluorescent nanoparticle uptake for brain tumor visualization. *Neoplasia* 8: 302–11.

Valtonen, S., Timonen, U., Toivanen, P. et al. 1997. Interstitial chemotherapy with carmustine-loaded polymers for high-grade gliomas: a randomized double-blind study. *Neurosurgery* 41: 44–8; discussion 48–9.
van Landeghem, F.K., Maier-Hauff, K., Jordan, A. et al. 2009. Post-mortem studies in glioblastoma patients treated with thermotherapy using magnetic nanoparticles. *Biomaterials* 30: 52–7.
Varallyay, P., Nesbit, G., Muldoon, L.L. et al. 2002. Comparison of two superparamagnetic viral-sized iron oxide particles ferumoxides and ferumoxtran-10 with a gadolinium chelate in imaging intracranial tumors. *AJNR Am J Neuroradiol* 23: 510–9.
Veiseh, O., Sun, C., Gunn, J. et al. 2005. Optical and MRI multifunctional nanoprobe for targeting gliomas. *Nano Lett* 5: 1003–8.
Vescovi, A.L., Galli, R., and Reynolds, B.A. 2006. Brain tumour stem cells. *Nat Rev Cancer* 6: 425–36.
von Eckardstein, K.L., Patt, S., Kratzel, C. et al. 2005. Local chemotherapy of F98 rat glioblastoma with paclitaxel and carboplatin embedded in liquid crystalline cubic phases. *J Neuro-Oncol* 72: 209–15.
Vredenburgh, J.J., Desjardins, A., Herndon, J.E., II et al. 2007. Bevacizumab plus irinotecan in recurrent glioblastoma multiforme. *J Clin Oncol* 25: 4722–9.
Winer, J.L., Kim, P.E., Law, M. et al. 2011. Visualizing the future: enhancing neuroimaging with nanotechnology. *World Neurosurg* 75: 626–37; discussion 618–9.
Yang, H. 2010. Nanoparticle-mediated brain-specific drug delivery, imaging, and diagnosis. *Pharm Res* 27: 1759–71.
Yoshida, J., Mizuno, M., Fujii, M. et al. 2004. Human gene therapy for malignant gliomas (glioblastoma multiforme and anaplastic astrocytoma) by in vivo transduction with human interferon beta gene using cationic liposomes. *Hum Gene Ther* 15: 77–86.

26 Tumor-Targeted Nanodrugs for Treatment of Primary Brain Tumors

Joseph T. Cheatle and Michele R. Aizenberg

CONTENTS

INTRODUCTION

Primary brain tumors are diverse, exceedingly difficult to treat, and potentially devastating neoplasms. Central nervous system (CNS) tumors represent about 2% of all cancers, with an average annual rate of 14.4 per 100,000 persons (Fisher et al. 2007). Brain tumors are generally divided between gliomas and nongliomas. The most common gliomas are astrocytic tumors, oliogodendritic tumors, and ependymomas. The highest-grade lesions—glioblastomas—have a nearly universal dismal prognosis. Brain tumors are unique in their ability to infiltrate along white matter tracts, making it nearly impossible to fully resect the tumor. In addition, CNS tumors are protected by the blood–brain barrier (BBB). The standard of care for many primary brain tumors includes surgery, chemotherapy, and radiation treatment. Current chemotherapeutic treatments are admittedly inadequate and often carry unwanted and even life-threatening side effects. Research is being conducted on diverse fronts to combat these neoplasms. Nanodrugs may provide more efficacious treatments than those currently available by their unique ability to reach primary brain tumors at sufficient concentrations to adequately target the tumor without severe systemic toxicity. Nanodrugs are designed to provide a mechanism where the drug can be released predictably and, therefore, more efficiently and safely, with decreased adverse effects and, subsequently, target-specific tumor-killing effects.

The BBB is a complex system that severely limits the ability of most substances to enter the highly regulated environment bathing the CNS. Endothelial cells form tight junctions that—combined with the multitude of enzymes, receptors, transporters, and efflux pumps constituting the multidrug resistance (MDR) pathway—form the BBB. Once a molecule successfully penetrates the BBB, it can distribute within the entirety of the CNS. In designing drugs to cross the BBB, of utmost importance are either sufficient lipophilic properties, to allow passive diffusion across the BBB, or other properties, to target specific transcytosis mechanisms (Lee et al. 2001). Although many high-grade primary brain tumors have disrupted the BBB, as seen by their contrast enhancement on imaging studies, this does not mean that other components of the BBB, including MDR and intratumoral drug resistance mechanisms, are completely inactivated. For adequate treatment of many primary brain tumors, which are commonly infiltrative, new drugs must target not only tumor-specific transporters and markers at the visible primary tumor site but also nonvisualized infiltrative areas (Juillerat-Jeanneret 2008).

The BBB is inherently incompetent in some brain tumors and therefore allows for passive and nonspecific targeted therapy. Gaps between adjacent endothelial cells in such tumors—some as large as 80 nM—allow certain molecules that would otherwise not permeate through the BBB to reach the tumor in relatively higher concentrations and more specifically than the normal brain parenchyma, which remains protected by the intact BBB (Moghimi et al. 2005; Rabinow 2004; Shenoy et al. 2005; Torchilin 2005). This passive targeting is active specifically in tumors with increased angiogenesis and subsequent breakdown of endothelial tight junctions. Increased permeability is not seen in primary brain tumors without angiogenesis. In addition, permeability is diminished in areas of necrosis, which have decreased tissue perfusion. Regardless, BBB permeability in many primary brain tumors far exceeds that of the normal brain and provides a mechanism by which a tumor can be targeted with some specificity.

Many mechanisms of nanodrug delivery that are currently being investigated are more sophisticated than passive targeting of tumors. Many of these nanodelivery systems have in common the following: (1) a relatively hydrophilic shell; (2) a hydrophobic core; (3) the ability to bind ligand substrates to the shell to allow for substrate-modulated binding to the specific target; (4) an imaging marker to track the distribution of the nanodrug; and (5) the ability to store high concentrations of drug in the relatively protected core. These nanoparticles are often crafted to circulate for prolonged periods, to be sufficiently small to cross the BBB with relative ease, and to avoid multidrug resistance mechanisms.

Scavenger mechanisms for removing nanoparticles from the circulation primarily involve the reticuloendothelial system (RES). Opsonization, the binding of plasma proteins to a nanoparticle, results in recognition of the nanoparticle by the RES in the liver, spleen, and marrow, and in removal of circulating nanoparticles from the circulation. Poly(ethylene glycol) (PEG), a hydrophilic and inert polymer, provides a steric barrier to the surface of the nanoparticle that reduces opsonization. Nanoparticles with PEGylation are referred to as stealth nanoparticles because they are less likely to be detected by the RES when compared to non-PEGylated nanoparticles (Li and Huang 2010). Stealth nanoparticles circulate for extended periods and therefore result in increased tissue distribution (Olivier 2005).

The targeting of a drug to a specific tumor may be approached either by direct targeting or by pretargeting methods. Direct targeting methods utilize a targeting ligand attached directly to a nanoparticle. In contrast, when utilizing pretargeting methods, the ligand is administered before the drug-loaded carrier is delivered. The ligand is intended to be concentrated in the target tumor tissue. The nanoparticle is administered, attaches to the ligand, and concentrates in the pretargeted tissue. The pretargeting method results in a longer circulation time for the nanoparticle compared to a direct target nanoparticle (Harding et al. 1997). More rapid clearance occurs with a direct target nanoparticle because its attached ligand, such as an antibody, results in changed physiochemical properties of the nanoparticle (Pulkkinen et al. 2008).

Adding a molecule to the surface of a nanoparticle may target the nanoparticle to a specific ligand but may produce other effects as well. Coating of a nanoparticle by proteins, such as antibodies, reduces the number of other ligands that may be bound to its surface since the size of a single protein may approach the size of the nanoparticle's core. A small number of bound molecules prevent a favorable binding equilibrium with the ligand of the target cell (Juillerat-Jeanneret 2008; Soni et al. 2005). In addition, proteins may introduce immunogenicity that reduces the effectiveness of subsequent treatments. Binding of a positively charged ligand to the nanoparticle enables interaction with the negatively charged cell membrane to allow for adsorption-mediated endocytosis (Lu et al. 2007).

NANOPARTICLE DESIGN

Nanoparticles used to deliver drugs are submicroscopic colloidal systems that act as drug delivery vehicles. The structure of the nanoparticle defines the kinetics of drug release. The target drug can be dissolved, entrapped, encapsulated, or covalently bound to the nanoparticle (Figure 26.1). The nanoparticle surface can be modified to assist in efficient and selective drug delivery to target tissues and increased cellular uptake (Cengelli et al. 2006; Dubowchik and Walker 1999; Panyam and Labhasetwar 2004; Savic et al. 2003). One modification is PEG polymerized on the surface of a nanoparticle. PEG is highly water-soluble and can be combined or copolymerized with other polymers, allowing for control of mechanical properties and degradation rate, decreased toxicity, and immune surveillance (Lee et al. 2003; Lu and Chen 2004). The bonding between the nanoparticle and the target drug can be altered for selective release of the drug in the target tissue (Reents et al. 2002; Schoenmakers et al. 2004). Nanoparticles can be internalized into cells via fluid-phase endocytosis, receptor-mediated endocytosis, or phagocytosis (Juillerat-Jeanneret 2008).

Polymeric nanoparticles are colloidal particles encased in a biodegradable polymer (Allen and Cullis 2004; Beduneau et al. 2007; Duncan 2003; Feng et al. 2004; Ferrari 2005; Juillerat-Jeanneret 2006; Rabinow 2004). In vitro experiments with 29 different neural and nonneural cell lines found a more than 10-fold increase in targeted cytotoxicity with nanoparticles than with drugs alone (Feng et al. 2004). Polymers are versatile because they can be crafted as either nanoparticles, nanospheres, or nanocapsules. Polymer-coated nanoparticles were effective in transporting peptides across the BBB and in providing a sustained release of peptides in the target tissue for an extended time (Kreuter et al. 2003). Polymeric nanocapsules can be designed to have degradation times that are specific for a certain target tissue's milieu (Na and Bae 2002).

Polymeric micelles are aggregates of amphiphilic copolymers with a hydrophobic core that is shielded from the aqueous environment (Oh et al. 2004). This nanoparticle compound can deliver water-insoluble drugs smaller than 50 nM (Sahoo and Labhasetwar 2003). The hydrophobic drug is encapsulated in the hydrophobic core and can be delivered intravenously because of the aqueous solubility provided by the hydrophilic outer shell. This hydrophilic layer also increases circulation time. The outer layer can be modified with target ligands conjugated to the micelle surface.

Hydrated nontoxic phospholipids can form small, closed vesicles called liposomes. These liposomes are small (smaller than 100 nm) unilamellar vesicles encapsulated by a single lipid layer. These vesicles are high-potency carriers that can entrain several tens of thousands of drug molecules (Cavalcanti 2003). Stealth liposomes are made by attaching PEG to the surface of liposomes (Harris and Chess 2003). Liposomes that incorporate DNA can be used for gene transfer (Ogris and Wagner 2002).

Dendrimers are macromolecules with a central hydrophobic core and a highly branching three-dimensional structure. The structure derives from polymerized growth from a central core molecule with subsequent outward expansion by a series of polymerization reactions. The size of the dendrimer can be controlled exactly (Haag 2004; Tomalia and Frechet 2002; Tomalia et al. 1985). The branched macromolecule provides hundreds of terminal groups for conjugation with numerous chemical moieties.

Several additional classes of nanoparticles, including hydrogels, fullerenes, and nanoshells can be utilized. Hydrogels are polymers that form highly hydrophobic three-dimensional networks, are permeable, and can incorporate drug molecules. Water causes their molecular structure to swell

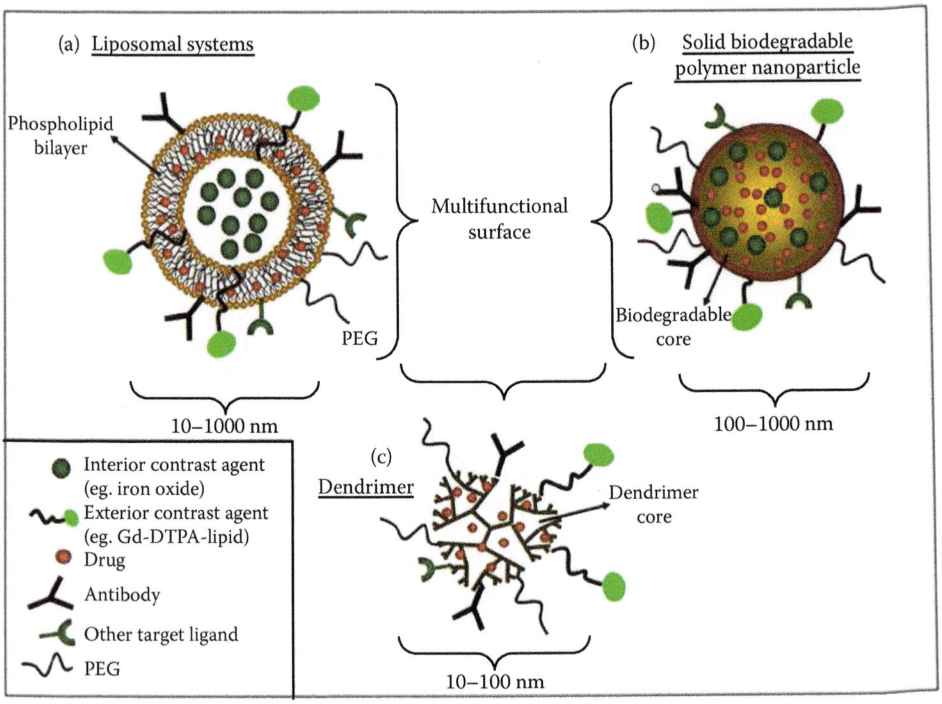

FIGURE 26.1 Schematic of typical nanoparticles that can be used for the delivery of chemotherapeutic agents in neuro-oncology. (a) Liposomal systems (b) solid biodegradable nonparticles, polymers (c) macromolecular dendrimer complexes. PEG, polyethylene glycol, Gd-DTPA, gadolinium diethylenetriaminepentaacetic acid. (From Fahmy, T.M. et al., *AAPS J* 9(2):E171–80, 2007. With permission.)

but not to dissolve (Leary et al. 2006). Fullerenes are carbon 60-based molecules with the potential to form a backbone to enable the formation of more complex nanostructures (Leary et al. 2006). Nanoshells are gold-layered dielectric nanoparticles with optical properties that can be controlled by varying the relative size of their constituent layers. Gold-layered molecules can be used as agents to mediate the thermal ablation of cancerous tissues (Hirsch et al. 2003).

PRECLINICAL APPLICATIONS

Doxorubicin

There have been extensive efforts to improve the efficacy of chemotherapeutic agents in treating malignant brain tumors, including the development of carmustine wafer implants (Lin and Kleinberg 2008), metronomic temozolomide (Clarke et al. 2009), convection-enhanced delivery, and multimodality treatments. The drug doxorubicin exhibits a cytotoxic effect on glioma cells and prolongs survival after intratumoral administration (Stan et al. 1999; Walter et al. 1995). Clinical use of unbound doxorubicin is limited by a high incidence of cardiac toxicity (Kreuter 2004). The ATP-binding casette transporter, a membrane transport protein, reduces intracellular concentrations of doxorubicin, attenuates drug effect on tumor cells, and limits the clinical efficacy of doxorubicin (von Holst et al. 1990). Doxorubicin does not normally pass the BBB but has been shown to be transported across it after binding to nanoparticles (Garcia-Garcia et al. 2005; Gelperina 2006; Kreuter 2006). Poly(butyl cyanoacrylate) (PBCA) coated with polysorbate 80, a highly biodegradable compound, has been used to encapsulate doxorubicin; the resulting nanoparticle can be administered systemically to provide long-term remission of glioblastoma in animal models (Steiniger et al. 2004). Uptake of PBCA nanoparticles coated with polysorbate 80 appears to occur by receptor-mediated endocytosis (Alyaudtin et al. 2001; Kreuter et al. 2002, 2003; Ramge et al. 2000; Steiniger et al. 2004). The encapsulation of doxorubicin by PBCA/polysorbate 80 also results in significantly reduced toxic effects on the heart and spermatogenic epithelium (Gelperina et al. 2002; Pereverzeva et al. 2007, 2008). Doxorubicin bound to PBCA nanoparticles and subsequently coated with polysorbate 80 diminished the proliferation of malignant glioma cells, tumor size, vascular density, necrosis, and microvascular proliferation in experimental rats (Hekmatara et al. 2009). Doxorubicin-loaded nanoparticles coated with polysorbate 80 led to a 40% cure rate in a glioma-infiltrated rat model in which untreated animals died uniformly 2 weeks after implantation (Gelperina et al. 2000). Doxorubicin-loaded nanoparticles have also been coated with poloxamine 908 and poloxamer 188. Poloxamine 908 imparted stealth characteristics that decreased hepatic uptake, prolonged circulation of nanoparticles, and secondarily enhanced target drug delivery to the tumor site (Moghimi and Hunter 2000). In contrast, poloxamer 188 was not an effective stealth agent and did not enhance the delivery of the target drug across the BBB (Moghimi 1995). Another nanoparticle formulation may greatly improve CNS drug delivery. Poly(methoxy-PEG cyanoacrylate-*co*-*n*-hexadecylcyanoacrylate) (PEGylated PHDCA) nanoparticles can penetrate the BBB more extensively than all other tested types of nanoparticles without modifying BBB permeability (Calvo et al. 2001).

Many drugs, including loperamide (Alyautdin et al. 1997), tubocurarine (Alyautdin et al. 1998), dalargin (Schroeder et al. 2000), and kyotorphin (Schroeder et al. 1998), have been successfully incorporated into PBCA nanoparticles coated with polysorbate 80. The actual mechanism by which surfactant-coated PBCA nanoparticles cross the BBB has not been completely elucidated. One mechanism is through the interaction of apolipoprotein A-I (ApoA-I) with the BBB. ApoA-I adsorbs to nanoparticles coated with PBCA, presumably because polysorbate 80 mimics low-density lipoproteins that promote absorption of ApoA-I. ApoA-I promotes endocytosis or transcytosis (Abbott and Romero 1996; Ramge et al. 2000) and interacts with scavenger receptor class B type I located on the BBB (Kreuter et al. 2002). Coating PBCA nanoparticles with polysorbate 80 promotes the uptake of a drug such as doxorubicin through the BBB and reduces drug elimination by avoiding transmembrane pumps that are components of the BBB (Petri et al. 2007). Nanoparticles loaded with the analgesic dalargin or loperamide require apolipoprotein B or E to cross the BBB (Kreuter et al. 2002). The role of ApoA-I in hepatic uptake is controversial, with one study reporting decreased hepatic uptake (Gessner et al. 2001) but with other pharmacokinetic studies reporting no effect on hepatic uptake (Ambruosi et al. 2005, 2006; Gulyaev et al. 1999; Panzenboeck et al. 2002). The drug loaded in the nanoparticle affects the distribution of the nanoparticle because drug payload affects the surface charge of the nanodrug. Empty PBCA nanoparticles have surface charges, body distribution, and BBB penetration different from those of doxorubicin-loaded PBCA particles (Ambruosi et al. 2005; Brigger et al. 2004; Panzenboeck et al. 2002). The infiltrating margin of primary brain tumors contains cells that are shielded from chemotherapy by a functioning BBB. To provide effective therapy for primary brain tumors, nanoparticles must be able to penetrate an intact BBB to eliminate tumor cells that infiltrate well beyond the core of the tumor.

Paclitaxel

Paclitaxel is an anticancer agent with in vitro activity against many tumor cell lines, including gliomas (Liebmann et al. 1993; Schiff et al. 1979); however, its poor solubility, narrow therapeutic window, and P-glycoprotein (P-gp) substrate specificity severely limit its widespread clinical application (Fellner et al. 2002; Heimans et al. 1994; Pulkkinen et al. 2008; Rowinsky et al. 1989; Sparreboom et al. 1997). Paclitaxel-loaded stealth nanoparticles have been targeted to cancer cells using avidin–biotin technology to overcome these problems. This targeting system is achieved by preliminary administration of biotinylated targeting ligands, followed by administration of avidin, and, finally, administration of a biotinylated drug or carrier. The avidin–biotin bond is the strongest known protein–ligand interaction (Green 1975). This technology has been applied for drug targeting, imaging technologies, and cancer therapy (Boerman et al. 2003; Breitz et al. 1999; Cremonesi et al. 1999; De La Fuente et al. 1997; Goodwin and Meares 2001; Hoya et al. 2001; Knox et al. 2000; Nobs et al. 2006; Paganelli et al. 1998; Rodriguez et al. 2007). Paclitaxel loaded in biotinylated nanoparticles and polylactic acid (PLA)/PEG–biotin polymers have been constructed and targeted against glioma cell line biotinylated transferrin receptors. Transferrin has been previously used for brain tumor targeting (Gijsens et al. 2002; Huwyler et al. 1996; Ishida et al. 2001; Maruyama et al. 2004; Shi et al. 2001). This targeting system resulted in a markedly increased antitumor activity of nanoparticle-bound paclitaxel compared to the form of paclitaxel currently in use (Pulkkinen et al. 2008).

Topoisomerase I Inhibitors

Camptothecin, a plant alkaloid, targets nuclear enzyme topoisomerase I, preventing its ability to transiently cleave and rejoin a DNA strand, which is essential for transcription and replication. Camptothecin is rapidly inactivated clinically by hydrolysis. Solid lipid nanoparticles loaded with camptothecin provided higher and more sustained brain camptothecin levels than the naked drug (Yang et al. 1999). Camptothecin-loaded nanoparticles significantly increased the survival of glioma-implanted rats (Cirpanli et al. 2011). The cytotoxicity of other topoisomerase inhibitors, including irinotecan and topotecan, increased when they were encapsulated in liposomes and administered by convection-enhanced delivery (Krauze et al. 2007; Saito et al. 2006; Yamashita et al. 2007).

Methotrexate

Methotrexate is a hydrophilic compound with poor CNS penetration that is commonly used as a chemotherapeutic agent. Methotrexate is generally administered intrathecally because of its inability to effectively cross the BBB. Loading methotrexate onto polysorbate 80-coated polyalkylcyanoacrylate nanoparticles increases the concentration of methotrexate in the cerebrospinal fluid and brain parenchyma compared to methotrexate administered alone and to methotrexate loaded onto nanoparticles without a polysorbate 80 coating (Gao and Jiang 2006).

Magnetic and Metallic Nanoparticles

Magnetic nanoparticles may be utilized to carry drugs across the BBB (McBain et al. 2008). Glioma cells actively take up these nanoparticles in vitro. Magnetic nanoparticles can infiltrate the extravascular interstitial space of gliomas after internal carotid injection (Alexiou et al. 2007; Pulfer et al. 1999). Magnetic paclitaxel nanoparticles (MPNPs) in glioma-bearing rats resulted in higher paclitaxel content in the brain tissue (compared with free paclitaxel) and in twofold prolongation of survival (Zhao et al. 2010). MPNPs can direct drug selectively to the tumor and avoid delivery to normal surrounding tissue. Magnetic nanoparticles have clinical uses for imaging as well (Shapiro et al. 2010).

Anisotropic Au–Ag nanoparticles were tested against several human solid tumor cell lines. Nanorods exposed to bovine serum albumin and conjugated with aptamers showed specificity for tumor cells. The absorption of light by Au–Ag nanorods releases heat and kills 93% of exposed cells (Huang et al. 2008). Nanoparticles can cross the BBB and attach to tumor cells through targeting molecules, suggesting that treating solid tumors with Au–Ag nanorods may be generalized to the treatment of glioma. Although interstitial treatment can be provided by laser irradiation (Vogl et al. 1999), other types of energy sources have the potential to deliver thermal energy through the intact skull to nanoparticles within primary brain tumors. These include radiofrequency (Gazelle et al. 2000; Mirza et al. 2001), microwave (Seki et al. 1999), and focused ultrasound energy (Jolesz and Hynynen 2002).

Nonsteroidal Anti-Inflammatory Drugs

Increased expression of cyclooxygenase-2 (COX-2) is associated with angiogenesis and resistance to many cytotoxic chemotherapy drugs (Dandekar et al. 2005; Kesari et al. 2005). Indomethacin, a nonsteroidal anti-inflammatory drug (NSAID), inhibits COX-2 and may impede malignant glioma growth by suppressing angiogenesis and drug resistance. In vitro studies showed a twofold increase in the cytotoxicity of glioma cell lines with indomethacin nanocapsules compared to indomethacin in solution. In addition, indomethacin-loaded nanocapsules resulted in significant antiproliferative effects without inducing necrotic cell death (Bernardi et al. 2008). A low dose of indomethacin nanocapsules utilized in an in vivo rat model inhibited the proliferation and viability of glioma cells, lowered the mitotic index, reduced the growth of implanted gliomas, and significantly increased survival (Bernardi et al. 2009). Nanocapsule delivery increased the intracerebral concentration of indomethacin to much higher levels than were possible with the drug alone (Bernardi et al. 2009). Normal hippocampal cells survived a dose of indomethacin nanocapsules 10-fold higher than that which inhibited glioma growth (Bernardi et al. 2008).

Targeted Radiodrug Delivery

Boron compounds, sodium borocaptate (BSH), and boronophenylalanine (BPA) are being investigated as potential therapeutic agents for brain tumors (Yang et al. 2008; Yokoyama et al. 2006). Epidermal growth factor receptor (EGFR) is an attractive therapeutic target because it is often amplified in glioblastoma multiforme (GBM) but remains weakly expressed or undetectable in normal brain tissue (Friedman and Bigner 2005; Sauter et al. 1996; Schwechheimer et al. 1995; Yang et al. 2008). BSH encapsulated in a nickel liposome conjugated with an antibody targeted against EGFR had an affinity for tumor tissue versus normal brain tissue in mice (Feng et al. 2009). Such effectiveness of immunoliposomal targeting with this agent depends on the degree of inherent BBB breakdown.

Polycefin

Nanoparticles are being studied as nonviral vectors that deliver DNA to tumor cells and provide gene therapy for treatment of primary brain tumors. Modified silica nanoparticles are as effective as viral vectors in gene delivery and have fewer toxic effects (Bharali et al. 2005). Polycefin is a nanoconjugate for the targeted delivery of antisense oligonucleotides and monoclonal antibodies specific for gliomas. The polycefin nanoparticle has a backbone of poly(β-L-malic acid), which is conjugated to the following: (1) antisense oligonucleotides that target the α4 and β chains of laminin-8; (2) oligonucleotide-releasing disulfide units; (3) antitransferrin receptor antibody for tumor targeting; (4) two valine pH-sensitive membrane-disrupting units; (5) PEG; and (6) fluorescence dye (Lee et al. 2006). Laminins are basement membrane proteins that participate in cell differentiation, migration, and proliferation (Hallmann et al. 2005). Laminin-8, a vascular basement membrane component, is overexpressed in GBM (Ljubimova et al. 2001, 2004). Polycefin, with a tandem configuration of antibodies compared to a single antibody complex, improved the efficiency of nanoparticle uptake in murine glioma models (Fujita et al. 2007). This nanoconjugate specifically targeted only glioma tissue with remarkable specificity, with very minimal uptake by kidney and liver cells in mice models (Ljubimova et al. 2008). Polycefin reduced the incorporation of laminin-8 into the basement membrane, reduced tumor microvessel density, and increased survival in human glioblastoma-bearing rats (Fujita et al. 2006). Another transferrin-conjugated nanoparticle based on poly(D,L-lactide-*co*-glycolide) has been patented with sustained-release characteristics, a biodegradable polymer, and a transferring ligand conjugated to the functional group that provides a backbone similar to polycefin (Labhasetwar and Sahoo 2010).

Nucleotides

Anticancer genes, antisense oligonucleotides, and small interfering RNA (siRNA) molecules may play a role in the treatment of brain tumors. Nucleotides are unstable molecules that require a carrier for transport—a role that nanoparticles play well. Liposomes have been studied as an encapsulating nanocarrier for viral vectors to transmit small plasmids. Virus encapsulation has an advantage in that it reduces immunogenicity, protects against in vivo inactivation, and prolongs in situ residency time (Ren et al. 2003). Epithelial cell adhesion molecules (EpCAMs) can be targeted by nanoparticles, and bound nanoparticles rapidly release their content. Antisense oligonucleotides can be carried by nanoparticles to malignant glioma cells to halt the production of bcl-2 and bcl-xl and to sensitize them to doxorubicin (Winkler et al. 2009). Carbon nanotubes have several favorable characteristics for the delivery of nucleotides. They appear to stabilize siRNA (Kateb et al. 2007), may prevent lysosomal degradation (Kateb et al. 2007), are taken up by tumor-associated macrophages, and result in a transient inflammatory cell infiltration that is nontoxic (VanHandel et al. 2009).

Virogel is a drug delivery nanoparticle. This nanoparticle mimics a virus in size, morphology, and action, and consists of a capsid-like protein capsule that is capable of cell-specific entry, toxin release, cell lysis, and migration to neighboring cells. Such nanoparticles may be able to transport nucleotides or drugs through the BBB and to deliver their payload to target cells while avoiding some inherent risks of viral transport. Folic acid is the ligand used to target this patented molecule (Bae et al. 2009). The nanoparticle has a hydrophobic polymeric core of poly(L-histidine-*co*-phenylalanine), an inner shell of PEG, and an outer shell of bovine serum albumin.

EMERGING CLINICAL APPLICATIONS

There are currently two nanoparticles for treating various cancers approved by the Food and Drug Administration. Doxil (Centocor Ortho Biotech Inc., Horsham, PA) is a stealth liposomal nanoparticle that encapsulates doxorubicin. It is approved to treat the following: (1) recurrent ovarian cancer after failure of platinum-based chemotherapy; (2) acquired immune deficiency syndrome (AIDS)-related Kaposi's sarcoma after failure of prior systemic chemotherapy or intolerance to such therapy; and (3) multiple myeloma, when given in combination with bortezomib and after receiving at least one prior therapy. The other approved nanoparticle is Abraxane, an albumin bound form of paclitaxel that treats breast cancer after failure of combination chemotherapy for metastatic disease or relapse within 6 months of adjuvant chemotherapy. Prior therapy should have included an anthracycline unless clinically contraindicated.

There are 87 currently registered clinical trials that mention nanoparticles (U.S. National Institutes of Health 2011). A phase I trial to treat high-grade gliomas is currently recruiting patients. The patients must have histologically proven intracranial malignant glioma that contains the wild-type or heterozygous UGT1A1*28 gene. Thirty-six patients are to be enrolled and will receive the drug nanoliposomal CPT-11 (liposomal irinotecan). The primary goals are to study the safety and pharmacokinetics of the drug for 1–2 years and to determine the maximum tolerated dose. Another phase I trial that utilizes lapatinib and paclitaxel albumin-stabilized nanoparticle formulation to treat solid tumors, including brain and CNS tumors, is underway. The hypothesis is that lapatinib will work synergistically with the paclitaxel nanoparticle to increase antitumor efficacy. The study will enroll 15 patients, measure the maximum tolerated dose of a 2-day pulse of lapatinib given prior to paclitaxel nanoparticles, and yield preliminary data for antitumor efficacy. Another currently registered trial is a diagnostic study utilizing ferrimoxytol, a superparamagnetic iron oxide nanoparticle, to assess differences in the effects of antiangiogenic chemotherapy and steroid on tumor imaging (U.S. National Institutes of Health 2011).

A 15-person trial of CALAA-01 (an siRNA), using a nanoparticle delivery system, has commenced to treat metastatic melanoma. The nanoparticle contains a targeting ligand, PEG, siRNA designed to reduce the expression of the M2 subunits of ribonucleotide reductase, and a linear cyclodextrin polymer backbone. Tumor biopsies from melanoma patients demonstrated successful uptake of nanoparticles and reduced amounts of ribonucleotide reductase (RRM2) mRNA and protein compared to predosing tissue. This suggests that nanoparticle-encapsulated siRNA can inhibit gene expression specifically by an RNAi mechanism and not produce an interferon response (Davis et al. 2010). The potential role of siRNA nanoparticles in treating CNS neoplasms requires further study.

CONCLUSION

Nanoparticles constitute a diverse and potentially life-saving technology that may soon be utilized to treat primary brain malignancies. Several drugs that are known to be antineoplastic agents may be encapsulated in nanoparticles to promote their delivery through the BBB.

Nanoparticles are being aggressively studied for use in humans; however, the possible risks to human health have not been elucidated. Other questions remain to be answered: How will the immune system react to these very small structures? Can nanoparticles act like haptens, potentially evoking unpredictable autoimmune effects? On the other hand, nanoparticles may go undetected by the immune system, gaining access to unintended tissues and resulting in toxicity to normal structures. Presently, biotoxicity studies are elucidating the interaction of nanoparticles with the human body and hopefully will yield strategies to design nanoparticles of increased efficacy and reduced toxicity (Invernici et al. 2011).

The search for more effective and less toxic treatments for primary brain tumors is driven by the limited effect of the present treatment. The studies reviewed in this chapter reveal a new treatment paradigm for primary brain tumors using tumor-targeted nanodrugs. Undoubtedly, more research is needed to evaluate and develop the use of nanodrugs. As we explore potential applications for these agents, we will come to better understand their pharmacokinetics in valid animal models. Human trials can be performed with select nanodrugs that appear to be promising candidates based on preclinical testing. There is optimism that the unique targeting properties of nanodrugs will lead to advances in the treatment of primary brain tumors to improve prognosis and to reduce the suffering of affected patients.

REFERENCES

Abbott, N.J., and Romero, I.A. 1996. Transporting therapeutics across the blood–brain barrier. *Mol Med Today* 2: 106–13.

Alexiou, C., Jurgons, R., Seliger, C. et al. 2007. Delivery of superparamagnetic nanoparticles for local chemotherapy after intraarterial infusion and magnetic drug targeting. *Anticancer Res* 27: 2019–22.

Allen, T.M., and Cullis, P.R. 2004. Drug delivery systems: entering the mainstream. *Science* 303: 1818–22.

Alyautdin, R.N., Petrov, V.E., Langer, K. et al. 1997. Delivery of loperamide across the blood–brain barrier with polysorbate 80-coated polybutylcyanoacrylate nanoparticles. *Pharm Res* 14: 325–8.

Alyaudtin, R.N., Reichel, A., Lobenberg, R. et al. 2001. Interaction of poly(butylcyanoacrylate) nanoparticles with the blood–brain barrier in vivo and in vitro. *J Drug Targeting* 9: 209–21.

Alyautdin, R.N., Tezikov, E.B., Ramge, P. et al. 1998. Significant entry of tubocurarine into the brain of rats by adsorption to polysorbate 80-coated polybutylcyanoacrylate nanoparticles: an in situ brain perfusion study. *J Microencapsulation* 15: 67–74.

Ambruosi, A., Khalansky, A.S., Yamamoto, H. et al. 2006. Biodistribution of polysorbate 80-coated doxorubicin-loaded [^{14}C]-poly(butyl cyanoacrylate) nanoparticles after intravenous administration to glioblastoma-bearing rats. *J Drug Targeting* 14: 97–105.

Ambruosi, A., Yamamoto, H., and Kreuter, J. 2005. Body distribution of polysorbate-80 and doxorubicin-loaded [^{14}C]poly(butyl cyanoacrylate) nanoparticles after i.v. administration in rats. *J Drug Targeting* 13: 535–42.

Bae, Y.H., Lee, E.S., Kim, D.Y. et al. 2009. Drug delivery vehicle that mimics viral properties. US Patent 2009/0060955.

Beduneau, A., Saulnier, P., and Benoit, J.P. 2007. Active targeting of brain tumors using nanocarriers. *Biomaterials* 28: 4947–67.

Bernardi, A., Braganhol, E., Jager, E. et al. 2009. Indomethacin-loaded nanocapsules treatment reduces in vivo glioblastoma growth in a rat glioma model. *Cancer Lett* 281: 53–63.
Bernardi, A., Frozza, R.L., Jager, E. et al. 2008. Selective cytotoxicity of indomethacin and indomethacin ethyl ester-loaded nanocapsules against glioma cell lines: an in vitro study. *Eur J Pharmacol* 586: 24–34.
Bharali, D.J., Klejbor, I., Stachowiak, E.K. et al. 2005. Organically modified silica nanoparticles: a nonviral vector for in vivo gene delivery and expression in the brain. *Proc Natl Acad Sci U S A* 102: 11539–44.
Boerman, O.C., van Schaijk, F.G., Oyen, W.J. et al. 2003. Pretargeted radioimmunotherapy of cancer: progress step by step. *J Nucl Med* 44: 400–11.
Breitz, H.B., Fisher, D.R., Goris, M.L. et al. 1999. Radiation absorbed dose estimation for ^{90}Y-DOTA-biotin with pretargeted NR-LU-10/streptavidin. *Cancer Biother Radiopharm* 14: 381–95.
Brigger, I., Morizet, J., Laudani, L. et al. 2004. Negative preclinical results with stealth nanospheres-encapsulated doxorubicin in an orthotopic murine brain tumor model. *J Controlled Release* 100: 29–40.
Calvo, P., Gouritin, B., Chacun, H. et al. 2001. Long-circulating PEGylated polycyanoacrylate nanoparticles as new drug carrier for brain delivery. *Pharm Res* 18: 1157–66.
Cavalcanti, A. 2003. Assembly automation with evolutionary nanorobots and sensor-based control applied to nanomedicine. *IEEE Trans Nanotechnol* 2: 82–7.
Cengelli, F., Maysinger, D., Tschudi-Monnet, F. et al. 2006. Interaction of functionalized superparamagnetic iron oxide nanoparticles with brain structures. *J Pharmacol Exp Ther* 318: 108–16.
Cirpanli, Y., Allard, E., Passirani, C. et al. 2011. Antitumoral activity of camptothecin-loaded nanoparticles in 9L rat glioma model. *Int J Pharm* 403: 201–6.
Clarke, J.L., Iwamoto, F.M., Sul, J. et al. 2009. Randomized phase II trial of chemoradiotherapy followed by either dose-dense or metronomic temozolomide for newly diagnosed glioblastoma. *J Clin Oncol* 27: 3861–7.
Cremonesi, M., Ferrari, M., Chinol, M. et al. 1999. Three-step radioimmunotherapy with yttrium-90 biotin: dosimetry and pharmacokinetics in cancer patients. *Eur J Nucl Med* 26: 110–20.
Dandekar, D.S., Lopez, M., Carey, R.I. et al. 2005. Cyclooxygenase-2 inhibitor celecoxib augments chemotherapeutic drug-induced apoptosis by enhancing activation of caspase-3 and -9 in prostate cancer cells. *Int J Cancer* 115: 484–92.
Davis, M.E., Zuckerman, J.E., Choi, C.H. et al. 2010. Evidence of RNAi in humans from systemically administered siRNA via targeted nanoparticles. *Nature* 464: 1067–70.
De La Fuente, E.K., Dawson, C.A., Nelin, L.D. et al. 1997. Biotinylation of membrane proteins accessible via the pulmonary circulation in normal and hyperoxic rats. *Am J Physiol* 272: L461–70.
Dubowchik, G.M., and Walker, M.A. 1999. Receptor-mediated and enzyme-dependent targeting of cytotoxic anticancer drugs. *Pharmacol Ther* 83: 67–123.
Duncan, R. 2003. The dawning era of polymer therapeutics. *Nat Rev Drug Discov* 2: 347–60.
Fellner, S., Bauer, B., Miller, D.S. et al. 2002. Transport of paclitaxel (Taxol) across the blood brain barrier in vitro and in vivo. *J Clin Invest* 110: 1309–18.
Feng, B., Tomizawa, K., Michiue, H. et al. 2009. Delivery of sodium borocaptate to glioma cells using immunoliposome conjugated with anti-EGFR antibodies by ZZ-His. *Biomaterials* 30: 1746–55.
Feng, S.S., Mu, L., Win, K.Y. et al. 2004. Nanoparticles of biodegradable polymers for clinical administration of paclitaxel. *Curr Med Chem* 11: 413–24.
Ferrari, M. 2005. Cancer nanotechnology: opportunities and challenges. *Nat Rev Cancer* 5: 161–71.
Fisher, J.L., Schwartzbaum, J.A., Wrensch, M. et al. 2007. Epidemiology of brain tumors. *Neurol Clin* 25: 867–90, vii.
Friedman, H.S., and Bigner, D.D. 2005. Glioblastoma multiforme and the epidermal growth factor receptor. *N Engl J Med* 353: 1997–9.
Fujita, M., Khazenzon, N.M., Ljubimov, A.V. et al. 2006. Inhibition of laminin-8 in vivo using a novel poly(malic acid)-based carrier reduces glioma angiogenesis. *Angiogenesis* 9: 183–91.
Fujita, M., Lee, B.S., Khazenzon, N.M. et al. 2007. Brain tumor tandem targeting using a combination of monoclonal antibodies attached to biopoly(beta-L-malic acid). *J Controlled Release* 122: 356–63.
Gao, K., and Jiang, X. 2006. Influence of particle size on transport of methotrexate across blood brain barrier by polysorbate 80-coated polybutylcyanoacrylate nanoparticles. *Int J Pharm* 310: 213–9.
Garcia-Garcia, E., Andrieux, K., Gil, S. et al. 2005. Colloidal carriers and blood–brain barrier (BBB) translocation: a way to deliver drugs to the brain? *Int J Pharm* 298: 274–92.
Gazelle, G.S., Goldberg, S.N., Solbiati, L. et al. 2000. Tumor ablation with radio-frequency energy. *Radiology* 217: 633–46.
Gelperina, S. 2006. Brain delivery by nanoparticles. In *Nanoparticle Technology for Drug Delivery, Drugs, and the Pharmaceutical Sciences*, ed. U.B. Kompella, 273–318. New York: Marcel Dekker.

Gelperina, S.E., Khalansky, A.S., Skidan, I.N. et al. 2002. Toxicological studies of doxorubicin bound to polysorbate 80-coated poly(butyl cyanoacrylate) nanoparticles in healthy rats and rats with intracranial glioblastoma. *Toxicol Lett* 126: 131–41.

Gelperina, S.E., Smirnova, Z.S., Khalansky, A.S. et al. 2000. Chemotherapy of brain tumours using doxorubicin bound to polysorbate 80-coated nanoparticles. *Proceedings of the 3rd World Meeting APV/APGI, Berlin*, 441–3.

Gessner, A., Olbrich, C., Schroder, W. et al. 2001. The role of plasma proteins in brain targeting: species dependent protein adsorption patterns on brain-specific lipid drug conjugate (LDC) nanoparticles. *Int J Pharm* 214: 87–91.

Gijsens, A., Derycke, A., Missiaen, L. et al. 2002. Targeting of the photocytotoxic compound AlPcS4 to HeLa cells by transferrin conjugated PEG-liposomes. *Int J Cancer* 101: 78–85.

Goodwin, D.A., and Meares, C.F. 2001. Advances in pretargeting biotechnology. *Biotechnol Adv* 19: 435–50.

Green, N.M. 1975. Avidin. *Adv Protein Chem* 29: 85–133.

Gulyaev, A.E., Gelperina, S.E., Skidan, I.N. et al. 1999. Significant transport of doxorubicin into the brain with polysorbate 80-coated nanoparticles. *Pharm Res* 16: 1564–9.

Haag, R. 2004. Supramolecular drug-delivery systems based on polymeric core–shell architectures. *Angew Chem Int Ed Engl* 43: 278–82.

Hallmann, R., Horn, N., Selg, M. et al. 2005. Expression and function of laminins in the embryonic and mature vasculature. *Physiol Rev* 85: 979–1000.

Harding, J.A., Engbers, C.M., Newman, M.S. et al. 1997. Immunogenicity and pharmacokinetic attributes of poly(ethylene glycol)-grafted immunoliposomes. *Biochim Biophys Acta* 1327: 181–92.

Harris, J.M., and Chess, R.B. 2003. Effect of PEGylation on pharmaceuticals. *Nat Rev Drug Discov* 2: 214–21.

Heimans, J.J., Vermorken, J.B., Wolbers, J.G. et al. 1994. Paclitaxel (Taxol) concentrations in brain tumor tissue. *Ann Oncol* 5: 951–3.

Hekmatara, T., Bernreuther, C., Khalansky, A.S. et al. 2009. Efficient systemic therapy of rat glioblastoma by nanoparticle-bound doxorubicin is due to antiangiogenic effects. *Clin Neuropathol* 28: 153–64.

Hirsch, L.R., Stafford, R.J., Bankson, J.A. et al. 2003. Nanoshell-mediated near-infrared thermal therapy of tumors under magnetic resonance guidance. *Proc Natl Acad Sci U S A* 100: 13549–54.

Hoya, K., Guterman, L.R., Miskolczi, L. et al. 2001. A novel intravascular drug delivery method using endothelial biotinylation and avidin–biotin binding. *Drug Deliv* 8: 215–22.

Huang, Y.F., Sefah, K., Bamrungsap, S. et al. 2008. Selective photothermal therapy for mixed cancer cells using aptamer-conjugated nanorods. *Langmuir* 24: 11860–5.

Huwyler, J., Wu, D., and Pardridge, W.M. 1996. Brain drug delivery of small molecules using immunoliposomes. *Proc Natl Acad Sci U S A* 93: 14164–9.

Invernici, G., Cristini, S., Alessandri, G. et al. 2011. Nanotechnology advances in brain tumors: the state of the art. *Recent Pat Anticancer Drug Discov* 6: 58–69.

Ishida, O., Maruyama, K., Tanahashi, H. et al. 2001. Liposomes bearing polyethyleneglycol-coupled transferrin with intracellular targeting property to the solid tumors in vivo. *Pharm Res* 18: 1042–8.

Jolesz, F.A., and Hynynen, K. 2002. Magnetic resonance image-guided focused ultrasound surgery. *Cancer J* 8 (Suppl 1): S100–12.

Juillerat-Jeanneret, L. 2006. Critical analysis of cancer therapy using nanomaterials. In *Nanomaterials for Cancer Therapy*, ed. C. Kumar, 199–232. Weinheim: Wiley-VCH.

Juillerat-Jeanneret, L. 2008. The targeted delivery of cancer drugs across the blood–brain barrier: chemical modifications of drugs or drug-nanoparticles? *Drug Discov Today* 13: 1099–106.

Kateb, B., Van Handel, M., Zhang, L. et al. 2007. Internalization of MWCNTs by microglia: possible application in immunotherapy of brain tumors. *Neuroimage* 37 (Suppl 1): S9–17.

Kesari, S., Ramakrishna, N., Sauvageot, C. et al. 2005. Targeted molecular therapy of malignant gliomas. *Curr Neurol Neurosci Rep* 5: 186–97.

Knox, S.J., Goris, M.L., Tempero, M. et al. 2000. Phase II trial of yttrium-90-DOTA-biotin pretargeted by NR-LU-10 antibody/streptavidin in patients with metastatic colon cancer. *Clin Cancer Res* 6: 406–14.

Krauze, M.T., Noble, C.O., Kawaguchi, T. et al. 2007. Convection-enhanced delivery of nanoliposomal CPT-11 (irinotecan) and PEGylated liposomal doxorubicin (Doxil) in rodent intracranial brain tumor xenografts. *Neuro-Oncology* 9: 393–403.

Kreuter, J. 2004. Influence of the surface properties on nanoparticle-mediated transport of drugs to the brain. *J Nanosci Nanotechnol* 4: 484–8.

Kreuter, J. 2006. Nonparticulate carriers for drug delivery to the brain. In *Nanoparticles as Drug Carriers*, ed. V.P. Torchilin, 527–47. London: Imperial College Press.

Kreuter, J., Ramge, P., Petrov, V. et al. 2003. Direct evidence that polysorbate-80-coated poly(butylcyanoacrylate) nanoparticles deliver drugs to the CNS via specific mechanisms requiring prior binding of drug to the nanoparticles. *Pharm Res* 20: 409–16.

Kreuter, J., Shamenkov, D., Petrov, V. et al. 2002. Apolipoprotein-mediated transport of nanoparticle-bound drugs across the blood–brain barrier. *J Drug Targeting* 10: 317–25.

Labhasetwar, V.D., and Sahoo, S.K. 2010. Transferrin-conjugated nanoparticles for increasing efficacy of a therapeutic agent. US Patent 2010/0015051 A1.

Leary, S.P., Liu, C.Y., and Apuzzo, M.L. 2006. Toward the emergence of nanoneurosurgery: Part III. Nanomedicine: targeted nanotherapy, nanosurgery, and progress toward the realization of nanoneurosurgery. *Neurosurgery* 58: 1009–26; discussion 1009–26.

Lee, B.S., Fujita, M., Khazenzon, N.M. et al. 2006. Polycefin, a new prototype of a multifunctional nanoconjugate based on poly(beta-L-malic acid) for drug delivery. *Bioconjugate Chem* 17: 317–26.

Lee, G., Dallas, S., Hong, M. et al. 2001. Drug transporters in the central nervous system: brain barriers and brain parenchyma considerations. *Pharmacol Rev* 53: 569–96.

Lee, K.B., Yoon, K.R., Woo, S.I. et al. 2003. Surface modification of poly(glycolic acid) (PGA) for biomedical applications. *J Pharm Sci* 92: 933–7.

Li, S.D., and Huang, L. 2010. Stealth nanoparticles: high density but sheddable PEG is a key for tumor targeting. *J Controlled Release* 145: 178–81.

Liebmann, J.E., Cook, J.A., Lipschultz, C. et al. 1993. Cytotoxic studies of paclitaxel (Taxol) in human tumour cell lines. *Br J Cancer* 68: 1104–9.

Lin, S.H., and Kleinberg, L.R. 2008. Carmustine wafers: localized delivery of chemotherapeutic agents in CNS malignancies. *Expert Rev Anticancer Ther* 8: 343–59.

Ljubimova, J.Y., Fugita, M., Khazenzon, N.M. et al. 2004. Association between laminin-8 and glial tumor grade, recurrence, and patient survival. *Cancer* 101: 604–12.

Ljubimova, J.Y., Fujita, M., Khazenzon, N.M. et al. 2008. Nanoconjugate based on polymalic acid for tumor targeting. *Chem Biol Interact* 171: 195–203.

Ljubimova, J.Y., Lakhter, A.J., Loksh, A. et al. 2001. Overexpression of alpha4 chain-containing laminins in human glial tumors identified by gene microarray analysis. *Cancer Res* 61: 5601–10.

Lu, W., Wan, J., She, Z. et al. 2007. Brain delivery property and accelerated blood clearance of cationic albumin conjugated PEGylated nanoparticle. *J Controlled Release* 118: 38–53.

Lu, Y., and Chen, S.C. 2004. Micro and nano-fabrication of biodegradable polymers for drug delivery. *Adv Drug Deliv Rev* 56: 1621–33.

Maruyama, K., Ishida, O., Kasaoka, S. et al. 2004. Intracellular targeting of sodium mercaptoundecahydrododecaborate (BSH) to solid tumors by transferrin–PEG liposomes, for boron neutron-capture therapy (BNCT). *J Controlled Release* 98: 195–207.

McBain, S.C., Yiu, H.H., and Dobson, J. 2008. Magnetic nanoparticles for gene and drug delivery. *Int J Nanomed* 3: 169–80.

Mirza, A.N., Fornage, B.D., Sneige, N. et al. 2001. Radiofrequency ablation of solid tumors. *Cancer J* 7: 95–102.

Moghimi, S.M. 1995. Mechanisms regulating body distribution of nanospheres conditioned with pluronic and tetronic block-copolymers. *Adv Drug Deliv Rev* 16: 183–93.

Moghimi, S.M., and Hunter, A.C. 2000. Poloxamers and poloxamines in nanoparticle engineering and experimental medicine. *Trends Biotechnol* 18: 412–20.

Moghimi, S.M., Hunter, A.C., and Murray, J.C. 2005. Nanomedicine: current status and future prospects. *FASEB J* 19: 311–30.

Na, K., and Bae, Y.H. 2002. Self-assembled hydrogel nanoparticles responsive to tumor extracellular pH from pullulan derivative/sulfonamide conjugate: characterization, aggregation, and adriamycin release in vitro. *Pharm Res* 19: 681–8.

Nobs, L., Buchegger, F., Gurny, R. et al. 2006. Biodegradable nanoparticles for direct or two-step tumor immunotargeting. *Bioconjugate Chem* 17: 139–45.

Ogris, M., and Wagner, E. 2002. Tumor-targeted gene transfer with DNA polyplexes. *Somatic Cell Mol Genet* 27: 85–95.

Oh, K.T., Bronich, T.K., and Kabanov, A.V. 2004. Micellar formulations for drug delivery based on mixtures of hydrophobic and hydrophilic pluronic block copolymers. *J Controlled Release* 94: 411–22.

Olivier, J.C. 2005. Drug transport to brain with targeted nanoparticles. *NeuroRx* 2: 108–19.

Paganelli, G., Orecchia, R., Jereczek-Fossa, B. et al. 1998. Combined treatment of advanced oropharyngeal cancer with external radiotherapy and three-step radioimmunotherapy. *Eur J Nucl Med* 25: 1336–9.

Panyam, J., and Labhasetwar, V. 2004. Sustained cytoplasmic delivery of drugs with intracellular receptors using biodegradable nanoparticles. *Mol Pharm* 1: 77–84.

Panzenboeck, U., Balazs, Z., Sovic, A. et al. 2002. ABCA1 and scavenger receptor class B, type I, are modulators of reverse sterol transport at an in vitro blood–brain barrier constituted of porcine brain capillary endothelial cells. *J Biol Chem* 277: 42781–9.

Pereverzeva, E., Treschalin, I., Bodyagin, D. et al. 2007. Influence of the formulation on the tolerance profile of nanoparticle-bound doxorubicin in healthy rats: focus on cardio- and testicular toxicity. *Int J Pharm* 337: 346–56.

Pereverzeva, E., Treschalin, I., Bodyagin, D. et al. 2008. Intravenous tolerance of a nanoparticle-based formulation of doxorubicin in healthy rats. *Toxicol Lett* 178: 9–19.

Petri, B., Bootz, A., Khalansky, A. et al. 2007. Chemotherapy of brain tumour using doxorubicin bound to surfactant-coated poly(butyl cyanoacrylate) nanoparticles: revisiting the role of surfactants. *J Controlled Release* 117: 51–8.

Pulfer, S.K., Ciccotto, S.L., and Gallo, J.M. 1999. Distribution of small magnetic particles in brain tumor-bearing rats. *J Neuro-Oncol* 41: 99–105.

Pulkkinen, M., Pikkarainen, J., Wirth, T. et al. 2008. Three-step tumor targeting of paclitaxel using biotinylated PLA–PEG nanoparticles and avidin–biotin technology: formulation development and in vitro anticancer activity. *Eur J Pharm Biopharm* 70: 66–74.

Rabinow, B.E. 2004. Nanosuspensions in drug delivery. *Nat Rev Drug Discov* 3: 785–96.

Ramge, P., Unger, R.E., Oltrogge, J.B. et al. 2000. Polysorbate-80 coating enhances uptake of polybutylcyanoacrylate (PBCA) nanoparticles by human and bovine primary brain capillary endothelial cells. *Eur J Neurosci* 12: 1931–40.

Reents, R., Jeyaraj, D.A., and Waldmann, H. 2002. Enzymatically cleavable linker groups in polymer-supported synthesis. *Drug Discov Today* 7: 71–6.

Ren, H., Boulikas, T., Lundstrom, K. et al. 2003. Immunogene therapy of recurrent glioblastoma multiforme with a liposomally encapsulated replication-incompetent Semliki forest virus vector carrying the human interleukin-12 gene—a phase I/II clinical protocol. *J Neuro-Oncol* 64: 147–54.

Rodriguez, J.A., Helguera, G., Daniels, T.R. et al. 2007. Binding specificity and internalization properties of an antibody–avidin fusion protein targeting the human transferrin receptor. *J Controlled Release* 124: 35–42.

Rowinsky, E.K., Burke, P.J., Karp, J.E. et al. 1989. Phase I and pharmacodynamic study of Taxol in refractory acute leukemias. *Cancer Res* 49: 4640–7.

Sahoo, S.K., and Labhasetwar, V. 2003. Nanotech approaches to drug delivery and imaging. *Drug Discov Today* 8: 1112–20.

Saito, R., Krauze, M.T., Noble, C.O. et al. 2006. Convection-enhanced delivery of Ls-TPT enables an effective, continuous, low-dose chemotherapy against malignant glioma xenograft model. *Neuro-Oncology* 8: 205–14.

Sauter, G., Maeda, T., Waldman, F.M. et al. 1996. Patterns of epidermal growth factor receptor amplification in malignant gliomas. *Am J Pathol* 148: 1047–53.

Savic, R., Luo, L., Eisenberg, A. et al. 2003. Micellar nanocontainers distribute to defined cytoplasmic organelles. *Science* 300: 615–8.

Schiff, P.B., Fant, J., and Horwitz, S.B. 1979. Promotion of microtubule assembly in vitro by Taxol. *Nature* 277: 665–7.

Schoenmakers, R.G., van de Wetering, P., Elbert, D.L. et al. 2004. The effect of the linker on the hydrolysis rate of drug-linked ester bonds. *J Controlled Release* 95: 291–300.

Schroeder, U., Schroeder, H., and Sabel, B.A. 2000. Body distribution of ^{3}H-labelled dalargin bound to poly(butyl cyanoacrylate) nanoparticles after i.v. injections to mice. *Life Sci* 66: 495–502.

Schroeder, U., Sommerfeld, P., Ulrich, S. et al. 1998. Nanoparticle technology for delivery of drugs across the blood–brain barrier. *J Pharm Sci* 87: 1305–7.

Schwechheimer, K., Huang, S., and Cavenee, W.K. 1995. EGFR gene amplification—rearrangement in human glioblastomas. *Int J Cancer* 62: 145–8.

Seki, T., Wakabayashi, M., Nakagawa, T. et al. 1999. Percutaneous microwave coagulation therapy for patients with small hepatocellular carcinoma: comparison with percutaneous ethanol injection therapy. *Cancer* 85: 1694–702.

Shapiro, B., Dormer, K., and Rutel, I.B. 2010. A two-magnet system to push therapeutic nanoparticles. *AIP Conf Proc* 1311: 77–88.

Shenoy, V.S., Vijay, I.K., and Murthy, R.S. 2005. Tumour targeting: biological factors and formulation advances in injectable lipid nanoparticles. *J Pharm Pharmacol* 57: 411–22.

Shi, N., Zhang, Y., Zhu, C. et al. 2001. Brain-specific expression of an exogenous gene after i.v. administration. *Proc Natl Acad Sci U S A* 98: 12754–9.

Soni, V., Kohli, D.V., and Jain, S.K. 2005. Transferrin coupled liposomes as drug delivery carriers for brain targeting of 5-florouracil. *J Drug Targeting* 13: 245–50.
Sparreboom, A., van Asperen, J., Mayer, U. et al. 1997. Limited oral bioavailability and active epithelial excretion of paclitaxel (Taxol) caused by P-glycoprotein in the intestine. *Proc Natl Acad Sci U S A* 94: 2031–5.
Stan, A.C., Casares, S., Radu, D. et al. 1999. Doxorubicin-induced cell death in highly invasive human gliomas. *Anticancer Res* 19: 941–50.
Steiniger, S.C., Kreuter, J., Khalansky, A.S. et al. 2004. Chemotherapy of glioblastoma in rats using doxorubicin-loaded nanoparticles. *Int J Cancer* 109: 759–67.
Tomalia, D.A., Baker, H., Dewald, J. et al. 1985. A new class of polymers: starburst-dendritic macromolecules. *Polym J* 17: 117–32.
Tomalia, D.A., and Frechet, J.M.J. 2002. Discovery of dendrimers and dendritic polymers: a brief historical perspective. *Journal of Polymer Science Part A Polymer Chemistry* 40: 2719–28.
Torchilin, V.P. 2005. Recent advances with liposomes as pharmaceutical carriers. *Nat Rev Drug Discov* 4: 145–60.
U.S. National Institutes of Health. 2011. ClinicalTrials.gov. http://www.clinicaltrials.gov/.
VanHandel, M., Alizadeh, D., Zhang, L. et al. 2009. Selective uptake of multi-walled carbon nanotubes by tumor macrophages in a murine glioma model. *J Neuroimmunol* 208: 3–9.
Vogl, T.J., Mack, M.G., Muller, P.K. et al. 1999. Interventional MR: interstitial therapy. *Eur Radiol* 9: 1479–87.
von Holst, H., Knochenhauer, E., Blomgren, H. et al. 1990. Uptake of adriamycin in tumour and surrounding brain tissue in patients with malignant gliomas. *Acta Neurochir (Wien)* 104: 13–6.
Walter, K.A., Tamargo, R.J., Olivi, A. et al. 1995. Intratumoral chemotherapy. *Neurosurgery* 37: 1128–45.
Winkler, J., Martin-Killias, P., Pluckthun, A. et al. 2009. EpCAM-targeted delivery of nanocomplexed siRNA to tumor cells with designed ankyrin repeat proteins. *Mol Cancer Ther* 8: 2674–83.
Yamashita, Y., Krauze, M.T., Kawaguchi, T. et al. 2007. Convection-enhanced delivery of a topoisomerase I inhibitor (nanoliposomal topotecan) and a topoisomerase II inhibitor (PEGylated liposomal doxorubicin) in intracranial brain tumor xenografts. *Neuro-Oncology* 9: 20–8.
Yang, S.C., Lu, L.F., Cai, Y. et al. 1999. Body distribution in mice of intravenously injected camptothecin solid lipid nanoparticles and targeting effect on brain. *J Controlled Release* 59: 299–307.
Yang, W., Wu, G., Barth, R.F. et al. 2008. Molecular targeting and treatment of composite EGFR and EGFRvIII-positive gliomas using boronated monoclonal antibodies. *Clin Cancer Res* 14: 883–91.
Yokoyama, K., Miyatake, S., Kajimoto, Y. et al. 2006. Pharmacokinetic study of BSH and BPA in simultaneous use for BNCT. *J Neuro-Oncol* 78: 227–32.
Zhao, M., Liang, C., Li, A., Chang, J., et al. 2010. Magnetic paclitaxel nanoparticles inhibit glioma growth and improve the survival of rats bearing glioma xenografts. *Anticancer Res* 30: 2217–23.

27 Application of Nanotechnology in the Diagnosis and Treatment of Brain Metastasis

Joel C. Sunshine, Zachary D. Guss, Jay J. Pillai, and Michael Lim

CONTENTS

INTRODUCTION

Despite decades of research and incremental improvements, optimal strategies for managing brain metastases remain elusive. Surgery and radiation comprise the cornerstone of current treatment for brain metastases, and clinical context determines individualized treatments.

Surgery and radiation possess distinct attributes for combating metastases. Surgery is well suited for addressing solitary lesions or symptomatic lesions because it can remove a large mass of tumor and relieve mass effect when present. However, the recurrence rates of tumors are high in the absence of adjuvant radiation, and the role of surgery is somewhat limited in patients with multiple metastases or extensive leptomeningeal diseases in which intrathecal chemotherapy plays a significant role.

Radiation can be further divided into whole-brain radiation therapy (WBRT) and radiosurgery. WBRT applies radiation relatively indiscriminately to the brain, which is advantageous for inhibiting the growth of micrometastases that cannot be detected by current imaging technology. However, the sequelae associated with radiating large expanses of normal parenchyma are considerable and limit the radiation dose that can be delivered in a single session. In particular, cognitive impairment, such as dementia, is a worrisome complication of this treatment (DeAngelis et al. 1989). Despite these drawbacks, WBRT is effective in extending survival for these patients. A study by Nussbaum et al. (1996) revealed that patients who received WBRT survived for 6–7 months compared to the natural history of the disease (5 months).

Radiosurgery, despite its mechanistic similarities to WBRT, functions more closely to conventional surgery. By focusing on discrete lesions rather than large expanses of the brain, radiosurgery can apply

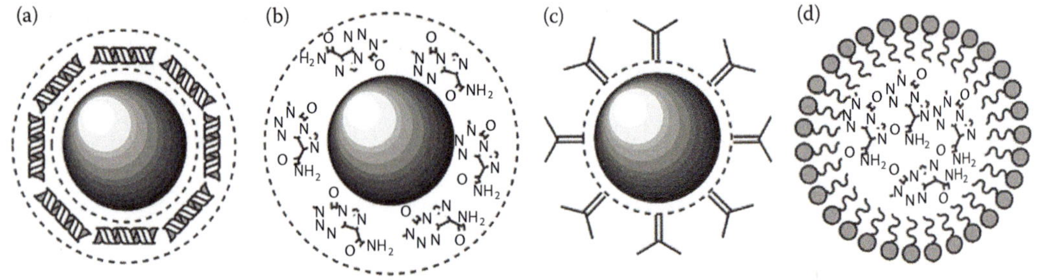

FIGURE 27.1 Gold and SPIO nanoparticles can be coated with polymers or lipids, can be used to deliver nucleic acids (a) or cytotoxic drugs (b), and can be conjugated to antibodies (c) that target the BBB (e.g., anti-transferrin receptor antibody), metastatic cancer (e.g., anti-HER3 antibody), or both. Polymer particles and liposomes (d) can be used to deliver a variety of drugs to the CNS and can be conjugated to antibodies to target metastatic cancer.

large doses of radiation in a single fraction and spare normal tissue. This is advantageous for controlling known metastases but does not treat micrometastases as does WBRT. It avoids the complications of surgery, such as cerebrospinal fluid leaks and surgical wound infections, and can target multiple lesions.

Despite this diverse array of therapeutic strategies, there is considerable room for progress. Nanotechnology is an exciting discipline that can augment existing strategies and open new therapeutic avenues. In the following sections, we will discuss the current state of this field as it applies to brain metastases and highlight areas of nanotechnology research that can be adapted to combat brain metastases.

DRUG DELIVERY

Ideally, a nanodelivery drug complex designed to reach brain metastases would be administered either orally or intravenously and would cross the blood–brain barrier (BBB) to selectively target malignant cells in the central nervous system (CNS) with minimal systemic side effects. Once they have reached their desired location within the CNS, nanocomplexes can perform many functions, such as acting as imaging contrast agents, releasing drugs at target tissues, or allowing intracellular delivery of drugs (Figure 27.1).

There is a wide range of materials available in nanoparticle formulations, ranging from inorganic nanoparticles—magnetic (superparamagnetic iron oxide [SPIO]), metallic (gold, silver), ceramic (silica, alumina), and quantum dots—to organic nanoparticles (dendrimers, liposomes, and polymeric nanoparticles), each with varying advantages and disadvantages (Table 27.1). We will focus

TABLE 27.1
Characteristics of Candidate Nanoparticles

Particle	Size	Features
SPIO/USPIO	>50 to <50 nm	Well tolerated; iron resorbed into iron stores; easily manufactured; useful for imaging properties (T_2 MRI contrast agent), hysteresis heating, and as scaffold for drug delivery; must be coated for solubility in water
Gold	10–100 nm	Chemically inert; relatively nontoxic; readily modifiable with thiol-containing groups; useful as imaging agent, for photothermal therapy, and as scaffold for drug delivery
PBCA	>200 nm	Endocytosed; readily coated for good BBB penetration
Liposomes	~100 nm	Can conjugate to antibodies for targeting; can deliver many small-molecule drugs
PEI/PEG nanogels	<100 nm	Intracellular delivery of nucleic acids
aAPC	Several micrometers	Ex vivo T-cell cancer stimulation

on magnetic, metallic, and organic nanoparticles and their potential applications for diagnosing and treating brain metastases.

MATERIAL PROPERTIES OF NANOPARTICLES

SPIO Nanoparticles

SPIO nanoparticles come in two main forms: magnetite (Fe_3O_4) and the oxidized form maghemite (γFe_2O_3) (Tang et al. 2007). SPIO nanoparticles have been mostly used clinically as magnetic resonance imaging (MRI) contrast agents, resulting in focal signal loss on T_2-weighted imaging due to substantial local magnetic susceptibility effects (Tang et al. 2007). They are classified by size, with particles >50 nm classified as SPIO nanoparticles and with particles <50 nm classified as ultrasmall SPIO (USPIO) nanoparticles (Tang et al. 2007). General advantages that allow tumor enhancement by circulating SPIO nanoparticles include accumulation at the tumor site by enhanced permeability and retention effect, which results from the combination of leaky tumor vasculature and the absence of a lymphatic system in the expanding tumor mass (Phillips et al. 2010; Tang et al. 2010).

SPIO nanoparticles can also be coated with polymers or inorganic metals and surface-functionalized, which both reduce the propensity for SPIO nanoparticles to aggregate and cause emboli in the presence of a magnetic field and allow addition of targeting moieties and other bioactivities to the nanoparticles (Tang et al. 2007). SPIO nanoparticles are generally well tolerated, are metabolized, and enter endogenous iron reserves via the hematopoietic pathway (Tang et al. 2007).

SPIO nanoparticles possess both therapeutic and diagnostic potentials. As mentioned before, SPIO nanoparticles can be surface-functionalized with a variety of molecules such as antibodies or small interfering RNA. In addition, when an alternating magnetic field is applied, SPIO nanoparticles can cause local hyperthermia due to hysteresis heating (Tang et al. 2007). Research in this area is ongoing to develop particles that effectively combine the MRI properties of SPIO nanoparticles with enhanced magnetic hyperthermia properties. Recently, one group showed that by doping manganite SPIO nanoparticles with silver, the Curie temperature reached the hyperthermia range while maintaining MRI characteristics (Melnikov et al. 2009).

Gold Nanoparticles

Bulk gold is colored yellow gold but can assume a variety of colors (from red to brown) when formed into nanoparticles, depending on size. This is due to surface plasmon resonance (SPR), a phenomenon resulting from conduction band electrons interacting with incident light (Kelly et al. 2003). Gold nanoparticles come in many shapes, including nanospheres, nanoshells, and nanorods. Gold nanospheres can be synthesized by reducing hydrogen tetrachloroaurate ($HAuCl_4$) in water using a variety of reducing agents (Turkevich et al. 1951). Gold nanorods can also be produced by multiple processes (Kim et al. 2002; Sau and Murphy 2004) and have been complexed into multifunctional, multicomponent nanorods for gene delivery applications (Salem et al. 2003). Since SPR is size-dependent, gold nanorods have two SPR bands. This property is tunable based on size, allowing the selection of appropriate absorption bands for imaging applications.

Another size-dependent property of nanogold is that 15-nm nanospheres can passively permeate the BBB and accumulate in the brain (Sonavane et al. 2008). This appears to be size-dependent, possibly due to the geometry of the BBB itself, since the basement membrane is covered by astrocytic end-feet processes that are separated from the endothelial basement membrane by around 20 nm (Sonavane et al. 2008).

Nanogold possesses a variety of advantages for use in nanoparticle formulations. Bulk gold is chemically inert and relatively nontoxic, and has great potential as an imaging agent and for photothermal therapy (Alkilany and Murphy 2010). Gold nanoparticles, such as nanorods or nanorod-in-shell nanostructures, are excitable by near-infrared light. When excited, they produce intense localized heat that is able to destroy tumor cells. Near-infrared light can penetrate into brain tissue to a depth of between 2 and 3 mm (Stolik et al. 2000); therefore, this technique would be most useful during surgery as an adjuvant to resection.

In addition, the surface of colloidal gold is readily modified by molecules containing thiol groups, allowing rapid and diverse modulation of surface properties and conjugation of a variety of cargoes and targeting agents to its surface.

NANOPARTICLE-BASED TARGETING OF BRAIN METASTASES

Crossing the BBB

Poly(alkylcyanoacrylate) (PBCA) nanoparticles coated with polysorbate 80 have been used to deliver a range of drugs (Alyautdin et al. 1997; Gulyaev et al. 1999; Kreuter 2001; Steiniger et al. 2004) across the BBB into the brain parenchyma via intravenous injection. The mechanism for the endocytosis of these nanoparticles has been proposed to be due to the surface adsorption of apolipoproteins B and E on the surface of the nanoparticles, thereby enabling the nanoparticles to be taken up by brain capillary endothelial cells via receptor-mediated endocytosis (Kreuter 2001). However, simply adsorbing polysorbate 80 onto any nanoparticle surface is not always sufficient to permit the nanoparticles to cross the BBB, and drugs adsorbed onto the surface of these nanoparticles dissociate rapidly when the nanoparticles are administered intravenously (Olivier 2005). As such, it appears likely that the PBCA nanoparticles actually caused nonspecific permeabilization of the BBB (Olivier 2005).

Solid lipid nanoparticles (SLNs) are also able to cross the BBB. Additionally, they can deliver superparamagnetic SPIO nanoparticles to the CNS circulation (Peira et al. 2003). Therefore, SLNs could serve as CNS MRI contrast agents for superior visualization of brain metastases.

Nanogels consisting of cross-linked polyethylene glycol (PEG) and polyethyleneimine (PEI) can enable a 15-fold accumulation of oligonucleotides in the brain after 1 h, with a concurrent twofold decrease in accumulation in the liver and spleen when compared with freely administered oligonucleotides (Vinogradov et al. 2004).

Liposomes conjugated to PEG, which is then conjugated to an antibody, are referred to as PEGylated immunoliposomes. In a rat model, conjugation of anti-transferrin receptor antibody to liposomes allowed passage through the BBB and resulted in daunomycin being accumulated in and delivered to the brain (Huwyler et al. 1996). This is especially powerful because maximal brain delivery occurred with around 30 antibodies/immunoliposomes; since each liposome can deliver tens of thousands of small-molecule drugs, conjugation of the antibodies to immunoliposomes increased the carrying capacity of the antibody by four log orders in magnitude (Huwyler et al. 1996). In addition, gene delivery using these PEGylated immunoliposomes showed specific expression of the delivered genes in the brain (Shi et al. 2001).

Targeting Brain Metastases

A commonly used strategy outside the brain involves addition of targeting ligands to nanoparticles to promote specific localization and uptake of the nanoparticles at the cellular level (rather than crossing the BBB, as with PEGylated immunoliposomes). There are many examples that exploit this principle to target metastases outside the brain, for example, attaching YIGSR peptide from laminin was used to target lung metastases (Sarfati et al. 2011), and alendronate was coupled to poly(D,L-lactide-*co*-glycolic) acid (PLGA) nanoparticles to target breast cancer bone metastases (Salerno et al. 2010).

One study compared brain metastases from primary breast carcinoma to primary tumors and found that HER3, in particular, and other members of the HER family were significantly overexpressed in brain metastases when compared to primary tumor controls (Da Silva et al. 2010).

Melanotransferrin (MTf) was identified as a key component that allows metastatic melanoma to migrate across the BBB and to form brain metastases; it is presented on the surface (Rolland et al. 2009). Systemic administration of anti-MTf antibody in a metastatic melanoma mouse model reduced the formation of brain metastases (Rolland et al. 2009). Nanoparticles conjugated with anti-HER3 or anti-MTf antibody might be promising agents for their dual functionality—disrupting the targeted pathway and targeting a functionalized nanoparticle to the cell surface.

CELL-BASED TARGETING OF BRAIN METASTASES

One of the most innovative approaches to achieving a targeted delivery of drugs and other molecules to brain metastases involves using engineered neural and mesenchymal stem cells (MSCs) as delivery agents. Neural stem/progenitor cells (NSPCs) show extensive tropism for pathological areas in the CNS (Aboody et al. 2000). When implanted into experimental models of intracranial gliomas in adult rodents, NSPCs can distribute themselves rapidly throughout the tumor and migrate to widely expanding and aggressively advancing tumor cells (Aboody et al. 2000). When implanted intracranially at distant sites from the tumor (normal tissue, contralateral hemisphere, or cerebral ventricles), NSPCs migrate through normal tissue, targeting the tumor cells, including human glioblastomas (Aboody et al. 2000). Moreover, when implanted outside the CNS intravascularly, NSPCs can target an intracranial tumor (Aboody et al. 2000).

This impressive targeting has also been demonstrated successfully for melanoma brain metastases (Aboody et al. 2006). Based on an in vivo model for melanoma metastasis, NSPCs that stably express cytosine deaminase (CD), which converts the prodrug 5-fluorocytosine (5-FC) into the active chemotherapeutic agent 5-fluorouracil (5-FU), were implanted intracranially and followed 3 days later by systemic 5-FC treatment, resulting in a significant (71%) reduction in tumor burden (Aboody et al. 2006). Another group modified human neural stem cells (NSCs) to secrete anti-HER2 immunoglobulin molecules as a tool to target and attack metastatic breast cancer in the brain. They showed that anti-HER2-secreting NSCs exhibit preferential tropism to tumor cells and can deliver antibodies to human breast carcinoma xenografts in mice (Frank et al. 2009). A variety of genes have also been delivered to primary brain tumors in mice using stem cells: NSPCs and MSCs, including interleukin (IL) 2 (Nakamura et al. 2004), IL-4 (Benedetti et al. 2000), IL-12 (Ehtesham et al. 2002), CD and interferon-β (Lee et al. 2009), and a secretable trimeric tumor necrosis factor–related apoptosis-inducing ligand (stTRAIL) (Kim et al. 2008), among others.

NSC-based targeting can also be diversely applied to nanotechnology. MSCs and NSPCs could serve as delivery vehicles for contrast agents, gold and iron oxide nanoparticles, and nonviral gene or nucleic acid nanoparticles. A few days before surgical resection of metastases, injecting a cell-based technology impregnated with iron oxide nanoparticles allows cancer detection, since SPIO nanoparticles can act as MRI contrast agents and potentially as radiation adjuvants via magnetic hyperthermia; loading the cells with gold nanorods would permit near-infrared-based thermotherapy localized only to malignant cells. This technique (impregnating cells with SPIO nanoparticles, delivering the cells, and detecting the SPIO nanoparticles with MRI) has been used successfully to track MSCs to lung metastases in vivo (Loebinger et al. 2009).

NANOPARTICLE-BASED ACTIVATION OF THE IMMUNE SYSTEM TO TREAT BRAIN METASTASES

Another novel approach to microparticle/nanoparticle-based immune activation is the concept of artificial antigen-presenting cells (aAPCs). aAPC involves the coupling of a tumor-specific peptide in major histocompatibility complex (MHC) (to interact with and stimulate the T-cell receptor) and a costimulatory signal (to induce proper activation in response to antigen presentation) to a variety of microparticle or nanoparticle surfaces (Steenblock et al. 2009). Magnetic microparticles that can be used for ex vivo stimulation and removed magnetically before the cells are administered have been developed and show strong stimulation and good antitumor effect both in vitro and in vivo (Durai et al. 2009; Oelke et al. 2003; Ugel et al. 2009). In addition, biodegradable particles that release cytokines and present both peptide-in-MHC and a costimulatory signal have been developed (Steenblock and Fahmy 2008). At this point, 4- to 5-μm particles appear to be most effective in inducing a strong immune response but are only useful for ex vivo applications due to the potential for particles of that size to be trapped in and occlude microvessels (Steenblock et al. 2009). However, there is ongoing work to develop nanoparticles of similar efficacy (to take advantage of the pharmacological benefits of nanoparticles) and the ability to deliver them systemically.

CONCLUSION

Nanotechnology holds promise for managing metastases to the brain, although research is still at the preliminary stage. Magnetic, metallic, and organic nanoparticles each have unique characteristics that favor their use in select clinical scenarios. Nanotechnology can be effective for both diagnosing and treating brain metastases. In particular, it offers promise for crossing of the BBB, targeted drug therapy, activation of the host immune system, and tumor hyperthermia. These novel approaches could potentially improve patient outcomes alone or in combination with conventional modalities such as surgery, radiation, or chemotherapy. Future progress will reveal the practical value of these exciting possibilities.

REFERENCES

Aboody, K.S., Brown, A., Rainov, N.G. et al. 2000. Neural stem cells display extensive tropism for pathology in adult brain: evidence from intracranial gliomas. *Proc Natl Acad Sci U S A* 97: 12846–51.

Aboody, K.S., Najbauer, J., Schmidt, N.O. et al. 2006. Targeting of melanoma brain metastases using engineered neural stem/progenitor cells. *Neuro-Oncology* 8: 119–26.

Alkilany, A.M., and Murphy, C.J. 2010. Toxicity and cellular uptake of gold nanoparticles: what we have learned so far. *J Nanopart Res* 12: 2313–33.

Alyautdin, R.N., Petrov, V.E., Langer, K. et al. 1997. Delivery of loperamide across the blood–brain barrier with polysorbate 80-coated polybutylcyanoacrylate nanoparticles. *Pharm Res* 14: 325–8.

Benedetti, S., Pirola, B., Pollo, B. et al. 2000. Gene therapy of experimental brain tumors using neural progenitor cells. *Nat Med* 6: 447–50.

Da Silva, L., Simpson, P.T., Smart, C.E. et al. 2010. HER3 and downstream pathways are involved in colonization of brain metastases from breast cancer. *Breast Cancer Res* 12: R46.

DeAngelis, L.M., Delattre, J.Y., and Posner, J.B. 1989. Radiation-induced dementia in patients cured of brain metastases. *Neurology* 39: 789–96.

Durai, M., Krueger, C., Ye, Z. et al. 2009. In vivo functional efficacy of tumor-specific T cells expanded using HLA-Ig based artificial antigen presenting cells (aAPC). *Cancer Immunol Immunother* 58: 209–20.

Ehtesham, M., Kabos, P., Kabosova, A. et al. 2002. The use of interleukin 12-secreting neural stem cells for the treatment of intracranial glioma. *Cancer Res* 62: 5657–63.

Frank, R.T., Edmiston, M., Kendall, S.E. et al. 2009. Neural stem cells as a novel platform for tumor-specific delivery of therapeutic antibodies. *PLoS One* 4: e8314.

Gulyaev, A.E., Gelperina, S.E., Skidan, I.N. et al. 1999. Significant transport of doxorubicin into the brain with polysorbate 80-coated nanoparticles. *Pharm Res* 16: 1564–9.

Huwyler, J., Wu, D., and Pardridge, W.M. 1996. Brain drug delivery of small molecules using immunoliposomes. *Proc Natl Acad Sci U S A* 93: 14164–9.

Kelly, K.L., Coronado, E., Zhao, L.L. et al. 2003. The optical properties of metal nanoparticles: the influence of size, shape, and dielectric environment. *J Phys Chem B* 107: 668–77.

Kim, F., Song, J.H., and Yang, P. 2002. Photochemical synthesis of gold nanorods. *J Am Chem Soc* 124: 14316–7.

Kim, S.M., Lim, J.Y., Park, S.I. et al. 2008. Gene therapy using TRAIL-secreting human umbilical cord blood-derived mesenchymal stem cells against intracranial glioma. *Cancer Res* 68: 9614–23.

Kreuter, J. 2001. Nanoparticulate systems for brain delivery of drugs. *Adv Drug Deliv Rev* 47: 65–81.

Lee, D.H., Ahn, Y., Kim, S.U. et al. 2009. Targeting rat brainstem glioma using human neural stem cells and human mesenchymal stem cells. *Clin Cancer Res* 15: 4925–34.

Loebinger, M.R., Kyrtatos, P.G., Turmaine, M. et al. 2009. Magnetic resonance imaging of mesenchymal stem cells homing to pulmonary metastases using biocompatible magnetic nanoparticles. *Cancer Res* 69: 8862–7.

Melnikov, O.V., Gorbenko, O.Y., Markelova, M.N. et al. 2009. Ag-doped manganite nanoparticles: new materials for temperature-controlled medical hyperthermia. *J Biomed Mater Res A* 91: 1048–55.

Nakamura, K., Ito, Y., Kawano, Y. et al. 2004. Antitumor effect of genetically engineered mesenchymal stem cells in a rat glioma model. *Gene Ther* 11: 1155–64.

Nussbaum, E.S., Djalilian, H.R., Cho, K.H. et al. 1996. Brain metastases. Histology, multiplicity, surgery, and survival. *Cancer* 78: 1781–8.

Oelke, M., Maus, M.V., Didiano, D. et al. 2003. Ex vivo induction and expansion of antigen-specific cytotoxic T cells by HLA-Ig-coated artificial antigen-presenting cells. *Nat Med* 9: 619–24.

Olivier, J.C. 2005. Drug transport to brain with targeted nanoparticles. *NeuroRx* 2: 108–19.

Peira, E., Marzola, P., Podio, V. et al. 2003. In vitro and in vivo study of solid lipid nanoparticles loaded with superparamagnetic iron oxide. *J Drug Targeting* 11: 19–24.

Phillips, M.A., Gran, M.L., and Peppas, N.A. 2010. Targeted nanodelivery of drugs and diagnostics. *Nano Today* 5: 143–59.

Rolland, Y., Demeule, M., Fenart, L. et al. 2009. Inhibition of melanoma brain metastasis by targeting melanotransferrin at the cell surface. *Pigment Cell Melanoma Res* 22: 86–98.

Salem, A.K., Searson, P.C., and Leong, K.W. 2003. Multifunctional nanorods for gene delivery. *Nat Mater* 2: 668–71.

Salerno, M., Cenni, E., Fotia, C. et al. 2010. Bone-targeted doxorubicin-loaded nanoparticles as a tool for the treatment of skeletal metastases. *Curr Cancer Drug Targets* 10: 649–59.

Sarfati, G., Dvir, T., Elkabets, M. et al. 2011. Targeting of polymeric nanoparticles to lung metastases by surface-attachment of YIGSR peptide from laminin. *Biomaterials* 32: 152–61.

Sau, T.K., and Murphy, C.J. 2004. Seeded high yield synthesis of short Au nanorods in aqueous solution. *Langmuir* 20: 6414–20.

Shi, N., Zhang, Y., Zhu, C. et al. 2001. Brain-specific expression of an exogenous gene after i.v. administration. *Proc Natl Acad Sci U S A* 98: 12754–9.

Sonavane, G., Tomoda, K., and Makino, K. 2008. Biodistribution of colloidal gold nanoparticles after intravenous administration: effect of particle size. *Colloids Surf B* 66: 274–80.

Steenblock, E.R., and Fahmy, T.M. 2008. A comprehensive platform for ex vivo T-cell expansion based on biodegradable polymeric artificial antigen-presenting cells. *Mol Ther* 16: 765–72.

Steenblock, E.R., Wrzesinski, S.H., Flavell, R.A. et al. 2009. Antigen presentation on artificial acellular substrates: modular systems for flexible, adaptable immunotherapy. *Expert Opin Biol Ther* 9: 451–64.

Steiniger, S.C., Kreuter, J., Khalansky, A.S. et al. 2004. Chemotherapy of glioblastoma in rats using doxorubicin-loaded nanoparticles. *Int J Cancer* 109: 759–67.

Stolik, S., Delgado, J.A., Perez, A. et al. 2000. Measurement of the penetration depths of red and near infrared light in human "ex vivo" tissues. *J Photochem Photobiol B* 57: 90–3.

Tang, C., Russell, P.J., Martiniello-Wilks, R. et al. 2010. Nanoparticles and cellular carriers—allies in cancer imaging and cellular gene therapy? *Stem Cells* 28: 1686–702.

Tang, M., Russell, P.J., and Khatri, A. 2007. Magnetic nanoparticles: prospects in cancer imaging and therapy. *Discov Med* 7: 68–74.

Turkevich, J., Stevenson, P.C., and Hillier, J. 1951. A study of the nucleation and growth processes in the synthesis of colloidal gold. *Discuss Faraday Soc* 11: 55–75.

Ugel, S., Zoso, A., De Santo, C. et al. 2009. In vivo administration of artificial antigen-presenting cells activates low-avidity T cells for treatment of cancer. *Cancer Res* 69: 9376–84.

Vinogradov, S.V., Batrakova, E.V., and Kabanov, A.V. 2004. Nanogels for oligonucleotide delivery to the brain. *Bioconjugate Chem* 15: 50–60.

28 Nanotechnology in Extra-Axial and Pituitary Tumors

Aimee Two, J. Peter Gruen, and Gabriel Zada

CONTENTS

INTRODUCTION

Extra-axial tumors of the central nervous system (CNS) originate outside the brain and spinal cord parenchyma and typically result in symptoms due to mass effect on surrounding neurovascular structures. This group of lesions is most commonly composed of meningiomas, schwannomas, metastatic lesions, dermoid/epidermoid tumors, bone tumors, chordomas and chondrosarcomas, and pituitary adenomas. While many of these neoplasms are benign with regard to inherent biology, local invasion and recurrences may occur frequently, thus presenting significant treatment challenges for both patients and practitioners. Surgical resection has historically been the primary means to treat these tumors, resulting in histological confirmation of tumor type, decreased mass effect, and potential for long-term remission or cure.

For many extra-axial tumors, such as meningiomas, pituitary adenomas, chordomas, and epidermoid tumors, the ability to achieve gross total surgical resection is an important aspect of the operation and has been correlated with long-term surgical remission or cure. In functional pituitary adenomas, the ability to maximally resect adenoma tissue while preserving normal pituitary gland function often results in surgical remission of endocrinopathies, such as acromegaly and Cushing's disease, and avoidance of hypopituitarism. However, surgery carries risks, which must always be weighed against potential benefits before proceeding with any invasive treatment. Depending on tumor pathology, additional treatment strategies, including chemotherapy and radiation, may be included as adjunctive modalities for a variety of extra-axial tumors but also carry their own side effects. For these reasons, emphasis has recently been placed on developing newer approaches to diagnosing, visualizing, and maximizing resection of these tumors. With recent advances in nanotechnology, targeting specific tumor cells may become more precise, thereby preserving normal neurological tissue and helping to provide more focused treatments that minimize adverse effects. The remainder of this chapter focuses on the application of current and developing nanotechnology techniques to various diagnostic and treatment areas, including neuroimaging, medical treatment, radiation therapy and radiosurgery, and surgical intervention, and how their use can improve outcomes for extra-axial brain tumors.

NANOTECHNOLOGY AND NEUROIMAGING

Advanced neuroimaging plays an important role in detecting and localizing brain tumors. Current imaging techniques, such as computed tomography and magnetic resonance imaging (MRI) scans, exploit different properties of cerebral tumors, such as their density and water content, to allow gross visualization and differentiation of tumor characteristics. Although these primary neuroimaging modalities have revolutionized the neurosciences and the ability to safely diagnose and resect these lesions, they often fail to provide finer structural information that would aid in more precisely differentiating neoplastic from non-neoplastic cells and in determining tumor type and biological behavior based on imaging features alone. Advances in nanotechnology may help overcome many of these shortcomings. Quantum dots and magnetic nanoparticles, for instance, have been explored as potential imaging agents with the ability to bind and label specific tumor cells. These labels can aid in more accurately distinguishing tumor cells from surrounding brain parenchyma and local inflammatory cells in both preoperative and postoperative settings, as well as in defining resectable margins intraoperatively. Quantum dots are nanoparticles that consist of a polymer coating surrounding a central core. This central core absorbs white light and re-emits it several nanoseconds later at a specific wavelength as monochromatic light with extremely bright fluorescence. The polymer coating on these nanoparticles enables targeting molecules, such as monoclonal antibodies, to be attached to them. When these molecules are directed against ligands or receptors known to be located specifically on cells of certain tumors, these particles can be injected systemically into patients and subsequently used as contrast agents for tumor detection. Several studies have already investigated the potential of quantum dots to act as probes for cancer cells. In one such study, quantum dots designed to target Her2, a molecule often overexpressed in breast cancer cells, successfully bound to and labeled the intended Her2 targets in these cancerous cells, proving their potential use as tumor markers (Wu et al. 2003). These findings also suggest an extremely useful role for quantum dots intraoperatively. With proper imaging tools, such as a fluorescent filter attached to an operative microscope, quantum dots may be helpful in guiding surgical resections by making cancerous cells visible and thereby potentially allowing more selective and thorough surgical resections.

Magnetic nanoparticles are also proving to be helpful in tumor imaging. Like quantum dots, magnetic nanoparticles are useful as contrast agents when injected systemically. Instead of using fluorescent labeling, this form of nanotechnology utilizes monocrystalline iron oxide nanoparticles (MIONs) that appear hyperintense on MRI to highlight intracranial neoplasms. When conjugated with antibodies directed against tumor cells, these particles will selectively demarcate cancerous cells. MIONs can also be used to demonstrate evidence of residual tumor after surgical resection. In one study, MIONs were injected intravenously into male rats 2 weeks after malignant glioma cells had been stereotactically implanted into their basal ganglia (Knauth et al. 2001). The rats then underwent open craniotomy, during which only a portion of their tumors was resected. Postoperative MRI scans revealed residual tumor signal as expected, but with no new hyperintensities indicating surgically induced changes (Knauth et al. 2001). Had gadolinium been used instead of MIONs, intense enhancement of surgery-induced changes would have prevented distinguishing between these changes and the residual tumor. Therefore, these nanoparticles may play a role not only in postoperative imaging but also in intraoperative imaging as the use of intraoperative MRI becomes more widespread.

Another benefit of magnetic nanoparticles is that they may be more sensitive than gadolinium in detecting malignant tumors. In one study of ultrasmall superparamagnetic iron oxide (USPIO) nanoparticles, patients with tumors known to enhance initially on MRI scans underwent scans after injections of both gadolinium and USPIO. Several patients had regions of increased enhancement representing either extended regions of tumor or new lesions altogether. Follow-up scans obtained 5 months later revealed gadolinium enhancement at the sites that had previously only enhanced with USPIO nanoparticles, suggesting that USPIO may demonstrate increased sensitivity to tumor detection on MRI compared to gadolinium (Neuwelt et al. 2004).

Although quantum dots and magnetic nanoparticles can potentially be used in the future to detect all extra-axial tumors, their use in current neurosurgical protocols remains limited given that it requires knowledge of a biomarker specific for each extra-axial tumor. One major limitation in this paradigm is that current research has failed to uncover such novel and specific biomarkers for extra-axial tumors. Nonetheless, there may still be a current role for these molecules in detecting extra-axial metastasis once biomarkers from the primary tumor are identified. Assuming that such biomarkers would label cells not otherwise found in the CNS, any cell in the CNS labeled by these particles could be assumed to be malignant.

Another possible strategy for the current application of such nanoparticles to current neurosurgery practices takes advantage of the fact that certain receptors are up-regulated in cancerous cells. If the difference in fluorescence intensity between cancerous cells and noncancerous cells is great enough to be detectable, this information could be used to differentiate cancerous cells from their normal counterparts. For example, studies on the expression pattern of growth hormone secretagogue receptors showed that this receptor is greatly overexpressed on the surface of several types of functional pituitary adenomas (Korbonits et al. 1998; Skinner et al. 1998). In theory, with more receptors able to bind quantum dots, adenoma cells should emit more fluorescence than normal pituitary cells, making them more distinguishable from normal pituitary cells and potentially easier to resect.

PHARMACOLOGIC INTERVENTIONS

Advances in nanotechnology have also led to new developments in drug delivery systems that can be used for improved medical therapies for tumors. These medical therapies can be used alone or in conjunction with surgical resection to achieve tumor control. Most chemotherapeutic drugs are administered orally or intravenously, where they enter the systemic circulation and exert widespread effects on both cancerous and noncancerous cells throughout the body. The toxicity of these drugs on normal cells can result in devastating side effects. Further complicating the issue of systemic drug delivery is the presence of membrane barriers that protect certain organs by only allowing the extravasation of a limited number of substances with certain properties through these barriers. By using more specific targeting systems that allow drugs to act on cell-specific markers, as well as advances in time-induced release, nanotechnology can greatly enhance the efficacy and safety of antitumor therapy.

Polymeric nanoparticles, polymeric micelles, liposomes, and dendrimers represent the main classes of nanoparticles with potential for improved delivery of pharmacologic agents to extra-axial tumors. Polymeric nanoparticles are colloid particles that carry drugs on or within a degradable polymer matrix. Their small size allows them to penetrate small capillaries, providing increased drug accumulation at their target sites and permitting them to pass through capillaries of the blood–brain barrier (Desai et al. 1997; Thomas and Klibanov 2003). These biodegradable polymers can also provide sustained drug release at target sites over extended periods of time. Their capsules can be manipulated so that they break down at predetermined rates and release drugs only in specific environments. Polymeric micelles have similar properties; these molecules have a hydrophobic core where drugs can be stored and protected from the environment by the molecule's hydrophilic mantle. Furthermore, these nanoparticles have increased circulation time because of their small size and hydrophilic shell, which assist them in evading the reticuloendothelial system (Jones and Leroux 1999; Torchilin 2001). An added benefit of these micelles is their ability to be modified by conjugating target ligands to their surface so that the micelles preferentially bind to tumor cells and release drugs at these sites. Liposomes share these benefits of cellular targeting and decreased early degradation. These vesicles consist of a lipid layer that can store drugs in their central aqueous space or in the lipid layer itself. Dendrimers are the final group of nanoparticles used for drug delivery to tumor sites. These macromolecules have cavities in their core where drugs can be stored. Their surfaces are densely packed with highly branched terminal groups that serve different purposes, including attaching to specific tumor ligands or receptors and binding fluorescent molecules. Dendrimers containing both these types of terminal groups enable simultaneous treatment and visualization of tumors, which is important for research on antitumor therapies (Balzani et al. 2003).

Using these nanoparticles as drug delivery systems offers great potential for treating extra-axial tumors. Similar to nanotechnology advances in tumor imaging, however, this technology is currently limited by the availability of novel biomarkers for extra-axial tumors to which these nanoparticles can be directed.

Drug implant systems are another area of drug delivery that has been impacted by advancements in nanotechnology. Unlike previous microchips containing only a single well for drug storage and with no means to control drug release, more modern microchips contain multiple drug reservoirs whose contents can be released based on cues from external stimuli. In one such microchip model, gold membrane anodes are used to seal individual reservoirs so that drugs are released from each reservoir when a current is applied between the gold membrane and an embedded cathode (Santini et al. 1999). This technology provides precise control over drug release; if different drugs are loaded in the reservoirs, differential release of different drugs at various times can be achieved. Other implantable devices currently under investigation use stimuli such as heat, pH, or electrical charges to control drug release (Ichimura et al. 2000). Ideally, in the future, these devices will also have sensors and feedback systems to measure the concentrations of drug biomarkers and to release appropriate amounts of these drugs in response to current levels, making them fully autonomous. The fact that these devices are implanted directly at the tumor site avoids issues such as drug targeting and systemic side effects. They can be used as monotherapy or adjunctive measures in combination with surgical resection. Extra-axial tumors, such as meningiomas, schwannomas, and bone tumors, are ideally suited for treatment with these devices. Their location outside the brain parenchyma enables these chips to be placed relatively safely without damaging brain tissue.

STEREOTACTIC RADIOSURGERY AND NANOTECHNOLOGY

Compared to the time of its development more than 50 years ago, stereotactic radiosurgery has come to play a significant role in treating several types of tumors. The treatment of meningiomas, pituitary adenomas, and schwannomas, in particular, has changed greatly because stereotactic radiosurgery has improved the ability to selectively target these frequently radiosensitive tumors. In a recent study by Kondziolka et al. (2008), the long-term tumor control rates of meningiomas after surgical resection, followed by stereotactic radiosurgery, were reported as higher than 85% over a 10-year follow-up period (Zada et al. 2010). While these results indicate good long-term results, these rates could potentially be improved even further by combining nanosurgical techniques and radiosurgery (Leary et al. 2006). As an example, nanoshells are gold-layered dielectric nanoparticles that readily extravasate into the abnormal vasculature of tumor tissue on account of their small size. These particles can be filled with antitumor agents. After systemic injection, they preferentially extravasate into and bind tumor tissue, where they remain biologically inert until activated by a source of thermal or ionizing ablation, such as radiosurgery. With this combined approach to tumor treatment, higher levels of remission and lower levels of required radiation may be possible, thereby potentially decreasing the toxicity associated with radiosurgery. This application would be especially beneficial in treating pituitary adenomas and meningiomas of the sellar and suprasellar regions, where the proximity of the optic nerve, tract, and chiasm often limits the amount of radiation that these tumors can receive.

NANONEUROSURGICAL TOOLS AND EXTRA-AXIAL TUMOR RESECTION

The development of imaging modalities and implantable devices that function at the nanoscale level would have very limited applications unless surgical tools were also available to permit surgical manipulation at this scale. Current surgical tools and equipment designed for microsurgery are much too bulky to be used at the nanoscale level. Nanotools, such as nanoknives, nanotweezers, and nanoscalpels, are currently under development; these would allow precise manipulation of tissue at the micron-level scale. In addition, lasers such as the femtosecond laser system can ablate single cells with only minimal damage to surrounding structures (Maxwell et al. 2005). When combined with imaging techniques that identify individual tumor cells, total tumor resection with maximal preservation of normal tissue becomes an ever greater reality.

CONCLUSIONS

The field of nanoneurosurgery holds exciting promise in contributing to the future treatment of extra-axial brain tumors. Through nanotechnology's progress in advanced neuroimaging, drug delivery devices, and surgical tools, the ability to not only better treat but also potentially cure these frequently benign tumors is becoming a reality. In addition, less emphasis may be placed on aggressive surgical resection, realizing that adjunctive nanotechnology techniques, such as nanoparticles for drug therapy or stereotactic radiosurgery, are even more effective. In turn, less aggressive resections may reduce postoperative neurological deficits, minimize the extent of surgical approaches, and ultimately contribute to shorter operative time and faster postoperative recovery.

REFERENCES

Balzani, V., Ceroni, P., Maestri, M. et al. 2003. Luminescent dendrimers. Recent advances. *Top Curr Chem* 228: 159–91.

Desai, M.P., Labhasetwar, V., Walter, E. et al. 1997. The mechanism of uptake of biodegradable microparticles in Caco-2 cells is size dependent. *Pharm Res* 14: 1568–73.

Ichimura, K., Oh, S.K., and Nakagawa, M. 2000. Light-driven motion of liquids on a photoresponsive surface. *Science* 288: 1624–6.

Jones, M., and Leroux, J. 1999. Polymeric micelles—a new generation of colloidal drug carriers. *Eur J Pharm Biopharm* 48: 101–11.

Knauth, M., Egelhof, T., Roth, S.U. et al. 2001. Monocrystalline iron oxide nanoparticles: possible solution to the problem of surgically induced intracranial contrast enhancement in intraoperative MR imaging. *AJNR Am J Neuroradiol* 22: 99–102.

Kondziolka, D., Mathieu, D., Lunsford, L.D. et al. 2008. Radiosurgery as definitive management of intracranial meningiomas. *Neurosurgery* 62: 53–8; discussion 58–60.

Korbonits, M., Jacobs, R.A., Aylwin, S.J. et al. 1998. Expression of the growth hormone secretagogue receptor in pituitary adenomas and other neuroendocrine tumors. *J Clin Endocrinol Metab* 83: 3624–30.

Leary, S.P., Liu, C.Y., and Apuzzo, M.L. 2006. Toward the emergence of nanoneurosurgery: Part III. Nanomedicine: targeted nanotherapy, nanosurgery, and progress toward the realization of nanoneurosurgery. *Neurosurgery* 58: 1009–26; discussion 1009–26.

Maxwell, I., Chung, S., and Mazur, E. 2005. Nanoprocessing of subcellular targets using femtosecond laser pulses. *Med Laser Appl* 20: 193–200.

Neuwelt, E.A., Varallyay, P., Bago, A.G. et al. 2004. Imaging of iron oxide nanoparticles by MR and light microscopy in patients with malignant brain tumours. *Neuropathol Appl Neurobiol* 30: 456–71.

Santini, J.T. Jr., Cima, M.J., and Langer, R. 1999. A controlled-release microchip. *Nature* 397: 335–8.

Skinner, M.M., Nass, R., Lopes, B. et al. 1998. Growth hormone secretagogue receptor expression in human pituitary tumors. *J Clin Endocrinol Metab* 83: 4314–20.

Thomas, M., and Klibanov, A.M. 2003. Conjugation to gold nanoparticles enhances polyethylenimine's transfer of plasmid DNA into mammalian cells. *Proc Natl Acad Sci U S A* 100: 9138–43.

Torchilin, V.P. 2001. Structure and design of polymeric surfactant-based drug delivery systems. *J Controlled Release* 73: 137–72.

Wu, X., Liu, H., Liu, J. et al. 2003. Immunofluorescent labeling of cancer marker Her2 and other cellular targets with semiconductor quantum dots. *Nat Biotechnol* 21: 41–6.

Zada, G., Pagnini, P.G., Yu, C. et al. 2010. Long-term outcomes and patterns of tumor progression after gamma knife radiosurgery for benign meningiomas. *Neurosurgery* 67: 322–8; discussion 328–9.

29 Using Nanoparticles in Diagnosis and Treatment of CNS Infection

Zion Zibly, Cody D. Schlaff, and John D. Heiss

CONTENTS

In Thasus, the wife of Dealces, who was lodged upon the Plain, from sorrow was seized with an acute fever, attended with chills... no sleep; bowels irritable, but passed nothing... urine thin and scanty; to the touch of the hand the fever was slight; coldness of the extremities... On the fourteenth, breathing rare, large at intervals... On the twentieth, much talk, and again became composed; loss of speech; respiration hurried. On the twenty-first she died. Her respiration throughout was rare and large; she was totally insensible... phrenitis [what is now commonly described as meningitis or encephalitis.]

—The Epidemics, Section III, Constitution 2, Section 17, Case XV of Hippocrates
(Hippocrates and Adams 1886)

INTRODUCTION

The existence and devastation of central nervous system (CNS) infections have plagued and bewildered physicians since they were first described over two millennia ago in works such as *The Epidemics* of Hippocrates. CNS infections are characterized by inflammation of various components of the nervous system [i.e., brain, meninges, vasculature, cerebrospinal fluid (CSF), and ventricles]. Hematogenous or local invasion of the CNS introduces a nidus of bacteria that develops into an active infection if the immune response is attenuated by host or bacterial factors. Multiplication of the pathogen, subsequent inflammatory response within the brain parenchyma and blood vessels, and microvascular thrombosis can permanently damage the nervous system. CNS infections may rapidly become life-threatening if they are not diagnosed and treated promptly. Despite numerous advances in medical treatment (new antimicrobial and anti-inflammatory agents) over the past century, CNS infections often require prolonged and sometimes complex treatment regimens. Preferred are antibiotics that have sufficient blood–brain barrier (BBB) permeability to reach the CNS target and bactericidal activity against the microorganism—requirements that are becoming progressively more difficult to satisfy in an era with widespread development of multidrug-resistant (MDR) microorganisms.

Since current conventional treatment modalities will eventually become ineffective as microorganisms become resistant to them, new antimicrobial agents and drug delivery methods that can overcome drug resistance mechanisms must be developed to treat future CNS infections. Successful development of nanoparticles as drug carriers for oncologic applications suggests that antibiotic formulations using nanoparticles might increase CNS transport of antibiotics that otherwise have poor BBB penetration. These nanoconjugates are a new family of drugs that can be designed to target a specific microorganism and to avoid multidrug resistance.

In this chapter, we will provide an overview of CNS infections and current treatment methods. The role of the BBB in providing immune surveillance, preventing CNS infection, and limiting access of drugs to the CNS will be discussed. The state-of-the-art in research into the possible role of nanoparticles in treating CNS infection will be reviewed. Finally, steps required to develop and apply nanomedicine to clinical CNS infections will be outlined.

PATHOGENESIS

Most CNS infections are caused by bacteria, viruses, fungi, or protozoa. In addition, some CNS infections are the result of misfolded proteins (prions) or parasitic worms (helminthes). CNS infections can originate by a variety of routes. Bacteria, fungi, protozoa, and worms usually gain access to the CNS by hematogenous (arterial or venous) delivery or direct extension from an adjacent anatomical structure. Bacteria may alternatively be introduced through traumatic opening of the CNS (penetrating trauma, open skull fracture) or bacterial seeding during a surgical procedure. Viruses usually invade the CNS through the axons of peripheral or cranial nerves. Most pathogens initially multiply at extraneural sites and then enter the CNS by crossing the BBB.

The four leading types of CNS infection are as follows: (1) viral meningitis, in which viral infection induces an inflammatory process within the meningeal layers covering the brain and spinal cord and produces the most frequent form of aseptic (nonbacterial) meningitis; (2) septic meningitis, in which bacterial infection of the meninges creates a robust inflammatory response within the meninges and CSF; (3) encephalitis, an infectious process restricted to brain tissue (parenchyma) in which nonviral forms may lead to brain abscess formation; and (4) meningoencephalitis, an infection of the brain parenchyma and meninges (Scheld et al. 2004).

EPIDEMIOLOGY AND PATIENT GROUPS

Diagnosis and proper treatment of CNS infections constitute a medical emergency because delay may result in disabling neurological deficits (Venes 2001; Rantakallio et al. 1986). Meningitis of bacterial and fungal origin requires prompt initiation of antimicrobial therapy. Fever and meningismus after a surgical operation on the brain or spine may signal the onset of meningitis. Diagnostic evaluation of patients with signs of meningitis may disclose noninfectious causes of aseptic meningitis, such as an inflammatory reaction within the leptomeninges to the presence of cancer cells that have spread to the CNS from a systemic malignancy.

Bacterial meningitis has a high mortality rate, ranging from 3% to 21%, depending on the responsible bacterium (Sakushima et al. 2011; Schuchat et al. 1997; van de Beek et al. 2004). In the United States, there are approximately 25,000 cases annually (Roos 2005; Thigpen et al. 2011). Bacterial meningitis has an annual incidence of 4–6 cases per 100,000 persons (van de Beek et al. 2006). Incidence rates are higher in developing countries. The incidence is much greater in African countries encompassing the "meningitis belt," a region of 21 countries from Senegal to Ethiopia (Roos 2005). The World Health Organization (WHO) reported that more than 400 million people live within the "meningitis belt." In a recent 15-year period (1996–2010), 800,000 cases were reported there, with 10% resulting in death.

Meningococcal and pneumococcal (*Streptococcus pneumoniae*) meningitis makes up nearly 80% of bacterial meningitis cases (Schuchat et al. 1997; van de Beek et al. 2006). Meningococcal (*Neisseria meningitidis*) meningitis has the highest mortality rate. *Hemophilus influenza* (Phifer's bacillus or *Bacillus influenza*) was formerly a frequent cause of infantile meningitis in the Western world; however, after the introduction of protein-conjugated vaccines for *H. influenza* in 1988, its incidence dropped from 100 cases per 100,000 in 1990 to 0.3 cases per 100,000 in 1996 (Roos 2005; Schuchat et al. 1997).

Patients with the greatest susceptibility to CNS infections are those with diseases or on treatments that create cell-mediated immunity defects such as: (1) acquired immune deficiency syndrome (AIDS); (2) Hodgkin's disease; (3) drug regimens that prevent graft rejection in organ transplant recipients; and (4) daily corticosteroid therapy for autoimmune diseases. Among immunocompromised patients, the most common causative agents are *S. pneumoniae*, *Listeria monocytogenes*, and Gram-negative Enterobacteriaceae. The underlying disease and subsequent treatment, duration of immunosuppression, and type of immune abnormality all play major roles in determining the risk for acquiring bacterial meningitis (Rubin and Young 2002; Scheld et al. 2004).

The primary immune defects that most frequently increase the risk of bacterial meningitis are as follows: (1) impaired cell-mediated immunity resulting from a defect in T lymphocytes and other mononuclear phagocytes; (2) impaired neutrophil activity resulting from their insufficient number or decreased function; and (3) defects in humoral immunity (e.g., antibody response). Deficits in humoral immunity may arise after splenectomy. Impaired nutritional status from chronic disease can result in general suppression of the immune system and increased susceptibility to CNS infections (Armstrong 1984). Open epithelial/skin barriers also increase the risk of sepsis leading to meningitis. Nosocomial (acquired in a healthcare setting) bacterial meningitis usually results from invasion of commonly found bacteria (*Staphylococcus aureus* and *Staphylococcus epidermidis*) and Gram-negative bacilli (Roos 2005; Scheld et al. 2004), often in the context of placing a foreign body such as a CSF shunt or a deep-brain stimulator (DBS) system (Forward et al. 1983; Piacentino et al. 2011; Schoenbaum et al. 1975). Approximately 60%–70% of all meningitis cases after neurosurgical procedures are caused by Gram-negative bacilli, most commonly *Klebsiella pneumoniae*, *Acinetobacter baumanii*, and *Escherichia coli*. Craniotomies for trauma or tumor resection are also more likely to be associated with infection of these organisms.

Aseptic meningitis refers to the clinical syndrome of meningeal inflammation in which bacterial pathogens cannot be identified in the CSF. There are many causes of aseptic meningitis, although the vast majority of cases are of viral origin. Enteroviruses account for nearly 85%–95% of all aseptic meningitis cases (Rotbart 2000). These viruses are acquired primarily through fecal–oral contamination, where the virus, once swallowed, binds to specific receptors on enterocytes. Other causes of aseptic meningitis include fungal, parasitic, noninfectious, and even undetected bacterial infections. Although aseptic meningitis implies a benign clinical course and the absence of long-term sequelae, some causes are associated with substantial morbidity and recurrent episodes (Rotbart 2000; Scheld et al. 2004; Wallgren 1951).

DIAGNOSIS AND IMAGING

The classical clinical presentation of meningitis includes fever, neck stiffness, headache, and altered mental status. However, the diagnosis should be suspected even in the absence of some of these elements. In a prospective study of Dutch patients infected with community-acquired bacterial meningitis, at least two of the four signs of infection were present in 95% of patients (van de Beek et al. 2004). Other clinical presentations may include photophobia, vomiting, and lethargy. As the disease progresses, consciousness levels deteriorate to the level of coma (Roos 2005; van de Beek et al. 2004). Presentation of petechial rash is suggestive of *N. meningitidis*. Focal neurological symptoms, such as paresis, aphasia, and cranial nerve palsy, are found in approximately 15% of patients. The diagnosis is usually confirmed by analysis of CSF obtained via lumbar puncture. Polymorphonuclear CSF pleocytosis and hypoglycorrhachia (low CSF glucose) signal septic meningitis, and Gram stain and culture usually identify the causative organism.

Computed tomography (CT) scans often prove useful in detecting the extent and severity of a bacterial meningitis infection. Contrast-enhanced CT may help predict impending neurological sequelae and reveal meningeal, cisternal, or parenchymal enhancement (Mathews et al. 1989). The common abnormalities seen on CT that arise secondary to bacterial meningitis are brain edema, hydrocephalus, cerebritis, abscess, subdural empyema or effusion, ventriculitis, and infarction. The pathogenic bacterium may compromise the BBB, resulting in abnormal contrast enhancement. Other imaging techniques, such as magnetic resonance imaging (MRI), magnetic resonance angiography (MRA), or transcranial Doppler sonography (TCD), may also be employed to provide a more complete understanding of the extent and effects of the infection (Enzmann 1984).

BBB AND CSF

What we now call the BBB was first described by Paul Ehrlich in 1885 after a series of in vivo experiments. His work with his student, Edwin Goldman, led to the discovery of this barrier between the CNS and the systemic circulation (Ehrlich 1885; Hartz and Bauer 2010). Lewandowsky (1900) described Ehrlich's findings in German as a *Bluthirnschranke*, which in English translates to "blood–brain barrier" (Hartz and Bauer 2010).

The BBB is a complex structure. An associated structure is the blood–CSF (BCSF) barrier, which prevents the passage of most molecules from the systemic circulation to the CSF. In the last decade, research has improved our understanding of the histology and functions of the BBB and the BCSF barrier. Although these structures were originally thought to function principally as impregnable barriers to the movement and exchange of different molecules between the blood, CSF, and brain, it has become increasingly clear that they retain the ability to transport certain molecules into the CNS and CSF. The BBB prevents most systemically circulating drugs from entering the brain. Drugs that can penetrate the BBB still have much lower levels (1%–40%) of active drug in the CSF than in the systemic circulation. A study of a comprehensive medicinal chemistry database showed that less than 5% of the drugs have a CNS effect (Iqbal et al. 2011).

The BBB is a highly active, selective, and dynamic structure (Deeken and Loscher 2007; Donelli et al. 1992; Neuwelt et al. 1986, 1999). Exposure of the BBB to different conditions and stimuli, such as infection, inflammation, ischemia, and malignancy, alters its properties and increases its permeability, allowing greater entrance of certain molecules into the CNS. In ischemic stroke and traumatic brain injury, BBB permeability is increased as a consequence of the disease process. In multiple sclerosis, increased permeability to immune factors may precipitate tissue injury and further inflammatory changes in the brain.

In addition to the BBB's function as a static structure that prevents molecules from passing from the systemic circulation to the brain, it also has an immunologic role. The BBB maintains the isolated environment (sanctuary site) of the brain, separating the brain's immune system from the peripheral immune system. It also keeps the environment within the brain sterile by preventing various microorganisms, such as bacteria and viruses, from reaching the brain (Hartz and Bauer 2010). The relative inaccessibility of the CNS to circulating drugs and components of the immune system becomes an impediment to treatment when the CNS develops a bacterial, viral, or parasitic infection.

The ability of a drug to cross the BBB depends on several physicochemical properties, including size, electric charge, and lipophobicity (Boockvar et al. 2011). In addition to the anatomical features of the BBB that make it a highly selective gateway for entrance, it has been recently revealed that this structure contains numerous membrane-based energy-dependent transporters (ATP-binding cassettes) that control the molecular milieu of the brain and CSF by refluxing molecules from the

CNS to the blood against their concentration gradient. New discoveries suggest that these ATP-dependent pumps actively reduce the effects of drugs that penetrate the CNS or CSF. For instance, antiepileptic drug resistance is partly caused by elevated expression of these pumps (Aronica et al. 2003, 2004). The BBB also contains integrated enzymes within its endothelium to inactivate xenobiotics and to limit their passage from the systemic circulation to the brain (Hartz and Bauer 2010).

Pharmacological and mechanical methods can be used to increase BBB permeability. For example, intra-arterial administration of osmotic agents, such as mannitol and urea, may open the barrier (Kroll and Neuwelt 1998). However, the brief duration of barrier opening and the need for arterial catheterization make it impractical for treating most infections. Another method for increasing BBB permeability is via the systemic delivery of cytotoxic drugs. Unfortunately, the toxicity of most of these drugs counteracts any beneficial effects achieved by opening the BBB. Recently, it was also shown that using high-intensity focused ultrasound (HIFU) can disrupt the BBB and enhance drug delivery to the brain. Methods other than BBB opening were used to increase levels of antimicrobial agents in the CNS and CSF, including intrathecal injection, dose augmentation, and drug solubility modification. Unfortunately, they may result in side effects, including headache, drug-induced fever, vomiting, and neurotoxicity. The limitations and side effects of current treatment have spurred efforts to develop novel agents, including nanoparticles of both synthetic and biological origin, as potential carriers and transporters of hydrophilic and sterically unfavorable molecules across the BBB.

NANOMEDICINE IN CNS INFECTIONS

The application of nanotechnology to CNS diseases has, until recently, focused primarily on the treatment of brain tumors, neurodegenerative disorders such as Alzheimer's and Parkinson's diseases, and acute ischemic stroke (Wong et al. 2011). A multitude of nanoparticle formulations was created, ranging from gold and other transition metal carriers, to organic polymeric carriers, to already existing biological nanoparticles (e.g., baculovirus, adenoassociated virus, etc.).

Some of the strategies utilized to introduce nanoparticles into the CNS to treat brain malignancies and neurodegenerative disorders can be applied to treat CNS infections. For the remaining discussion, we will begin by reviewing the nanotechnology being developed for a more sensitive detection of CNS human immunodeficiency virus (HIV) infection. Then, we will focus on the various nanotechnology-based methods to treat HIV infection and their associated neurological effects, keeping in mind that many of the methods explained here may also be used pharmacologically against other CNS infections such as meningitis, encephalitis, and transmissible spongiform encephalopathy (TSE; prion disease). Finally, we will discuss some of the mechanisms that nanoparticles can exploit to gain entrance into the brain.

NANOTECHNOLOGY IN THE DETECTION OF CNS INFECTIONS

In regions of the world with high rates of HIV infection, particularly developing countries, reliable and cost-effective viral detection methods are critically important. In developed countries, the gold standard detection method is based on flow cytometry, which is expensive and technically demanding. Nanotechnology-based techniques have a high degree of sensitivity and specificity, can multiplex, and do not require enzymes for functioning (Nowacek and Gendelman 2009). Additionally, they can be portable and do not require highly technical skills. An accurate and uncomplicated method for detecting the virus is essential for early diagnosis of HIV infection and for tracking therapeutic response and disease recurrence.

A promising revolutionary new amplification and detection technique is the bio-barcode assay (Hill and Mirkin 2006). This technique can identify exceptionally low numbers of proteins and short oligonucleotide molecules without using polymerase chain reaction (PCR). Two types of particles are involved: (1) magnetic microparticles (MMPs), which contain recognition elements for the target of interest (e.g., antibodies and aptamers) that are covalently attached to the surface; and (2) gold nanoparticles (Au-NP), which contain another recognition agent that can form a "sandwich" around the target with MMPs. Au-NPs also contain barcodes. After the two types of particles are incubated with the target and after the sandwich is formed, a magnetic field is used to localize and collect the sandwiched targets, while the remaining unbound Au-NPs are washed away. A dithiothreitol (DTT) solution at high temperature is then used to liberate Au-NPs. Once released, the barcodes can be analyzed by scanometric analysis (such as by microarray analysis using more nanoparticle probes) or in situ (if the barcodes carry a detectable marker, such as fluorophore, chemiluminescent probe, or redox-active moiety). A variety of factors (ease of Au-NP fabrication, greater oligonucleotide binding capabilities, stability under a variety of conditions, and optical properties) make this technique extremely advantageous for detecting very small quantities of proteins and nucleic acids on the order of attomolar (10^{-18} M) and zeptomolar (10^{-21} M), respectively. When modified to detect HIV-1 p24, the bio-barcode assay was approximately 150 times more sensitive to p24 detection with nearly 100% sensitivity and specificity when compared to the traditional enzyme-linked immunosorbent assay (Hill and Mirkin 2006).

As previously mentioned, in developed countries, the standard method for enumerating $CD4^+$ T lymphocytes has been flow cytometry. The severity of immune system compromise increases as the number of these lymphocytes in the blood declines. The inability to test CD4 HIV immune function and to establish disease severity in developing countries has been described as the "missing link" in managing HIV^+ patients (Nowacek and Gendelman 2009). The optical requirements for counting $CD4^+$ T lymphocytes are significantly reduced when semiconductor nanocrystals, known as quantum dots (QDs), are integrated into microfluidic systems. The QD method has shown good correlation ($R^2 = 0.97$) with flow cytometry measurements and is an attractive alternative to other methods because it is simpler, more cost-effective, and just as accurate. QDs can be integrated into portable microfluid-based lymphocyte capture and detection systems that can isolate and count $CD3^+$ $CD4^+$ lymphocyte subpopulations from whole blood (Jokerst et al. 2008; Santra et al. 2005).

A final novel HIV detection method incorporates the integration of metallic nanoparticles, such as Au-NPs, into carbon nanotubes (CNTs). CNTs and Au-NP-modified electrodes are particularly interesting because of their low-resistance ohmic contacts (in the case of CNTs), in addition to their high surface area, favorable electronic properties, ease of biomolecule attachment, and electrocatalytic effects. Most importantly, like the bio-barcode assay, this method provides ultrasensitive detection. In the case of HIV, a gold electrode surface with thiolated single-walled CNT/Au-NP modified with ferrocene–pepstatin provided an estimated detection limit of 0.8 pM for HIV-1 protease (HIV-1 PR). Previous ferrocene–peptide conjugates bound to gold electrodes enabled the identification and screening of HIV-1 PR under physiological conditions with a limit of detection of 1 nM (Mahmoud et al. 2008).

Nanoparticles in Delivering Treatment for CNS Infections

The properties of nanomaterials and nanotechnology provide unique ways to inhibit viral mechanisms such as invasion, growth, and replication. Several strategies using viral and nonviral nanoparticles have been developed. Metallic (such as gold and silver) nanoparticles have been extensively evaluated to treat various maladies (Figure 29.1). Au-NPs are attractive for nanomedicine, particularly for treating CNS infections, because of their biocompatibility, ease of construction, and ability to readily conjugate molecules to their surfaces. Silica nanoparticles, baculovirus, and transactivating transcriptor (TAT)-conjugated polymers have all been used to deliver HIV drugs or gene therapy vectors to the brain (Bharali et al. 2005; Liu et al. 2008; Rao et al. 2008). However, a crucial limiting factor with these modalities is that the biodistribution of particles after systemic administration varies greatly and that a limited number of cell types take up most of the particles. Intrathecal or intraparenchymal administration of nanoparticles can increase CNS delivery of nanoparticles (Shriver et al. 2009).

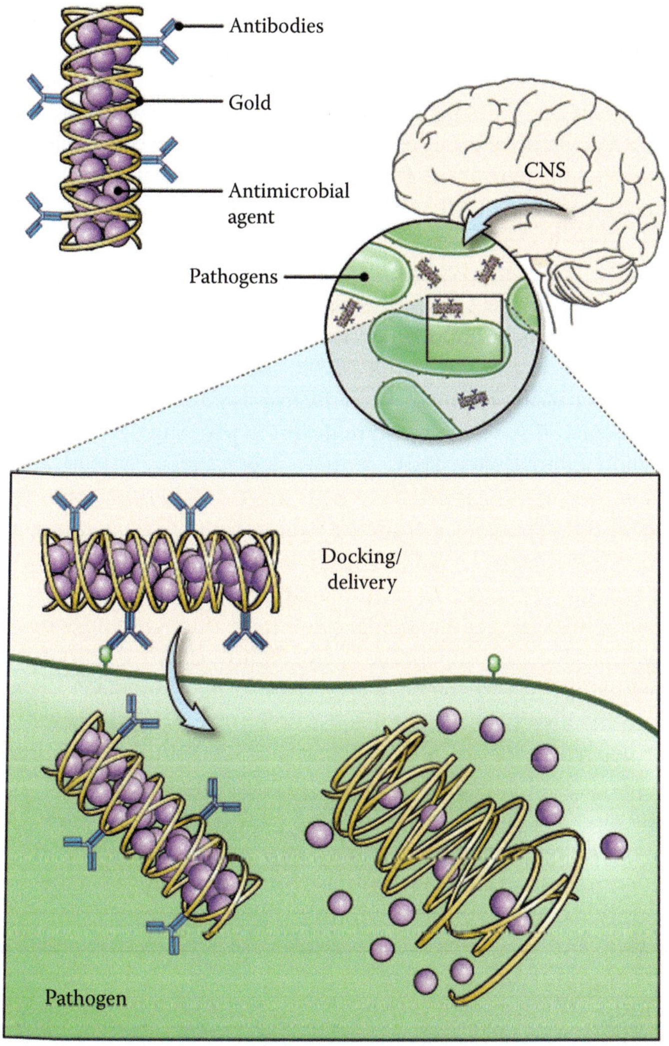

FIGURE 29.1 Au-NP cage with attached antibodies carries the antimicrobial agent to the pathogen. The cage enters the pathogen after the attachment of the antibody to the cell membrane. The antimicrobial agent is released within (or around) the pathogen to selectively eradicate the pathogen.

The in vivo distribution of viral nanoparticles depends on the intrinsic properties of the virus such as surface charge, size, and the interaction of viral proteins with molecules expressed in target cells (Shriver et al. 2009). In vitro silver nanoparticles were shown to directly interact with HIV in a size-dependent manner; only particles within the size range of 1–10 nm attached to the virus. Based on the regular spatial relationships of attached nanoparticles and exposed sulfur-bearing residues, HIV gp120 would provide attractive sites for interaction; the suggested mechanism of interaction occurs through preferential binding to HIV-1 gp120 glycoprotein knobs (Nowacek and Gendelman 2009).

The method of administration can play a role in directing nanoparticles to different tissues in the body. For instance, stereotactic injection of iron oxide-conjugated baculovirus into the cerebral ventricle of rodent results in localization of particles in cells of the choroid plexus (Raty et al. 2006). Intravenous injection of cationic albumin-conjugated PEGylated nanoparticles into mice results in accumulation of particles in the periventricular region and third ventricle. On the other hand, adenovirus and herpes virus, two viral vectors commonly used for gene therapy, require osmotic disruption or direct injection into the CNS to cross the BBB. Once across the BBB, these particles are taken up by neurons (Shriver et al. 2009).

Cowpea mosaic virus (CPMV), a plant virus, has been previously explored as a nanoparticle delivery system for therapeutic agents targeting cancer and neurodegeneration. CPMV is a member of the genus *Comovirus* and is part of the *Picornaviridae* superfamily. It is an attractive candidate for CNS infection treatment because CPMV has excellent bioavailability, is nontoxic even at high doses, and distributes to multiple organs after injection. Moreover, the reactive lysines and cysteines on the surface of the viral capsid can be conjugated with fluorescent dyes and metals (such as gadolinium for imaging studies) or even with peptide ligands for targeted drug delivery. Studies conducted to understand the biodistribution of CPMV in the CNS found that in mice infected with neurotropic mouse hepatitis virus (MHV), CPMV localized mainly to the CNS endothelium in areas that contained intact BBB. However, in inflammatory lesions with macrophage/microglial infiltration and IgG presence, CPMV localized in the brain parenchyma. In an in vitro model of the BBB, CPMV was rapidly internalized. These studies suggest that CPMV particles might be developed as a vehicle for delivering drugs across the BBB to treat CNS infections (Shriver et al. 2009).

Solid lipid nanoparticles are another attractive option for delivering therapeutics to the brain (Chattopadhyay et al. 2008; Kaur et al. 2008; Wong et al. 2011). Ranging in size from 1 to 1000 nm (<1 μm), solid lipid nanoparticles readily cross the BBB owing to their lipidic nature. Solid lipid nanoparticles are usually dispersed as a colloidal solution in water or in an aqueous surfactant solution; they have the potential to carry lipophilic or hydrophilic drugs. Additionally, their small size prolongs their circulation time in the blood. Furthermore, solid lipid nanoparticles do not exhibit burst effects, therefore resulting in controlled release of the drug from the particle. Because of their lipidic nature, these nanoparticles are bioacceptable and biodegradable, making them less toxic than polymeric nanoparticles. Solid lipid nanoparticles enable controlled release of an incorporated drug for up to several weeks. They provide high drug payload for both hydrophilic and hydrophobic drugs. The relationship between the carrier and the drug to be delivered to the brain is critical. Usually, there is an optimal lipid/surfactant ratio that yields the most favorable loading and entrapment efficiency, stability, uptake, transport, and release of each drug (Nowacek and Gendelman 2009). This optimal ratio, however, is based on the structure, size, charge, and degree of lipophilicity of the particles. The best lipid/surfactant composition ratio of a certain drug must be determined to optimize drug delivery (Nowacek and Gendelman 2009). Moreover, coating or attaching ligands to their surface can improve drug targeting.

Surfactant coating allows nanoparticles to preferentially absorb apolipoprotein E, apolipoprotein B, or both from the bloodstream and to mimic lipoprotein particles that are endocytosed by low-density lipoprotein receptors on the endothelial linings of brain blood capillaries (Chattopadhyay et al. 2008). Solid lipid nanoparticles, especially those in the range of 120–200 nm, if given orally, are not readily taken up by cells of the reticuloendothelial system (RES) and therefore bypass liver and spleen filtration. The various formulations of solid lipid nanoparticles are stable for up to 3 years, are biodegradable, and do not require organic solvents. Furthermore, feasible, large-scale, cost-effective production and sterilization techniques are available, yielding excellent reproducibility (Kaur et al. 2008).

In delivering HIV antiretroviral therapies, particularly protease inhibitors, which show fairly low brain permeability, solid lipid nanoparticles are promising. In a human brain endothelial cell line representative of the BBB, hCMEC/D3, solid lipid nanoparticles approximately 167 nm in diameter were formulated to deliver atazanavir, an azapeptide inhibitor of HIV-1 PR that would normally be too large (>500 kDa) to cross the BBB (Chattopadhyay et al. 2008). Atazanavir is a highly selective HIV-1 PR inhibitor with a mean EC_{50} of 2–5 nM in the absence of human serum, as measured in a variety of laboratory and clinical HIV-1 isolates grown in various cell lines. Compared to free drug, incorporating tritium-labeled atazanavir into nanoparticles resulted in a significantly higher accumulation of atazanavir by the endothelial cell monolayer. Additionally, cytotoxicity was seen only with nanoparticles delivering concentrations of atazanavir higher than 200 nM, a dose 50 times higher than the EC_{50} of atazanavir in vitro. Furthermore, solid lipid nanoparticles had controlled release of atazanavir, with an initial burst of 17% within the first hour and a gradual release up to 46% after 24 h.

Cationic solid lipid nanoparticles are similar to solid lipid nanoparticles. The only difference is that cationic lipid nanoparticles are modified with a positively charged lipid on their shells. Positively charged nanoparticles can improve drug loading and cellular uptake as a result of electrostatic interactions (Nowacek and Gendelman 2009). The complex structural organization of cationic solid lipid nanoparticles can improve the pharmacokinetics of drug entrapment and release. These particles have received considerable interest as a novel treatment modality for MDR infections. A common CNS infection in immunosuppressed patients, especially those diagnosed with late-stage HIV infections, is cryptococcal meningitis. Cryptococcal meningitis results from an opportunistic infection of the encapsulated yeast *Cryptococcus neoformans*. With the increasing number of *C. neoformans* isolates resistant to standard treatments (fluconazole and amphotericin B), in addition to the poor CSF penetration of standard antimicrobial agents, new antimicrobial peptides encapsulated in self-assembling cationic micelles have been created and standard antimicrobial agents have been entrapped in polysorbate 80-coated polymeric nanoparticles (Xu et al. 2011).

Polymeric nanoparticles have various mechanisms of uptake into the brain: (1) enhanced retention in brain capillaries with absorption to capillary walls, resulting in a high-concentration gradient across the BBB; (2) opening of tight junctions due to the presence of nanoparticles; and (3) transcytosis of nanoparticles through the endothelium. Coating these particles with polysorbate has been reported to improve their brain bioavailability by (1) inhibiting the efflux system, particularly P-glycoprotein (P-gp); (2) inducing endocytosis of polymeric nanoparticles as a result of their facilitated interaction with endothelial cells of the BBB; (3) inducing solubilization of endothelial cell membrane lipids; and (4) increasing membrane fluidization due to the surfactant effects of polysorbates. However, various problems are associated with using polymeric nanoparticles for drug delivery to the brain. For example, residual contaminants (such as organic solvents, polymer aggregates, toxic monomers, and degradation products) of the production process remain. Other problems include expensive production methods, lack of large-scale production methods, and unsuitable sterilization techniques. Therefore, solid lipid nanoparticles are generally preferred to polymeric nanoparticles for drug transport.

Antimicrobial peptides can affect a broad spectrum of microorganisms, including bacteria, fungi, parasites, enveloped viruses, and even cancer cells. These peptides act as an essential part of innate host resistance and serve as a first-line defense against infections. Crucial to circumventing multidrug resistance is to find therapeutics, such as these peptides, that have mechanisms of action entirely distinct from those of other antimicrobial agents. The efficacy of these self-assembling nanoparticles was evaluated in vitro and in vivo in a rabbit model. Amphiphilic cholesterol-conjugated G_3R_6TAT (CG_3R_6TAT) peptides were self-assembled into core/shell cationic nanoparticles containing TAT anchoring molecules on the shell and having a diameter of approximately 177 nm. TAT peptide is a protein transduction domain from the transcriptional activator TAT protein of HIV-1 that is commonly used to assist proteins and QDs across the BBB. These peptides are popular nanoparticle conjugates because they can pass through biological membranes without relying on transporters or receptor-mediated endocytosis (Rao et al. 2008). In determining the minimum inhibitory concentration (MIC) of these particles, they found that, in 12 yeast isolates, concentrations between 1.0 and 8.1 μmol/L inhibited growth and induced vacuolar changes, cell wall breakdown, and release of cytoplasm. This value was only slightly higher than that of amphotericin B, but threefold higher than that of fluconazole. The nanoparticles successfully crossed the BBB through absorptive endocytosis and resulted in a significant and continuous drop in CSF yeast counts compared to controls. There was no significant difference in CSF yeast counts between treatment with the nanoparticle and treatment with amphotericin B. Also, yeast counts were significantly reduced in the brain parenchyma after 11 days of treatment with CG_3R_6TAT nanoparticles but were not significantly different after treatment with amphotericin. Despite being equally effective as amphotericin B, CG_3R_6TAT nanoparticles, unlike amphotericin B, do not cause significant damage to liver or kidney function and do not interfere with serum electrolyte balance. The mechanism of action of these nanoparticles is not fully understood, but one hypothesis is that, unlike conventional antimicrobial agents that target specific metabolic enzymes and thereby selectively induce drug resistance, nanoparticles containing antimicrobial peptides may target membrane-targeting pore-forming mechanisms that may disrupt the cell wall and induce lysis. This mechanism is inherently more difficult for microbes to circumvent.

Monoclonal antibodies against specific receptors may also help target nanoparticles to the brain (Wong et al. 2011). For instance, antibodies such as OX26 and 8D3, which target the transferrin receptor, have been studied in detail. The transferrin receptor is highly expressed in brain tissue and endothelial cells of BBB microvessels. This common receptor has previously been exploited to enhance the brain delivery of the chemotherapeutic agent daunomycin (Daunorubicin) and of DNA plasmids in rats and mice, and may prove useful for targeting the treatment of CNS infection. However, no transferrin antibodies that target the human CNS have been identified.

Another receptor that is highly expressed at the BBB interface is insulin receptor. As with monoclonal antibodies against the transferrin receptor, insulin receptor antibodies conjugated to nanoparticles have been studied primarily in cancers, particularly in glioma. In addition to the insulin receptor being highly expressed on the BBB interface, it is also highly expressed on the plasma membrane of glioma cells, thus making it a target to treat glioma. These receptors can be targeted with specific monoclonal antibodies, but further research is required to develop species-specific targeting agents. For example, mouse–antihuman antibody 83-14 binds to insulin receptors of larger Old World primates (e.g., Rhesus monkey) but not to insulin receptors of humans and New World primates.

In delivering treatment to the CNS, the pharmacological modality must have some mechanism to circumvent the BBB and the BCSF barrier. Nanoparticles can use a multitude of methods to traverse the BBB and to enter the brain, such as efflux transport inhibition, paracellular transport enhancement, BBB disruption, cell-mediated delivery, receptor-mediated absorptive endocytosis, and many more. The method employed depends on the chemical and physical properties of the nanoparticle formulations. Membrane-associated efflux transporters hamper the accumulation of lipophilic drugs in the brain (Loscher and Potschka 2005). The combined actions of membrane-associated efflux transporters (e.g., P-gp, MDR proteins, and organic anion transporting polypeptides) that are found in brain capillary endothelial cells and astrocytic end feet enforce the highly selective permeability properties of the BBB and cause the rapid efflux of xenobiotics from the CNS. They impact the pharmacological action of most drugs by affecting drug absorption, disposition, and elimination from the body. Additionally, these transporters can play a crucial role in the emergence of multidrug resistance. Inhibiting or bypassing these transporters is key to introducing lipophilic drugs into the brain for various CNS infections (Loscher and Potschka 2005).

Inhibition of efflux transportation can be achieved pharmacologically. Using nanoparticle carriers with encapsulated efflux transporter inhibitors can limit systemic side effects and significantly enhance accumulation of nanoparticles/drugs in the brain. In managing high blood pressure, the delivery of liposomes encapsulated with PSC833, an MDR1 (P-gp) blocker, produced a 4.6-fold increase in the brain accumulation of verapamil in nonhuman primates (Lee et al. 2006). Inhibition of efflux transporters has also shown efficacy for chemoresistant cancers; thus, it may have potential applications for increasing the accumulation of antimicrobial drugs in the brain, particularly in treating MDR microorganisms.

Another example of the efficacy of efflux transport inhibition in increasing drug transport across the BBB is the modulation of P-gp using Pluronic P85. P85 is an amphiphilic block copolymer with two chains of hydrophilic polyethylene oxide groups flanking a hydrophobic polypropylene oxide group. In a SCID model of HIV-1 encephalitis created by inoculation with HIV-1-infected monocyte-derived macrophages, P85 greatly increased BBB permeability and improved penetration of the HIV drugs saquinavir, zidovudine, lamivudine, and nelfinavir. The presence of P85 on the plasma membrane has been shown to decrease the microviscosity of the membrane, thus disrupting lipid rafts (where viral infection and HIV-1 budding occur) and influencing receptor distribution. Using nanoparticles that inhibit or affect the efflux of drugs can prevent resistance to antiretroviral therapy (Nowacek and Gendelman 2009; Spitzenberger et al. 2007).

As previously mentioned, tight junctions between capillary endothelial cells limit the paracellular transport of molecules across the BBB; therefore, for these molecules to enter the brain, they must use transcellular membrane permeability pathways. There are several methods of BBB disruption that expand interstitial spaces and can be used to bypass transcellular membrane pathways. The first method is pharmacological. Vasoactive agents, such as peptidase inhibitors, angiotensin II, and bradykinin, can temporarily and reversibly increase the permeability of blood vessels. At low doses, antimicrotubule agents can also be used to transiently open tight junctions. Increased paracellular transport of nanoparticles across the BBB leads to accumulation of nanoparticles in the CNS (Wong et al. 2011).

Altering serum osmolarity is a second method that expands interstitial spaces. It has been applied extensively to treat malignant gliomas. A hypertonic solution shrinks the size of endothelial cells by osmosis, resulting in expanded interendothelial distance. Mannitol, a high-osmotic-pressure solution, was given to transiently open interendothelial distance and to increase the delivery of bevacizumab to tumor tissue via superselective intra-arterial infusions (Boockvar et al. 2011). Osmolaric alteration can also be used to increase the delivery of various other agents, including antiretroviral therapies, to the brain. However, the hyperosmotic approach carries the risk of neurological damage, particularly with high doses or prolonged infusions. Despite this risk, BBB opening is an attractive method for increasing the accumulation of nanoparticles in the brain.

A third method that has recently gained popularity in BBB disruption and has been used to deliver HIV antiretroviral-containing nanoparticles is electromagnetic interference. Electromagnetic interference increases permeability in a number of different tissues, including the brain (Nowacek and Gendelman 2009; Salford et al. 1994; Wong et al. 2011). This is usually achieved through the application of amplitude-modulated (AM) and frequency-modulated (FM) waves of all types (square, sinusoidal, and triangle). Increased permeability is accompanied by increased surface softness after exposure. All types of AM and FM waves increase permeability; however, not all electromagnetic interference has the same effect. Square AM waves had the largest effect, increasing the permeability of the BBB to nanoparticles by more than 14 times without increasing the permeability of free unencapsulated drug (Kuo and Kuo 2008). Electromagnetic interference has been used with various types of nanoparticle carriers. After applying electromagnetic interference, polymeric (polybutylcyanoacrylate and methylmethacrylate–sulfopropylmethacrylate) and solid lipid nanoparticles carrying saquinavir (an anti-HIV drug) demonstrated increased passage across an in vitro model of the BBB (Calvo et al. 2001; Kuo and Kuo 2008). Solid lipid nanoparticles showed the highest permeability coefficient and the highest number of particles that were taken up by human brain microvascular endothelial cells.

The brain mounts an immune response to microbiological invasion, resulting in inflammation and other host reactions. A key component of this response is the recruitment of mononuclear phagocytes, such as dendritic cells and macrophages, to the site of inflammation. In fact, entry of macrophages is one way in which HIV-1 enters the brain, commonly resulting in HIV-1-associated neurocognitive disorders. The formation of multinucleated giant cells from the mononuclear phagocytic lineage serves as the principal reservoir of the virus within the CNS (Dou et al. 2007). Research is being pursued to exploit this mechanism of recruitment of mononuclear phagocytes to CNS inflammation to improve the delivery of therapy to a site of CNS infection.

Acting as Trojan horses, mononuclear phagocytes can incorporate various types of microsized and nanosized particles (e.g., liposomes, solid lipid nanoparticles, cationic nanoparticles, polymeric nanoparticles, etc.) (Dou et al. 2006, 2009). Different combinations of polymers and surface ligands (modifiers) can influence the rate and amount of uptake by phagocytes. Furthermore, different blends of nanoparticles may be more effective in targeting or avoiding these cells. Currently, most nanoparticles designed for cell-mediated delivery contain a single drug, but attempts are being made to encapsulate multiple drugs into a single nanoparticle. Because many CNS infections are associated with regional brain inflammation, mononuclear phagocytes are a natural vehicle owing to their innate ability to migrate to sites of inflammation through diapedesis and chemotaxis. Additionally, these cells can phagocytose, produce cytotoxic compounds, and release them by exocytosis. They also endocytose colloidal nanoparticles, although mononuclear phagocytes appear to have a preference for charged compounds rather than neutral nanoparticles.

Using mononuclear phagocytes for cell-mediated delivery of nanoparticles assists drugs that usually cannot cross cellular membranes to reach target cells and tissues. One way in which nanoparticles were shown to be transported by mononuclear phagocytes is by endosomal transport. Nanoparticles are engulfed by mononuclear phagocytes into primary endosomes, where they are sorted to either recycling or secondary endosomes. Once within the secondary endosomes, the acidic environment causes the surface charge on the nanoparticles to change from anionic to cationic, resulting in release into the cytoplasm. Importantly, during HIV infection, the same cells that traffic the antiretroviral nanoparticles also traffic the virus. Thus, cell-mediated trafficking delivers the nanoparticles to the same cellular compartments where the virus replicates. For this delivery mechanism to be effective clinically, however, several crucial questions must be answered. The first and most obvious question is whether the mononuclear phagocytes are even able to cross the BBB when loaded with drug; the second question is whether these cells can then reach the disease-affected regions of the brain. One research study showed that loading mononuclear phagocytes with antiretroviral nanoparticles did not decrease their ability to cross an in vitro model of the BBB. In an in vivo model, most macrophages accumulated in organs other than the brain after adoptive transfer, although a significant portion of cells did localize to the brain and were able to persist and release drug into extracellular space (Nowacek and Gendelman 2009).

Another transport mechanism for delivering nanoparticles to the CNS is receptor-mediated endocytosis. Brain endothelial cells are readily equipped for this process. Brain endothelial cells bind and uptake molecules at the luminal surface via receptor-mediated endocytosis and then release them via exocytosis on the abluminal surface, thus providing a means for transporting nanoparticles (and other molecules) through the brain endothelial cytoplasm. Attaching surface molecules to nanoparticles can induce this pathway to transport antiretroviral conjugated nanoparticles across the BBB. A popular surface molecule that is known to induce this pathway in endothelial cells is HIV TAT. The principal function of the TAT protein is to increase viral transcription by binding to cellular factors and affecting phosphorylation. Confocal microscopic visualization of TAT-conjugated nanoparticles showed that these particles crossed the BBB and entered brain tissue despite having sizes of 100–200 nm (Liu et al. 2008). TAT conjugation also allowed transport of QDs to the brain parenchyma (Santra et al. 2005; Schwarze et al. 1999). Additionally, drug levels within the brain were shown to dramatically increase with administration of TAT-conjugated antiretroviral nanoparticles. Compared to free ritonavir, TAT-conjugated nanoparticles with encapsulated ritonavir exhibited drug release and produced brain drug levels 800-fold higher 2 weeks after administration. However, this modality has two significant limitations: neurotoxicity and immunogenicity.

Nanoparticles are being developed and evaluated for their ability to enhance the transport of drugs to the CNS to treat bacterial CNS infections. Filamentous bacteriophage is a nanoparticle that is well-suited for drug transport because of its 3000 copies of the p8 major coat protein, which are potential sites for conjugation of antibacterial drugs (Yacoby et al. 2006). The bacteriophage can in turn be conjugated to an antibody that recognizes *S. aureus* or *E. coli*. Alternatively, its phage coat can be genetically engineered to display a targeting peptide or protein that specifically binds to these pathogenic bacteria (Vaks and Benhar 2011a). In vitro testing demonstrated that antibody-conjugated nanoparticles loaded with an antibiotic prodrug of chloramphenicol specifically bound to the pathogen and unloaded the prodrug, which assumed its active form and inhibited bacterial growth. This nanoparticle platform has also undergone in vivo testing in mice and has proved to be safe and nonimmunogenic (Vaks and Benhar 2011b).

Intranasal administration is a final unique method for CNS drug delivery that has primarily been explored to treat and prevent neurodegenerative diseases and to administer morphine (Wong et al. 2011). Compared to intravenous injection of nimodipine microemulsions to treat neurodegenerative diseases, intranasal delivery showed a threefold increase in uptake by the olfactory bulb (Zhang et al. 2004). Also, CSF concentrations of nimodipine were significantly higher after intranasal injection. Similar findings were also seen with the intranasal administration of morphine, providing a solid rationale for testing this modality for other maladies such as CNS infections (Betbeder et al. 2000). The brain and the nasal cavity are connected by the olfactory nerve system (which terminates in the olfactory neuroepithelium) and the trigeminal nerve. Exploiting these two nerve systems bypasses the BBB and allows direct nose-to-brain drug delivery, thereby reducing potential systemic toxicity. However, this route of administration is notoriously inefficient. It has been suggested that improved nanoformulations with conjugated mucoadhesives may increase the efficiency of drug uptake into the brain. Agents such as chitosan, a compound that interacts with junctional complexes or lectins, have been explored to increase the uptake of drugs across the nasal epithelium and into the brain. Further development of intranasal drug delivery is necessary to explore its possible role in treating CNS infections (Wong et al. 2011).

CONCLUSION

There are a variety of potential combinations of carrier systems, conjugates, formulations, and delivery mechanisms that can be used to apply nanotechnology to develop and deliver novel treatments for CNS infections. Specific treatments utilizing nanoparticles must be tailored to take into account the route of administration, cellular target, disease, and drug being carried. A broad application of nanomedicine to CNS infections will require a better understanding of the relationships between nanoparticles, surface ligands, and payload drugs, and the dynamics and properties of the BBB and the BCSF barrier.

REFERENCES

Armstrong, D. 1984. Central nervous system infections in the immunocompromised host. *Infection* 12 (Suppl 1): S58–S64.

Aronica, E., Gorter, J.A., Jansen, G.H. et al. 2003. Expression and cellular distribution of multidrug transporter proteins in two major causes of medically intractable epilepsy: focal cortical dysplasia and glioneuronal tumors. *Neuroscience* 118: 417–29.

Aronica, E., Gorter, J.A., Ramkema, M. et al. 2004. Expression and cellular distribution of multidrug resistance-related proteins in the hippocampus of patients with mesial temporal lobe epilepsy. *Epilepsia* 45: 441–51.

Betbeder, D., Sperandio, S., Latapie, J.P. et al. 2000. Biovector nanoparticles improve antinociceptive efficacy of nasal morphine. *Pharm Res* 17: 743–8.

Bharali, D.J., Klejbor, I., Stachowiak, E.K. et al. 2005. Organically modified silica nanoparticles: a nonviral vector for in vivo gene delivery and expression in the brain. *Proc Natl Acad Sci U S A* 102: 11539–44.

Boockvar, J.A., Tsiouris, A.J., Hofstetter, C.P. et al. 2011. Safety and maximum tolerated dose of superselective intraarterial cerebral infusion of bevacizumab after osmotic blood–brain barrier disruption for recurrent malignant glioma. Clinical article. *J Neurosurg* 114: 624–32.

Calvo, P., Gouritin, B., Brigger, I. et al. 2001. PEGylated polycyanoacrylate nanoparticles as vector for drug delivery in prion diseases. *J Neurosci Methods* 111: 151–5.

Chattopadhyay, N., Zastre, J., Wong, H.L. et al. 2008. Solid lipid nanoparticles enhance the delivery of the HIV protease inhibitor, atazanavir, by a human brain endothelial cell line. *Pharm Res* 25: 2262–71.

Deeken, J.F., and Loscher, W. 2007. The blood–brain barrier and cancer: transporters, treatment, and Trojan horses. *Clin Cancer Res* 13: 1663–74.

Donelli, M.G., Zucchetti, M., and D'Incalci, M. 1992. Do anticancer agents reach the tumor target in the human brain? *Cancer Chemother Pharmacol* 30: 251–60.

Dou, H., Destache, C.J., Morehead, J.R. et al. 2006. Development of a macrophage-based nanoparticle platform for antiretroviral drug delivery. *Blood* 108: 2827–35.

Dou, H., Grotepas, C.B., McMillan, J.M. et al. 2009. Macrophage delivery of nanoformulated antiretroviral drug to the brain in a murine model of neuroAIDS. *J Immunol* 183: 661–9.

Dou, H., Morehead, J., Destache, C.J. et al. 2007. Laboratory investigations for the morphologic, pharmacokinetic, and anti-retroviral properties of indinavir nanoparticles in human monocyte-derived macrophages. *Virology* 358: 148–58.

Ehrlich, P. 1885. *Das Sauerstoff-Bedurfnis des Organismus: eine farbenanalytische Studie*. Berlin: Hirschward.

Enzmann, D.R. 1984. *Imaging of Infections and Inflammations of the Central Nervous System: Computed Tomography, Ultrasound, and Nuclear Magnetic Resonance*. New York: Raven Press.

Forward, K.R., Fewer, H.D., and Stiver, H.G. 1983. Cerebrospinal fluid shunt infections. A review of 35 infections in 32 patients. *J Neurosurg* 59: 389–94.

Hartz, A.M., and Bauer, B. 2010. Regulation of ABC transporters at the blood–brain barrier: new targets for CNS therapy. *Mol Interv* 10: 293–304.

Hill, H.D., and Mirkin, C.A. 2006. The bio-barcode assay for the detection of protein and nucleic acid targets using DTT-induced ligand exchange. *Nat Protoc* 1: 324–36.

Hippocrates and Adams, F. 1886. *The Genuine Works of Hippocrates. Wood's Library of Standard Medical Authors*. New York: W. Wood and Company.

Iqbal, U., Abulrob, A., and Stanimirovic, D.B. 2011. Integrated platform for brain imaging and drug delivery across the blood–brain barrier. *Methods Mol Biol* 686: 465–81.

Jokerst, J.V., Floriano, P.N., Christodoulides, N. et al. 2008. Integration of semiconductor quantum dots into nano-bio-chip systems for enumeration of CD4+ T cell counts at the point-of-need. *Lab Chip* 8: 2079–90.

Kaur, I.P., Bhandari, R., Bhandari, S. et al. 2008. Potential of solid lipid nanoparticles in brain targeting. *J Controlled Release* 127: 97–109.

Kroll, R.A., and Neuwelt, E.A. 1998. Outwitting the blood–brain barrier for therapeutic purposes: osmotic opening and other means. *Neurosurgery* 42: 1083–99; discussion 1099–100.

Kuo, Y.C., and Kuo, C.Y. 2008. Electromagnetic interference in the permeability of saquinavir across the blood–brain barrier using nanoparticulate carriers. *Int J Pharm* 351: 271–81.

Lee, Y.J., Maeda, J., Kusuhara, H. et al. 2006. In vivo evaluation of P-glycoprotein function at the blood–brain barrier in nonhuman primates using [^{11}C]verapamil. *J Pharmacol Exp Ther* 316: 647–53.

Lewandowsky, M. 1900. Zur Lehre von der Cerebrospinalflussigkeit. *Z Klin Med* 40: 480–94.

Liu, L., Guo, K., Lu, J. et al. 2008. Biologically active core/shell nanoparticles self-assembled from cholesterol-terminated PEG-TAT for drug delivery across the blood–brain barrier. *Biomaterials* 29: 1509–17.

Loscher, W., and Potschka, H. 2005. Role of drug efflux transporters in the brain for drug disposition and treatment of brain diseases. *Prog Neurobiol* 76: 22–76.

Mahmoud, K.A., Hrapovic, S., and Luong, J.H. 2008. Picomolar detection of protease using peptide/single walled carbon nanotube/gold nanoparticle-modified electrode. *ACS Nano* 2: 1051–7.

Mathews, V.P., Kuharik, M.A., Edwards, M.K. et al. 1989. Dyke Award. Gd-DTPA-enhanced MR imaging of experimental bacterial meningitis: evaluation and comparison with CT. *AJR Am J Roentgenol* 152: 131–6.

Neuwelt, E.A., Abbott, N.J., Drewes, L. et al. 1999. Cerebrovascular biology and the various neural barriers: challenges and future directions. *Neurosurgery* 44: 604–8; discussion 608–9.

Neuwelt, E.A., Specht, H.D., and Hill, S.A. 1986. Permeability of human brain tumor to 99mTc-glucoheptonate and 99mTc-albumin. Implications for monoclonal antibody therapy. *J Neurosurg* 65: 194–8.

Nowacek, A., and Gendelman, H.E. 2009. NanoART, neuroAIDS and CNS drug delivery. *Nanomedicine (London)* 4: 557–74.

Piacentino, M., Pilleri, M., and Bartolomei, L. 2011. Hardware-related infections after deep brain stimulation surgery: review of incidence, severity and management in 212 single-center procedures in the first year after implantation. *Acta Neurochir (Wien)* 153: 2337–41.

Rantakallio, P., Leskinen, M., and von Wendt, L. 1986. Incidence and prognosis of central nervous system infections in a birth cohort of 12,000 children. *Scand J Infect Dis* 18: 287–94.

Rao, K.S., Reddy, M.K., Horning, J.L. et al. 2008. TAT-conjugated nanoparticles for the CNS delivery of anti-HIV drugs. *Biomaterials* 29: 4429–38.

Raty, J.K., Liimatainen, T., Wirth, T. et al. 2006. Magnetic resonance imaging of viral particle biodistribution in vivo. *Gene Ther* 13: 1440–6.

Roos, K.L. 2005. *Principles of Neurologic Infectious Diseases*. New York: McGraw-Hill Medical Publishing Division.

Rotbart, H.A. 2000. Viral meningitis. *Semin Neurol* 20: 277–92.

Rubin, R.H., and Young, L.S. 2002. *Clinical Approach to Infection in the Compromised Host*, 4th ed. New York: Kluwer Academic/Plenum.

Sakushima, K., Hayashino, Y., Kawaguchi, T. et al. 2011. Diagnostic accuracy of cerebrospinal fluid lactate for differentiating bacterial meningitis from aseptic meningitis: a meta-analysis. *J Infect* 62: 255–62.

Salford, L.G., Brun, A., Sturesson, K. et al. 1994. Permeability of the blood–brain barrier induced by 915 MHz electromagnetic radiation, continuous wave and modulated at 8, 16, 50, and 200 Hz. *Microsc Res Tech* 27: 535–42.

Santra, S., Yang, H., Stanley, J.T. et al. 2005. Rapid and effective labeling of brain tissue using TAT-conjugated CdS:Mn/ZnS quantum dots. *Chem Commun (Camb)* 3144–6.

Scheld, W.M., Whitley, R.J., and Marra, C.M. 2004. *Infections of the Central Nervous System*, 3rd ed. Philadelphia, PA: Lippincott Williams & Wilkins.

Schoenbaum, S.C., Gardner, P., and Shillito, J. 1975. Infections of cerebrospinal fluid shunts: epidemiology, clinical manifestations, and therapy. *J Infect Dis* 131: 543–52.

Schuchat, A., Robinson, K., Wenger, J.D. et al. 1997. Bacterial meningitis in the United States in 1995. Active Surveillance Team. *N Engl J Med* 337: 970–6.

Schwarze, S.R., Ho, A., Vocero-Akbani, A. et al. 1999. In vivo protein transduction: delivery of a biologically active protein into the mouse. *Science* 285: 1569–72.

Shriver, L.P., Koudelka, K.J., and Manchester, M. 2009. Viral nanoparticles associate with regions of inflammation and blood brain barrier disruption during CNS infection. *J Neuroimmunol* 211: 66–72.

Spitzenberger, T.J., Heilman, D., Diekmann, C. et al. 2007. Novel delivery system enhances efficacy of antiretroviral therapy in animal model for HIV-1 encephalitis. *J Cereb Blood Flow Metab* 27: 1033–42.

Thigpen, M.C., Whitney, C.G., Messonnier, N.E. et al. 2011. Bacterial meningitis in the United States, 1998–2007. *N Engl J Med* 364: 2016–25.

Vaks, L., and Benhar, I. 2011a. Antibacterial application of engineered bacteriophage nanomedicines: antibody-targeted, chloramphenicol prodrug loaded bacteriophages for inhibiting the growth of *Staphylococcus aureus* bacteria. *Methods Mol Biol* 726: 187–206.

Vaks, L., and Benhar, I. 2011b. In vivo characteristics of targeted drug-carrying filamentous bacteriophage nanomedicines. *J Nanobiotechnol* 9: 58.

van de Beek, D., de Gans, J., Spanjaard, L. et al. 2004. Clinical features and prognostic factors in adults with bacterial meningitis. *N Engl J Med* 351: 1849–59.

van de Beek, D., de Gans, J., Tunkel, A.R. et al. 2006. Community-acquired bacterial meningitis in adults. *N Engl J Med* 354: 44–53.

Venes, D. 2001. *Taber's Cyclopedic Medical Dictionary, 19th Edition.* Philadelphia: F.A. Davis Co.

Wallgren, A. 1951. [Etiology of meningoencephalitis in children, especially the syndrome of acute aseptic meningitis]. *Acta Paediatr* 40: 541–65.

Wong, H.L., Wu, X.Y., and Bendayan, R. 2011. Nanotechnological advances for the delivery of CNS therapeutics. *Adv Drug Deliv Rev* 64: 686–700.

Xu, N., Gu, J., Zhu, Y. et al. 2011. Efficacy of intravenous amphotericin B-polybutylcyanoacrylate nanoparticles against cryptococcal meningitis in mice. *Int J Nanomed* 6: 905–13.

Yacoby, I., Shamis, M., Bar, H. et al. 2006. Targeting antibacterial agents by using drug-carrying filamentous bacteriophages. *Antimicrob Agents Chemother* 50: 2087–97.

Zhang, Q., Jiang, X., Jiang, W. et al. 2004. Preparation of nimodipine-loaded microemulsion for intranasal delivery and evaluation on the targeting efficiency to the brain. *Int J Pharm* 275: 85–96.

30 Applications of Nanotechnology in Movement Disorders

Patrick Pezeshkian, Karlo J. Lizarraga, Alessandra Gorgulho, and Antonio De Salles

CONTENTS

INTRODUCTION

Nanotechnology is defined as the science involved in the design, synthesis, characterization, and application of materials and devices whose smallest functional organization in at least one dimension is on the nanometer scale or is one billionth of a meter (e.g., a sodium atom is about 0.2 nm long). Materials and devices engineered at the nanometer scale imply a controlled manipulation of individual constituent molecules and atoms on how they are arranged to form a macroscopic substrate, bringing together the collective expertise of traditional sciences such as chemistry, physics, materials science, and biology (Silva 2004). Nanotechnology designs materials to interact with cells and tissues at the subcellular level with a high degree of specificity, giving a unique opportunity to develop new solutions for treating movement disorders by stimulating, responding to, and interacting with specific target sites to induce physiological responses while striving to minimize side effects (Andrews 2007).

The delivery of drugs to the brain to treat movement disorders remains highly challenging. Conventional drug delivery systems that are currently employed to treat various movement disorders fall short in providing the cytoarchitecture restoration and connection patterns essential for functional recovery in movement disorders, in part due to the limitations posed by the blood–brain barrier (BBB). Nanotechnology may yield a possible solution to many of the challenges in the conventional treatment of movement disorders by providing a targeted drug delivery system and by enhancing the bioavailability and efficacy of various drugs and other bioactive agents used to treat movement disorders.

This chapter will provide a concise review of current and future applications of nano-enabled delivery systems to treat movement disorders, in particular Parkinson's disease (PD), and will explore the applications of nanotechnology in clinical neuroscience to develop innovative therapeutic modalities targeted for movement disorders.

POTENTIAL NANOSTRUCTURES FOR TREATMENT OF AND PROPHYLAXIS AGAINST MOVEMENT DISORDERS

The majority of nanotechnological drug delivery systems used to treat neurodegenerative disorders are in the form of polymeric nanoparticles. Polymeric nanoparticles are promising for movement disorders such as PD since they can pass through cellular tight junctions, cross the BBB, achieve a high drug-loading capacity, and be targeted toward mutagenic proteins in the disorder. The features of these nanosystems in central nervous system (CNS)-targeted drug delivery are as follows: (1) their chemical properties can be easily modified to achieve cell-specific and selective drug delivery; (2) they can control the targeted delivery of drugs; (3) they increase the bioavailability and efficacy of incorporated drugs by masking physicochemical characteristics, thereby increasing drug transfer across the BBB; (4) they protect incorporated drugs against enzymatic degradation; and (5) they have fewer side effects.

Fullerene Derivatives as Neuroprotectors against Reactive Oxygen Species Damage

Initial results on the use of fullerenes as neuroprotectors in animal models of a group of neurologic diseases in which free radical-mediated damage is implicated as a causative force demonstrate protective effects on diseases such as ischemic stroke, multiple sclerosis, PD, and Alzheimer's disease (Kim et al. 2008).

Fullerenes were experimented with in PD models to determine if they have any effect on disease progression. In a rat model, carboxyfullerene directly infused into the substantia nigra prevented regional iron-induced lipid peroxidation and oxidative injury and inhibited the loss of dopaminergic nerve terminals in the striatum (Lin et al. 1999). These early results in an animal model demonstrate that PD is among the many neurodegenerative diseases for which fullerenes may be efficacious. Furthermore, fullerenols, which are hydroxyl-functionalized fullerene derivatives, possess antioxidant and free radical scavenger properties that can reduce glutamate-, *N*-methyl D-aspartate (NMDA)-, 2-amino-3-(3-hydroxy-5-methyl-isoxazol-4-yl)propanoic acid (AMPA)-, and kainite-induced excitotoxicity and apoptosis, which are implicated in many neurodegenerative disorders (Lin et al. 1999). The mechanism of fullerenol-mediated neuroprotection is thought to be partly due to inhibition of glutamate channels (Lin et al. 1999).

If further animal studies continue to show beneficial effects and if toxicity studies demonstrate that fullerenes are sufficiently safe for human use, future clinical studies in humans may be feasible to test their efficacy in neurodegenerative disorders, including movement disorders such as PD.

Nanoparticles for Target-Specific Drug Delivery

By assisting in passage across the BBB, nanoparticles can serve as vehicles for the efficient delivery of drugs that are otherwise unable to be delivered at therapeutic levels to target sites in the brain owing to the selective permeability of the BBB.

Solid lipid nanoparticles (SLNs) are among the best characterized lipid-based nanoscale compounds developed for brain tumor drug delivery, and this same technology can potentially be used to deliver therapeutic drugs, such as dopaminergic or anticholinergic agents, to target sites in the brain to treat movement disorders. These nanoparticles are prepared by hot or cold high-pressure homogenization or microemulsion of solid physiological lipids (Muller et al. 2000), with adjustment of the synthetic technique to accommodate the chemical properties of the drug incorporated into the SLNs. By encapsulating the drugs and by separating them from blood plasma, the lipid matrix of the SLN protects the drugs from degradation and greatly reduces the systemic distribution of the drugs, thereby reducing their systemic side effects. Furthermore, selective unloading of drugs within target brain tissues can be controlled by modifying the surface coat of the SLN and the constituent lipids of the particle itself (Wissing et al. 2004). By achieving high drug bioavailability to target tissue, SLNs can minimize the therapeutic dose of the drug and, consequently, its systemic toxicity. In addition, since they are composed of physiological lipids, SLNs are nontoxic. However, further characterization of SLNs in in vitro and in vivo model systems is required before their clinical use in movement disorders can be explored.

Non-lipid-based drug delivery nanoparticles, which are colloidal particles consisting of the matrices of synthetic biocompatible and biodegradable polymers, can also be used for drug delivery to specific targets in the brain. These nanoparticles can be chemically modified to absorb, trap, or bond to drugs. Similar to lipid-based nanoparticles, they can isolate their payload from the physiological environment to prevent degradation and toxicity, and they can be designed to promote favorable drug delivery kinetics to target sites in the brain (Wissing et al. 2004).

POTENTIAL NANOSYSTEMS FOR TREATING MOVEMENT DISORDERS

Nonviral Vectors as Vehicles for Delivery of Genes

The development of viral and nonviral gene transfer vectors can make gene therapy for treating movement disorders such as PD clinically relevant. The safety of the vector is an important concern for any type of gene therapy. When viruses are used as transfection agents, there is a risk of excessive immune response, mutagenesis at the site of insertion, or both. One of the foci of nanotechnology is the development and use of nonviral vectors to deliver genes safely and efficiently.

The delivery of genes in vivo involves using genetic materials such as DNA, RNA, and oligonucleotides that can inhibit undesirable gene expression or synthesize therapeutic proteins. For gene therapy to be effective, the genetic payload must be delivered to the targeted cell or tissue and then transported to the nucleus of the cell for phenotypic expression. Ultrafine silica nanoparticles—organically modified silica (ORMOSIL) nanoparticles that have been amino-functionalized—can bind and protect plasmid DNA from enzymatic digestion and were used as nonviral vectors for in vitro transfection (Klejbor et al. 2007). ORMOSIL nanoparticles, measuring approximately 30 nm, are relatively easy to produce on a large scale, and their surfaces can be readily modified during synthesis (Roy et al. 2003, 2005). For instance, adding cationic groups to the surface of ORMOSIL nanoparticles enhances binding with negatively charged plasmids for successful carriage inside the cells. This binding protects the DNA structure from environmental insult during the process involved in in vivo transfer. Furthermore, the process of loading the nanoparticles with DNA for transfection is considerably simpler than the production of transfectable material and its encapsulation in viral particles.

Recent work has established the feasibility of using ORMOSIL nanoparticles as a nonviral vector for efficient in vivo gene delivery (Bharali et al. 2005). ORMOSIL nanoparticles that were surface-functionalized with amino groups for DNA binding were prepared and characterized. These nanoparticles, complexed with plasmid DNA encoding for enhanced green fluorescent protein (EGFP), were then stereotactically injected into the mouse ventral midbrain and lateral ventricle, allowing fluorescent visualization of the extensive transfection of neuronal-like cells in the substantia nigra and in areas surrounding the lateral ventricle. No ORMOSIL-based toxicity was observed 4 weeks after transfection; the efficiency of transfection equaled or exceeded that obtained in studies using a viral vector (Bharali et al. 2005).

Studies such as these demonstrated that ORMOSIL nanoparticles can be effectively used to introduce genes into the dopaminergic cells of the substantia nigra pars compacta (Bharali et al.

2005). This approach should allow modeling of PD by (1) transfecting with the mutant α-synuclein gene; (2) blocking fibroblast growth factor (FGF) and glia-derived growth factor signaling with dominant-negative receptor mutants; or (3) knocking down parkin gene activity using antisense or small interfering RNA technology (Goldberg et al. 2003; Masliah et al. 2001; McBride and Kordower 2002; Tooyama et al. 1994). ORMOSIL-mediated transfections of midbrain dopaminergic neurons also would allow testing of diverse gene therapeutic strategies for other disorders involving dopaminergic neurons.

Nano-enabled immunoliposomes are yet another type of nonviral vector that may be used for gene replacement therapy in PD. PEGylated immunoliposomes containing antibodies directed to insulin or transferrin receptors have been used successfully as carriers for gene replacement therapy in a PD model (Pardridge 2005). The mechanism by which immunoliposomes cross the BBB is not fully known; however, it has been hypothesized that they do so via binding to capillary luminal membrane receptors, fusion of the liposomes with several vesicular pits into a large vesicle, and phagocytosis of this vesicle into the abluminal membrane border (Cornford and Cornford 2002).

Nanoparticle-Based Gene Therapy

The novel idea that DNA plasmids can be condensed into nanoparticles for delivery to the brain to stop or prevent neurodegeneration has been assessed in animal models of PD by several investigators. Yurek et al. (2009), using condensed DNA plasmids expressing specific genes shown to be lacking in their animal PD model, transfected specific target cells and repaired defective genes.

In this study, the authors used a nonviral technique to transfer the gene encoding for glial cell line-derived neurotrophic factor (GDNF) to striatal cells. Plasmid DNA encoding for GDNF was compacted into DNA nanoparticles (DNPs) and then injected into the denervated striatum of rats with unilateral 6-hydroxydopamine lesions. Sham controls were injected with saline. One week later, experimental animals received either a ventral mesencephalic (VM) tissue chunk graft or a cell suspension VM graft implanted into the denervated striatum. Grafts were allowed to integrate for 4–6 weeks; during this period, spontaneous and drug-induced motor activities were monitored. A 16-fold increase in the number of surviving transfected cells within tissue chunk grafts placed into the striatum pretreated with plasmid glial cell line-derived neurotrophic factor (pGDNF) DNPs was noted when compared to grafts placed into the striatum pretreated with saline. Similarly, a sevenfold increase in the number of transfected cells within cell suspension grafts placed into the striatum treated with pGDNF DNPs was observed when compared to cell suspension grafts placed into the saline-dosed striatum. Behavioral observations involved significant improvement in rotational scores and spontaneous forepaw usage of the affected forelimb in grafted animals receiving prior treatment with compacted pGDNF DNPs when compared to grafted animals receiving saline control pretreatment. Data analysis for protein, morphological, and behavioral measures suggested that compacted pGDNF DNPs injected into the striatum can result in transfected cells overexpressing GDNF protein at levels that provide neurotrophic support for grafted embryonic dopamine (DA) neurons.

The study not only represents an encouraging first step in developing a promising new approach to PD therapy but also provides a platform for translating a variety of new gene therapy agents into human clinical trials for numerous neurodegenerative disorders.

Nanofibers as Stem Cell Therapy

Strategies in neural tissue engineering focus on developing scaffolds that artificially generate cellular microenvironments that favor regeneration. Some recent advances using such scaffolds to promote regeneration within the CNS, particularly in conjunction with stem cells, have generated encouraging results. One such advancement is the use of hydrogel scaffolds, which are polymer-based biodegradable nanofiber scaffolds that have extensive neural tissue engineering applications (Batchelor and Howells 2003).

Polymer-based biodegradable nanofibers have been engineered to prepare a scaffold for holding stem cells that can be injected into the brain at the site requiring nerve regeneration, such as the substantia nigra pars compacta in PD, potentially allowing stem cells to repair damaged nerves rapidly and effectively. Stem cells can be imbedded into the scaffold externally or once the scaffold is implanted. The cells adhere to the scaffold, which forms a bridge in the target tissue; with time, the scaffold erodes and is naturally eliminated from the body, leaving the newly regenerated neurons intact (Nisbet et al. 2008).

Another scaffolding technology for neural regeneration that holds a lot of potential are self-assembling peptide nanofibers. Silva et al. (2004) and Webber et al. (2010) developed a peptide amphiphile (PA) molecule, which, when introduced into an ionic environment, self-assembles into cylindrical nanofibers. These PA molecules can be injected in liquid form into a desired location in the body, such as the substantia nigra pars compacta in PD patients, where they self-assemble into cylindrical nanofibers 6–8 nm in diameter, similar in dimension to native extracellular matrix proteins. The hydrophilic group in these cylindrical nanofibers can be selected to be a bioactive compound, such as neuroactive epitope from laminin. These PA nanofibers functionalized with the neuroactive epitope from laminin were tested for their ability to aid in axon regeneration after spinal cord injury in the mouse model. When the spinal cord of mice (which was lesioned so that the mice could no longer use their hind legs) was injected with the abovementioned PA molecules into the lesion site, reduced cell death, increased number of oligodendroglia, and increased regeneration of both motor and sensory spinal cord tracts were measured. This regeneration was also functionally manifested by the mice' use of their hind legs at 9 weeks compared to control animals, which had range of movement but no functional movement in the hind limbs. This finding suggests the potential use of such nanoparticles in spinal cord injury patients as a way to reverse trauma-induced paralysis. The same concept can potentially be applied to PD to assist in regenerating dopaminergic neurons in the substantia nigra pars compacta by providing a scaffold on which transplanted dopaminergic stem cells could grow.

Carbon Nanotube and Nanowire Biosensors

Microelectrodes have a wide range of neurosurgical applications, including deep brain stimulation (DBS) for movement and psychiatric disorders, and spinal and peripheral nerve stimulation for pain, with increasing applications in neuromodulation. However, they have limitations since, with time, the efficacy of neural stimulation therapies may be limited by the degradation of the electrode and the response of the neural tissue to the implanted foreign material. Degeneration of the neuroelectrode, usually composed of either platinum or iridium oxide, occurs via electrochemical erosion of the electrode itself (Agnew et al. 1986) and delamination of the electrode coat. Furthermore, the target being stimulated develops a frequency-dependent, stimulation-induced depression of neuronal excitability or persistent refractory state as a result of chronic high-frequency microstimulation. Micromotion, including movement of tissue away from the electrode and motion of the electrode itself, can change the initial positioning and further reduce electrode efficacy. Certain properties of nanomaterials may help avoid some of these common neuroelectrode problems.

Carbon nanotubes, consisting of nanoscale graphite hollow tubules with lengths ranging on the order of hundreds of nanometers to several micrometers, are among the nanomaterials that can be used as molecular-scale electrochemical biosensors. In addition, their ability to carry a high current density despite their small size and chemical stability gives carbon nanotubes excellent electronic properties (Yang et al. 2007). Carbon nanotubes also have great mechanical strength and can withstand large angle flexural strains without breaking (Falvo et al. 1997). The latter is beneficial for the implantation phase of neuroelectrodes and for resisting breakage from micromotions that occur in the tissue once implanted. Since conventional microelectrodes are significantly larger than individual neurons, the resulting interface between the microelectrode and surrounding neurons is two-dimensional. Nanoelectrodes, by being more similar in dimension to a neuron, would enable a finer resolution, thereby potentially enabling a three-dimensional interface between electrodes and neural tissue (Nguyen-Vu et al. 2006). Furthermore, carbon nanotubes can be functionalized with bioactive molecules, providing an additional avenue for design features. Nanoscale wires can also enhance neural stimulation by delivering smaller amounts of current, resulting in less damage to local tissue and less damping of the excitability of nearby neurons with time.

The applicability of nanotechnology can be extended to the use of wireless implantable biosensors that may help treat PD patients (Li et al. 2005). Biosensors comprise carbon nanotubes and nanowires, which are grown and organized into arrays that are combined to make carbon nanochips that are biocompatible and are not rejected by the human body as a foreign object. Once inside the body, nanochips can perform several functions, such as sensing and monitoring the release of DA produced in specific areas of the brain.

The carbon nanotube-based biosensor developed by Li et al. (2005) was shown to record loss of DA and to stimulate activity between neurons and neurites, thus generally monitoring and controlling DA levels in the brain. In addition to sensing DA release, the biosensor can also communicate with an organic polymer-based sensor attached to an area of the body in which a tremor occurs. The signal from the implanted sensor can control and direct the motion of the area of the body on which the exterior sensor is attached (Li et al. 2005).

In summary, the limitations of traditional DBS using macroelectrodes include the relative large size of current electrodes, the lack of monitoring of local brain electrical activity and DA levels, and the open-loop nature of the stimulation. Therefore, exquisite improvement in spatial resolution, temporal resolution, and sensitivity of monitoring and stimulation is achieved by reducing the size of monitoring and by stimulating electrodes to the nanoscale level (Li and Andrews 2007). Carbon nanofiber nanoelectrode technology offers the possibility of trimodal arrays for monitoring electrical activity, measuring DA levels, and making precise stimulations (Li and Andrews 2007). DBS can then become a closed-loop system, guided by changes in brain electrical activity, DA levels, or both (Lovat et al. 2005). Such a concept can offer a generic electrical–neural interface that can be applicable to a variety of neurodegenerative disorders, including PD.

CONCLUSION

Nanotechnology is currently showing great potential for providing new therapeutic modalities to limit and reverse the pathophysiology of movement disorders. It offers the following possibilities: functional regeneration of damaged neurons; localized and specific delivery of drugs, genes, and cells across the BBB; molecular guidance for reestablishment of circuits; and monitoring of drug and neurotransmitter levels at specific anatomical locations. Nanoscale molecules will widen the scope of therapeutic action beyond modifying transmitter function to include stem cell and gene therapies that could offer more selective targeting.

The challenges associated with nanotechnology applications in movement disorders are numerous. Nanotechnology offers intervention at the molecular level, and any desired cellular signaling pathway can be targeted. This specificity may be thought of as sufficient, but there is a need for even greater specificity. Furthermore, there is a need for technologies that can multitask and carry out a diverse set of specific cellular and physiological functions, such as targeting multiple receptors or ligands. Widespread cellular heterogeneity and multidimensional cellular interactions in the spatial and temporal domains underlie the nervous system's anatomical and functional wiring that is the basis of its extremely complex information processing. This processing is disrupted in disease states, and it is important not to cause further disruptions by applying an interacting nanodevice since the resultant side effects could be unpredictable and severe.

The development and application of nanotechnology to treat movement disorders will require the cooperation of basic and clinical researchers to completely understand the pathophysiology of movement disorders and to develop specifically targeted nanosystems, along with neurologists and neurosurgeons, to investigate the clinical advantages and disadvantages of its use.

REFERENCES

Agnew, W.F., Yuen, T.G., McCreery, D.B. et al. 1986. Histopathologic evaluation of prolonged intracortical electrical stimulation. *Exp Neurol* 92: 162–85.

Andrews, R.J. 2007. Neuroprotection at the nanolevel: Part I. Introduction to nanoneurosurgery. *Ann N Y Acad Sci* 1122: 169–84.

Batchelor, P.E., and Howells, D.W. 2003. CNS regeneration: clinical possibility or basic science fantasy? *J Clin Neurosci* 10: 523–34.

Bharali, D.J., Klejbor, I., Stachowiak, E.K. et al. 2005. Organically modified silica nanoparticles: a nonviral vector for in vivo gene delivery and expression in the brain. *Proc Natl Acad Sci U S A* 102: 11539–44.

Cornford, E.M., and Cornford, M.E. 2002. New systems for delivery of drugs to the brain in neurological disease. *Lancet Neurol* 1: 306–15.

Falvo, M.R., Clary, G.J., Taylor, R.M. II et al. 1997. Bending and buckling of carbon nanotubes under large strain. *Nature* 389: 582–4.

Goldberg, M.S., Fleming, S.M., Palacino, J.J. et al. 2003. Parkin-deficient mice exhibit nigrostriatal deficits but not loss of dopaminergic neurons. *J Biol Chem* 278: 43628–35.

Kim, Y.O., Kim, H.J., Kim, S.K. et al. 2008. Neuroprotective effects of hydroxyfullerene in rats subjected to global cerebral ischemia. *Mol Cell Toxicol* 4: 218–23.

Klejbor, I., Stachowiak, E.K., Bharali, D.J. et al. 2007. ORMOSIL nanoparticles as a non-viral gene delivery vector for modeling polyglutamine induced brain pathology. *J Neurosci Methods* 165: 230–43.

Li, J., and Andrews, R.J. 2007. Trimodal nanoelectrode array for precise deep brain stimulation: prospects of a new technology based on carbon nanofiber arrays. *Acta Neurochir Suppl* 97: 537–45.

Li, J., Ng, H.T., and Chen, H. 2005. Carbon nanotubes and nanowires for biological sensing. In *Protein Nanotechnology: Protocols, Instrumentation, and Applications*, ed. T. Vo-Dinh; in *Methods in Molecular Biology*, ed. J.M. Walker, 191–223. New York: Humana Press.

Lin, A.M., Chyi, B.Y., Wang, S.D. et al. 1999. Carboxyfullerene prevents iron-induced oxidative stress in rat brain. *J Neurochem* 72: 1634–40.

Lovat, V., Pantarotto, D., Lagostena, L. et al. 2005. Carbon nanotube substrates boost neuronal electrical signaling. *Nano Lett* 5: 1107–10.

Masliah, E., Rockenstein, E., Veinbergs, I. et al. 2001. Beta-amyloid peptides enhance alpha-synuclein accumulation and neuronal deficits in a transgenic mouse model linking Alzheimer's disease and Parkinson's disease. *Proc Natl Acad Sci U S A* 98: 12245–50.

McBride, J.L., and Kordower, J.H. 2002. Neuroprotection for Parkinson's disease using viral vector-mediated delivery of GDNF. *Prog Brain Res* 138: 421–32.

Muller, R.H., Mader, K., and Gohla, S. 2000. Solid lipid nanoparticles (SLN) for controlled drug delivery—a review of the state of the art. *Eur J Pharm Biopharm* 50: 161–77.

Nguyen-Vu, T.D., Chen, H., Cassell, A.M. et al. 2006. Vertically aligned carbon nanofiber arrays: an advance toward electrical–neural interfaces. *Small* 2: 89–94.

Nisbet, D.R., Crompton, K.E., Horne, M.K. et al. 2008. Neural tissue engineering of the CNS using hydrogels. *J Biomed Mater Res B Appl Biomater* 87: 251–63.

Pardridge, W.M. 2005. The blood–brain barrier: bottleneck in brain drug development. *NeuroRx* 2: 3–14.

Roy, I., Ohulchanskyy, T.Y., Bharali, D.J. et al. 2005. Optical tracking of organically modified silica nanoparticles as DNA carriers: a nonviral, nanomedicine approach for gene delivery. *Proc Natl Acad Sci U S A* 102: 279–84.

Roy, I., Ohulchanskyy, T.Y., Pudavar, H.E. et al. 2003. Ceramic-based nanoparticles entrapping water-insoluble photosensitizing anticancer drugs: a novel drug-carrier system for photodynamic therapy. *J Am Chem Soc* 125: 7860–5.

Silva, G.A. 2004. Introduction to nanotechnology and its applications to medicine. *Surg Neurol* 61: 216–20.

Silva, G.A., Czeisler, C., Niece, K.L. et al. 2004. Selective differentiation of neural progenitor cells by high-epitope density nanofibers. *Science* 303: 1352–5.

Tooyama, I., McGeer, E.G., Kawamata, T. et al. 1994. Retention of basic fibroblast growth factor immunoreactivity in dopaminergic neurons of the substantia nigra during normal aging in humans contrasts with loss in Parkinson's disease. *Brain Res* 656: 165–8.

Webber, M.J., Kessler, J.A., and Stupp, S.I. 2010. Emerging peptide nanomedicine to regenerate tissues and organs. *J Intern Med* 267: 71–88.

Wissing, S.A., Kayser, O., and Muller, R.H. 2004. Solid lipid nanoparticles for parenteral drug delivery. *Adv Drug Deliv Rev* 56: 1257–72.

Yang, W., Thordarson, P., Gooding, J.J. et al. 2007. Carbon nanotubes for biological and biomedical applications. *Nanotechnology* 18: 412–6.

Yurek, D.M., Flectcher, A.M., Kowalczyk, T.H. et al. 2009. Compacted DNA nanoparticle gene transfer of GDNF to the rat striatum enhances the survival of grafted fetal dopamine neurons. *Cell Transplant* 18: 1183–96.

31 Nanomedicine in Demyelinating Disease
Application to Diagnosis and Therapy in Multiple Sclerosis

Emily C. Leibovitch and Steven Jacobson

CONTENTS

INTRODUCTION

Multiple sclerosis (MS) is the most prevalent inflammatory, demyelinating disease involving the central nervous system (CNS) (Noseworthy et al. 2000). Lesions within myelinated regions of the CNS, mediated by infiltrating T cells and autoantibodies, are a hallmark of the disease (Keegan and Noseworthy 2002) and lead to progressive neurologic dysfunction (Hartung and Kieseier 2010). Although the etiological agent(s) of MS remains unknown, the heterogeneous disease course suggests a host of complex interactions between immunologic, genetic, and environmental factors (Sospedra and Martin 2005).

The blood–brain barrier (BBB) poses a challenge for translating research initiatives into clinically effective therapeutics to manage MS (Pardridge 2002). The barrier function of the lining microvessel endothelial cells is multifaceted, reinforced by tight junctions with minimal pinocytosis and by drug efflux systems [such as permeability-glycoprotein (Pgp)] (Tsuji and Tamai 1999) and ectoenzymes on the plasma membrane. As shown in Figure 31.1, Pgp is among the multiple transport systems expressed on BBB endothelial cells that impede molecular passage across the BBB. Together, these physical, enzymatic, and efflux functions severely restrict macromolecular passage

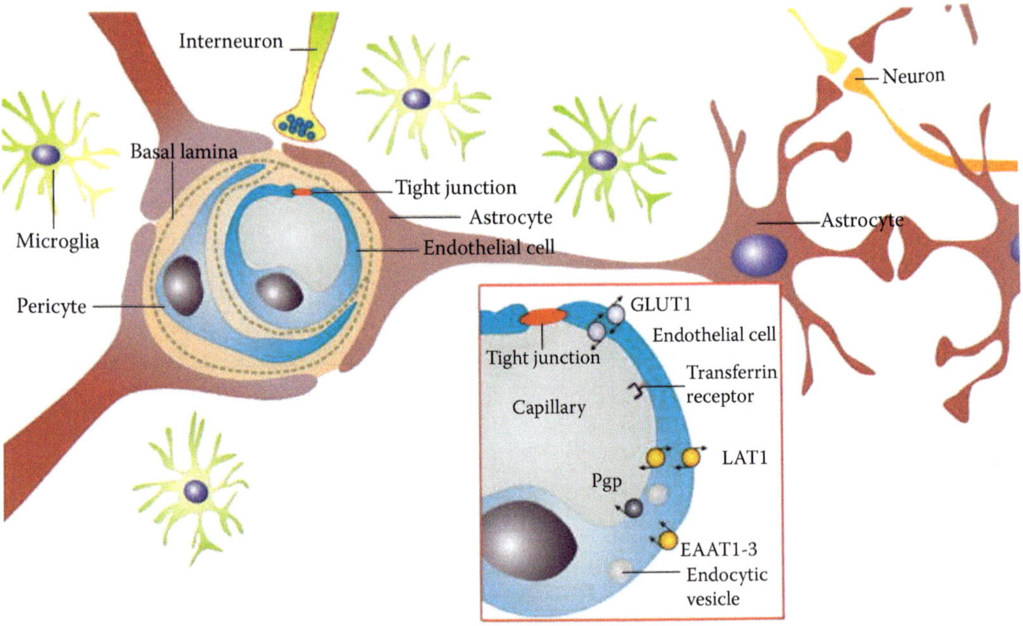

FIGURE 31.1 The blood–brain barrier (BBB) is composed of endothelial cells that form tight junctions and express various transport systems that exclude a variety of molecules. These transport systems include: permeability glycoprotein (Pgp), glucose transporter (GLUT1), large amino acid transporter (LAT1), excitatory amino acid transporters (EAAT1–3), and transferrin receptor. (Reprinted with permission from Kabanov, A.V., Gendelman, H.E., *Prog. Polym. Sci.*, 32, 1054–82, 2007.)

to the brain (Kabanov and Gendelman 2007). Only highly lipid-soluble molecules can penetrate the BBB, which excludes more than 98% of small-molecular-weight drugs and almost all large-molecular-weight drugs (Pardridge 2002). Consequently, in managing diseases such as MS, drugs are often delivered to the CNS at high doses, a strategy that increases the likelihood of successful delivery, but also side effects (Murthy 2007).

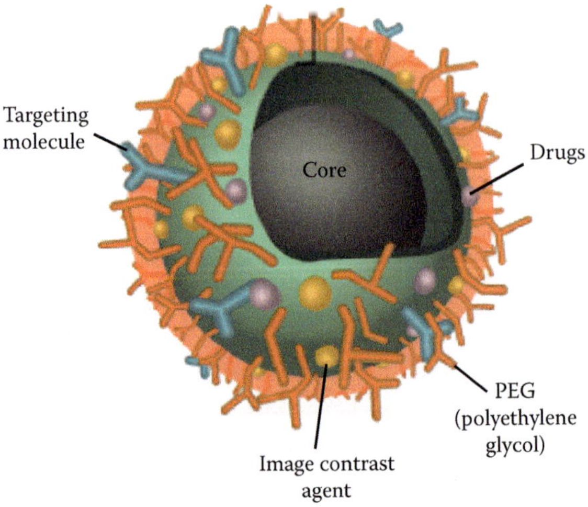

FIGURE 31.2 A multifunctional nanoparticle (NP). The core can be solid or aqueous, whereas the surface can be functionalized with targeting molecules, drugs, or contrast agents. (Adapted and reprinted with permission from McNeil, S.E., *J. Leukoc. Biol.*, 78, 585–94, 2005.)

The past decade has seen rapid advances in the field of nanomedicine, the integration of nanotechnology and medicine. Nanoparticles (NPs), synthetic polymers with dimensions in the nanoscale (generally less than 100 nm) (Murthy 2007), are at the forefront of drug delivery and imaging applications because of the versatility in their core and surface compositions. Figure 31.2 presents a multifunctional NP, with surface modifications that include targeting molecules, drugs, and image contrast agents. Each modification will be discussed in further detail.

NPs can cross the BBB, likely by receptor-mediated transcytosis and/or adsorptive transcytosis. These routes of entry rely on interactions between NPs and moieties or receptors on the luminal surface of BBB endothelial cells (Yang 2010). Utilizing NPs to circumvent the barrier effects of the BBB holds great promise for understanding and treating disorders of the CNS. This chapter will cover NP modifications mostly developed in rodents with experimental autoimmune encephalitis (EAE), a widely used animal model in MS research (Furlan et al. 2009), in addition to NP applications that may advance current goals for MS treatment. Although discussed in the context of MS, these NPs have broad applicability to other CNS demyelinating diseases.

ENCAPSULATION OF MS THERAPEUTICS

NP encapsulation has the potential to enhance many aspects of drug delivery. Therapeutic agents such as steroids, peptides, and oligonucleotides can be captured in a precipitating NP during synthesis. Encapsulation protects the targeted molecule from degradation en route to the CNS, effectively increasing its bioavailability (Kabanov and Gendelman 2007). Although lipophilic NPs can cross the BBB by transcytosis or by permeating the tight junctions of BBB epithelial cells (Bondi et al. 2010), their surfaces can be further functionalized to promote stability and solubility, and enable targeting to specific cells. Moreover, NPs can act as drug carriers in a sustained release system because their biocompatible polymer matrix degrades with controlled and modifiable kinetics (Vinogradov et al. 2004).

LIPOSOMES

Liposomes consist of an aqueous compartment surrounded by a lipid bilayer. These vesicles are optimal for drug delivery because lipophilic drugs can be incorporated into the lipid bilayer, and water-soluble drugs can be incorporated into the aqueous center (Kabanov and Gendelman 2007). Liposome surfaces are often modified with water-soluble polyethylene glycol (PEG) to increase their half-life in circulation, partly by reducing the adhesion of opsonizing plasma proteins. Nanoscale liposomes are successful CNS drug delivery vehicles because they are readily phagocytosed by microglia, astrocytes, and macrophages (Kabanov and Gendelman 2007). A macrophage-based NP drug delivery system (illustrated in Figure 31.3) is advantageous for MS and other CNS diseases with an inflammatory component because macrophages traffic along chemokine gradients, in the process directing NP-encapsulated drugs to inflammatory foci.

Intravenous (i.v.) injection of methylprednisolone, an anti-inflammatory glucocorticosteroid, is a common treatment for MS relapses. In a 2003 study, Schmidt et al. (2003) encapsulated prednisolone in 90–100 nm radiolabeled liposomes and evaluated the efficacy versus free drug in a rat model of adoptive transfer EAE. ^{3}H-labeled liposomes localized to the inflamed CNS, and higher prednisolone concentrations were detected for longer periods in rats administered liposome-encapsulated drug compared to rats administered free drug. Strikingly, liposome-encapsulated prednisolone was more effective than a 5-fold higher dose of free prednisolone in terms of reducing T cell and macrophage infiltration, inducing T cell apoptosis, restoring BBB integrity, and ameliorating the clinical course of EAE (Schmidt et al. 2003). Using NPs such as liposomes to enhance the delivery efficacy of current MS drugs may allow for dose reductions and/or decreased side effects.

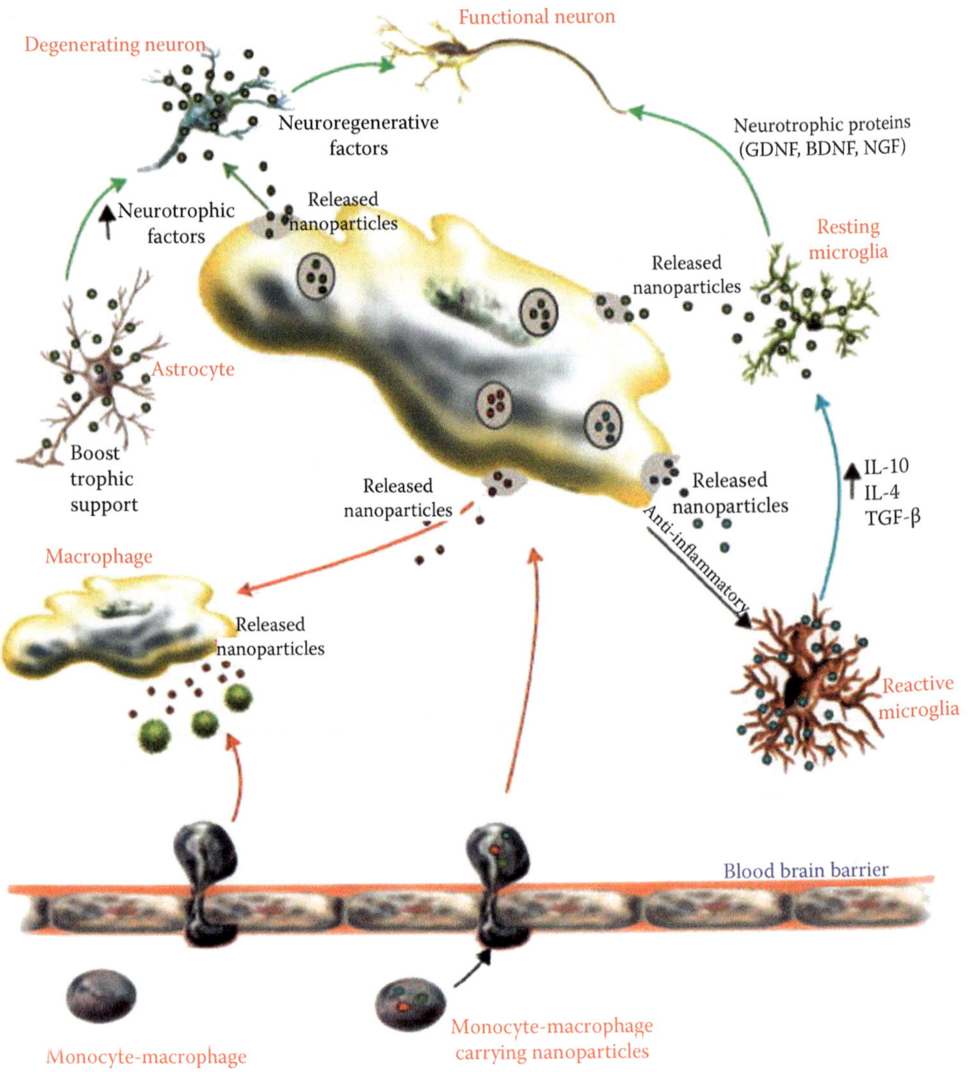

FIGURE 31.3 A macrophage-based NP drug delivery system for treating CNS disorders. Upon entering the brain, macrophages with phagocytosed NPs traffic to sites of inflammation along chemokine gradients. As NP dissolves, the encapsulated therapeutic is released to nearby cells, such as other macrophages, microglia, astrocytes, or neurons. Drugs or other therapeutic agents could act to dampen neurotoxic microglial activation and influence neuronal viability, among other restorative effects on CNS homeostasis. (Adapted and reprinted with permission from Kabanov, A.V., Gendelman, H.E., *Prog. Polym. Sci.*, 32, 1054–82, 2007.)

Solid Lipid NPs

Solid lipid nanoparticles (SLNs) recently became popular for drug delivery across the BBB. They are synthesized by emulsifying the drug and surfactant with molten lipid. Upon cooling, the result is a lipid drug delivery vehicle that is solid at room temperature (Muller et al. 2000). In a rat model of EAE, Bondi et al. (2010) encapsulated the neuroprotective agent Riluzole in SLNs and compared the efficacy to free drug. Riluzole was administered intraperitoneally (i.p.) 7 days after EAE induction, preceding the onset of clinical symptoms. Incorporating the drug into SLN particles reduced the accumulation of Riluzole in off-target organs. Riluzole was found at higher concentrations in the brain and lower concentrations in the bloodstream of rats treated with the SLN-drug particle compared to rats treated with free drug. Moreover, rats in the SLN-drug particle group displayed a delayed onset of EAE clinical symptoms. These results suggest that SLN-mediated drug delivery may result in reduced systemic toxicity, correlating with enhanced efficacy for a given dose (Bondi et al. 2010).

Aquasomes

Aquasomes have been proposed as peptide delivery vehicles because their surface coating stabilizes the biologically active conformation of proteins (Umashankar et al. 2010). Aquasomes consist of a solid nanocrystalline ceramic core coated with carbohydrates, which creates a glossy, aqueous film that adsorbs proteins with negligible denaturation. Trehalose is reportedly an optimal coating because it is inert toward protein surfaces (Umashankar et al. 2010) and prevents aggregation of lyophilized recombinant human serum albumin (Han et al. 2007). Aquasomes represent a possible delivery vehicle for peptides in MS therapy. In 2006, Warren et al. published the results of a double-blinded, placebo-controlled phase II clinical trial of MBP-8298, a synthetic polypeptide corresponding to an immunodominant epitope of human MBP (myelin basic protein). MBP-8298 was administered i.v. for 6 months to 32 primary progressive MS (PPMS) and secondary progressive MS (SPMS) patients. Although anti-MBP antibody levels were lowered in many MBP-8298-treated patients, this did not correlate with measurable clinical benefit(s) (Warren et al. 2006). Peptide therapies for MS may benefit from the stabilizing effects of aquasome delivery (Paolicelli et al. 2009).

Dendrimers

Dendrimers are highly branched NPs (approximately 5–10 nm) increasingly used as scaffolds in drug delivery because dendrimer multivalency allows for high drug payload and/or multifunctionality (Dai et al. 2010). Polyamidoamine (PAMAM)-modified dendrimers are commonly investigated as drug carriers. In this type of dendrimer, PAMAM tendrils radiate from a hydrophobic core, resulting in a globular structure with the capacity to accommodate a high density of ligands (Fahmy et al. 2007). Microglial cells activated with LPS readily phagocytose dendrimers, with peak intracellular concentrations occurring after 1–2 h of exposure (Wang et al. 2009). Dai et al. (2010) demonstrated that FITC labeled PAMAM dendrimers, in the absence of any targeting ligands, were phagocytosed by activated microglia and astrocytes in a rabbit model of neuroinflammation. Using PAMAM dendrimers to deliver anti-inflammatory agents to activated CNS cells may mitigate the pathological effects mediated by microglial activation.

Dendrimers can be targeted to specific cells when appropriately functionalized. Fahmy et al. encapsulated the antimitogenic drug Doxorubicin (Dox) in a PEGylated PAMAM dendrimer and modified the surface with antibodies or MHC molecules. When a biotinylated anti-CD3 antibody was attached, the drug-loaded dendrimer effectively targeted T cell populations. Similar dendrimer modifications have applications as immunostimulatory or immunosuppressive tools (Fahmy et al. 2007) in MS. Acute relapses of MS and optic neuritis, which is a common occurrence among MS patients, are treated with high-dose glucocorticosteroids (e.g., methylprednisolone). Wüst et al. (2008) reported that peripheral T cells are key targets of the glucocorticoid actions of dexamethasone in a mouse model of EAE. Drug-loaded dendrimers can also be targeted to B cells, a population increasingly recognized as central to MS immunopathogenesis. If surface-modified dendrimers can direct a drug precisely to its site of action, then a lower dose may be used to achieve the same (and possibly even greater) efficacy with fewer side effects resulting from interactions between the drug and non-target cells.

OLIGONUCLEOTIDE DELIVERY

DNA Vaccines

Plasmid DNA vaccines, which aim to induce tolerance in MS by encoding for one or multiple myelin antigen(s) (Fissolo et al. 2010), show great potential for MS therapy. In a rat model of EAE,

Wefer et al. (2004) demonstrated that immunization with DNA encoding MOG_{91-108} elicited protection against a subsequent challenge with MOG_{91-108} peptide in complete Freund adjuvant, possibly through a mechanism of IFNβ induction. There are two reported clinical trials of DNA vaccines in MS patients. Both evaluate the clinical outcomes of relapsing–remitting MS (RRMS) and SPMS patients injected intramuscularly (i.m.) with BHT-3009, which encodes the full-length MBP. In the 2007 trial, patients were vaccinated for 9 weeks, and although BHT-3009 reduced the number and volume of gadolinium-enhancing lesions for some patients, this did not achieve statistical significance (Bar-Or et al. 2007). In the 2008 trial, patients were vaccinated for 44 weeks and despite a statistically significant reduction in gadolinium-enhancing lesions, BHT-3009 vaccination failed to achieve significance on other measures of clinical outcome, such as relapse rate (Garren et al. 2008).

Immunological data support BHT-3009 as an effective DNA vaccine for MS treatment although statistical significance was not reached for all clinical trial measures. BHT-3009-vaccinated patients exhibited decreased proliferation of IFNγ-producing CD4 cells specific for epitopes MBP, proteolipid protein, and myelin oligodendrocyte glycoprotein (MOG). In a follow-up study, BHT-3009 vaccination was associated with significant decreases in MBP-specific CSF antibody titers (Bar-Or et al. 2007). These encouraging results suggest the need for more efficient methods of DNA vaccine delivery (Fissolo et al. 2010), such as NP-mediated delivery. Biodistribution studies show that only a small fraction of DNA injected i.m. survives to transfect target cells (Donnelly et al. 2005). NP encapsulation would protect DNA from extracellular degradation and can be functionally enhanced through surface conjugations. Furthermore, NPs composed of biodegradable polymers provide a scaffold for DNA release, with modifiable kinetics according to core and surface compositions (Xiang et al. 2010).

NPs of approximately 40 nm reportedly elicit strong CD8 T cell and antibody responses against a DNA-encoded antigen (Minigo et al. 2007; Xiang et al. 2010). DNA adsorption onto cationic polymer coatings (such as poly-L-lysine) further strengthens DNA vaccine immunogenicity, likely resulting from increased antigen presenting cell uptake (Xiang et al. 2010) due to electrostatic interactions between the DNA (Minigo et al. 2007), and the negatively charged target cell membrane (Xiang et al. 2010). In a 2007 study, Minigo et al. (2007) demonstrated that DNA complexed with a poly-L-lysine-coated NP elicited greater cellular and humoral immune responses in mice than naked DNA or even DNA complexed with uncoated NPs. Currently formulated DNA vaccines might have improved efficacy in clinical trials if complexed with NPs for delivery.

Antisense Oligonucleotides

Several recent studies have implicated microRNAs (miRs) in the pathogenesis of MS. MicroRNA expression profiles differ between MS patients and controls, and between active and inactive lesions (Junker et al. 2011). Du et al. (2009) found that miR-326 is significantly higher in the peripheral blood lymphocytes of patients with RRMS than in age-matched controls or patients with neuromyelitis optica. MiR-326 expression in CD4 cells correlated with clinical relapses, and is 9-fold up-regulated in active MS lesions compared to normal brain matter (Junker et al. 2009). Knocking down miR326 in EAE mice reduced the number of T_H17 cells and resulted in more clinically severe EAE, whereas overexpressing it increased the number of T_H17 cells and resulted in a more clinically severe EAE (Du et al. 2009).

Targeting miRNA by delivering a cognate antisense sequence (anti-miR) represents a novel therapeutic strategy for regulating pathogenic gene expression (Junker et al. 2011). This may be especially relevant for MS, in which the expression of a particular miR correlates with disease severity. The challenges associated with delivering antisense miRs, namely, adequate delivery to the target organ and off-target effects (Junker et al. 2011), may be largely surmountable with NP-mediated delivery.

Nanogels and interfering NPs (iNOPs) are candidate NPs for in vivo targeting of pathogenic miRs. Cationic nanogels are hydrophilic polymer gels composed of cross-linking PEG and PEI (polyethylenimine) chains that can bind and encapsulate negatively charged oligonucleotides, such as antisense miRs (Vinogradov et al. 2004). In 2004, Vinogradov et al. demonstrated that nanogels successfully delivered encapsulated oligonucleotides across the BBB, with CNS target specificity similar to other therapeutic-NP complexes. After i.v. injection with the oligonucleotide–nanogel complexes, there was increased oligonucleotide concentration in the brain and decreased levels in the liver and spleen compared to freely delivered oligonucleotides (Vinogradov et al. 2004). In a 2011 study, Su et al. demonstrated the effective silencing of liver-expressed miR122 in mice after an i.v. injection with a chemically modified anti-miR122 complexed with a dendrimer-based NP, iNOP-7. The NP iNOP-7 is functionalized with lysine and lipid chains and displayed favorable pharmacokinetics and immune tolerance in a mouse model (Baigude et al. 2007), suggesting its capacity to deliver anti-miRs in a clinical setting (Su et al. 2011). Given the high degree of branching enabling dendrimer multifunctionality, it is likely that these NPs can accommodate sufficient modifications to optimize their passage across the BBB to deliver therapeutic anti-miRs to treat MS.

NEUROPROTECTION TO COMPLEMENT MS DRUG THERAPY

Because axonal degeneration correlates with disability in MS, neuroprotection can be considered a complementary therapeutic approach for MS treatment (Nowacek et al. 2009). Microglial cells, activated by infection and inflammation, are recruited to lesions in inflammatory demyelinating pathologies. Microglial activation leads to secretion of proinflammatory molecules and reactive oxygen species (ROS) that may exacerbate neuronal damage during an inflammatory response (Dai et al. 2010). Microglia and macrophages within active CNS lesions express high levels of inducible nitric oxide synthase (iNOS) (De Groot et al. 1997). There is elevated iNOS, nitric oxide (NO), peroxynitrate, and superoxide in the CNS of MS patients (Calabrese et al. 2002), suggesting a role for these damaging oxidants in the immunopathogenesis of MS. Importantly, oligodendrocytes are susceptible to NO and can be lysed by the levels produced by activated microglial cells (Witherick et al. 2010). Pre-oligodendrocytes are more sensitive to oxidative stress than mature oligodendrocytes, suggesting that microglial activation leading to proinflammatory cytokine and ROS production may hinder remyelination and repair (Witherick et al. 2010).

Carbon-60 (C_{60}) fullerenes are three-dimensional arrays of evenly spaced carbon atoms, arranged in a pattern comparable to the rhombuses of a soccer ball (Silva 2005). A class of neuroprotective compounds is based on C_{60} fullerenes, as functionalized derivatives, for example, hydroxylated fullerenes or fullerenols, and reportedly possess antioxidant properties. In 2003, Hirsch et al. modified a C_{60} core with malonate ester dendrimers to increase the particle's solubility and surface area. This modification resulted in increased antioxidant activity, relative to an unmodified C_{60} fullerenol, because water-soluble fullerenol derivatives are highly reactive with oxygen free radicals (Gilmore et al. 2008).

ABS-75 is a water-soluble fullerene derivative functionalized with an NMDA receptor antagonist, in order to combine antioxidant and anti-excitotoxic activities. This NP was demonstrated to slow EAE disease progression in mice immunized with MBP. Mice injected i.p. with ABS-75 exhibited decreased oxidative injury, less monocyte recruitment, and reduced axon and myelin loss in the spinal cord (Basso et al. 2008). Despite reports that fullerenes induce lipid peroxidation and aggregate with cell surface membranes, this can be minimized with surface modifications (Gilmore et al. 2008).

Schubert et al. (2006) reported that NPs composed of cerium (CeO_2) and yttrium (Y_2O_3) oxides protect neurons from oxidative stress-induced cell death in vitro, likely by ROS reduction. Interestingly, such neuroprotection was not accomplished with equivalent concentrations of the nonparticle oxides and salts, and the protective effect was independent of NP size between 6 and 1000 nm (Schubert et al. 2006). NPs can also encapsulate radicals that mediate neuroprotective effects (Yang 2010). Kizelsztein et al. (2007) developed nanoliposomes to deliver tempamine (TMN), a stable radical with antioxidant and proapoptotic and/or pronecrotic activities (Wasserman et al. 2007). This

TMN–NP complex effectively ameliorated the clinical symptoms of EAE in both chronic and acute mouse models. Nanoliposome conjugation extended the half-life of TMN in plasma to more than 10 h, whereas free TMN was cleared from plasma in 1 h (Kizelsztein et al. 2009). Promoting neuroprotection by minimizing oxidative stress in the CNS microenvironment may be an important underpinning for the success of MS therapy.

MAGNETIC NPs

Magnetic resonance imaging (MRI) provides distinctive spatial resolution and tissue contrast without ionizing radiation (Vellinga et al. 2008) and is a widely used imaging technique for investigating neurologic disease (van Kasteren et al. 2009). Gadolinium is a contrast agent routinely used to visualize BBB breakdown, an important assessment for the diagnosis of MS (Vellinga et al. 2008). Gadolinium-enhancing lesions may reflect a relatively advanced pathology, because inflammation is likely to precede and/or fuel BBB breakdown (McAteer et al. 2007). Contrast agents enabling the visualization of inflammatory dynamics independent of BBB breakdown could allow for earlier detection and diagnosis of MS. Functionalized magnetic NPs are emerging as the tools to enable such early detection because they augment MRI sensitivity. Magnetic NPs can act as both diagnostic agents and drug carriers, and are at the intersection of theragnostics, the union of therapeutics and diagnostics (Shubayev et al. 2009).

Iron Oxide NPs

Ultrasmall superparamagnetic particles composed of iron oxide (USPIO) are often used as MRI contrast agents. The iron oxide core decreases the T1 and T2 relaxation times of water, resulting in an increased signal on T1-weighted images and a reduced signal on T2-weighted gradient-echo images (Corot et al. 2006). USPIOs are phagocytosed by mononuclear phagocytic cells, which facilitates visualization of macrophage infiltration and provide complementary information to gadolinium enhancement for disease state assessment (Vellinga et al. 2008). The USPIO ferumoxtran-10 is currently used to image phagocytic cells in MS patients (Dousset et al. 2006). Vellinga et al. investigated the USPIO particle SHU555C, which is smaller than ferumoxtran-10 (25 nm compared to 30 nm) and has a shorter in vivo half-life (6–8 h compared to 24–30 h). SHU555C enhancement occurred more frequently and was longer lasting than gadolinium enhancement. In some patients, SHU555C enhancement even preceded gadolinium enhancement (Vellinga et al. 2008). Very small superparamagnetic iron oxide particles (VSOPs) represent another category of NP contrast agents. Tysiak et al. (2009) demonstrated that VSOPs locate to neuroinflammatory foci in a murine model of EAE and were immediately visible as T2* hypointensity. VSOPs delineated all gadolinium-enhancing lesions, and 13 of 55 inflammatory foci were visualized exclusively by VSOPs. Using superparamagnetic NPs to detect macrophage infiltration into the CNS may provide insight into the early pathophysiology of MS and other demyelinating diseases (Vellinga et al. 2008).

Functionalized Magnetic NPs

As with previously discussed NPs, magnetic NP surfaces can also be functionalized. McAteer et al. (2007) developed a 1-μm iron oxide particle conjugated to a VCAM-1 antibody that detected VCAM-1 expression with high specificity in a model of mouse brain inflammation. VCAM-1 and its cognate ligand, $\alpha_4\beta_1$ integrin, mediate leukocyte recruitment and transmigration across the BBB (McAteer et al. 2007). Leukocyte trafficking into the CNS is reflected in the characteristic leukocytic infiltrates of MS lesions (Noseworthy et al. 2000). Inhibitors that bind $\alpha_4\beta_1$ to block its interaction with VCAM-1 provide clinical benefit for MS patients, and may reduce enhancing lesions and episodes of clinical relapse (Polman et al. 2006).

Magnetic NPs with iron oxide cores can be conjugated to antibodies such as anti-VCAM-1, or other endovascular targets implicated in BBB breakdown that could indicate the beginning of a neuroinflammatory and/or demyelinating pathology. Cayrol et al. (2008) applied a proteomics approach to lipid rafts isolated from human BBB epithelial cells to identify additional BBB adhesion molecules. They reported that CD166 (or ALCAM, activated leukocyte cell adhesion molecule) is up-regulated during neuroinflammation (Cayrol et al. 2008). CD166 may, therefore, represent another antibody target for decreasing peripheral leukocyte migration into the CNS (Cayrol et al. 2008).

Veiseh et al. (2009) designed an iron oxide NP coated with PEGylated chitosan and functionalized with a targeting ligand and the fluorophore Cy5.5. Although this particular NP targeted a brain tumor ligand, the design is conceptually applicable to other CNS pathologies because this NP entered the CNS without compromising BBB integrity when injected i.v. (Veiseh et al. 2009). van Kasteren et al. (2009) designed an MRI-visible glyconanoparticle to detect the endothelial markers E-/P-selectin (CD62E/CD62P) in a rat model of MOG-EAE. E- and P-selectin are up-regulated during injury or disease and are involved in leukocyte homing to inflammatory sites (von Andrian and Mackay 2000). This glyconanoparticle was coated with sialyl-Lewis^{x}, the cognate glycan ligand for CD62, and delineated chronically activated brain endothelium adjacent to the site of a focal MS-like lesion in an EAE rat model (van Kasteren et al. 2009). Functionalizing magnetic NP surfaces can provide additional dimensions to their utility as imaging contrast agents.

Stem Cell Labeling

Mesenchymal stem cells (MSCs) may dampen neuroinflammation by secreting superoxide dismutase 3 (Kemp et al. 2010) and promoting decreased iNOS expression (Lanza et al. 2009), thereby reducing microglial activation. MSCs may differentiate into glial lineages such as oligodendrocytes, or promote neurogenesis in the brain parenchyma, effectively mitigating the neurodegenerative arm of MS (Karussis and Kassis 2008). When nonhematopoietic MSCs were transplanted in a murine model of EAE, the cells migrated to CNS white matter lesions, correlating with reduced immune cell infiltrates and smaller areas of demyelination (Kassis et al. 2008). Gordon et al. delivered (i.v.) bone marrow-derived human MSCs to EAE mice 6 days post-MOG_{35-55} injection and reported a significant reduction in clinical disease severity. MSC-treated EAE mice exhibited reduced white matter spinal cord lesions (size and absolute number) relative to controls. The injected MSCs localized to gray and white matter throughout the spinal cord and likely promoted reparative effects in the demyelinated regions (Gordon et al. 2010).

NPs can be used to track transplanted MSCs to ensure proper homing, survival, and function upon differentiation in the target tissue (Strauer and Kornowski 2003). Reddy et al. (2010) successfully tracked transplanted MSCs using SPIO NPs coated with chitosan. Chitosan is a biocompatible, biodegradable polymer prepared by the *N*-deacetylation of chitin. NP endocytosis did not alter MSC surface marker expression, proliferation, or chondrogenic (cartilage-forming) differentiation capacity. The NP-labeled MSCs were tracked for 2 weeks in vivo in a rabbit model of brain ischemia (Reddy et al. 2010).

CONCLUSION

In conclusion, incorporating NPs into laboratory and clinical settings may profoundly impact current diagnostic and treatment strategies for MS and other CNS demyelinating diseases. NPs are advantageous therapeutic delivery vehicles owing to their variable core compositions, which can accommodate a wide range of therapeutics as illustrated in Figure 31.4. NPs ferrying drugs and contrast agents across the BBB will likely facilitate more efficacious and precise drug delivery and disease assessment by MRI. Concerns regarding NP toxicity and in vivo pharmacokinetics can be largely addressed through core and surface modifications.

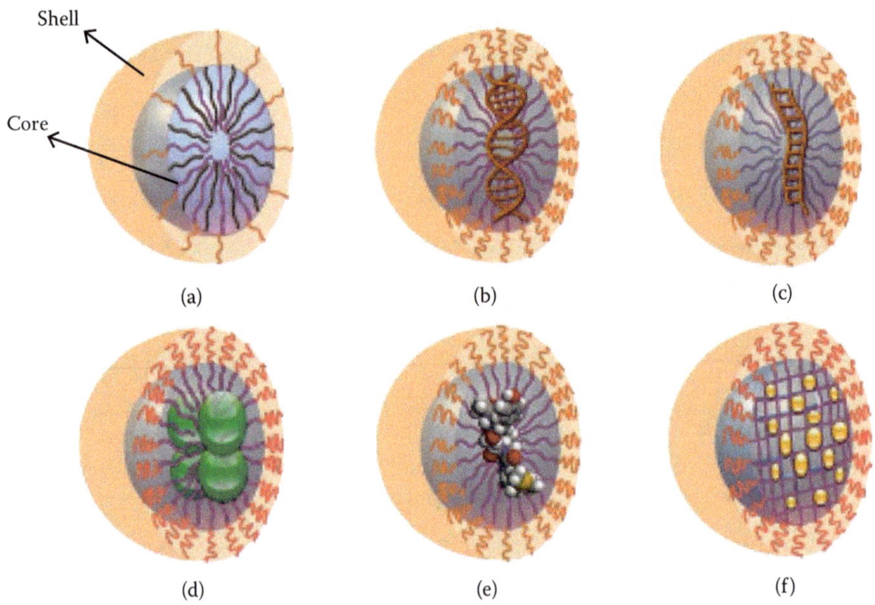

FIGURE 31.4 NP cores can be composed of neutralized polyions. (a) Complexes of synthetic polyions facilitate encapsulation of a variety of molecules, such as (b) polycation and plasmid DNA; (c) polycation and siRNA; (d) polycation or polyanion and protein; or (e) peptide. The core ionic chains can also cross-link to form polymer networks that facilitate the incorporation of (f) small polypeptides or charged hydrophobic drugs. (Reprinted with permission from Gilmore, J.L. et al., *J. Neuroimmune Pharmacol.*, 3, 83–94, 2008.)

REFERENCES

Baigude, H., McCarroll, J., Yang, C.S. et al. 2007. Design and creation of new nanomaterials for therapeutic RNAi. *ACS Chem Biol* 2: 237–41.

Bar-Or, A., Vollmer, T., Antel, J. et al. 2007. Induction of antigen-specific tolerance in multiple sclerosis after immunization with DNA encoding myelin basic protein in a randomized, placebo-controlled phase 1/2 trial. *Arch Neurol* 64: 1407–15.

Basso, A.S., Frenkel, D., Quintana, F.J. et al. 2008. Reversal of axonal loss and disability in a mouse model of progressive multiple sclerosis. *J Clin Invest* 118: 1532–43.

Bondi, M.L., Craparo, E.F., Giammona, G. et al. 2010. Brain-targeted solid lipid nanoparticles containing riluzole: preparation, characterization and biodistribution. *Nanomedicine (Lond)* 5: 25–32.

Calabrese, V., Scapagnini, G., Ravagna, A. et al. 2002. Nitric oxide synthase is present in the cerebrospinal fluid of patients with active multiple sclerosis and is associated with increases in cerebrospinal fluid protein nitrotyrosine and *S*-nitrosothiols and with changes in glutathione levels. *J Neurosci Res* 70: 580–7.

Cayrol, R., Wosik, K., Berard, J.L. et al. 2008. Activated leukocyte cell adhesion molecule promotes leukocyte trafficking into the central nervous system. *Nat Immunol* 9: 137–45.

Corot, C., Robert, P., Idee, J.M. et al. 2006. Recent advances in iron oxide nanocrystal technology for medical imaging. *Adv Drug Deliv Rev* 58: 1471–504.

Dai, H., Navath, R.S., Balakrishnan, B. et al. 2010. Intrinsic targeting of inflammatory cells in the brain by polyamidoamine dendrimers upon subarachnoid administration. *Nanomedicine (Lond)* 5: 1317–29.

De Groot, C.J., Ruuls, S.R., Theeuwes, J.W. et al. 1997. Immunocytochemical characterization of the expression of inducible and constitutive isoforms of nitric oxide synthase in demyelinating multiple sclerosis lesions. *J Neuropathol Exp Neurol* 56: 10–20.

Donnelly, J.J., Wahren, B., and Liu, M.A. 2005. DNA vaccines: progress and challenges. *J Immunol* 175: 633–9.

Dousset, V., Brochet, B., Deloire, M.S. et al. 2006. MR imaging of relapsing multiple sclerosis patients using ultra-small-particle iron oxide and compared with gadolinium. *AJNR Am J Neuroradiol* 27: 1000–5.

Du, C., Liu, C., Kang, J. et al. 2009. MicroRNA miR-326 regulates TH-17 differentiation and is associated with the pathogenesis of multiple sclerosis. *Nat Immunol* 10: 1252–9.

Fahmy, T.M., Schneck, J.P., and Saltzman, W.M. 2007. A nanoscopic multivalent antigen-presenting carrier for sensitive detection and drug delivery to T cells. *Nanomedicine* 3: 75–85.

Fissolo, N., Montalban, X., and Comabella, M. 2010. DNA-based vaccines for multiple sclerosis: current status and future directions. *Clin Immunol* 142: 76–83.

Furlan, R., Cuomo, C., and Martino, G. 2009. Animal models of multiple sclerosis. *Methods Mol Biol* 549: 157–73.

Garren, H., Robinson, W.H., Krasulova, E. et al. 2008. Phase 2 trial of a DNA vaccine encoding myelin basic protein for multiple sclerosis. *Ann Neurol* 63: 611–20.

Gilmore, J.L., Yi, X., Quan, L. et al. 2008. Novel nanomaterials for clinical neuroscience. *J Neuroimmune Pharmacol* 3: 83–94.

Gordon, D., Pavlovska, G., Uney, J.B. et al. 2010. Human mesenchymal stem cells infiltrate the spinal cord, reduce demyelination, and localize to white matter lesions in experimental autoimmune encephalomyelitis. *J Neuropathol Exp Neurol* 69: 1087–95.

Han, Y., Jin, B.S., Lee, S.B. et al. 2007. Effects of sugar additives on protein stability of recombinant human serum albumin during lyophilization and storage. *Arch Pharm Res* 30: 1124–31.

Hartung, H.P., and Kieseier, B.C. 2010. Atacicept: targeting B cells in multiple sclerosis. *Ther Adv Neurol Disord* 3: 205–16.

Junker, A., Hohlfeld, R., and Meinl, E. 2011. The emerging role of microRNAs in multiple sclerosis. *Nat Rev Neurol* 7: 56–9.

Junker, A., Krumbholz, M., Eisele, S. et al. 2009. MicroRNA profiling of multiple sclerosis lesions identifies modulators of the regulatory protein CD47. *Brain* 132: 3342–52.

Kabanov, A.V., and Gendelman, H.E. 2007. Nanomedicine in the diagnosis and therapy of neurodegenerative disorders. *Prog Polym Sci* 32: 1054–82.

Karussis, D., and Kassis, I. 2008. The potential use of stem cells in multiple sclerosis: an overview of the preclinical experience. *Clin Neurol Neurosurg* 110: 889–96.

Kassis, I., Grigoriadis, N., Gowda-Kurkalli, B. et al. 2008. Neuroprotection and immunomodulation with mesenchymal stem cells in chronic experimental autoimmune encephalomyelitis. *Arch Neurol* 65: 753–61.

Keegan, B.M., and Noseworthy, J.H. 2002. Multiple sclerosis. *Annu Rev Med* 53: 285–302.

Kemp, K., Gray, E., Mallam, E. et al. 2010. Inflammatory cytokine induced regulation of superoxide dismutase 3 expression by human mesenchymal stem cells. *Stem Cell Rev* 6: 548–59.

Kizelsztein, P., Ovadia, H., Garbuzenko, O. et al. 2009. Pegylated nanoliposomes remote-loaded with the antioxidant tempamine ameliorate experimental autoimmune encephalomyelitis. *J Neuroimmunol* 213: 20–5.

Lanza, C., Morando, S., Voci, A. et al. 2009. Neuroprotective mesenchymal stem cells are endowed with a potent antioxidant effect in vivo. *J Neurochem* 110: 1674–84.

McAteer, M.A., Sibson, N.R., von Zur Muhlen, C. et al. 2007. In vivo magnetic resonance imaging of acute brain inflammation using microparticles of iron oxide. *Nat Med* 13: 1253–8.

McNeil, S.E. 2005. Nanotechnology for the biologist. *J Leukoc Biol* 78: 585–94.

Minigo, G., Scholzen, A., Tang, C.K. et al. 2007. Poly-L-lysine-coated nanoparticles: a potent delivery system to enhance DNA vaccine efficacy. *Vaccine* 25: 1316–27.

Muller, R.H., Mader, K., and Gohla, S. 2000. Solid lipid nanoparticles (SLN) for controlled drug delivery—a review of the state of the art. *Eur J Pharm Biopharm* 50: 161–77.

Murthy, S.K. 2007. Nanoparticles in modern medicine: state of the art and future challenges. *Int J Nanomedicine* 2: 129–41.

Noseworthy, J.H., Lucchinetti, C., Rodriguez, M. et al. 2000. Multiple sclerosis. *N Engl J Med* 343: 938–52.

Nowacek, A., Kosloski, L.M., and Gendelman, H.E. 2009. Neurodegenerative disorders and nanoformulated drug development. *Nanomedicine (Lond)* 4: 541–55.

Paolicelli, D., Direnzo, V., and Trojano, M. 2009. Review of interferon beta-1b in the treatment of early and relapsing multiple sclerosis. *Biologics* 3: 369–76.

Pardridge, W.M. 2002. Drug and gene delivery to the brain: the vascular route. *Neuron* 36: 555–8.

Polman, C.H., O'Connor, P.W., Havrdova, E. et al. 2006. A randomized, placebo-controlled trial of natalizumab for relapsing multiple sclerosis. *N Engl J Med* 354: 899–910.

Reddy, A.M., Kwak, B.K., Shim, H.J. et al. 2010. In vivo tracking of mesenchymal stem cells labeled with a novel chitosan-coated superparamagnetic iron oxide nanoparticles using 3.0T MRI. *J Korean Med Sci* 25: 211–9.

Schmidt, J., Metselaar, J.M., Wauben, M.H. et al. 2003. Drug targeting by long-circulating liposomal glucocorticosteroids increases therapeutic efficacy in a model of multiple sclerosis. *Brain* 126: 1895–904.

Schubert, D., Dargusch, R., Raitano, J. et al. 2006. Cerium and yttrium oxide nanoparticles are neuroprotective. *Biochem Biophys Res Commun* 342: 86–91.
Shubayev, V.I., Pisanic, T.R., 2nd, and Jin, S. 2009. Magnetic nanoparticles for theragnostics. *Adv Drug Deliv Rev* 61: 467–77.
Silva, G.A. 2005. Nanotechnology approaches for the regeneration and neuroprotection of the central nervous system. *Surg Neurol* 63: 301–6.
Sospedra, M., and Martin, R. 2005. Immunology of multiple sclerosis. *Annu Rev Immunol* 23: 683–747.
Strauer, B.E., and Kornowski, R. 2003. Stem cell therapy in perspective. *Circulation* 107: 929–34.
Su, J., Baigude, H., McCarroll, J. et al. 2011. Silencing microRNA by interfering nanoparticles in mice. *Nucleic Acids Res* 39: e38.
Tsuji, A., and Tamai, I.I. 1999. Carrier-mediated or specialized transport of drugs across the blood–brain barrier. *Adv Drug Deliv Rev* 36: 277–90.
Tysiak, E., Asbach, P., Aktas, O. et al. 2009. Beyond blood brain barrier breakdown—in vivo detection of occult neuroinflammatory foci by magnetic nanoparticles in high field MRI. *J Neuroinflammation* 6: 20.
Umashankar, M.S., Sachdeva, R.K., and Gulati, M. 2010. Aquasomes: a promising carrier for peptides and protein delivery. *Nanomedicine* 6: 419–26.
van Kasteren, S.I., Campbell, S.J., Serres, S. et al. 2009. Glyconanoparticles allow pre-symptomatic in vivo imaging of brain disease. *Proc Natl Acad Sci U S A* 106: 18–23.
Veiseh, O., Sun, C., Fang, C. et al. 2009. Specific targeting of brain tumors with an optical/magnetic resonance imaging nanoprobe across the blood–brain barrier. *Cancer Res* 69: 6200–7.
Vellinga, M.M., Oude Engberink, R.D., Seewann, A. et al. 2008. Pluriformity of inflammation in multiple sclerosis shown by ultra-small iron oxide particle enhancement. *Brain* 131: 800–7.
Vinogradov, S.V., Batrakova, E.V., and Kabanov, A.V. 2004. Nanogels for oligonucleotide delivery to the brain. *Bioconjug Chem* 15: 50–60.
von Andrian, U.H., and Mackay, C.R. 2000. T-cell function and migration. Two sides of the same coin. *N Engl J Med* 343: 1020–34.
Wang, B., Navath, R.S., Romero, R. et al. 2009. Anti-inflammatory and anti-oxidant activity of anionic dendrimer-*n*-acetyl cysteine conjugates in activated microglial cells. *Int J Pharm* 377: 159–68.
Warren, K.G., Catz, I., Ferenczi, L.Z. et al. 2006. Intravenous synthetic peptide MBP8298 delayed disease progression in an HLA class II-defined cohort of patients with progressive multiple sclerosis: results of a 24-month double-blind placebo-controlled clinical trial and 5 years of follow-up treatment. *Eur J Neurol* 13: 887–95.
Wasserman, V., Kizelsztein, P., Garbuzenko, O. et al. 2007. The antioxidant tempamine: in vitro antitumor and neuroprotective effects and optimization of liposomal encapsulation and release. *Langmuir* 23: 1937–47.
Wefer, J., Harris, R.A., and Lobell, A. 2004. Protective DNA vaccination against experimental autoimmune encephalomyelitis is associated with induction of IFNbeta. *J Neuroimmunol* 149: 66–76.
Witherick, J., Wilkins, A., Scolding, N. et al. 2010. Mechanisms of oxidative damage in multiple sclerosis and a cell therapy approach to treatment. *Autoimmune Dis* 2011: 164608.
Wust, S., van den Brandt, J., Tischner, D. et al. 2008. Peripheral T cells are the therapeutic targets of glucocorticoids in experimental autoimmune encephalomyelitis. *J Immunol* 180: 8434–43.
Xiang, S.D., Selomulya, C., Ho, J. et al. 2010. Delivery of DNA vaccines: an overview on the use of biodegradable polymeric and magnetic nanoparticles. *Wiley Interdiscip Rev Nanomed Nanobiotechnol* 2: 205–18.
Yang, H. 2010. Nanoparticle-mediated brain-specific drug delivery, imaging, and diagnosis. *Pharm Res* 27: 1759–71.

32 Applications of Nanotechnology in Degenerative Disorders of the Brain

Ayaz M. Khawaja and John D. Heiss

CONTENTS

INTRODUCTION

Dementia is characterized by global cognitive impairment, including memory dysfunction, personality changes, and apraxia. Dementia has many possible etiologies, but most are progressively degenerative in nature. Specific etiologies of dementia manifest different phenotypes at initial presentation. The spectrum of the phenotypic manifestations may sometimes be sufficiently specific to clinically diagnose the type of dementia, such as normal pressure hydrocephalus (NPH) and Wernicke's encephalopathy, and accompanying radiographic evidence can exclude other types, such as vascular dementia. Onset of most types of dementia is chronic, although more rapid onset of symptoms is typical of spongiform encephalopathy and Huntington's chorea, and become the basis for clinical diagnosis in these conditions. The presence of an infectious trigger in the patient's history and laboratory investigations may lead to recognizing the type of dementia, such as a declining

CD4+ (helper T cell) count in HIV encephalopathy. When there is a specific cause such as hypothyroidism or vitamin B_{12} deficiency, correction of the deficiency may alleviate the resulting dementia. Lumbar puncture may lead to dramatic improvement in NPH patients (Hakim and Adams 1965). However, more often than not, the exact diagnosis may still not be elucidated despite a very detailed history and physical examination, appropriate laboratory investigations, and radiographic analysis, although this evaluation excludes the types of dementia that have a well-defined cause. Similar pathogenic mechanisms appear to be responsible for the most common types of dementia.

Novel diagnostic methods using nanotechnology may increase the sensitivity of the detection of amyloid β-derived diffusible ligands within the cerebrospinal fluid (CSF) and may be valuable in diagnosing Alzheimer's disease (AD). Significant work is being done to characterize nanoparticles (NPs) with enhanced ability to cross the blood–brain barrier (BBB) and carry therapeutic agents such as drugs or genes. Other modalities such as nanoneuromodulation may be used to assess neuronal activity in vivo and provide feedback to deep brain stimulation (DBS) devices to modulate this activity. Given the diverse etiologies of dementia, treatments must be designed that counter specific disease mechanisms. Nanotechnology is being integrated into treatments ranging from pharmacotherapy to functional nanoneurosurgery.

NANODIAGNOSTICS

Amyloid-derived diffusible ligands (ADDLs) have been proposed as endogenous neurotoxins in the pathophysiology of AD; their presence is correlated with disease progression (Klein 2002). The corresponding binding proteins are primarily located in frontal cortex and hippocampus, accounting for a possible receptor-mediated pathogenesis (Klein 2002). Georganopoulou et al. developed an ultrasensitive biobarcode assay to measure extremely low concentrations of ADDLs in the CSF of patients with dementia. The approach uses NPs that carry DNA and are tagged with ADDL-complementary antibodies to measure ADDL concentration. The specifics of the method are detailed in Chapter 5 ("Peptide and Protein-Based Nanoparticles").

A nanoscale optical biosensor based on localized surface plasmon resonance (LSPR) spectroscopy was also developed to detect the concentration of ADDLs (Haes et al. 2005). Surface plasmon resonance is a phenomenon applied to valence electrons in the atom. For metal NPs, the incident electromagnetic radiation induces an oscillation in the valence electrons (Haes et al. 2005; Jensen et al. 1999; Kreibig et al. 1997) that results in photon absorption and a change in the LSPR spectrum that can be measured with transmission extinction spectroscopy (Haes et al. 2005; Jensen et al. 1999, 2000; Kreibig et al. 1997). Silver NPs were used as a component of the optical nanosensor, whose extinction spectrum is very sensitive to the nanoenvironment of the nanosensor (Haes and Van Duyne 2002). The NPs are bound to polyclonal anti-ADDL antibody before being exposed to ADDL antigen. The associated extinction spectra accurately predict concentrations between 1 fmol and 100 nmol of ADDL. This method was used to measure ADDLs in *postmortem* brain extracts and in CSF from AD patients and normal controls, with confirmation of raised concentrations of ADDL in AD patients (Haes et al. 2005).

Chikae et al. (2008) reported a unique method of detecting amyloid beta-protein (Aβ) based on protein–saccharide interactions. Gold NPs on a screen-printed strip were coated with a monolayer of acetylenyl groups; azide-terminated sialic acid (a monosaccharide) was then conjugated to the acetylenyl groups. Interaction of sialic acid with tyrosine groups on Aβ resulted in a peak oxidation current whose signal intensity was concentration-dependent. This technique was very sensitive and cost-effective in determining Aβ concentration (Chikae et al. 2008). Kang et al. (2009) developed an immunosensor system using scanning tunneling microscopy that detected concentrations of Aβ as low as 10 fg. Other investigators attempted to measure CSF levels of tau protein, another biomarker of AD. In one study, binding of tau protein to gold NPs functionalized with anti-tau antibody resulted in increased two-photon Rayleigh scattering intensity and the detection of tau levels as low as 1 pg/mL (Neely et al. 2009).

NANOIMAGING

Detection of amyloid plaques, through imaging of NPs tagged with contrast agents or fluorescent probes, is being developed for early identification of AD (Brambilla et al. 2011a). The utility of iron oxide NPs conjugated with Aβ1-40 peptide (a peptide that binds amyloid-beta with high affinity) was explored in transgenic mice that were genetically modified to express Aβ plaques and mimic AD (Wadghiri et al. 2003). The NPs were delivered in vivo together with mannitol to transiently open the BBB. Magnetic resonance (MR) microimaging detected Aβ plaques; the density of plaques seen on imaging correlated well with the density of plaques observed by histology (Wadghiri et al. 2003). Synthetic combination of a cholinesterase inhibitor and fluorophore resulted in a compound that strongly fluoresced, inhibited cholinesterase, and bound robustly to amyloid plaques (Elsinghorst et al. 2009). Hartig et al. (2010) injected NPs labeled with this compound directly into the hippocampi of triple-transgenic mice and demonstrated their very specific binding to Aβ plaques by confocal laser scanning microscopy. Intracerebral injection of thioflavins tagged to NP constructs also resulted in specific binding to Aβ plaques in transgenic mice (Siegemund et al. 2006). Conjugation of Aβ peptide to quantum dots (QD), NPs that emit photons when excited (Khawaja 2011), enabled the study of kinetics of Aβ oligomerization in vitro in cultured microglia using real-time 3-D slit-scanning confocal microscopy (Tokuraku et al. 2009). Conjugation of Aβ peptides to a cobalt–platinum–gold NP construct having a high magnetization value (Brambilla et al. 2011a; Choi et al. 2008) resulted in contrast changes on T2-weighted MR imaging that correlated with the concentration of peptides, permitting analysis of the evolution of aggregation of peptides (Brambilla et al. 2011a; Choi et al. 2008). Finally, radiolabeled clioquinol, a copper–zinc chelator shown to decrease amyloid deposition in an in vivo animal model (Cherny et al. 2001), was conjugated to polymeric NPs and could penetrate the BBB and bind amyloid in vivo (Roney et al. 2009). NPs could potentially be used to detect amyloid using positron emission tomography (PET) and single-photon emission computed tomography (SPECT) imaging.

NANOTHERAPEUTICS

DENDRIMERS

Dendrimers are macromolecular structures comprising repetitive units of monomers linked to a common core. They have highly branched, symmetric, repetitive, and regular structures. Dendrimers permit an incredible degree of structural manipulation because of the multivalency of their surface functional groups. Additionally, dendrimers in solution tend to be regular in terms of size, shape, and mass (monodisperse). Dendrimers can clear prions from an in vitro cell culture of neuroblastoma cells (Lim et al. 2010). The dendrimers degraded the prions and inhibited the conversion of normal prion proteins into misfolded configurations. Quaternization, which minimizes the peripheral cationic charge on the dendrimer and reduces its toxicity, decreased the efficacy of prion clearance (Lim et al. 2010). Increasing the number of dendrimer branches (otherwise known as increasing generation number) and surface cationic charge increased efficiency in destabilizing scrapie prion protein aggregates from in vitro cell cultures (Cordes et al. 2007). Solassol et al. (2004) demonstrated in an in vivo murine model that cationic phosphorus-containing dendrimers could clear 80% of scrapie prion protein stores from the spleen.

An increasing number of publications report that molecules sequestering Aβ can attenuate Aβ-induced neurotoxicity (Drouet et al. 1999). Dendrimers inhibit aggregation of amyloid fibrils in vitro (Aβ 1–28) (Klajnert et al. 2006). Amyloid fibrils normally cluster around the sialic acid residues on cell surfaces. Patel et al. (2007) prepared a sialic acid functionalized dendrimer construct that attenuated Aβ neurotoxicity in an in vitro human neuroblastoma cell culture. Sialic acid functionalized on dendrimers better sequestered Aβ compared to sialic acid residues alone (Patel et al. 2007). The KLVFF hydrophobic sequence of Aβ is important for formation of β-sheet structures and prevents aggregation of Aβ into fibrils when bound to the corresponding sequence on Aβ (Hilbich et al. 1992; Tjernberg et al. 1997). Chafekar et al. (2007) subsequently functionalized first-generation dendrimers with either one (K1) or four KLVFF residues (K4). K4 was much more potent than K1 in inhibiting Aβ aggregation and in disintegrating existing amyloid aggregates because of its multivalent display of KLVFF residues (Chafekar et al. 2007). Heparin exhibits properties that promote Aβ aggregation (McLaurin et al. 1999) because of the structural homology of a specific segment between the two proteins (Klajnert et al. 2006; Mahfoud et al. 2002).

Gene Therapy

Nanotechnology has enabled the development of nonviral vectors for gene therapy. ORMOSIL (organically modified silica) NPs were used experimentally as nonviral vectors for transferring genes that result in the development of animal models of neurodegenerative disorders (Klejbor et al. 2007). The mouse model of polyglutamine disorders was successfully developed by tagging ORMOSIL NPs with plasmids expressing 127 glutamine repeats and injecting them directly into a lateral cerebral ventricle in vivo. Immunocytochemistry in this mouse model revealed increased GFAP+ astrocytes throughout the brain as well as increased ventricular size. Such changes mimic those that occur in Huntington's disease, a polyglutamine disorder. These animal models can be used to evaluate the utility of gene silencing. Small-interfering RNA (siRNA) belongs to the class of double-stranded RNAs that function in gene silencing. Staber et al. (2006) injected lipid NPs tagged with siRNAs directly into the lateral ventricle and noticed an 80% decrease in huntington protein through the silencing effect (Staber et al. 2006). This effect was accomplished at much a lower dose of vector compared to adeno-associated viral vectors harboring small hairpin RNAs.

Gonzalez-Barrios et al. (2006) tested the utility of a neurotensin polyplex NP carrier system in an animal model of hemiparkinsonism. The investigators transfected the substantia nigra of mice in vivo with the hGDNF (human glial cell derived neurotrophic factor) gene, whose expression was confirmed with RT-PCR analysis. GDNF had a neurotrophic effect on embryonal dopaminergic cells and induced anatomical, biochemical and functional recovery in the mice. In a mouse model of Parkinson's disease (PD), DNA NPs complexed with plasmid encoding GDNF were injected into the striatum, and their permissive neurotrophic effects were compared to a control group receiving saline injections (Yurek et al. 2009). One week later, ventral mesencephalic tissues were grafted into the denervated striatum of mice belonging to both groups (GDNF- and saline-injected). The GDNF group had a 16-fold increase in cell count within the implanted tissue compared to controls. The GDNF mice also exhibited a significant functional improvement (Yurek et al. 2009). Huang et al. (2010) further developed this system of introducing GDNF by modifying NPs with surface lactoferrin to enhance their BBB penetration and allow their delivery via intravenous rather than intrastriatal injections. In a rat model of rotenone-induced Parkinsonism, multiple intravenous injections of these modified NPs resulted in greater and more sustained GDNF expression than that provided by single injection. Multiple injections also resulted in greater functional improvement and increased striatal neurotransmitter levels. These studies suggest that gene therapy using GDNF-expressing plasmids delivered via NPs might play a potential role in the future treatment of PD.

Stem Cell Therapy

Stem cell therapy is a potential therapeutic option for PD (Parati et al. 2003). Clinical studies in PD patients demonstrated that mesencephalic embryonic tissue can be implanted into the striatum, restore lost neurons, increase dopamine release, and improve functional outcome (Lindvall and Hagell 2002). Neural progenitor cells in the grafted tissue can differentiate into mature neurons, establish a dendritic and axonal network, and functionally integrate with the surrounding neural tissue (Englund et al. 2002). Lack of adequate quantities of stem cells, development of dyskinesia, and variable outcomes are some of several problems associated with stem cell therapy (Lindvall and Hagell 2002). Biodegradable scaffold material made of self-assembling nanofibers containing the neurite-promoting laminin epitope IKVAV (isoleucine–lysine–valine–alanine–valine) may improve the effect of stem cell transplantation by providing a "pathway" for nerves to regenerate into and by directing selective differentiation of neural stem cells into mature neurons and not astrocytes (Silva et al. 2004). Hydrogel scaffolds were also used to enable growth and differentiation of implanted cells (Nisbet et al. 2008).

Drug Delivery and Therapy

Oral administration of the dopamine precursor, L-DOPA, is the current standard of care for PD. Dopamine does not have a transporter to penetrate the BBB, whereas L-DOPA is transported across the BBB by LAT-1 (large amino acid transporter, type 1) (Kageyama et al. 2000). Response fluctuations and dyskinetic side effects are associated with L-DOPA, and alternative treatments that more consistently restore function and avoid side effects are needed.

In PD, the activity of aromatic L-amino acid decarboxylase (AADC) (an enzyme that converts the precursor L-DOPA to dopamine in the CNS) declines, thereby limiting L-DOPA's ability to provide dopamine to the striatum (Forsayeth et al. 2006). Forsayeth et al. (2006) tested a gene therapy approach using adeno-associated virus carrying cDNA of human AADC in Parkinsonian monkeys. The investigators used PET, L-DOPA response, and AADC activity to evaluate the relationship of these markers to vector dose and found a linear relationship between vector dose and AADC activity. They established a threshold dose above which the PET and behavioral response significantly improved. The same group later translated this approach to a Phase I clinical trial of patients with moderate-to-severe PD, administering AADC gene therapy through bilateral intraputaminal convection-enhanced infusions of the viral vector and evaluating its effect on neurological function and PET tracer uptake (Eberling et al. 2008). There was no control group, and the trial was not blinded. The therapy was safe and resulted in about 30% increase in PET uptake and modest clinical improvement. The same group later examined MRI and PET data from the clinical study and MRI and immunohistochemical data from the nonhuman primate studies and found that hyperintensities on T2 weighted MR images occurred in areas of increased PET uptake in the clinical studies and in areas of immunohistochemical evidence of AADC in nonhuman primate studies (Valles et al. 2010). The group concluded that T2-weighted MRI can effectively follow the distribution of the administered AADC gene. The size of the AAV is 20 nm (Flotte 2000), and therefore at the nanoscale, suggests that nonbiologic NP gene-therapy vectors may be delivered similarly.

Rivastigmine is an inhibitor of acetylcholinesterase and butyrylcholinesterase aimed at augmenting cholinergic levels to alleviate the cognitive and memory decline seen in AD. Rivastigmine has poor penetration through the BBB and attempts to increase brain cholinergic levels by escalating drug dose results in peripheral cholinergic side effects (Brambilla et al. 2011a; Gauthier et al. 2010). Polysorbate 80-coated polymeric NPs were a delivery platform for rivastigmine in an in vivo animal model using intravenous administration (Wilson et al. 2008a). There was a 3.8-fold increase in rivastigmine transfer to the brain with NP formulation compared to free drug. The increase is purported to be due to the internalization of the NPs by endothelial cells through the LDL receptor and subsequent transcytosis (Brambilla et al. 2011a; Wilson et al. 2008a). The same NP construct delivered tacrine, an acetylcholinesterase inhibitor, with a subsequent 4-fold increase in its uptake compared to free drug (Wilson et al. 2008b).

Mifepristone, an anti-progesterone and anti-glucocorticoid agent, has a favorable effect on decreasing the rate of cognitive decline in mild-to-moderate AD (Belanoff et al. 2002). Using PLGA NPs enhanced the bioavailability of mifepristone in an in vivo animal model, although the purpose of the study did not address efficacy in AD (He et al. 2007).

Several agents were characterized by virtue of their interaction with Aβ, either as inhibitors of Aβ aggregation into fibrils and plaques, or facilitators of disintegration of established Aβ fibrils and plaques (Brambilla et al. 2011a; Cizas et al. 2010). Copolymeric NiPAM/BAM (*N*-polyisopropylacrylamide/*N*-*tert*-butylacrylamide) NPs inhibited fibrillation of Aβ, mainly by affecting the nucleation process (Cabaleiro-Lago et al. 2008). Effect on fibrillation depended on the surface character of NPs: fluorinated, hydrogenated, sulfonated, and sulfated polymeric NPs all induced a disordered state in Aβ, delayed oligomerization, and decreased cytotoxicity (Rocha et al. 2008; Saraiva et al. 2010). Cabaleiro-Lago et al. (2010) demonstrated the effect of increasing surface area of amino-modified polystyrene NPs, which are not biodegradable, on decreasing the rate of fibrillation of Aβ and the effect of altering NP concentration on fibrillation kinetics. The aggregation kinetics with Aβ were further elaborated by Brambilla et al. (2010) with biodegradable pegylated polymeric NPs using a novel method based on capillary electrophoresis with laser-induced fluorescence detection (CE-LIF). Confocal microscopy confirmed binding between the NPs and Aβ, and displayed large aggregate formations at high NP concentrations. The same group used the CE-LIF technique to evaluate the aggregation kinetics of NPs functionalized with selegeline with the Aβ protein and the role played by PEG chains (Le Droumaguet et al. 2011). Important NP characteristics responsible for optimal interaction with Aβ include smaller size and greater surface negative charge, which can be accomplished using NPs stabilized with sodium cholate (Brambilla et al. 2011b). These pegylated polycyanoacrylate NPs were shown to cross the BBB, and therefore offer a promising therapeutic modality for AD (Calvo et al. 2001; Kim et al. 2007). Sterically stabilized pegylated phospholipid nanomicelles are another pegylated NP construct demonstrated in vitro in human neuroblastoma cells to enhance the alpha-helicity of $A\beta_{1-42}$. It inhibits the conversion of the alpha-helix form to β-pleated sheets and significantly lowers neurotoxicity in vitro (Pai et al. 2006). The probable mechanism includes shielding the hydrophobic domains of Aβ protein by the hydrophobic residues on the pegylated NP, which impairs interactions between monomeric Aβ proteins, thereby inhibiting aggregation (Brambilla et al. 2011a; Pai et al. 2006). Another example of preventing amyloid protein aggregation involves nanogels. Nanogels are NPs comprising a hydrophilic polymer network with a high degree of loading capacity and stability (Brambilla et al. 2011a). Cholesterol-bearing pullulan hydrogels (CHPs) were constructed by Ikeda et al. (2006) using a polysaccharide pullulan backbone and partially substituted cholesterol. CHPs and amino-CHPs were used in vitro in PC12 cell culture. The hydrogels incorporated the Aβ peptide and impaired aggregation into fibrils. A greater hydrogel concentration increased conformational changes of Aβ protein into β sheets and later alpha-helices. Amino-CHPs had a more remarkable effect (Ikeda et al. 2006). The CHP hydrogel NPs were also shown in in vitro primary cortical and microglial cell cultures, to bind both $A\beta_{1-42}$ monomers and oligomers (Boridy et al. 2009). Such binding decreased the neurotoxicity of the Aβ oligomers; the unmodified CHPs were nontoxic. Such a construct could possibly be used to complement other therapeutic approaches to AD (Boridy et al. 2009).

Miscellaneous Examples

Oxidative stress has an important role in the pathogenesis of AD, and there is accumulating evidence that transitional metal ions are involved in modulating this stress. Metal chelating molecules diminish the ability of Aβ to catalyze oxidation reactions, showing the role of chelators in abolishing the release of free radicals (Praticò et al. 2002; Sayre et al. 2000). In a single-blind prospective study of AD patients, those receiving intramuscular desferoxamine had a significant decrease in rate of decline of activities of daily living compared to the group receiving no treatment (Crapper McLachlan et al. 1991). The poor transport of metal chelators across the BBB could potentially be circumvented by conjugating them to NPs. The copper chelator D-penicillamine was conjugated covalently to NPs through disulfide or thioether linkage, with subsequent release achieved using reducing agents such as dithiothreitol (Cui et al. 2005). In an in vitro model, the NP-chelator solubilized copper-Aβ_{1-42} aggregates. NP-chelator conjugates developed by other investigators inhibited aggregate Aβ protein formation in vitro in human cortical cells (Liu et al. 2009). Additionally, the NPs reduced the neurotoxicity of Aβ aggregates because of oxidative stress.

β-Secretase significantly contributes to the generation of Aβ peptides and hence, the pathogenesis of AD (Rajendran et al. 2008). Therefore, β-secretase provides a potential therapeutic target for addressing AD. The inhibitor characterized by Rajendran et al. (2008) provides binding to the active site and the subcellular portion of the enzyme. This was achieved by attaching the inhibitor to a sterol moiety that provided anchoring to the cellular membrane and also steered the inhibitor toward the endosomes. Enzyme activity decreased in both cell culture and in vivo to a greater extent compared to free inhibitor. Epigallocatechin-3-gallate (EGCG) was shown to up-regulate alpha-secretase activity, which is important to promote the non-amyloidogenic processing pathway of amyloid precursor protein (Smith et al. 2010). When nanolipidic particles are conjugated with EGCG, the construct increases the alpha-secretase activity by 91% in vitro and increases oral bioavailability in vivo twice over that of free EGCG (Smith et al. 2010). Another target includes the EP(4) receptor which, when stimulated by prostagladin E_2, increases Aβ production, by activating gamma-secretase (Hoshino et al. 2012). Oral administration of the EP(4) antagonist AE3-208 to transgenic mice resulted in decreased Aβ generation and improved cognitive function. NPs are potentially useful in improving the bioavailability of β-secretase inhibitors and EP(4) antagonists (Hoshino et al. 2012). These therapeutic strategies could complement other therapeutic options in AD.

Curcumins have multiple therapeutic effects as a result of their antioxidant, anti-inflammatory, antiviral, and antitumor activities (Osawa 2007; Aggarwal et al. 2007). Tetrahydrocurcumin, a metabolite of curcumin, has antioxidant properties both in vitro and in vivo (Osawa 2007). Curcumin reduced cognitive decline in an in vivo animal model of AD (Ringman et al. 2005). Curcumin conjugated with poly(*n*-butylcyanoacrylate) NP enhanced half-life in circulation and increased the area under the concentration–time curve (Sun et al. 2010). The area under the curve in the brain was 2.5 times greater for curcumin-NP compared to free curcumin. The same NP construct was used by other investigators and conjugated to ApoE3 ligand for targeting LDL receptors of the BBB (Mulik et al. 2010). The NP construct produced greater reduction in amyloid-neurotoxicity than free curcumin in neuroblastoma cell culture. ApoE3, which also has antioxidant properties, was synergistic with curcumin in reducing amyloid neurotoxicity in the same experimental setup (Mulik et al. 2010).

Chitosan NPs can carry dopamine, increase its transport rate across an in vitro model of the BBB, and reduce its neurotoxicity in cell culture (Trapani et al. 2011). Intraperitoneal injections of dopamine NPs in vivo in mice dose-dependently increase striatal dopamine (Trapani et al. 2011). Other investigators used chitosan NPs for targeted immunotherapy against AD by surface conjugating them with the anti-amyloid antibody IgG4.1 (Ch-Ig-NP) (Agyare et al. 2008). Uptake of the Ch-Ig-NP in different brain regions was about 10 times higher than that of a control NP that was coated with albumin.

NANONEUROPROTECTION

Fullerenols

Fullerenols, otherwise known as "buckyballs," were first discovered in 1985 by Curl, Kroto, and Smalley, who eventually went on to win the 1996 Nobel Prize in Chemistry (Khawaja 2011). Fullerenes are three-dimensional structures of evenly spaced carbon atoms that assume a spherical shape (Khawaja 2011). These structures possess antioxidant properties that reduce free radical-mediated damage arising from glutamate excitotoxicity (Dugan et al. 1997). It also diminishes Aβ_{1-42} protein-related apoptosis in cells, a process normally mediated through glutamate activation. In transgenic mice carrying the G93A superoxide dismutase gene that is responsible for familial amyotrophic lateral sclerosis, fullerenes reduced apoptosis and functional decline (Dugan et al. 1997).

Fibrillary β-amyloid (Aβ) can be incorporated into the lipid bilayer of cellular membranes, mediating influx of extracellular calcium (Huang et al. 2000), which alters homeostasis and induces apoptosis (Arispe et al. 1994). Fullerenols inhibited Aβ-mediated apoptosis by interacting with the lipid bilayer in a manner that specifically inhibited the influx of extracellular calcium but did not affect the inositol triphosphate-mediated release of intracellular calcium (Huang et al. 2000).

Other Neuroprotective Agents

Glutathione is a tripeptide with antioxidant properties involved in scavenging oxygen, superoxide, and hydroxyl free radicals (Masella et al. 2005). It exists in two physiological states: GSH (reduced) and GSSG (oxidized). Williams et al. (2009) prepared a self-assembling NP construct based on glutathione (GSH) and polyethylene glycol diacrylate. The investigators tested its utility in vitro in human neuroblastoma cells and found that it effectively relieved oxidative stress produced by hydrogen peroxide. Other investigators created a PLGA NP construct tagged with superoxide dismutase and found that it effectively protected human neurons in vitro from the oxidative stress of hydrogen peroxide (Reddy et al. 2008). Ferulic acid (a phenolic compound found in plant cell walls that has strong anti-inflammatory and antioxidant properties; Ozaki 1992) decreased reactive oxygen generation and other apoptotic factors in human neuroblastoma cell culture exposed to Aβ (Picone et al. 2009). Tagging ferulic acid with NPs further relieved the oxidative stress by increasing drug stability (Picone et al. 2009).

Drug Therapy for Neuroprotection

The neurotoxicity of HIV infection includes neuronal apoptosis, particularly in the basal ganglia, and AIDS dementia (Gray et al. 2001; Khawaja 2011). Apoptosis is not the only pathological event resulting in dementia, as other factors including microglial activation and axonal damage are also involved. Early aggressive highly active antiretroviral therapy (HAART) regimen may prevent the cognitive decline and neurological and neuropsychological manifestations of CNS AIDS, indicating that treatment may reverse some of the stages in the pathogenesis of AIDS dementia (Gray et al. 2001; Sacktor et al. 2000). However, some cases may be associated with failure attributable either to "burnt-out" CNS damage or inadequate penetration of HAART through the BBB (Gray et al. 2001). The HIV virus has a predilection for affecting macrophages and creating multinucleated giant cells, which are a pathognomonic finding of HIV infection on histopathological analysis (Budka 1989). Specific macrophage targeting was demonstrated by Schafer et al. (1992) in vitro in a macrophage culture acquired from a perinatally affected HIV+ child, using NPs composed of polyalkylcyanoacrylate, polymethylacrylate, and albumin. A high phagocytic intake was observed for 200-nm-sized NPs, with infected macrophages exhibiting a higher phagocytic intake compared to their noninfected counterparts. In another in vitro macrophage culture, compared to the aqueous form, saquinavir bound to NPs had significantly higher phagocytic uptake and reduction in HIV antigen production across most concentrations ranging from 0.1 nmol to 100 nM (Bender et al. 1996). QDs conjugated to transferrin were utilized in transferring saquinavir across the BBB in an in vitro BBB setup (Mahajan et al. 2010). Brain microvascular endothelial cells efficiently took up the QDs, and HIV replication was inhibited in mononuclear cells. Transferrin enabled the uptake of the functionalized QDs through transferrin receptors and subsequent transcytosis through the in vitro BBB.

Implanting a biodegradable composite intracranial nano-enabled device (CIND) within the frontal lobe is a proposed method for introducing HAART drugs directly into the brain (Khan et al. 2010). This strategy is intended to reduce the dose of drug and systemic toxicity that result from systemic administration. Under laboratory testing, release of the drug azidothymidine (AZT) from the CIND had an initial lag phase followed by steady delivery without any "burst effect" (Khan et al. 2010).

NANOMANIPULATION

Nanomanipulation revolves around the idea of performing surgery or manipulating tissues at the nanometer scale through the use of surgical "nanorobots" (Khawaja 2011). The concept involves introducing an external device within the patient and manipulating it for diagnostic and/or therapeutic purposes. The idea behind using such devices is not unique to nanotechnology. Robotic surgery on a macroscale is being increasingly adopted for urologic procedures such as prostatectomy. The use of endoscopic "capsules" is also being attempted in gastroenterology to diagnose the cause and specific site of gastrointestinal (GI) bleeding in the region between upper and lower GI tracts. The capsule can be remotely manipulated and takes multiple images from within the GI tract. It is usually used after conventional endoscopic procedures, angiography, and nuclear studies fail to reveal the site of bleeding.

Application of nanomanipulation in neurosurgery is beginning to be explored. A rapidly vibrating micropipette (100 Hz) with less than 1-μm-wide tip has been used to dissect the dendrites of rat hippocampal CA1 cells (Kirson and Yaari 2000). The procedure resulted in complete disconnection of the dendrites, with cell viability maintained. To study nerve regeneration in an animal model, Yanik et al. (2004) devised a precise method of severing nerves using femtosecond laser surgery in roundworms. Functional regeneration occurred in the severed nerves, and this technique would potentially be useful to study nerve regeneration in mammals. The dynamics of such nanoneurosurgery with a femtosecond laser were elaborated through imaging within *Caenorhabtis elegans* nematode worms (Santos et al. 2010).

The concept of nanomanipulation is broad and can be used to identify applications in monitoring and modulating neuronal activity in vivo using carbon nanotubes and nanowires (detailed in the next section). One example of nanomanipulation is controlling the growth and differentiation of stem cells implanted into the striatum using customized self-assembling nanofibers (Silva et al. 2004).

An example to illustrate the principle of "remote manipulation" includes using gold NPs attached to amyloid protein and subsequently activated by low-gigahertz electromagnetic fields (Kogan et al. 2005). The energy released by the gold NPs dissolves the amyloid aggregates (Kogan et al. 2005).

NANONEUROMODULATION

CARBON NANOFIBERS

Nanoneuromodulation depends on modalities, or tools, derived from nanotechnology that have the potential for stimulating and monitoring neurons in vivo (Khawaja 2011). Brain implant devices are now commonly used to treat a variety of neurological disorders such as essential tremor, PD, and obsessive compulsive disorder (Eberle et al. 2011). However, such brain implants can be complicated by electrode failure prompting another surgical procedure to replace the implant and electrode (Voges et al. 2006). These implants are not entirely biocompatible and may elicit an immune and fibrotic response leading to device failure (Polikov et al. 2005). Additionally undesirable, they deliver a large current through a direct electrical contact and stimulate a large field of neurons (Agnew et al. 1986; Mohy-Ud-Din et al. 2010). In vitro studies showed that carbon nanofibers with diameters less than 100 nm and high surface energy have significantly less astrocytic adherence and glial scar tissue formation than carbon nanofibers with diameters greater than 100 nm and low surface energy (McKenzie et al. 2004). Carbon nanofiber technology could also improve the precision of electrical stimulation over present clinical implants and allow simultaneous monitoring of electrical activity and transmitter levels (Li and Andrews 2007). This would enable closed-loop control of electrical stimulation guided by fluctuations in brain electrical activity and neurotransmitter levels (Li and Andrews 2007). In vitro studies using vertically aligned carbon nanofiber neural chips implanted with animal hippocampal slice cultures demonstrated dual mode recordings of neuroelectrophysiological and neurochemical signals (Zhe et al. 2010). These arrays can be conveniently modified to control host tissue response (Zhe et al. 2010).

Carbon Nanotubes

Carbon nanotube structures were also investigated in in vivo animal studies to provide neuronal stimulation (Mohy-Ud-Din et al. 2010). These structures were modified so they could be activated through optical stimulation. It was found that 50-nm^2 formulations of single- and multi-walled carbon nanotubes stimulated 40 and 12 μm^2 areas of brain, respectively, thereby significantly limiting the area of brain stimulated (Mohy-Ud-Din et al. 2010). Carbon nanotubes were used to measure concentrations of serotonin and dopamine in both in vitro and in vivo conditions (Wang et al. 2003). Carbon nanotubes were also used to analyze concentrations of potential neurotoxins in the CSF of patients with PD (Zhang et al. 2004). These patients had elevated concentration in the CSF of *N*-methylsalsolinol (NM-Sal), an endogenous MPTP-like neurotoxin. This finding was confirmed in other studies (Maruyama et al. 1996; Naoi et al. 2002). The increased level of NM-Sal in PD patients arose from increased activity of neutral *N*-methyltransferase specific for salsonilol (Naoi et al. 1997, 1998); salsonilol concentration was the same in patients and controls (Zhang et al. 2004).

Nanowires

Nanowires are structures only a few hundreds of nanometers thick that could potentially be used to monitor brain electrical activity (Khawaja 2011). Platinum nanowires were used to study neuronal activity in animal studies, guiding the wires into the CNS through the vascular system, thereby eliminating artifacts seen with surface electrodes (Llinas et al. 2005). Polymer nanowires undergo a change in the direction of the leading edge in response to electrical fields. They have smaller diameters than platinum nanowires and are biodegradable (Llinas et al. 2005). The latter property can be potentially useful in the nanoneuromodulation of CNS disorders such as PD.

CONCLUSIONS

AD and PD are common neurodegenerative diseases that are the focus of much nanotechnology research. These presently incurable disorders are managed with drug therapy, and in some patients with PD, DBS. Nanotechnology may lead to improved treatment of these diseases by identifying additional potential therapeutic targets based on disease mechanisms and by fine-tuning DBS therapy. Targeting more than one step in a disease mechanism may result in synergistic effects that improve functional parameters to a greater degree than that with monotherapy. Much in vitro and in vivo research has been performed to study the toxicity and therapeutic efficacy of multiple nanotechnology-based modalities for AD and PD, but few clinical trials have been initiated. Fewer research studies have been conducted to develop nanotechnology-based therapies for more rapidly progressing types of dementia, including prion disease and Huntington's disease. Nanotechnology offers unique opportunities to develop novel diagnostic tools and therapies that could lead to improved care of patients with degenerative diseases of the CNS.

REFERENCES

Aggarwal, B. B., Surh, Y.-J., and Shishodia, S. (eds) 2007. *The Molecular Targets and Therapeutic Uses of Curcumin in Health and Disease. Advances in Experimental Medicine and Biology*, 407–23. New York: Springer.

Agnew, W.F., Yuen, T.G.H., McCreery, D.B. et al. 1986. Histopathologic evaluation of prolonged intracortical electrical stimulation. *Experimental Neurology* 92: 162–85.

Agyare, E., Curran, G., Ramakrishnan, M. et al. 2008. Development of a smart nano-vehicle to target cerebrovascular amyloid deposits and brain parenchymal plaques observed in Alzheimer's disease and cerebral amyloid angiopathy. *Pharmaceutical Research* 25: 2674–84.

Arispe, N., Pollard, H.B., and Rojas, E. 1994. β-Amyloid Ca^{2+}-channel hypothesis for neuronal death in Alzheimer disease. *Molecular and Cellular Biochemistry* 140: 119–25.
Belanoff, J., Jurik, J., Schatzberg, L. et al. 2002. Slowing the progression of cognitive decline in Alzheimer's disease using Mifepristone. *Journal of Molecular Neuroscience* 19: 201–06.
Bender, A.R., von Briesen, H., Kreuter, J. et al. 1996. Efficiency of nanoparticles as a carrier system for antiviral agents in human immunodeficiency virus-infected human monocytes/macrophages in vitro. *Antimicrobial Agents and Chemotherapy* 40: 1467–71.
Boridy, S., Takahashi, H., Akiyoshi, K. et al. 2009. The binding of pullulan modified cholesteryl nanogels to Aβ oligomers and their suppression of cytotoxicity. *Biomaterials* 30: 5583–91.
Brambilla, D., Le Droumaguet, B., Nicolas, J. et al. 2011a. Nanotechnologies for Alzheimer's disease: diagnosis, therapy, and safety issues. *Nanomedicine-Nanotechnology Biology and Medicine* 7: 521–40.
Brambilla, D., Souguir, H., Nicolas, J. et al. 2011b. Colloidal properties of biodegradable nanoparticles influence interaction with amyloid-β peptide. *Journal of Biotechnology* 156: 338–40.
Brambilla, D., Verpillot, R., Taverna, M. et al. 2010. New method based on capillary electrophoresis with laser-induced fluorescence detection (CE-LIF) to monitor interaction between nanoparticles and the amyloid-β peptide. *Analytical Chemistry* 82: 10083–9.
Budka, H. 1989. Human immunodeficiency virus (HIV)-induced disease of the central nervous system: pathology and implications for pathogenesis. *Acta Neuropathologica* 77: 225–36.
Cabaleiro-Lago, C., Quinlan-Pluck, F., Lynch, I. et al. 2008. Inhibition of amyloid beta protein fibrillation by polymeric nanoparticles. *Journal of the American Chemical Society* 130: 15437–43.
Cabaleiro-Lago, C., Quinlan-Pluck, F., Lynch, I. et al. 2010. Dual effect of amino modified polystyrene nanoparticles on amyloid β protein fibrillation. *ACS Chemical Neuroscience* 1: 279–87.
Calvo, P., Gouritin, B., Chacun, H. et al. 2001. Long-circulating PEGylated polycyanoacrylate nanoparticles as new drug carrier for brain delivery. *Pharmaceutical Research* 18: 1157–66.
Chafekar, S.M., Malda, H., Merkx, M. et al. 2007. Branched KLVFF tetramers strongly potentiate inhibition of beta-amyloid aggregation. *Chembiochem* 8: 1857–64.
Cherny, R.A., Atwood, C.S., Xilinas, M.E. et al. 2001. Treatment with a copper–zinc chelator markedly and rapidly inhibits β-amyloid accumulation in Alzheimer's disease transgenic mice. *Neuron* 30: 665–76.
Chikae, M., Fukuda, T., Kerman, K. et al. 2008. Amyloid-β detection with saccharide immobilized gold nanoparticle on carbon electrode. *Bioelectrochemistry* 74: 118–23.
Choi, J.S., Choi, H.J., Jung, D.C. et al. 2008. Nanoparticle assisted magnetic resonance imaging of the early reversible stages of amyloid beta self-assembly. *Chemical Communications* 2008: 2197–9.
Cizas, P., Budvytyte, R., Morkuniene, R. et al. 2010. Size-dependent neurotoxicity of beta-amyloid oligomers. *Archives of Biochemistry and Biophysics* 496: 84–92.
Cordes, H., Boas, U., Olsen, P. et al. 2007. Guanidino- and urea-modified dendrimers as potent solubilizers of misfolded prion protein aggregates under non-cytotoxic conditions. Dependence on dendrimer generation and surface charge. *Biomacromolecules* 8: 3578–83.
Crapper McLachlan, D.R., Dalton, A.J., Kruck, T.P. et al. 1991. Intramuscular desferrioxamine in patients with Alzheimer's disease. *Lancet* 337: 1304–8.
Cui, Z., Lockman, P.R., Atwood, C.S. et al. 2005. Novel D-penicillamine carrying nanoparticles for metal chelation therapy in Alzheimer's and other CNS diseases. *European Journal of Pharmaceutics and Biopharmaceutics* 59: 263–72.
Drouet, B., Pincon-Raymond, M., Chambaz, J. et al. 1999. Laminin 1 attenuates beta-amyloid peptide Abeta(1–40) neurotoxicity of cultured fetal rat cortical neurons. *Journal of Neurochemistry* 73: 742–9.
Dugan, L.L., Turetsky, D.M., Du, C. et al. 1997. Carboxyfullerenes as neuroprotective agents. *Proceedings of the National Academy of Sciences of the United States of America* 94: 9434–9.
Eberle, W., Penders, J., and Yazicioglu, R.F. 2011. Closing the loop for deep brain stimulation implants enables personalized healthcare for Parkinson's disease patients. *Conference Proceedings of IEEE Engineering in Medicine and Biology Society* 2011: 1556–8.
Eberling, J.L., Jagust, W.J., Christine, C.W. et al. 2008. Results from a phase I safety trial of hAADC gene therapy for Parkinson disease. *Neurology* 70: 1980–3.
Elsinghorst, P.W., Hartig, W., Goldhammer, S. et al. 2009. A gorge-spanning, high-affinity cholinesterase inhibitor to explore beta-amyloid plaques. *Organic & Biomolecular Chemistry* 7: 3940–6.
Englund, U., Björklund, A., Wictorin, K. et al. 2002. Grafted neural stem cells develop into functional pyramidal neurons and integrate into host cortical circuitry. *Proceedings of the National Academy of Sciences of the United States of America* 99: 17089–94.
Flotte, T.R. 2000. Size does matter: overcoming the adeno-associated virus packaging limit. *Respiratory Research* 1: 16–8.

Forsayeth, J.R., Eberling, J.L., Sanftner, L.M. et al. 2006. A dose-ranging study of AAV-hAADC therapy in Parkinsonian monkeys. *Molecular Therapy* 14: 571–7.

Gauthier, S., Juby, A., Dalziel, W. et al. 2010. Effects of rivastigmine on common symptomatology of Alzheimer's disease (EXPLORE). *Current Medical Research and Opinion* 26: 1149–60.

Gonzalez-Barrios, J.A., Lindahl, M., Bannon, M.J. et al. 2006. Neurotensin polyplex as an efficient carrier for delivering the human GDNF gene into nigral dopamine neurons of hemiparkinsonian rats. *Mol Ther* 14: 857–65.

Gray, F., Adle-Biassette, H., Chretien, F. et al. 2001. Neuropathology and neurodegeneration in human immunodeficiency virus infection. Pathogenesis of HIV-induced lesions of the brain, correlations with HIV-associated disorders and modifications according to treatments. *Clinical Neuropathology* 20: 146–55.

Haes, A.J., Chang, L., Klein, W.L. et al. 2005. Detection of a biomarker for Alzheimer's disease from synthetic and clinical samples using a nanoscale optical biosensor. *Journal of the American Chemical Society* 127: 2264–71.

Haes, A.J., and Van Duyne, R.P. 2002. A nanoscale optical blosensor: sensitivity and selectivity of an approach based on the localized surface plasmon resonance spectroscopy of triangular silver nanoparticles. *Journal of the American Chemical Society* 124: 10596–604.

Hakim, S., and Adams, R.D. 1965. The special clinical problem of symptomatic hydrocephalus with normal cerebrospinal fluid pressure. Observations on cerebrospinal fluid hydrodynamics. *Journal of Neurological Sciences* 2: 307–27.

Hartig, W., Kacza, J., Paulke, B.R. et al. 2010. In vivo labelling of hippocampal beta-amyloid in triple-transgenic mice with a fluorescent acetylcholinesterase inhibitor released from nanoparticles. *European Journal of Neuroscience* 31: 99–109.

He, W., Horn, S.W., and Hussain, M.D. 2007. Improved bioavailability of orally administered mifepristone from PLGA nanoparticles. *International Journal of Pharmaceutics* 334: 173–8.

Hilbich, C., Kisterswoike, B., Reed, J. et al. 1992. Substitutions of hydrophobic amino-acids reduce the amyloidogenicity of Alzheimer's disease beta-A4 peptides. *Journal of Molecular Biology* 228: 460–73.

Hoshino, T., Namba, T., Takehara, M. et al. 2012. Improvement of cognitive function in Alzheimer's disease model mice by genetic and pharmacological inhibition of the EP4 receptor. *Journal of Neurochemistry* 120: 795–805.

Huang, H.M., Ou, H.C., Hsieh, S.J. et al. 2000. Blockage of amyloid beta peptide-induced cytosolic free calcium by fullerenol-1, carboxylate C60 in PC12 cells. *Life Sciences* 66: 1525–33.

Huang, R., Ke, W., Liu, Y. et al. 2010. Gene therapy using lactoferrin-modified nanoparticles in a rotenone-induced chronic Parkinson model. *Journal of the Neurological Sciences* 290: 123–30.

Ikeda, K., Okada, T., Sawada, S.-I. et al. 2006. Inhibition of the formation of amyloid β-protein fibrils using biocompatible nanogels as artificial chaperones. *FEBS Letters* 580: 6587–95.

Jensen, T., Kelly, L., Lazarides, A. et al. 1999. Electrodynamics of noble metal nanoparticles and nanoparticle clusters. *Journal of Cluster Science* 10: 295–317.

Jensen, T.R., Malinsky, M.D., Haynes, C.L. et al. 2000. Nanosphere lithography: tunable localized surface plasmon resonance spectra of silver nanoparticles. *Journal of Physical Chemistry B* 104: 10549–56.

Kageyama, T., Nakamura, M., Matsuo, A. et al. 2000. The 4F2hc/LAT1 complex transports L-DOPA across the blood–brain barrier. *Brain Research* 879: 115–21.

Kang, D.Y., Lee, J.H., Oh, B.K. et al. 2009. Ultra-sensitive immunosensor for beta-amyloid (1–42) using scanning tunneling microscopy-based electrical detection. *Biosensors & Bioelectronics* 24: 1431–6.

Khan, R.A., Harilall, S., Pillay, V. et al. 2010. An intracranial neurocompatible nanocomposite polymeric device for the chronic management of AIDS dementia complex. *Nanotech* 3: 361–3.

Khawaja, A.M. 2011. The legacy of nanotechnology: revolution and prospects in neurosurgery. *Int J Surg* 9: 608–14.

Kim, H., Gil, S., Andrieux, K. et al. 2007. Low-density lipoprotein receptor-mediated endocytosis of PEGylated nanoparticles in rat brain endothelial cells. *Cellular and Molecular Life Sciences* 64: 356–64.

Kirson, E.D., and Yaari, Y. 2000. A novel technique for micro-dissection of neuronal processes. *J Neurosci Methods* 98: 119–22.

Klajnert, B., Cortijo-Arellano, M., Cladera, J. et al. 2006. Influence of dendrimer's structure on its activity against amyloid fibril formation. *Biochemical and Biophysical Research Communications* 345: 21–8.

Klein, W.L. 2002. Abeta toxicity in Alzheimer's disease: globular oligomers (ADDLs) as new vaccine and drug targets. *Neurochemistry International* 41: 345–52.

Klejbor, I., Stachowiak, E.K., Bharali, D.J. et al. 2007. ORMOSIL nanoparticles as a non-viral gene delivery vector for modeling polyglutamine induced brain pathology. *Journal of Neuroscience Methods* 165: 230–43.

Kogan, M.J., Bastus, N.G., Amigo, R. et al. 2005. Nanoparticle-mediated local and remote manipulation of protein aggregation. *Nano Letters* 6: 110–15.
Kreibig, U., Gartz, M., and Hilger, A. 1997. Mie resonances: Sensors for physical and chemical cluster interface properties. *Berichte der Bunsen-Gesellschaft-Physical Chemistry Chemical Physics* 101: 1593–604.
Le Droumaguet, B., Souguir, H., Brambilla, D. et al. 2011. Selegiline-functionalized, PEGylated poly(alkyl cyanoacrylate) nanoparticles: Investigation of interaction with amyloid-β peptide and surface reorganization. *International Journal of Pharmaceutics* 416: 453–60.
Li, J., and Andrews, R.J. 2007. Trimodal nanoelectrode array for precise deep brain stimulation: prospects of a new technology based on carbon nanofiber arrays. *Acta Neurochirurgica Supplement* 97: 537–45.
Lim, Y.B., Mays, C.E., Kim, Y. et al. 2010. The inhibition of prions through blocking prion conversion by permanently charged branched polyamines of low cytotoxicity. *Biomaterials* 31: 2025–33.
Lindvall, O., and Hagell, P. 2002. Role of cell therapy in Parkinson disease. *Neurosurgical Focus* 13: e2.
Liu, G., Men, P., Kudo, W. et al. 2009. Nanoparticle–chelator conjugates as inhibitors of amyloid-β aggregation and neurotoxicity: A novel therapeutic approach for Alzheimer disease. *Neuroscience Letters* 455: 187–90.
Llinas, R.R., Walton, K.D., Nakao, M. et al. 2005. Neuro-vascular central nervous recording/stimulating system: using nanotechnology probes. *Journal of Nanoparticle Research* 7: 111–27.
Mahajan, S.D., Roy, I., Xu, G.X. et al. 2010. Enhancing the delivery of anti retroviral drug "Saquinavir" across the blood brain barrier using nanoparticles. *Current HIV Research* 8: 396–404.
Mahfoud, R., Garmy, N., Maresca, M. et al. 2002. Identification of a common sphingolipid-binding domain in Alzheimer, prion, and HIV-1 proteins. *Journal of Biological Chemistry* 277: 11292–6.
Maruyama, W., Abe, T., Tohgi, H. et al. 1996. A dopaminergic neurotoxin, (*R*)-*N*-methylsalsolinol, increases in parkinsonian cerebrospinal fluid. *Annals of Neurology* 40: 119–22.
Masella, R., Di Benedetto, R., Vari, R. et al. 2005. Novel mechanisms of natural antioxidant compounds in biological systems: Involvement of glutathione and glutathione-related enzymes. *Journal of Nutritional Biochemistry* 16: 577–86.
McKenzie, J.L., Waid, M.C., Shi, R. et al. 2004. Decreased functions of astrocytes on carbon nanofiber materials. *Biomaterials* 25: 1309–17.
McLaurin, J., Franklin, T., Zhang, X.Q. et al. 1999. Interactions of Alzheimer amyloid-beta peptides with glycosaminoglycans—effects on fibril nucleation and growth. *European Journal of Biochemistry* 266: 1101–10.
Mohy-Ud-Din, Z., Woo, S.H., Kim, J.H. et al. 2010. Optoelectronic stimulation of the brain using carbon nanotubes. *Annals of Biomedical Engineering* 38: 3500–8.
Mulik, R.S., Mönkkönen, J., Juvonen, R.O. et al. 2010. ApoE3 mediated poly(butyl) cyanoacrylate nanoparticles containing curcumin: study of enhanced activity of curcumin against beta amyloid induced cytotoxicity using in vitro cell culture model. *Molecular Pharmaceutics* 7: 815–25.
Naoi, M., Maruyama, W., Akao, Y. et al. 2002. Dopamine-derived endogenous *N*-methyl-(*R*)-salsolinol: its role in Parkinson's disease. *Neurotoxicology and Teratology* 24: 579–91.
Naoi, M., Maruyama, W., Matsubara, K. et al. 1997. A neutral *N*-methyltransferase activity in the striatum determines the level of an endogenous MPP+-like neurotoxin, 1,2-dimethyl-6,7-dihydroxyisoquinolinium ion, in the substantia nigra of human brains. *Neuroscience Letters* 235: 81–4.
Naoi, M., Maruyama, W., Nakao, N. et al. 1998. (*R*)Salsolinol *N*-methyltransferase activity increases in parkinsonian lymphocytes. *Annals of Neurology* 43: 212–6.
Neely, A., Perry, C., Varisli, B. et al. 2009. Ultrasensitive and highly selective detection of Alzheimer's disease biomarker using two-photon Rayleigh scattering properties of gold nanoparticle. *ACS Nano* 3: 2834–40.
Nisbet, D.R., Crompton, K.E., Horne, M.K. et al. 2008. Neural tissue engineering of the CNS using hydrogels. *Journal of Biomedical Materials Research Part B: Applied Biomaterials* 87B: 251–63.
Osawa, T. 2007. Nephroprotective and hepatoprotective effects of curcuminoids. Review. *Advances in Experimental Medicine and Biology* 595: 407–23.
Ozaki, Y. 1992. Antiinflammatory effect of tetramethylpyrazine and ferulic acid. *Chemical & Pharmaceutical Bulletin* 40: 954–6.
Pai, A.S., Rubinstein, I., and Onyuksel, H. 2006. PEGylated phospholipid nanomicelles interact with beta-amyloid(1–42) and mitigate its beta-sheet formation, aggregation and neurotoxicity in vitro. *Peptides* 27: 2858–66.
Parati, E.A., Bez, A., Ponti, D. et al. 2003. Neural stem cells. Biological features and therapeutic potential in Parkinson's disease. *Journal of Neurosurgcial Sciences* 47: 8–17.
Patel, D.A., Henry, J.E., and Good, T.A. 2007. Attenuation of beta-amyloid-induced toxicity by sialic-acid-conjugated dendrimers: role of sialic acid attachment. *Brain Research* 1161: 95–105.

Picone, P., Bondi, M.L., Montana, G. et al. 2009. Ferulic acid inhibits oxidative stress and cell death induced by Ab oligomers: improved delivery by solid lipid nanoparticles. *Free Radical Research* 43: 1133–45.

Polikov, V.S., Tresco, P.A., and Reichert, W.M. 2005. Response of brain tissue to chronically implanted neural electrodes. *Journal of Neuroscience Methods* 148: 1–18.

Praticò, D., Uryu, K., Sung, S. et al. 2002. Aluminum modulates brain amyloidosis through oxidative stress in APP transgenic mice. *The FASEB Journal* 16:1138–40.

Rajendran, L., Schneider, A., Schlechtingen, G. et al. 2008. Efficient inhibition of the Alzheimer's disease β-secretase by membrane targeting. *Science* 320: 520–3.

Reddy, M.K., Wu, L., Kou, W. et al. 2008. Superoxide dismutase-loaded PLGA nanoparticles protect cultured human neurons under oxidative stress. *Applied Biochemistry and Biotechnology* 151: 565–77.

Ringman, J.M., Frautschy, S.A., Cole, G.M. et al. 2005. A potential role of the curry spice curcumin in Alzheimer's disease. *Current Alzheimer Research* 2: 131–6.

Rocha, S., Thunemann, A.F., Pereira Mdo, C. et al. 2008. Influence of fluorinated and hydrogenated nanoparticles on the structure and fibrillogenesis of amyloid beta-peptide. *Biophysical Chemistry* 137: 35–42.

Roney, C.A., Arora, V., Kulkarni, P.V. et al. 2009. Nanoparticulate radiolabelled quinolines detect amyloid plaques in mouse models of Alzheimer's disease. *International Journal of Alzheimer's Disease* 2009: 481031.

Sacktor, N.C., Skolasky, R.L., Lyles, R.H. et al. 2000. Improvement in HIV-associated motor slowing after antiretroviral therapy including protease inhibitors. *Journal of Neurovirology* 6: 84–8.

Santos, S.I., Mathew, M., and Loza-Alvarez, P. 2010. Real time imaging of femtosecond laser induced nanoneurosurgery dynamics in *C. elegans*. *Optics Express* 18: 364–77.

Saraiva, A.M., Cardoso, I., Saraiva, M.J. et al. 2010. Randomization of amyloid-beta-peptide(1–42) conformation by sulfonated and sulfated nanoparticles reduces aggregation and cytotoxicity. *Macromolecular Bioscience* 10: 1152–63.

Sayre, L.M., Perry, G., Harris, P.L.R. et al. 2000. In situ oxidative catalysis by neurofibrillary tangles and senile plaques in Alzheimer's disease. *Journal of Neurochemistry* 74: 270–9.

Schafer, V., von Briesen, H., Andreesen, R. et al. 1992. Phagocytosis of nanoparticles by human immunodeficiency virus (HIV)-infected macrophages: A possibility for antiviral drug targeting. *Pharmaceutical Research* 9: 541–6.

Siegemund, T., Paulke, B.R., Schmiedel, H. et al. 2006. Thioflavins released from nanoparticles target fibrillar amyloid beta in the hippocampus of APP/PS1 transgenic mice. *International Journal of Developmental Neuroscience* 24: 195–201.

Silva, G.A., Czeisler, C., Niece, K.L. et al. 2004. Selective differentiation of neural progenitor cells by high-epitope density nanofibers. *Science* 303: 1352–5.

Smith, A., Giunta, B., Bickford, P.C. et al. 2010. Nanolipidic particles improve the bioavailability and α-secretase inducing ability of epigallocatechin 3 gallate (EGCG) for the treatment of Alzheimer's disease. *International Journal of Pharmaceutics* 389: 207–12.

Solassol, J., Crozet, C., Perrier, V. et al. 2004. Cationic phosphorus-containing dendrimers reduce prion replication both in cell culture and in mice infected with scrapie. *Journal of General Virology* 85: 1791–9.

Staber, P.D., Vargeese, C., Martins, I.H. et al. 2006. 88. Synthetic siRNAs for Huntington's disease therapy. *Molecular Therapy* 13: S36–S37.

Sun, M., Gao, Y., Guo, C. et al. 2010. Enhancement of transport of curcumin to brain in mice by poly(*n*-butylcyanoacrylate) nanoparticle. *Journal of Nanoparticle Research* 12: 3111–22.

Tjernberg, L.O., Lilliehook, C., Callaway, D.J.E. et al. 1997. Controlling amyloid beta-peptide fibril formation with protease-stable ligands. *Journal of Biological Chemistry* 272: 12601–5.

Tokuraku, K., Marquardt, M., and Ikezu, T. 2009. Real-time imaging and quantification of amyloid-beta peptide aggregates by novel quantum-dot nanoprobes. *PLoS One* 4: e8492.

Trapani, A., De Giglio, E., Cafagna, D. et al. 2011. Characterization and evaluation of chitosan nanoparticles for dopamine brain delivery. *International Journal of Pharmaceutics* 419: 296–307.

Valles, F., Fiandaca, M.S., Eberling, J.L. et al. 2010. Qualitative imaging of adeno-associated virus serotype 2-human aromatic L-amino acid decarboxylase gene therapy in a phase I study for the treatment of Parkinson disease. *Neurosurgery* 67: 1377–85.

Voges, J., Waerzeggers, Y., Maarouf, M. et al. 2006. Deep-brain stimulation: long-term analysis of complications caused by hardware and surgery—experiences from a single centre. *Journal of Neurology, Neurosurgery & Psychiatry* 77: 868–72.

Wadghiri, Y.Z., Sigurdsson, E.M., Sadowski, M. et al. 2003. Detection of Alzheimer's amyloid in transgenic mice using magnetic resonance microimaging. *Magnetic Resonance in Medicine* 50: 293–302.

Wang, Z.H., Liang, Q.L., Wang, Y.M. et al. 2003. Carbon nanotube-intercalated graphite electrodes for simultaneous determination of dopamine and serotonin in the presence of ascorbic acid. *Journal of Electroanalytical Chemistry* 540: 129–34.

Williams, S.R., Lepene, B.S., Thatcher, C.D. et al. 2009. Synthesis and characterization of poly(ethylene glycol)-glutathione conjugate self-assembled nanoparticles for antioxidant delivery. *Biomacromolecules* 10: 155–61.

Wilson, B., Samanta, M.K., Santhi, K. et al. 2008a. Poly(*n*-butylcyanoacrylate) nanoparticles coated with polysorbate 80 for the targeted delivery of rivastigmine into the brain to treat Alzheimer's disease. *Brain Research* 1200: 159–68.

Wilson, B., Samanta, M.K., Santhi, K. et al. 2008b. Targeted delivery of tacrine into the brain with polysorbate 80-coated poly(*n*-butylcyanoacrylate) nanoparticles. *European Journal of Pharmaceutics and Biopharmaceutics* 70: 75–84.

Yanik, M.F., Cinar, H., Cinar, H.N. et al. 2004. Neurosurgery: Functional regeneration after laser axotomy. *Nature* 432: 822.

Yurek, D., Fletcher, A., Kowalczyk, T. et al. 2009. Compacted DNA nanoparticle gene transfer of GDNF to the rat striatum enhances the survival of grafted fetal dopamine neurons. *Cell Transplantation* 18: 1183–96.

Zhang, W., Wan, F., Xie, Y. et al. 2004. Amperometric determination of (*R*)-salsolinol, (*R*)-*N*-methylsalsolinol and monoamine neurotransmitters with liquid chromatography using functionalized multi-wall carbon nanotube modified electrode in Parkinson's patients' cerebrospinal fluid. *Analytica Chimica Acta* 512: 207–14.

Zhe, Y., McKnight, T.E., Ericson, M.N. et al. 2010. Vertically aligned carbon nanofiber neural chip for interfacing with neurological system. *Nano/Molecular Medicine and Engineering (NANOMED), 2010 IEEE 4th International Conference on*, 188–91.

33 Brain–Machine Interface and Rehabilitation

Martin Pham, Alexander Taghva, and Ali R. Rezai

CONTENTS

BACKGROUND

Brain–machine interface (BMI) technology has the potential to improve the quality of life and restore function in patients with severe motor disabilities. Motor disability is seen in a variety of chronic conditions including multiple sclerosis, Parkinson's disease, amyotrophic lateral sclerosis, as well as acute onset conditions with chronic disability such as stroke and traumatic brain and spinal cord injuries. Although there are a variety of assistive devices to augment their existing capabilities, there is a disconnect between these devices and patient intention and patient action. Devices in existence today, such as walkers and wheelchairs, primarily compensate for motor weakness by providing structural support; however, they do not decode patient desires for movement into augmented motor capabilities.

BMI systems allow people to interact with their environment by linking different patterns of brain activity with specific tasks. This interaction occurs in lieu of normal neuromuscular outputs, without activating any muscle or peripheral nerve. This has already been demonstrated as a proof of concept with driving a robot or wheelchair (Del et al. 2009), operating prosthetic devices (Muller-Putz et al. 2005, 2006; Obermaier et al. 2003; Pfurtscheller et al. 2000a), selecting letters from a virtual keyboard (Birbaumer et al. 1999; Obermaier et al. 2003; Scherer et al. 2004; Sellers et al. 2006a), Internet browsing (Bensch et al. 2007; Karim et al. 2006), navigating virtual realities (Bayliss 2003; Leeb et al. 2007a, 2007b), and playing games (Millan et al. 2010).

NANOTECHNOLOGY

There are both noninvasive and invasive methods to BMI technology that will be discussed in this chapter. The invasive forms of surgically implanted BMI devices may be the preferred approach for high-resolution neural prosthetic control. For this to happen, electrode arrays must be developed that can provide stable recordings of neuron populations for long periods. Electrode technology using microwire arrays (Caminiti et al. 1990; Fetz 1999; Nicolelis et al. 2003; Williams et al. 1999), the Utah array (Maynard et al. 1999, Rousche and Normann 1998), or ceramic thin-film electrodes (Moxon et al. 2004a) have all been used for multineuron recordings, but the loss of neural signal recordings over time hinders their potential for long-term use. It has been proposed that this loss of signal is due to the formation of a nonconductive glial scar along the electrode from damage during insertion (Cui et al. 2001). This, in effect, creates a physical barrier between the electrode and healthy cells and hinders the ability to record neuronal activity (Fitch et al. 1999). Using nanoscale silicon substrates to create an active matrix modification on the surface of ceramic thin-film electrodes has led to decreased glial adhesion and enhanced neurite outgrowth formation, without interfering with the ability to record single neurons (Moxon et al. 2004a). This nanoporous surface is thought to enhance the interaction of neural tissue by mimicking the biologic topology of its surrounding surface interactions, which is accustomed to nanoscale structures such as microtubules and proteins (Moxon et al. 2004a). By mimicking the biologic topology and surface interaction of cells, nanotechnology opens the door for further development of biocompatible electrodes that can be chronically implanted into the human cortex.

Carbon nanotubes have received much interest with regard to their use in implantable electrical devices within the nervous system as well (Iijima 1991). In all intracortically implanted BMI applications, the final contact between the neural tissue and a neural prosthetic device is that of the implanted metal electrode itself (Campbell et al. 1990, 1991; Cheung et al. 2007; Jensen et al. 2006; Loeb et al. 1995; Moxon et al. 2004b; Pellinen et al. 2005). Naturally, the type and texture of the metal as well as its area of exposure determine its sensitivity and mechanical fragility. An ideal electrode would be chemically stable and be sensitive to neural signals or electrical stimulation (Cogan et al. 2004; Loeb et al. 1995). Metal electrodes coated with carbon nanotubes are chemically inert, biocompatible, and have high charge transfer characteristics (Keefer et al. 2008). They have an increased sensitivity in recording neuronal activity, decreased susceptibility to electrical noise, and have the ability to record local field potentials, multiunit activity, and neuronal spiking simultaneously from a single electrode (Keefer et al. 2008). The extremely broad resolution and biocompatibility of carbon nanotube-coated electrodes may one day make them the ideal conduit for brain–machine technologies.

BMI TECHNOLOGIES

EEG-Based Technologies

The noninvasive approach to BMI technology relies on the recording of electroencephalogram (EEG) signals at the scalp (Figure 33.1). Although the surface electrodes on the skin can deliver extremely stable signals, skin, fatty tissue, bone, and dura attenuate these signals and limit them to about 100 Hz in frequency (Stieglitz et al. 2009). This approach has worked in selecting letters on a keyboard (Pfurtscheller et al. 2000b), moving a cursor on a computer screen (Birbaumer 2006), and controlling an electrical stimulation system for grasp function (Pfurtscheller et al. 2003; Stieglitz et al. 2009).

Several types of EEG-based BMI technologies have been described based on the particular EEG feature used to translate the user's intent. In the first type, a BMI will focus on a P300 event-related brain potential. This is based on a positive potential over a central cortical area that occurs roughly 300 ms after a stimulus (Donchin et al. 2000; Sellers et al. 2006b). By arranging visual stimuli such as letters or numbers in matrices that flash in succession, a P300 event will be produced on the row and column of the user's focused attention. The BMI system can then determine the user's desired selection. Communication with this method can occur at rates of 20–30 bits/min, which translates into roughly 2–4 words/min using predictive word processing software (Lenhardt et al. 2008; Serby et al. 2005). A second type of EEG-based BMI is based on 8–12 Hz (μ) and 18–26 Hz (β) rhythms originating from the sensorimotor cortex. Because these sensorimotor oscillations change with movement, sensation, or motor imagery, users can learn to control these rhythms to communicate on-screen or to operate a device (Daly and Wolpaw 2008). Three-dimensional cursor control (McFarland et al. 2010) as well as power wheelchair navigation (Hema et al. 2011) have

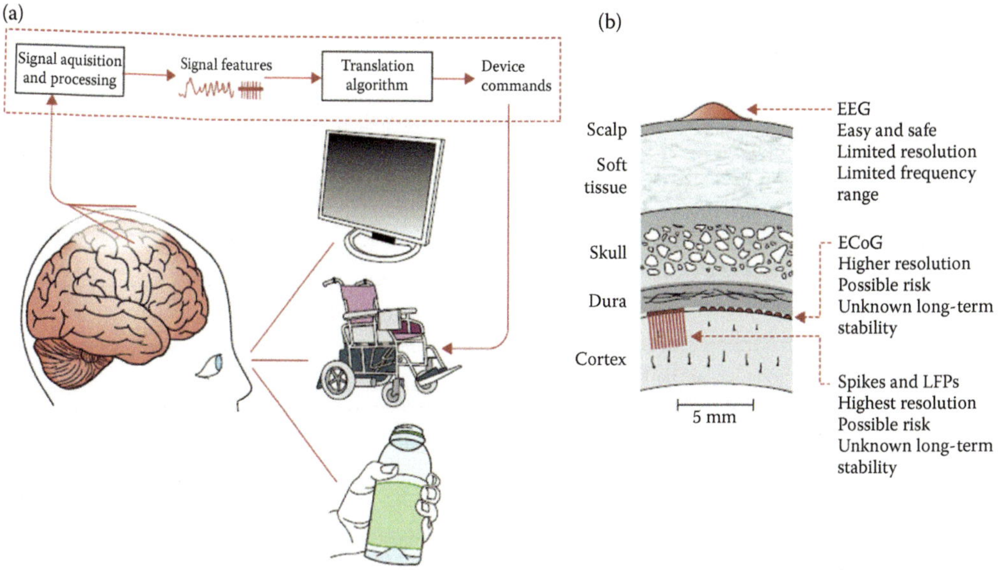

FIGURE 33.1 Overview of a BMI system. (a) Design and operation of a BMI system. Electrophysiological signals indicating brain activity are obtained from scalp, cortical surface, or within the brain and are analyzed to derive particular signal features (such as amplitudes of event-related potentials, EEG rhythms, or firing rates of single neurons). These features are translated into commands that operate an output device, such as a word processing program, a wheelchair, or a neuroprosthetic limb. (b) Recording locations for electrophysiological signals used by BMI systems. EEG activity is recorded using electrodes on scalp. ECoG activity is recorded using electrodes on cortical surface. Action potentials from single neurons or LFPs are recorded using electrode arrays implanted in motor cortex or in other brain areas. (Adapted from Daly, J.J., Wolpaw, J.R., *Lancet Neurol.*, 7, 1032–43, 2008.)

been described using this method. The third type of EEG-based BMI can be used for slow cortical potentials (SCPs) as well. SCPs are thought to be associated with preparatory depolarization of an underlying cortical network or with cortical inhibition (Daly and Wolpaw 2008). Control of SCPs to induce positive or negative voltage shifts have been used for basic word processing and Internet browsing (Birbaumer et al. 1999, 2000; Kubler et al. 2001).

From a clinical perspective, EEG-based BMI devices are valuable because no surgical intervention is required for their use. They can be initiated at any point in time during the patient's recovery and rehabilitation. Although there is some attenuation of signal based on the external nature of the scalp attachments, this resolution is sufficient to establish meaningful communication between the user and their environment (Stieglitz et al. 2009).

ECoG-Based Technologies

Electrocorticogram (ECoG) arrays implanted on the cortical surface for presurgical monitoring in epilepsy patients have also been used to record evoked potentials and rhythms (Leuthardt et al. 2004). In addition to μ and β rhythms, higher frequency 30–200 Hz (γ) sensorimotor oscillations can be distinguished for use in BMI devices (Figure 33.2). The use of motor imagery to influence ECoG rhythm amplitudes for on-screen cursor control can be learned within only a few minutes of training (Hinterberger et al. 2008; Schalk et al. 2008). Combined with a superior topographical resolution and the absence of any signal contamination from cranial muscle electromyographic or electrooculographic activity, ECoG-based BMI may offer significantly better communication and control than their EEG-based BMI counterparts (Daly and Wolpaw 2008).

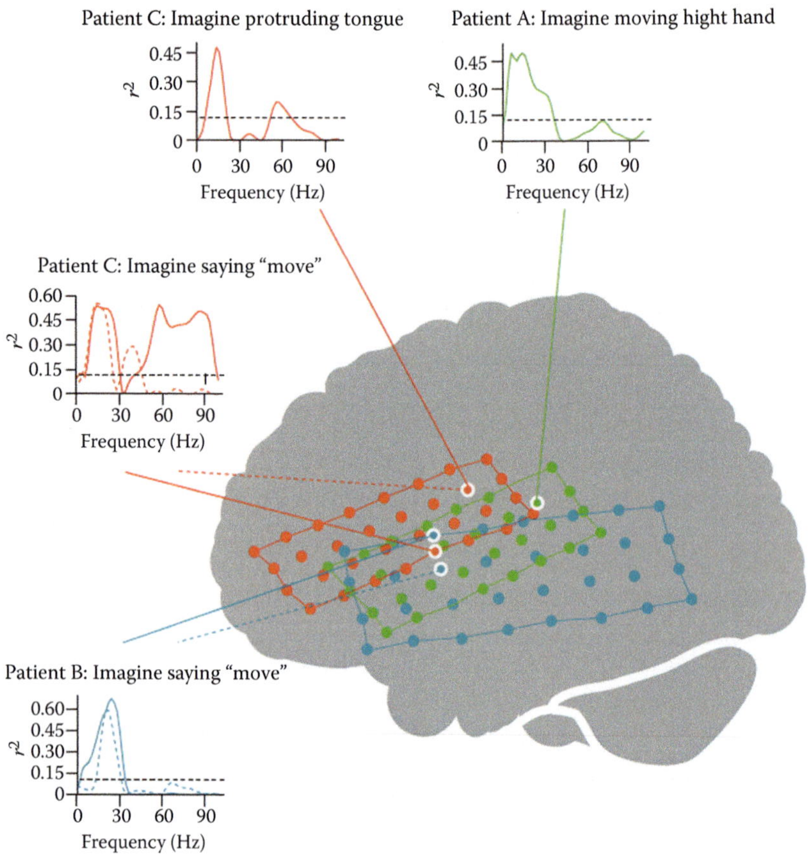

FIGURE 33.2 An ECoG-based BMI. ECoG control of vertical cursor movement using specific motor imagery to move the cursor up and rest (i.e., no imagery) to move it down. Electrodes used for control are circled and spectral correlations of their respective ECoG activity with target location (i.e., top or bottom of screen) are shown. Specific imagined actions used are indicated. Significant levels of control achieved with different kinds of imagery are apparent (dashed horizontal lines indicate significance at $p = 0.01$). Solid and dotted r^2 spectra shown for patients B and C correspond to sites indicated by dotted and solid line locators, respectively (Adapted from Daly, J.J., Wolpaw, J.R., *Lancet Neurol.*, 7, 1032–43, 2008.)

Thus far, research using ECoG-based BMI devices has been limited to short-term studies on patients who were temporarily implanted with electrode arrays before epilepsy surgery (Leuthardt et al. 2004; Schalk et al. 2008). Practical use of these devices will depend on developing fully implantable systems that can function safely and reliably over many years (Daly and Wolpaw 2008; Stieglitz et al. 2009).

Intracortical-Based Technologies

Placement of intracortical multielectrode arrays within the motor cortex can record both single neuron activity as well as local field potentials. User modulation of single neuron activity has been implemented to control cursors in one, two, and three dimensions (Chapin et al. 1999; Musallam et al. 2004; Serruya et al. 2002; Taylor et al. 2002, 2003). Local field potentials that indicate nearby synaptic activity can be used similarly for cursor control in multiple dimensions as well (Pesaran et al. 2002). By identifying the neuronal or synaptic activity associated with the user's limb movements, cursor control and neural prostheses can be linked to intuitive cortical activity that would be comparable to the user's actual limb movements (Andersen et al. 2010; Scherberger et al. 2005). When applied to word processing, optimization analyses demonstrated performances exceeding that of 6.5 bits/s, or 15 words/min (Gilja et al. 2011; Santhanam et al. 2006).

There are several challenges associated with the using intracortical-based BMI devices. Long-term safety and stability of the multielectrode arrays within the brain parenchyma still need to be fully addressed. Electrode-tissue recording failure due to electrode encapsulation, materials failure, or micromotion are obstacles to creating arrays that can offer stable recordings over decades (Gilja et al. 2011). Although sensorimotor EEG-based BMI systems have been shown to be equal to that of intracortical BMI with regard to the speed and accuracy of cursor control (Hochberg et al. 2006; Wolpaw and McFarland 2004), the increased resolution of intracortical arrays may support a more robust system on which to base the use of dexterous prosthetic limbs, particularly those that would operate in four dimensions (the fourth dimension being velocity) (Figure 33.3) (Velliste et al. 2008).

Perhaps the most sophisticated BMI is the cortical–cortical prosthesis. In theory, these prosthetics may recapitulate the function of a damaged brain area by replacing that area with a similarly functioning silicon-based computer chip (Figure 33.4). In order for this type of prosthetic to function, several conditions must be met: (1) the prosthetic must be able to both receive and transmit data by electrical signals to surrounding neuron populations; (2) the prosthetic must be biocompatible; and (3) the prosthetic must be able to efficiently compute the functions normally performed by that brain region. A proof of concept of this type of prosthetic has been described for recapitulation of function of the rat hippocampus. A multiple-input, multiple-output nonlinear dynamic model of spike-train to spike-train transformations was previously formulated for hippocampal–cortical prostheses; this model agrees well with experimentally observed output from

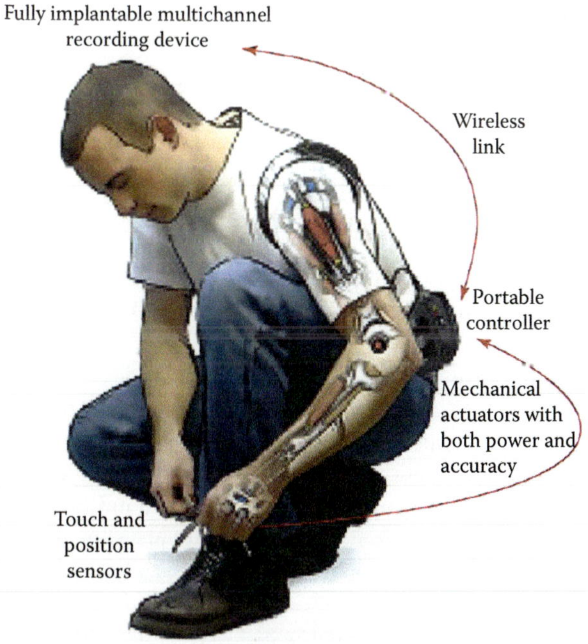

FIGURE 33.3 How a fully implantable BMI could restore limb mobility in paralyzed subjects or amputees. Although details of this system have to be worked out through future research, it is clear that BMI for human clinical applications should be encased in the patient's body as much as possible. Wireless telemetry offers a viable solution. The prosthesis not only should have functionality of the human arm in terms of power and accuracy of actuators, but also should be equipped with sensors of touch and position from which signals can be transmitted back to subject's brain. (Adapted from Lebedev, M.A., Nicolelis, M.A.L., *Trends Neurosci.*, 29, 536–46, 2006.)

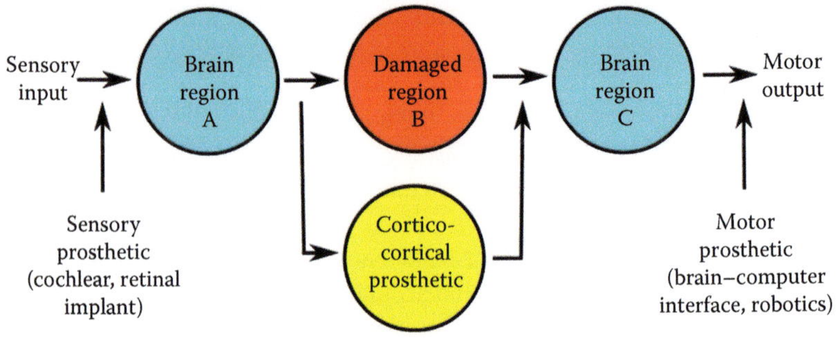

FIGURE 33.4 Schematic layout of different neural prosthetics and their insertion points. Different neural prosthetic types and their point of insertion along a schematized neural circuit.

hippocampal neurons (Song et al. 2007). Two primary barriers must be overcome to make these implants clinically applicable. First, at this time, a priori knowledge of the cortical region to be replaced must be obtained before the creation of the implant. This knowledge may not exist for a patient before their injury or loss of function. Second, selection of neuron inputs and outputs can be computationally expensive; however, methods to decrease this complexity are described (Taghva et al. 2011).

REHABILITATION

BMI technologies have been utilized for motor recovery in the stroke population. This was based on the framework of activity-dependent central nervous system (CNS) plasticity with BMI technologies used as a motor relearning intervention (Daly and Wolpaw 2008; Nudo 2006; Wolpaw and Tennissen 2001). Activity-dependent plasticity results in learning that can change motor function in the injured nervous system. If interventions are targeted successfully, this can result in restoration of more normal motor function. However, repetitive abnormal movements can solidify or worsen abnormal motor function as well based on this framework of plasticity (Carmichael et al. 2001; Dancause et al. 2005; Stroemer et al. 1995; Teasell et al. 2005).

Currently, the standard approach to improving motor function in stroke and brain-injured victims is through peripheral repetitive motion of the upper and lower limbs during physical and occupational therapy. This approach uses the repetitive motion to influence CNS activity in an activity-dependent way to improve motor function. BMI technology aims to improve motor function by either training users to modulate their brain signals directly or by using an assistive device for movement to induce CNS plasticity through sensory input and feedback (Figure 33.5) (Daly and Wolpaw 2008).

The modulation of brain signals as a conditioning regimen can influence EEG, ECoG, and single neuron activity (Carmena et al. 2003; Fetz 1969; Leuthardt et al. 2004; Taylor et al. 2002; Wolpaw and McFarland 2004). Mechanisms of neuronal plasticity such as synaptogenesis, axonal sprouting, and neurite outgrowth have been shown in both animals and humans (Cramer et al. 1997, 2006; Stroemer et al. 1995; Teasell et al. 2005). Neuronal changes have also been shown to occur in regions either distant or contralateral to the infarct as well and include neuronal hyperexcitability, reorganization of cortical sensory and motor maps, and alterations of intra- and interhemispheric connections to the motor regions of the brain (Frost et al. 2003; Napieralski et al. 1996). Enhanced motor activity results in improvement in particular features of EEG activity as well (Napieralski et al. 1996). With studies showing that stroke survivors can learn to control certain aspects of their

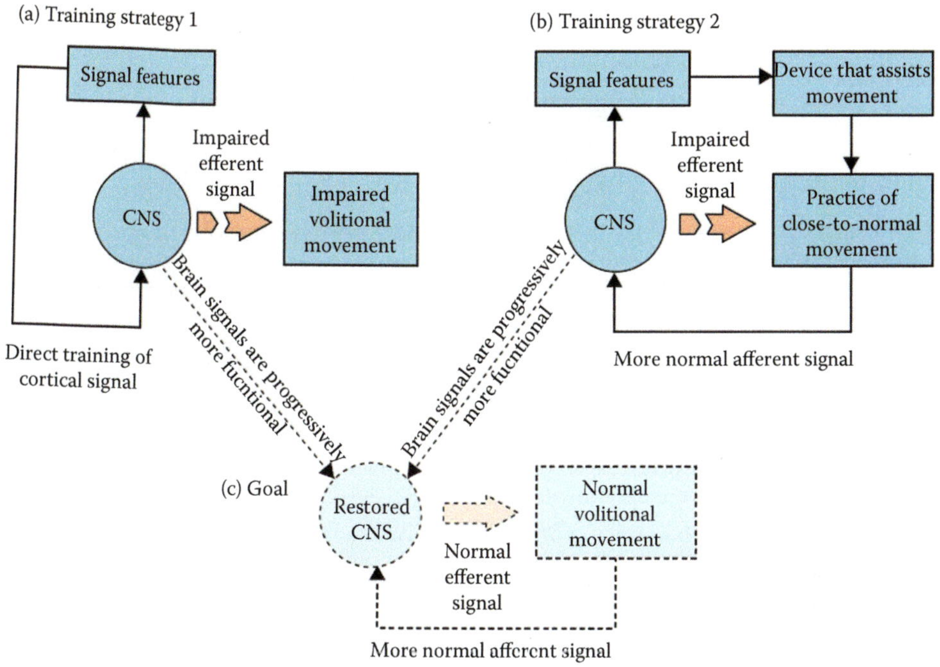

FIGURE 33.5 Two BMI-based training strategies to encourage and guide CNS plasticity to improve motor function. (a) This training strategy translates specific features of brain activity into an action (e.g., cursor movement) and uses that action as feedback to train patients to produce more normal brain activity. The hypothesis is that plasticity that produces this more normal activity will also restore more normal CNS function and will therefore improve motor control. (b) This training strategy uses specific features of brain activity to activate a device that assists movement that can compensate for patient's impaired neuromuscular control during motor tasks. The hypothesis is that, by improving motor function, this assistance will produce sensory input that induces CNS plasticity to restore more normal motor control. (c) The first strategy aims to normalize brain activity with the expectation that this will be accompanied by improved motor function, whereas the second strategy uses brain activity to assist practice of more normal neuromuscular control with the expectation that the more normal sensory input produced by the better motor function will induce plasticity that improves neuromuscular control. (Adapted from Daly, J.J., Wolpaw, J.R., *Lancet Neurol.*, 7, 1032–43, 2008.)

EEG activity using BMI protocols, motor function involving the same areas producing the signal may be improved (Birbaumer and Cohen 2007; Buch et al. 2008).

Using brain signals to activate an assistive movement device using BMI technology may also help improve motor function. By providing a means to practice motor movement that is as close to normal as possible, assistive movement devices may help guide sprouting axons to their appropriate cortical regions (Carmichael and Chesselet 2002). One study demonstrated that by attaching a flexible hand prosthesis to a paralyzed hand, users learned how to open and close the prosthesis by increasing or decreasing EEG activity over the lesioned hemisphere (Figure 33.6). This showed that clinical improvements were possible from ipsilaterally lesioned hemispheres using a prosthesis (Birbaumer and Cohen 2007). EEG-based control of robot-assisted movement using motor imagery was also used in stroke survivors to examine gains in upper limb rehabilitation (Ang et al. 2009). This study showed significant gains in motor improvement using BMI-based robotic rehabilitation, providing evidence of BMI technologies in restoring motor control of upper limbs in stroke victims.

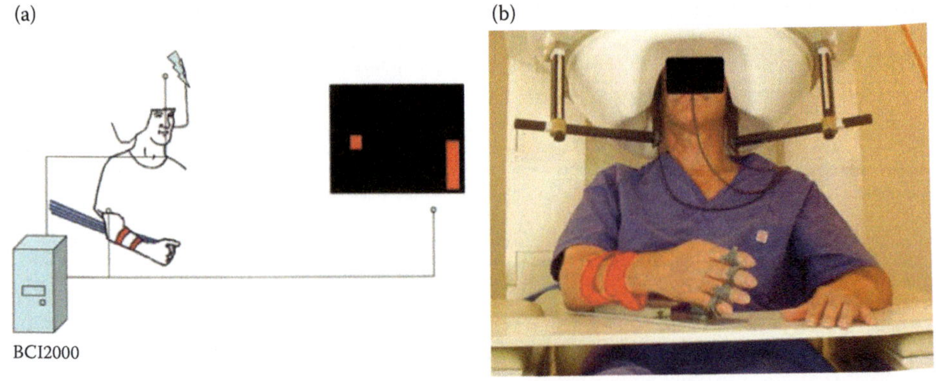

FIGURE 33.6 Brain–orthosis interface. (a) Patients observe their SMR (sensory motor rhythm) activity represented by red cursor on a screen (right part) after instructed to increase (top goal bar) or decrease (lower goal bar) SMR. Decrease closes and increase opens the hand (right) in five steps depending on SMR 2 s before. (b) Stroke patient with hand in orthosis during training. Note that patients not only receive visual feedback from feedback screen but also from watching and feeling their own paralyzed hand moving. (Adapted from Birbaumer, N., *Psychophysiology*, 43, 517–32, 2006.)

REFERENCES

Andersen, R.A., Hwang, E.J., and Mulliken, G.H. 2010. Cognitive neural prosthetics. *Annu Rev Psychol* 61: 169–90, C1–3.

Ang, K.K., Guan, C., Chua, K.S. et al. 2009. A clinical study of motor imagery-based brain–computer interface for upper limb robotic rehabilitation. *Conf Proc IEEE Eng Med Biol Soc* 2009: 5981–4.

Bayliss, J.D. 2003. Use of the evoked potential P3 component for control in a virtual apartment. *IEEE Trans Neural Syst Rehabil Eng* 11: 113–6.

Bensch, M., Karim, A.A., Mellinger, J. et al. 2007. Nessi: an EEG-controlled web browser for severely paralyzed patients. *Comput Intell Neurosci* 2007: 71863.

Birbaumer, N. 2006. Breaking the silence: brain–computer interfaces (BCI) for communication and motor control. *Psychophysiology* 43: 517–32.

Birbaumer, N., and Cohen, L.G. 2007. Brain–computer interfaces: communication and restoration of movement in paralysis. *J Physiol* 579: 621–36.

Birbaumer, N., Ghanayim, N., Hinterberger, T. et al. 1999. A spelling device for the paralysed. *Nature* 398: 297–8.

Birbaumer, N., Kubler, A., Ghanayim, N. et al. 2000. The thought translation device (TTD) for completely paralyzed patients. *IEEE Trans Rehabil Eng* 8: 190–3.

Buch, E., Weber, C., Cohen, L.G. et al. 2008. Think to move: a neuromagnetic brain–computer interface (BCI) system for chronic stroke. *Stroke* 39: 910–17.

Caminiti, R., Johnson, P.B., Burnod, Y. et al. 1990. Shift of preferred directions of premotor cortical cells with arm movements performed across the workspace. *Exp Brain Res* 83: 228–32.

Campbell, P.K., Jones, K.E., Huber, R.J. et al. 1991. A silicon-based, three-dimensional neural interface: manufacturing processes for an intracortical electrode array. *IEEE Trans Biomed Eng* 38: 758–68.

Campbell, P.K., Jones, K.E., and Normann, R.A. 1990. A 100 electrode intracortical array: structural variability. *Biomed Sci Instrum* 26: 161–5.

Carmena, J.M., Lebedev, M.A., Crist, R.E. et al. 2003. Learning to control a brain–machine interface for reaching and grasping by primates. *PLoS Biol* 1: E42.

Carmichael, S.T., and Chesselet, M.F. 2002. Synchronous neuronal activity is a signal for axonal sprouting after cortical lesions in the adult. *J Neurosci* 22: 6062–70.

Carmichael, S.T., Wei, L., Rovainen, C.M. et al. 2001. New patterns of intracortical projections after focal cortical stroke. *Neurobiol Dis* 8: 910–22.

Chapin, J.K., Moxon, K.A., Markowitz, R.S. et al. 1999. Real-time control of a robot arm using simultaneously recorded neurons in the motor cortex. *Nat Neurosci* 2: 664–70.

Cheung, K.C., Renaud, P., Tanila, H. et al. 2007. Flexible polyimide microelectrode array for in vivo recordings and current source density analysis. *Biosens Bioelectron* 22: 1783–90.
Cogan, S.F., Guzelian, A.A., Agnew, W.F. et al. 2004. Over-pulsing degrades activated iridium oxide films used for intracortical neural stimulation. *J Neurosci Methods* 137: 141–50.
Cramer, S.C., Nelles, G., Benson, R.R. et al. 1997. A functional MRI study of subjects recovered from hemiparetic stroke. *Stroke* 28: 2518–27.
Cramer, S.C., Shah, R., Juranek, J. et al. 2006. Activity in the peri-infarct rim in relation to recovery from stroke. *Stroke* 37: 111–5.
Cui, X., Lee, V.A., Raphael, Y. et al. 2001. Surface modification of neural recording electrodes with conducting polymer/biomolecule blends. *J Biomed Mater Res* 56: 261–72.
Daly, J.J., and Wolpaw, J.R. 2008. Brain–computer interfaces in neurological rehabilitation. *Lancet Neurol* 7: 1032–43.
Dancause, N., Barbay, S., Frost, S.B. et al. 2005. Extensive cortical rewiring after brain injury. *J Neurosci* 25: 10167–79.
Del, R.M.J.J., Galan, F., Vanhooydonck, D. et al. 2009. Asynchronous non-invasive brain-actuated control of an intelligent wheelchair. *Conf Proc IEEE Eng Med Biol Soc* 2009: 3361–4.
Donchin, E., Spencer, K.M., and Wijesinghe, R. 2000. The mental prosthesis: assessing the speed of a P300-based brain–computer interface. *IEEE Trans Rehabil Eng* 8: 174–9.
Fetz, E.E. 1969. Operant conditioning of cortical unit activity. *Science* 163: 955–8.
Fetz, E.E. 1999. Real-time control of a robotic arm by neuronal ensembles. *Nat Neurosci* 2: 583–4.
Fitch, M.T., Doller, C., Combs, C.K. et al. 1999. Cellular and molecular mechanisms of glial scarring and progressive cavitation: in vivo and in vitro analysis of inflammation-induced secondary injury after CNS trauma. *J Neurosci* 19: 8182–98.
Frost, S.B., Barbay, S., Friel, K.M. et al. 2003. Reorganization of remote cortical regions after ischemic brain injury: a potential substrate for stroke recovery. *J Neurophysiol* 89: 3205–14.
Gilja, V., Chestek, C.A., Diester, I. et al. 2011. Challenges and opportunities for next-generation intracortically based neural prostheses. *IEEE Trans Biomed Eng* 58: 1891–9.
Hema, C.R., Paulraj, M.P., Yaacob, S. et al. 2011. Asynchronous brain machine interface-based control of a wheelchair. *Adv Exp Med Biol* 696: 565–72.
Hinterberger, T., Widman, G., Lal, T.N. et al. 2008. Voluntary brain regulation and communication with electrocorticogram signals. *Epilepsy Behav* 13: 300–6.
Hochberg, L.R., Serruya, M.D., Friehs, G.M. et al. 2006. Neuronal ensemble control of prosthetic devices by a human with tetraplegia. *Nature* 442: 164–71.
Iijima, S. 1991. Helical microtubes of graphitic carbon. *Nature* 354: 56–8.
Jensen, W., Yoshida, K., and Hofmann, U.G. 2006. In-vivo implant mechanics of flexible, silicon-based ACREO microelectrode arrays in rat cerebral cortex. *IEEE Trans Biomed Eng* 53: 934–40.
Karim, A.A., Hinterberger, T., Richter, J. et al. 2006. Neural internet: web surfing with brain potentials for the completely paralyzed. *Neurorehabil Neural Repair* 20: 508–15.
Keefer, E.W., Botterman, B.R., Romero, M.I. et al. 2008. Carbon nanotube coating improves neuronal recordings. *Nat Nanotechnol* 3: 434–9.
Kubler, A., Neumann, N., Kaiser, J. et al. 2001. Brain–computer communication: self-regulation of slow cortical potentials for verbal communication. *Arch Phys Med Rehabil* 82: 1533–9.
Lebedev, M.A., and Nicolelis, M.A.L. 2006. Brain–machine interfaces: past, present and future. *Trends Neurosci* 29: 536–46.
Leeb, R., Friedman, D., Muller-Putz, G.R. et al. 2007a. Self-paced (asynchronous) BCI control of a wheelchair in virtual environments: a case study with a tetraplegic. *Comput Intell Neurosci* 79642.
Leeb, R., Lee, F., Keinrath, C. et al. 2007b. Brain–computer communication: motivation, aim, and impact of exploring a virtual apartment. *IEEE Trans Neural Syst Rehabil Eng* 15: 473–82.
Lenhardt, A., Kaper, M., and Ritter, H.J. 2008. An adaptive P300-based online brain–computer interface. *IEEE Trans Neural Syst Rehabil Eng* 16: 121–30.
Leuthardt, E.C., Schalk, G., Wolpaw, J.R. et al. 2004. A brain–computer interface using electrocorticographic signals in humans. *J Neural Eng* 1: 63–71.
Loeb, G.E., Peck, R.A., and Martyniuk, J. 1995. Toward the ultimate metal microelectrode. *J Neurosci Methods* 63: 175–83.
Maynard, E.M., Hatsopoulos, N.G., Ojakangas, C.L. et al. 1999. Neuronal interactions improve cortical population coding of movement direction. *J Neurosci* 19: 8083–93.
McFarland, D.J., Sarnacki, W.A., and Wolpaw, J.R. 2010. Electroencephalographic (EEG) control of three-dimensional movement. *J Neural Eng* 7: 036007.

Millan, J.D., Rupp, R., Muller-Putz, G.R. et al. 2010. Combining brain–computer interfaces and assistive technologies: state-of-the-art and challenges. *Front Neurosci* 4: 161.

Moxon, K.A., Kalkhoran, N.M., Markert, M. et al. 2004a. Nanostructured surface modification of ceramic-based microelectrodes to enhance biocompatibility for a direct brain–machine interface. *IEEE Trans Biomed Eng* 51: 881–9.

Moxon, K.A., Leiser, S.C., Gerhardt, G.A. et al. 2004b. Ceramic-based multisite electrode arrays for chronic single-neuron recording. *IEEE Trans Biomed Eng* 51: 647–56.

Muller-Putz, G.R., Scherer, R., Pfurtscheller, G. et al. 2005. EEG-based neuroprosthesis control: a step towards clinical practice. *Neurosci Lett* 382: 169–74.

Muller-Putz, G.R., Scherer, R., Pfurtscheller, G. et al. 2006. Brain–computer interfaces for control of neuroprostheses: from synchronous to asynchronous mode of operation. *Biomed Tech (Berl)* 51: 57–63.

Musallam, S., Corneil, B.D., Greger, B. et al. 2004. Cognitive control signals for neural prosthetics. *Science* 305: 258–62.

Napieralski, J.A., Butler, A.K., and Chesselet, M.F. 1996. Anatomical and functional evidence for lesion-specific sprouting of corticostriatal input in the adult rat. *J Comp Neurol* 373: 484–97.

Nicolelis, M.A., Dimitrov, D., Carmena, J.M. et al. 2003. Chronic, multisite, multielectrode recordings in macaque monkeys. *Proc Natl Acad Sci U S A* 100: 11041–6.

Nudo, R.J. 2006. Mechanisms for recovery of motor function following cortical damage. *Curr Opin Neurobiol* 16: 638–44.

Obermaier, B., Muller, G.R., and Pfurtscheller, G. 2003. "Virtual keyboard" controlled by spontaneous EEG activity. *IEEE Trans Neural Syst Rehabil Eng* 11: 422–6.

Pellinen, D., Moon, T., Vetter, R. et al. 2005. Multifunctional flexible parylene-based intracortical microelectrodes. *Conf Proc IEEE Eng Med Biol Soc* 5: 5272–5.

Pesaran, B., Pezaris, J.S., Sahani, M. et al. 2002. Temporal structure in neuronal activity during working memory in macaque parietal cortex. *Nat Neurosci* 5: 805–11.

Pfurtscheller, G., Guger, C., Muller, G. et al. 2000a. Brain oscillations control hand orthosis in a tetraplegic. *Neurosci Lett* 292: 211–4.

Pfurtscheller, G., Muller, G.R., Pfurtscheller, J. et al. 2003. 'Thought'—control of functional electrical stimulation to restore hand grasp in a patient with tetraplegia. *Neurosci Lett* 351: 33–6.

Pfurtscheller, G., Neuper, C., Guger, C. et al. 2000b. Current trends in Graz Brain–Computer Interface (BCI) research. *IEEE Trans Rehabil Eng* 8: 216–9.

Rousche, P.J., and Normann, R.A. 1998. Chronic recording capability of the Utah Intracortical Electrode Array in cat sensory cortex. *J Neurosci Methods* 82: 1–15.

Santhanam, G., Ryu, S.I., Yu, B.M. et al. 2006. A high-performance brain–computer interface. *Nature* 442: 195–8.

Schalk, G., Miller, K.J., Anderson, N.R. et al. 2008. Two-dimensional movement control using electrocorticographic signals in humans. *J Neural Eng* 5: 75–84.

Scherberger, H., Jarvis, M.R., and Andersen, R.A. 2005. Cortical local field potential encodes movement intentions in the posterior parietal cortex. *Neuron* 46: 347–54.

Scherer, R., Muller, G.R., Neuper, C. et al. 2004. An asynchronously controlled EEG-based virtual keyboard: improvement of the spelling rate. *IEEE Trans Biomed Eng* 51: 979–84.

Sellers, E.W., Krusienski, D.J., McFarland, D.J. et al. 2006a. A P300 event-related potential brain–computer interface (BCI): the effects of matrix size and inter stimulus interval on performance. *Biol Psychol* 73: 242–52.

Sellers, E.W., Kubler, A., and Donchin, E. 2006b. Brain–computer interface research at the University of South Florida Cognitive Psychophysiology Laboratory: the P300 Speller. *IEEE Trans Neural Syst Rehabil Eng* 14: 221–4.

Serby, H., Yom-Tov, E., and Inbar, G.F. 2005. An improved P300-based brain–computer interface. *IEEE Trans Neural Syst Rehabil Eng* 13: 89–98.

Serruya, M.D., Hatsopoulos, N.G., Paninski, L. et al. 2002. Instant neural control of a movement signal. *Nature* 416: 141–2.

Song, D., Chan, R.H., Marmarelis, V.Z. et al. 2007. Nonlinear dynamic modeling of spike train transformations for hippocampal–cortical prostheses. *IEEE Trans Biomed Eng* 54: 1053–66.

Stieglitz, T., Rubehn, B., Henle, C. et al. 2009. Brain–computer interfaces: an overview of the hardware to record neural signals from the cortex. *Prog Brain Res* 175: 297–315.

Stroemer, R.P., Kent, T.A., and Hulsebosch, C.E. 1995. Neocortical neural sprouting, synaptogenesis, and behavioral recovery after neocortical infarction in rats. *Stroke* 26: 2135–44.

Taghva, A., Song, D., Hampson, R.E. et al. 2011. Determination of relevant neuron–neuron connections for neural prosthetics using time-delayed mutual information: tutorial and preliminary results. *World Nuerosurgery*, Nov. 4 Epub ahead of print.

Taylor, D.M., Tillery, S.I., and Schwartz, A.B. 2002. Direct cortical control of 3D neuroprosthetic devices. *Science* 296: 1829–32.

Taylor, D.M., Tillery, S.I., and Schwartz, A.B. 2003. Information conveyed through brain-control: cursor versus robot. *IEEE Trans Neural Syst Rehabil Eng* 11: 195–9.

Teasell, R., Bayona, N.A., and Bitensky, J. 2005. Plasticity and reorganization of the brain post stroke. *Top Stroke Rehabil* 12: 11–26.

Velliste, M., Perel, S., Spalding, M.C. et al. 2008. Cortical control of a prosthetic arm for self-feeding. *Nature* 453: 1098–101.

Williams, J.C., Rennaker, R.L., and Kipke, D.R. 1999. Long-term neural recording characteristics of wire microelectrode arrays implanted in cerebral cortex. *Brain Res Brain Res Protoc* 4: 303–13.

Wolpaw, J.R., and McFarland, D.J. 2004. Control of a two-dimensional movement signal by a noninvasive brain–computer interface in humans. *Proc Natl Acad Sci U S A* 101: 17849–54.

Wolpaw, J.R., and Tennissen, A.M. 2001. Activity-dependent spinal cord plasticity in health and disease. *Annu Rev Neurosci* 24: 807–43.

34 Nanobioelectronics for Noninvasive Neuronal Therapy

Evangelia Hondroulis, Jairo Nelson, and Chen-Zhong Li

CONTENTS

INTRODUCTION

The effect of exogenous electric fields on physiology and their possible relationship to the treatment of diseases has been a source of interest for researchers for years (Barker et al. 1984; Ieran et al. 1990; Sandyk 1997). In 1855, Guillaume Duchenne discovered that alternating current leads to electrotherapeutic triggering of muscle contractions, thus spurring the use of electrical energy as medical treatment. Since then, the use of electric fields in medicine, also known as electrotherapy, has become popular in fields such as gene and cellular therapies (Golzio et al. 2004; Song et al. 2004; Teissie et al. 2008) and has even progressed to clinical trials for drug delivery (Fantozzi et al. 2010).

Cells have intrinsic electrical properties that fuel this behavior, the most prevalent stemming from the cellular membrane consisting of a lipid bilayer riddled with various transmembrane proteins creating ionic channels responsible for the survival of the cell. Because of these channels, the membrane can be considered to have mostly capacitive properties with the inside of the cell more resistive creating a membrane potential as shown in Figure 34.1. The field of bioelectronics emerged as a result of the convergence of biology and electronics. Understanding these electrical properties of cells and how they are affected under different stimuli, we can find out more information on how the cell functions. Neural cells, already electrically active, behave differently under external stimuli, and thus have been the focus of extensive research studies for neuronal therapy and the treatment of various neurodegenerative diseases.

Knowledge of these electrical properties of cells has led to the development of the field of bioelectronics. Bioelectronics is the application of electronics to biology and medicine and can be broken down into two categories. Physically interfacing electronic devices with biological systems has led to technologies such as the cardiac pacemaker, implantable electrical bone growth simulators, deep brain simulators, and electrical nerve simulation (Nowak et al. 2011). The other aspect of bioelectronics is electronics for both detection and characterization of biological materials, such as on the cellular and subcellular levels.

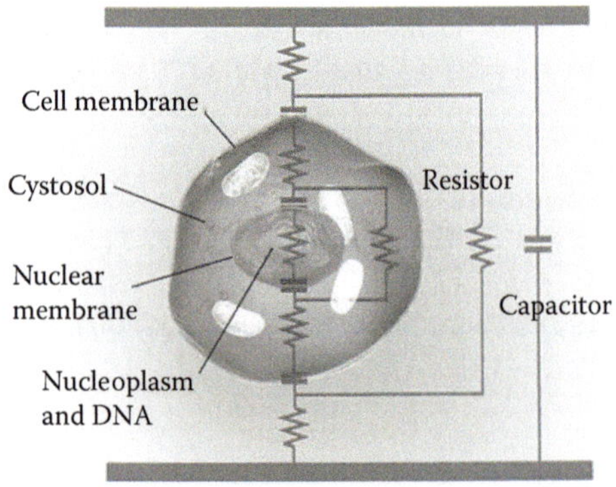

FIGURE 34.1 A cell can be thought of as a circuit made up of capacitors and resistors with the membrane acting like a capacitor and the cytosol as a resistor. (From Schoenbach, K.H., Nuccitelli, R., and Beebe, S.J. Zap. http://www.biotele.com/bioelectric.htm. With permission.)

Nanotechnology has great potential benefits in rapidly developing fields such as biomedical engineering, drug delivery, environmental health, pharmaceutical industries, and even electronics and communication technologies. For instance, in the healthcare field, nanomaterials are being considered in the development of new drugs and new therapies for disease control and improving the quality of life (Bianco et al. 2005; Faraji and Wipf 2009; Slowing et al. 2007). More recently, nanomaterials have been used in tissue engineering and medical imaging, leading to improved diagnostics and new therapeutic treatments (Harrison and Atala 2007; Kim et al. 2008; Shi 2009). The combination of electrotherapy and nanotechnology has tremendous potential for neural therapy.

ELECTROMAGNETIC FIELDS

Before delving into the applications of electromagnetic fields on tissues, we must first review exactly what a magnetic field is and how it is produced. It is important to understand that everything has a magnetic field, and the reason for this will be provided once a few basic principles of electromagnetism have been explained.

Whenever someone mentions the word magnetic, it is common to think of a bar magnet with a north and a south pole as seen in Figure 34.2. These magnets are typically made of metals, such as

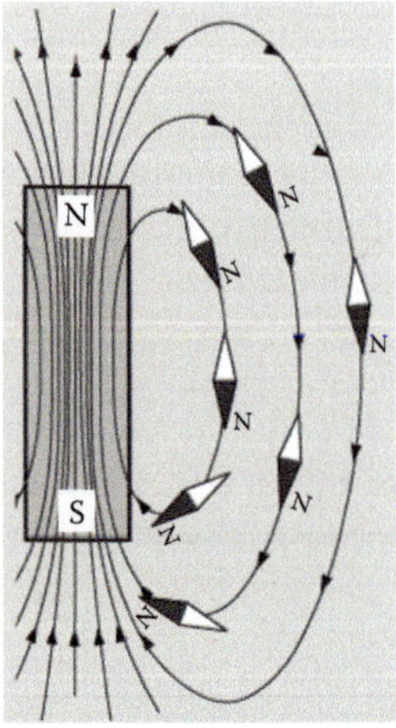

FIGURE 34.2 Effect of electron spin states in bar magnet material produce magnetism, and the resulting flux is denoted by arrow directions. Internet source: http://hyperphysics.phy-astr.gsu.edu/hbase/magnetic/elemag.html.

iron, that have electrons in the valence shells of their atomic orbitals. These electrons have a spin state that can be oriented in one of two directions: an up-spin and a down-spin. Typically, atoms have variations in electron spin states, and placing a magnet close to that substance can influence the direction of the spin of their electrons. However, permanent magnets have the spin of their electrons locked in a single state (up or down, but not both) and this state cannot be influenced by other magnets. This property is what gives a magnet its magnetic field and its resulting force on other magnets and certain materials with electrons in their own valence shells. As seen in the bar magnet of Figure 34.2, arrows are shown leaving the northern pole of the magnet and reentering at the southern pole. This direction of magnetic flow (or flux) is dictated by the common spin state that all the electrons share in that material.

What Causes a Magnetic Field to Form?

To understand why magnetic fields are formed, a deeper understanding of what is occurring at the atomic level is required. This deeper understanding is crucial for proper design of electromagnetic devices for treatment and imaging.

The best depiction of how a magnetic field is formed would be that of a conducting wire with electricity flowing through it. A magnetic field originates from the movement of charged particles, such as electrons, along the axis of the wire. When charged particles flow uniformly in a certain direction an electrical current is formed (measured in amperes). The direction of this current determines the direction in which the magnetic field will be projected and follows "The Right Hand Rule" as shown in Figure 34.3. The basis of the right hand rule is that the thumb is on the same axis as the wire and points in the direction in which the current is flowing. The resulting magnetic field of this current will loop around the wire and point in the direction of your fingers as shown in Figure 34.3.

The intensity of the magnetic field (denoted as B_0) is governed by the following equation:

$$B_0 = \frac{\mu_0 I}{2\pi r}$$

where I is the current and denotes the rate at which electrons move through the wire, μ_0 is the permeability of the material and is a mathematical representation of "how easily" a magnetic flux can affect a material within its field, and r is the radius from the wire the magnetic field expands. Notice that the denominator of the equation contains the factors for a circular area, $2\pi r$. This is important for two reasons. First, it proves that the magnetic field is circular in shape when it is produced from a current holding wire. Second, because the radius variable is in the denominator, an increase in the

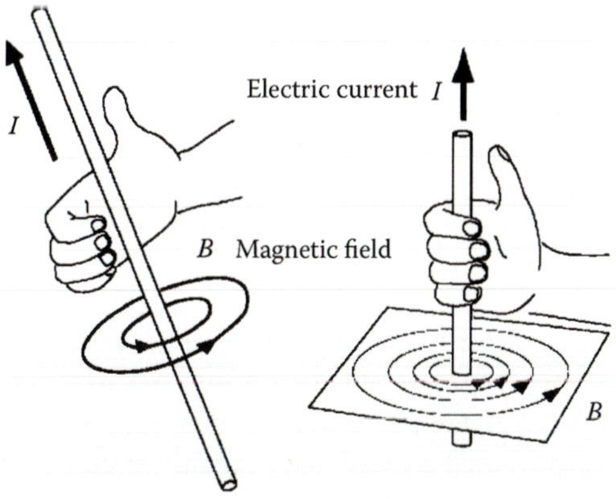

FIGURE 34.3 Propagation of magnetic field caused by flow of current in the wire. Flow of current (I) is in the direction of the thumb, and the magnetic field (B) propagates in the direction the fingers as they move toward the palm of the hand. Internet source: http://hyperphysics.phy-astr.gsu.edu/hbase/magnetic/magcur.html#c2.

radius will result in a decrease of the magnetic flux B_0. This implies that the farther away from the wire, the weaker the electromagnetic field (flux) will be. Factors such as this are of great importance when designing medical applications using these electromagnetic fields.

When creating a magnetic field by inducing a current flow in a wire, it is important to be able to change the intensity of the field to perform a useful action. For instance, when a patient is placed within an MRI machine, the operator must be careful to increase the field intensity in increments to begin the scan and decrease the intensity incrementally once the scan is complete. If the field is collapsed too fast at the end of the scan, a current can be induced in parts of the body that can damage tissues. The heart is a great example of this. The autorhythmic cells of the AV and SV nodes within the heart serve as pacemakers by setting the rate of atrioventricular contractions. Action potentials occur automatically when voltage triggered I_f channels open when the autorhythmic cells repolarize to –60 mV. The induction of intense enough currents from a magnetic field within these cells can temporarily disrupt the rhythm of the heart and in extreme cases lead to cardiac arrest. For these reasons, it is important to control the intensity of the EMF, especially when applying it to a biological system. This is done in two ways: (1) by increasing the current as discussed earlier and (2) coiling wires together so that their magnetic fields sum, much like electrical action potentials can sum at ganglia to increase their effect.

CELLULAR ELECTRICAL PROPERTIES

One must consider the electronic properties of the cells and tissues when designing electrotherapy treatments. Cells have electrostatic forces due to the charged ions within the cell creating an electric field creating the cell's permittivity, the ability to form an electric field, and its conductivity, the ability to conduct an electric current. Cells also have electromotive forces to maintain a required potential difference such as voltage, current, resistance, and capacitance. All these properties are found in developing, wounded, normal, and diseased tissues, and are the center of various technologies such as dielectrophoresis, electroporation, and impedance spectroscopy.

Cells also control the exchange of electrically charged ions across their membrane creating a membrane potential with different amounts of electrical charges inside and outside the cell. Thus, the structure of the cellular membrane plays an important role in the electronic properties of the cell. The cell membrane creates the boundary between the regulated internal environment of the cell and the external environment. The membrane is constructed from molecules with long fatty chains with electrically charged head groups in contact with the surrounding water allowing it to be water soluble. Within the membrane structure are various types of proteins responsible for a variety of cellular functions. The integral proteins, passing through the cell membrane, and the peripheral proteins, lining the inside and outside of the membrane, give the cell control of motion, transport of molecules, and adhesion to surfaces. The integral proteins are the ones that form pumps that actively transport ions across the cell membrane developing the transmembrane potential. Because the membrane is semipermeable to certain ions, a concentration gradient forms causing a potential to develop across the membrane as defined by the Nernst equation:

$$E_s = \frac{RT}{F} \ln \frac{[S]_2}{[S]_1}$$

where E_s is the equilibrium potential, $[S]_1$ and $[S]_2$ represent the concentrations of the ion S on each side of the membrane, R is the gas constant (8.315 J K^{-1} mol^{-1}), and F is Faraday's constant (9.648×10^4 C mol^{-1}). Concerning the cell membrane, the ions of interest are K^+, Na^+, Ca^{2+}, and Cl^- with potassium and chloride creating negative and sodium and calcium creating positive intracellular potential with respect to the extracellular potential. For most electrically active cells, there is a higher concentration of K^+ ions on the inside of the cell and a higher concentration of Na^+ and Cl^- on the outside, creating an overall negative transmembrane potential. Changes in the cell's membrane potential cause action potentials, electrical signals that send messages around the body to perform the appropriate reaction.

ELECTROTHERAPY FOR NEURAL TISSUES

It is important to understand how neural tissues interact with induced electromagnetic fields. In 1980, Cohen et al. validated the innate magnetic field strengths found in the human body ranging between 10^{-10} and 10^{-5} Gauss (G). The reason for this traces back to the properties of magnetism, which state that the flow of a charged particle will have a corresponding magnetic field propagating along the axis of its motion. Because all materials have atoms containing electrons (except for H^+), all materials made of atoms must have some electromagnetic footprint no matter how small. Cohen's group also found that the stronger fields were found in the brain and limbs with relative field strength of about 0.1 μG/cm (Cohen et al. 1980). The weaker fields were found around the torso, which makes sense when considering the conditions required for a magnetic field: movement of charge. The reason for the difference in field strength is attributable to the functions and properties of neurons.

The continuous firing of the neurons in the brain creates an overall magnetic field in a person. The brain has more neurons condensed in a single region than any other part of the body. These neurons fire countless times throughout the day. The constant depolarization and repolarization of these neurons tightly packed together can cause regions in the brain where the magnetic fields sum up or cancel out. The resulting electromagnetic pattern of a healthy individual brain varies depending on the arrangement of their neural network and on what neural pathways depolarize at any given time. Because the direction of the magnetic field depends on the direction of the flow of current, as neurons carry electrical messages to and from the brain (efferent and afferent signals, respectively), the projection of the field along the axis of hyperpolarization causes some canceling of waves and strengthening of magnetic waves in other regions. Therefore, by performing any activity, by having a thought, by feeling an emotion, or by reading, the magnetic patterns of the brain slightly change as the result of a change in the direction of the overall current flow. Figure 34.4 shows the magnetic field direction of a person at rest. The figures show the head and the forearms with their overall

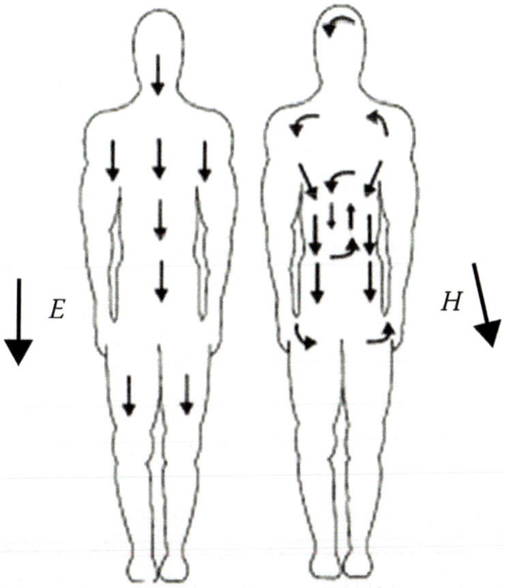

FIGURE 34.4 Effects of subjecting an entire body to an external magnetic field (*H*) can induce amplified magnetic flux within the body as denoted by arrows in the person to the right. When subjected to external magnetic field (*H*), a secondary electric field (*E*) and current can be produced within the body as shown on the left. (From Hatoum, N. Electromagnetic interfaces (EMI) and their effect on the nervous system. http://www.biotele.com/EMI.htm. With permission.)

magnetic flux. Both pictures contain the direction of the field a normal person exhibits and how applying an external magnetic field can influence the innate field of the body.

By controlling the direction, intensity, and frequency of an applied field to neural tissue, one can change the rates of depolarization and repolarization, allowing for suppression or stimulation of certain behaviors. Current research is using this phenomenon integrated with nanotechnology for many biomedical applications for neural regeneration, the treatment of neurodegenerative diseases and cancer.

NANOELECTRONICS FOR NERVE REGENERATION

Functional recovery of peripheral nerves in the body after damage is often poor and thus is a major focus of research in regenerative therapy. After a peripheral nerve is severed, there is only regeneration on the side of the lesion still connected to the cell body resulting in loss of sensory and motor function. There is limited success of reattaching completely to the detached section. Numerous in vivo studies that have been performed prove that the poor regeneration occurs when there is a larger distance between the two severed sections of the nerve (Terzis and Smith 1990; Sunderland 1978; Kim et al. 2003). Traditional methods of treatment for nerve repair are surgical excision of scarred nerve segments and graft repair usually with loss of regenerative ability of the nerve and loss of function at the donor nerve graft site (Kim et al. 2003). To improve on the grafting technique, various types of artificial nerve conduits were developed to stimulate nerve growth. Then brief electrical pulses were shown to promote axonal outgrowth. Initially, Hoffman et al. (1952) provided electrical stimuli to the severed spinal cords and noticed reinnervation and axon growth near the nerve root. Since then, applications of electrical stimuli were performed directly on the peripheral nerves (Nix 1983; Pockett 1985) with noticeable improved functional reinnervation.

With the influx of nanotechnology to medicine, advances are being made toward new technologies for nerve regeneration. Nanofibrous scaffolds have attracted a lot of attention for a broad range of biomedical applications particularly for the replacement of structurally or physiologically deficient tissues. The combination of nanotechnology and electrotherapy has led to the development of electrically conductive polymers that incorporate electrical signals directly into biomaterials that would provide the desired electrical field while the nerve tissue regenerates. Scaffold materials produced from nanofibers offer a large surface area that can support cell growth with the electrical stimuli to promote neuronal cell growth. Ghasemi-Mobarakeh et al. (2009) created electrospun conductive nanofibrous scaffolds for nerve cells that showed enhanced cell proliferation and neurite outgrowth compared to the scaffolds without electrical stimulation. The nanosized fibers act as physical guides for regenerating nerve fibers and enhance and direct the nerve growth.

Nanofibers are also being used to magnetically align collagen fibers to direct nerve growth. Collagen is a component of the cell's extracellular matrix, and it is a popular choice as a material in tissue engineering applications. It has been shown that in strong magnetic fields, the collagen fibers start to align in the direction of the field, and when seeded with neuronal cells such as Schwann cells and dorsal root ganglions, neurite extension was observed (Biazar et al. 2010; Eguchi et al. 2003).

Although treatments for damage to the peripheral nervous system may benefit from tissue engineering strategies that optimize neural and Schwann cell function, the treatments for damage to the central nervous system need to be approached differently because of the lack of tissue regeneration in the damaged area from the formation of glial scars. Magnetic nanoparticles are on the forefront of this research as they may be used to create mechanical tension that stimulates the growth and elongation of axons of neurons in the central nervous system (De Silva et al. 2007).

Synthetic nerve guidance channels (Figure 34.5) integrated with nanomaterials to reduce glial scar formation and promote axon growth and tissue regeneration are also being developed to promote regeneration (Seil 2010).

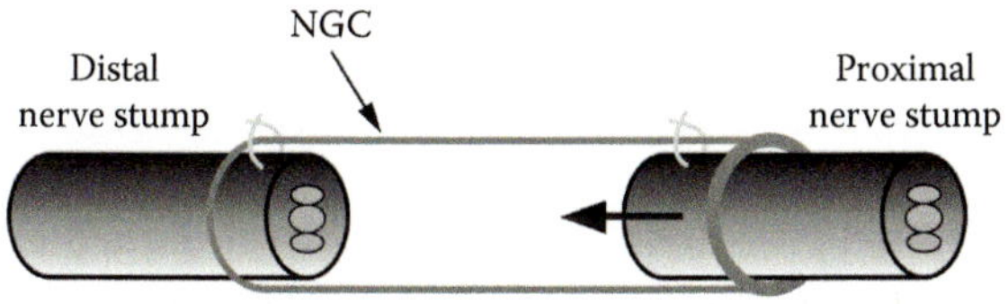

FIGURE 34.5 Illustration of a nerve guidance channel connecting two severed ends of a nerve. (Source: www.che.utexas.edu/~schmidt/overview/nerveguide.html.)

NANOELECTRONICS FOR NEURODEGENERATIVE DISEASES

Neurodegenerative disorders that lead to dementia, such as Parkinson's disease (PD) and Alzheimer's disease (AD), have been the current focus of much research. The common denominator for these diseases is the loss of neurons, and nanotechnology is being used to support and promote neurite growth.

PARKINSON'S DISEASE

PD is caused by the degradation of neurons within the midbrain leading to a decrease in levels of dopamine causing abnormal control of motion. The current forms of treatment involve supplementing the brain with more dopamine and inhibiting the enzymes that break down dopamine, with both methods involving traditional drugs such as Levodopa and Carbidopa. Advances in nanotechnology prove beneficial as carbon nanotubes (CNTs) are being researched for targeted drug delivery to the disease site, and in particular, they can be electrically stimulated causing the CNTs to dilate and provide a controlled release of the drugs (Abidian 2006).

Deep brain stimulation has been another potential treatment of the disabling motor symptoms of Parkinson's disease by using implanted electrodes. For example, Medtronic (a biomedical company) has produced the "Activa Parkinson's Control System," a subcutaneous neural stimulator made of four metallic electrodes that serve to stimulate the basal nuclei responsible for refining skeletal muscle movements of the body. The currents produced by the device are very small and yet have the similar effects as the drug treatments; however, this device is invasive, because it requires the electrodes to be placed under the scalp, pulled down the neck under the skin to a signal generator that is surgically implanted to the chest. CNTs are being used to replace the bulkier microelectrodes for deep brain stimulation (Kabanov and Gendelman 2007).

ALZHEIMER'S DISEASE

Nanotechnology is also promising for the diagnosis and treatment of AD. One of the major issues with diagnosing the disease is its mistaken identity. Most of the early stage symptoms are overlooked as normal age-related or stress-related changes, and ultimately the disease is only diagnosed well after the neural tissue begins to deteriorate. One avenue that biomedical researchers are taking for potential early diagnosis of AD is the detection of biomarkers for the plaque forming amyloid-β (Aβ) peptide. Nanoelectronic sensors have the potential to detect the presence of biomolecules, cells, or microbes at extremely low concentrations and thus would prove very useful for early diagnosis of AD. For example, Prabhulkar et al. (2012) designed an electrochemical immunosensor capable of simultaneously detecting multiple Aβ isoforms using the intrinsic electroactivity of Aβ.

Gold nanoparticles in conjunction with weak microwaves are also being considered for the removal of the Aβ plaques. Kogan et al. (2006) produced gold nanoparticles functionalized with an antibody that selectively binds to the Aβ peptide and under microwaves with energy levels six times smaller than those from standard cell phones, and noticed degradation in the Aβ fibrils with no damage to the surrounding healthy cells.

NANOELECTRONICS FOR CANCER THERAPY

Current advancements in cancer treatments such as surgery, radiotherapy, and chemotherapy have proven to be successful; however, these options have many side effects. Surgery is the best chance of cure for many cancers that have not spread; however, as with all surgeries, there are complications. Radiotherapy, the use of ionizing radiation to damage the DNA of exposed cells, kills cancer cells in and around a specific tumor; however, it also kills surrounding healthy cells, producing deleterious side effects known as radiation toxicities. Chemotherapy, the use of drugs to destroy cancer cells, is beneficial in destroying cancer cells that may have spread; however, like radiotherapy; it also kills the surrounding healthy cells.

Living cells are associated with electrical characteristics including resting membrane potential, ionic current flow, resistance, capacitance, permittivity, and conductivity and are thus responsive to and even generate electric fields and currents. The idea of classifying cancers by their electrical properties was first proposed by Fricke and Morse in 1926. Recently, there has been interest in the use of electrotherapy as a treatment for cancer because of the different electrical properties. Current treatments for cancer have much potential; however, a major limitation in these treatments is the negative side effects that occur. Electrotherapy for cancer treatment is a very promising alternative because it eliminates the toxic chemicals and possible immunogenic responses in the host tissue. Kirson et al. (2007) demonstrated that low intensity, intermediate frequency (100 kHz–1 MHz) alternating electric fields have detrimental effects on glioblastoma cells in vivo with the theory that the electric fields disrupt the normal mitotic process of proliferating cells.

CONCLUSION

The electrical properties of biological materials have been the focus of biomedical research for decades. Advancements in the fields of nanotechnology and electrotherapy, and their progeny field, nanobioelectronics, can greatly benefit areas of neuroregenerative medicine and the treatment of neurodegenerative diseases. Limiting factors in the current treatments can be overcome such as reducing the formation of scar tissue formed during the traditional nerve graft procedure. Nanobioelectronics has the potential to revolutionize current treatments in neurosurgery.

REFERENCES

Abidian, M.R., Kim, D.H., Martin, D.C. 2006. Conducting-polymer nanotubes for controlled drug release. *Adv Mater* 18: 405–9.

Barker, A.T., Dixon, R.A., Sharrard, W.J. et al. 1984. Pulsed magnetic field therapy for tibial non-union. Interim results of a double-blind trial. *Lancet* 5;1(8384): 994–6.

Bianco, A., Kostarelos, K., Prato, M. 2005. Applications of carbon nanotubes in drug delivery. *Curr Opin Chem Biol* 9(6): 674–9.

Biazar, E., Khorasani, M.T., Zaeifi, D. 2010. Nanotechnology for peripheral nerve regeneration. *Int J Nanomed* 1(1): 1–23.

Cohen, D. Palti, Y., Cuffin, N. et al. 1980. Magnetic fields produced by steady currents in the body. *Proc Natl Acad Sci U S A* 77(3): 1447–51.

De Silva, M.N., Almeida, M.V., Goldberg, J.L. 2007. Developing super-paramagnetic nanoparticles for central nervous system axon regeneration. *Nanotech* 2: 791–4.

Eguchi Y., Ogiue-Ikeda M., Ueno S. 2003. Control of orientation of rat Schwann cells using an 8-T static magnetic field. *Neurosci Lett* 351(2): 130–2.

Fantozzi, F., Arturoni, E., Barbucci, R. 2010. The effects of the electric fields on hydrogels to achieve antitumoral drug release. *Bioelectrochemistry* 78(2): 191–5.

Faraji, A. H., Wipf, P. 2009. Nanoparticles in cellular drug delivery. *Bioorg Med Chem* 17(8): 2950–62.

Fricke, H., Morse, S. 1926. The electric capacity of tumors of the breast. *J Cancer Res* 10: 340–76.

Ghasemi-Mobarakeh, L., Prabhakaran, M.P., Morshed, M., Nasr-Esfahani, M.H., and Ramakrishna, S. 2009. Electrospun poly (ε-caprolactone)/gelatin nanofibrous scaffolds for nerve tissue engineering. *Biomaterials* 29: 4532.

Golzio, M., Rols, M.P., Teissie, J. 2004. In vitro and in vivo electric field-mediated permeabilization, gene transfer, and expression. *Methods* 33(2): 126–35.

Harrison, B.S., Atala, A. 2007. Carbon nanotube applications for tissue engineering. *Biomaterials* 28(2): 344–53.

Hatoum, N. Electromagnetic interfaces (EMI) and their effect on the nervous system. http://www.biotele.com/EMI.htm.

Hoffman, H. 1952. Acceleration and retardation of the process of axon-sprouting in partially denervated muscles. *Aust J Exp Bioi Med Sci* 30: 541–66.

Ieran, M., Zaffuto, S., Bagnacani, M., Annovi, M., Moratti, A., Cadossi, R. 1990. Effect of low frequency pulsing electromagnetic fields on skin ulcers of venous origin in humans: a double-blind study. *J Orthop Res* 8(2): 276–82.

Kabanov, A.V., Gendelman, H.E. 2007. Nanomedicine in the diagnosis and therapy of neurodegenerative disorders. *Prog Polym Sci* 32: 1054–82.

Kim, D.H., Cho, Y.J., Tiel, R.L. et al. 2003. Outcomes of surgery in 1019 brachial plexus lesions treated at Louisiana State University Health Sciences Center. *J Neurosurg* 98: 1005–16.

Kim, J., Lee, J.E., Lee, S. H. et al. 2008. Designed fabrication of a multifunctional polymer nanomedical platform for simultaneous cancer-targeted imaging and magnetically guided drug delivery. *Adv Mater* 20(3): 478–83.

Kirson, E.D., Dbaly, V., Tovaryš, F. et al. 2007. Alternating electric fields arrest cell proliferation in animal tumor models and human brain tumors. *Proc Natl Acad Sci* 104(24): 10152–7.

Kogan, M.J., Bastus, N.G., Grillo-Bosch, R.D. et al. 2006. Nanoparticle-mediated local and remote manipulation of protein aggregation. *Nano Let.* 6: 110–15.

Nix, W., Hopf, H. 1983. Electrical stimulation of regenerating nerve and its effect on motor recovery. *Brain Res* 272: 21–25.

Nowak, K., Mix, E., Gimsa, J. et al. 2011. Optimizing a rodent model of Parkinson's disease for exploring the effects and mechanisms of deep brain stimulation. *Parkinson's Disease* 2001: 414682.

Pockett, S., Gavin, R.M. 1985. Acceleration of peripheral nerve regeneration after crush injury in the rat. *Neurosci Lett* 59: 221–24.

Prabhulkar, S., Piatyszek, R., Cirrito, J.R. et al. 2012. Microbiosensor for Alzheimer's disease diagnostics: detection of amyloid beta biomarkers. *J Neurochem* 122: 374–81.

Sandyk, R. 1997. Therapeutic effects of alternating current pulsed electromagnetic fields in multiple sclerosis. *J Altern Complement Med* 3(4): 365–86.

Schoenbach, K.H., Nuccitelli, R., Beebe, S.J. Zap. http://www.biotele.com/bioelectric.htm.

Seil, J.T., Webster, T.J. 2010. Electrically active nanomaterials as improved neural tissue regeneration scaffolds. *Wiley Interdiscip Rev Nanomed Nanobiotechnol* 2(6): 635–47.

Shi, D. 2009. Integrated multifunctional nanosystems for medical diagnosis and treatment. *Adv Funct Mater* 19(21): 3356–73.

Slowing, I.I., Trewyn, B.G., Giri, S. et al. 2007. Mesoporous silica nanoparticles for drug delivery and biosensing applications. *Adv Funct Mater* 17(8): 1225–36.

Song, L., Chau, L., Sakamoto, Y. et al. 2004. Electric field-induced molecular vibration for noninvasive, high-efficiency DNA transfection. *Mol Ther* 9: 607–16.

Sunderland, S. 1978. *Nerve and Nerve Injuries*, 2nd edn. New York: Churchill Livingstone.

Teissie, J., Escoffre, J.M., Rols, M.P. et al. 2008. Time dependence of electric field effects on cell membranes. A review for a critical selection of pulse duration for therapeutical applications. *Radiol Oncol* 42: 196–206.

Terzis, J.K., Smith, K.L. 1990. *The Peripheral Nerve: Structure, Function and Reconstruction*. New York: Raven Press.

35 In Vitro and In Vivo Techniques to Assess Neurotoxicity of Nanoparticles

Thomas Birngruber, Eleonore Fröhlich, Frank Sinner, and Thomas R. Pieber

CONTENTS

INTRODUCTION

The basic idea of nanotoxicology is that the small size of nanoparticles (NPs) is the main cause for their adverse biological effects. Many materials that are cell compatible in bulk form (e.g., polystyrene used for cell culture) may become cytotoxic in sizes ≤20 nm (Fröhlich et al. 2009). NPs can be used as fluorescent biological labels—such as drug and gene delivery agents, biodetectors for pathogens, probes for DNA structure, for separation and purification of biological molecules and cells—and in tissue engineering, tumor destruction via heating, and magnetic resonance imaging (MRI) contrast enhancement (Bhaskar et al. 2010). In the central nervous system (CNS), the blood–brain barrier (BBB) functions as a protective shield against harmful substances such as viruses and toxins but also against potentially harmful NPs circulating in the bloodstream. The BBB consists of tight junctions between endothelial cells around the capillary lumen (Reese and Karnovsky 1967) (Figure 35.1) protecting the brain but also preventing potential CNS therapeutic agents from entering the brain (Miller 2002; Pardridge 2005). Some NPs can cross the BBB and have been identified as possible transporter systems to carry CNS active drugs across it. BBB permeability and toxicity of drug candidates and NPs are important parameters during CNS drug development and must be considered at an early stage of pharmaceutical development. The

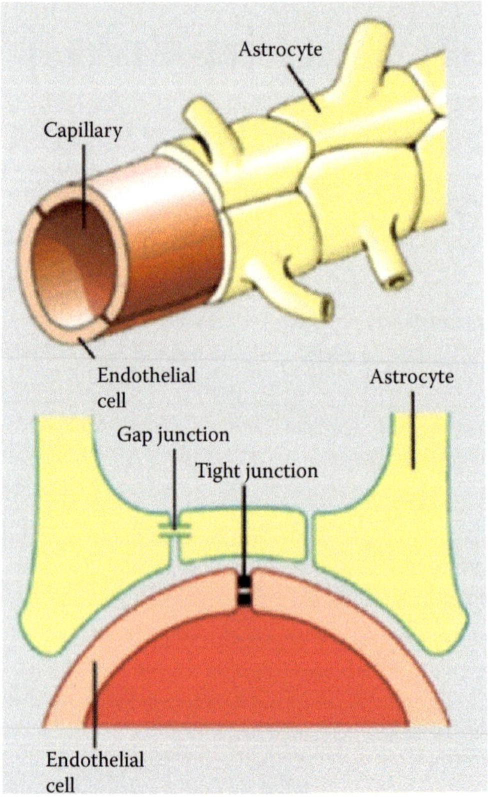

FIGURE 35.1 The BBB consists of tight junctions on the vascular endothelial cells in the brain. (Reprinted with permission from Speckman, E.-J. et al., *Physiologie*, Urban & Fischer, Munich, 2008.)

ability of NPs to cross the BBB may also be useful for detection and therapy of brain pathologies. Whether NPs are used in terms of a pharmaceutical formulation or are taken up unintentionally from the surroundings as a result of pollution, the question of how NPs interact with brain cells is highly interesting.

NPs entering the brain come into contact with five brain-specific cell types: neurons, astrocytes, oligodendrocytes, ependymal cells, and microglial cells (Figure 35.2). Astrocytes and microglial cells are the most important populations for evaluating neurotoxicity in vitro based on their role in the brain (transmission of information, protection of neurons, and regulation of the immune system) and their quantitative representation around brain neurons. Nano-sized materials such as single-wall carbon nanotubes have been reported to cause oxidative stress in neurons leading to apoptosis and necrosis (Zhang et al. 2010) and to decrease viability not only in neurons but also in glial cells (Belyanskaya et al. 2009). Cells treated with nano-sized titanium dioxide have been reported to show intracellular accumulations of reactive oxygen species (ROS) leading to apoptosis, necrosis, and cell cycle arrest (Liu et al. 2010; Wu et al. 2010). U87 astrocytoma cells can suffer from cytotoxic effects after exposure to titanium dioxide, zinc oxide, and magnesium oxide NPs (Lai et al. 2008), whereas iron oxide NPs can interfere with the adherence of astrocytes to the substratum (Au et al. 2007).

In this chapter, we provide an overview of currently available in vitro and in vivo screening techniques for assessing the cytotoxicity of NPs in general and in relation to brain-specific cells.

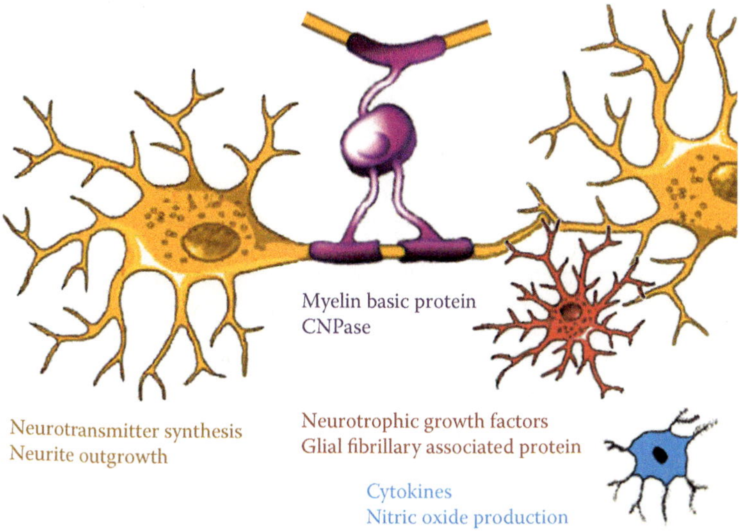

FIGURE 35.2 Important cell types in the brain and their main functions that could be used for specific neurotoxicity testing. For assessing neurons (yellow), the synthesis of neurotransmitters (e.g., tyrosine hydroxylase for DOPA synthesis) and neurite outgrowth are useful parameters. Function of oligodendrocytes (pink) can be assessed by expression of myelin basic protein and 2′,3′-cyclic nucleotide 3′-phosphodiesterase (CNPase). Status of astrocytes (red) can be monitored by neurotrophic growth factor expression and glial fibrillary acidic protein expression. Changes in microglia cells (blue) can be studied in cytokine and nitric oxide production.

IN VITRO TECHNIQUES

Basal Cytotoxicity Screening

Screening for basal cytotoxicity is the first toxicological assessment in drug development, and it can also be used to measure nanotoxicity (Table 35.1). Basal cytotoxicity screening assays (bCSAs) are used to identify any negative influence on cell metabolism and viability by monitoring cell functions common to all cell types such as proliferation, organelle function, and cell death. Basal cytotoxicity screening assays have been used for drug compound testing for many years to identify any potentially damaging effects. In these assays, IC_{50} values (compound concentration causing the half-maximal effect) in human cell lines show a strong correlation with LD_{50} values in blood (compound concentration death in 50% of exposed subjects) (Shrivastava et al. 1992). For bCSAs, well-characterized commercially available cell lines are used because they show higher reproducibility of growth and stimuli response than primary cells, which have to be freshly isolated from animals or humans for each experiment. Primary cells can also vary in quality because of the isolation procedure and donor differences; furthermore, they cannot be maintained in culture for a prolonged period. On the other hand, primary cells retain more cell-specific characteristics than cell lines and are therefore used when cell-specific functions are assessed. Most bCSAs are based on determining cell number (cell counting), cellular products (proteins, DNA), enzyme activities, and organelle function (Table 35.1). bCSAs are also used to detect disruption of membrane integrity, induction of apoptosis, inhibition of proliferation, and oxidative stress.

Assays for *membrane integrity* use either detection of intracellular enzymes (e.g., lactate dehydrogenase, adenylate kinase) in the supernatant of cultured cells or the uptake of cell-impermeable dyes into the cells. For dye uptake, the fluorescent dyes propidium iodide, ethidium

TABLE 35.1
Detection Principles for Most Commonly Used Basal Cytotoxicity Assays

	Basic Parameters
Cell number	Coulter counter, CASY Technology, xCELLigence®
Protein content	Sulforhodamine B staining, leucine incorporation
DNA content	Nuclear dyes, especially PicoGreen[a]
Metabolism	ATP content
Enzyme activity	Dehydrogenase activity (formazan bioreduction), esterase activity
Organelle function	Neutral Red uptake
	Membrane Integrity
Dye exclusion	Lactate dehydrogenase, adenylate kinase
Dye uptake	Propidium iodide, ethidium homodimer-1, 7 Z-AAD
	Apoptosis
Membrane changes	Phosphatidylserine exposure: Annexin V-binding, increase in membrane permeability: Yo-PRO-1 uptake
Mitochondrial changes	Mitochondrial membrane potential, cytochrome C translocation
Caspase activation	Caspase 3/7 activation
Cytokeratin changes	Cytokeratin 18 fragmentation
DNA changes	Reparation: Detection of poly(ADP-ribose)polymerase activity or protein, condensation and degradation: terminal deoxynucleotidyl transferase-dUTP nick end labeling (TUNEL) assay
	Proliferation
Incorporation of labeled nucleotides	^{3}H-thymidine, immunocytochemical detection of BrdU
	Production of Reactive Oxygen and Oxidative Cell Damage
Cellular oxidative stress	Glutathione, catalase, superoxide dismutase, hemoxigenase-1 expression
Reactive oxygen detection	Rhodamine 123, lucigenin, dihydrodichlorofluorescein, dihydroethidium, cytochrome *C*
Detection of oxidation products	Lipid peroxidation (malondialdehyde and 4 hydroxynonenal), DNA (8-hydroxy deoxyguanine)

[a] There are many other nuclear dyes, e.g., Hoechst, SYTO and TOTO dyes to stain nuclei but PicoGreen has the advantage of binding only to double-strand DNA and not to single-strand DNA and RNA and therefore detects only the nuclear volume.

homodimer-1 (Figure 35.3), 7-AAD, TO-PRO-3, TOTO-3, SYTOX dyes, all binding to DNA, are used. Alternatively, simple color detection with trypan blue and eosin can be performed.

Apoptosis can be identified in many different ways. On the one hand, morphological changes can be used (e.g., cell shrinkage, chromatin condensation, membrane blebbing, cell fragmentation, formation of apoptotic bodies, lysis or phagocytosis of apoptotic bodies). On the other hand, biochemical changes can quantify changes in the mitochondrial membrane, caspase activation, and plasma membrane permeability, or detect DNA strand breaks and cytoskeleton breakdown. Usually, more than one assay is performed because apoptotic events do not occur continuously, and one specific assay at a certain time point may not detect the total apoptotic effect: for example, at different time points increased caspase activity indicating apoptosis can be noted at varying levels (Figure 35.4a). Almost all phases of apoptosis can be detected using cytokeratin 18 fragmentation where signals are present over a longer period (Figure 35.4b).

Proliferation is evaluated by incorporating either fluorescent or radioactive-labeled nucleotides. Radioactive thymidine uptake is still regarded as the gold standard to assess cell proliferation and

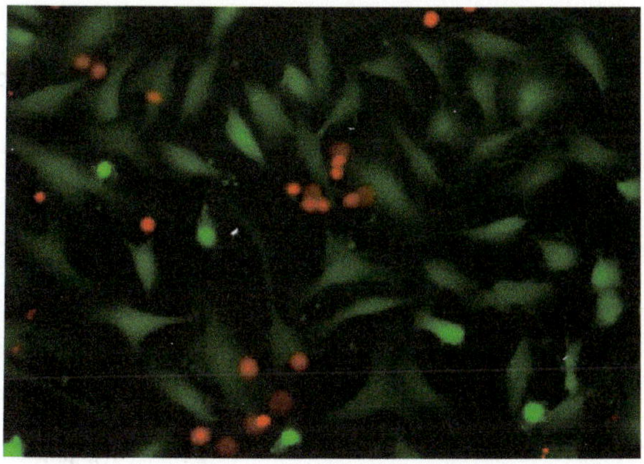

FIGURE 35.3 Detection of living and dead cells according to metabolization of calcein acetoxymethyl ester (green) for vitality and ethidium homodimer-1 (red) uptake for loss of membrane integrity. Upon exposure to 20 nm carboxyl polystyrene particles, several cells (red) show loss of membrane integrity.

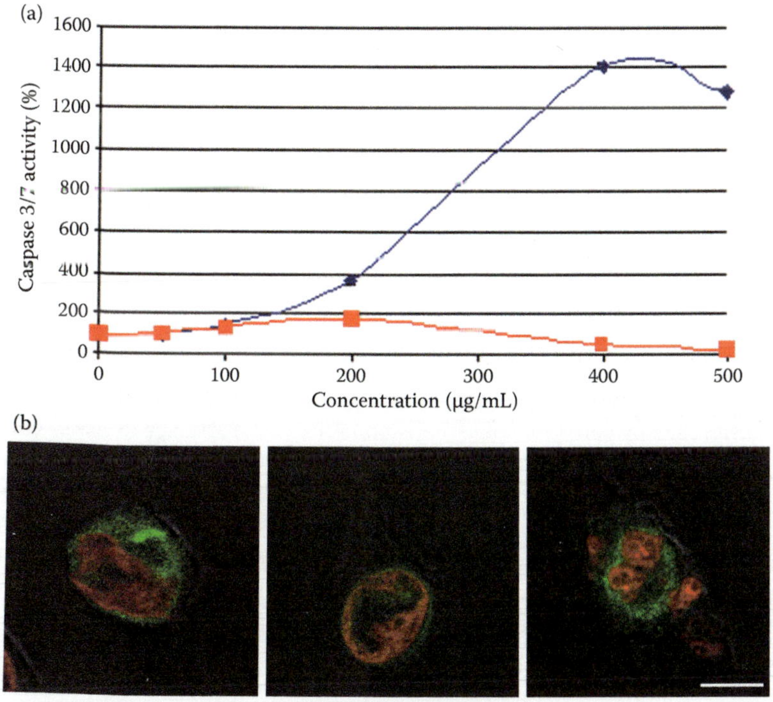

FIGURE 35.4 Assessment of apoptosis by caspase 3/7 activation (a) and by cytokine fragmentation (b). (a) Caspase 3/7 activation after 4 h (blue curve) is much higher than after 24 h (red curve) in the presence of 20 nm carboxyl polystyrene particles. The different shapes of the two curves show the time dependency of apoptotic events. (b) Immunocytochemical detection of cytokeratin fragmentation by M30 antibody (green) identifies cells in various phases of apoptosis. The staining is present at the early phase of apoptosis when only a few nuclear changes are seen (left); at the stage where nuclear condensation has occurred (middle); and at the late stage, when fragmentation of the nucleus is present (right). Nuclei are stained with propidium iodide (red). Scale bar: 10 μm.

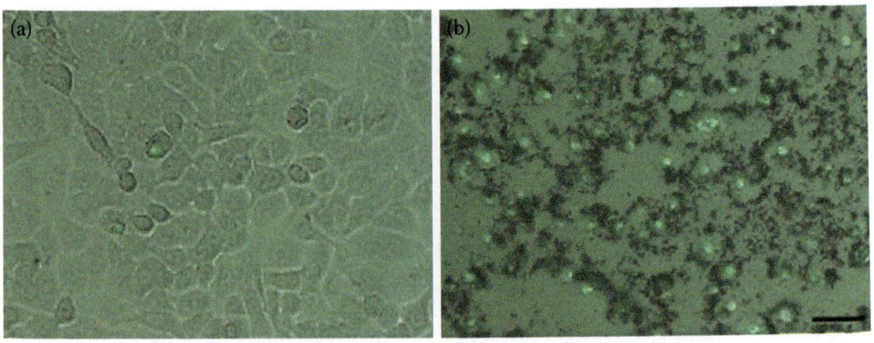

FIGURE 35.5 Detection of oxidative stress by DCF (green) in A549 cells exposed to high concentrations of quartz particles (500 μg/mL). Pictures of fluorescence and bright field are merged. Oxidative stress seen as green fluorescent cells in untreated cultures (a) is rare. Upon exposure to quartz particles (b), oxidative stress in cells is markedly increased. Quantification of the fluorescent signal using fluorescent readers is complicated by fluorescence quenching caused by the particles.

it is also the more sensitive technique, but fluorescent detection of BrdU by antibodies is frequently used because it avoids the need for radioactive material.

Oxidative stress, one of the key modes of action of NPs, can be studied by (1) detecting intracellular radicals, mainly superoxides; (2) converting substrates such as dichlorodihydrofluorescein, dihydroethidium, cytochrome *C*, lucigenin, and luminal; (3) molecules involved in the oxidative stress reaction (glutathione, superoxide dismutase, catalase, hemoxigenase-1); and (4) detecting cellular oxidation products (malondialdehyde, 4-hydroxynonenal and 8-hydroxy deoxyguanine). However, in all oxidative stress assays, quantification in the presence of high NP concentrations can be difficult (Figure 35.5).

To assess cytotoxicity of NPs, adherent cells are usually preferred to cells growing in suspension, mainly because NPs are most often designed for cancer-cell treatment and regenerative medicine where epithelial cells and fibroblasts, both adherently growing, are the most relevant cell types. Many NPs stick to these cells and influence the readout, for instance, by color absorption. A major advantage during NP cytotoxicity testing is that adherent cells can be washed and NPs can be removed. Generally, a 96-well plate with 4- and 24-h exposure is adequate for testing. In addition to the usual controls (blank, growth control, solvent control), several wells should be reserved to test for NP interference with assay compounds and readout (NPs in the absence of cells) but also for NP interference by cell adherence (cells and NPs in the absence of assay compounds). For interference control, NPs are used as positive and negative controls because, in contrast to conventional drugs, NPs may cause problems in bCSAs by interfering with color, chemical reactivity, light scattering, and physical absorption of assay compounds. Carbon nanotubes and mesoporous silica particles are especially known to produce false positive or false negative results in most bCSAs (Casey et al. 2008; Fisichella et al. 2009; Monteiro-Riviere et al. 2009; Worle-Knirsch et al. 2006).

The MTT assay, a frequently used screening assay based on the intracellular reduction of 3-(4,5-dimethylthiazol-2-yl)-2,5-diphenyltetrazolium bromide to a colored product, is especially affected by NP interference (Worle-Knirsch et al. 2006). Other assays based on esterase activity such as calcein acetoxymethyl ester (AM) metabolization can be combined with membrane integrity assessment (Figure 35.3) to obtain additional information from a single experiment.

Specific Cytotoxicity Screening

Some NPs impair cell function at concentrations that do not decrease viability (Fröhlich et al. 2010). Specific cytotoxicity screening assays (sCSAs) address the effects of NPs on specific cell functions.

Potential specific cytotoxicity assays for brain cells are summarized in Table 35.2. In the brain, NPs can have cytotoxic effects on brain-specific cell types. For *neurons*, synthesis of transmitters and specific proteins (neuron-specific enolase, S-100) as well as neurite formation are relevant parameters. *Astrocytes* play an important role in brain structure (forming the BBB) and brain metabolism (creating the optimal milieu for neuron and oligodendrocyte function) and in the repair of brain lesions (glial scars). They produce a variety of peptides such as calcitonin gene-related peptide, brain-derived neurotrophic factor (BDNF), somatostatin, and vasoactive intestinal peptide. Upon activation, astrocytes increase expression of their main intermediary filament, glial fibrillary acidic protein (GFAP), which can serve as a toxicity parameter. *Oligodendrocytes* produce myelin for electrical isolation of axons in the brain. Production of myelin basic protein serves as an indicator of oligodendrocyte damage in vivo (Follett et al. 2004). The main functions of microglial cells include production of cytokines (mainly IFN-γ, TNF-α, IL-8, IL-1α and IL-1β) and chemokines (MCP-1), phagocytosis, and oxidative burst (H_2O_2 and NO secretion). Because NPs may induce inflammation and inflammatory processes in the brain by microglia activation and may promote neurodegenerative diseases associated with microglia degeneration (such as prion disease, schizophrenia, and Alzheimer's disease), assessing microglia activation is of primary importance.

Certain cell lines can be used for specific neurotoxicity screening; these cell lines include: glioma C6 cells for oligodendrite function (myelinization); PC-12 (pheochromocytoma cells) for neurotransmitter synthesis and release; NSC34 (hybrid murine neuronal cell line) for axonal transport, neurofilament organization, and action potential generation; and N1E115 (murine neuroblastoma cells) for differentiation (Harry et al. 1998). As markers for astrocyte function, GFAP and BDNF production of U251 (human glioma) cells can be studied (Kashima et al. 1993). Myelin basic protein expression as an indicator for oligodendrocyte function is lost in many cell lines but expressed upon differentiation in the MO3.13 (human oligodendroglial) cell line (Buntinx et al. 2003). Screening for microglia activation during inflammation is often performed in the murine microglial cell line BV2, which is a good model for microglial cells in mice but may be less representative for human microglia (Henn et al. 2009). Screening for inflammatory effects using a human, non-microglial cell line is another option to assess activation of microglia. For brain-targeted NPs, increased cytokine secretion (IL-1b) and ROS levels in microglial cells exposed to silica particles can also be relevant (Choi et al. 2010).

Specific changes in cell function and cell damage have also been reported after exposure to NPs in vivo, for instance, astrocyte swelling and neuronal degeneration after administration of silver NPs (Tang et al. 2009) or depletion of antioxidant capacity in addition to changes in the neurotransmitter release pattern after intraperitoneal application of TiO_2 NPs to mice (Ma et al. 2010). Intraperitoneal injection of Al_2O_3 particles caused glia cell activation according to increased ED1 in microglia, nesting in progenitor cells, and GFAP staining in astrocytes (Li et al. 2009). These data suggest that assessing brain-targeted NPs has to include not only basal but also specific cytotoxicity assays.

NPs may also interfere with sCSAs: the secretion of protein can be underestimated because many NPs adsorb proteins. This could lead to an underestimation of the inflammatory response

TABLE 35.2
Cell Types Used in Specific Neurotoxicity Assays

Astrocytes	BDNF, GFAP expression (e.g., in U251 cells)
Oligodendrocytes	Major basic protein (e.g., MO3.13 cells)
Neurons	TH immunoreactivity (e.g., in PC-12 cells)
	Neurite outgrowth (e.g., in PC-12, SY-SY5Y cells)
	Axonal transport (e.g., in NSC-34 cells)
Microglia	IL-1 secretion, NO production, phagocytosis (e.g., in BV2, EOC-2 cells)

based on cytokine secretion. Interference of NPs with color, fluorescence, and light scattering of specific CSAs can similarly occur for basic CSAs.

IN VIVO TECHNIQUES

In vivo studies on pharmacokinetics (PK) and pharmacodynamics (PD) are a reliable source of information during drug development and toxicological studies, and are the last step before a clinical trial. Target-tissue toxicity is difficult to predict accurately in vitro but is a special focus of in vivo studies in addition to the ability of NPs to cross the BBB (PK) and the tissue response to the presence of NPs (PD). Several non-invasive and invasive methods are available to evaluate PK and PD of NPs in the brain:

- Histopathology
- Imaging techniques: positron emission tomography (PET), MRI, and magnetic resonance spectroscopy (MRS)
- Microdialysis (MD) and open flow microperfusion (OFM)
- Behavioral tests

HISTOPATHOLOGY

Histopathology remains the gold standard of neurotoxicity testing. There is a constant search for better neuropathological techniques for in vivo studies. The first step after brain extraction is tissue fixation. In routine studies, immersion fixation, paraffin embedding, and staining with hematoxylin and eosin (H&E) are performed. Perfusion fixation has the benefit of faster distribution of fixation media but if perfusion pressure is too high, blood vessels can rupture and cause artifacts. Sections from different regions of the brain should be made because composition and cell type varies among gray and white matter, cerebrum, cerebellum, and brain stem. Regions around the third and fourth ventricles are of special interest because they are supplied by blood vessels with a higher permeability allowing a higher exchange rate between blood and brain. Specific regions of the brain are prepared in consecutive sections, a tricky task demanding a skilled operator. Metal templates or molds for slicing provide assistance in a standardized production of sections. After tissue sectioning, a number of staining and immunohistochemical methods can be applied: GFAP is used to detect glial cells in paraffin-embedded sections in rodents and humans. S100 protein is used to detect neuronal cells such as Schwann cells, glial cells, and neurons. The presence of cytokines such as IL-1α, IL-1β, IL-8, IFN-γ, and TNF-α mainly provides information on the inflammatory status of the tissue and therefore toxic reactions. These markers bridge the gap between in vivo and in vitro tests because they are used in both. Additionally, in vivo sections contain histological information on cell necrosis, apoptosis, reactive gliosis, and other possible reactions to NPs.

Histological analysis of brain tissue requires brain tissue extraction at the end of the experiment and therefore sacrificing the subject. Time-resolved experiments necessitate the sacrifice of several animals per time point, which results in high animal usage.

IMAGING TECHNIQUES

Positron Emission Tomography

Positron emission tomography is a noninvasive technique that can be used to measure NP-induced changes in the brain metabolism of humans and animals (Brooks et al. 1984; Josserand et al. 2006). The subject with the incorporated marker is positioned in a detector that counts emission of gamma photons emitted by the tracer. A two-dimensional (2-D) or 3-D image of the brain can be generated in real time by applying computer-supported imaging techniques (Dingemanse et al. 1988; Hilbert

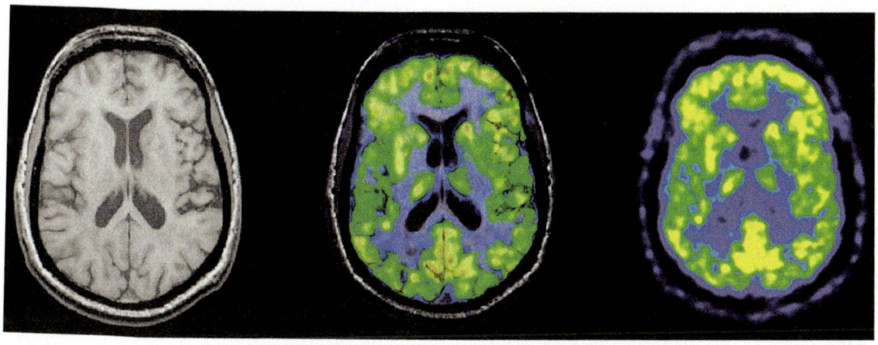

FIGURE 35.6 MR image (left) and PET image (right) and are merged to an image incorporating anatomical and metabolic information (center). Siemens Reference Number: somed200701-01.

and Battista 1991). Altered glucose metabolism, for instance, can be recorded by (18F)fluoro-deoxy glucose. PET provides high sensitivity and exact time resolution; in addition, pharmacological compartment modeling can be used to assess PK. PET scans are increasingly combined with other imaging techniques such as MRI or CT to acquire both anatomical and metabolic information (Figure 35.6). This combination has many potential applications for brain imaging. Limitations of PET arise from the high cost of the cyclotrons necessary to produce short-lived radionuclides for PET scanning and the need for a specially adapted on-site chemical-synthesis apparatus to produce radiopharmaceuticals. Because no distinction can be made between parent compound and metabolites, directly monitoring a toxicological reaction to NPs is limited.

MRI and MRS

MRI is a sensitive, non-invasive technique for investigating a range of clinical pathologies such as multiple sclerosis, stroke, and brain tumors (Ewing et al. 2003; Patlak et al. 1983). Contrast agents make it easier to identify changes in blood perfusion and morphology. MRS can differentiate a wide range of chemical substances in the body but has the disadvantage of poor spatial resolution compared to MRI. A combination of both technologies can provide information about the distribution of NPs as well as information about PD during toxicological analyses. Similar to PET scans, MRI and MRS are very cost-intensive methods, which limit their applicability as screening methods.

Analytical Techniques

Microdialysis

Intracerebral MD is an invasive technique for the continuous measurement of toxicity markers in extracellular tissue space. A cylindrical, semipermeable membrane is implanted into the target area in the brain and perfused with an artificial physiological solution, the perfusate (Elmquist and Sawchuk 1997). The perfusate is ideally closely matched to the ionic composition and the pH value of the extracellular fluid (ECF) of the brain to avoid changes in brain physiology that could interfer with the BBB (de Lange et al. 2000). The membrane allows substance exchange between the ECF and perfusate according to concentration gradients and diffusion parameters. The perfusate inside the membrane is continuously exchanged—the standard setup consists of an active inflow, driven by a syringe pump; the outflow is passive and the perfusate (enriched with components from ECF) is collected in a vial, which is replaced periodically. A constant perfusate flow through the membrane allows continuous, parallel measurement of several markers and metabolites, and also allows differentiation between parent compounds and metabolites. Toxicity markers such as the above-mentioned S100B and cytokines are accessible by MD. Continuous measurement reduces the number of animals used and statistical problems such as intersubject variability. MD allows the simultaneous and time-resolved determination of cerebral ECF and plasma levels, which enables kinetic and dynamic monitoring of substances. To investigate a specific brain region, the MD probe can be implanted stereotactically with high local accuracy; several probes can be implanted.

A drawback of MD implantation of the probe damages cerebral capillaries and consequently affects BBB function. Different time spans (hours up to days) for BBB reestablishment have been reported (de Lange et al. 1997; Groothuis et al. 1998; Westergren et al. 1995). A short time span between probe implantation and start of measurement may result in decreased local blood flow (Benveniste et al. 1987). Moreover, the membrane is the limiting factor in the exchange between tissue and perfusate because substances larger than the pore size of the membrane and lipophilic substances cannot pass the membrane and are subsequently excluded from the analysis (Rosenbloom et al. 2005). Finally, accurately determining the relationship between concentrations of the drug of interest in the solution that exits the dialysis probe and concentrations in the tissue (recovery) can be difficult. The in vitro recovery is inaccurate because in vitro results usually exceed in vivo measurements. There are, however, two methods for circumventing the recovery problem: no-net flux, which delivers a reliable outcome but is very time consuming, and retro dialysis, where a substance with similar diffusion behavior is added to the perfusate and the loss to the tissue of this substance is detected.

Open Flow Microperfusion (OFM)

OFM is a recently developed method with a working principle similar to MD (Bodenlenz et al. 2005; Schaupp et al. 1999), in that it offers the possibility of continuous measurement in healed tissue with an intact BBB. The main difference is that OFM incorporates no membrane but, rather, an exchange area with macroscopic perforations of 100 μm or larger (Figure 35.7) that allow the sampling of substances from tissues without limitations due to size, protein binding, charge, or lipophilicity (Ellmerer et al. 2000; Ikeoka et al. 2010; Pachler et al. 2007). Many CNS-active substances are lipophilic because the BBB allows the transcellular diffusion of small lipophilic substances. Therefore, of sampling lipophilic substances with OFM offers many benefits over MD. OFM requires two pumps operated at the same flow rate, for injection and withdrawal of the perfusate, since the open

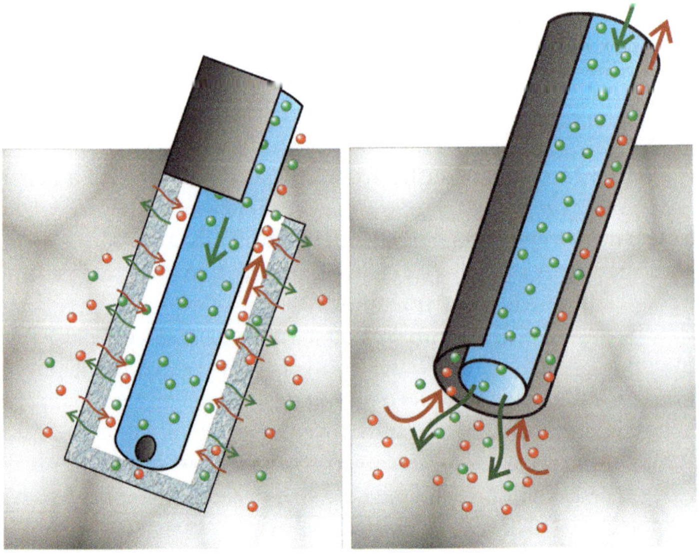

FIGURE 35.7 Working principle of MD (left) and OFM (right): MD uses a membrane as tissue interface. OFM provides a more direct contact to the cerebral tissue.

structure provides less protection for internal microfluids compared to the MD membrane. OFM overcomes the limitation of MD membrane occlusion and pore narrowing that can arise through membrane exposure to proteins (Rosenbloom et al. 2006). Further, it shows no relevant formation of a trauma layer, gliosis, or scar tissue around the probe, which has been observed in MD (Benveniste 1989). The absence of these limitations with OFM allows a long healing period to be used between OFM implantation and measurement start; therefore, full BBB recovery can be expected. Moreover, repeated measurements can be performed over several days without change in measurement performance, thereby promoting OFM use in long-term toxicity testing. The OFM probe can also be used to introduce NPs directly into the brain, bypassing the BBB, and allowing separate testing of transport across the BBB and nanotoxicity in brain tissue.

Behavioral Tests

One aspect of brain toxicology is the change in the behavior of animals. Two types of behavioral tests can be made: stimulus-oriented and internally generated behavior tests (Hodgson 2004). Stimulus-oriented behavior involves conditioned and unconditioned experimental setups in which the response to a particular stimulus is tracked. The conditioned test involves training to perform an action in order to receive a reward or avoid pain. Internally generated behavioral tests involve monitoring animal behavior in experimental situations such as social behavior, circadian behavior, and exploratory behavior. The performance of subjects treated with NPs is compared to that of a control group. Groups must be matched in terms of age, sex, species, and housing conditions. Combined with other techniques, behavioral variations represent an integrated whole organism response that is mostly nonspecific but sensitive and can increase our understanding of the toxicological effects of NPs in the CNS.

CONCLUSIONS

For a good assessment of the toxic effects of NPs in the brain, both in vitro and in vivo studies are needed. High-throughput testing in the early stages of pharmaceutical development demands in vitro methods for screening. Cytotoxicity screening methods can assess a large number of samples but they lack the complexity of in vivo systems. Preclinical in vivo testing is more time consuming and more expensive than in vitro testing, but it represents the distribution, metabolic, and elimination processes occurring in the body in a far better manner.

For in vivo assessment, histopathology is still the gold standard, but analytical methods such as MD and OFM offer the advantages of continuous measurement and monitoring of PK/PD in a single subject. Because OFM does not suffer from MD's limitations (i.e., a narrow measurement spectrum and limited application time), it allows measurement of high molecular markers for cell damage, such as lactate dehydrogenase, and of cytokines for identifying the inflammatory response to NPs. Noninvasive imaging techniques in development and, to some extent, available, but their high costs impede their widespread use as preclinical tests.

In vivo models provide more reliable information than in vitro models, and reflect an experimental situation closer to that of clinical trials. Target organ toxicity is a question that is particularly difficult to answer accurately in vitro. Even for organs such as the liver, in which primary cells and tissue slices are commercially available, in vitro assays often cannot reliably predict in vivo dose-response effects (Kramer et al. 2007). A close consensus of in vitro and in vivo assessments is a prerequisite for efficient pharmaceutical development. In vivo models can be used to validate in vitro models and to verify outcomes of in vitro studies. In applying NPs to the brain, this means selecting brain-relevant cell lines such as astrocytes, brain-endothelial cells, neurons, and microglia, and choosing adequate assays such as lactate dehydrogenase leakage, S100, BDNF, and cytokines, thereby allowing a direct comparison of in vitro and in vivo results.

REFERENCES

Au, C., Mutkus, L., Dobson, A. et al. 2007. Effects of nanoparticles on the adhesion and cell viability on astrocytes. *Biol Trace Elem Res* 120: 248–56.

Belyanskaya, L., Weigel, S., Hirsch, C. et al. 2009. Effects of carbon nanotubes on primary neurons and glial cells. *Neurotoxicology* 30: 702–11.

Benveniste, H. 1989. Brain microdialysis. *J Neurochem* 52: 1667–79.

Benveniste, H., Drejer, J., Schousboe, A. et al. 1987. Regional cerebral glucose phosphorylation and blood flow after insertion of a microdialysis fiber through the dorsal hippocampus in the rat. *J Neurochem* 49: 729–34.

Bhaskar, S., Tian, F., Stoeger, T. et al. 2010. Multifunctional nanocarriers for diagnostics, drug delivery and targeted treatment across blood–brain barrier: perspectives on tracking and neuroimaging. *Part Fibre Toxicol* 7: 3.

Bodenlenz, M., Schaupp, L.A., Druml, T. et al. 2005. Measurement of interstitial insulin in human adipose and muscle tissue under moderate hyperinsulinemia by means of direct interstitial access. *Am J Physiol Endocrinol Metab* 289: E296–300.

Brooks, D.J., Beaney, R.P., Lammertsma, A.A. et al. 1984. Quantitative measurement of blood–brain barrier permeability using rubidium-82 and positron emission tomography. *J Cereb Blood Flow Metab* 4: 535–45.

Buntinx, M., Vanderlocht, J., Hellings, N. et al. 2003. Characterization of three human oligodendroglial cell lines as a model to study oligodendrocyte injury: morphology and oligodendrocyte-specific gene expression. *J Neurocytol* 32: 25–38.

Casey, A., Herzog, E., Lyng, F.M. et al. 2008. Single walled carbon nanotubes induce indirect cytotoxicity by medium depletion in A549 lung cells. *Toxicol Lett* 179: 78–84.

Choi, J., Zheng, Q., Katz, H.E. et al. 2010. Silica-based nanoparticle uptake and cellular response by primary microglia. *Environ Health Perspect* 118: 589–95.

de Lange, E.C., Danhof, M., de Boer, A.G. et al. 1997. Methodological considerations of intracerebral microdialysis in pharmacokinetic studies on drug transport across the blood–brain barrier. *Brain Res Brain Res Rev* 25: 27–49.

de Lange, E.C., de Boer, A.G., and Breimer, D.D. 2000. Methodological issues in microdialysis sampling for pharmacokinetic studies. *Adv Drug Deliv Rev* 45: 125–48.

Dingemanse, J., Sollie, F.A., Breimer, D.D. et al. 1988. Pharmacokinetic modeling of the anticonvulsant response of oxazepam in rats using the pentylenetetrazol threshold concentration as pharmacodynamic measure. *J Pharmacokinet Biopharm* 16: 203–28.

Ellmerer, M., Schaupp, L., Brunner, G.A. et al. 2000. Measurement of interstitial albumin in human skeletal muscle and adipose tissue by open-flow microperfusion. *Am J Physiol Endocrinol Metab* 278: E352–6.

Elmquist, W.F., and Sawchuk, R.J. 1997. Application of microdialysis in pharmacokinetic studies. *Pharm Res* 14: 267–88.

Ewing, J.R., Knight, R.A., Nagaraja, T.N. et al. 2003. Patlak plots of Gd-DTPA MRI data yield blood–brain transfer constants concordant with those of 14C-sucrose in areas of blood–brain opening. *Magn Reson Med* 50: 283–92.

Fisichella, M., Dabboue, H., Bhattacharyya, S. et al. 2009. Mesoporous silica nanoparticles enhance MTT formazan exocytosis in HeLa cells and astrocytes. *Toxicol In Vitro* 23: 697–703.

Follett, P.L., Deng, W., Dai, W. et al. 2004. Glutamate receptor-mediated oligodendrocyte toxicity in periventricular leukomalacia: a protective role for topiramate. *J Neurosci* 24: 4412–20.

Fröhlich, E., Kueznik, T., Samberger, C. et al. 2010. Size-dependent effects of nanoparticles on the activity of cytochrome P450 isoenzymes. *Toxicol Appl Pharmacol* 242: 326–32.

Fröhlich, E., Samberger, C., Kueznik, T. et al. 2009. Cytotoxicity of nanoparticles independent from oxidative stress. *J Toxicol Sci* 34: 363–75.

Groothuis, D.R., Ward, S., Schlageter, K.E. et al. 1998. Changes in blood–brain barrier permeability associated with insertion of brain cannulas and microdialysis probes. *Brain Res* 803: 218–30.

Harry, G.J., Billingsley, M., Bruinink, A. et al. 1998. In vitro techniques for the assessment of neurotoxicity. *Environ Health Perspect* 106 Suppl 1: 131–58.

Henn, A., Lund, S., Hedtjarn, M. et al. 2009. The suitability of BV2 cells as alternative model system for primary microglia cultures or for animal experiments examining brain inflammation. *Altex* 26: 83–94.

Hilbert, J.M., and Battista, D. 1991. Quazepam and flurazepam: differential pharmacokinetic and pharmacodynamic characteristics. *J Clin Psychiatry* 52 Suppl: 21–6.

Hodgson, E. (ed.). 2004. *Book A Textbook of Modern Toxicology*. Hoboken, NJ: John Wiley and Sons.

Ikeoka, D., Pachler, C., Korsatko, S. et al. 2010. Interleukin-6 produced in subcutaneous adipose tissue is linked to blood pressure control in septic patients. *Cytokine* 50: 284–91.
Josserand, V., Pelerin, H., de Bruin, B. et al. 2006. Evaluation of drug penetration into the brain: a double study by in vivo imaging with positron emission tomography and using an in vitro model of the human blood–brain barrier. *J Pharmacol Exp Ther* 316: 79–86.
Kashima, T., Tiu, S.N., Merrill, J.E. et al. 1993. Expression of oligodendrocyte-associated genes in cell lines derived from human gliomas and neuroblastomas. *Cancer Res* 53: 170–5.
Kramer, J.A., Sagartz, J.E., and Morris, D.L. 2007. The application of discovery toxicology and pathology towards the design of safer pharmaceutical lead candidates. *Nat Rev Drug Discov* 6: 636–49.
Lai, J.C., Lai, M.B., Jandhyam, S. et al. 2008. Exposure to titanium dioxide and other metallic oxide nanoparticles induces cytotoxicity on human neural cells and fibroblasts. *Int J Nanomed* 3: 533–45.
Li, X., Zheng, H., Zhang, Z. et al. 2009. Glia activation induced by peripheral administration of aluminum oxide nanoparticles in rat brains. *Nanomedicine* 5: 473–79.
Liu, S., Xu, L., Zhang, T. et al. 2010. Oxidative stress and apoptosis induced by nanosized titanium dioxide in PC12 cells. *Toxicology* 267: 172–7.
Ma, L., Liu, J., Li, N. et al. 2010. Oxidative stress in the brain of mice caused by translocated nanoparticulate TiO_2 delivered to the abdominal cavity. *Biomaterials* 31: 99–105.
Miller, G. 2002. Drug targeting. Breaking down barriers. *Science* 297: 1116–8.
Monteiro-Riviere, N.A., Inman, A.O., and Zhang, L.W. 2009. Limitations and relative utility of screening assays to assess engineered nanoparticle toxicity in a human cell line. *Toxicol Appl Pharmacol* 234: 222–35.
Pachler, C., Ikeoka, D., Plank, J. et al. 2007. Subcutaneous adipose tissue exerts proinflammatory cytokines after minimal trauma in humans. *Am J Physiol Endocrinol Metab* 293: E690–6.
Pardridge, W.M. 2005. The blood–brain barrier: bottleneck in brain drug development. *NeuroRx* 2: 3–14.
Patlak, C.S., Blasberg, R.G., and Fenstermacher, J.D. 1983. Graphical evaluation of blood-to-brain transfer constants from multiple-time uptake data. *J Cereb Blood Flow Metab* 3: 1–7.
Reese, T.S., and Karnovsky, M.J. 1967. Fine structural localization of a blood–brain barrier to exogenous peroxidase. *J Cell Biol* 34: 207–17.
Rosenbloom, A.J., Ferris, R.L., Sipe, D.M. et al. 2006. In vitro and in vivo protein sampling by combined microdialysis and ultrafiltration. *J Immunol Methods* 309: 55–68.
Rosenbloom, A.J., Sipe, D.M., and Weedn, V.W. 2005. Microdialysis of proteins: performance of the CMA/20 probe. *J Neurosci Methods* 148: 147–53.
Schaupp, L., Ellmerer, M., Brunner, G.A. et al. 1999. Direct access to interstitial fluid in adipose tissue in humans by use of open-flow microperfusion. *Am J Physiol* 276: E401–8.
Shrivastava, R., Delomenie, C., Chevalier, A. et al. 1992. Comparison of in vivo acute lethal potency and in vitro cytotoxicity of 48 chemicals. *Cell Biol Toxicol* 8: 157–70.
Speckmann, E.-J., Hescheler, J., and Kohling, R. 2008. *Physiologie*. 5th ed. Munich: Urban & Fischer.
Tang, J., Xiong, L., Wang, S. et al. 2009. Distribution, translocation and accumulation of silver nanoparticles in rats. *J Nanosci Nanotechnol* 9: 4924–32.
Westergren, I., Nystrom, B., Hamberger, A. et al. 1995. Intracerebral dialysis and the blood–brain barrier. *J Neurochem* 64: 229–34.
Worle-Knirsch, J.M., Pulskamp, K., and Krug, H.F. 2006. Oops they did it again! Carbon nanotubes hoax scientists in viability assays. *Nano Lett* 6: 1261–8.
Wu, J., Sun, J., and Xue, Y. 2010. Involvement of JNK and P53 activation in G2/M cell cycle arrest and apoptosis induced by titanium dioxide nanoparticles in neuron cells. *Toxicol Lett* 199: 269–76.
Zhang, Y., Ali, S.F., Dervishi, E. et al. 2010. Cytotoxicity effects of graphene and single-wall carbon nanotubes in neural phaeochromocytoma-derived PC12 cells. *ACS Nano* 4: 3181–6.

36 Nanorobotics for Neurosurgery

Robbin Miranda, Gill Pratt, and Geoffrey S. F. Ling

CONTENTS

INTRODUCTION

Nanorobotics is the field devoted to developing robots in the nanometer scale (Abhilash 2010). The term 'nanorobot' conjures an image of a tiny abiotic device that scurries about, performing autonomous functions at the atomic or molecular level. In reality, however, nanorobotics describes a broader, more encompassing field of nanotechnology, including macrorobotic tools for interaction, manipulation, and/or construction of objects at a nanoscale resolution, as well as devices that are themselves at the nanoscale level and/or constructed of nanoscale components. Within this latter category, the term "nanorobots" is used to describe a wealth of devices, including those developed through top-down approaches, such as engineered biological organisms or biomolecules, as well as through bottom-up approaches, including construction of devices through the assembly of individual nanoscale components. Given the recent emergence of the field of nanorobotics, the distinction between nanorobots and non-robotic nanotechnology is not yet well defined; for instance, engineered viral vectors for delivery of drugs and imaging agents to tumor cells have been described as nanorobots, whereas encapsulation of nanoparticles for targeted delivery (described elsewhere in this volume) is generally not considered under the field of nanorobotics. In the current chapter, we focus on topics that have previously been described within the field of nanorobotics, and specifically those geared toward advancing surgical procedures (and neurosurgery, where applicable). However, the reader should remain aware of the potential overlap of "robotic" and "non-robotic" approaches within the field of nanotechnology. As these techniques continue to emerge, it is anticipated that such distinctions will be better defined.

TOOLS FOR NANOSCALE SURGICAL OPERATIONS

Femtosecond Laser Systems

The use of femtosecond laser pulses, which last only one quadrillionth of a second, has enabled precise manipulation of single cells and subcellular structures, with spatial resolution in the nanometer range (Leary et al. 2006). Studies have demonstrated the ability to dissect or ablate individual subcellular organelles without compromising the cell's viability, and to transfect cells with foreign DNA (Leary et al. 2006). Femtosecond laser technology enables fundamental investigations of nervous system function through its ability to sever individual neuronal fibers (Leary et al. 2006). Such technology was also proposed for axonal repair through the substitution of healthy donor axons (Leary et al. 2006).

Nanoneedles

Nanoscale surgical operations were demonstrated on living cells using atomic force microscopy tips sharpened to a diameter of 200–300 nm (Leary et al. 2006). The nanoneedle penetrated cellular and nuclear membranes and inserted various molecules into cellular nuclei without destroying the cells (Leary et al. 2006).

Nanotweezers

Nanotechnology has enabled tweezer-like capabilities through multiple approaches. The first general-purpose nanotweezer was developed in 1999 by Kim and Lieber and consisted of two bundles of multiwalled carbon nanotubes (Freitas 2005). Carbon nanotubes attached to electrodes open and close upon application of voltage, and were used to manipulate subcellular structures within a living cell (Leary et al. 2006). DNA binding peptides were also used as nanotweezers that open and close upon modulation of the solution pH between neutral and acidic values (Sharma et al. 2008). Additionally, for optical tweezers, light energy has been used to immobilize and manipulate molecules such as nucleic acids and actin filaments (Leary et al. 2006).

Manufacturing for Tissue Repair in Central and Peripheral Nervous Systems

Trauma, stroke, and degeneration are common examples of injury to neurons, glia, nerves, and support structures. Effective techniques to repair such injuries would be a major clinical advance. Traditional surgical and medical approaches have had limited benefit. Self-assembling peptide nanofiber scaffolds are a recent innovation and were shown to permit axons to grow through a site of injury thereby allowing restoration of function such as vision (Ellis Behnke et al. 2006). Peptide amphiphiles can be introduced in a site needing repair. There, they will self-assemble into cylindrical nanofibers, similar in dimension to endogenous extracellular matrix (Collins et al. 2010). In an ionic environment, these amphiphiles will orient such that their hydrophobic ends are in the center and their hydrophilic ends are on the surface (Webber et al. 2010). The hydrophilic end can be engineered with a bioactive compound such as isoleucine–lysine–valine–alanine–valine (IKVAV). IKVAV can promote neural stem cells to become neurons as opposed to glia (Silva et al. 2004). Carbon nanotubes are also used as scaffolds and when combined with appropriate DNA or protein, can support tissue regeneration (Collins et al. 2010; Elder et al. 2008; Khawaja 2011).

Although not yet nanosized, several miniaturized robotic systems for image-guided surgeries are in use. The SpineAssist device, developed by Mazor Robotics Ltd., allows surgeons to plan a spinal procedure in 3D. Designed to reduce the need for computed tomography (CT) scanning during spinal implant procedures and improve placement accuracy, the device has been used in more than 2000 surgeries worldwide, which have placed tens of thousands of implants in patients. In May 2011, SpineAssist received a European CE Mark for use in brain surgery. Likewise, the StealthStation guidance system, developed by Medtronic Surgical Navigation Technologies, collects images from traditional imaging sources [x-ray, CT, magnetic resonance imaging (MRI), and ultrasound] and transforms them into 3D images, which can then be used by the surgeon to plan the surgical procedure as well as monitor the surgery in real time. The StealthStation system is used for brain, spinal, orthopedic, and ear/nose/throat surgeries.

NANOSCALE ROBOTIC DEVICES FOR AUTOMATED SURGICAL PROCEDURES

Unlike macroscale robots, nanorobots are governed by molecular quantum mechanics and are subject to positional uncertainties inherent to quantum mechanics (Ummat et al. 2006). Given the difficulty of working with and constructing such small devices, techniques such as scanning electron microscopy and atomic force microscopy are being leveraged to provide researchers with visual and haptic feedback on the molecular structure of nanoscale devices. Additionally, virtual reality techniques are being explored as a more immersive means of using such feedback (Ummat et al. 2006).

Engineered Biological Organisms

One top-down approach to creating new nanoscale artificial devices is to manipulate natural nanoscale biological systems such as viruses and bacteria (Freitas 2005). In the mid-1990s, combinatorial genetic engineering of bacteriophage T7 produced T7 mutants that replicated twice as fast as wild type T7 (Freitas 2005). In the early 2000s, chemical modification of viruses enabled the addition of tumor-targeting peptides onto the viral coat, as well as the incorporation of contrast agents such as gadolinium atoms into the viral capsid, resulting in MRI-detectable tumor-targeting particles (Freitas 2005). More recently, viral vectors were engineered to deliver cancer-fighting drugs directly to tumor cells. An ongoing clinical trial at the University of California San Diego's Moores Cancer Center is investigating the efficacy of Toca 511, a retroviral replicating vector, engineered to selectively transfect a cytosine deaminase gene into cancer cells. Over time, cancer cells expressing the transfected gene convert orally administered Toca FC (flucytosine) into the anticancer drug 5-fluorouracil, enabling targeted killing of the cancer cells with minimal toxicity to healthy cells (University of California, San Diego Health Sciences 2012).

Using self-replicating organisms as nanorobots has both advantages and limitations. On one hand, the ability of nanorobots to self-replicate would reduce manufacturing costs and could streamline the administration of nanorobots for clinical applications. However, uncontrolled self-replication is also a major safety concern, and unless mechanisms are used to halt replication and destroy or remove the nanorobots after completion of their intended function, using self-replicating nanorobots for in vivo medical applications will be greatly limited.

Engineered Biomolecules

An alternative approach to engineering complete nanoscale biological systems is to engineer biomolecules to exhibit novel functional properties. Recently, a DNA nanorobot was developed to autonomously transport molecular payloads to cells (Douglas et al. 2012). Construction of the 35 × 35 × 45 nm hexagonal barrel-shaped nanorobot was facilitated by the DNA origami method, in which a single strand of DNA was folded into the desired shape through the interaction with 196 oligonucleotide staple strands. Upon interaction with either one or two specific types of proteins, the nanorobot uses a DNA aptamer-based lock mechanism to reconfigure its shape in a spring-like fashion, exposing surfaces that were previously enclosed within the device. This mechanism of "opening" the nanorobot enables loading molecular payloads onto the newly exposed surfaces.

In an experiment used to test this nanorobot function, the robots were loaded with fluorescently labeled antibody fragments. When the robots' aptamer-based lock mechanism encountered specific "key" antigens located on human leukocyte cells, the resulting reconfiguration of the nanorobots exposed the fluorescent label, allowing fluorescent imaging of the nanorobot-bound leukocyte cells.

A limitation in using biomolecules such as DNA is the inability to create complex nanodevices required to develop precise medical tools (Cavalcanti et al. 2009b).

Hybrid Biotic/Abiotic Nanorobots

Biological molecules and organisms have also been engineered with nonbiological parts to create hybrid nanoscale systems. The first artificial hybrid nanomotor was reported in 1999 through the addition of metal-binding amino acid residues to the enzyme ATPase, which naturally contains a protein "rotor" whose rotation is controlled via electrochemical reactions with the molecule (Freitas 2005). A device to blend fluids twice as fast as diffusion alone was developed by affixing *Serratia marcesens* bacteria, which contain naturally moving appendages called fimbriae, onto tiny beads (Freitas 2005; Pearson 2003). When incorporated within tiny fluid-filled tubes, the bacterial fimbriae set the beads in motion to blend the fluids.

ABIOTIC NANOROBOTIC COMPONENTS

In addition to the top-down approach of modifying existing organisms and biomolecules to exhibit new nanoscale functional properties, bottom-up approaches have also been considered for developing atomic or molecular devices, machines, and circuits through the assembly of nanoscale components. A number of nanorobotic components were theoretically designed and tested through molecular dynamics simulations, but to date have either not yet or have only recently begun to be physically constructed. Such components include molecular bearings, molecular gears, and molecular motors.

The first reported theoretical nanobearings were published in 1992 by Drexler and Merkle and included a 2.2-nm-diameter, 206-atom overlap-repulsion bearing design, consisting of carbon, silicon, oxygen, and hydrogen, as well as a 4.8-nm-diameter, 2808-atom strained-shell sleeve bearing (Drexel 1992; Freitas 2005). More recent nanobearing designs consisted of double-walled carbon nanotubes, in which the outer tube functions as a rotating sleeve (Shenai et al. 2010).

Drexler and Merkle also designed the first molecular gear, a 4.3 × 4.4-nm, 3557-atom planetary gear consisting of 12 moving parts with a rapidly rotating central shaft and slowly rotating peripheral output shaft (Drexel 1992; Freitas 2005). In 2009, scientists from A*STAR's Institute of Materials Research and Engineering developed the first controllable molecular gear (Manzano et al. 2009). The rotation of the 1.2-nm, single-molecule hexa-*t*-butyl-pyrimidopentaphenylbenzene rotor was controlled with the tip of a scanning tunneling microscope while the molecule was mounted on a single-atom axis.

In addition to the top-down bioengineering approach to create a molecular motor as described above, gas-powered, abiotic molecular motors were also proposed and conceptualized by Drexler and Merkle (Drexel 1992; Freitas 2005). Depending on how the 6165-atom device is controlled, it could either serve as a pump for neon gas atoms or convert neon gas into rotary power (Freitas 2005). In 2003, a 550-nm nanomotor was developed that used electrical current to trigger tortional oscillations of a gold rotor (Freitas 2005).

POWERING NANOROBOTIC DEVICES

In some cases, autonomous nanorobots may not require a power source, as they could rely on natural circulation mechanisms in vivo, such as blood flow, to reach their target destinations. Moreover, the presence of imaging agents attached to the nanorobots may mitigate the need for active data transmission by the nanorobots. However, in other cases, untethered nanorobots may require a power source for self-propulsion, data transmission, and/or other activities.

Engineered organisms may entail a natural source of power, such as the motion of bacterial fimbriae described above. Hybrid or abiotic nanorobots could also be biomimetically designed to metabolize local glucose and oxygen to generate power for self-propulsion (Patel et al. 2006). The use of electrical actuators, such as electromagnetic, piezoelectric, electrostatic, electrothermal, and complementary metal oxide semiconductor (CMOS) technologies has also been addressed (Cavalcanti et al. 2009a, 2009b). Monitoring and controlling power transmission can be facilitated through radiofrequency-based telemetry procedures (Cavalcanti et al. 2009a, 2009b).

In addition to enabling movement of nanorobots through a built-in capacity for self-propulsion, an alternative method of facilitating nanorobotic movement is through interaction of the nanorobots with an electromagnetic field. A robotic platform was developed that uses real-time feedback from MRI to control the navigation of untethered magnetic carriers, nanorobots, and magnetotactic bacteria carrying sensory or therapeutic agents through blood vessels (Martel et al. 2009). As part of a project called "MR-Sub" (Magnetic Resonance Submarine), a magnetic resonance propulsion force was used to navigate a ferromagnetic bead in real time through the carotid artery of a living swine. The researchers had previously demonstrated the ability to propel and steer a single bacterium carrying a 3-μm bead along a preplanned path.

The activity of engineered nanobiomaterials can also be controlled with light. A technique called "optogenetics" entails the use of viral vectors containing the genetic blueprint of one or more naturally occurring or engineered microbial opsins—ion channels that respond to specific wavelengths of light (Deisseroth 2011). The viral constructs can be targeted to specific types of cells. When the constructs are injected into the brain, the target neurons incorporate the opsin genes and express the light-sensitive opsins within their cellular membranes. Real-time activation and inhibition of target neurons were demonstrated in the awake, behaving nonhuman primate upon exposure of the neurons to pulses of light at specific wavelengths (Diester et al. 2011).

In powering nanorobots in vivo, heat dissipation must be carefully regulated, especially during the use of large numbers of nanorobots so as not to destroy healthy biological tissue (Patel et al. 2006). However, once the nanoparticles reach their targets, controlled heat generation, for instance, through the exposure of magnetic nanoparticles to an alternating magnetic field, can be used to selectively destroy adverse targets such as tumor cells (Martel et al. 2009).

BIOMIMETIC NANOROBOTIC DEVICES

Although not yet physically developed, Freitas has proposed a theoretical design for a 1-μm, 18-billion-atom artificial mechanical red blood cell, or "respirocyte" (Freitas 2005). The device would mimic the biological actions of natural red blood cells but could store and transport 236 times as much gas per unit volume. An addition of 1 L of a respirocyte suspension could oxygenate a patient's tissues for up to 4 h in the case of heart or respiratory failure, and in the field of neurosurgery could also lead to enhancement of neurovascular procedures. Similarly, artificial white blood cells, or "microbivores," were proposed by Freitas to enhance the destruction of microbiological pathogens in the bloodstream, exhibiting greater lifetime capacity and phagocytic efficiency compared to natural white blood cells (Freitas 2005).

Investigators at the University of Southern California's Viterbi School of Engineering developed a design for an artificial neural synapse consisting of a carbon nanotube transistor circuit (Friesz et al. 2007). The modeled synapse roughly emulates the biological functions of neurotransmitters as well as ion channels and pumps, and takes into account the temporal summation of postsynaptic potentials to generate simulated approximations of biological action potentials. The synaptic design was recently fabricated as an analog circuit in a laboratory environment (Joshi et al. 2011). Such biomimetic devices could one day be implanted in the brain to compensate for lost neuronal function due to neurodegeneration or acquired brain injury. An advantage of using carbon nanotubes over silicon for neural implants is that carbon nanotubes do not provoke adverse immunological responses in biological tissue. They can also be arranged into three-dimensional conformations, an important factor in the ability to mimic the large-scale connectivity naturally found within the brain (Joshi et al. 2011).

FUTURE CONSIDERATIONS FOR NANOROBOTIC NEUROSURGICAL APPLICATIONS

The use of nanorobots has been proposed to improve upon a microelectromechanical system (MEMS) device for repairing damaged nerves (Hogg and Sretavan 2005). The MEMS device methodology, which was demonstrated in cell culture, uses enzymes to expose the axon at the end of a damaged host nerve and the axon of a healthy donor nerve. The axonal membranes of the host–graft axon pair are then fused together through application of an electric field, and the process is repeated for additional host–graft axon pairs. Hogg and Sretavan proposed that using nanorobots released from the MEMS device could reduce the required length of axon exposure from 100 to 10 μm, resulting in greater nerve preservation and quicker healing. Moreover, they suggested that the nanorobots could detect viable axons, sense axonal membrane chemicals, move together pairs of host and graph axons of the same functional type (e.g., sensory vs. motor), and test the functionality of repaired axons.

Another proposed neurosurgical application of nanorobotics is using them for intracranial pressure monitoring and detecting intracranial nitric oxide synthase protein overexpression signals in the early stages of aneurysm development (Cavalcanti et al. 2009a, 2009b; Leary et al. 2006). Such nanorobots could utilize CMOS technologies for power supply, chemical sensing, and active telemetry, transmitting data and receiving commands wirelessly to and from an external device such as a mobile phone (Cavalcanti et al. 2009a, 2009b). Data collected by the nanorobot could trigger medical intervention before hemorrhage occurs. Nanorobots could also be designed to independently increase or decrease drainage of cerebrospinal fluid based on detected changes in intracranial pressure, or could be coupled with intraventricular catheters to eliminate debris and prevent shunt occlusion (Leary et al. 2006).

Leary et al. (2006) speculated that nanorobots equipped with optical tweezers and femtosecond laser systems may one day be used to target defective genes and eliminate them from every cell in the human body. They may even be able to repair certain genetic mutations via intracellular genetic nanoneurosurgery (Leary et al. 2006). Furthermore, Elder and associates (2008) foresee a time when there will be molecular manufacturing at the atomic scale and controlled 3D nano-assembly resulting in nanorobots that will be used for modifying neural connections or promoting neural regeneration.

CONCLUSION

The emerging field of nanorobotics offers significant advancements in medical diagnostics and treatments. Developments in nanorobotics are dissolving traditional boundaries between biotic and abiotic devices and utilizing advantages of both to optimize functional interactions in vivo for clinical diagnosis and repair. Although challenges remain in transitioning these developments into practical applications for clinical use, recent and current research efforts are providing foundations for the development of nanorobotic devices with the potential to revolutionize the field of neurosurgery. Recent developments in nanorobotics are leading to the emergence of autonomous nanoscale robots for brain imaging and repair, as well as tools to aid the neurosurgeon in visualizing neural features and performing precise surgical manipulations at the nanoscale level. Ongoing research efforts continue to shape the field of nanorobotics and its transition from science fiction to reality.

DISCLAIMER

The opinions expressed herein belong solely to the authors. They do not represent and should not be interpreted as being those of or endorsed by the U.S. Army, the Defense Advanced Research Projects Agency, the Department of Defense, or any other branch of the federal government.

DISCLOSURES

The authors have no disclosures to make.

ACKNOWLEDGMENTS

The authors gratefully acknowledge the technical and editorial help of Nicole Draghic in the preparation of this manuscript.

REFERENCES

Abhilash, M. 2010. Nanorobots. *Int J Pharm Bio Sci* 1: 1–10.

Cavalcanti, T.A., Shirinzadeh, B., Fukuda, T. et al. 2009a. Errata: update to "Nanorobot for brain aneurysm". *Int J Robot Res* 28: 927.

Cavalcanti, T.A., Shirinzadeh, B., Fukuda, T. et al. 2009b. Nanorobot for brain aneurysm. *Int J Robotic Res* 28: 558–70.

Collins, K.I., Orringer, D.A., and Patil, P.G. 2010. Nanotechnology in neurosurgery. *J Nanotechnol Eng Med* 1: 034001-1–01-15.

Deisseroth, K. 2011. Optogenetics. *Nat Methods* 8: 26–9.

Diester, I., Kaufman, M.T., Mogri, M. et al. 2011. An optogenetic toolbox designed for primates. *Nat Neurosci* 14: 387–97.

Douglas, S.M., Bachelet, I., and Church, G.M. 2012. A logic-gated nanorobot for targeted transport of molecular payloads. *Science* 335: 831–4.

Drexel, K.E. 1992. *Nanosystems—Molecular Machinery, Manufacturing, and Computation*. New York: Wiley.

Elder, J.B., Liu, C.Y., and Apuzzo, M.L. 2008. Neurosurgery in the realm of 10(-9): Part 2. Applications of nanotechnology to neurosurgery—present and future. *Neurosurgery* 62: 269–84; discussion 284–5.

Ellis-Behnke, R.G., Liang, Y.X., You, S.W. et al. 2006. Nano neuro knitting: peptide nanofiber scaffold for brain repair and axon regeneration with functional return of vision. *Proc Natl Acad Sci U S A* 103: 5054–9.

Freitas, R.A. 2005. Current status of nanomedicine and medical robotics. *J Comput Theor Nanosci* 2: 1–25.

Friesz, A.K., Parker, A.C., Zhou, C. et al. 2007. A biomimetic carbon nanotube synapse circuit. Presented at the *Biomed. Eng. Soc. (BMES) Annu. Fall Meet.*, Los Angeles, CA, Sep. 26–29.

Hogg, T., and Sretavan, D.W. 2005. Controlling tiny multi-scale robots for nerve repair. *Proceedings of the American Association of Artificial Intelligence*, pp. 1286–91.

Joshi, J., Zhang, J., Wang, C. et al. 2011. A biomimetic fabricated carbon nanotube synapse for prosthetic applications. *Life Science Systems and Applications Workshop (LiSSA), 2011 IEEE/NIH*, pp. 139–42.

Khawaja, A.M. 2011. The legacy of nanotechnology: revolution and prospects in neurosurgery. *Int J Surg* 9: 608–14.

Leary, S.P., Liu, C.Y., and Apuzzo, M.L. 2006. Toward the emergence of nanoneurosurgery: Part III. Nanomedicine: targeted nanotherapy, nanosurgery, and progress toward the realization of nanoneurosurgery. *Neurosurgery* 58: 1009–26; discussion 1009–26.

Manzano, C., Soe, W.H., Wong, H.S. et al. 2009. Step-by-step rotation of a molecule-gear mounted on an atomic-scale axis. *Nat Mater* 8: 576–9.

Martel, S., Felfoul, O., Mathieu, J.B. et al. 2009. MRI-based medical nanorobotic platform for the control of magnetic nanoparticles and flagellated bacteria for target interventions in human capillaries. *Int J Robot Res* 28: 1169–82.

Patel, G.M., Patel, G.C., Patel, R.B. et al. 2006. Nanorobot: a versatile tool in nanomedicine. *J Drug Target* 14: 63–7.

Pearson, H. 2003. Construction bugs find tiny work. *Nature News*, http://www.nature.com/news/2003/030711/full/news030707-9.html.

Sharma, G., Rege, K., Budil, D.E. et al. 2008. Reversible pH-controlled DNA-binding peptide nanotweezers: an in-silico study. *Int J Nanomed* 3: 505–21.

Shenai, P.M., Ye, J., and Zhao, Y. 2010. Sustained smooth dynamics in short-sleeved nanobearings based on double-walled carbon nanotubes. *Nanotechnology* 21: 495303.

Silva, G.A., Czeisler, C., Niece, K.L. et al. 2004. Selective differentiation of neural progenitor cells by high-epitope density nanofibers. *Science* 303: 1352–5.

Ummat, A., Sharma, G., Mavroidis, C. et al. 2006. Bio-nano-robotics: state of the art and future challenges. In *Tissue Engineering and Artificial Organs*, ed. M.L. Yarmush and M. Toner, *The Biomedical Engineering Handbook*, ed. J. D. Bronzino. Boca Raton, FL: CRC Press.

University of California, San Diego Health Sciences. 2012. Treating brain cancer with viral vector. *Science Daily*.

Webber, M.J., Kessler, J.A., and Stupp, S.I. 2010. Emerging peptide nanomedicine to regenerate tissues and organs. *J Intern Med* 267: 71–88.

37 Imaging and Tracking Therapeutic Cells with Nanoparticles

Gang Niu and Joseph A. Frank

CONTENTS

CELL-BASED THERAPY

Cell replacement therapy and gene transfer to the diseased or injured brain have provided the basis for developing potentially powerful new therapeutic strategies (Carpenter et al. 2009; Kim and de Vellis 2009). Neurological disorders such as Parkinson's disease, Huntington's disease, amyotrophic lateral sclerosis (ALS), Alzheimer's disease, multiple sclerosis (MS), stroke, and spinal cord injury are caused by loss of neurons and glial cells in the brain or spinal cord. Thus, neural regeneration has been one of the most actively developing fields in regenerative medicine. In addition to neural stem cells (NSCs) and neural progenitor cells (NPCs), embryonic stem cells (ESCs), adult bone marrow-derived cells including hematopoietic stem cells (HSCs), bone marrow stromal cells (BMSCs) or mesenchymal stem cells (MSCs), and endothelial progenitor cells (EPCs) have also been extensively studied to find the best source for tissue regeneration (Schoenhard and Hatzopoulos 2010). Recently, induced pluripotent stem (iPS) cells have been developed that may have the potential to treat central nervous system (CNS) diseases.

Malignant brain tumors resist effective treatment and remain lethal despite extensive surgical excision, radiotherapy, and chemotherapy. Recent studies have shown that NSCs or NPCs exhibit tumor-tropic properties and have the capacity to carry therapeutic genes selectively to brain tumors, such as malignant gliomas, medulloblastomas, and metastasis (Aboody et al. 2000; Dickson et al. 2007; Kim et al. 2006). It has been suggested that infused NSCs are able to home, target, and deliver genes to tumors in the CNS (Thu et al. 2009). Immortalized human NSCs have emerged as a highly efficient source of cells that can be genetically manipulated ex vivo for gene transfer into the CNS. Once transplanted into damaged brain, these cells survive, integrate into host tissues, and differentiate into both neurons and glial cells (Kim 2007). Another appealing strategy is cell-based immunotherapy that induces the immune system to mount a potent antitumor response or that uses harvested cells genetically engineered with a suicide gene (Yaghoubi et al. 2009). Immune targeting has the potential to result in more selective killing of cancer cells and in reduced toxicity to normal cells compared to standard therapies (Disis et al. 2009; Eksioglu et al. 2010; Kapp et al. 2009; Shablak et al. 2009; Sutlu and Alici 2009). Generally, cell-based cancer immunotherapy is derived from the ex vivo manipulation of various therapeutically useful cell types, such as lymphocytes, dendritic cells (DCs), and natural killer (NK) cells, followed by local or systemic administration to cancer patients (Aldrich et al. 2010).

Currently, our knowledge on the biodistribution and pharmacokinetics of the therapeutic cells relies mainly on histological examination of tissue samples obtained at biopsies/autopsy. Mechanisms underlying the migration and roles the therapeutic cells themselves played in therapy have not been fully elucidated. Thus, it is critical to develop sensitive and reproducible noninvasive methods for real-time assessment of the therapeutic process and the refinement of current cell-based therapy (Budde and Frank 2009; Pham et al. 2009).

MOLECULAR IMAGING

Molecular imaging enables noninvasive visualization of molecular and cellular function in living organisms using advanced probes. Clinical molecular imaging techniques, including nuclear medicine, positron emission tomography (PET), and magnetic resonance imaging (MRI), can diagnose disease earlier and may contribute to improved therapeutic outcome by optimizing the methods used in conducting preclinical and clinical trials of novel therapies (Table 37.1) (Arbab and Frank 2008; Weissleder and Mahmood 2001). Successful application of cell-based therapies requires the ability to monitor transplanted cell survival and integration noninvasively and dynamically with high temporal and spatial resolution (Lee et al. 2008). Molecular imaging has been extensively adapted to effectively track cell homing and function in target tissues by prelabeling cell products with intrinsic or exogenous probes. Intrinsic cell labeling is usually achieved by ex vivo transduction of cells with imaging reporter genes (Kang and Chung 2008). This approach allows for reliable and near-stable visualization of cellular trafficking, proliferation, and function at the target site. Moreover, depending on transduction efficiency, tagging cells with a reporter gene may provide a general solution for long-term monitoring of genetically engineered cells that contain a suicide gene. However, genetic modification of stem cells may affect their "stemness" characteristics, and there is concern that transduced cells may undergo malignant transformation (Seggewiss et al. 2006). In addition, the sensitivity of this technique is determined by the reporter gene expression level in the transfected cells and the pharmacokinetics of the ligands used. For cell labeling with exogenous probes, radioactive, fluorescent paramagnetic or superparamagnetic agents can be incorporated into cells according to standardized protocols (Bulte and Kraitchman 2004; Michalet et al. 2005). One potential limitation of cell labeling with agents such as superparamagnetic iron oxide nanoparticles (SPIONs) or quantum dots (QDs) is the transfer of the exogenous label to tissue macrophages in areas of damage and inflammation (Pawelczyk et al. 2009), thereby confounding the interpretation of noninvasive cellular imaging or histology (Jiang et al. 2005; Zeng et al. 2007). However, because genetic reporters may also be incorporated into macrophages, it may be necessary to use multiple probes or stains on histological examination to correctly identify and differentiate the implanted or infused cells from tissue macrophages (Pawelczyk et al. 2008, 2009).

There are a variety of nanoparticles with unique physical properties that can be used as exogenous agents to directly label cells. The commonly used nanomaterials for cell labeling include QDs (Chan and Nie 1998), carbon nanotubes (Balasubramanian and Burghard 2005), and superparamagnetic nanoparticles (Frank et al. 2003). The unique physicochemical properties of nanoparticles allow visualization of cells in vivo using different molecular imaging modalities such as MRI, radionuclide imaging [PET and single-photon emission computed tomography (SPECT)] and optical imaging (Willmann et al. 2008). The unique optical and/or magnetic properties of the nanoparticles permit noninvasive, accurate, and real-time cell tracking (Frangioni 2008). The prolonged intracellular retention of these nanomaterials makes it possible to conduct longitudinal cell tracking in vivo.

TABLE 37.1
Comparison of Different Imaging Modalities

Technique	Resolution	Sensitivity	Depth	Imaging Agents	Application	Main Characteristics	Clinical Translation
MRI	10–100 μm	10^{-3} to 10^{-4} M	No limit	Gadolinum, manganese, iron oxides nanoparticles	Anatomy, physiology, metabolism, cellular, molecular	Versatile, high soft tissue contrast, long imaging time	Yes
CT	25–50 μm	Low	No limit	Iodine	Anatomy, physiology	High spatial resolution (lung and bone)	Yes
Ultrasound	50 μm	Low	<5 cm	Microbubbles	Anatomy, physiology	Vascular interventional	Yes
PET	1–2 mm (small animal scanner); 3–5 mm (clinical system)	10^{-11} to 10^{-12} M	No limit	Positron emitting radioisotopes labeled compounds	Physiology, metabolism, molecular, cellular	Versatile, cyclotron needed	Yes
SPECT	1–2 mm (small animal scanner), 5 mm (clinical system)	10^{-10} to 10^{-11} M	No limit	γ-ray emitting radioisotopes labeled compounds	Physiology, metabolism, molecular, cellular	Imaging with different isotopes simultaneously	Yes
Fluorescence imaging	2–3 mm	~10^{-15} M	<1 cm	NIR fluorochromes (QDs)	Physiology, molecular, cellular	Rapid screening of molecular event	Limited
BLI	2–3 mm	~10^{-15} M	1 cm	D-Luciferin, coelenterazine	Cellular, molecular	Gene expression	No
Intravital microscopy	1 μm	High	400 μm	Fluorochromes	Physiology, molecular, cellular	Limited depth	Limited

Note: MRI, magnetic resonance imaging; PET, positron emission computed tomography; CT, computed tomography; BLI, bioluminescence imaging; NIR, near infrared; SPECT, single photon emission computed tomography.

MAGNETIC RESONANCE IMAGING

Among the different imaging approaches, MRI is particularly attractive because of its superior spatial resolution and lack of ionizing radiation (Anderson and Frank 2007; Weissleder et al. 1997). The high temporal and spatial resolution of clinically relevant MRI is greater than that of nuclear and optical imaging, allowing for detailed morphological images without in-depth limitations. Moreover, physiological and metabolic measures can be obtained with multiple MRI techniques providing additional characterization of the target tissue. To date, MRI has also been widely exploited for molecular/cellular imaging because of its ability to localize labeled cells within tissues, which can serve as a basis for optimizing dose and dose schedule of cell products in early phase cellular therapy trials.

Labeling stem cells or immune cells for detection with MRI is usually achieved by loading the cells in vitro with either paramagnetic contrast agents [i.e., gadolinium (Gd) chelates], experimental or clinically approved SPIONs, or perfluorocarbon nanoparticles. SPIONs have been extensively used because of their unique properties including biocompatibility, high signal contrast, particularly on T_2^*-weighted images, large surface area for functional modification, and easy histological detectability (Perl's iron staining) or electronic microscopy. SPIONs are usually coated with dextran or another carbohydrate (e.g., polyethylene glycol) for improved biocompatibility, reduced nonspecific protein binding, and prolonged circulation in biological systems. Ferumoxides (Feridex, Bayer HealthCare) is a dextran-coated SPION that was approved by the U.S. Food and Drug Administration in 1996 and is sold in Europe under the name Endorem (Guerbet) (Bulte 2009). Ferucarbotran (Resovist) is the second clinically approved superparamagnetic iron oxide developed for contrast-enhanced MRI of the liver in Europe (Reimer and Balzer 2003). To improve the efficiency of endocytosis in cells, transfection agents such as poly-L-lysine (Arbab et al. 2003; Liu et al. 2010), protamine sulfate (Arbab et al. 2004), or cationic liposome (Arbab et al. 2005) have often been used to facilitate endosomal uptake of the SPIONs. Another strategy involves surface modifications of SPIONs through linking to targeting moieties such as HIV Tat peptide or monoclonal antibodies (Ahrens et al. 2003; Bulte and Kraitchman 2004; Lewin et al. 2000; Zhang et al. 2007). Magnetic labeling of cells with SPIONs has little or no effect on proliferation, metabolic expression profiles, reactive oxygen species formation, apoptosis rates, or differentiation in the majority of cells studied (Arbab et al. 2005; Pawelczyk et al. 2006).

SPIONs have been used extensively to track the migration of stem cells, therapeutic immune cells, and many other cell types in animal models (Budde and Frank 2009). Walczak et al. (2008) labeled MSCs with ferumoxides mixed with the transfection agent poly-L-lysine and visualized the delivery of these labeled MSCs with MRI after intra-arterial (IA), but not after intravenous (IV), injection in animal models of stroke. IA stem cell injection during acute stroke resulted in a high variability of cerebral engraftment (Walczak et al. 2008). Daldrup-Link et al. (2005) showed that the human NK cell line NK-92 can be efficiently labeled with ferumoxides and ferucarbotran using simple incubation, lipofection, and electroporation techniques. They demonstrated that accumulation of these labeled cells in murine tumors can be monitored in vivo with MR imaging. Arbab et al. (2006) labeled EPCs with ferumoxide–protamine sulfate (FePro) complexes and injected the cells into mice bearing xenografted glioma cells. Migration and incorporation of labeled EPCs into tumor neovasculature were detected as low signal intensity on MRI at the tumor periphery as early as 3 days after EPC administration in preformed tumors. Confocal microscopy showed incorporation into the neovasculature, and immunohistochemistry clearly demonstrated the maturation of the administered EPCs into endothelial cells. The same group also reported the feasibility of imaging the migration and incorporation of SPION-labeled sensitized splenocytes in an experimental 9L glioma brain tumor model (Arbab et al. 2007). In addition, using SPION labeling of sensitized cytotoxic T cells to glioma antigens, Arbab et al. (2010) could distinguish implanted gliomas from radiation-induced injury in the rat brain based on the hypointense regions on MRI that formed in the tumors but not in areas of parenchymal destruction. These results suggest that magnetically labeled cells could be used to differentiate radiation necrosis from recurrent tumor in patients who have previously received irradiation of a malignant glioma (Figure 37.1). Guzman et

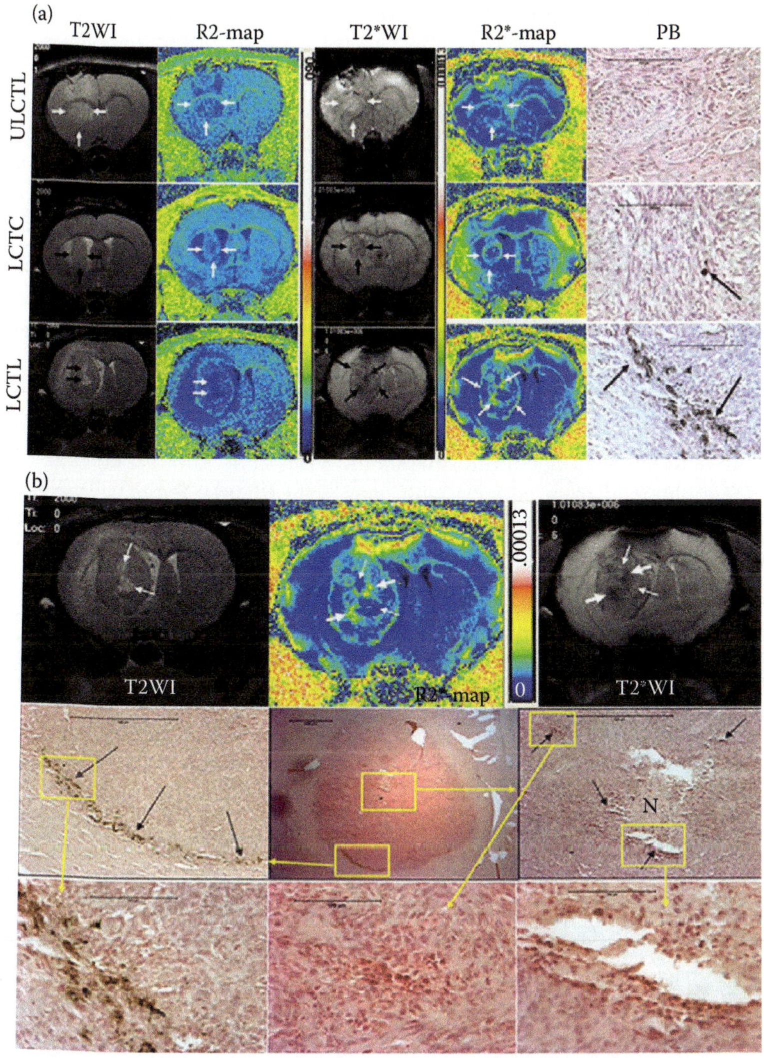

FIGURE 37.1 MRI and Prussian blue positive cells in tumors. (a) T2-weighted and T2*-weighted images and their corresponding R2 and R2* maps and DAB-enhanced Prussian blue staining from representative animals that received unlabeled CTL (ULCTL, upper row); labeled control T-cells (LCTC, middle row); and CTL (LCTL, lower row). Both T2W and T2*W images show well established tumors in the brain; however, low signal intensity areas were only seen in tumors that received LCTC and LCTL. Corresponding R2* maps show high-signal intensity areas. Animals that received LCTL show high-signal intensity areas both at the peripheral and central part of the tumors (arrows). Corresponding DAB-enhanced Prussian blue staining shows multiple Prussian blue positive cells in tumors that received LCTL (arrows). There are a few Prussian blue positive cells seen in a tumor that received LCTC (arrow). No definite Prussian blue positive cells were seen in a tumor that received ULCTL. Areas of necrosis can easily be identified by comparing T2WI and R2 maps (thick arrows) in a tumor that received LCTL (lower row). Bars on images measure 100 mm. (b) Detailed histological analysis of the tumor that received LCTL. Bars on images measure 100 mm. Upper panel: T2-weighted image (T2WI) and T2*-weighted image (T2*WI) show areas of necrosis (high signal intensity on T2WI and T2*WI and low-signal intensity on R2* map). Thin arrows show sites of necrosis, and thick arrows show possible site of accumulated iron-positive cells. Middle panel: representative histological section with similar tumor orientation (within the constraints of the experimental limitations, i.e., 1-mm thick MRI slices versus 10-mm thick histological section) show central necrosis (N) in the tumor with areas of iron-positive cells (black arrows) seen in the central and peripheral part of the tumor that received iron-labeled CTLs. Lower panel: enlarged view of the boxed areas. (Adapted with permission from Arbab, A.S. et al., *PLoS One*, 5, e9365, 2010.)

al. (2007) followed SPION-labeled NSC migration and site-specific differentiation in neonatal and adult rodent brains and in a brain injury model. Anderson et al. (2005) demonstrated the incorporation of SPION-labeled mouse endothelial progenitor cells into the growing neovasculature in a glioma model. Recently, Long et al. (2009) developed an in vivo labeling method based on cell-to-cell transfer of SPIONs from tumor cells to endogenous antigen-presenting cells. In this study, mice were immunized with a tumor cell-based vaccine that was irradiated and labeled with SPIONs. Antigen-presenting cells that had captured SPIONs can be quantitatively imaged over time as they accumulated in lymph nodes.

There are several clinical studies that monitored migration of magnetically labeled stem cells to target tissue by MRI (Callera and de Melo 2007; de Vries et al. 2005; Toso et al. 2008; Zhu et al. 2006). de Vries et al. harvested autologous DCs and dual-labeled the cells with either ferumoxides or ^{111}In oxine. The labeled DCs were coinjected intranodally in melanoma patients under ultrasound guidance as part of an early phase vaccine trial. The results demonstrated that in vivo MR tracking of magnetically labeled cells is feasible in humans for detecting very low numbers of DCs in conjunction with detailed anatomical information (de Vries et al. 2005). Zhu et al. (2006) cultured and labeled NSCs obtained from patients with open-head trauma. Patients received intracerebral injections of either ferumoxides-labeled or unlabeled cells around the injured tissue. Serial MRIs up to 7 weeks demonstrated migration of the magnetically labeled NSCs from injection sites as hypointense voxels (Figure 37.2). Neurologic complications were not reported in patients receiving labeled cells. In patients with spinal cord injury, autologous bone marrow CD34-positive cells labeled with magnetic beads home to the injury site for several weeks after infusion into the cerebral spinal fluid (Callera and de Melo 2007) on MRI. All these studies nonetheless provide strong evidence that tracking cells with MRI could be important clinically.

Although it is promising to use clinically approved SPIONs to label cells for cell trafficking studies, this approach has several limitations including sensitivity of the imaging technique, dilution of the exogenous label with cell division, and potential transfer of the magnetic label to tissue macrophages or bystander cells. For example, once the labeled cells die, SPIONs remain in and around dead cells or are engulfed by host macrophages that may lead to a misinterpretation of MRI findings. It is important to note that, although studies have shown that about 10–20% of macrophages may contain iron nanoparticles transferred from labeled cells in an area of inflammation, a similar number of macrophages have also been found to be positive for genetic reporters that were introduced into the implanted stem cells (Pawelczyk et al. 2008, 2009). Furthermore, MRI cell tracking is also more difficult in certain cases such as traumatic injury in which endogenous blood derivatives exhibit similar susceptibility artifacts as magnetically labeled cells.

Gd- and manganese-based nanoparticle formulations are possible experimental alternatives to SPIONs to label cells and are increasingly used as T_1-based MRI contrast agents. Modo et al. (2004) tracked the migration of intracerebrally injected stem cells with a combination Gd chelate–fluorescent label and detected the presence and migration of cells on T_2-weighted images at least 14 days after injection, with correlative findings by fluorescence microscopy. Gd-based agents with substantially greater T_1 relaxivities have been developed to label cells. Anderson et al. (2006) labeled MSCs with Gd fullerenol, which has a T_1 relaxivity 10-fold greater than that of Gd chelates, and detected an increase in signal intensity on T_1-weighted images obtained at 7 T after direct injection into the rat thigh. Although Gd-based contrast agents do not have the signal void problems associated with SPIONs, Gd-based contrast agents are much less sensitive, because they work by shortening the T_1 of the water protons in the tissue in which they accumulate (Sengar et al. 2009). Moreover, a negative effect of Gd-based agents on cell proliferation has been observed (Anderson et al. 2006; Modo et al. 2004); therefore, further evaluation is needed to ensure that there is no long-term toxicity and that the ability of the cells to repair damage has not been compromised (Brekke et al. 2007). There also remains a need to identify paramagnetic agents that exert strong T_1 effects and allow sufficient detection of cells versus surrounding tissues when there are low numbers of labeled cells or low concentrations of Gd (Anderson et al. 2006).

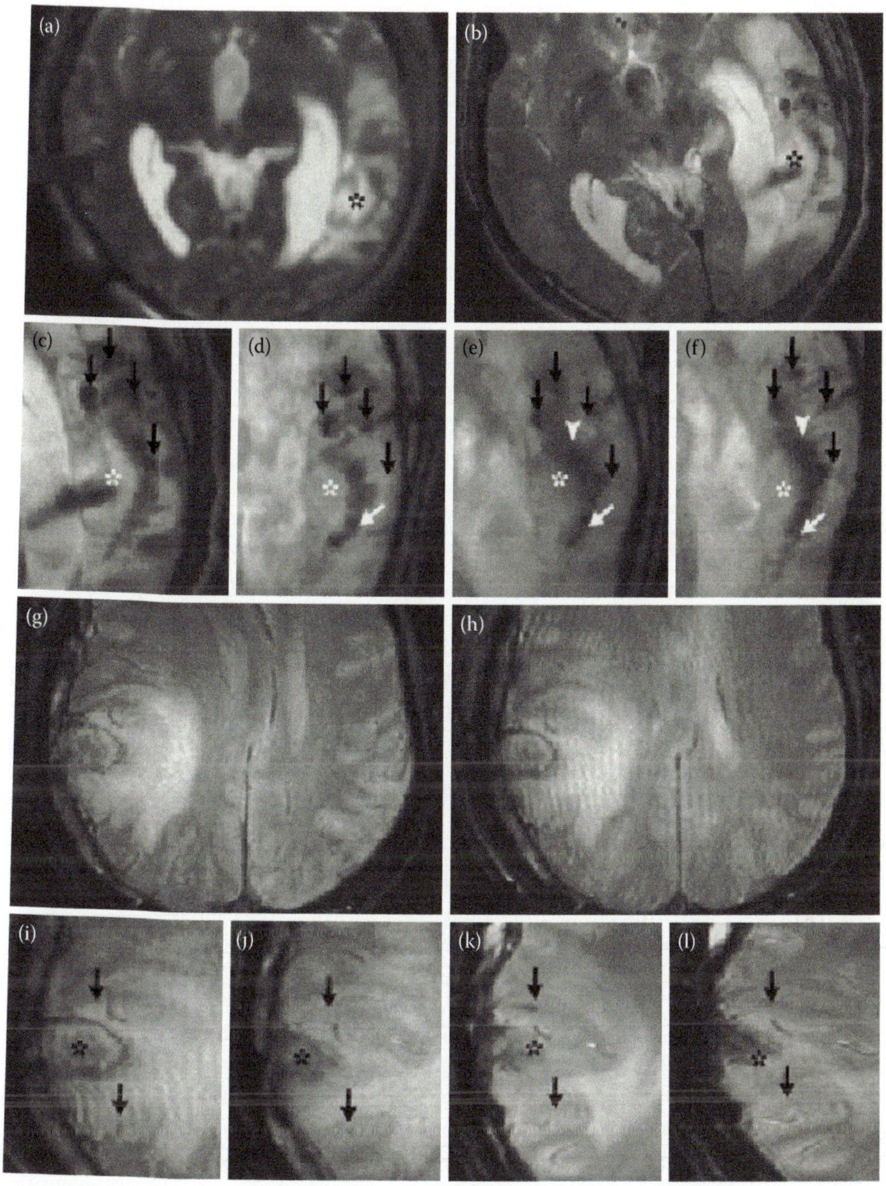

FIGURE 37.2 MRI scans from the patient with traumatic brain injury who received neural stem cells labeled with iron oxide nanoparticles (a–f) and the patient who received unlabeled cells (g–l). The scan obtained before the implantation of the labeled neural stem cells (a) did not show a pronounced hypointense signal around the TBI lesion (asterisk) in left temporal lobe, whereas circular areas of hypointense signal were visible at injection sites 1 day after implantation (b). Magnified images are shown in panels c–f. Four hypointense signals (black arrows) were observed at injection sites around the lesion on day 1 (c), day 7 (d), day 14 (e), and day 21 (f). On day 7 (d), dark signals (white arrow) posterior to the lesion were observed, indicating presence of labeled cells. By day 14 (e), hypointense signals at injection sites had faded, and another dark signal (white arrowhead) had appeared and spread along the border of damaged brain tissue. By day 21 (f), the dark signal had thickened and extended further along the lesion (white arrow). Scans in g and h, from the patient who underwent implantation of unlabeled cells, were obtained on days 0 and 1, respectively; magnified views in i, j, k, and l were obtained on days 1, 7, 14, and 21, respectively. A slightly hypointense signal is present around injection sites in i–l. In these panels, black arrows indicate the hypointense signal, and asterisks indicate the lesion. (Adapted with permission from Zhu, J. et al., *N. Engl. J. Med.*, 355, 2376–8, 2006.)

OPTICAL IMAGING

Optical imaging is one of the most widely used molecular imaging techniques in preclinical settings, although translation of this technique to the clinic is limited by poor tissue penetration of fluorescent labels. By labeling cells with fluorescent probes, optical imaging is an effective means of noninvasively serially tracking cells, thereby providing insight into cell migration to the target site (Sutton et al. 2008). Fluorescence imaging probes can be broadly classified as conventional fluorophores (e.g., organic dyes, fluorescent proteins), QDs, and hybrid architectures combining one or more of these emitters in an inert matrix. Organic fluorophores are widely used for a variety of straightforward shorter term cell labeling applications because of their commercial availability, low cost, and easy handling (Boddington et al. 2008). However, these fluorphores are often subjected to photobleaching and/or quenching and may be sensitive to changes in pH and chemical degradation.

QDs are inorganic fluorescent semiconductor nanoparticles with unique optical features including high quantum yields, high molar excitation coefficients, resistance to photobleaching and chemical degradation, broad excitation and narrow emission spectra, long lifetimes, and large stroke shift (Kim et al. 2008). Moreover, the emission wavelength of QDs is size-dependent and can be fine-tuned to cover a wide range of visible and near-infrared (NIR) spectra by controlling the QD size and composition (Resch-Genger et al. 2008). QDs are preferred over organic fluorophores for multicolor cell labeling (Chan and Nie 1998); because of the nonlinear two-photon absorption cross section and the fluorescence quantum yield provided by QD, there is increased light that is suitable for imaging deep structures. Recently, numerous cell-based applications were discovered for QDs (Noh et al. 2008). Slotkin et al. (2007) reported that direct labeling of neural stem and progenitor cells (NSPCs) was achieved using ultrasound guided biomicroscopy injection of 620 nm COOH-conjugated QDs into the parenchyma of the ventral telencephalon of mouse embryos. QDs were found in principal cell types generated by NSPCs at substantial distances away from the initial site of injection. In another study, Lim et al. (2009a) labeled human NK cells with antibody-coated QDs for fluorescence imaging. The labeled NK cells showed a therapeutic effect similar to that of unlabeled NK cells. QDs labeled NK cells injected directly into the tumors could be visualized using NIR optical imaging.

Recently, up-conversion nanoparticles have been developed as new generation imaging probes. Unlike common QDs, up-conversion nanoparticles exhibit anti-Stokes emission upon low levels of irradiation in the NIR spectral region where biological molecules are optically transparent. Idris et al. (2009) reported the use of up-conversion fluorescent nanoparticles for tracking transplanted cells in a living mouse model of cryo-injured hindlimb. They found that labeling the cells with the particles permitted noninvasive imaging of cells in the living subject for at least 1 week after delivery. In addition to QDs and up-conversion nanoparticles, several other imaging nanoplatforms, such as silica nanoparticles, carbon single wall nanotubes and gold nanorods have been used for cell labeling (Basiruddin et al. 2010; Huang et al. 2005; Mooney et al. 2008). However, in vivo application of these nanoparticles needs further investigation.

NUCLEAR IMAGING

Nuclear imaging techniques, including PET and SPECT are the primary choices for molecular imaging because of their inherent high sensitivity (Dobrucki and Sinusas 2010). Using radionuclides that decay with positron emission, PET imaging leads to high-acquisition efficiency and results in high-quality three-dimensional images or maps of functional processes within the body. SPECT imaging detects γ-rays and the major advantage of SPECT imaging is that it can be used for simultaneous imaging of multiple radionuclides emitting γ-rays with different energies. Thus, SPECT can potentially allow simultaneous detection of multiple biologic events with multiple isotopes, which is not possible with PET. However, compared with PET, SPECT has lower sensitivity and spatial resolution and limited ability to perform follow-up studies. Both imaging methods have been widely used in cellular imaging after labeling cells intrinsically with reporter genes or extrinsically with ^{111}In oxine or ^{18}FDG (Ponomarev 2009).

MULTIMODALITY IMAGING

Ideally, cell labeling and imaging would gather information at both anatomic and functional levels with high spatial and temporal resolution. Unfortunately, no single modality is sufficient to provide all these details because each modality has its own limitations as shown in Table 37.1. For example, MRI can provide three-dimensional tomography but is limited by low target sensitivity, whereas PET and optical imaging have good sensitivity but suffer from low spatial resolution or tissue penetration. To capitalize on the strengths of different imaging methods, multimodality imaging has become attractive for both preclinical and clinical studies (Chen 2006; Pichler et al. 2008b, 2010). Multimodality imaging assembles anatomical, functional, and molecular information by combining images from different modalities taken at the same time point. This strategy combines the strengths of different modalities and yields a hybrid imaging platform (i.e., MRI/PET or PET/CT) with characteristics superior to those of any of its constituents considered alone (Iagaru et al. 2009; Pichler et al. 2008a). Coregistration of MRI images with PET provides the anatomic landscape for localizing the molecular information (Pichler et al. 2010). Multimodality imaging can also be achieved after the cells are directly labeled with multifunctional imaging probes (Chen et al. 2008, 2009; Cormode et al. 2008; Hsu et al. 2007; Lee and Chen 2009; Wang et al. 2010; Xie et al. 2010). NPs having large surface areas allow multiple functional moieties to be incorporated for multimodality molecular imaging (Choi et al. 2007; Tallury et al. 2008).

Lim et al. (2009b) synthesized a novel multifunctional probe for both ^{19}F-based MRI and NIR optical imaging, with perfluorooctyl bromide (PFOB) and IRDye800 selected as ^{19}F MR imaging nanoprobes and NIR fluorescent dyes, respectively. The ^{1}H-based MR provides a whole-body image, whereas the ^{19}F-based MR shows only signals generated from the injected DCs labeled with IRDye800-coated PFOB nanoemulsions. Several groups developed ion sensing nanoparticle formulation composed of a superparamagnetic iron oxide core encapsulated within a porous silica shell for multimodality imaging (Lu et al. 2007; Patel et al. 2010). The latter could be readily anchored with ligands capable of coordinating with fluorescent dyes or positron-emitting metals. This hybrid nanoparticle has great potential for use in cell tracking with both MRI and optical/PET functions (Patel et al. 2010). Park et al. (2010) described silica-coated magnetic nanoparticles incorporating rhodamine B isothiocyanate, MNPs@SiO_2(RITC) for multimodal visualization of human MSCs in vivo. The transplanted hMSCs were successfully detected using MR and optical imaging, indicating that MNPs@SiO_2(RITC) can be applied for multimodal cell labeling and tracking. Shi et al. (2010) recently developed bifunctional anionic Eu^{3+}-doped Gd_2O_3 hybrid nanoparticles as a luminescent and T_1-weighted MRI contrast agent for stem cell labeling. The endosomal uptake of these nanoparticles into hMSCs was confirmed by confocal laser scanning microscopy after a 2-h incubation. Although multimodality imaging has the potential to broaden our horizons in the emerging field of molecular imaging, the novel imaging technology may not be clinically routine before its impact on diagnostic accuracy has been proven, and before its effect on subsequent therapeutic management and cost-effectiveness has been considered and validated (Pichler et al. 2010).

CONCLUSION AND PERSPECTIVES

Using different imaging modalities and probes, molecular imaging has spurred the development of exciting new cell-based therapeutic strategies. Tracking and monitoring transplanted cells using nanoparticles has yielded promising results both in vitro and in vivo. Incorporation of noninvasive imaging techniques and cellular labeling will enable novel treatment trials to be designed and lead to the development of strategies to optimize cell dosage, timing, and frequency to maximize therapeutic response. SPIONs have emerged as the most widely adopted cell labeling probes for

T_2-based MR imaging. Gd- and manganese-based nanoparticle formulations are increasingly used as T_1-based MRI contrast agents but have limitations with regard to sensitivity and potential toxicity. For optical imaging, QDs remain the most useful cell labeling probe in experimental studies (Bhirde et al. 2011).

Most probes and cell labeling methods have been studied in either an in vitro environment or in preclinical models. In order to be translated into clinical applications, comprehensive biosafety studies related to toxicity and immunogenicity will be required. Most studies have shown that in the short term, cell labeling with various nanoparticles, including QDs, SPIONs, and nanotubes, did not alter cell viability, differentiation, and function. However, systemic toxicity is the least understood feature of nanoparticles and needs to be addressed in detail before they can be applied clinically. Once introduced into systemic circulation, toxicity can occur from the NPs themselves or from the associated components of the NPs that might be released during their degradation in vivo. Traditionally, QDs are composed of cadmium salts that can release soluble cadmium after cellular labeling leading to toxicity; some reports suggest that surface modification of the QDs can address this issue (Gao et al. 2010; Rzigalinski and Strobl 2009). Toxicity of the NPs can also be influenced by several other factors such as concentration, hydrodynamic size, surface charge and type of coatings, and route of administration (Holgate 2010). Lastly, the ability to label cells at the bench may not easily translate into labeling clinical cell products in an appropriate manufacturing practice facility.

Quantifying the number of labeled therapeutic cells in target tissues is important for optimizing dose and timing of cellular therapy. However, to our knowledge there is no optimized technique available to determine the number of administered labeled cells in vivo (Arbab and Frank 2008). With multiple functional nanoplatforms as imaging agents, multimodality imaging seeks to overcome the limits of current molecular diagnostics and permit diagnosis as well as to develop cell-based therapy. Future investigation is needed to address sensitivity and determine the detection thresholds for the different modalities. For example, PET is a highly sensitive imaging modality requiring introduction of only a trace amount of probes, whereas a relatively large number of probes is needed for MRI. Moreover, the design of multifunctional probes needs to be optimized because functional group modification of the probes during conjugation may change their chemical properties. Fundamentally understanding the molecular interactions between the components of multifunctional probes at the nanoscale is critical for assembling these components into efficacious nanocomposite particles. The influence of one structural component on the performance of the others must be carefully investigated to ensure synergy in the integrated design and the utility of the probe for molecular imaging in the CNS.

ACKNOWLEDGMENTS

This work was supported by NIH Intramural Research Program. G.N. acknowledges an Imaging Sciences Training Fellowship jointly funded by the Radiology and Imaging Sciences Department of NIH Clinical Center and by the Intramural Research Program, NIBIB, NIH.

REFERENCES

Aboody, K.S., Brown, A., Rainov, N.G. et al. 2000. Neural stem cells display extensive tropism for pathology in adult brain: evidence from intracranial gliomas. *Proc Natl Acad Sci U S A* 97: 12846–51.

Ahrens, E.T., Feili-Hariri, M., Xu, H. et al. 2003. Receptor-mediated endocytosis of iron-oxide particles provides efficient labeling of dendritic cells for in vivo MR imaging. *Magn Reson Med* 49: 1006–13.

Aldrich, J.F., Lowe, D.B., Shearer, M.H. et al. 2010. Vaccines and immunotherapeutics for the treatment of malignant disease. *Clin Dev Immunol* 2010: 697158.

Anderson, S.A., and Frank, J.A. 2007. MRI of mouse models of neurological disorders. *NMR Biomed* 20: 200–15.

Anderson, S.A., Glod, J., Arbab, A.S. et al. 2005. Noninvasive MR imaging of magnetically labeled stem cells to directly identify neovasculature in a glioma model. *Blood* 105: 420–5.

Anderson, S.A., Lee, K.K., and Frank, J.A. 2006. Gadolinium-fullerenol as a paramagnetic contrast agent for cellular imaging. *Invest Radiol* 41: 332–8.

Arbab, A.S., Bashaw, L.A., Miller, B.R. et al. 2003. Characterization of biophysical and metabolic properties of cells labeled with superparamagnetic iron oxide nanoparticles and transfection agent for cellular MR imaging. *Radiology* 229: 838–46.

Arbab, A.S., and Frank, J.A. 2008. Cellular MRI and its role in stem cell therapy. *Regen Med* 3: 199–215.

Arbab, A.S., Janic, B., Jafari-Khouzani, K. et al. 2010. Differentiation of glioma and radiation injury in rats using in vitro produce magnetically labeled cytotoxic T-cells and MRI. *PLoS One* 5: e9365.

Arbab, A.S., Pandit, S.D., Anderson, S.A. et al. 2006. Magnetic resonance imaging and confocal microscopy studies of magnetically labeled endothelial progenitor cells trafficking to sites of tumor angiogenesis. *Stem Cells* 24: 671–8.

Arbab, A.S., Rad, A.M., Iskander, A.S. et al. 2007. Magnetically-labeled sensitized splenocytes to identify glioma by MRI: a preliminary study. *Magn Reson Med* 58: 519–26.

Arbab, A.S., Yocum, G.T., Kalish, H. et al. 2004. Efficient magnetic cell labeling with protamine sulfate complexed to ferumoxides for cellular MRI. *Blood* 104: 1217–23.

Arbab, A.S., Yocum, G.T., Rad, A.M. et al. 2005. Labeling of cells with ferumoxides–protamine sulfate complexes does not inhibit function or differentiation capacity of hematopoietic or mesenchymal stem cells. *NMR Biomed* 18: 553–9.

Balasubramanian, K., and Burghard, M. 2005. Chemically functionalized carbon nanotubes. *Small* 1: 180–92.

Basiruddin, S.K., Saha, A., Pradhan, N. et al. 2010. Functionalized gold nanorod solution via reverse micelle based polyacrylate coating. *Langmuir* 26: 7475–81.

Bhirde, A., Xie, J., Swierczewska, M. et al. 2011. Nanoparticles for cell labeling. *Nanoscale* 3: 142–53.

Boddington, S., Henning, T.D., Sutton, E.J. et al. 2008. Labeling stem cells with fluorescent dyes for non-invasive detection with optical imaging. *J Vis Exp* 14: e686.

Brekke, C., Morgan, S.C., Lowe, A.S. et al. 2007. The in vitro effects of a bimodal contrast agent on cellular functions and relaxometry. *NMR Biomed* 20: 77–89.

Budde, M.D., and Frank, J.A. 2009. Magnetic tagging of therapeutic cells for MRI. *J Nucl Med* 50: 171–4.

Bulte, J.W. 2009. In vivo MRI cell tracking: clinical studies. *AJR Am J Roentgenol* 193: 314–25.

Bulte, J.W., and Kraitchman, D.L. 2004. Monitoring cell therapy using iron oxide MR contrast agents. *Curr Pharm Biotechnol* 5: 567–84.

Callera, F., and de Melo, C.M. 2007. Magnetic resonance tracking of magnetically labeled autologous bone marrow CD34+ cells transplanted into the spinal cord via lumbar puncture technique in patients with chronic spinal cord injury: CD34+ cells' migration into the injured site. *Stem Cells Dev* 16: 461–6.

Carpenter, M.K., Frey-Vasconcells, J., and Rao, M.S. 2009. Developing safe therapies from human pluripotent stem cells. *Nat Biotechnol* 27: 606–13.

Chan, W.C., and Nie, S. 1998. Quantum dot bioconjugates for ultrasensitive nonisotopic detection. *Science* 281: 2016–8.

Chen, K., Li, Z.B., Wang, H. et al. 2008. Dual-modality optical and positron emission tomography imaging of vascular endothelial growth factor receptor on tumor vasculature using quantum dots. *Eur J Nucl Med Mol Imaging* 35: 2235–44.

Chen, K., Xie, J., Xu, H. et al. 2009. Triblock copolymer coated iron oxide nanoparticle conjugate for tumor integrin targeting. *Biomaterials* 30: 6912–9.

Chen, X. 2006. Multimodality imaging of tumor integrin alphavbeta3 expression. *Mini Rev Med Chem* 6: 227–34.

Choi, J.H., Nguyen, F.T., Barone, P.W. et al. 2007. Multimodal biomedical imaging with asymmetric single-walled carbon nanotube/iron oxide nanoparticle complexes. *Nano Lett* 7: 861–7.

Cormode, D.P., Skajaa, T., van Schooneveld, M.M. et al. 2008. Nanocrystal core high-density lipoproteins: a multimodality contrast agent platform. *Nano Lett* 8: 3715–23.

Daldrup-Link, H.E., Meier, R., Rudelius, M. et al. 2005. In vivo tracking of genetically engineered, anti-HER2/neu directed natural killer cells to HER2/neu positive mammary tumors with magnetic resonance imaging. *Eur Radiol* 15: 4–13.

de Vries, I.J., Lesterhuis, W.J., Barentsz, J.O. et al. 2005. Magnetic resonance tracking of dendritic cells in melanoma patients for monitoring of cellular therapy. *Nat Biotechnol* 23: 1407–13.

Dickson, P.V., Hamner, J.B., Burger, R.A. et al. 2007. Intravascular administration of tumor tropic neural progenitor cells permits targeted delivery of interferon-beta and restricts tumor growth in a murine model of disseminated neuroblastoma. *J Pediatr Surg* 42: 48–53.

Disis, M.L., Bernhard, H., and Jaffee, E.M. 2009. Use of tumour-responsive T cells as cancer treatment. *Lancet* 373: 673–83.

Dobrucki, L.W., and Sinusas, A.J. 2010. PET and SPECT in cardiovascular molecular imaging. *Nat Rev Cardiol* 7: 38–47.

Eksioglu, E.A., Eisen, S., and Reddy, V. 2010. Dendritic cells as therapeutic agents against cancer. *Front Biosci* 15: 321–47.

Frangioni, J.V. 2008. New technologies for human cancer imaging. *J Clin Oncol* 26: 4012–21.

Frank, J.A., Miller, B.R., Arbab, A.S. et al. 2003. Clinically applicable labeling of mammalian and stem cells by combining superparamagnetic iron oxides and transfection agents. *Radiology* 228: 480–7.

Gao, J., Chen, K., Xie, R. et al. 2010. In vivo tumor-targeted fluorescence imaging using near-infrared non-cadmium quantum dots. *Bioconjug Chem* 21: 604–9.

Guzman, R., Uchida, N., Bliss, T.M. et al. 2007. Long-term monitoring of transplanted human neural stem cells in developmental and pathological contexts with MRI. *Proc Natl Acad Sci U S A* 104: 10211–6.

Holgate, S.T. 2010. Exposure, uptake, distribution and toxicity of nanomaterials in humans. *J Biomed Nanotechnol* 6: 1–19.

Hsu, A.R., Cai, W., Veeravagu, A. et al. 2007. Multimodality molecular imaging of glioblastoma growth inhibition with vasculature-targeting fusion toxin VEGF121/rGel. *J Nucl Med* 48: 445–54.

Huang, D.M., Hung, Y., Ko, B.S. et al. 2005. Highly efficient cellular labeling of mesoporous nanoparticles in human mesenchymal stem cells: implication for stem cell tracking. *FASEB J* 19: 2014–6.

Iagaru, A., Mittra, E., Yaghoubi, S.S. et al. 2009. Novel strategy for a cocktail 18F-fluoride and 18F-FDG PET/CT scan for evaluation of malignancy: results of the pilot-phase study. *J Nucl Med* 50: 501–5.

Idris, N.M., Li, Z., Ye, L. et al. 2009. Tracking transplanted cells in live animal using upconversion fluorescent nanoparticles. *Biomaterials* 30: 5104–13.

Jiang, Q., Zhang, Z.G., Ding, G.L. et al. 2005. Investigation of neural progenitor cell induced angiogenesis after embolic stroke in rat using MRI. *Neuroimage* 28: 698–707.

Kang, J.H., and Chung, J.K. 2008. Molecular–genetic imaging based on reporter gene expression. *J Nucl Med* 49 Suppl 2: 164S–79S.

Kapp, M., Rasche, L., Einsele, H. et al. 2009. Cellular therapy to control tumor progression. *Curr Opin Hematol* 16: 437–43.

Kim, B.Y., Jiang, W., Oreopoulos, J. et al. 2008. Biodegradable quantum dot nanocomposites enable live cell labeling and imaging of cytoplasmic targets. *Nano Lett* 8: 3887–92.

Kim, S.K., Kim, S.U., Park, I.H. et al. 2006. Human neural stem cells target experimental intracranial medulloblastoma and deliver a therapeutic gene leading to tumor regression. *Clin Cancer Res* 12: 5550–6.

Kim, S.U. 2007. Genetically engineered human neural stem cells for brain repair in neurological diseases. *Brain Dev* 29: 193–201.

Kim, S.U., and de Vellis, J. 2009. Stem cell-based cell therapy in neurological diseases: a review. *J Neurosci Res* 87: 2183–200.

Lee, S., and Chen, X. 2009. Dual-modality probes for in vivo molecular imaging. *Mol Imaging* 8: 87–100.

Lee, Z., Dennis, J.E., and Gerson, S.L. 2008. Imaging stem cell implant for cellular-based therapies. *Exp Biol Med (Maywood)* 233: 930–40.

Lewin, M., Carlesso, N., Tung, C.H. et al. 2000. Tat peptide-derivatized magnetic nanoparticles allow in vivo tracking and recovery of progenitor cells. *Nat Biotechnol* 18: 410–4.

Lim, Y.T., Cho, M.Y., Noh, Y.W. et al. 2009a. Near-infrared emitting fluorescent nanocrystals-labeled natural killer cells as a platform technology for the optical imaging of immunotherapeutic cells-based cancer therapy. *Nanotechnology* 20: 475102.

Lim, Y.T., Noh, Y.W., Kwon, J.N. et al. 2009b. Multifunctional perfluorocarbon nanoemulsions for (19)F based magnetic resonance and near-infrared optical imaging of dendritic cells. *Chem Commun (Camb)* 6952–4.

Liu, G., Yang, H., Zhang, X.M. et al. 2010. MR imaging for the longevity of mesenchymal stem cells labeled with poly-L-lysine–resovist complexes. *Contrast Media Mol Imaging* 5: 53–8.

Long, C.M., van Laarhoven, H.W., Bulte, J.W. et al. 2009. Magnetovaccination as a novel method to assess and quantify dendritic cell tumor antigen capture and delivery to lymph nodes. *Cancer Res* 69: 3180–7.

Lu, C.W., Hung, Y., Hsiao, J.K. et al. 2007. Bifunctional magnetic silica nanoparticles for highly efficient human stem cell labeling. *Nano Lett* 7: 149–54.

Michalet, X., Pinaud, F.F., Bentolila, L.A. et al. 2005. Quantum dots for live cells, in vivo imaging, and diagnostics. *Science* 307: 538–44.

Modo, M., Mellodew, K., Cash, D. et al. 2004. Mapping transplanted stem cell migration after a stroke: a serial, in vivo magnetic resonance imaging study. *Neuroimage* 21: 311–7.

Mooney, E., Dockery, P., Greiser, U. et al. 2008. Carbon nanotubes and mesenchymal stem cells: biocompatibility, proliferation and differentiation. *Nano Lett* 8: 2137–43.
Noh, Y.W., Lim, Y.T., and Chung, B.H. 2008. Noninvasive imaging of dendritic cell migration into lymph nodes using near-infrared fluorescent semiconductor nanocrystals. *FASEB J* 22: 3908–18.
Park, K.S., Tae, J., Choi, B. et al. 2010. Characterization, in vitro cytotoxicity assessment, and in vivo visualization of multimodal, RITC-labeled, silica-coated magnetic nanoparticles for labeling human cord blood-derived mesenchymal stem cells. *Nanomedicine* 6: 263–76.
Patel, D., Kell, A., Simard, B. et al. 2010. Cu^{2+}-labeled, spion loaded porous silica nanoparticles for cell labeling and multifunctional imaging probes. *Biomaterials* 31: 2866–73.
Pawelczyk, E., Arbab, A.S., Chaudhry, A. et al. 2008. In vitro model of bromodeoxyuridine or iron oxide nanoparticle uptake by activated macrophages from labeled stem cells: implications for cellular therapy. *Stem Cells* 26: 1366–75.
Pawelczyk, E., Arbab, A.S., Pandit, S. et al. 2006. Expression of transferrin receptor and ferritin following ferumoxides-protamine sulfate labeling of cells: implications for cellular magnetic resonance imaging. *NMR Biomed* 19: 581–92.
Pawelczyk, E., Jordan, E.K., Balakumaran, A. et al. 2009. *In vivo* transfer of intracellular labels from locally implanted bone marrow stromal cells to resident tissue macrophages. *PLoS One* 4: e6712.
Pham, W., Kobukai, S., Hotta, C. et al. 2009. Dendritic cells: therapy and imaging. *Expert Opin Biol Ther* 9: 539–64.
Pichler, B.J., Judenhofer, M.S., and Pfannenberg, C. 2008a. Multimodal imaging approaches: PET/CT and PET/MRI. *Handb Exp Pharmacol* 109–32.
Pichler, B.J., Kolb, A., Nagele, T. et al. 2010. PET/MRI: paving the way for the next generation of clinical multimodality imaging applications. *J Nucl Med* 51: 333–6.
Pichler, B.J., Wehrl, H.F., and Judenhofer, M.S. 2008b. Latest advances in molecular imaging instrumentation. *J Nucl Med* 49 Suppl 2: 5S–23S.
Ponomarev, V. 2009. Nuclear imaging of cancer cell therapies. *J Nucl Med* 50: 1013–6.
Reimer, P., and Balzer, T. 2003. Ferucarbotran (resovist): a new clinically approved res-specific contrast agent for contrast-enhanced MRI of the liver: properties, clinical development, and applications. *Eur Radiol* 13: 1266–76.
Resch-Genger, U., Grabolle, M., Cavaliere-Jaricot, S. et al. 2008. Quantum dots versus organic dyes as fluorescent labels. *Nat Methods* 5: 763–75.
Rzigalinski, B.A., and Strobl, J.S. 2009. Cadmium-containing nanoparticles: perspectives on pharmacology and toxicology of quantum dots. *Toxicol Appl Pharmacol* 238: 280–8.
Schoenhard, J.A., and Hatzopoulos, A.K. 2010. Stem cell therapy: pieces of the puzzle. *J Cardiovasc Transl Res* 3: 49–60.
Seggewiss, R., Pittaluga, S., Adler, R.L. et al. 2006. Acute myeloid leukemia is associated with retroviral gene transfer to hematopoietic progenitor cells in a rhesus macaque. *Blood* 107: 3865–7.
Sengar, R.S., Spokauskiene, L., Steed, D.P. et al. 2009. Magnetic resonance imaging-guided adoptive cellular immunotherapy of central nervous system tumors with a T1 contrast agent. *Magn Reson Med* 62: 599–606.
Shablak, A., Hawkins, R.E., Rothwell, D.G. et al. 2009. T cell-based immunotherapy of metastatic renal cell carcinoma: modest success and future perspective. *Clin Cancer Res* 15: 6503–10.
Shi, Z., Neoh, K.G., Kang, E.T. et al. 2010. Bifunctional Eu(3+)-doped Gd(2)O(3) nanoparticles as a luminescent and T(1) contrast agent for stem cell labeling. *Contrast Media Mol Imaging* 5: 105–11.
Slotkin, J.R., Chakrabarti, L., Dai, H.N. et al. 2007. *In vivo* quantum dot labeling of mammalian stem and progenitor cells. *Dev Dyn* 236: 3393–401.
Sutlu, T., and Alici, E. 2009. Natural killer cell-based immunotherapy in cancer: current insights and future prospects. *J Intern Med* 266: 154–81.
Sutton, E.J., Henning, T.D., Pichler, B.J. et al. 2008. Cell tracking with optical imaging. *Eur Radiol* 18: 2021–32.
Tallury, P., Payton, K., and Santra, S. 2008. Silica-based multimodal/multifunctional nanoparticles for bioimaging and biosensing applications. *Nanomedicine (Lond)* 3: 579–92.
Thu, M.S., Najbauer, J., Kendall, S.E. et al. 2009. Iron labeling and pre-clinical MRI visualization of therapeutic human neural stem cells in a murine glioma model. *PLoS One* 4: e7218.
Toso, C., Vallee, J.P., Morel, P. et al. 2008. Clinical magnetic resonance imaging of pancreatic islet grafts after iron nanoparticle labeling. *Am J Transplant* 8: 701–6.
Walczak, P., Zhang, J., Gilad, A.A. et al. 2008. Dual-modality monitoring of targeted intraarterial delivery of mesenchymal stem cells after transient ischemia. *Stroke* 39: 1569–74.

Wang, P., He, J., Wang, P.N. et al. 2010. Poly(*n*-isopropylacrylamide)-coated multifunctional nanoparticles for cell tracking. *Photomed Laser Surg* 28: 201–5.
Weissleder, R., Cheng, H.C., Bogdanova, A. et al. 1997. Magnetically labeled cells can be detected by mr imaging. *J Magn Reson Imaging* 7: 258–63.
Weissleder, R., and Mahmood, U. 2001. Molecular imaging. *Radiology* 219: 316–33.
Willmann, J.K., van Bruggen, N., Dinkelborg, L.M. et al. 2008. Molecular imaging in drug development. *Nat Rev Drug Discov* 7: 591–607.
Xie, J., Chen, K., Huang, J. et al. 2010. PET/NIRF/MRI triple functional iron oxide nanoparticles. *Biomaterials* 31: 3016–22.
Yaghoubi, S.S., Jensen, M.C., Satyamurthy, N. et al. 2009. Noninvasive detection of therapeutic cytolytic T cells with 18F-FHBG pet in a patient with glioma. *Nat Clin Pract Oncol* 6: 53–8.
Zeng, L., Hu, Q., Wang, X. et al. 2007. Bioenergetic and functional consequences of bone marrow-derived multipotent progenitor cell transplantation in hearts with postinfarction left ventricular remodeling. *Circulation* 115: 1866–75.
Zhang, C., Jugold, M., Woenne, E.C. et al. 2007. Specific targeting of tumor angiogenesis by RGD-conjugated ultrasmall superparamagnetic iron oxide particles using a clinical 1.5-T magnetic resonance scanner. *Cancer Res* 67: 1555–62.
Zhu, J., Zhou, L., and XingWu, F. 2006. Tracking neural stem cells in patients with brain trauma. *N Engl J Med* 355: 2376–8.

38 Micro- and Nanotechnology in Nerve Repair

Ali C. Ravanpay and Michel Kliot

CONTENTS

INTRODUCTION

Rodin's most formative lesson to Camille Claudel was to "see in depths and not in surfaces." Deconstructing and manipulating surfaces into smaller building units have defined the evolution of not only fine arts but also medicine in the twentieth century. Because of several decades of deconstructionist research, current understanding of disease processes expands beyond the gross anatomic changes produced by trauma or chronic illness and encompasses cellular and subcellular elements involved in various pathologies. Looking ahead, the most exciting and challenging frontiers of tomorrow's medicine are defined by applying this understanding to manipulate individual cells and their parts to reconstruct damaged organs.

In neuroscience, the tradition of seeing beyond surfaces dates back to the early twentieth century, when Ramon y Cajal produced stunning images of complex neuronal networks (Albright et al. 2001). Since then, neuroscientists have come a long way by studying the origins and life cycle of neurons. Similar to other areas of the biosciences, more recently, with the availability of micro- and nano-scale tools, the goal of neuroscience has become to manipulate and repair damaged neuronal connectivity and function.

Neuronal injury, which frequently results in neuronal death, remains a devastating medical condition. In the United States alone, 12,000 new cases of spinal cord injury occur annually (National Spinal Cord Injury Statistical Center 2011). About one in 60 people live with some form of paralysis (Fox et al. 2008). Unlike other organs, regeneration in the nervous system is a rare and challenging process for multiple reasons. Despite recent research identifying small pockets of neuronal stem cells, neuronal regeneration after injury in humans is extremely rare and almost nonexistent. To overcome the devastating consequences of neuronal loss, several areas of research have evolved over the past decades. One area has focused on understanding neural stem cell biology and possible transplantation (Peljto and Wichterle 2011). In parallel, research has focused on studying mechanisms of cell and axonal death secondary to injury, and developing ways of avoiding or delaying it in an attempt to afford sufficient time for axonal repair. Unlike other organs, the challenge of repair in the nervous system is not just limited to replacing dead or injured cell populations. Equally important is the requirement for reestablishing appropriate connections within existing circuits.

The peripheral nervous system has offered a unique system, both in vivo and in vitro, not only to study the mechanisms of neuronal and axonal injury but also to explore various repair mechanisms. Peripheral nerves are known to regenerate and reconnect with their distal fragments after injury; however, this is an exceedingly slow process approximating 2 mm/day (de Ruiter et al. 2009). Several challenges to successfully complete this process include ensuring survival of the neuronal cell body in the spinal cord or dorsal root ganglia; reapproximating the nerve endings after a transection injury; regenerating axons over long distances with many branch points; and finally survival, and successful reinnervation, of the target end organs such as muscle and sensory receptors. A functional consequence of the generally slow process of axonal growth and the mechanical interference of connective tissue injury is the prolonged absence of retrograde input to the neuronal cell body, which can result in neuronal degeneration and target muscle atrophy.

These challenges in axonal repair and recovery have spawned two dynamic areas of research. One area has focused on devising methods to physically reconnect the severed axonal ends rapidly and effectively, which in turn would prevent subsequent retrograde, and target death. Second, is the ability to maintain and support the neuronal cell body and remnant distal axons to prevent cell death until axonal repair is achieved and signaling is reestablished.

With the advent of micro- and nano-scale technologies, which allow manipulating neurons and their processes, new approaches were proposed to facilitate axonal repair through approximation, alignment, and fusion.

AXONAL REPAIR

The earliest reference to suturing severed axons dates back to Paul of Aegina in the seventh century. Eleven centuries later, French surgeons Laugier and Nelaton performed the first nerve suture in humans (Sretavan et al. 2005). Since then and despite decades of surgically manipulating injured axonal ends, the outcomes remain suboptimal. Although in recent years, there has been a strong interest in exploiting pockets of potential stem cells within the central nervous system to replace injured neurons, a pragmatic focus of nerve repair remains inducing axonal regrowth and reconnectivity. The ability to reconnect severed neuronal connections allows for rapid repair and helps to circumvent the secondary and frequently irreversible atrophy of target organs. The current surgical approach to axonal repair is fairly crude and involves manually aligning nerve stumps under the microscope and using sutures or fibrin glue to reapproximate the severed ends. Despite using current magnification tools, this procedure remains imprecise and imperfect, leading to suboptimal outcomes. It should be noted that with current surgical methods, nerve ends are approximated and sutured, but there is no specific intervention at the scale of individual axons that form the basis of electrical transmission. Nerve repair today still depends on axon regeneration. Surgically manipulating nerves or ends of nerves that are millimeters in scale is highly unlikely to be effective in specifically aligning and rejoining individual axons that are microns in diameter.

Recently, with the development of micro- and nano-scale technologies, exciting and innovative approaches were proposed to enable surgical intervention at the level of individual axons and to reconnect injured axonal ends. These novel approaches may be deconstructed into three simple steps: first, removal of unhealthy edges of axons immediately after injury or axonal trimming; second, effective alignment of the axonal edges; and third, the actual fusion of axonal ends. Micro- and nano-scale technologies offer new possibilities to visualize and manipulate axonal edges and endings with the potential of revolutionizing axonal repair.

AXONAL TRIMMING

The first step in interventional axonal repair is trimming unhealthy and injured axonal edges. The nano-knife is a miniaturized cutting tool produced using silicon microfabrication and microassembly techniques and has been used both in vitro and in vivo (Wheeler et al. 1999). The cutting edge of the nano-knife, effectively the apex of a pyramid, forms a nano-scale cutting surface with 20-nm radius of curvature, which corresponds to the diameter of a single microtubule. This degree of precision is very important to avoid damaging the remainder of the axon (Figure 38.1). The nano-knife has been tested extensively to incise both myelinated and unmyelinated axons in vitro. Furthermore, this tool was used to make precise repeated cuts of individual axons in vivo in sciatic nerve samples of mice (Chang et al. 2007) (Figure 38.2).

In addition to a mechanical nano-knife, laser microbeams have also been used to perform high-precision cutting (Berns 2007). Not surprisingly, the microbeam involves very brief pulses to avoid

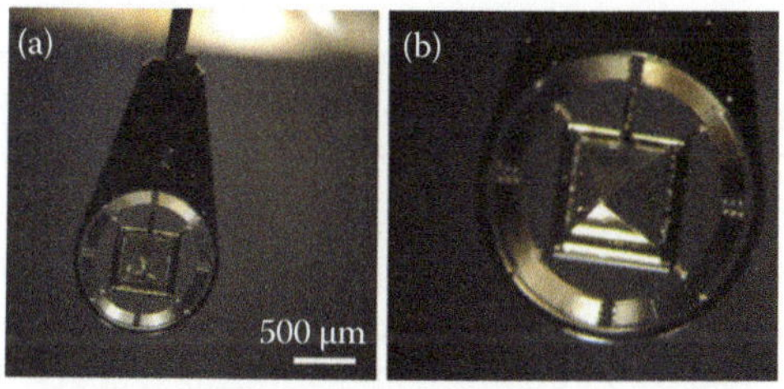

FIGURE 38.1 Photos showing microfabricated nano-knife for axon surgery. (a) The nano-knife consists of a pyramidal-shaped silicon nitride cutting shell mounted on a four-point silicon suspension for optimal device compliance during use. (b) Higher magnification view showing apex of pyramidal-shaped cutting shell that serves as cutting edge. Cutting edges with 20 nm radius of curvature can be achieved. (Photo courtesy of Mynosys Cellular Devices.)

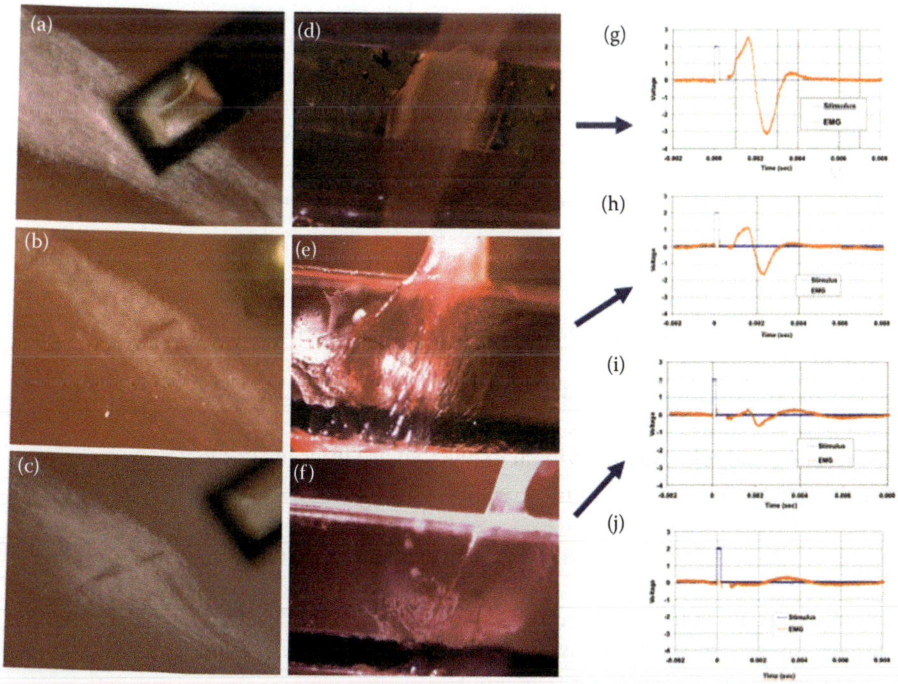

FIGURE 38.2 Axon nano-knife used in a mouse model of sciatic nerve surgery. (a–c) Images of axon nano-knife use, and resultant cuts in a branch of the sciatic nerve. (d–f) Progressive paring away of axons from a sciatic nerve branch using axon nano knife. (g–i) EMG recorded distally from the surgical site showing progressive decrease in EMG amplitude with nerve cutting. (j) Elimination of EMG signal after last axons were cut. Note that a single axon nanoknife was used for the entire cutting sequence, indicating that microdevices for axon surgery are physically robust.

temperature-related damage to adjacent structures. Photosensitization was used in conjunction with laser microbeam technology to reduce the necessary length of time and the unwanted potential damage to the axonal edge (Guo et al. 2008).

AXONAL ALIGNMENT

Using microelectromechanical systems (MEMS) to construct micro-scale tools has made reconnecting axonal edges possible. In the "direct repair" model, the injured edges of the severed axons are trimmed and subsequently reapproximated and then fused in a process similar to molecular splicing (Chang et al. 2008).

Dielectrophoresis (DEP) is an effective noncontact mechanism of aligning axonal edges. In a nonhomogeneous electric field, the gradient of field strengths will result in movement of objects in one direction (Gascoyne and Vykoukal 2002). This property has been exploited in vitro to align isolated sciatic nerve axons (Sretavan et al. 2005). This approach is limited by the magnitude of applied voltage because high voltage damages cell membranes, and from distances no longer than 30–40 mm. DEP has been demonstrated to induce two cells to approach each other within an electric field and form an end-on-end configuration (Zimmermann 1986; Neil and Zimmermann 1993). Furthermore, the effect of DEP on cell and cellular membrane integrity has also been studied and it does not appear to affect cell viability (Fuhr et al. 1994).

AXONAL FUSION

To generate functional continuity, the aligned edges of the damaged neurons ought to be fused. Cellular fusion is not a novel concept in biology and has most commonly formed hybridomas in immunology to mass-produce specific antibodies (Chiarella and Fazio 2008). Three technologies that potentially can be used for axonal fusion are microelectrofusion, chemical fusion, and laser-induced membrane fusion. In microelectrofusion, electric pulses create a temporary breakdown in axonal membrane. Through a stochastic process, the micropores, which occur at points of contact between the neighboring axons, will reseal and fuse the two axonal ends (Neil and Zimmermann 1993). Sretavan et al. (2005) have demonstrated the feasibility of microelectrofusion in culture by showing spread of GFP signal from a GFP-expressing axon to another nonexpressing axon after microelectrofusion of the two fusion partners (Figure 38.3).

An essential tool for developing and validating this strategy of direct axon repair via membrane fusion is an in vitro platform to perform high "throughput" experiments for identifying and testing a wide range of electrofusion parameters and post-fusion treatments to maximize success and yield of the repair process. For such a platform, microtechnology plays yet another critical role by providing the means to spatially micropattern cultured axons in neatly arranged arrays for easy and rapid experimental manipulations (Figure 38.4). Although many similar methods for such cell and axon micropatterning have long been explored (Branch et al. 1998; Chang and Sretavan 2008; Corey et al. 1996), Chang and Sretavan recently developed and demonstrated a robust and inexpensive micropatterning method able to maintain axons in desired configurations with high compliance for several weeks in culture (Wheeler et al. 1999). Known as mPLACeR micropatterning, this method involves depositing high-resolution micropatterns of cell adhesive material on a cell-repellant background. Narrow, micron-wide lanes can easily be laid out to guide individual axons, providing a large number of identically rendered, isolated targets for nano-knife cutting and electrofusion. Subsequently, because of the longevity of the substrate engineering, the desired micropatterning of cells and axons can be maintained long enough to permit observing and analyzing the functional recovery of individual axons after the repair process. In addition to its role in developing nerve repair strategies, this inexpensive platform can also be readily manufactured for diverse variety of experiments and assays supporting basic research and drug discovery.

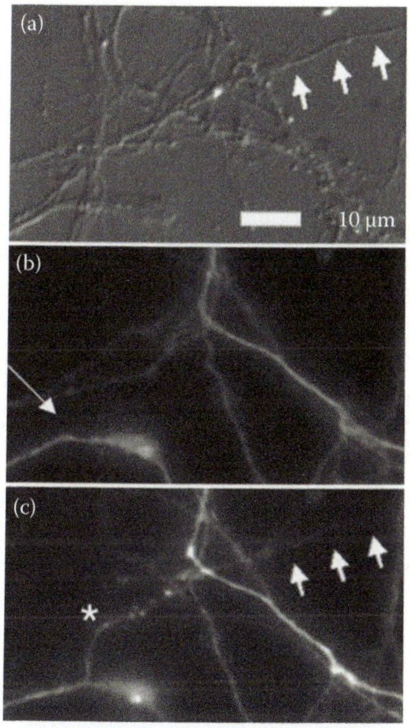

FIGURE 38.3 Example of axon splicing mediated by microelectrofusion. (a) Brightfield view of GFP containing and non-GFP containing axons. Arrows point to one axon that originally does not contain GFP. (b) Fluorescence image of the field shown in (a). Arrow points to location where microelectrofusion was applied subsequent to acquisition of this image. (c) Approximately 9 min after fusion. Asterisk indicates GFP filling of an axon that originally did not contain GFP, demonstrating the formation of physical and cytoplasmic continuity with a GFP-containing fusion partner. GFP eventually spread to fill in the portion of the axon indicated by arrows. (Photo courtesy of Wesley Chang and David Sretavan, UCSF.)

Second, chemical approaches, which may be easier and more economical, have been proposed to induce axonal fusion. In one approach, agents with membrane disrupting properties such as polyethylene glycol (PEG) were applied to the ends of severed spinal cord white matter strips followed by manually aligning the proximal and distal ends (Branch et al. 1998). The investigators reported dramatic reestablishment of axon functional continuity. The results, if verified by other groups, would be a very significant advance in neural repair. However, given the potential toxicity of PEG, the millions of axons and other cell types present in the spinal cord, and the mismatch of scale between whole spinal cords and individual axons, it is difficult to imagine that this manually applied method is reliable and promotes stable, functional connectivity at the level of single axons that are the essential cellular units for electrical conduction in the nervous system. Rather, a substantially different approach involving systematic and scale-appropriate surgical intervention on individual axons is needed.

To date, laser-induced membrane fusion has not been used for axon fusion but has been investigated for cell fusion. It achieved a 38% rate of hybridoma formation in one study (Ohkohchi et al. 2000). In addition to these modalities, studies are ongoing to explore and harness biologically occurring mechanisms of membrane fusion such as fusion of neurotransmitter-containing vesicles with the presynaptic membrane during signal transmission (Sudhof and Rothman 2009). This process involves the SNARE molecular complex, whose components are being studied in an attempt to use this structure for axonal fusion. Enveloped viruses use a similar mechanism to fuse with and gain entry into their host cells (White et al. 2008).

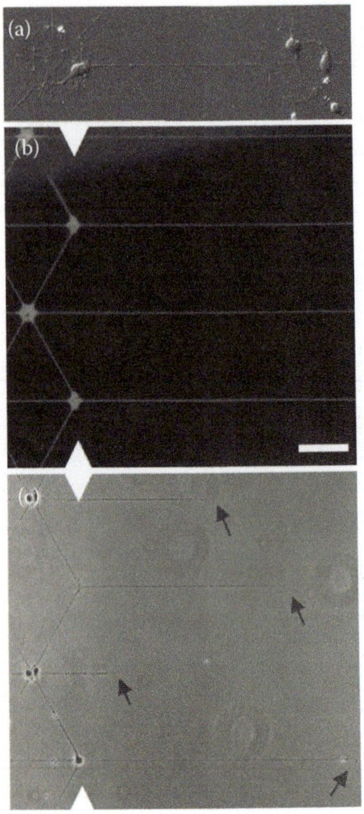

FIGURE 38.4 (a) With surface micropatterning developed by Chang and Sretavan (2008), neuronal cell bodies can be precisely positioned, whereas individual axons can be isolated and directed along specific paths. (b) Lattice micropatterns (left of white triangles) are intended to receive neurons and guide axonal outgrowths into long parallel lanes to the right. (Bright areas are cell adhesive.) (c) Individual axons from cortical neurons seeded on the left project axons (black arrows) into these lanes (scale = 100 μm). By using a lattice micropattern at the seeding area and limiting cell seeding density, it is possible to maximize occupancy rate of lanes while isolating individual axons.

An alternative approach, proposed by Kameswaren and colleagues, uses continuous mechanical tension to generate elongated axonal tracts of up to 10 cm in length. These tracts are subsequently used not only to link severed axonal edges, but also to serve as scaffolding for additional regeneration (Kameswaran et al. 2008).

In addition to mechanical approaches specifically proposed for fusing axonal ends, novel techniques in other areas such as vascular neurosurgery used to approximate vascular ends under very high magnification, may be applied to bring axonal edges together. For instance, recently, Matsumura et al. (2011) proposed using 30–50× magnification in "supermagnified" operations to create microvascular anastamoses. This and similar approaches, which help to better visualize smaller surgical fields and improve skills in manipulating microscopic units, in combination with robotic computer controlled micromanipulation that overcomes human limitations in dexterity and speed, could be applied to fuse delicate axonal edges with higher precision (Matsumura et al. 2011).

AXONAL REPAIR MICRODEVICE

The approaches proposed here require new operative instruments. The operative fields involved in manipulating axonal edges will be far smaller than a conventional neurosurgical operative field, even under an operative microscope. Furthermore, trimming the injured axonal edges, their alignment, and fusion require new technologies, which would allow constant monitoring of electrical activity and exquisite mechanical stability. Prototypes have been proposed and generated to complete parts or most of the repair mechanism.

A miniature axonal nano-knife has already been constructed and used to excise sciatic nerve axons of laboratory mice. This tool had a built-in recording electrode, which allowed monitoring of nerve activity and confirming complete transection of the nerve. This was an early attempt at applying one of the steps of the axonal repair paradigm. Certainly, the eventual goal of axonal repair microsurgery is combining all three steps presented here including trimming, aligning, and fusing within one functional structure. A prototype of such a platform has been constructed, which contains both the microcutting device and the fusion electrode arrays (Chang et al. 2010). Fabricating and using such microtools will make it necessary to modify the current neurosurgical operative suite and field. Although at present the use of the intraoperative microscope allows handling of small nerves, the proposed approaches here involve manipulating single axonal ends.

While axonal edges are being manipulated and approximated, it is necessary to promote the survival of the neuronal cell body and both the proximal and the distal axonal segments. Federici and Boulis harnessed the naturally existing retrograde axonal transport properties to introduce genetically modified vectors into neuronal cell bodies. This approach can be used to introduce neurotrophic survival factors into injured neurons and help promote their survival while the axonal segments are being fused (Federici and Boulis 2009). Similarly, distal axonal segments could be maintained by applying inhibitors of such factors as calpains, which are known to initiate a proteolytic cascade in response to neuronal injury resulting in irreversible damage (Vosler et al. 2008).

CONCLUSION

Despite a century of investigation and experimentation, the field of neuroscience is still struggling with the fundamental challenge of neuronal repair after injury. Multiple areas of research have evolved over the past several decades to wrestle this albatross. Aside from exploiting stem cell biology with the goal of directly replacing dead or dying neurons, peripheral nerves have been used as a model system for axonal rescue and repair. This latter area of research relies on cutting edge technologies in biology and engineering to promote the survival of cell bodies of injured neurons, their proximal and distal axonal segments, and to visualize and manipulate axonal ends at micro- and nano-scales to promote axonal fusion with the goal of generating physiologically functioning neurons. These goals resulted in producing micro- and nanosurgical tools such as micro-knives for trimming axonal edges, as well as scaffoldings and MEM-constructed tools for aligning and approximating injured axonal ends and chemical, electrical, or laser mediated fusion of axonal edges. The ability to reconnect severed or damaged axons in place of neuronal or axonal regeneration, enables rapid repair of neuronal circuitry that in turn circumvents other complications such as inappropriate innervation or irreversible end organ damage. For this intervention to be effective, two fundamental criteria ought to be met. First, the fused axons need to be electrophysiologically and biochemically active with the ability to transmit electrophysiologic signals, and have the biochemical continuity of a healthy axon with functioning axonal transport. Additionally, a minimum number of these axons, approximately 10–20%, ought to be fused to achieve acceptable functionality. Needless to say, the progress described here highlights the nodal points of recent discovery and at the same time underlines the importance of continued research.

REFERENCES

Albright, T.D., Jessell, T.M., Kandell, E.R. et al. 2001. Progress in the neural sciences in the the century after Cajal (and the mysteries that remain). *Annals of the New York Academy of Sciences* 929: 11–40.

Berns, M.W. 2007. A history of laser scissors (microbeams). *Methods in Cell Biology* 82: 1–58.

Branch, D.W., Corey, J.M., Weyhenmeyer, J.A. et al. 1998. Microstamp patterns of biomolecules for high-resolution neuronal networks. *Medical & Biological Engineering & Computing* 36: 135–41.

Chang, W.C., Hawkes, E., Keller, C.G. et al. 2010. Axon repair: surgical application at a subcellular scale. *Wiley Interdisciplinary Reviews. Nanomedicine and Nanobiotechnology* 2: 151–61.

Chang, W.C., Hawkes, E.A., Kliot, M. et al. 2007. *In vivo* use of a nanoknife for axon microsurgery. *Neurosurgery* 61: 683–91; discussion 691–2.

Chang, W.C., Kliot, M., and Sretavan, D.W. 2008. Microtechnology and nanotechnology in nerve repair. *Neurological Research* 30: 1053–62.

Chang, W.C., and Sretavan, D.W. 2008. Novel high-resolution micropatterning for neuron culture using poly-lysine adsorption on a cell repellant, plasma-polymerized background. *Langmuir* 24: 13048–57.

Chiarella, P., and Fazio, V.M. 2008. Mouse monoclonal antibodies in biological research: strategies for high-throughput production. *Biotechnology Letters* 30: 1303–10.

Corey, J.M., Wheeler, B.C., and Brewer, G.J. 1996. Micrometer resolution silane-based patterning of hippocampal neurons: critical variables in photoresist and laser ablation processes for substrate fabrication. *IEEE Transactions on Bio-medical Engineering* 43: 944–55.

de Ruiter, G.C., Spinner, R.J., Yaszemski, M.J. et al. 2009. Nerve tubes for peripheral nerve repair. *Neurosurgery Clinics of North America* 20: 91–105, vii.

Federici, T., and Boulis, N.M. 2009. Invited review: Festschrift edition of neurosurgery peripheral nervous system as a conduit for delivering therapies for diabetic neuropathy, amyotrophic lateral sclerosis, and nerve regeneration. *Neurosurgery* 65: A87–92.

Fox, M.H., Rowland, J.L., Froehlich-Grobe, K. et al. 2008. Determining paralysis prevalence in the United States. *Disability and Health Journal* 1: 172–9.

Fuhr, G., Glasser, H., Muller, T. et al. 1994. Cell manipulation and cultivation under a.c. electric field influence in highly conductive culture media. *Biochimica et Biophysica Acta* 1201: 353–60.

Gascoyne, P.R., and Vykoukal, J. 2002. Particle separation by dielectrophoresis. *Electrophoresis* 23: 1973–83.

Guo, S.X., Bourgeois, F., Chokshi, T. et al. 2008. Femtosecond laser nanoaxotomy lab-on-a-chip for *in vivo* nerve regeneration studies. *Nature Methods* 5: 531–3.

Kameswaran, N., Cullen, D.K., Pfister, B.J. et al. 2008. A novel neuroprosthetic interface with the peripheral nervous system using artificially engineered axonal tracts. *Neurological Research* 30: 1063–7.

Matsumura, N., Hayashi, N., Kamiyama, H. et al. 2011. Microvascular anastomosis at 30–50x magnifications (super-microvascular anastomosis) in neurosurgery. *Surgical Neurology International* 2: 6.

National Spinal Cord Injury Statistical Center. 2011. Spical cord injury facts and figures at a glance. Birmingham, AL.

Neil, G.A., and Zimmermann, U. 1993. Electrofusion. *Methods in Enzymology* 220: 174–96.

Ohkohchi, N., Itagaki, H., Doi, H. et al. 2000. New technique for producing hybridoma by using laser radiation. *Lasers in Surgery and Medicine* 27: 262–8.

Peljto, M., and Wichterle, H. 2011. Programming embryonic stem cells to neuronal subtypes. *Current Opinion in Neurobiology* 21: 43–51.

Sretavan, D.W., Chang, W., Hawkes, E. et al. 2005. Microscale surgery on single axons. *Neurosurgery* 57: 635–46; discussion 635–46.

Sudhof, T.C., and Rothman, J.E. 2009. Membrane fusion: grappling with SNARE and SM proteins. *Science* 323: 474–7.

Vosler, P.S., Brennan, C.S., and Chen, J. 2008. Calpain-mediated signaling mechanisms in neuronal injury and neurodegeneration. *Molecular Neurobiology* 38: 78–100.

Wheeler, B.C., Corey, J.M., Brewer, G.J. et al. 1999. Microcontact printing for precise control of nerve cell growth in culture. *Journal of Biomechanical Engineering* 121: 73–8.

White, J.M., Delos, S.E., Brecher, M. et al. 2008. Structures and mechanisms of viral membrane fusion proteins: multiple variations on a common theme. *Critical Reviews in Biochemistry and Molecular Biology* 43: 189–219.

Zimmermann, U. 1986. Electrical breakdown, electropermeabilization and electrofusion. *Reviews of Physiology, Biochemistry and Pharmacology* 105: 176–256.

39 Nanotechnology
Applications to Pediatric Neurosurgery

Giancarlo Suffredini and Robert F. Keating

CONTENTS

INTRODUCTION

The pediatric neurosurgeon confronts a variety of complex neurologic disorders, which include brain tumors, vascular malformations, congenital spine deformities, hydrocephalus, and bony deformities of the skull. Nanotechnology is an emerging field that has the potential to improve the treatment and care provided by the pediatric neurosurgeon for a multitude of conditions. These improvements may result from using: (1) smart catheters resistant to bacterial colonization; (2) nanosensors for precisely detecting pathologic changes in intraventricular pressure or in ventricular size; (3) nanosensors to measure cortical vascular oxygenation; (4) novel targeting strategies for tumors and vascular malformations; and (5) superior biomaterials with decreased immunogenicity.

APPLICATIONS IN HYDROCEPHALUS

Smart Catheters: Reduction of Bacterial Colonization on Ventricular Catheters

Bacterial adherence to implantable medical devices, such as ventriculoperitoneal (VP) shunts, has long been recognized as a substantial risk factor in the development of shunt-related infections (Kandasamy et al. 2011; Livni et al. 2004). Prophylactic measures, which include impregnating shunt tubing with antibiotics, have recently been investigated and noted to reduce early infection rates after both initial and revision shunt placement (Kandasamy et al. 2011; Sciubba et al. 2008). The bacteria most frequently associated with VP shunt infection are *Staphylococcus epidermidis*, followed by *Staphylococcus aureus* (Kontny et al. 1993; Livni et al. 2004; Quigley et al. 1989). These bacteria secrete a polysaccharide "biofilm" that promotes adherence of their cell walls to each other and to the surface of the shunt (Younger et al. 1987). The thick polysaccharide biofilm prevents penetration of antibiotics and effectively shields the interior of the colony, making this matrix of bacteria within a polysaccharide biofilm particularly resistant to antibiotics. Nanotechnologies that attempt to prevent infection of ventricular catheters by impeding bacterial colonization are currently under investigation. These include coating the surfaces of catheters with bactericidal nanoparticles containing silver, superparamagnetic iron oxide, and selenium. To date, silver nanoparticles have seen the most scrutiny. Other strategies are under current investigation that prevent adherence of bacteria to the catheter by making its surface superhydrophilic. These include coating the catheter with silica nanoparticle-coated graphene oxide and altering the surface nanotopography of the catheter with nanoetching (Machado et al. 2010; Webster 2010).

Silver nanoparticles, ranging in size from 1 to 100 nm, are formed from silver atoms metallically bonded together in clusters (Chaloupka et al. 2010). Silver nanoparticles are produced from silver nitrate using a reducing agent such as $NaBH_4$ or UV light photoreduction (Chaloupka et al. 2010; Choi et al. 2008; Sato-Berru 2009). Silver nanoparticles have different physicochemical properties from bulk silver including high catalytic capabilities and the ability to generate reactive oxygen species (Limbach et al. 2007). The bactericidal activity of silver nanoparticles was recently found to be superior to silver ions and silver chloride colloids (Choi et al. 2008) (Figures 39.1 and 39.2). Proteomic analysis identified targeting of the bacterial cell membrane and subsequent dissipation of proton motive force as a possible mechanism for the strong bactericidal properties of silver

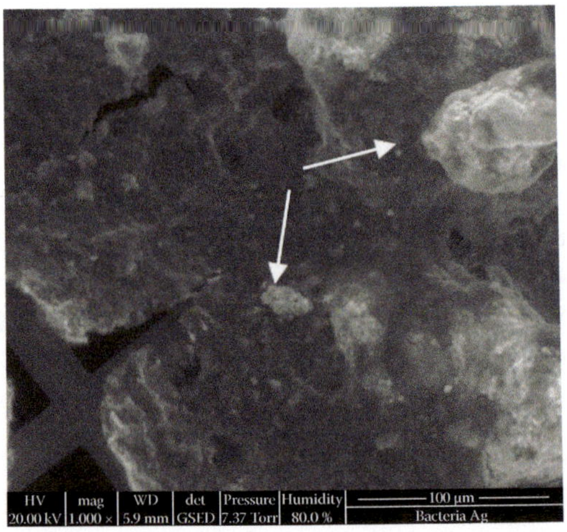

FIGURE 39.1 Silver nanoparticles adsorbed to enriched nitrifying culture on a copper grid using an environmental scanning electron microscope. Arrows show aggregated silver nanoparticles that attached to microbial cells or embedded in microbial extracellular polymeric substances. Bar size: 100 μm. (From Choi, O. et al., *Water Res.*, 42, 3066–74, 2008. With permission.)

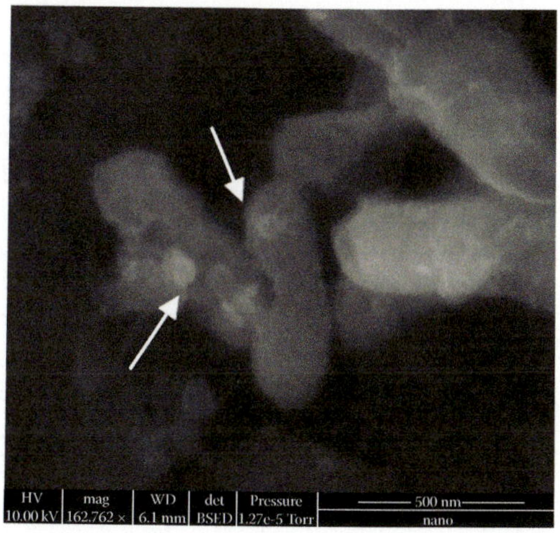

FIGURE 39.2 Silver nanoparticles adsorbed to enriched nitrifying culture using a high-speed backscattered electron detector. Arrows show spherical or hexagon types of Ag NPs that attached to the microbial cells, probably causing cell wall pitting. Bar size: 500 nm. (From Choi, O. et al., *Water Res*., 42, 3066–74, 2008. With permission.)

nanoparticles (Lok et al. 2006). Silver nanoparticles in aqueous solution may also release silver ions, which are also biologically active with strong bactericidal properties (Panacek et al. 2006).

Neurosurgical catheters impregnated with nanosilver particles have been manufactured to reduce catheter-associated infections (Galiano et al. 2008; Roe et al. 2008). In vitro data suggest that silver ions from nanoparticles are released locally around the impregnated catheter. Silver ion concentration within the cerebrospinal fluid (CSF) is undetectable (Galiano et al. 2008). Recently, a pilot clinical cohort study addressed the efficacy of nanosilver-impregnated ventricular catheters in preventing catheter-associated ventriculomeningitis (CAV) associated with external CSF drainage (Lackner et al. 2008). A group of 19 patients who received nanosilver-impregnated catheters were compared retrospectively to a group of 20 patients who had received standard catheters. None of the patients who received the nanosilver-impregnated catheters developed CAV, whereas five of the patients with standard catheters developed CAV and positive CSF cultures. These findings support further clinical investigation via prospective randomized trials to better evaluate the clinical safety and efficacy of nanosilver-impregnated catheters.

Improving Catheter Design

A novel ventricular catheter using a tip involving a membrane without any direct opening and able to filter the CSF may have the potential to reduce CSF shunt obstruction. When CSF flows through the membrane, tissue (choroid plexus, blood cells, tumor cells, ependymal tissue), and protein cannot cross the membrane, thus preventing this catheter being occluded by tissue invasion and in turn preventing obstruction downstream from clogging due to protein precipitation, coagulation or flocculation (Sevrain 2010).

Pressure Monitoring Devices

The substantial failure rate (20–50%) of hydrocephalus shunt valves that open only in response to elevated intracranial pressure has motivated research for alternative treatment approaches. A novel impedance sensor designed to monitor and control cerebral ventricular volume was recently described (Linninger et al. 2009). This sensor uses differences in electrical conductivity between CSF and brain tissue for volumetric measurement of the lateral ventricles. The sensor can permit active control of ventricular volume through feedback principles. The volume sensor (consisting of platinum/iridium electrodes surrounded by a silicon conductance catheter) feeds information about voltage drops to a microcontroller unit, which in turn sends the information to a micropump. The micropump receives a control signal from the microcontroller to maintain the desired ventricular volume (Linninger et al. 2009). Experiments using this volume-sensing technique have been conducted with safe levels of current stimulation without causing neuron activation (Gilad et al. 2007). Future long-term in vivo animal studies are planned to test the biocompatibility of the sensor.

APPLICATIONS IN CEREBRAL TRAUMA

Monitoring Devices for Oxygenation and Metabolism

Recently, two photon-enhanced microscopy of a phosphorescent nanoprobe consisting of platinum porphyrin–coumarin-343 provided accurate and rapid measurement of cortical vascular pO_2 at depths reaching 250 μm with high temporal resolution (Lebedev et al. 2009; Sakadzic et al. 2010). The core of the nanoprobe consists of highly phosphorescent platinum porphyrin encapsulated and protected by a poly(arylglycine) dendrimer. The coumarin-343 units attached to the periphery of the dendrimer and captured and transfered two-photon excitation energy to the porphyrin. The signal for the analytical pO_2 determination is formed by the phosphorescence decay lifetime, which is inversely proportional to the pO_2. High-molecular-weight polyethylene glycol is attached to the remaining termini on the dendrimers to protect against interference from proteins and macromolecules, such as albumin (Sakadzic et al. 2010).

APPLICATIONS IN CEREBRAL ARTERIOVENOUS MALFORMATION

Arteriovenous malformations (AVMs) are a major cause of nontraumatic intracranial hemorrhage in childhood (Schisano and Franco 1982). Major treatments for AVMs are surgery, embolization, and stereotactic radiosurgery, alone or in combination. Complete elimination of the nidus is considered a requirement for a cure (Lee et al. 2009). Growing understanding of the vascular biology and endothelial profiles of AVMs is increasing and may provide insight into how nanotechnology may be used to enhance minimally invasive treatments for this condition (Alexander and Tolbert 2006; Sammons et al. 2011).

Endothelial Targets

Differential molecule expression on AVM endothelium as compared to normal brain endothelial cells has been well characterized over the past decade. Several members of the vascular endothelial growth factor (VEGF) family, including VEGF-A, VEGF-B, VEGF-C and VEGF-D, and thrombomodulin are most prominently up-regulated in AVM endothelium compared to normal brain endothelial cells (Sammons et al. 2011). These biomarkers are attractive candidates for vascular therapy targets. Gold nanoparticles can be functionalized with biological moieties to target specific ligands or receptors such as those overexpressed in AVMs. Gold nanoparticles targeted to vascular endothelium produce low-energy, short-range photoelectrons during external beam radiotherapy, boosting

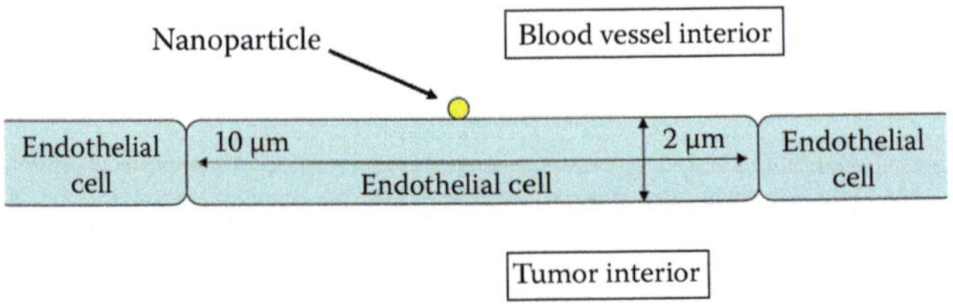

FIGURE 39.3 Simplified model of endothelial cell layer between intravascular cavity and tumor cells. Gold nanoparticles are attached to the vascular side of endothelium. (From Berbeco, R.I. et al., *Int. J. Radiat. Oncol. Biol. Phys.*, 81, 270–6, 2010. With permission.)

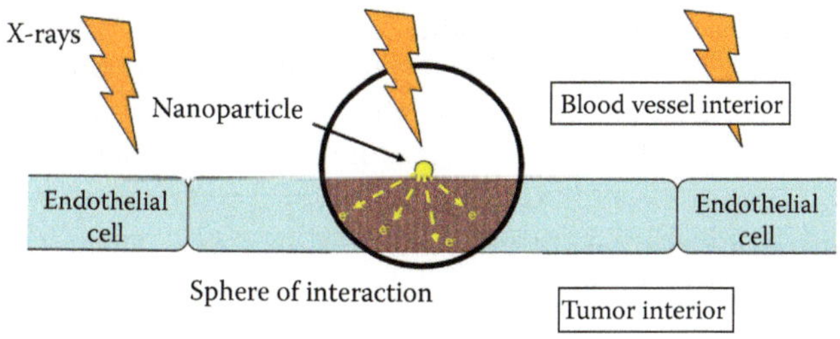

FIGURE 39.4 Range of photoelectrons generated within gold nanoparticles (AuNPs) is shown as a "sphere of interaction" with the nanoparticle at center. Extra dose deposited in nearest endothelial cell by AuNP photoelectron emissions (shaded region) is used to calculate dose enhancement. (From Berbeco, R.I. et al., *Int. J. Radiat. Oncol. Biol. Phys.*, 81, 270–6, 2010. With permission.)

the dose from 20% to 340% to the targeted endothelium while maintaining established dose constraints (Berbeco et al. 2011) (Figures 39.3 and 39.4). Targeted radiotherapy may lead to enhanced minimally invasive treatment strategies for AVMs.

Quantum Dots and Imaging

Advances in nanotechnology, specifically those in semiconductor nanocrystals, show the potential to dramatically improve real-time in vivo molecular imaging, bridging the gap between traditional in vivo imaging and immunohistochemistry (Taghva et al. 2010). Specifically, quantum dots designed to target proteins expressed on endothelial cells may soon be able to generate real-time color images of discrete components of an AVM because of the different endothelial cell expression proteins present in the feeding artery, nidus, and draining veins (Taghva et al. 2010).

Quantum dots can be administered by injection and excited in vivo using a long wavelength. Their resulting fluorescence can be recognized using a sensitive charge-captured device (Cai and Chen 2008). Quantum dots can be engineered by modifying surface properties to improve solubility, sensitivity, specificity, and visualization of target tissue. Quantum dot–gadolinium wedge nanoparticles functionalized with Annexin A5 (a molecule that recognizes phosphatidylserine) have been created to provide contrast on MRI in regions of apoptosis (Prinzen et al. 2007). Recently, simultaneous multicolor fluorescence imaging of five different lymphatic basins using five near-infrared quantum dots was demonstrated in a mouse model (Kobayashi et al. 2007).

APPLICATIONS IN TETHERED CORD SYNDROME

Iatrogenic tethered cord syndrome can result a from scar that forms after intradural spinal surgery and attaches the spinal cord to the dorsal dura (Lew and Kothbauer 2007). Numerous techniques have been explored to avoid bridging scar formation including using Gore-Tex and silastic membranes for adhesion prevention (Aliredjo et al. 1999; Boop and Chadduck 1991; Inoue et al. 1994), which to date have failed to provide significant long-term efficacy in preventing recurrent tethering.

Biomaterials with Decreased Immunogenicity

Recent in vitro studies identified the role of topographic nanoscale features as potent modulators of cellular behavior through the onset of focal adhesion formation (Biggs et al. 2010) (Figure 39.5). Specifically, nanoscale topographies affect modulation of integrin-mediated cellular adhesion; topographic modification influences cellular function. Important findings in these studies show decreased cellular adhesion with increasing nanoprotrusion height (greater than 70 nm), whereas reduced height of nanoprotrusions (less than 50 nm) increased the frequency of focal adhesion (Dalby et al. 2002; Sjostrom et al. 2009). Other topographic nanoscale details such as feature

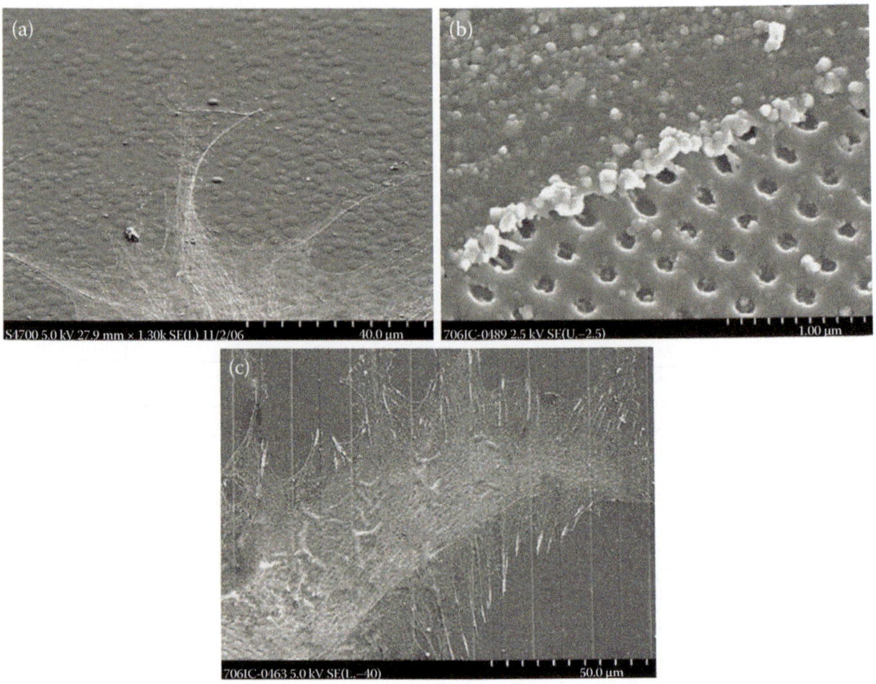

FIGURE 39.5 Nanoscale topographical features influence cellular spreading and focal adhesion formation. (a) Nanoprotrusion with microscale *x*–*y* dimensions and a *z* dimension <70 nm increases cellular spreading. Nanoisland topography increases cellular spreading by providing tactile stimuli. (b) Immunogold labeling of focal adhesions (electron-dense clusters) in adherent cells allows the visualization of cell–substratum interactions. Nanoscale pits >70 nm in diameter perturb integrin clustering forcing adhesion formation to occur at the interpit regions. (c) Focal adhesions as visualized by scanning electron microscopy and immunogold labeling indicate that grooves with *z* dimensions down to a minimum 30–40 nm can induce adhesion alignment to groove orientation. (From Biggs, M.J. et al., *Nanomedicine*, 6, 619–33, 2010. With permission.)

diameter and edge–edge spacing are also defining factors in regulating integrin clustering and adhesion formation (Sjostrom et al. 2009). Modulation of topographic nanoscale features shows great promise in creating biomembranes that avoid bridging scar formation.

TARGETING CANCER: PEDIATRIC APPLICATIONS

Based on data from the Central Brain Tumor Registry of the United States from 2000 to 2004, tumors of neuroepithelial tissue origin (gliomas) account for almost 75% of all primary childhood CNS tumors and include: astrocytoma, brain stem glioma, ependymoma, and optic nerve glioma.

Chlorotoxin-Labeled Nanoparticles and Gene Therapy

Chlorotoxin is a 36-amino acid peptide purified from the venom of *Leiurus quinquestriatus*, a large scorpion found in Israel. It binds with high affinity to membrane-bound matrix metalloproteinase-2 endopeptidase, which is preferentially up-regulated in tumors of neuroectodermal origin including glioma and medulloblastoma (Deshane et al. 2003; Lyons et al. 2002). When attached to iron oxide particles, chlorotoxin can localize to the perinuclear region, making it a promising targeting molecule for gene delivery to treat glioma (Kievit et al. 2010).

A multifunctional nanoprobe constructed by coating iron oxide nanoparticles with covalently bound polyethylene glycol functionalized with chlorotoxin can target the vast majority of gliomas;

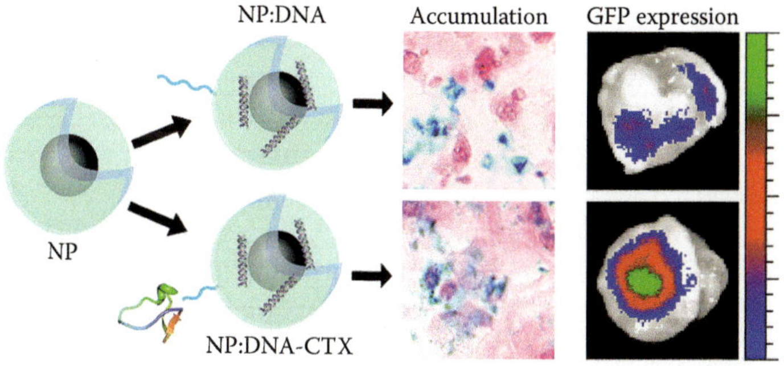

FIGURE 39.6 Chlorotoxin-labeled nanoparticles show significantly increased cellular uptake as measured by gene expression, compared to nanoparticles without chlorotoxin. (From Kievit, F.M. et al., *ACS Nano*, 4, 4587–94, 2010. With permission. Copyright © 2010 American Chemical Society.)

it is detectable both by magnetic resonance and optical imaging (Kievit et al. 2010; Veiseh et al. 2005). Because of their magnetic susceptibility and fluorescence properties, these probes have the potential to be used for imaging during glioma resection in real time and to correlate preoperative diagnostic images with intraoperative pathology at the cellular level (Veiseh et al. 2005). In addition, in the current era of intraoperative imaging using MRI, tremendous potential exists to couple this with intraoperative imaging post-tumor resection to ascertain the presence of any residual disease.

A recent in vivo study compared the efficacy of (1) chlorotoxin-labeled nanoparticles loaded with DNA and (2) nanoparticles loaded with DNA (without chlorotoxin) for gene therapy in a C6 glioma mouse xenograft model (Kievit et al. 2010). The different types of nanoparticles accumulated equally at the tumor site, presumably because of the enhanced retention and permeability effect. However, uptake of the chlorotoxin-labeled nanoparticles by glioma cells was significantly greater than unlabeled nanoparticles, as measured by gene expression (Figure 39.6).

Magnetic Fluid Hyperthermia

Magnetic fluid hyperthermia (MFH) is a form of thermotherapy where magnetic nanoparticles are used to locally heat tumor tissue to temperatures of 42–46°C causing tumor cell lysis or apoptosis (Krishnan 2010). Superparamagnetic iron oxide nanoparticles are well suited for this application because they generate heat via relaxation processes when they are placed in an alternating magnetic field (Gneveckow et al. 2004). MagForce Nanotechnologies® has developed superparamagnetic iron oxide nanoparticles with an aminosilane coating and an alternating magnetic field applicator (MFH 300F) formulated specifically for MFH.

Recently, this company completed a prospective, single-arm, two-center phase II study evaluating the safety and efficacy of MFH in conjunction with percutaneous irradiation to treat recurrent glioblastoma multiforme (Maier-Hauff et al. 2011). Fifty-nine patients with recurrent glioblastoma multiforme received direct intratumoral injection of an aqueous dispersion of aminosilane-coated superparamagnetic iron nanoparticles. The ensuing hyperthermia treatment consisted of six semi-weekly sessions with each session lasting 1 h. The overall survival from the initial date of tumor diagnosis was 23.2 months compared to 14.6 months in the reference group, and overall survival from the date of tumor recurrence was 13.4 months compared to 6.2 months in the reference group (Maier-Hauff et al. 2011).

Plasmonic Nanobubbles as Surgical Tools

Combining the photothermal properties of plasmonic gold nanoparticles with the mechanical and optical properties of transient vapor nanobubbles has led to the development of a novel tunable surgical microtool (Lukianova-Hleb et al. 2010). When a plasmonic gold nanoparticle is transiently heated with a short laser pulse, a nanometer-sized vapor bubble is created that expands and then collapses within a nanosecond. This expansion creates a localized, nonthermal, mechanical impact (Lukianova-Hleb et al. 2010, 2011). Plasma nanobubbles create optical scattering proportional to their size, and this effect can potentially guide ablative treatments in real time. Gold nanoparticles can be functionalized with biological moieties that target specific tumor markers, ligands, and receptors, which allow plasmonic nanobubbles to act locally in and around tumor cells. This therapy has the potential to greatly improve the quality of minimally invasive therapies for a wide range of tumors.

Arsenicals and Hedgehog-Related Tumors

Arsenic trioxide (ATO) is currently FDA-approved as a second-line therapy for acute promyelocytic leukemia in patients who do not respond or relapse on transretinoic acid therapy. Known targets of ATO include promyelocytic leukemia–retinoic acid receptor alpha (PML-RAR) fusion protein, NF-κB, thioredoxin reductase, and JNK (c-JUN N-terminal kinase) (Beauchamp et al. 2011; Cavigelli et al. 1996; Kapahi et al. 2000; Lu et al. 2007; Mann et al. 2006). Additionally, ATO has been shown to inhibit the hedgehog-glioma associated oncogene homologue (Hh-GLI) signaling pathway by directly binding to GLI (Beauchamp et al. 2011). Hedgehog was originally discovered for its prominent role in the development of multiple processes in *Drosphilia*. Dysregulation of Hedgehog has been well characterized in basal cell carcinoma and in medulloblastoma (Teglund and Toftgard 2010). A major challenge in treating children and young adults with therapeutics that inhibit Hedgehog signaling are the serious side effects including profound and permanent defects in bone structure (Kimura et al. 2008). Advances in molecular targeting such as those demonstrated with chlorotoxin specificity in tumors of neuroectodermal origin may minimize many of the serious and dramatic off-target effects of ATO therapy.

FUTURE DIRECTIONS

Combination drugs for tumors of neuroectodermal origin, such as chlorotoxin-labeled nanoparticles for gene delivery that also contain a superparamagnetic iron oxide core enabling thermotherapy, could potentially be delivered intravenously and yield two distinct treatment modalities: gene therapy and MFH. There are endless possibilities for engineering nanoparticles with unique tumor-targeting features and modes of action. Nanoconjugates that selectively kill tumor cells in in vitro studies require further testing in preclinical studies to assess their selectivity of action, toxicity, and pharmacokinetics. Promising agents will move forward to clinical studies. The ultimate role of each nanodrug in treating diseases of the pediatric nervous system will be determined by its safety and efficacy, as is currently the case with molecular pharmaceuticals. Similarly, devices that incorporate nanotechnology to reduce the risk of catheter-associated infection or postoperative scarring will require further testing.

Contemporary, state-of-the art neurosurgery demands increasing sophistication with respect to pursuing minimally invasive approaches to the patient. Nowhere is this better seen than in the treatment of children with neurosurgical disease. The spectrum of neurosurgical conditions, such as congenital disorders that may prompt surgery, differs significantly between children and adults. Pediatric patients are smaller in size, and it is not uncommon for them to have less pronounced neurological deficits than adult patients. Thus, in turn, minimally invasive surgery is being developed to provide intervention with reduced impact on normal structures in the growing child. Whether one is approaching treatment of hydrocephalus, a brain or spinal cord tumor, vascular malformation, intracranial hypertension, or spinal dysraphism, nanotechnology is poised to play a pivotal role in the future care of these patients.

REFERENCES

Alexander, M.J., and Tolbert, M.E. 2006. Targeting cerebral arteriovenous malformations for minimally invasive therapy. *Neurosurgery* 59: S178–83; discussion S3–13.

Aliredjo, R.P., de Vries, J., Menovsky, T. et al. 1999. The use of Gore-Tex membrane for adhesion prevention in tethered spinal cord surgery: technical case reports. *Neurosurgery* 44: 674–7; discussion 677–8.

Beauchamp, E.M., Ringer, L., Bulut, G. et al. 2011. Arsenic trioxide inhibits human cancer cell growth and tumor development in mice by blocking Hedgehog/GLI pathway. *J Clin Invest* 121: 148–60.

Berbeco, R.I., Ngwa, W., and Makrigiorgos, G.M. 2011. Localized dose enhancement to tumor blood vessel endothelial cells via megavoltage X-rays and targeted gold nanoparticles: new potential for external beam radiotherapy. *Int J Radiat Oncol Biol Phys* 81: 270–6.

Biggs, M.J., Richards, R.G., and Dalby, M.J. 2010. Nanotopographical modification: a regulator of cellular function through focal adhesions. *Nanomedicine* 6: 619–33.

Boop, F.A., and Chadduck, W.M. 1991. Silastic duraplasty in pediatric patients. *Neurosurgery* 29: 785–7; discussion 788.

Cai, W., and Chen, X. 2008. Preparation of peptide-conjugated quantum dots for tumor vasculature-targeted imaging. *Nat Protoc* 3: 89–96.

Cavigelli, M., Li, W.W., Lin, A. et al. 1996. The tumor promoter arsenite stimulates AP-1 activity by inhibiting a JNK phosphatase. *EMBO J* 15: 6269–79.

Chaloupka, K., Malam, Y., and Seifalian, A.M. 2010. Nanosilver as a new generation of nanoproduct in biomedical applications. *Trends Biotechnol* 28: 580–8.

Choi, O., Deng, K.K., Kim, N.J. et al. 2008. The inhibitory effects of silver nanoparticles, silver ions, and silver chloride colloids on microbial growth. *Water Res* 42: 3066–74.

Dalby, M.J., Yarwood, S.J., Riehle, M.O. et al. 2002. Increasing fibroblast response to materials using nanotopography: morphological and genetic measurements of cell response to 13-nm-high polymer demixed islands. *Exp Cell Res* 276: 1–9.

Deshane, J., Garner, C.C., and Sontheimer, H. 2003. Chlorotoxin inhibits glioma cell invasion via matrix metalloproteinase-2. *J Biol Chem* 278: 4135–44.

Galiano, K., Pleifer, C., Engelhardt, K. et al. 2008. Silver segregation and bacterial growth of intraventricular catheters impregnated with silver nanoparticles in cerebrospinal fluid drainages. *Neurol Res* 30: 285–7.

Gilad, O., Horesh, L., and Holder, D.S. 2007. Design of electrodes and current limits for low frequency electrical impedance tomography of the brain. *Med Biol Eng Comput* 45: 621–33.

Gneveckow, U., Jordan, A., Scholz, R. et al. 2004. Description and characterization of the novel hyperthermia- and thermoablation-system MFH 300F for clinical magnetic fluid hyperthermia. *Med Phys* 31: 1444–51.

Inoue, H.K., Kobayashi, S., Ohbayashi, K. et al. 1994. Treatment and prevention of tethered and retethered spinal cord using a Gore-Tex surgical membrane. *J Neurosurg* 80: 689–93.

Kandasamy, J., Dwan, K., Hartley, J.C. et al. 2011. Antibiotic-impregnated ventriculoperitoneal shunts—a multi-centre British Paediatric Neurosurgery Group (BPNG) study using historical controls. *Childs Nerv Syst* 27: 575–81.

Kapahi, P., Takahashi, T., Natoli, G. et al. 2000. Inhibition of NF-kappa B activation by arsenite through reaction with a critical cysteine in the activation loop of Ikappa B kinase. *J Biol Chem* 275: 36062–6.

Kievit, F.M., Veiseh, O., Fang, C. et al. 2010. Chlorotoxin labeled magnetic nanovectors for targeted gene delivery to glioma. *ACS Nano* 4: 4587–94.

Kimura, H., Ng, J.M., and Curran, T. 2008. Transient inhibition of the Hedgehog pathway in young mice causes permanent defects in bone structure. *Cancer Cell* 13: 249–60.

Kobayashi, H., Hama, Y., Koyama, Y. et al. 2007. Simultaneous multicolor imaging of five different lymphatic basins using quantum dots. *Nano Lett* 7: 1711–6.

Kontny, U., Hofling, B., Gutjahr, P. et al. 1993. CSF shunt infections in children. *Infection* 21: 89–92.

Krishnan, K.M. 2010. Biomedical nanomagnetics: a spin through possibilities in imaging, diagnostics, and therapy. *IEEE Trans Magn* 46: 2523–58.

Lackner, P., Beer, R., Broessner, G. et al. 2008. Efficacy of silver nanoparticles-impregnated external ventricular drain catheters in patients with acute occlusive hydrocephalus. *Neurocrit Care* 8: 360–5.

Lebedev, A.Y., Cheprakov, A.V., Sakadzic, S. et al. 2009. Dendritic phosphorescent probes for oxygen imaging in biological systems. *ACS Appl Mater Interfaces* 1: 1292–304.

Lee, B.B., Lardeo, J., and Neville, R. 2009. Arterio-venous malformation: how much do we know? *Phlebology* 24: 193–200.

Lew, S.M., and Kothbauer, K.F. 2007. Tethered cord syndrome: an updated review. *Pediatr Neurosurg* 43: 236–48.

Limbach, L.K., Wick, P., Manser, P. et al. 2007. Exposure of engineered nanoparticles to human lung epithelial cells: influence of chemical composition and catalytic activity on oxidative stress. *Environ Sci Technol* 41: 4158–63.

Linninger, A., Basati, S., Dawe, R. et al. 2009. An impedance sensor to monitor and control cerebral ventricular volume. *Med Eng Phys* 31: 838–45.

Livni, G., Yuhas, Y., Ashkenazi, S. et al. 2004. In vitro bacterial adherence to ventriculoperitoneal shunts. *Pediatr Neurosurg* 40: 64–9.

Lok, C.N., Ho, C.M., Chen, R. et al. 2006. Proteomic analysis of the mode of antibacterial action of silver nanoparticles. *J Proteome Res* 5: 916–24.

Lu, J., Chew, E.H., and Holmgren, A. 2007. Targeting thioredoxin reductase is a basis for cancer therapy by arsenic trioxide. *Proc Natl Acad Sci U S A* 104: 12288–93.

Lukianova-Hleb, E., Hu, Y., Latterini, L. et al. 2010. Plasmonic nanobubbles as transient vapor nanobubbles generated around plasmonic nanoparticles. *ACS Nano* 4: 2109–23.

Lukianova-Hleb, E.Y., Koneva, II, Oginsky, A.O. et al. 2011. Selective and self-guided micro-ablation of tissue with plasmonic nanobubbles. *J Surg Res* 166: e3–13.

Lyons, S.A., O'Neal, J., and Sontheimer, H. 2002. Chlorotoxin, a scorpion-derived peptide, specifically binds to gliomas and tumors of neuroectodermal origin. *Glia* 39: 162–73.

Machado, M.C., Cheng, D., Tarquinio, K.M. et al. 2010. Nanotechnology: pediatric applications. *Pediatr Res* 67: 500–4.

Maier-Hauff, K., Ulrich, F., Nestler, D. et al. 2011. Efficacy and safety of intratumoral thermotherapy using magnetic iron-oxide nanoparticles combined with external beam radiotherapy on patients with recurrent glioblastoma multiforme. *J Neurooncol* 103: 317–24.

Mann, K.K., Davison, K., Colombo, M. et al. 2006. Antimony trioxide-induced apoptosis is dependent on SEK1/JNK signaling. *Toxicol Lett* 160: 158–70.

Panacek, A., Kvitek, L., Prucek, R. et al. 2006. Silver colloid nanoparticles: synthesis, characterization, and their antibacterial activity. *J Phys Chem B* 110: 16248–53.

Prinzen, L., Miserus, R.J., Dirksen, A. et al. 2007. Optical and magnetic resonance imaging of cell death and platelet activation using annexin a5-functionalized quantum dots. *Nano Lett* 7: 93–100.

Quigley, M.R., Reigel, D.H., and Kortyna, R. 1989. Cerebrospinal fluid shunt infections. Report of 41 cases and a critical review of the literature. *Pediatr Neurosci* 15: 111–20.

Roe, D., Karandikar, B., Bonn-Savage, N. et al. 2008. Antimicrobial surface functionalization of plastic catheters by silver nanoparticles. *J Antimicrob Chemother* 61: 869–76.

Sakadzic, S., Roussakis, E., Yaseen, M.A. et al. 2010. Two-photon high-resolution measurement of partial pressure of oxygen in cerebral vasculature and tissue. *Nat Methods* 7: 755–9.

Sammons, V., Davidson, A., Tu, J. et al. 2011. Endothelial cells in the context of brain arteriovenous malformations. *J Clin Neurosci* 18: 165–70.

Sato-Berru, R.e.a. 2009. Silver nanoparticles synthesized by direct photoreduction of metal salts. Application in surface-enhanced Raman spectroscopy. *J. Raman Spectrosc* 376–80.

Schisano, G., and Franco, A. 1982. Primary non traumatic intracranial hemorrhage: 139 consecutive cases treated at an emergency regional hospital. *J Neurosurg Sci* 26: 199–204.

Sciubba, D.M., Noggle, J.C., Carson, B.S. et al. 2008. Antibiotic-impregnated shunt catheters for the treatment of infantile hydrocephalus. *Pediatr Neurosurg* 44: 91–6.

Sevrain, L.C. 2010. Anti-clogging ventricular catheter for cerebrospinal fluid drainage. http://www.freepatentsonline.com/y2010/0222732.html.

Sjostrom, T., Dalby, M.J., Hart, A. et al. 2009. Fabrication of pillar-like titania nanostructures on titanium and their interactions with human skeletal stem cells. *Acta Biomater* 5: 1433–41.

Taghva, A., Khalessi, A.A., Kim, P.E. et al. 2010. From atom to brain: applications of molecular imaging to neurosurgery. *World Neurosurg* 73: 477–85.

Teglund, S., and Toftgard, R. 2010. Hedgehog beyond medulloblastoma and basal cell carcinoma. *Biochim Biophys Acta* 1805: 181–208.

Veiseh, O., Sun, C., Gunn, J. et al. 2005. Optical and MRI multifunctional nanoprobe for targeting gliomas. *Nano Lett* 5: 1003–8.

Webster, T.J. 2010. Assessment of nanomodified endotracheal tubes in a benchtop airway model. *Open Conf Proc Jl* 1: 224–9.

Younger, J.J., Christensen, G.D., Bartley, D.L. et al. 1987. Coagulase-negative staphylococci isolated from cerebrospinal fluid shunts: importance of slime production, species identification, and shunt removal to clinical outcome. *J Infect Dis* 156: 548–54.

40 Food and Drug Administration Regulatory Framework for Nanodrugs and Nanodevices and the Combination

Vicky Yamamoto, Giancarlo Suffredini, John D. Heiss, David Mishelevich, Grant Palmer, Warren S. Grundfest, and Babak Kateb

CONTENTS

INTRODUCTION

Responsibility for the regulation of medical devices traveled through a number of different agencies before the Food and Drug Administration (FDA) became a standalone organization in 1931. FDA has had a long legislative history with multiple legislative acts contributing to its current structure. In 1883, the US Department of Agriculture (USDA) formed a new position of Chief Chemist of Bureau of Chemistry. The first comprehensive Food and Drug Law or the "Wiley Act" was passed in 1906, which was enforced by this bureau. In 1927, Food, Drug and Insecticide Administration (FDIA) headed by Walter G. Campbell assumed responsibility for the 1906 act; his department was renamed as the Food and Drug Administration (Paradise et al. 2009).

In order to better understand the regulatory framework and the process of the FDA approval in this area, we need to have a better understanding of legal definitions of drugs, medical devices, biologics, and combination products.

Food, Drug and Cosmetic Act defines drug (21 U.S.C. § 321(p)(2)) as "(A) articles recognized in the official United State Pharmacopoeia, official Homoeopathic Pharmacopoeia of the United States, or official National Formulary, or any supplement to any of them; and (B) articles intended for use in the diagnosis, cure, mitigation, treatment, or prevention of diseases in man or other animal; and (C) articles (other than food) intended to affect the structure or any function of the body of man or other animals; and (D) articles for use as component of any article specified in clause (A), (B), or (C)."

A medical device is defined as "an instrument, apparatus, any component, implement, machine, contrivance, implant, in vitro reagent, or other similar or related article, including any component part, or accessory, which is (1) recognized in the official National Formulary, or the United States Pharmacopeia, or any supplement to them, (2) intended for use in the diagnosis of disease or other conditions, or in the cure, mitigation, treatment, or prevention of disease, in man or other animals, or (3) intended to affect the structure or any function of the body of man or other animals, and which does not achieve its primary intended purposes through chemical action within or on the body of man or other animals and which is not dependent upon being metabolized for the achievement of its primary intended purposes" [21 U.S.C. § 321(h)].

Biologics is defined as "a virus, therapeutic serum, toxin, antitoxin, vaccine, blood, blood component or derivative, allergenic product, or analogous product, or arsphenamine or derivative of arsphenamine (or any other trivalent organic arsenic compound), applicable to the prevention, treatment, or cure of a disease or condition of human beings" [42 U.S.C. § 262(i)].

Combination product is a single product containing both a drug and a medical device, a drug and a biologic, and medical device and a biologic, or all three; Center for Drug Evaluation and Research (CDER), Center for Biologics Evaluation Research (CBER), and Center for Devices and Radiological Health (CDRH) categorize and review combination products according to the primary mode of action (e.g., a product with the principle mode of action of a drug is classified as a drug) [21 C.F.R. § 3.2 (e)].

FDA is one of the 11 agencies of the Department of Health and Human Services (HHS). FDA's CDRH is responsible for medical device review and CBER review device and products associated with biological specimens. FDA's CDER job is to evaluate new drugs prior to their release for distribution and marketing public. CDER's evaluation protects patients and doctors by informing them as to the potential risks and benefits of approved products. This includes the likelihood of complications and potential for allergic reactions in addition to the risk-benefit. The results of the clinical trials are used to develop these analysis and reports (FDA website; Johnson 2012).

510(K) AND PMA PROCESS, DEVICE CLASSIFICATION, AND MARKETING

The CDRH review process begins with the evaluation of the device design, its intended use, and in vitro test results. The FDA requires that all information from discovery or creation of the device through animal trials be available to evaluate the performance of the device (Eifler and Thaxton 2011). After the preclinical phase, the manufacturer of the device is requiring to file an Investigational Device Exemption (IDE) application with the FDA. Once the IDE is approved, the investigator may choose to compare the device with the predicate device and apply for a 510(k) approval, which is based on a regulation. If the FDA determines that the device poses a significant risk or does not meet the predicate requirements, then the device must continue along the IDE pathway for a premarket approval (PMA) application.

By comparison there are different paths for medical devices:

1. PMA: The PMA pathway generally applies to devices, with significant risk or the potential to cause life-threatening injuries. Devices such as heart valves, pacemakers, and almost all implants require collection of safety and efficacy data through clinical trials leading to PMA application. Several technologies have become so common that FDA allows for 510(k) approval in specialized categories including angioplasty balloons and surgical staplers.

2. 510(k): In 510(k) applications, the manufacturer must demonstrate that the proposed new device is similar in both construction and application to an existing device (a predicate device) that has already been approved for the US market. The 510(k) process is faster and less costly than the corresponding PMA process. Once approval is granted either by 510(k) or PMA, the manufacturer is required to keep detailed records of the manufacture, sale, and use of the product in the market place. Perhaps more importantly, the manufacturing is required to keep track of all product-related adverse events and report them to the FDA.

By act of the US Congress, the FDA is now required to charge device manufacturer fees every time a device is submitted for approval. Prices for small businesses are considerably lower than those of large manufacturers and the prices for the 510(k) is lower than PMA. In 2010, the average large company PMA cost approximately $218,000 ($55,000 for smaller companies), while the average for a 510(k) was reported to be $4000 ($2000 for smaller businesses).

Based on the Medical Device Amendments of 1976 (MDA. PL. 94-295), FDA categorized all medical devices to three classes:

- *Class I, low risk:* For example, surgical gloves require registration only unless 510(k) is specifically needed to evaluate novel materials or biocompatibility. This category falls under general controls for safety. Many of the class I devices are exempt from the Quality System Regulations (QSR) or premarket announcements.
- *Class II, moderate risk:* Special controls are needed to lower risk. For example, specific labeling, performance standards, and postmarket examination are enforced. A clear example of class II devices is powered wheelchair, which requires 510(k) and IDE for evaluation of new models. Fifteen percent of all devices belong to this category.
- *Class III, high risk:* Metal-on-metal joint implants, pedicle screw spine, pacemakers, heart valves, silicone gel–filled breast implants, and implantable brain devices are subjected to the PMA process. In the same category, there are devices, such as dental implants, that are allowed into the market through a 510(k) process. Determination of risk is the critical factor in categorizing a new product (Johnson 2012).

More than 4000 510(k) notifications are filed annually versus 40 original PMA applications (Johnson 2012). Good manufacturing practices (GMPs), device listing, establishment of registration, labeling, and premarket notification are the five elements of the general controls (Johnson 2012).

To begin the process, manufacturers must assess the risk category for their device. Devices in class I generally do not require presubmission meetings with the FDA. The 510(k) with FDA is submitted and the agency has 90 days to respond to the applicant. The FDA may approve the product or ask for more information. The 90-day clock starts again with the submission of the additional data and with the response to the FDA. For classes II and III devices, presubmission meetings are generally recommended. This alerts the FDA and allows the manufacturer to discuss its plans for moving forward [PMA vs. 510(k)] with the regulators at the agency. Applicants must provide a detailed plan for collecting the necessary data and the evaluation of risk and benefits. After the documents are submitted to the FDA, the following may be determined by the FDA:

1. Request more information or testing.
2. The device is similar to already approved predicate(s).
3. The device is Not Substantially Equivalent (NSE), which prohibits marketing until the device safety and efficacy profile is known via a PMA process.
4. The device is exempted from the 510(k) process.

The manufacturer has 30 days to respond and a 90-day window to withdraw the application (Johnson 2012). However, this process is much more complex when it comes to the PMA (Johnson 2012).

FDA has established the PMA rout in order to ensure that a device has sufficient performance and safety. Clinical data are necessary for this application. Preclinical assessments should be done prior to initiation of the trial, and investigators need an IDE. Institutional Review Board (IRB) and IDE approvals are needed prior to the initiation of the clinical trial. Single site trials can be conducted with the approval of the local IRB. However, discussions with the FDA should take place before expanding the trial beyond the single site, and IDE is generally required before multiple sites are involved.

PMA application needs

1. Research and clinical data supporting the intended application
2. Device instructions for use
3. Device manual
4. Packaging and labeling
5. Software and hardware validation
6. Publications reviewing the disease to be treated and the performance of the similar devices already in the market
7. Manufacturing process and material to be used

For PMAs, FDA usually conducts the administrative review within 45–60 days and responds to the application 75 days after administrative review (Johnson 2012). CDRH has the option to hold a joint committee meeting to discuss the fate of an application if FDA could not label the device as approved, approvable with modification, or nonapprovable (Johnson 2012).

PRECLINICAL CONSIDERATIONS

The regulations that apply to nanomaterials are significantly different from those that apply to regular materials. Evaluation of nanomaterials must include characterization of their size, shape, biodistribution, tendency to agglomerate, stability and physiologic conditions, routes of elimination, and shelf life. Standardization of production with online characterization to ensure particle size and shape consistency is required.

Preclinical evaluation of nanomaterials for use as drugs depends on the material properties at the nanoscale. This broad range of nanomaterial requires many diverse technologies and characterization techniques. Assessing the biologic impact of these materials using in vitro systems can be difficult. Critical factors for selecting nanoparticles for clinical use are (Eifler and Thaxon 2011)

1. Size
2. Charge
3. Hydrophobicity
4. Targeting capability
5. Biomaterials

Rational design of the nanoparticles ultimately has to be based on applicability, adaptability, ease of manufacturing, and functionality. In this phase, investigators should think about how to test for biological activity, toxicology, immunogenicity, biocompatibility, biodistributions, biodegradability, as well as tracking. Deactivation (permanent vs. biodegradable) of nanodrugs and devices is a major challenge to the therapeutic and diagnostic use of nanodrugs.

In the preclinical stage, in vivo trials in animal models must demonstrate the efficacy of the nanodrug to perform its intended function. The safety of the drug must be assessed in terms of its biodistribution, potential to cause side effects, and potential to cause harm to other organs or systems. In this regard, the drugs that are biodegradable appear to have an advantage over those that are not.

The pharmacological interactions of the nanodrug with other drugs must be explored and the dose response must be clearly defined. Predicting the potential interactions, side effects, and their frequency is a critical part in the design of preclinical and clinical trials. These predictions are based in part on prior experience with similar drug responses observed in in vitro models and the results of in vivo animal trials. Input from a broad range of experts including chemists, toxicologists, clinicians, biostatisticians, cell biologists, pharmacologists, bioengineers, and others who are familiar with the drug and device development process leads to the development of a rational and well thought-out research plan. Once this plan is developed, it should be discussed with the appropriate regulatory officials at the FDA. Building a multidisciplinary team and using their expertise to develop the evaluation of a regime for a new drug can save time and cost and speed up the FDA approval process. These multidisciplinary teams help to focus the research on critical issues and eliminate unnecessary investigations.

A discussion at the discovery stage with the FDA official enables investigators to identify challenges to FDA approval for nanodevices and nanodrugs or their combinations and to provide suggestions that could help scientists to obtain FDA approval. All discussions among the groups of collaborators, lab members, and FDA have to be properly documented with the time, date, and location of the discussion, so that progress could be backtracked and milestones could be documented as major decisions take place over years or decades.

Thus, the preclinical evaluation of a drug is meant to determine if the drug is reasonably safe for initial use in humans, to allow investigators to put all questions in regard to the research design and experimentation on the table, and to try to come up with a clear roadmap ending in a clear pathway for regulatory approval. If this process takes place properly and methodically, the data will support the application.

The preclinical or discovery stage of drugs and devices usually takes, on average, 6.6 years focusing on efficacy and toxicity assessments (Eifler and Thaxon 2011). The clinical trials of drugs are usually conducted in three phases.

Phase I clinical trials are designed to examine the safety of the drug and usually take 2–3 years. Phase II clinical trials are studies designed to determine the metabolism and pharmacologic actions of drugs, the side effects associated with increasing doses, and early evidence of effectiveness in humans. Study subjects may include healthy participants and/or patients. This phase is typically short (6–12 months), and 20–100 volunteers could be assessed in this stage (Eifler and Thaxon 2011). Phase II clinical trials are controlled clinical studies conducted to evaluate the effectiveness of the drug for a particular indication in patients with the disease or condition under study and to determine the common short-term side effects and risks. More than 100 patients are usually registered in this phase with experimental studies extended from 12 to 24 months. Phase III clinical trials are expanded, controlled, and randomized trials that are undertaken after preliminary evidence drug efficacy is established. These larger trials are designed to identify potential side effects and complications resulting from the use of the drug. They can also explore dosimetry and applicability in a range of patients. Duration of this study usually takes a few years because the total number of patients may exceed 1000 across multiple trial centers (ClinicalTrials.gov 2008).

If the drug appears to be safe and effective based on the available clinical data, the manufacturer must collate and analyze all relevant information as part of the IND submission package. Then the manufacturer may apply for New Drug Application (NDA) approval. With NDA approval, the drug can be marketed for a specific use provided that the manufacturer is willing to conduct postmarket surveillance as required by the FDA. As a part of the approval, the FDA may request that further studies be conducted in a postmarketing phase (phase IV). The postmarketing phase is generally used to test the effectiveness and side effects of the drug in general use. It is estimated that the typical FDA submission process to develop a new drug takes between 10 and 12 years and costs about $1 billion (DiMasi et al. 2003).

Effects that Nanomaterials Can Impart on the Rest of a Formulation

Multiple functionalities, such as targeting, therapeutic, imaging, sensing, controlled delivery, and controlled degradation, can be added to nanoformulations to reduce undesirable systemic side effects and to increase the amount delivered to the targeted disease site (Fujita et al. 2006; Kateb et al. 2011). The individual components comprising these multifunctional nanoparticles may interact with other components in the formulation and result in a change in surface chemistry, release of a component, or other modification. Furthermore, when formulations are designed to degrade, that is, for controlled drug delivery, these changes need to be characterized and quantified. The components within a multifunctional delivery system should be characterized in how they interact with the body independently and with their counterparts. The development of technologies such as lab-on-a-chip (Richter et al. 2011) may make such high-throughput analysis possible.

BIOLOGICAL IMPACT ON PHYSICOCHEMICAL PROPERTIES

Nanobased formulations of existing drugs can result in remarkably decreased toxicities compared to their free-form drug as demonstrated with Aurimune (TNF-α bound to polyethylene glycol–coated nanosized colloidal gold) and Abraxane (nanoparticle albumin bound paclitaxel) (McNeil 2011b). There are certain cases where existing preclinical data could not support the clinical development of a nanotherapeutic. Nanobased platforms that reformulate previously approved drugs require thorough investigation and characterization because they may alter the pharmacokinetic profiles, biodistribution, and interactions with cellular components of these drugs and potentially create new toxicities. Toxicities associated with nanoparticles are generally mediated by their size, shape, and chemical composition (Maynard et al. 2011).

BIODISTRIBUTION STUDIES

The addition of a previously approved surface moiety to enhance the action of a nanotherapeutic may also dramatically change the biodistribution of the drug by expanding its access to previously inaccessible tissue (Hasadsri et al. 2009; Olivier et al. 1999; Vergoni et al. 2009). This reiterates the concept that new combinations of previously approved surface moieties may have much greater effects on the in vivo than the in vitro behavior of nanoparticles, making their physicochemical characterization highly important.

ANALYZING NANOPARTICLE SIZE

In 2004, the National Cancer Institute created the Nanotechnology Characterization Laboratory (NCL) to fund and centralize the development of standardization protocols to preclinically evaluate the safety and toxicity of emerging nanotherapeutics and other nanomaterials (Eifler and Thaxton 2011) (Table 40.1). The NCL has developed a three-phase standardized analytical cascade for testing the preclinical toxicology, pharmacology, and efficacy of nanotherapeutics. Physical characterization is the first phase of the cascade and includes assessment of properties such as size, size distribution, molecular weight, density, surface area, porosity, hydrophilicity, surface charge density, purity, sterility, surface chemistry, and stability (NCL 2012). The second phase is the in vitro characterization where the binding, internalization, and blood contact properties are assessed. In vivo characterization is the third phase of the assay cascade, in which NCL laboratory studies are designed to provide data that would support the pharmacology and toxicology section of an IND. Here, animal studies in rodents determine toxicity as well as absorption, distribution, metabolism, and excretion (ADME) profiles; establish the efficacy of the nanomaterial formulation; and provide the preliminary identification of target organs of acute toxicity.

TABLE 40.1
Objectives of NCL

1	Establish and standardize an analytical cascade for nanomaterial characterization.
2	Facilitate the clinical development and regulatory review of nanomaterials for cancer clinical trials.
3	Identify and characterize critical parameters related to nanomaterials' absorption, distribution, metabolism, excretion, and toxicity profiles using animal models.
4	Examine the biological and functional characteristics of multicomponent/combinatorial aspects of nanoscaled therapeutic, molecular, and clinical diagnostics, and detection platforms.
5	Engage and facilitate academic and industrial-based knowledge sharing of nanomaterial performance data and behavior resulting from preclinical testing (i.e., physical characterization, in vitro testing, and in vivo pharmacokinetics and toxicokinetics).
6	Interface with other nanotechnology efforts.

Source: Nanotechnology Characterization Laboratory. National Cancer Institute, Advanced Technology Program. Available at http://ncl.cancer.gov/about_mission.asp, 2012.

CHALLENGES IN CHARACTERIZATION: CONSISTENCY ACROSS LABORATORIES

A second preclinical concern is consistency among laboratories regarding the analytic methods used for particle characterization. Because the FDA does not have unique requirements to characterize nanobased therapeutics, techniques for drug characterization may vary across laboratories. For small molecule drugs, standardized techniques such as mass spectrometry, gas chromatography, and nuclear magnetic resonance can be processed in a high-throughput manner for molecular analysis and characterization (McNeil 2011a). However, nanoparticles often require more sophisticated techniques to characterize their size in solution such as dynamic light scattering, mass spectrometry, zeta potential measurement (surface charge), atomic force microscopy (topology), examination by transmission electron microscopy, scanning electron microscopy, and energy dispersive x-ray spectroscopy for elemental identification (McNeil 2011a). Because there is no current standard characterization methodology for nanoparticles, sponsors of nanobased drug candidates need to justify their methods used for characterization when submitting to the FDA.

CONSTRAINTS IN DEFINING NANOTECHNOLOGY, NANOMEDICINE, NANONEUROSCIENCE, AND NANONEUROSURGERY

The application of nanotechnology to medical diagnostics and therapeutics presents a problem to regulators because of the confusion and disagreement over the definition of "nanotechnology" (Bawa 2011). The definition proposed by the National Nanotechnology Initiative, "… Nanotechnology is science, engineering, and technology conducted at the nanoscale, which is about 1 to 100 nanometers…," is used by the Patent and Trade Office and loosely by the FDA, because the FDA has not adopted a formal definition (Bawa 2011). Obtaining a consensus definition of nanotechnology is important to identify candidate therapeutics that should have novel regulatory avenues.

In June 2011, the FDA released its "guidance for industry considering whether an FDA-regulated product involves the application of nanotechnology." The guidance documents describe the agency's current thinking on a topic and are intended as recommendations only and not as legally enforceable responsibilities. When deciding whether an FDA-regulated product contains nanomaterials or involves the application of nanotechnology, the FDA will consider two fundamental issues: (1) "Whether an engineered material or end product has at least one dimension in the nanoscale range or (2) whether an engineered material or end product exhibits properties or

TABLE 40.2
Characteristics of an FDA-Regulated Product that Involves Application of Nanotechnology

Engineered material/end product	To draw a distinction between biological or chemical substances that may naturally exist in small scales (such as proteins) and those particles that are deliberately manufactured to exhibit new size-dependent phenomena. The terms "engineered material" and "end product" are used to describe those particles that are deliberately engineered.
At least one dimension in the nanoscale range	In the size range of 1–100 nm, materials can exhibit new or altered physicochemical properties. This is applied as a first reference point in determining whether an FDA-regulated product contains nanomaterials or the application of nanotechnology.
Exhibits properties of phenomena that are attributable to its dimension(s)	Physicochemical properties exhibited at the nanoscale level enable many applications that can affect the safety, effectiveness, performance, quality, and public health impact of FDA-regulated products. These novel applications are not exhibited in the same material at the bulk level.
Size range up to 1 μm	Although a size range of 1–100 nm is commonly used in working definitions of nanotechnology, the size/dimension-dependent properties of nanomaterials can exist at scales greater than 100 nm. Of particular interest are structures such as agglomerates, aggregates, and functionalized nanoparticles.

phenomena, including physicochemical properties, or biological effects that are attributable to its dimension(s), even if these dimensions fall outside the nanoscale range (1–100 nm), up to one micrometer" (Nalubola and Pena 2011) (Table 40.2). These considerations are applicable to new products and possibly when manufacturing changes alter the dimensions, physical properties, or effects of an FDA-regulated product or any of its components. As new information becomes available, these are subject to change and refinement in future product specific-guidance documents (Nalubola and Pena 2011).

CASE STUDY

Doxil is the first nanodrug approved by the FDA. The drug has three important features:

1. The drug avoids being captured by the reticuloendothelial system (RES) due to its PEGylated nanoliposomes.
2. The drug could be released at the tumor because of its stable remote loading of doxorubicin, which is driven by a transmembrane ammonium sulfate gradient.
3. The drug has a liposome liquid bilayer composed of the high-Tm (53^{oc}) phosphatidylcholine and cholesterol (Barenholz 2012).

CALAA-01 is another good example of targeted therapy using nanoparticles to deliver siRNA. The drug was developed by Caltech scientists to be used as a component in systemic siRNA delivery. The chemical design of the nanoparticle was based on a multifunctional colloidal particle, a self-assembling cyclodextrin-containing polymer (CDP) core, and a targeting ligand. The nanodrug could escape from the endosome through an acidification mechanism (Eifler and Thaxton 2011). Tables 1.3 and 1.4 in Chapter 1 have lists of approved and currently under development nanodrugs.

The scalability and simplicity of the manufacturing process seemed to play a role in the design of these nanodrugs. However, these teams carefully planned their design by asking all the right questions. We have outlined design considerations in the preclinical and discovery stages. Their design included tailoring of proper chemical charge, hydrophobicity, targetability, and biomaterials, making these drugs not only a success but also a great classic model of drug design to study.

CONCLUSION

It is critical to acknowledge that the FDA is an evolving organization by nature because the organization deals with state of the art drugs and devices all the time. While there are some shortcomings in FDA actions by default, it is impossible to be perfect in developing policies for drug discovery and development while fields such as nanoneuroscience and nanoneurosurgery are continuously evolving. The FDA has to be much more open and technologically savvy in regard to its communication with the scientists. An insular system of communication and lack of access to FDA offices and officers creates a perception of lack of willingness by the US government to help scientists in this regard. FDA officers should present and speak at more conferences and participate in drafting manuscripts and book chapters like this in order to better educate scientists about FDA process, law, and regulations.

Scientists also play a major role in influencing policy decisions that foster successful drug and device development. While FDA is a major regulatory body, its power is derived from the people who elect officials to the United States Congress. Today, we must be more proactive in raising our voices and working more closely with our elected officials in order to ensure that laws and regulations are updated as advances are made in the field. One clear example of such interaction is Brain Mapping Day organized at the US Congress by the Society for Brain Mapping and Therapeutics (SBMT), which is devoted to educating policymakers and their staff about state-of-the-art neuroscience, stem cells, nanoneurosurgery, nanobioelectronics, image-guided therapy, and devices.

The following stages could be simultaneously pursued as a nanodevice or drug is developed (Tyner and Sadrieh 2011):

1. Choosing biomaterial: structure, composition, purity, size, morphology, and surface features
2. Determining the properties and effects: impact on characterization and proper instrumentation
3. Formulating product or characterization: stability, treatment effects, and size
4. Manufacturing characterization and product use: alteration of formulation and manufacturing

The initial preclinical steps and research design are absolutely critical. Early discussions with the FDA could make a huge difference in cost, duration, and success of the project. Individual scientists could take steps and prepare to ask the right questions based on the outline in this chapter, as well as other proper and recent resources, which could prepare them for the long journey of drug and device development and obtaining FDA approval.

The future of clinical application of nanoneuroscience and nanoneurosurgery drugs and devices, including development and discovery, hinges on a combination of smart planning, proper research design and strategy, and more importantly, the forward-looking policies put in place by the FDA, which can be shaped by advocacy from scientists, politicians, and the public.

REFERENCES

Barenholz, Y. 2012. Doxil—the first FDA-approved nano-drug: lessons learned. *J Control Release* 160: 117–34.
Bawa, R. 2011. Regulating nanomedicine—can the FDA handle it? *Curr Drug Deliv* 8: 227–34.
ClinicalTrials.gov. 2008. Glossary of Clinical Trials Terms. National Institutes of Health. Available at http://clinicaltrials.gov/ct2/info/glossary.
Cressey, D. 2010. Tiny traits cause big headaches. *Nature* 467: 264–5.
DiMasi, J.A., Hansen, R.W., and Grabowski, H.G. 2003. The price of innovation: new estimates of drug development costs. *J Health Econ* 22: 151–85.
Eifler, A.C., and Thaxton, C.S. 2011. Nanoparticle therapeutics: FDA approval, clinical trials, regulatory pathways, and case study. *Methods Mol Biol* 726: 325–38.

Force, N.T. 2007. Nanotechnology: a report of the U.S. Food and Drug Administration Nanotechnology Task Force. Availiable at: http://www.fda.gov/downloads/ScienceResearch/SpecialTopics/Nanotechnology/ucm110856.pdf.

Fujita, M., Khazenzon, N.M., Ljubimov, A.V. et al. 2006. Inhibition of laminin-8 in vivo using a novel poly(malic acid)-based carrier reduces glioma angiogenesis. *Angiogenesis* 9: 183–91.

Hamburg, M.A. 2012. FDA's approach to regulation of products of nanotechnology. *Science* 336: 299–300.

Hasadsri, L., Kreuter, J., Hattori, H. et al. 2009. Functional protein delivery into neurons using polymeric nanoparticles. *J Biol Chem* 284: 6972–81.

Johnson, J.A. 2012. FDA Regulation of Medical Devices, Congressional Research Report prepared for the members of the US Congress, Congressional Research Service, 7-5700, R42130.

Kateb, B., Chiu, K., Black, K.L. et al. 2011. Nanoplatforms for constructing new approaches to cancer treatment, imaging, and drug delivery: what should be the policy? *Neuroimage* 54(Suppl 1): S106–S124.

Maynard, A.D., Warheit, D.B., and Philbert, M.A. 2011. The new toxicology of sophisticated materials: nanotoxicology and beyond. *Toxicol Sci* 120: S109–29.

McNeil, S.E. 2011a. Challenges for nanoparticle characterization. *Methods Mol Biol* 697: 9–15.

McNeil, S.E. 2011b. Unique benefits of nanotechnology to drug delivery and diagnostics. *Methods Mol Biol* 697: 3–8.

Nalubola, R., and Pena, C. 2011. Considering Whether an FDA-Regulated Product Involves the Application of Nanotechnology Guidance for Industry. U.S. FDA, ed., Silver Spring, MD: U.S. Department of Health and Human Services.

Nanotechnology Characterization Laboratory. 2012. National Cancer Institute, Advanced Technology Program. Available at http://ncl.cancer.gov/about_mission.asp.

Olivier, J.C., Fenart, L., Chauvet, R. et al. 1999. Indirect evidence that drug brain targeting using polysorbate 80-coated polybutylcyanoacrylate nanoparticles is related to toxicity. *Pharm Res* 16: 1836–42.

Paradise, J., Tisdale, A.W., Hall, R.F., and Kokkoli, E. 2009. Evaluating oversight of human drug and medical device: a case study of the FDA and implications for nanobiotechnology. *J Law Med Ethics* (Winter): 598–624.

Richter, L., Charwat, V., Jungreuthmayer, C. et al. 2011. Monitoring cellular stress responses to nanoparticles using a lab-on-a-chip. *Lab Chip* 11: 2551–60.

Tyner, K., and Sadrieh, N. 2011. Considerations when submitting nanotherapeutics to FDA/CDER for regulatory review. *Methods Mol Biol* 697: 17–31.

Vergoni, A.V., Tosi, G., Tacchi, R. et al. 2009. Nanoparticles as drug delivery agents specific for CNS: in vivo biodistribution. *Nanomedicine* 5: 369–77.

41 Nanoneuroscience and Nanoneurosurgery

A Key Component of President Obama's Brain Mapping Initiative

Babak Kateb, Vicky Yamamoto, Peter J. Basser, Michael Roy, Lucien M. Levy, Jian Tajbakhsh, Gary K. Steinberg, Allyson C. Rosen, Keith L. Black, Charlie Teo, Kuldip Sidhu, Mitchel S. Berger, and Warren S. Grundfest*

CONTENTS

INTRODUCTION

President Obama said in his April 2nd speech "We have been a nation of dreamers and risk-takers; people who see what nobody else sees sooner than anybody else sees it. We do innovation better than anybody else—and that makes our economy stronger. When we invest in the best ideas before anybody else does, our businesses and our workers can make the best products and deliver the best services before anybody else." In the same event at the White House he followed up on the promise he made at his presidential state-of-the-union address and authorized $100 million in additional funding for neuroscience, directing it specifically toward brain mapping efforts (Figure 41.1).

This is a great achievement for the brain mapping community, which has developed and defined the field of Brain Mapping as: The study of the anatomy and function of the brain and spinal cord through the use of imaging (including intra-operative, Microscopic, Endoscopic and Multi-Modality

* Also the senior author.

FIGURE 41.1 President Obama unveiled the Brain Mapping Initiative before 100 top scientists at the East Wing of the White House on Tuesday April 2nd 2013. Chairman of the Board of SBMT and President of Brain Mapping Foundation, Babak Kateb, was amongst the attendees. (Courtesy of Brain Mapping Foundation.)

imaging), Immunohistochemistry, Molecular and optogenetics, Stem cell and Cellular Biology, Engineering (material, electrical and biomedical), Neurophysiology and Nanotechnology.

The Society of Brain Mapping and Therapeutics (SBMT) and the Brain Mapping Foundation (BMF) fully support President Obama's BRAIN Initiative (Brain Research through Advancing Innovative Neurotechnologies). It is our opinion that before significant funds are committed to speculative technologies, a complete, comprehensive and detailed survey of the key unresolved clinical research questions, and the most promising technologies, should be completed. The initial Brain Activity Mapping (BAM) initiative appeared to be based on a single Neuron article that sketches out a few such technologies, without providing a clear vision about how a BAM could be defined on an organism much more complex than a nematode (*Caenorhabditis elegans*) or fruit fly (*Drosophila*) (Paul Alivisatos et al. 2012a,b). If the ultimate goal is to understand how the human brain works, a more thorough survey of what is known and what can be measured in vivo is required, including all aspects of neuroimaging, nanotechnology, device application, cellular tracking and molecular genomics and photonics (Levy 2013).

Our view is that the nervous system is organized hierarchically over many spatiotemporal time scales, extending from nanometers to meters, and from microseconds to lifetimes. This demands a much more holistic approach to understanding what brain activity is, what it means, and how we can measure it. Organizing a visionary process that lays out critical problems in brain sciences and technical problems at the relevant spatial and temporal scales would serve to improve the quality of this project and prioritize the efficient distribution of limited funds. This could be done in various ways, such as by having high-profile talks and organizing Congresses such as the 10th Annual World

Brain Mapping Congress of the Society for Brain Mapping and Therapeutics (SBMT). This meeting incidentally took place at the Baltimore Convention Center shortly after the President Obama announcement. The meeting brought together more than top 500 scientists from across the globe, including leaders in the field of neuroscience and other relevant disciplines of science and engineering. Such meetings should engage policymakers, physicians and surgeons in order to address neurological disorders from all perspectives (www.WorldBrainMapping.Org).

Clinical input is essential to ensure that knowledge obtained from the BRAIN Initiative is translated into improved health outcomes. Simply mapping brain function and activity is not sufficient. We need to also address the therapeutic aspects of brain mapping to ensure that the knowledge acquired by functional and activity maps actually impacts treatment and healthcare delivery for patients with neurological disorders.

In this final chapter of the textbook we have identified items of relevance to national health issues and tried to address them through 3 major-encompassing programs, which could engage scientists across disciplines and enable translation, integration and commercialization of nanobioelectronics (nanotechnology, stem cell and cellular therapeutics, imaging, and devices). These novel proposed programs are: 1) National Alliance for NanoBioElectonics (NANBE), 2) National Network for Human Brain and Specimen Banks (NNHBSB), and 3) National Data Repository and Analysis for Neuroscience (NDRAN).

PROBLEM TO BE ADDRESSED

Neurologic and psychiatric disorders, including stroke, traumatic brain injury, brain tumors, spinal cord injuries, brain cancers, neurodegenerative disorders (Alzheimer's Disease, Parkinson's Disease, ALS, etc.), developmental disorders, epilepsy and depression, cost the US healthcare system near $400B annually (Rydell and Larson 2012). In order to improve the treatment of these devastating diseases, stimulate relevant research, reduce cost and disabilities, increase productivity and improve rehabilitation, a multidisciplinary approach that mobilizes all assets of the federal government, industry and academia is necessary. If properly designed, this investment in healthcare and research can simultaneously generate jobs and provide the stimulus for solving critical problems in healthcare.

One of the greatest challenges of the early 21st century is the translation and transfer of cutting-edge research and advances made in the nation's diverse research laboratories into the clinical setting. The SBMT and the Brain Mapping Foundation jointly propose the creation of a new Brain Mapping and Therapeutics program within BAMP in order to promote the development of new diagnostic and therapeutic approaches, as well as educational and preventive strategies to preserve and repair brain function. A focused, targeted, goal-oriented effort founded upon interagency and interdisciplinary collaboration can efficiently advance medical technology, fuel job creation (i.e., biotech spinoffs), and foster infrastructure development.

The proposed program would establish novel consortia involving federal agencies, academic centers, private foundations and industry *based on the two congressional initiatives by Congressmen Fattah and Moran*:

> "The Fattah Neuroscience Initiative (FNI) is an innovative, non-incremental policy initiative designed to make major progress in understanding the human brain by intensifying, in a collaborative fashion, federal research efforts across brain disease, disorder, injury, cognition and development. The initiative aims to coordinate Federal research across agencies and draw upon public-private partnerships and the world of academia. The initiative promotes research and discovery across brain cognition, development, disease and injury." and congressman Moran "The Committee encourages the Secretary of Defense to support multi-disciplinary research toward translational medicine that may provide better diagnostic tools and treatment outcomes for service members who suffer from traumatic brain injury, post-traumatic stress disorder, and other neurological diseases. With this support the Secretary of Defense will enable the creation of a nationwide scientific consortium aimed at integrating nanotechnology, stem cell, cellular therapy, medical imaging, electronic medical record, information technology, Brain

> Mapping, and medical device. The Committee encourages the Secretary of Defense to provide the capabilities necessary for researchers, scientists, surgeons, physicians, healthcare professionals, and patients to effectively communicate their findings and outcomes. With proper support, translational research outcomes would be augmented through real-time access to information and its integration between researchers, physicians, hospitals, and patients. The Committee encourages the Secretary of Defense to establish this program through the NAVY Bureau of Medicine and Surgery (BUMED). The Committee directs the Secretary of Defense to provide a report not later than 90 days after enactment of this act on a possible implementation of this program."

As opposed to current centers of excellence, the proposed consortia by the Brain Mapping Foundation and Congressman Moran would be: 1) larger and multi-institutional 2) focused upon multiple aspects of a disease, including medical, business, legal, and socioeconomic aspects rather than only single components 3) multifaceted in addressing the impact on overall public health, 4) attentive to job creation in addition to the advancement of science by providing bridge grants for the consortia, which could encourage biotech spinoffs, 5) able to increase investments in the US by encouraging up to 20% investment in the US in the area of Brain Mapping by international partners. These investments could be made by hiring postdocs in the US or by providing other forms of funding based on individual agreements required in the Research Announcement/Agreement. These initiatives are designed to significantly contribute to President Obama's Brain Mapping Initiative through the following key elements of the program, which would include:

The National Alliance for NanoBioElectonics (NANBE) in Brain Mapping

SBMT has developed a global network accessible to all neuroscientists, which could facilitate the translation of basic science and engineering into clinical trials. The National Alliance for NanoBioElectonics (NANBE) could similarly integrate diverse elements including nanotechnology, imaging and diagnostic devices, stem cell and cellular therapies, and genomics in order to develop novel diagnostic and therapeutic innovations that can protect and preserve the brain while simultaneously creating new technologies and new jobs. The NANBE could further foster international partnerships through a global alliance led by the US, which would do even more to create new jobs within the US. This program will be aimed at integration, translation and commercialization of already exciting technologies and treatment modalities and or technologies, which are under development. The alliance will function as a global hub for nanoneuroscisnce, which will help identify the best nanoplatform for imaging, diagnostics and the treatment of neurological disorders (Kateb et al. 2011).

National Network for Human Brain and Specimen Banks (NNHBSB)

The Brain Mapping Foundation (BMF) calls for the creation of a National Network for Human Brain and Specimen Banks (NNHBSB), which would work closely with existing organ, harvest teams and specimens banks across the country. This program could be implemented by the Department of the Health and Human Services (HHS). This database would include A) Postmortem data and B) Brain imaging data from living humans including patients with neurological disorders.

A. Postmortem Data: The NNHBSB could coordinate the harvest of donated brains from all organ donors and connect the specimens (including Brain, Cerebrospinal Fluid, eyes, spinal cord) with data from the electronic medical record of the patient to maximize what we are able to learn from the brains about conditions such as degenerative brain disease and traumatic brain injury. The bank would include both brains with known disease as well as ostensibly healthy brains. This national network would have far greater power to enhance our knowledge than existing institutional or regional programs.

B. Brain Imaging and Activity Maps: Brain imaging techniques have revolutionized the way we can study the human brain but in order to harness the potential to improve human health it is crucial to study both animals and living humans, both healthy and those with a variety of brain illnesses at the organ, tissue, cellular and subcellular levels (Levy 2013). Data from multiple modalities such as functional and structural MRI, DTI (HTDI, HFDTI, etc.), CT, Ultra-low field MRI, PET, noninvasive and surgically implanted brain and spinal cord stimulation devices can be compared and related to preclinical models of disease to provide the most powerful approach to understanding and advancing human brain health. Clinical applications of functional brain maps will require large populations to ensure the results sample a broad spectrum of the population cohort, both in health and disease. Longitudinal studies will be required, to examine aging, consequences of interventions and to establish stability of results. This in turn will require integration and standardization of methodology across multiple centers around the US, in order to perform the measurements in a useful time frame. This means attention must be focused on expanding multicenter brain mapping methodology such as has been developed by the Alzheimer's disease Neuroimaging Initiative and the FIRST Biomedical Imaging Research Network. Generating a broad series of brain maps is a truly multidisciplinary effort that will involve many scientists and clinicians. Next-generation brain mapping would highly benefit from integrating novel concepts such as cytomics, also referred to as spatial systems biology that is revolutionizing our knowledge about cellular interactions and signal transduction towards therapeutic intervention: by combining imaging-derived and genomic data in a cell-by-cell mode (Pierzchalski et al. 2008; Kriete 2005; Valet et al. 2003; Gertych et al. 2009; Tajbakhsh et al. 2011). Cytomics would apply bioinformatics to understand the relation between molecular composition, topology and functionality of cells in neural circuits. Translation of brain maps to studying and curing disease will require careful coordination of technology providers, basic scientists and clinicians with diverse viewpoints. The successful conclusion of this program will provide normative data that will be invaluable in studying the special needs of diverse patient groups. Therefore, helping with the development of customized and personalized medicine. To summarize: the program must be multimodal, multi-institutional, must include preclinical studies, and sample a broad demographic profile.

National Data Repository and Analysis for Neuroscience (NDRAN)

The Congressional Accountability Office documents that \$2.7B in research funding has been provided to the DoD to address Traumatic Brain Injury (TBI) and Psychological health over the past 5 years, while NIH has spent another \$405M. However, despite 3.1B+ in federal spending in the last 5 years on TBI and psychological health, we have not had a significant return on our investment. Progress in treating these disorders has been slow, and new therapies have yet to emerge.

In order to avoid duplication of effort, more closely monitor progress, identify particularly promising work and ensure efficient use of funds, we need to establish an electronic system that encompasses all areas of neuroscience and clearly documents the results. Funded research needs to be outcome-driven with attention to deliverables, timelines and cost-effectiveness. The NDRAN could help to achieve these aims. We believe that 3 different levels of access to the repository could be established.

1. Patient Education: One level for patients could educate them about the latest research linking to articles in Pubmed Health (http://www.ncbi.nlm.nih.gov/pubmedhealth/) and opportunities to participate in research (http://www.clinicaltrials.gov/).
2. Large Scale Research Collaboration: A second level would facilitate, incentivize, and target barriers to collaboration among researchers. For example, patient data donated to the

repository and development of tools to facilitate data quality and comparability across sites would be considered as accomplishments in evaluating research productivity. Bringing together data from diverse sources and studies will allow clinicians to base decisions on large, definitive, studies rather than small, inconsistent, studies.

3. Government Oversight of Public Health Needs, Service Delivery Efficiency, and Barriers to Moving Innovations to Clinical Care: The final level would allow multiple government agencies to access the data and systematically analyse, and produce publications to identify research gaps, barriers to rapid and efficient implementation of best clinical practices, and guide future funding. This level would also identify and consider burdensome regulatory barriers and delays that do not protect patient safety and which limit patient access to effective therapies and sensitive diagnostic tools.

One clear need is to have people working in different silos of neuroscience (usually characterized by working on problems in different "points" within the time and spatial parameter scales) be able to communicate and exchange information with each other, rather than remain in their scientific cliques, sometimes for their entire careers. To accomplish this neuroscience data repository and analysis it would be useful to bring diverse interest groups together under one umbrella; NDRAN is designed to accomplish this goal. NDRAN also could identify important life saving technologies such as High Definition Diffusion Tensor imaging, biophotonics, nano platforms, non-invasive multimodality imaging technologies (UV, IR, Ultra Low MRI, etc.) technologies and devices in order to introduce a new generation of life saving technologies, help reduce the cost of health care, and increase efficiency in delivering personalized medicine, which will be "outcome driven."

PURPOSE

As described previously, the SBMT and BMF proposed to the White House, NIH and DARPA (Defense Advance Research Program Agancy) 3 national strategic programs:

- National Alliance for NanoBioElectronics (NANBE)
- National Network for Human Brain and Specimen Banks (NNHBSB)
- National Data Repository and Analysis for Neuroscience (NDRAN)

The program would be established in order to:

1) Foster partnership in technology, science, and medicine through the development of national consortia.
2) Systemically and actively facilitate the integration, translation and commercialization of technologies and scientific advancement from various physical, social, and biological sciences into clinical medicine.
3) Establish consortia, scholarships and educational programs at all educational levels in order to encourage ingenuity and cross-disciplinary interaction, which could contribute to the development of future scientists, technologists and health professionals.
4) Promote education curricula addressing disease prevention. This novel program will contribute to US leadership in the global healthcare market place while building long term employment.
5) Establish brain mapping centers, minimally invasive and noninvasive technologies such as nanotechnology, biophotonics, ultra low MRI, portable CT, High definition Diffusion (HDTI) Tensor Imaging, simulations, cellular therapeutics/stem cell and devices, and these methods could help to map the brain and help physicians treat neurological disorders as well.

SIGNIFICANCE OF THE PROPOSED APPROACH AND ITS IMPACT ON THE FIELD OF BRAIN MAPPING

This 3-tiered proposal including NANBE, NNHBSB and NDRAN, is designed to achieve the following goals:

(1) Create jobs in the biotech sector through investment in research. This program will both directly and indirectly support technology, science, healthcare, medical research and related fields.
(2) Improve healthcare efficiency and cost-effectiveness through the identification of best practices and the promotion of personalized medicine, point of care diagnostics, electronic medical records and advance information technology such as grid technology. Grid technology allows hospitals to be connected to each other. This will reduce healthcare cost, increase efficiency and prevent medical errors.
(3) Stimulate small businesses through expanded SBIR and SBA programs.
(4) Promote disease prevention to reduce healthcare costs.
(5) Assemble multisite consortiums to advance the study of neurological diseases, including brain mapping, traumatic brain and spinal cord injuries, PTSD, neuro-oncology and neurodegenerative diseases. The programs will scientifically evaluate and apply the latest technologies in neuropharmacology, nanobioelectronics, nanoneurosurgery, nanoneuroscience, imageguided therapy, frontiers in stem cell research, therapeutics and regenerative medicine (Sidhu 2012), device, and nanotechnology.
(6) Translate medical research and discoveries to the social sciences including behavioral studies and neuro-economics.
(7) Foster education in neuroscience and related disciplines by supporting programs and scholarships at the K-12 such as SBMT kids corner, the high school, university such as SBMT student chapters, and postgraduate levels such as our fellowship programs to encourage cross-disciplinary thinking and interest in related fields.
(8) Advanced training for workers and professionals in the US.
(9) Promote US leadership in technology, healthcare policy, medical research, and science.
(10) Complement the missions of all participating agencies and allow them to be more productive by facilitating cross disciplinary activities and reducing wasteful spending, duplication and medical errors.

PROGRAMS AND METHODS

In order to assess the viability, novelty, and success of the proposed program, SBMT and BMF have contacted officials from many US government agencies. A multidisciplinary approach involving scientists, engineers, physicians, surgeons, legal scholars and economists is necessary to achieve the ultimate goals of the program. To implement this program SBMT and BMF propose to the White House Office of Science, Technology and Policy (OSTP) to assemble leading experts from, foundations, academia and industry. White House OSTP, NIH, NSF and DARPA along with other government and other nonprofit and industry partners will play a central role in coordinating these programs. Leading foundations and associations should be part of these unique consortia. These programs will be responsible for defining the programmatic goals, organizing the consortia, requesting proposals, distributing funds, and reporting back to the Congress and the executive branch through quarterly reports. This should foster rapid growth in healthcare-related jobs, the creation of a sustainable infrastructure, and continued US leadership in the healthcare sector.

Representatives from involved public and private partners and agencies will participate in an Oversight Committee to make sure that funds are spent properly and with transparency. The NANBE in Collaboration with National Institute of Health will request proposals for the development of the consortia. The proposals must include multiple centers and institutions within the US and explain in detail how funds will contribute to job creation in multiple fields using science, engineering, education and technology transfer. While this initiative is designed for national consortia, international participants whose country(ies) match the funds, effort and protect the intellectual property of US investigators are welcome to join this effort if and when they agree to invest at least up to 20% of their funds in the US. Participation from industry and private foundations is also welcomed provided that they too bring commensurate funds or efforts, and show how their participation will foster job creation using neurosciences.

This proposal will also promote scholarships, fellowships, and humanitarian work by physicians and scientists working in collaboration with other humanitarian organizations, government agencies and foundations. In conjunction with its government partners, this SBMT and BMF-proposed program will create the world's largest multidisciplinary research consortium for treatment of neurological diseases, with a focus on job creation and technology development by providing incentive/bridge grants for spinoffs and biotech formations.

CONCLUSION

It is our strong belief that no single foundation or organization should spearhead President Obama's Brain Mapping Initiative. Prominent organizations like ours have been on the forefront of brain mapping and therapeutics for decades. Therefore, the national policy must be shaped through an inclusive process, which is transparent and open to the public. Scientists at the grassroots level should be able to voice their opinion so that new ideas can be cultivated. Moreover, there is a great need for training a new generation of cross disciplinary scientists. In this regard, Brain Mapping Foundation is partnering with one of the prominent universities in the US to establish a doctoral program in NanoBioElectronics, which will train a new generation of scientists who could integrate nanotechnology, device, imaging and cellular/stem cell therapy. We also believe that standardization of the brain mapping field and technologies are the key to the advancement of the field. In this regard SBMT is launching the American Board of Brain Mapping, which will focus on certifying specialists (PhDs and MDs) in the field, in partnership with industry and pertinent governmental agencies.

The programs described in this chapter, which were also submitted to the White House in a "White Paper", are aimed at building national and global consortia, which could bring increased investments to domestic and international partners, and help to dramatically change the field of neuroscience by distinguishing between projects/areas that need resources and those that are less meritorious. These proposals address the "valley of death" by establishing a path extending from integration to translation and into commercialization. This could improve the efficiency and cost-effectiveness of research while enhancing the evaluation of neurological disorders and creating biotech jobs in the US. We strongly believe that funded research has to be outcome-based. Research policies should focus on how to avoid duplication, how to encourage camaraderie in research, and how to properly reward partnership and collaborations. This proposal could enable scientists to have open access to the work done by other scientists in real time, and help build consortia, which are not aimed at "competing amongst institutions but competing against the diseases," said Babak Kateb, Founding chairman of the Board of SBMT, President of Brain Mapping Foundation and Research Scientist at the Department of Neurosurgery at Cedars Sinai Medical Center, CA, USA.

REFERENCES

Alivisatos, A. P., Chun, M., Church, G. M., Deisseroth, K., Donoghue, J. P., Greenspan, R. J., McEuen, P. L., Roukes, M. L., Sejnowski, T. J., Weiss, P. S., and Yuste, R. 2012a. Brain activity map. Available at http://www.sciencemag.org/content/early/recent/7 March 2013/Page 2/10.1126/science.1236939, pp. 1–2.

Alivisatos, A. P., Chun, M., Church, G. M., Greenspan, R. J., Roukes, M. L., and Yuste, R. 2012b. The brain activity map project and the challenge of functional connectomics. *Neuron* 74:970–4.

Gertych, A., Wawrowsky, K. A., Lindsley, E., Vishnevsky, E., Farkas, D. L., and Tajbakhsh, J. 2009. Automated quantification of DNA demethylation effects in cells via 3D mapping of nuclear signatures and population homogeneity assessment. *Cytom A* 75(7):569–83.
Kateb, B., Chiu, K., Black, K. L., Yamamoto, V., Khalsa, B., Ljubimova, J. Y., Ding, H., Patil, R., Portilla-Arias, J. A., Modo, M., Moore, D. F., Farahani, K., Okun, M. S., Prakash, N., Neman, J., Ahdoot, D., Grundfest, W., Nikzad, S., and Heiss, J. D. 2011. Nanoplatforms for constructing new approaches to cancer treatment, imaging, and drug delivery: What should be the policy? *NeuroImage* 54:S106–24.
Kriete, A. 2005. Cytomics in the realm of systems biology. *Cytom A* 68(1):19–20.
Levy, L. M. 2013. Brain mapping project: Clinical aspects and role of neuroradiology. *AJNR Am J Neuroradiol.* Published online before print March 28, 2013, doi:10.3174/ajnr.A3587.
Pierzchalski, A., Robitzki, A., Mittag, A., Emmrich, F., Sack, U., O'Connor, J. E., Bocsi, J., and Tárnok, A. 2008. Cytomics and nanobioengineering. *Cytom B Clin Cytom* 74(6):416–26.
Rydell, C., and Larson, R. 2012. The critical role of neurologists in our health care system. Available at www.aan.com/go/advocacy/active/summaries.
Sidhu, K. S. 2012. *Frontiers in Pluripotent Stem Cells Research and Therapeutic Potentials; Bench-to-Bedside.* Bentham Science Publishers, Dubai, UAE. ISBN:978-1-60805-289-9, pp. 237.
Tajbakhsh, J., Gertych, A., Fagg, W. S., Hatada, S., and Fair, J. H. 2011. Early in vitro differentiation of mouse definitive endoderm is not correlated with progressive maturation of nuclear DNA methylation patterns. *PLoS One* 6(7):e21861.
United States Government Accountability Office. 2012. GAO-12-154, Defense Health Coordinating Authority Needed for Psychological Health and Traumatic Brain Injury Activities, pp. 1–53.
Valet, G. K., and Tárnok, A. 2003. Cytomics in predictive medicine. *Cytom B Clin Cytom* 53(1):1–3.

Index

Page numbers followed by f and t indicate figures and tables, respectively.

D

F

G

H

J

K

L

M

N

O

P

Q

R

S

T